Daniel S. Thomas

*Applications of Electronics*

*Bernard Grob*

INSTRUCTOR, RCA INSTITUTES, INC.

Author of *Basic Electronics*

*Basic Television*

*Milton S. Kiver*

Editor, *Electronic Packaging and*

*Production Magazine*

Author of *Color Television Fundamentals*

*Transistors*

Coauthor of *Transistor Laboratory Manual*

# *Applications of Electronics*

McGRAW-HILL BOOK COMPANY

*New York*

*St. Louis*

*San Francisco*

*London*

*Toronto*

*Sydney*

*Applications of Electronics*



LIBRARY OF CONGRESS CATALOG CARD NUMBER: 65-23208

ISBN 07-024930-X
1213 HDBP 765

# *Preface*

In *Applications of Electronics,* it is the aim of the authors to attempt to bring together in a single volume discussions of the principles and equipment that pertain to the many specialized areas of electronics and communications. Typical circuits and equipment are described for essentially all modern applications. These include amplifiers, oscillators, microwaves, power supplies, radio transmitters, radio receivers (including TV and FM with stereo multiplex), test instruments, industrial electronics, pulse circuits, radar, and electronics in military applications.

The book begins with a presentation of fundamental principles of tubes, transistors, amplifiers, oscillators, and rectifiers before applying these principles to circuits in commercial equipment. With this background the reader may then deal logically with such diverse subjects as superheterodyne receivers and industrial control circuits. Once the basic electronic circuits are understood, they can be shifted from subject to subject with confidence and understanding in many different applications.

A practical approach is employed in this book, which is designed specifically for technicians and servicemen. Mathematics is kept to a minimum. Knowledge of elementary electricity and electronics, as covered in *Basic Electronics* by Bernard Grob, is assumed, however.

The second edition continues the original concept of a practical survey course in electronics, but the material has been updated and expanded. Tunnel diodes and silicon controlled rectifiers are new topics in semiconductor discussions, and diode rectifiers are more fully covered. In addition to more coverage of transistors in Chap. 2, circuits throughout the book are shown with both tubes and transistors. Many photographs, with actual schematic diagrams, illustrate typical commercial equipment. In addition, simplified diagrams are used to help explain the main features of the circuits. The following topics have also been added: single-sideband transmission, FM stereo multiplex broadcasting, UHF antennas, and masers and lasers. Chap. 14, Pulse Circuits, has been completely revised, with more emphasis placed on multivibrators, blocking oscillators, and trigger circuits in general. Also, Chap. 15, Industrial Electronics, now presents a more logical development for different types of control circuits.

Teaching aids include self-examination questions and a summary at the end of each chapter. In addition, groups of chapters have an overall review with self-examination questions. These summaries also list supplementary references for additional information. The answer key for all the self-examination questions is at the back of the book. Each chapter also has a group of essay-type questions for more extensive testing, particularly on schematic diagrams. The book concludes with an extensive appendix of supplementary material, including FCC frequency allocations, logarithms and decibels, universal *RC* time constant curves, color coding of components, and the Morse code for telegraphy.

Special acknowledgment is due N. Buch, M. Goldstein, C. Nolan, E. J. Williamson, M. Whitmer, and the late J. R. Balton for their technical assistance in preparation of the manuscript. The authors are also indebted to the many organizations that provided photographs and operating manuals for their equipment. These credits are given with the illustrations.

Bernard Grob

Milton S. Kiver

# *Contents*

*Applications of Electronics*

# Chapter 1 Vacuum-tube amplifiers

This unit explains how any vacuum tube with a control grid can amplify its input signal. The tube and its circuit components then form an amplifier stage. Triodes, tetrodes, or pentodes can be used. Furthermore, we can have d-c amplifiers or a-c amplifiers for either r-f or a-f signal variations. The topics are as follows:



## *1·1 Types of amplifiers*

The amplifier is the foundation for practically all applications of electronics, including radio and television communications. Without this ability to amplify a small input signal and provide a much larger output signal, vacuum tubes, transistors, and electronic devices in general would

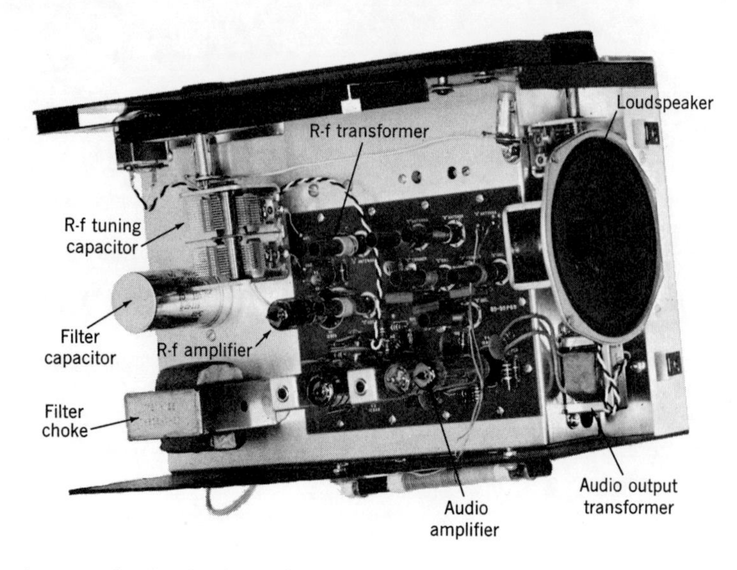


*Fig. 1 · 1 Radio receiver chassis, with r-f and a-f amplifiers. (Heathkit Corp.)*

have only limited application. All the forms of communications that we know could not exist and industrial control would have to rely on mechanical systems for its operation. Because the amplifier is so important, a detailed study will be made of the basic forms of amplifier circuits. First, vacuum-tube amplifiers are explained in this unit; then transistor amplifiers are described in the next unit. With these circuits for a foundation, we can consider the many applications of electronics based on amplifiers using tubes and transistors. Different names are assigned to amplifiers in order to describe a feature of the circuit or to indicate its function. For instance, an audio amplifier amplifies a-c signal voltages in the a-f range. An r-f stage amplifies r-f signal voltages. Because they deal with higher a-c signal frequencies, r-f amplifiers generally use pentodes instead of triodes. With their lower grid-plate capacitance, pentodes can provide more stable r-f gain. Pentodes can also be used for audio amplifiers. Triodes are preferable, however, for either a-f or r-f amplifiers where reduction of tube noise is important.

R-f amplifiers practically always employ tuned circuits. In the control-grid circuit, a tuned circuit can provide maximum r-f signal voltage at the resonant frequency; in the output circuit, parallel resonance allows maximum plate-load impedance at the desired frequency. As a result, the circuits are generally different for a-f and r-f amplifiers because the a-f stage must amplify a broad range of audio frequencies, while the r-f stage amplifies a band of frequencies centered about the resonant frequency of its tuned circuit.

For either r-f or a-f circuits, a stage can be considered a voltage amplifier or a power amplifier. Both amplify the input-signal voltage, but a power amplifier is required to supply appreciable signal current in the output rather than high voltage. A good illustration is the final audio amplifier that supplies signal to drive a loudspeaker. A speaker requires high current. Therefore the audio-output stage is a power amplifier. High values of output current are obtained by using a power tube, which is physically larger than a voltage amplifier tube, and by using a relatively low plate-load impedance. For example, where an a-f voltage amplifier may have a plate-load resistor of 500,000 ohms, an audio-output stage generally has an external plate-load impedance of 4,000 ohms. Different types of amplifiers can be seen in Fig. 1 • 1 showing the chassis of a radio receiver. In this application, r-f and a-f amplifiers are used for amplifying the desired a-c signal, with a power supply for the d-c voltages needed to operate the amplifier stages.

Before proceeding to the analysis of typical amplifier circuits, we can note the common practice of using the term *amplifier* to indicate both a single stage and a complete system with a number of amplifier stages. For instance, when you speak of a high-fidelity system as consisting of an amplifier, a record player, and a loudspeaker, you mean an amplifier system with several stages and probably including a d-c power supply. On the other hand, in this chapter the term amplifier is meant to indicate a single stage, which we can analyze in detail. Surprisingly, this dual meaning does not lead to any misunderstanding, as usually it is self-evident whether we are considering a single stage or an amplifier system.

## 1 • 2 *Amplifier requirements*

The primary function of an amplifier is to receive a small signal at its input and provide a much larger version of it at the output. In the voltage amplifier shown in Fig. 1 • 2, an input signal of 2 volts, peak value, applied to the grid circuit is amplified to provide 80 volts signal in the plate circuit. Therefore the amplitude of the signal voltage has been multiplied by the factor of 40. The same amplification applies to rms (root-mean-

*Fig. 1 • 2 The amplifier increases the amplitude of the 2-volt input signal to provide 80 volts output signal.*

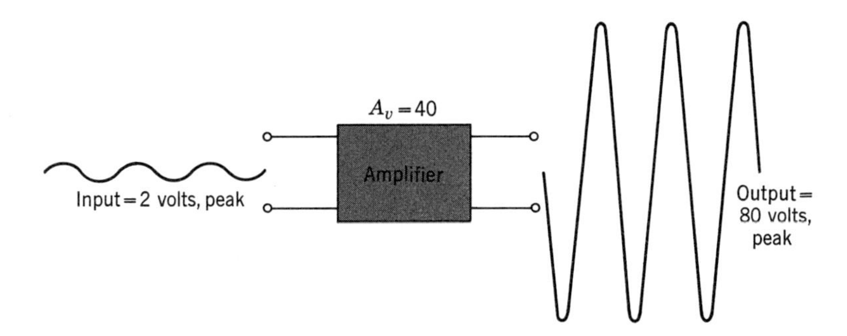

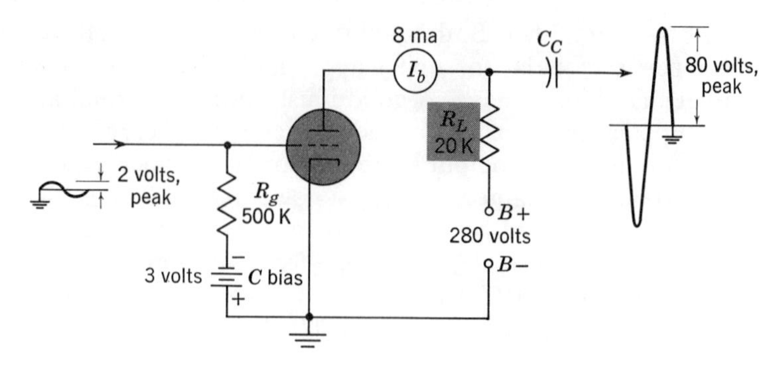


*Fig. 1 · 3 Circuit of a triode amplifier stage.*

square), peak, or peak-to-peak values of the signal. A typical circuit to provide this amplification is shown in Fig. 1 · 3. The principal component of the stage is the vacuum tube itself, which is the amplifier. A triode, a tetrode, or a pentode can be used.

**Plate load.** Resistor $R_L$ is placed in the plate circuit so that the plate current can produce a voltage drop outside the tube. When the voltage across the resistor ($i_bR_L$) varies, the net plate-cathode voltage also varies, since this equals the fixed B+ voltage minus the $i_bR_L$ drop. For this reason, the amplified signal-voltage output from plate to chassis ground has a polarity opposite the grid-signal–voltage input. Capacitor $C_c$ transfers the a-c component of the fluctuating d-c plate voltage, which is the amplified signal-voltage output, to the next stage while blocking the steady d-c component. Note that without this load resistance, the plate current can be varied by the grid voltage, but no amplified output voltage will be obtained.

**B+ voltage.** The B supply is required to produce plate current. The B+ voltage, applied between the plate and cathode, enables plate current to flow from the cathode to the plate, then through $R_L$ and the B supply, and finally back to the cathode again. This current can flow only in one direction. In essence, it is a fluctuating direct current, containing a strong d-c component on which is superimposed a smaller a-c (that is, signal) component. It is this a-c component[1] which produces the signal-voltage drop across $R_L$.

**Grid bias.** A low-voltage battery (commonly called a C battery) is inserted between the control grid and cathode. Its purpose is to bias the control grid negative with respect to the cathode. The bias is negative

[1] Because of the steady d-c level with a-c fluctuations, the following symbols are used to distinguish between the different components of current and voltage: capital $I$ or $E$ for the d-c axis; small $i$ or $e$ for instantaneous values of the variations; subscript $b$ for d-c values or $p$ for a-c values in the plate circuit, and subscript $c$ or $g$ in the control-grid circuit. The plate supply voltage is $E_{bb}$, which is often called simply B+.

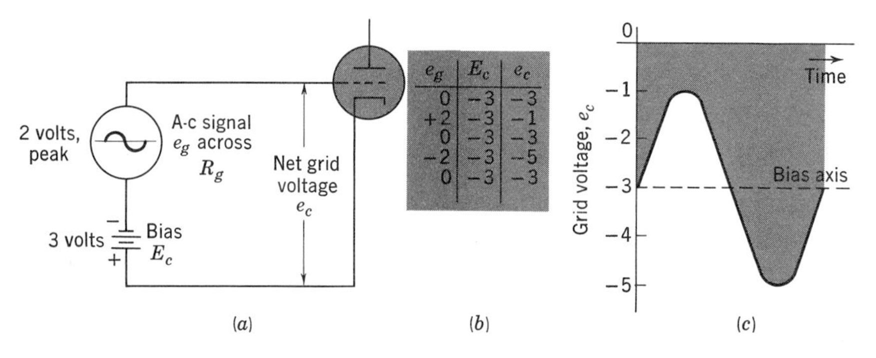


| $e_g$ | $E_c$ | $e_c$ |
|---|---|---|
| 0 | −3 | −3 |
| +2 | −3 | −1 |
| 0 | −3 | −3 |
| −2 | −3 | −5 |
| 0 | −3 | −3 |

*Fig. 1·4 How the bias and signal combine in the grid-cathode circuit. (a) D-c bias $E_c$ in series with a-c signal $e_g$. (b) Tabulation of peak and average values for net grid voltage $e_c$. (c) Graph of $e_c$ values.*

because the minus terminal of the C battery connects to the control grid, while the positive terminal reaches the cathode through the chassis ground. With the control grid biased negative, there is no grid current and no d-c voltage drop across the grid resistor $R_g$.

The bias voltage is large enough so that the input signal cannot make the net grid voltage positive, even during the positive half-cycle of the a-c voltage. As a result, equal variations of the a-c grid-signal voltage produce equal increases and decreases in the plate current. A grid resistor is inserted in the circuit in order to present a load to the incoming signal and to place this signal in series with the d-c grid bias.

**How the control-grid d-c bias and a-c signal combine.** The a-c signal voltage applied to the control grid is developed across the grid load resistor $R_g$. The C bias supply has a negligibly small impedance. Therefore the d-c bias voltage of the C supply and the a-c signal voltage across $R_g$ are in series with each other between grid and cathode (Fig. 1·4*a*). The resultant net grid-cathode voltage $e_c$, then, is the combination of the d-c grid voltage $E_c$ and the a-c grid signal $e_g$.

The bias is negative all the time. When the signal voltage is negative, it adds to the bias; when the signal is positive, it bucks the bias. In Fig. 1·4*b*, the net values of $e_c$ are tabulated for the zero and peak values of the a-c signal. Note that at no time does the combined grid voltage become positive with respect to the cathode.

**Grid drive.** The a-c signal input is called the *grid-drive voltage,* since it makes $e_c$ vary about its bias axis.

**Plate current.** With B+ voltage applied, plate current flows. When the signal drives the net grid voltage less negative, the plate current increases; when it drives the net grid voltage more negative, the plate current decreases. It is important to note, however, that the plate current itself is *not* the output-signal voltage. The plate-load resistor $R_L$ is needed to provide the amplified output-signal voltage.

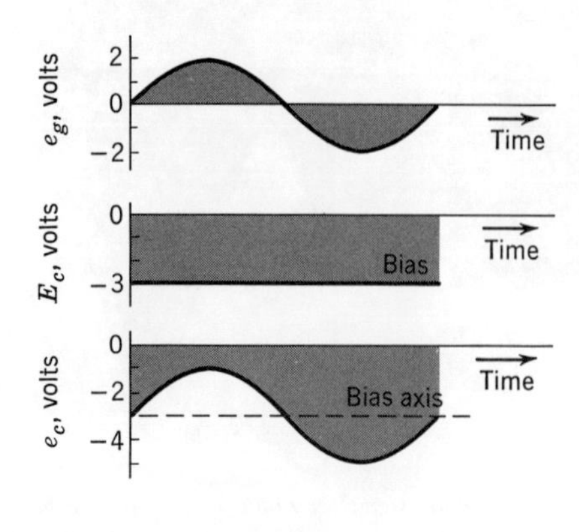


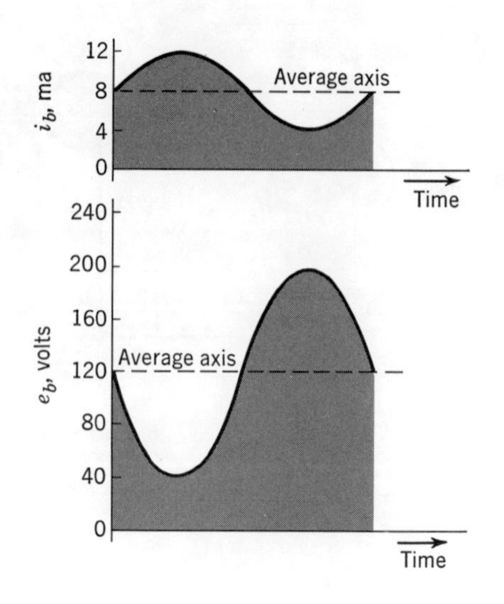


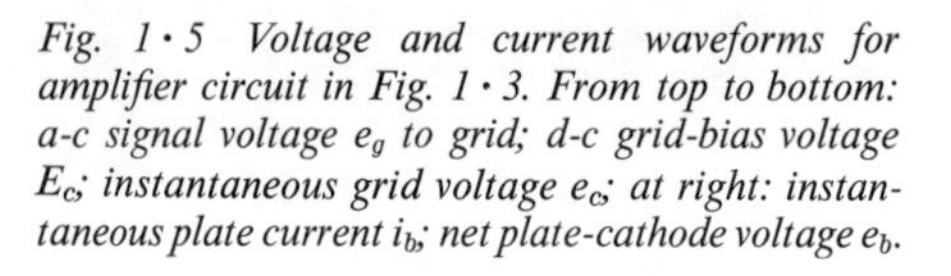

*Fig. 1·5 Voltage and current waveforms for amplifier circuit in Fig. 1·3. From top to bottom: a-c signal voltage $e_g$ to grid; d-c grid-bias voltage $E_c$; instantaneous grid voltage $e_c$; at right: instantaneous plate current $i_b$; net plate-cathode voltage $e_b$.*

## 1·3 Voltage gain

How the amplifier produces an output voltage greater than the input voltage is illustrated by the waveshapes in Fig. 1·5, corresponding to the current and voltage values listed in Table 1·1. These are calculated for the amplifier circuit in Fig. 1·3.

In Table 1·1, the vertical columns list the values of grid-signal voltage $e_g$, d-c bias $E_c$, and the combined grid voltage $e_c$. The corresponding values of plate current are listed also, together with the resultant voltage drop $i_bR_L$ across the 20,000-ohm $R_L$. Finally, the net plate-cathode voltage $e_b$ is given. The $e_b$ value equals the B+ of 280 volts minus the $i_bR_L$ voltage drop. The $i_bR_L$ value is subtracted from B+ because the voltage drop across $R_L$ reduces the plate voltage $e_b$.

The signal voltage $e_g$, given in the first column, has a peak voltage swing of 2 volts, positive and negative, from its zero axis. The d-c bias voltage $E_c$ stays steady at −3 volts. The combined grid voltage then varies between −1 volt and −5 volts.

*Table 1·1 Voltage and current values for amplifier in Fig. 1·3**

| $e_g$, volts | $E_c$, volts | $e_c$, volts | $i_b$, ma | $i_bR_L$, volts | $e_b$, volts $= 280 - i_bR_L$ |
|---|---|---|---|---|---|
| 0 | −3 | −3 | 8 | 160 | 120 |
| +2 | −3 | −1 | 12 | 240 | 40 |
| 0 | −3 | −3 | 8 | 160 | 120 |
| −2 | −3 | −5 | 4 | 80 | 200 |
| 0 | −3 | −3 | 8 | 160 | 120 |

* B+ = 280 volts, $R_L$ = 20,000 ohms.

In the next column are the corresponding values of plate current. The average value is 8 ma, set by the $-3$ volts bias. When the grid voltage varies 2 volts, the plate current changes by 4 ma, in this example. A grid-signal swing from $-3$ volts to $-1$ volt increases $i_b$ from 8 to 12 ma; a grid-voltage change from $-3$ to $-5$ volts lowers $i_b$ from 8 to 4 ma.

The column of voltage-drop values across $R_L$ simply equals the product $i_bR_L$. For the average $i_b$ value of 8 ma, determined by the bias, the voltage drop across the 20,000-ohm $R_L$ is 160 volts. These voltage values vary as $i_b$ is varied by the grid voltage.

Finally, the net plate-cathode voltages in the last column are calculated by subtracting each $i_bR_L$ voltage drop from the fixed B+ value of 280 volts. For the average $i_bR_L$ drop of 160 volts, as determined by the grid bias, the net plate-cathode voltage equals 280 – 160 volts, or 120 volts.

As the a-c signal drives the grid voltage $e_c \pm 2$ volts about the bias axis, the plate current swings $\pm$ 4 ma about the 8-ma axis. The plate-current variations follow the grid signal, increasing when $e_c$ becomes less negative and decreasing when $e_c$ becomes more negative. Because this is a variation of current, however, its amplitude cannot be compared with the input voltage. It is the varying net plate-cathode voltage $e_b$ that is the amplified output-signal voltage. The plate voltage has peak variations of 80 volts from its average axis. From 120 volts, $e_b$ swings down to 40 volts and up to 200 volts, corresponding to a swing of $\pm 80$ volts.

**Gain of the stage.** Compared with the grid-input swing of $\pm 2$ volts, the plate-voltage swing of $\pm 80$ volts is 40 times as great. This is the voltage gain of the stage. Its symbol is $A_v$:

$$A_v = \text{voltage gain} = \frac{\text{signal-voltage output}}{\text{signal-voltage input}} \qquad (1\cdot 1)$$

The output and input voltage amplitudes can be compared in rms, average, peak, or peak-to-peak values as long as the same measure is used for both. In this example,

$$A_v = \frac{\text{output}}{\text{input}} = \frac{\text{80 volts, peak}}{\text{2 volts, peak}} = 40$$

There is no unit for $A_v$, since it is a ratio of two voltages.

It should be noted that the voltage gain $A_v$ is not the same as the amplification factor $\mu$. The gain depends on the circuit, while the $\mu$ is a characteristic of the tube. These two factors are related, however, by the formula

$$\text{Gain} = \mu \frac{R_L}{R_L + r_p} \qquad (1\cdot 2)$$

where $R_L$ is the external plate-load resistance of the circuit, and $r_p$ is the internal plate resistance of the tube. The gain can be equal to $\mu$ or smaller but never greater. For example, with a tube having a $\mu$ of 80 and $r_p$ of 20,000 ohms when it is used in the circuit of Fig. 1·3 with an $R_L$ of 20,000 ohms,

$$A_v = \mu \frac{R_L}{R_L + r_p} = 80\left(\frac{20{,}000}{20{,}000 + 20{,}000}\right) = 80\left(\frac{20{,}000}{40{,}000}\right) = 80 \times \frac{1}{2} = 40$$

The higher the value of $R_L$ compared with $r_p$, the higher is the gain as it approaches the $\mu$ of the tube.

**Phase inversion.** The plate-voltage variations of $e_b$ in Fig. 1·5 have opposite polarity from the grid-voltage signal $e_g$. This phase inversion stems from the fact that the amplified output signal is produced by the varying voltage drop across $R_L$. As the $i_bR_L$ drop becomes greater with more current, the net plate-cathode voltage is smaller; when less plate current produces a smaller $i_bR_L$ voltage, the plate voltage rises. With a sine-wave signal, the amplified output voltage in the plate circuit can be considered 180° out of phase with the input voltage in the grid circuit.

**Fluctuating d-c plate voltage.** The amplifier plate voltage $e_b$ is a fluctuating d-c voltage with a steady d-c average value ($E_b$) and an a-c component ($e_p$) varying about this axis. If you measure from plate to cathode with a d-c voltmeter, it will read the average d-c value of 120 volts for $E_b$. An a-c voltmeter connected across the same two points reads the rms value of the a-c component. This value equals 0.707 × 80, or 56.56 volts here.

All values of plate voltage are positive, but the variations below the average can be considered negative with respect to the axis. When the steady d-c component is blocked by the coupling capacitor $C_c$, these variations below the axis provide the negative half-cycle of the a-c component coupled to the next circuit, while the variations above the axis correspond to the positive half-cycle. The a-c component is the desired output voltage, but there cannot be any signal output unless direct current flows to produce the fluctuating d-c plate voltage.

**Summary of amplifier operation.** The amplification can be considered in the following steps:

1. The grid signal varies the instantaneous values of grid voltage above and below the negative bias voltage axis.
2. The variations in grid voltage produce corresponding changes in plate current.
3. The plate-current variations vary the voltage drop across the plate-load resistor $R_L$.
4. The net plate-cathode voltage is equal to the B+ voltage minus the voltage drop across $R_L$.
5. Therefore the plate-cathode voltage varies. This varying voltage is the amplified output signal.

The output-signal voltage is an amplified duplicate of the grid-signal voltage, but with opposite polarity. For the one cycle illustrated in Fig. 1·5, every variation in grid voltage, whether at the peak or intermediate values, has a corresponding variation in plate voltage. The same variations are repeated every cycle. Therefore the variations in plate voltage provide an output signal with the same frequency as the input signal.

## *1 · 4 Methods of coupling*

This refers to the method used for connecting the desired output signal from one amplifier to the input circuit of the next stage. Where the desired output is an a-c signal variation, the problems in coupling include:

1. How to provide the best plate-load impedance for the a-c signal output.
2. How to supply d-c plate voltage for operation of the amplifier.
3. How to couple the a-c signal without too much loss of the desired frequency components.
4. How to block the d-c component of plate voltage from the grid circuit of the next stage, so that it can operate with its own grid-bias voltage.

How these problems are solved can be seen from the coupling circuits shown in Figs. 1 · 6 to 1 · 11.

**RC coupling.** This method is generally used with a-f voltage amplifiers because it provides the same amount of gain for a wide range of audio frequencies. In Fig. 1 · 6*a*, $R_L$ is the plate load in series with the plate and the B+ supply. Since $R_L$ presents the same resistance to all audio frequencies, the *RC* amplifier provides equal gain for different frequencies. This is called *uniform,* or *flat,* frequency response. As indicated in Fig. 1 · 6*b*, this a-f amplifier has a flat response from 100 to 5,000 cps.

The amplifier response is limited at low frequencies by coupling capacitor $C_{c_2}$. The reason why is the fact that capacitive reactance rises as the signal frequency drops. As a result, more of the output-signal voltage appears across the coupling capacitor and less across the following grid resistor $R_{g_2}$. At the opposite end of the frequency range, the amplifier

*Fig. 1 · 6 RC-coupled amplifier. (a) Circuit for a typical audio stage. $R_L$ is the resistance plate load and $C_{c_2}$ the output coupling capacitor. (b) Response curve showing gain of amplifier at different audio frequencies.*

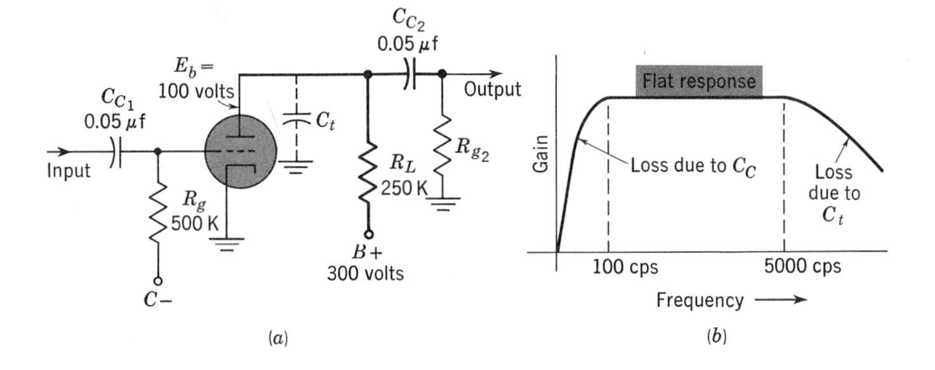

response decreases because of stray distributed capacitance in the plate circuit. This capacitance, labeled $C_t$ in Fig. 1·6*a*, is the total capacitance shunted across the plate-cathode circuit. Typical values for $C_t$ are 20 to 40 pf. At high frequencies, however, even this small amount of capacitance can effectively bypass signals around a high-valued $R_L$. As a matter of fact, when a circuit designer wishes to extend the frequency range of an amplifier, one of the first things he does is lower $R_L$. This makes the shunt capacitance $C_t$ less effective until the frequency rises to a higher value.

One disadvantage of the *RC* amplifier is the fact that the $I_bR_L$ voltage drop reduces the average plate voltage available for the tube. The amount of plate current that flows is determined by the voltage actually at the plate, with respect to cathode. In an *RC* amplifier, this average plate-cathode voltage is much less than the B-supply voltage. For example, if the average plate current in Fig. 1·6*a* is 0.8 ma with a 250,000-ohm $R_L$, the average $I_bR_L$ voltage drop equals 200 volts. The average plate-cathode voltage is then only 100 volts, equal to 300 volts minus the 200-volt $I_bR_L$ drop. Thus, $R_L$ cannot be made too high. Note the low wattage rating required for $R_L$. With an average voltage drop of 200 volts and an average current of 0.8 ma, its *EI* product equals 0.16 watt dissipated in heat. Typically, a ⅓- to 1-watt carbon resistor would be used here.

The function of the coupling capacitor is illustrated in Fig. 1·7, where *C* corresponds to $C_c$, and *R* is equivalent to $R_g$ in the *RC* amplifier. The voltage applied to this *RC* circuit is the fluctuating d-c plate voltage $e_b$. In

*Fig. 1·7 How an RC circuit couples the a-c component of $e_b$ across R, while the d-c component is blocked by C.*

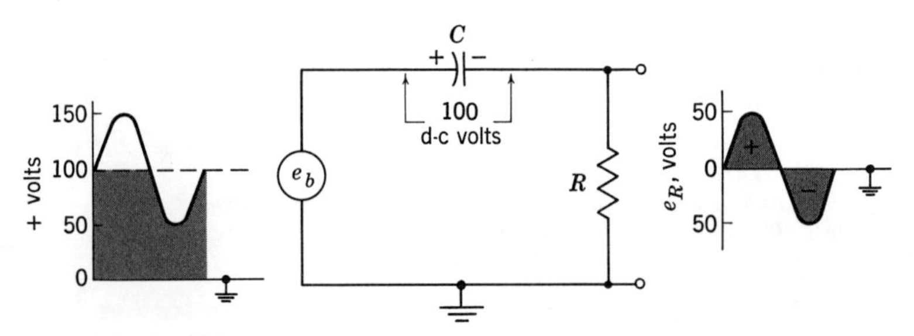


*Fig. 1·8 Impedance-coupled amplifier with audio choke as plate-load impedance. An r-f choke can be used instead for an r-f amplifier.*

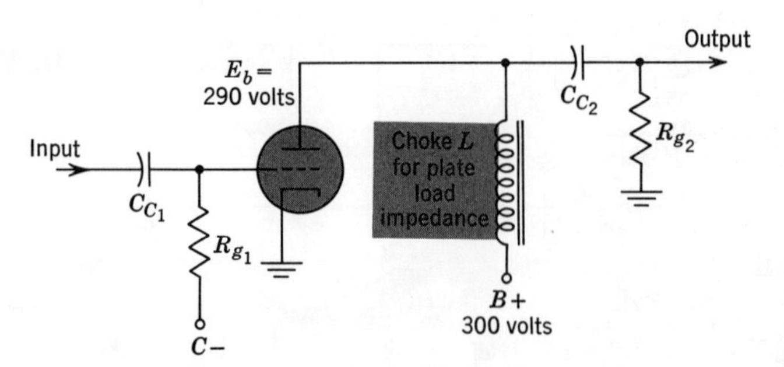

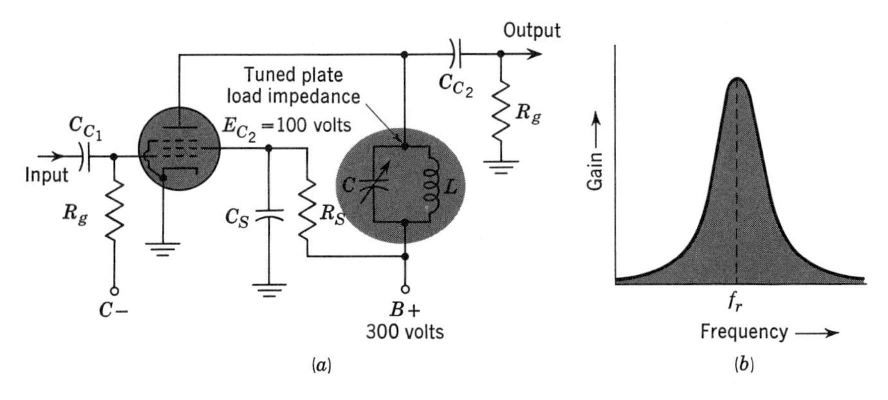


*Fig. 1·9 Single-tuned r-f amplifier. (a) Circuit. (b) Frequency response.*

this example, $e_b$ has an average value of 100 volts, varying from 150 to 50 volts as the signal varies.

Since the charge or discharge path is the same for the $RC$ circuit, $C$ soon charges to the average value of the applied voltage, which equals 100 volts in this example. With $C$ charged to 100 volts, when the applied voltage $e_b$ increases, it produces a charging current that develops a positive voltage across $R$; when $e_b$ decreases below 100 volts, $C$ discharges, and its discharge current produces a negative voltage across $R$. As a result, as $e_b$ varies from 50 volts above to 50 volts below the 100-volt axis, the increases produce the positive half-cycle of voltage $e_R$, while the decreases produce the negative half-cycle. The voltage across $R$ is an a-c voltage varying above and below zero, which is the chassis ground reference. The average d-c level of the fluctuating d-c input is the 100 volts across $C$. Therefore the steady d-c component is blocked as the voltage across $C$, while the a-c component is passed as the a-c voltage across $R$. This voltage is the amplified a-c signal to the control grid of the next stage.

**Impedance coupling.** The circuit shown in Fig. 1·8 uses an inductance as a choke for the plate-load impedance, instead of $R_L$ in the $RC$ amplifier. The coupling capacitor $C_c$ is still needed, however, to block the steady d-c component of plate voltage. Either an iron-core choke can be used for an audio amplifier or an air-core choke for an r-f amplifier. The advantage of a choke for the plate load is that it has low d-c resistance but can provide the high value of an a-c impedance required for high voltage gain.

**Single-tuned stage.** The r-f amplifier in Fig. 1·9 uses a single parallel-resonant circuit to provide the required plate-load impedance for the a-c signal. The d-c resistance of the r-f coil $L$ is negligible, and the average d-c plate voltage is practically equal to B+. However, the parallel $LC$ circuit provides a high impedance for r-f signal at the resonant frequency.

The response curve in $b$ is the same as the resonance curve of the tuned circuit. For parallel resonance, the impedance is maximum at $f_r$, decreasing to very low values at frequencies far off resonance. Since the gain of the

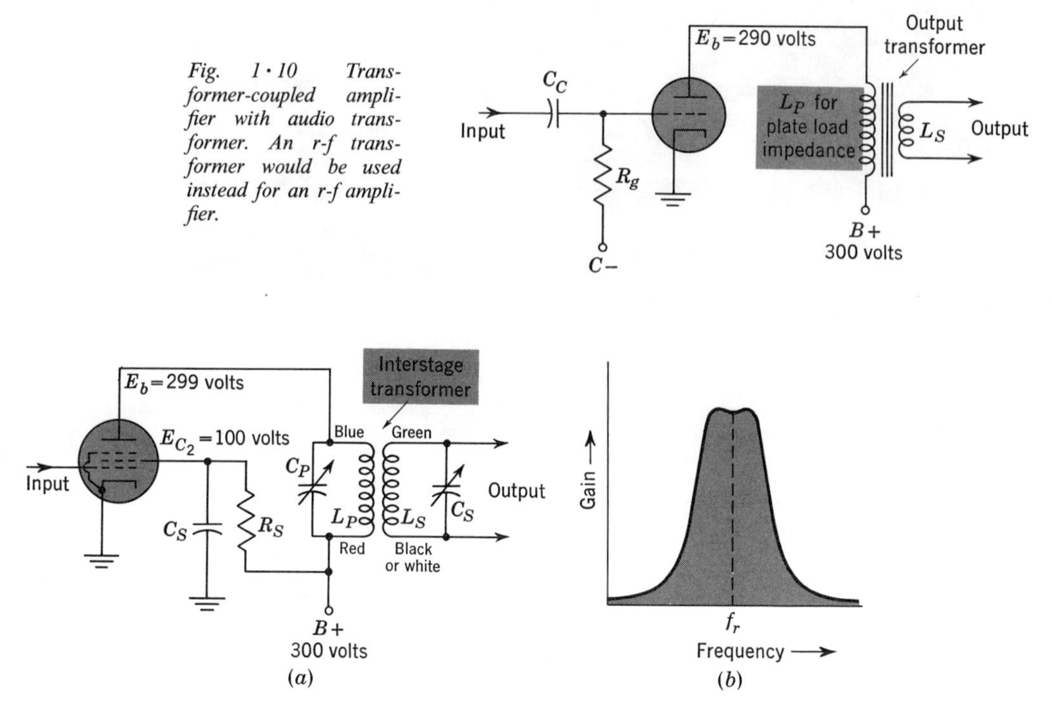


*Fig. 1·10 Transformer-coupled amplifier with audio transformer. An r-f transformer would be used instead for an r-f amplifier.*

*Fig. 1·11 Double-tuned r-f amplifier. (a) Circuit. (b) Frequency response.*

amplifier is proportional to the amount of plate-load impedance, the response of the amplifier for different frequencies corresponds to the response of the parallel-resonant circuit. That is, the tuned amplifier can provide gain only for frequencies at and near the resonant frequency. The bandwidth of the resonant response depends on the $Q$ of the $LC$ circuit.

Pentodes are generally used for r-f amplifiers. The only additional need of the pentode circuit is a positive d-c screen-grid voltage. In Fig. 1·9, $R_s$ is the screen-dropping resistor. The screen voltage is equal to the B+ voltage minus the $I_sR_s$ voltage drop, where $I_s$ is the screen-grid current. Assuming 4-ma screen current, the dropping resistor must be 50,000 ohms for a 200-volt $I_sR_s$ drop, with 100 volts left for the screen grid. $C_s$ is the screen-bypass capacitor. Its reactance must be low enough to bypass $R_s$ for the lowest frequency amplified in the stage.

**Transformer coupling.** Figure 1·10 shows a typical audio power-output-stage transformer coupled to a loudspeaker. The primary winding $L_p$ is the plate-load impedance. This arrangement has the advantage of low d-c resistance, resulting in a high value of $E_b$, which is important for maximum power output. The a-c impedance of the primary is usually 2,000 to 5,000 ohms, this value being high enough because the requirement is power output rather than voltage gain. The a-c component of the signal current in the primary $L_p$ induces the desired signal voltage in the secondary winding $L_s$ by transformer action. Since the secondary is isolated, the steady d-c component of the primary current is blocked. Therefore the secondary has a-c signal only.

**Double-tuned stage.** The circuit in Fig. 1·11*a* is a transformer-coupled amplifier, but both the primary and secondary of the r-f transformer are tuned. Because of the isolated secondary winding, no blocking capacitor is needed. The primary-tuned circuit $L_pC_p$ is the plate-load impedance for the pentode, as in the single-tuned stage. However, the tuned secondary affects the primary impedance by mutual coupling, generally causing the response of the double-tuned stage to have a greater bandwidth than a single-tuned circuit. This is demonstrated by the response curve in Fig. 1·11*b*. As in the single-tuned stage, only frequencies at or close to resonance can be amplified, since there is very little plate-load impedance for other frequencies.

*Table 1·2 Comparison of coupling methods*

| *Type of coupling* | *Plate load* | *D-c blocking by* | *Frequency response* | *D-c plate voltage* | *Applications* |
|---|---|---|---|---|---|
| *RC* | $R_L$ | $C_c$ | Flat for wide band of audio frequencies | Low because of $I_bR_L$ drop | Audio and wide-band voltage amplifiers |
| Impedance | $Z_L$ of choke | $C_c$ | Varies with $Z$ of choke | High because of low d-c drop across choke | A-f or r-f voltage amplifiers |
| Transformer | $Z_L$ of primary | Isolated secondary | Varies with $Z$ of transformer | High because of low d-c drop across primary | Audio power amplifiers |
| Single-tuned | $Z_L$ of resonant circuit at $f_r$ | $C_c$ | Same as resonance curve of single-tuned circuit | Practically equal to B+ | R-f amplifiers |
| Double-tuned | $Z_L$ of primary resonant circuit at $f_r$ | Isolated secondary | Same as resonance curve of double-tuned circuit | Practically equal to B+ | R-f amplifiers |

The color coding of the transformer leads is shown in Fig. 1 · 11 because it can be helpful in circuit tracing. The coding is for an *interstage transformer,* that is, a transformer that couples the signal from the plate of one stage to the control grid of the next. All interstage transformers, whether for a-f or for r-f amplifiers, are usually coded with a blue plate lead and a red lead for the return to B+ in the primary; in the secondary, the green wire connects to the grid of the next stage, while its return lead is coded either black or white.

**Summary of coupling circuits.** The chief characteristics of the different coupling methods are compared in Table 1 · 2. Only the *RC* amplifier has d-c plate voltage much lower than the B-supply voltage. In spite of this, *RC* coupling is generally used for audio voltage amplifiers because it is compact, economical, and can provide a good frequency response. Transformer coupling is almost always required for a power-output stage. For r-f amplifiers, tuned stages are used. The single-tuned stage generally has variable tuning, while the double-tuned stage is fixed-tuned. Typical components for these coupling circuits are shown in Fig. 1 · 12.

## *1 · 5 Classes of operation*

Another factor which governs the operation of amplifiers is the relative amount of d-c bias and a-c signal on the grid. On this basis, amplifiers can be classified class A, class B, or class C. When plate current flows for the full cycle of 360° of the input signal, the amplifier is operated class A. In class B operation, plate current flows for approximately 180°, or one-half the input cycle. Some applications use class AB operation, which is between class A and class B. For class C operation, plate current flows for appreciably less than 180°.

The applications for the different classes of operation are related to the permissible amount of distortion and the efficiency requirements. Distortion means that the plate current does not have exactly the same variations as the grid signal. Efficiency here is the ability to produce maximum a-c signal-power output with minimum plate dissipation of the d-c power from the B supply. Class A operation has the least distortion but is the most inefficient. At the opposite extreme, class C is most efficient but has the greatest distortion of plate current. Class B operation is between classes A and C in distortion and efficiency.

**Class A operation.** In class A operation, the grid is usually biased to about one-half the cutoff voltage or less. In addition, the peak amplitude of the a-c signal input must not drive the instantaneous grid voltage more negative than cutoff. Plate current will then flow for the full 360° of each cycle of the signal input. These requirements can be applied to the three examples of class A operation illustrated in Fig. 1 · 13. The example in *a* is for small signal amplitudes. Under this condition, the bias can be close to zero in order to utilize only the linear section of the transfer characteristic ($e_c$-$i_b$) curve. In *b*, the bias is made more negative to accommodate a stronger input signal. More grid drive allows more a-c signal output, but the dis-

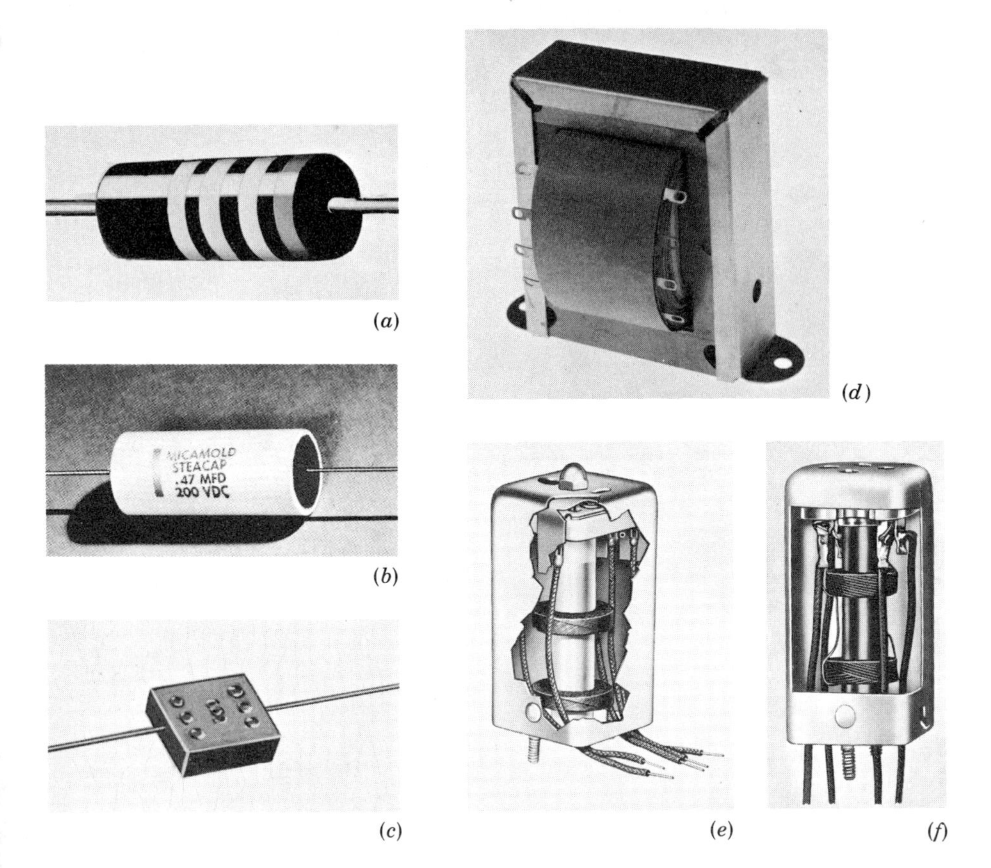


*Fig. 1·12 Typical components for different coupling methods. (a) Carbon-composition resistor for $R_L$. (b) 0.05-μf audio coupling capacitor. (c) 250-pf r-f coupling capacitor. (d) Audio-output transformer. (e) and (f) Double-tuned transformers for 455 kc.*

tortion increases. In *c*, the bias is the same, but even more grid signal is applied. Plate current still flows, however, for 360° of the input cycle.

Note the grid-voltage values in Fig. 1·13*c*. The peak of the negative half-cycle drives the grid voltage to −7 volts, which is just under the grid-cutoff voltage of −8 volts, so that plate current can flow. The peak of the positive half-cycle of the signal drives the grid 1 volt positive. Grid current flows when the control grid is driven positive with respect to the cathode, as indicated by the shaded portion of the grid signal. Plate current still flows, however, for positive control-grid voltage. In fact, this condition results in a very high plate current, approaching saturation with maximum plate current. Even with the grid driven positive, plate current still flows for the full 360° of the signal in *c*. This type of operation is indicated as class $A_2$, the subscript 2 indicating grid current. Class $A_1$ means there is no grid current. When no subscript is indicated, class $A_1$ is assumed.

Class A operation introduces the least distortion because every grid-voltage variation produces a corresponding plate-current variation. Average

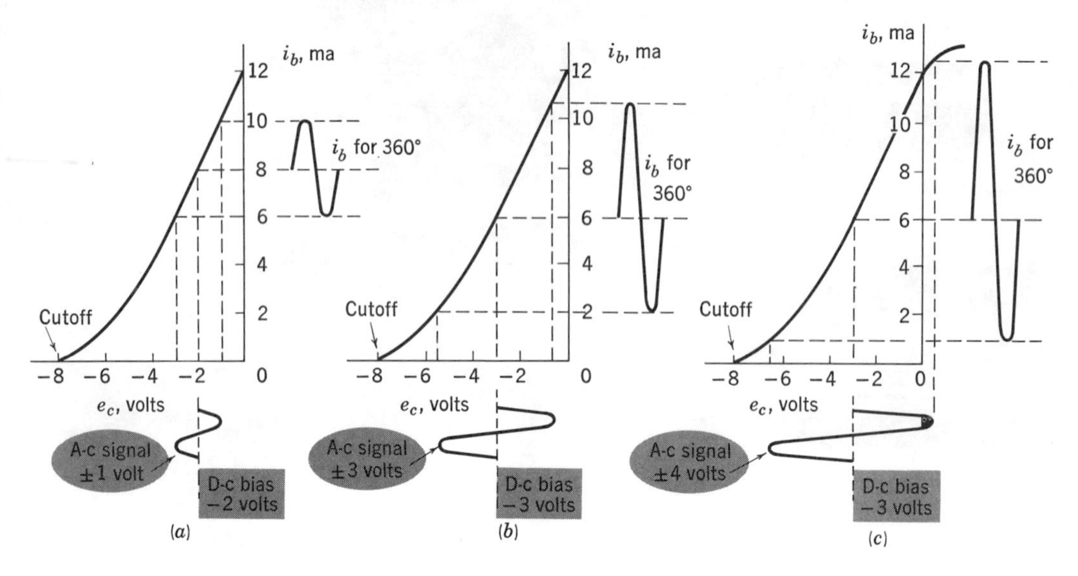


*Fig. 1 · 13 Class A operation with plate current for the full cycle of input signal. (a) Small-signal operation with bias close to zero. (b) Greater bias and drive but no grid current. (c) Class $A_2$ with grid current.*

plate current remains the same, with or without signal. In Fig. 1 · 13*a*, a d-c millimeter in the plate circuit would read an average plate current of 8 ma, corresponding to the bias of −2 volts, with or without grid drive if the operation were perfectly linear. For the example in Fig. 1 · 13*b*, the meter would read 6 ma, instead of 8 ma, because the bias has been increased from −2 to −3 volts. In Fig. 1 · 13*c*, the bias is still −3 volts, and the corresponding plate current is approximately the same 6 ma.

**Class B operation.** To operate the same tube class B, the negative bias is increased to the grid-cutoff value so that more input signal voltage can be applied (see Fig. 1 · 14). With the bias at cutoff, plate current flows only for the positive half-cycles of the a-c grid signal. This half-cycle is shaded to indicate the grid voltages that produce corresponding values of plate current. The negative half-cycle is cut off, however. Nothing happens during this negative half of the a-c input signal; it just does not affect the plate current. On the grid, the full cycle of signal voltage is present. In the plate circuit, though, only half-cycles of plate current flow. In effect, the class B operation corresponds to a half-wave rectifier. For audio amplifiers, this distortion would result in garbled sound. Class B audio amplifiers, however, use two tubes so that each supplies opposite half-cycles of the input signal. This dual arrangement is called a *push-pull* amplifier.

The advantage of class B operation is that more a-c signal output can be produced with less average plate current, compared with class A. This improved efficiency results from using more negative bias. Note in Fig. 1 · 14 that, without any signal input, the average direct plate current is zero.

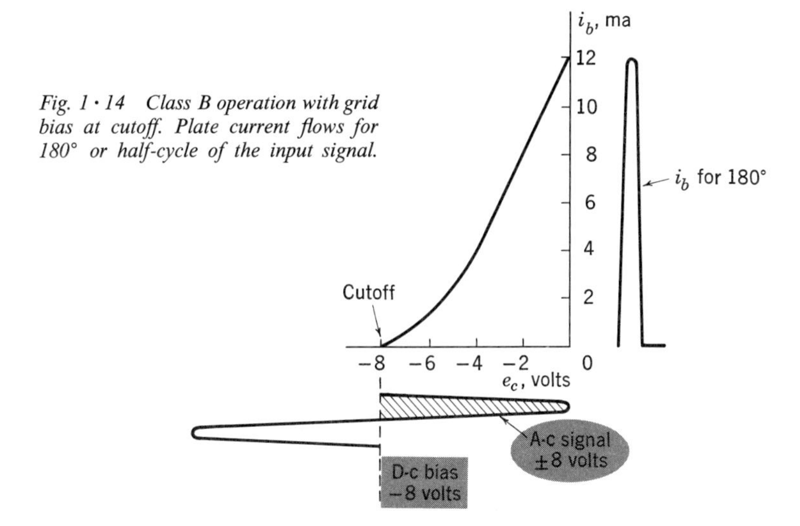


*Fig. 1·14 Class B operation with grid bias at cutoff. Plate current flows for 180° or half-cycle of the input signal.*

With signal, the average plate current increases to 7.6 ma, approximately. More a-c power output can be obtained because the alternating plate current swings between the extreme values of zero and the peak of 12 ma.

**Class C operation.** For class C operation, the bias is made more negative than cutoff, and more grid-voltage drive is applied (see Fig. 1·15). The result is plate current for less than one half-cycle. Typical class C operation is shown in Fig. 1·15; the bias is twice the grid-cutoff voltage and plate current flows for one-third the cycle, or 120°. This completely cuts off not only the full negative half-cycle of grid voltage, but part of the positive half-cycle as well. For the values shown here, the positive grid drive must be more than 8 volts to cancel enough of the 16-volt negative bias to reduce the grid voltage below cutoff so that plate current can flow.

The positive values of grid drive for one-third the signal cycle are shaded in Fig. 1·15 to indicate that they produce corresponding values of plate current. The plate current flows in 120° pulses. Starting at 0° in the

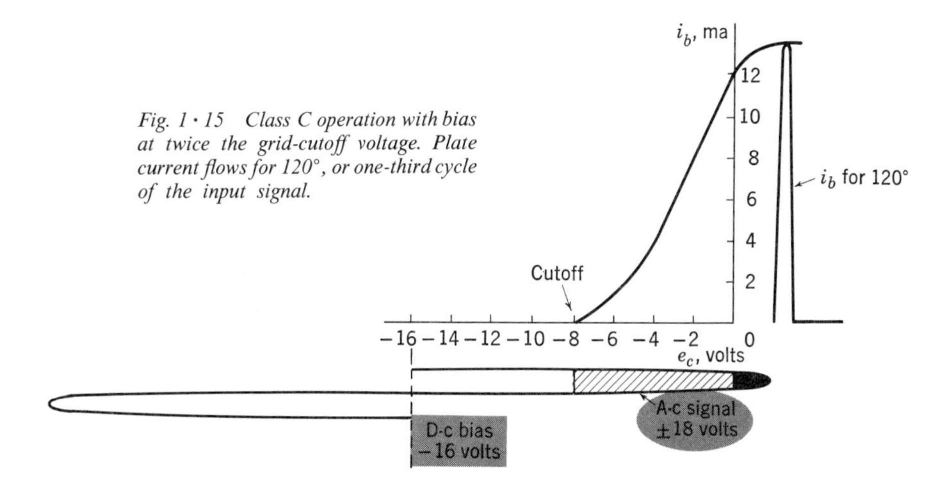


*Fig. 1·15 Class C operation with bias at twice the grid-cutoff voltage. Plate current flows for 120°, or one-third cycle of the input signal.*

full cycle of grid voltage, there is no current because the bias is past cutoff. After 30°, when the signal reaches one-half its peak positive amplitude, plate current starts to flow. During the next 120°, the plate current rises to a maximum and then drops to zero as the signal voltage drives the grid positive and then beyond cutoff. Therefore each full cycle of input signal produces a pulse of plate current for only one-third of the cycle.

Class C operation is generally used for r-f power amplifiers. With the high bias, a large amplitude of grid-signal drive can be used. Usually there is enough drive to produce grid current, as indicated by the shaded tip of the positive drive in Fig. 1 · 15. Class C operation provides maximum a-c power output with minimum average plate current.

The distortion is greatest for class C operation because the plate signal corresponds to only one-third the input cycle. For stages amplifying an r-f signal of constant amplitude, however, this distortion is no disadvantage. With an *LC* resonant circuit in the plate tuned to the signal frequency, the tuned circuit provides a full sine-wave cycle of output voltage for each pulse of plate current. Therefore, full cycles of amplified r-f output voltage result, with the same frequency as the grid signal.

## *1 · 6 Methods of bias*

In general, amplifier grid bias is a steady d-c voltage that makes the control grid negative with respect to the cathode so that the average grid voltage is negative. The a-c input signal then drives the instantaneous values of grid voltage above and below the bias value as an axis. As a result, the bias determines what portion of the grid-plate transfer characteristic will be effective in amplifying the grid signal. The amount of bias may range from −1 volt or less for class A operation with small signal input to −100 volts or more for class C power amplifiers. Specifically, the three main methods of supplying the required negative bias voltage are: (1) fixed bias from a battery or power supply, (2) cathode bias or self-bias from the d-c voltage produced by tube current flowing through a cathode resistor, and (3) grid-leak bias or signal bias from grid current

*Fig. 1 · 16 Fixed bias. (a) With C bias cell. (b) C voltage from a divider in the power supply.*

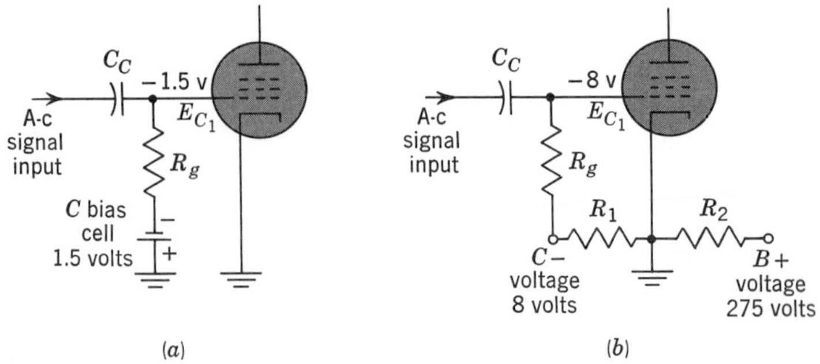

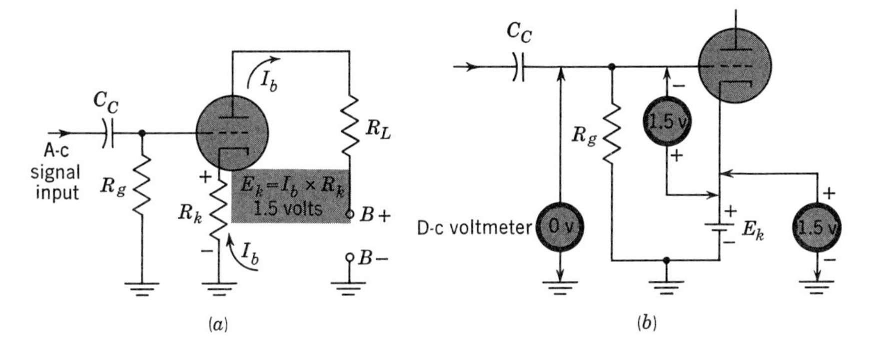


*Fig. 1 · 17 Cathode bias. (a) Actual circuit with d-c voltage drop across cathode-bias resistor $R_k$. (b) Equivalent arrangement of $E_k$ showing how it inserts a d-c bias voltage between grid and cathode.*

with an $R_gC_c$ coupling circuit. Any of the bias methods can be used for triodes or pentodes. These methods of bias are shown in Figs. 1 · 16 to 1 · 21.

## *1 · 7 Fixed bias*

As shown in Fig. 1 · 16*a*, fixed bias can be obtained with a dry cell. The bias cell can be extremely small in size, since it supplies practically no current. The negative side of the C bias voltage must have a d-c path to the control grid, while the positive side returns to cathode through the common-chassis connection. $R_g$ is used to prevent shorting the grid a-c input signal to chassis and cathode. It does not affect the bias voltage, however, because there is no grid current. $C_c$ couples the signal from the plate circuit of the preceding stage while blocking its d-c component; only the dry cell provides the required d-c bias. If you measure the voltage $E_c$ between control grid and cathode or chassis with a d-c voltmeter, it will indicate 1.5 volts, with the grid negative. This voltage is considered fixed because it is 1.5 volts with or without an a-c input signal and whether or not the tube is conducting plate current.

Another method of obtaining fixed bias is shown in Fig. 1 · 16*b*. Here a voltage divider in the power supply has a tap that supplies negative voltage with respect to chassis ground. Connecting the bottom end of $R_g$ to this negative tap brings the required fixed bias voltage to the control grid. The C voltage makes the grid negative as the cathode returns to chassis ground.

## *1 · 8 Cathode bias*

This method is used very often because the tube supplies its own bias. In Fig. 1 · 17*a*, cathode bias is obtained by inserting a resistor $R_k$ in the cathode-to-ground circuit, so that the returning cathode current produces an *IR* voltage drop. For a triode without grid current, the average cathode current equals the average plate current $I_b$. Since current can flow in only one direction, the $I_bR_k$ drop is a d-c voltage. Although the plate current varies, it has an average value that sets the bias. In addition, the average $I_b$ is the same with or without signal for class A amplifiers. The polarity of the voltage drop always makes the cathode side of $R_k$ more positive than its opposite end. The cathode side must be more positive because the grounded side is connected to B−, which is the most negative point in the circuit.

**The cathode voltage is negative grid bias.** The cathode voltage $E_k$ makes the cathode 1.5 volts more positive than the chassis. Since the grid returns to chassis through $R_g$, the cathode is also 1.5 volts more positive than the grid. It is assumed that no grid current flows, so that the voltage across $R_g$ equals zero. With the cathode 1.5 volts positive compared with the grid, the control grid is 1.5 volts more negative than the cathode. In terms of the a-c input voltage to the grid, the signal swings the grid voltage above and below the negative 1.5-volt bias axis as though this were negative bias in the grid-ground circuit rather than positive voltage in the cathode-ground circuit. The results are the same for either case. The equivalent arrangement is illustrated in Fig. 1 · 17*b*.

**Calculating the cathode bias.** The amount of cathode bias depends on the size of the cathode resistor $R_k$ and the amount of cathode current:

$$E_k = I_k R_k \qquad (1 \cdot 3)$$

or

$$R_k = \frac{E_k}{I_k} \qquad (1 \cdot 4)$$

For a triode, $I_k$ equals the average plate current $I_b$. For a pentode, $I_k$ is $I_b$ plus the screen-grid current $I_s$. If control-grid current flows, it must be added also. These formulas make it possible to calculate the amount of cathode bias for a given case or to determine the cathode resistance needed for the desired amount of bias. Typical examples are illustrated in Fig. 1 · 18. The values used for plate current and screen current of the tube are

*Fig. 1 · 18 Calculations for cathode bias. (a) Triode 3-ma $I_b$ is cathode current. (b) Pentode plate and screen currents are added for total cathode current of 40 ma.*

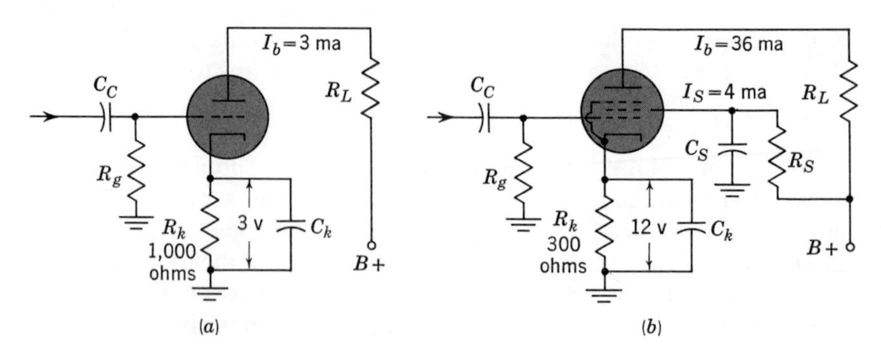

obtained from the operating characteristics given in the tube manual. Note that a normal voltmeter reading of the cathode d-c bias voltage indicates normal plate and screen currents.

**Cathode-bypass capacitor** $C_k$. With signal input and varying plate current, the voltage across $R_k$ is a fluctuating d-c voltage whose average value serves as the d-c bias. However, the a-c component of the cathode voltage affects the grid input signal. This effect is called *degeneration, negative feedback,* or *inverse feedback,* since the variations in cathode voltage cancel part of the a-c input signal. The result is less gain in the amplifier, although distortion is reduced also. The cathode-bias resistor is generally bypassed, as illustrated by $C_k$ in Figs. 1·18*a* and *b*, so that the cathode voltage will be a steady d-c bias. The presence of $C_k$ prevents degeneration in the grid-cathode circuit, and maximum gain is achieved. The required bypass capacitance can be calculated for a reactance one-tenth the value of $R_k$, or less, at the lowest signal frequency amplified by the stage.

**Applications of cathode bias.** Class A amplifiers generally use cathode bias, since plate current flows for the complete cycle. In addition, cathode bias is used as safety bias in power amplifiers. Should any trouble develop that would tend to produce excessive plate current, the bias would increase similarly and limit the current to a safe value.

## *1·9 Grid-leak bias*

This form of bias is called *signal bias* because it is obtained by rectifying the a-c signal input. Effectively, the grid-cathode circuit serves as a diode rectifier that charges the coupling capacitor to a d-c voltage proportional to the peak positive signal amplitude. There are two requirements. First, the signal must make the control grid positive with respect to cathode, so that grid current can flow. When there is no other bias, this condition is automatically satisfied, since every a-c signal has a positive half-cycle. The second requirement is an $R_gC_c$ coupling circuit with a time constant which is long compared with one cycle of the a-c input signal. This condition

*Fig. 1·19 How grid-leak bias is produced. (a) $C_c$ is charged by grid current when signal drives grid positive. (b) $E_c$ is d-c bias source in series with a-c input voltage.*

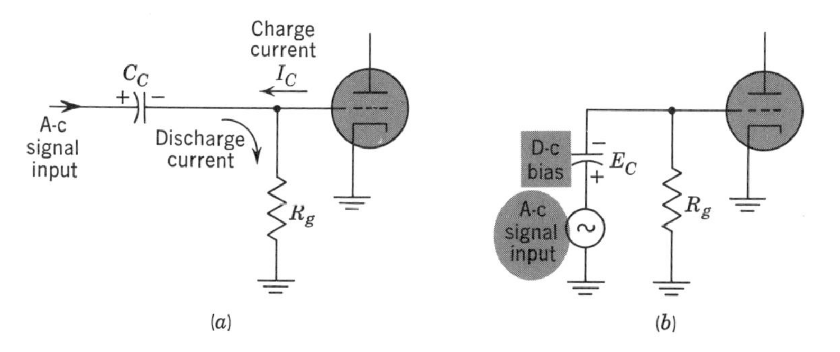

allows $C_c$ to remain charged at a steady d-c value. For a sine-wave signal, it means that $C_c$ must have low reactance compared with $R_g$.

Grid-leak bias can be produced in any amplifier with an $R_gC_c$ coupling circuit, where the signal drives the grid positive over part of the signal cycle.

**How $C_c$ produces grid-leak bias.** Figure 1 · 19*a* illustrates how grid-leak bias develops a steady d-c voltage across $C_c$ when grid current flows. When the grid is driven positive by the signal voltage, electrons flow from the cathode to the control grid in the tube and then accumulate on the grid side of $C_c$, repelling electrons from the other side of $C_c$. During the time when the grid is not positive, $C_c$ can discharge through $R_g$ to the cathode. This is the only possible path for the discharge current, since electrons cannot travel from grid to cathode in the tube. Now it is important to note that $C_c$ charges quickly when electrons travel from cathode to grid, but it discharges very slowly through the high resistance of $R_g$. After a few cycles of signal, $C_c$ has a net charge with its grid side negative. The resultant bias $E_c$ is a d-c voltage because it can have only one polarity. It can be considered a steady bias compared with the signal variations, since $E_c$ varies very little during one cycle of the a-c input.

The grid-leak bias action does not prevent $R_gC_c$ from acting as a coupling circuit for the a-c signal. This symmetrical charge and discharge through the same resistance of $R_g$, however, couples equal positive and negative half-cycles of voltage to the grid circuit. With grid-leak bias rectification, however, the charge and discharge paths are not equal. The grid current stores more charge in $C_c$ than is lost by discharge through $R_g$. This provides a net negative voltage across $C_c$, which serves as the bias. Therefore the a-c coupling and the d-c bias action due to grid current can be considered superimposed to provide the net result that the a-c signal varies the instantaneous grid voltage around the d-c bias voltage as an axis (Fig. 1 · 19*b*).

**The amount of grid-leak bias depends on signal amplitude.** The positive drive of the signal produces enough grid current to charge $C_c$ to a negative d-c voltage approximately 90 per cent of the peak positive amplitude. In effect, the negative grid bias makes the instantaneous grid voltage back off, so that only the peak of the signal can drive the grid positive, as illustrated in Fig. 1 · 20 for different amounts of signal voltage.

Starting at the top, there is no a-c signal input. No signal means no bias, since grid current cannot flow to charge $C_c$. When the a-c signal input has a peak amplitude of 1 volt, grid current flows. The resultant charging of $C_c$ produces some negative bias. After 10 to 20 cycles of signal, the negative bias voltage across $C_c$ stabilizes at a value of approximately 0.9 volt. The exact bias voltage is governed by the $R_gC_c$ time constant, but 90 per cent of the positive peak is a typical value. When the bias is less than this, enough grid current flows to increase the bias. If the bias should become too negative, no grid current could flow. Then $C_c$ would discharge through $R_g$, reducing the bias voltage. The bias voltage settles at a value where the grid current that flows at the positive peak of the signal

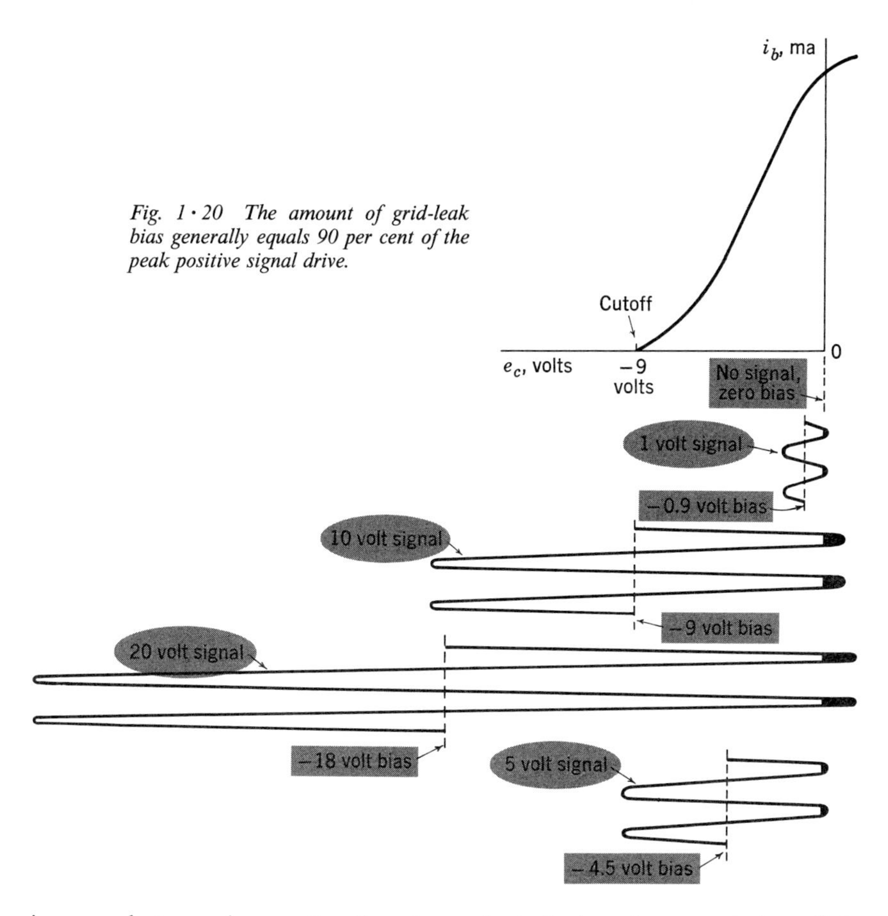


*Fig. 1 · 20 The amount of grid-leak bias generally equals 90 per cent of the peak positive signal drive.*

is enough to make up for the charge lost during the remainder of the cycle when $C_c$ discharges through $R_g$.

Once the bias has been established, grid current flows for only a small part of the signal cycle at the positive peak. Plate current can flow, however, either for the entire cycle or for part of the cycle, as determined by the grid-cutoff voltage of the tube. Therefore, the amplifier can operate class A, B, or C with grid-leak bias. These three classes of operation are illustrated in Fig. 1 · 20 for 1-volt, 10-volt, and 20-volt signals. The plate current may be slightly distorted for the small portion of the cycle when grid current flows, but this distortion is not serious for many applications.

**Applications of grid-leak bias.** Grid-leak bias can be used for class A, B, or C operation, voltage amplifiers and power amplifiers. When grid-leak bias is used with an *RC*-coupled amplifier, $R_gC_c$ provides the bias if the grid is driven positive. With transformer coupling, the *RC* grid-leak bias combination must be inserted, however, as shown in Fig. 1 · 21.

In *a*, $R_g$ and $C_c$ are in series, as they are in a coupling circuit, but here they are inserted just for grid-leak bias, since the transformer provides the required coupling from the previous stage.

In *b*, a parallel $R_gC_c$ combination is used. The parallel bank is in series with the grid side of the secondary, which is tuned for an r-f amplifier.

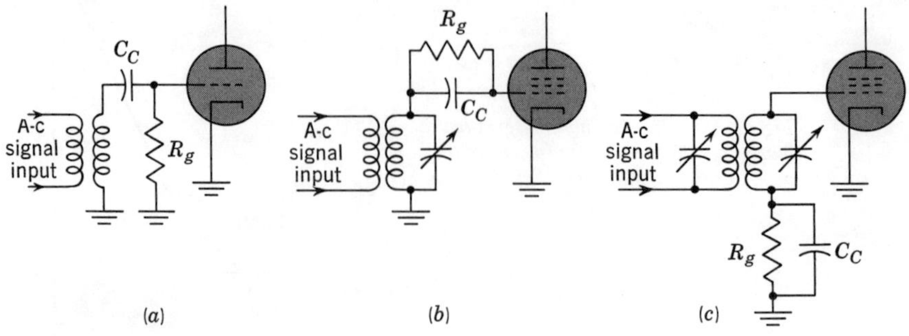

*Fig. 1·21 Grid-leak bias circuits. (a) RC coupling. (b) Transformer coupling with $R_gC_c$ on grid side. (c) Tuned coupling with $R_gC_c$ in return side.*

Grid-leak bias is produced the same way as with a series $R_gC_c$ combination. When the positive peak of the a-c input signal drives the grid positive, grid current flows to charge $C_c$. Between positive peaks, $C_c$ discharges slightly through its shunt $R_g$. With a long time constant for $R_gC_c$ compared with one cycle of signal, the result is a steady negative d-c bias voltage across $C_c$ approximately equal to the 90 per cent of the peak positive drive.

In *c*, the same $R_gC_c$ parallel bank for grid-leak bias is connected in the return side of the secondary. In this position, a d-c voltmeter can measure the bias across $R_gC_c$ without detuning the secondary circuit. It is often convenient to measure this voltage, since the amount of grid-leak bias indicates how much a-c input signal is applied to the circuit.

## *1·10 Contact bias*

Since the cathode is heated and placed close to a dissimilar metal electrode in the vacuum tube, a small potential difference called *contact potential* is generated between cathode and control grid. In diodes, a contact potential exists between plate and cathode. The amount of this open-circuit emf is between 0.5 and 1.0 volt. It is a d-c voltage with the cathode side positive. In small-signal applications, where this voltage provides sufficient bias, the contact potential is a convenient biasing method. Tubes frequently used with contact bias are the duodiode triodes, 6SQ7 and 6AT6, where the triode section is a low-level audio amplifier with contact bias of approximately $-1$ volt, as shown in Fig. 1·22. $R_g$ is 5 to 15 M to maintain the bias. A very high value of grid resistance is necessary to prevent shorting the contact potential.

## *1·11 Amplifiers in cascade*

When the plate signal output of one amplifier is coupled to the grid of the next stage, the two amplifiers are connected in *cascade*. The purpose of this arrangement is to multiply the gain, since each stage amplifies its input signal. As shown in Fig. 1·23, the first stage $V_1$ has a gain of 50, amplifying the 0.1-volt input signal to a 5-volt output signal, which is then

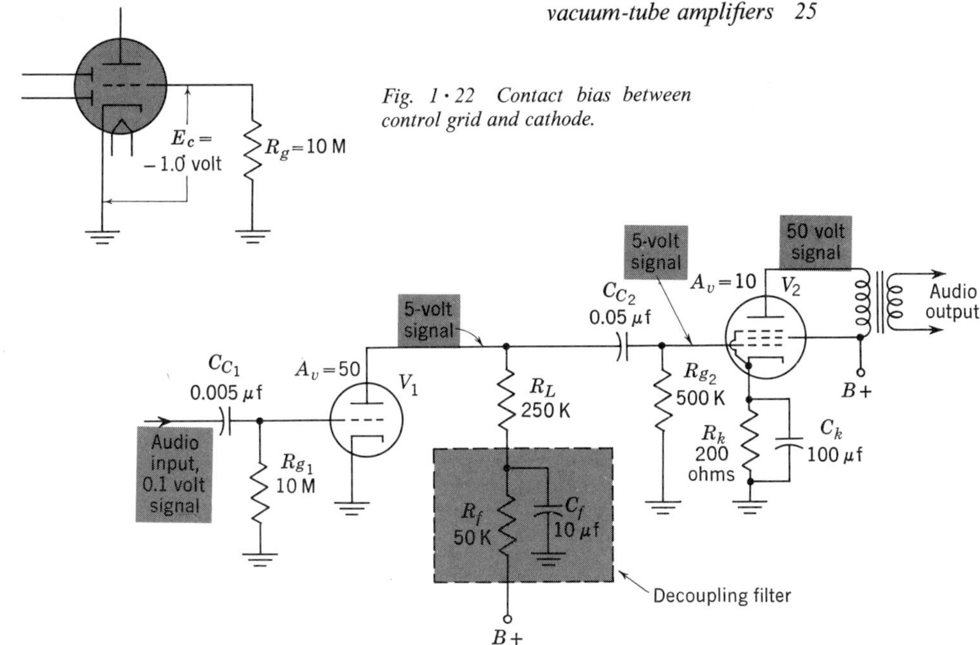


*Fig. 1 · 22 Contact bias between control grid and cathode.*

*Fig. 1 · 23 Two stages in cascade, with decoupling filter $R_fC_f$ in plate circuit of $V_1$.*

coupled to the grid of $V_2$. With a gain of 10, the output of $V_2$ is ten times the 5-volt input signal, or 50 volts. Although an $RC$-coupled amplifier is shown with a transformer-coupled amplifier here, any type of coupling can be used between cascaded stages. Also, more than two stages can be connected in cascade.

In Fig. 1·23, $V_1$ is an $RC$-coupled amplifier using contact bias. The amplified signal output is developed across the plate-load resistor $R_L$. Although in series in the plate circuit, $R_f$ is not part of the plate load, because it is bypassed by $C_f$. The $R_fC_f$ combination forms a decoupling filter to isolate the plate load $R_L$ from the B supply. The amplified output appearing across $R_L$ is coupled to the grid circuit of the transformer-coupled stage $V_2$. This stage has cathode bias provided by $R_k$, with $C_k$ large enough to bypass $R_k$ for the lowest signal frequency. In the plate circuit of $V_2$, the primary of the output transformer is the plate-load impedance for the amplifier, while the secondary couples the audio-output signal to the following circuit.

**Over-all gain.** The over-all gain of amplifiers in cascade is the product, not the sum, of the individual gain values for each stage. The reason can be seen from the numerical examples given in Fig. 1·23. The $V_1$ stage has 0.1-volt input signal and 5-volt output. Its gain is 50. With 5-volt input from $V_1$, the second stage has 50-volt output. Therefore its gain is 10. Thus, the over-all gain from the 0.1-volt input of $V_1$ to the 50-volt output of $V_2$ is 50/0.1, or 500. This is the product of the individual gain values of $50 \times 10$ for the two stages. With more than two stages, the over-all gain is still equal to the product of the individual gain values. For instance, four stages in cascade, where each has a gain of 20, provide an over-all gain of $20 \times 20 \times 20 \times 20$, or 160,000.

**Plate decoupling filter.** When amplifiers are connected in cascade, one or more stages may have a plate decoupling filter, such as $R_fC_f$ in Fig. 1·23. *Decoupling* means to separate or isolate two circuits for minimum effect between them. In the plate circuit, $R_f$ is actually an isolating resistor, separating $R_L$ from the impedance of the output filter capacitor in the B supply. $C_f$ is needed to bypass $R_f$, though, so that no signal voltage develops across $R_f$. Otherwise, $R_f$ would be part of the plate load, and there would be no decoupling. The need for decoupling arises when two or more cascaded stages are connected to a common B supply. The higher the gain, the more important is the decoupling.

## *1·12 Direct-coupled amplifiers*

As shown in Fig. 1·24, the direct-coupled amplifier is similar to an *RC*-coupled amplifier but without any coupling capacitors. As a result, d-c voltages as well as a-c voltages can be amplified because the plate of one stage is directly connected to the grid of the next. In Fig. 1·24, this connection makes the grid of $V_2$ +100 volts with respect to chassis ground, since the grid and plate are at the same potential. It is the potential difference between grid and cathode, however, that determines tube bias. Making the cathode of $V_2$ + 110 volts, therefore, provides 10-volt bias for $V_2$. Its grid is 10 volts negative to cathode, since 100 volts is 10 volts less positive than 110 volts.

**D-c potentials.** The problem with direct-coupled amplifiers is that each succeeding stage must be placed at a progressively higher potential, requiring a relatively high B-supply voltage. For the two stages in Fig. 1·24, the 350-volt supply is high enough, but each stage operates with only about one-half this amount for plate-supply voltage. The required positive voltages are obtained from a voltage divider across the B supply. In Fig. 1·24,

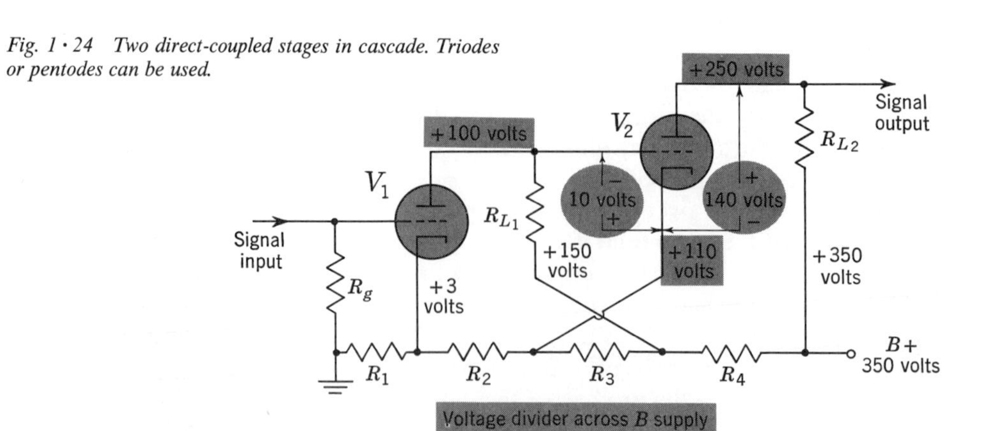


*Fig. 1·24 Two direct-coupled stages in cascade. Triodes or pentodes can be used.*

the divider provides positive voltages of 3, 110, 150, and 350 volts across $R_1$, $R_2$, $R_3$, and $R_4$, respectively.

**D-c voltage gain.** Direct-coupled amplifiers can be used for amplifying a d-c voltage. To illustrate, suppose that a dry cell is inserted in series with $R_g$ to decrease the grid voltage of $V_1$ from $-3$ to $-1.5$ volts. The plate current of $V_1$ goes up, decreasing the plate voltage. Assume that $E_b$ goes down from 100 to 94 volts. This 6-volt change makes the grid-cathode voltage on $V_2$ equal to 16 volts now. Then the plate current of $V_2$ decreases and its plate voltage rises. Assume an increase from 250 to 310 volts, which is a change of 60 volts. As a result of changing the d-c bias on the grid of $V_1$ by 1.5 volts, the d-c plate voltage of $V_2$ changes by 60 volts. Therefore the over-all gain is 60/1.5, or 40. The direct-coupled amplifier has amplified the d-c input voltage by a factor of 40.

There is not much point in amplifying the output voltage of a dry cell, since higher d-c voltages can be obtained more easily from a d-c source. There are applications, however, where the d-c voltage to be amplified is a control voltage, that is, a d-c voltage which indicates some desired characteristic in an automatic control system. For instance, rectifying and filtering an a-c signal may provide a d-c control voltage of 1 volt. However, 20 volts is needed for effective control. A direct-coupled amplifier could then be used to increase the amplitude of the d-c control voltage from 1 volt to 20 volts.

**Low-frequency response.** The other application of direct-coupled stages is for a resistance-loaded amplifier which has perfect low-frequency response down to 0 cps because there is no coupling capacitor. A-c signal voltage is amplified as well as d-c voltage. Furthermore, the high-frequency response is limited by the stray shunt capacitance, just as in an *RC*-coupled stage. Without the coupling capacitor to block d-c voltage, however, the d-c amplifier does not have this limitation on low-frequency response. For instance, a direct-coupled audio amplifier can have ideal response for the lowest audio frequencies.

## *1 · 13 Cathode-coupled stage*

Another variation in resistance-loaded amplifiers is the cathode-coupled stage shown in Fig. 1 · 25, where the load resistor for the output signal is in the cathode-to-ground circuit instead of the plate circuit. Input signal is

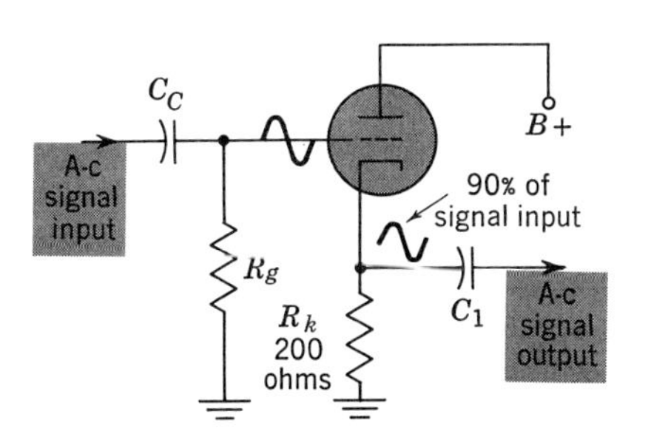


*Fig. 1 · 25 Cathode-coupled stage, or cathode follower. Either a triode or pentode can be used.*

applied to the grid circuit, as usual, to vary the plate current. There is no signal voltage in the plate circuit, however, because it has no load resistance. Still, the plate-signal current returning to cathode does develop a signal voltage across $R_k$. There is no cathode bypass because $R_k$ must provide the output-signal voltage. Its average d-c value serves as grid bias.

This stage is often called a *cathode follower* because the cathode voltage follows the grid-signal voltage. Both have the same polarity with respect to chassis ground, but the cathode-signal voltage is actually canceling part of the grid-signal voltage. If the signal drives the grid 10 volts less negative, and the cathode voltage goes 9 volts more positive, for example, the effective grid-cathode drive is only 1 volt. As a result, there is less output signal than input signal. The gain of a cathode-follower stage therefore must always be less than one. It is not difficult to make the gain close to one, though, so that the output can be 70 to 99 per cent of the input signal.

The purpose of using such a stage is not for its gain, but for its low output impedance. With a low resistance in the cathode circuit, such as the 200 ohms for $R_k$ in Fig. 1·25, the output can be coupled to a low-impedance load of 50 to 100 ohms without any appreciable loss in signal. Bear in mind that an amplifier with a high-resistance plate load cannot be coupled to such a low-impedance load without a drastic drop of voltage gain, which may be a greater loss than in a cathode follower. Furthermore, the cathode follower has very high input resistance in the grid circuit. One cathode-coupled stage can be used to advantage, therefore, as an isolating

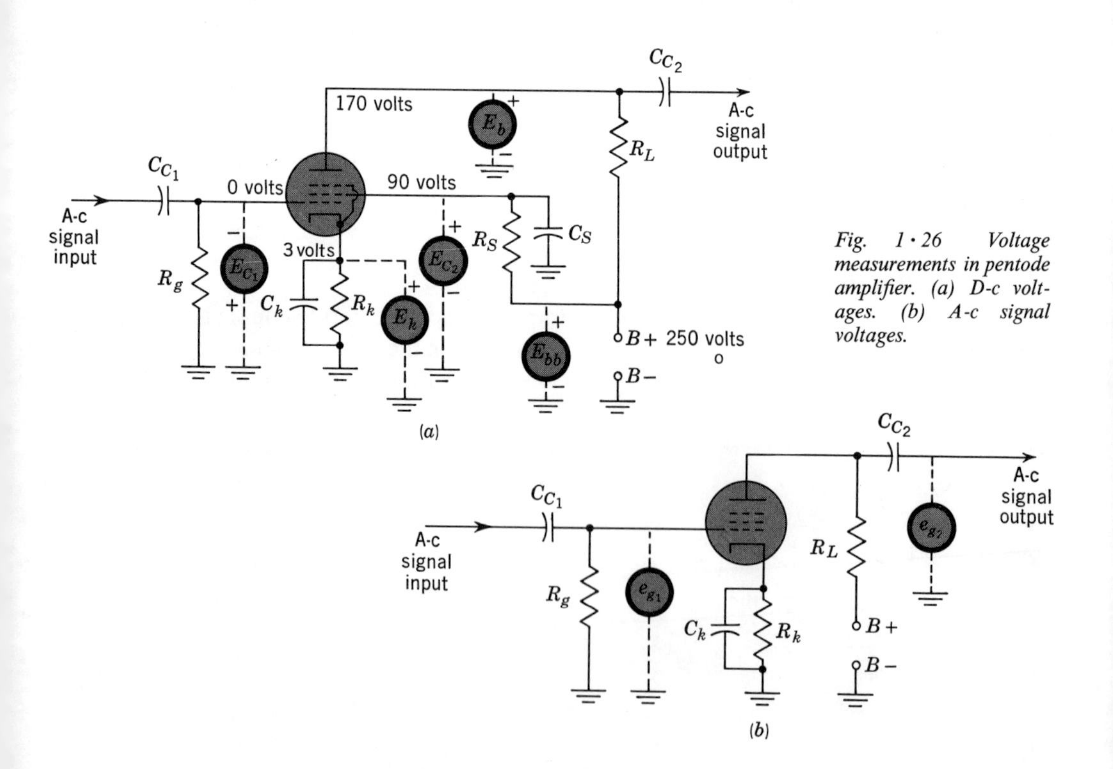


*Fig. 1·26 Voltage measurements in pentode amplifier. (a) D-c voltages. (b) A-c signal voltages.*

stage to present a high input resistance to the previous amplifier stage and low output resistance for a load that requires appreciable current. In short, the cathode follower is useful for isolation and impedance matching, although its voltage gain is less than one.

## *1 · 14 Direct and alternating voltages in the amplifier*

It is important to realize that the signal voltage in an amplifier is actually just the a-c component of a fluctuating d-c voltage. In the control-grid circuit, the grid voltage is a fluctuating negative d-c voltage with the a-c signal varying the instantaneous grid voltage above and below the steady d-c bias voltage. In the plate circuit, direct current must flow. The signal variations are the a-c component of the fluctuating direct plate current; the signal voltage is the a-c component of the fluctuating positive d-c plate voltage. Without steady d-c voltage supplied to the plate, there can be no direct plate current, and hence there is no signal. For a pentode, d-c voltage for the screen grid is also needed to produce plate current. The screen-grid circuit has no signal voltage, however, because of the screen-bypass capacitor. The suppressor grid is usually connected directly to cathode and also has no signal.

The d-c electrode voltages in the amplifier can be measured with a d-c voltmeter, as illustrated in Fig. 1 · 26. The measurements are made from each electrode to chassis ground, which is B−. With the ground lead of the meter connected to the chassis, the positive lead is placed at the desired pin to read its d-c voltage. Schematic diagrams generally give the normal d-c voltage at each pin with respect to chassis ground.

Starting with the cathode voltage $E_k$, a reading of +3 volts, for example, indicates normal cathode bias. Note that the d-c voltage from control grid to chassis equals zero, since it is assumed that there is no grid current. If there is grid-leak bias, the $E_{c_1}$ voltage reading will indicate the bias. It is best to use a vacuum-tube voltmeter (VTVM) for reading grid-leak bias voltages. In the plate circuit, the voltmeter from B+ to chassis reads the 250-volt B-supply voltage $E_{bb}$. At the plate the reading of +170 volts shows that there is an *IR* drop of 80 volts across $R_L$. Finally, the normal screen-grid voltage of +90 volts for $E_{c_2}$ results from the *IR* drop across $R_s$. These are all d-c voltages, but normal d-c electrode voltages are necessary for amplification of the a-c signal. When amplifier operation is checked, the d-c voltages are generally measured first because this is usually more convenient.

If it is desired to measure a-c signal voltage, an a-c voltmeter in the grid circuit will read the rms value of the a-c input voltage, as shown in Fig. 1 · 26*b* for the meter reading $e_{g_1}$. The amplified output can be measured after the coupling capacitor, as shown by the meter reading $e_{g_2}$. If necessary, the a-c signal-voltage output can be measured in the plate circuit also, but a coupling capacitor should be connected in series with the a-c voltmeter to block the d-c plate voltage. For measuring r-f signal voltage, an r-f voltmeter is necessary, which generally requires an additional r-f probe for the meter.

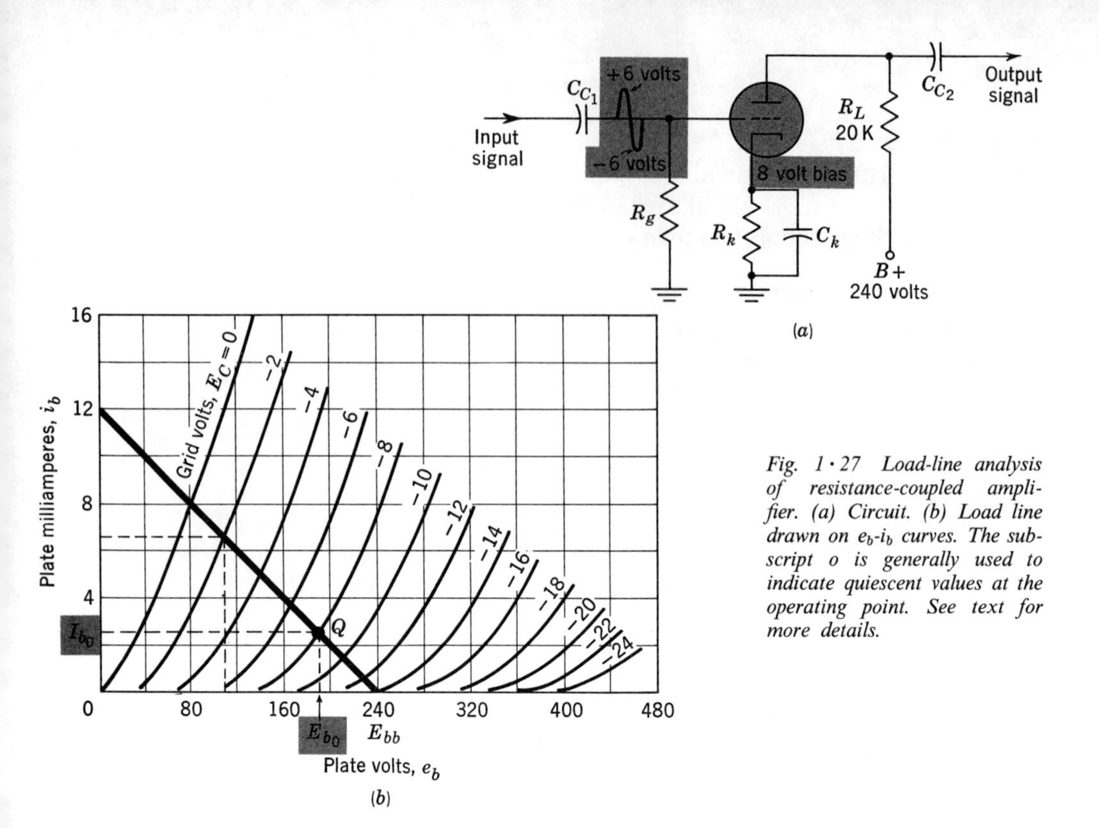


*Fig. 1·27 Load-line analysis of resistance-coupled amplifier. (a) Circuit. (b) Load line drawn on $e_b$-$i_b$ curves. The subscript o is generally used to indicate quiescent values at the operating point. See text for more details.*

## 1·15 Load line

The operation of a class A amplifier having a resistance plate load can be analyzed more exactly by the load-line method illustrated in Fig. 1·27. A typical circuit is shown in *a*, while *b* shows the load line for an $R_L$ value of 20,000 ohms. The load line is useful because it includes all values of plate voltage and current for the specified plate load. Although a triode is used for the example here, the same method applies to pentodes also.

To draw the load line on the family of plate characteristics, it is necessary to know only the B+ value and $R_L$ to determine the two end points. One point is at B+ on the horizontal axis, where $i_b$ is zero. This is one operating point because the plate voltage $e_b$ equals B+ when there is no plate current. The opposite point is the extreme value of maximum plate current, where the plate voltage is zero, because the $i_bR_L$ drop equals the B+ voltage. Therefore the current value on the vertical axis, where $e_b$ is zero, must be equal to B+/$R_L$. In this case, the maximum $i_b$ equals 240 volts/20,000 ohms, or 12 ma. Therefore, a straight line drawn between 12 ma on the vertical axis and 240 volts on the horizontal axis of the graph represents the load line. For any value of plate current, the corresponding plate voltage $e_b$ is on the load line, which takes into account the $i_bR_L$ voltage drop.

The load line shows how much bias is best for the amount of grid signal. In this example, −8-volt bias is chosen for a peak grid drive of 6 volts. The point where the load line crosses the curve for $e_c$ equal to −8 volts is the *Q* point, or the quiescent, or operating, point. The *Q* point is determined by the bias. Without signal input, the quiescent $I_{b_0}$ here is 2.5 ma with $E_{b_0}$

equal to 190 volts, approximately. With grid drive, the a-c signal varies the grid voltage between the peak $e_c$ values of $-2$ and $-14$ volts. When $e_c$ is $-2$ volts, the peak plate current equals 6.7 ma; plate voltage now reaches a minimum of 110 volts. The maximum negative drive, when $e_c$ is at $-14$ volts, makes $i_b$ practically zero, and $e_b$ equal to B+.

The peak-to-peak signal-voltage output in the plate circuit is the difference between the maximum and minimum values of $e_b$. Here this difference equals 240 − 110 volts, or 130 volts. In comparison to the grid-signal input, the output of 130 volts provides a gain of $^{130}/_{12}$, or approximately 10.8.

## *1·16 Troubles in amplifier circuits*

When there is no a-c signal output, or the output is weak or distorted, the trouble can be found by a check of the amplifier. It is assumed that the trouble has been localized to the amplifier by noting that the signal input is normal but the output is abnormal. General practice is to check the tube first, since tubes are the most common cause of trouble. If the tube is good, the following steps can be followed to find the trouble:

1. Measure the d-c electrode voltages, looking for readings that are too high or too low. The B+ voltage should be measured first, since all the other d-c voltages depend on this.
2. When one of the electrode voltages measures much too high or too low, check this circuit. After turning off the power, measure with an ohmmeter to find an open resistor or shorted capacitor.
3. If all the d-c electrode voltages are normal, the amplifier can still have an open coupling or bypass capacitor. A shorted capacitor generally affects the d-c voltage readings, but an open capacitor is best found by being temporarily bridged with another capacitor of about the same size.

When the defective component has been found, a new unit of the same value or close to it is soldered in its place. If a burned-out resistor is replaced, look for a shunt-bypass capacitor in the same circuit and replace it also, since a leaky or shorted capacitor can be the cause of the open resistor. A higher wattage rating than the original resistor is permissible for longer life, but use approximately the same resistance value. Similarly, capacitors with a higher d-c voltage rating than the original but approximately the same capacitance can be used. In high-frequency circuits, do not replace mica or ceramic capacitors with the rolled-foil type because of their inductance. When electrolytic capacitors are replaced, they must be connected in the correct polarity. These should also have about the same voltage rating as the original, since the forming voltage for electrolytic capacitors depends on the applied voltage.

**Plate voltage.** The two extreme cases are full B+ voltage at the plate and zero plate voltage. Both can prevent signal output, but for entirely different reasons. Zero plate voltage means that the d-c path for plate current to the B supply is open. Look for an open plate load or open decoupling resistor. Full B+ for plate voltage means the plate circuit is not open. First, it should be noted that in many amplifiers with a plate load that has very little d-c resistance, the plate voltage normally is practically equal to the full B voltage. With a resistance of 1,000 ohms or more in the plate circuit, however, full B+ at the plate means that there is no plate current. The plate voltage is connected, but without plate current there will be no *IR* voltage drop. The most common causes of no plate current are (1) defective tube, (2) open cathode circuit, (3) zero screen-grid voltage in a tetrode or pentode, and (4) control-grid bias more negative than cutoff.

**Screen-grid voltage.** When the screen-grid voltage is zero, there is no plate current and therefore no signal output. Generally, this is due to a shorted screen-bypass capacitor or an open screen-dropping resistor. What often happens is that the bypass capacitor shorts, causing excessive current through the screen resistor, which then burns open. This drops the screen-grid voltage to zero. If the screen-bypass capacitor should open, there would be loss of gain, and distortion, but the d-c voltage at the screen would be approximately normal.

**Cathode voltage.** With cathode bias, the cathode resistor can be open or its bypass capacitor may open or short. When the cathode bypass shorts, the cathode voltage becomes zero and the output becomes distorted. With an $R_gC_c$ coupling circuit in the grid, however, the signal input produces grid-leak bias, and there may be little difference in the output. If the cathode bypass opens, degeneration occurs with a resultant loss of gain. The most definite trouble here is an open cathode resistor. With the cathode circuit open, plate current cannot flow, and there is no output. The plate voltage now equals the full B+ value.

When the cathode-to-ground voltage is measured with a d-c voltmeter, the indication of an open cathode is a reading much higher than normal. If 2 volts is normal bias, for example, the voltmeter will read 10 to 50 volts for an open cathode because the high input resistance of the voltmeter closes the cathode circuit temporarily while connected for the reading. With this high resistance in the cathode circuit, enough plate current flows to produce a cathode-bias voltage almost equal to the grid-cutoff voltage.

**Grid-leak bias.** This negative d-c voltage at the grid indicates the amount of a-c signal input. No bias voltage in this case means no signal input from the preceding stage.

**Direct-coupled-amplifier voltages.** In checking direct-coupled amplifiers, remember that all the electrode voltages may be positive with respect to chassis ground as a normal condition. In measuring these voltages with a vacuum-tube voltmeter, always measure from the electrodes to chassis ground. Then subtract your readings to obtain the actual voltage between electrodes. If any one electrode has the wrong voltage, it will change all the other d-c voltages in the amplifier because of the direct coupling.

## SUMMARY

1. The main requirements of an amplifier circuit are d-c plate and screen voltages so that plate current can flow, an external plate-load impedance to develop output signal, and negative control-grid bias to set the operating point.

2. Voltage gain is the ratio of output-signal voltage to input-signal voltage. The gain is equal to or less than the $\mu$ of the tube but can never be more than the $\mu$, as indicated by formula (1 • 2).
3. Triodes or pentodes can be used for voltage amplifiers, which have a high-impedance load to provide maximum voltage gain; power amplifiers provide maximum signal current for a low-impedance load.
4. The main types of coupling circuits between amplifier stages are *RC* coupling, impedance coupling, and transformer coupling. For r-f amplifiers, tuned coupling circuits are generally used. Tuned stages amplify signal voltage just at the resonant frequency of the *LC* circuit. The characteristics of these coupling methods are summarized in Table 1 • 2.
5. The grid-plate transfer characteristic curve is a graph of grid voltages and corresponding plate-current values with fixed plate and screen voltage, as shown in Fig. 1 • 13. The value of negative grid voltage that results in no plate current is the grid-cutoff voltage.
6. Amplifier operation is classed as A, B, or C according to how the bias and signal swing utilize the grid-plate transfer characteristic curve. For class A, plate current flows 360°; for class B, 180°; for class C, less than 180°.
7. The main methods of bias to fix the operating point of the amplifier are (1) fixed bias from a C battery or from a negative tap on the power supply, (2) cathode bias, or self-bias from a resistor in the cathode circuit, (3) grid-leak bias, or signal bias from an *RC* grid-coupling circuit when grid current flows, and (4) contact bias, which is an open-circuit emf of approximately 1 volt between grid and cathode.
8. The cathode-bias voltage equals $I_k R_k$. Usually $R_k$ is bypassed with a shunt $C_k$ to prevent degeneration.
9. The amount of negative grid-leak bias voltage equals approximately 90 per cent of the peak positive grid-signal drive. No signal means no grid-leak bias.
10. When the plate signal output of one stage is coupled to the control grid of the next stage, the amplifiers are in cascade. This is the usual method of connecting amplifiers because it multiplies the gain. The over-all gain equals the product of the individual gain values for each stage.
11. Cascaded amplifiers often have decoupling filters, such as $R_f C_f$ in Fig. 1 • 23, to isolate the stages from the common impedance of the B supply. $R_f$ provides the isolation, while $C_f$ is its bypass, so that $R_f$ will not be part of the plate-load impedance for a-c signal.
12. In direct-coupled amplifiers, the plate of one stage is connected directly to the grid of the next stage. The positive plate voltage on the next grid is neutralized by a more positive cathode voltage, however, resulting in a net negative bias voltage from grid to cathode. Direct-coupled stages can amplify a-c signal voltage and d-c voltage. See Fig. 1 • 24.
13. The cathode follower, or cathode-coupled stage, has the load impedance for a-c signal in the cathode circuit, instead of the plate circuit. See Fig. 1 • 25.
14. The load line illustrated in Fig. 1 • 27 shows all values of $i_b$ and $e_b$ for a given value of plate-load resistance.
15. The troubles in amplifier stages are no output signal, weak output, or distorted output. To find the trouble, check the tubes and d-c voltages. If the d-c voltages are normal, check for an open capacitor by bridging each one with a similar capacitor.

## SELF-EXAMINATION

Here's a chance to find out how well you have learned the material in this chapter. Work the exercises, then check your answers against the Key at the back of this book. These exercises are for your self-testing only.

Each of the following incomplete statements is supplied with four possible answers. Circle the letter of the *one* which you think *best* completes the statement.

1. The external plate-load resistor is needed in an amplifier circuit to provide (*a*) screen-grid voltage; (*b*) signal variations in plate voltage; (*c*) higher $\mu$ for the tube; (*d*) the required amount of bias.
2. An amplifier has 10-mv rms signal input. The output signal is 1-volt rms. The voltage gain of the stage equals (*a*) 1; (*b*) 10; (*c*) 100; (*d*) 141.

3. A transformer-coupled audio-output stage operating class A, is a (*a*) voltage amplifier; (*b*) impedance-coupled stage; (*c*) power amplifier; (*d*) tuned amplifier.
4. An amplifier needs −6 volts on the grid to cut off the tube. Its cathode bias equals 2 volts. The a-c signal input is 1-volt rms. The class of operation is (*a*) A; (*b*) B; (*c*) AB; (*d*) C.
5. If the plate and screen-grid voltages are reduced, the control-grid cutoff voltage will (*a*) increase; (*b*) decrease; (*c*) stay the same; (*d*) depend on the amplitude of the a-c input signal.
6. A pentode has 5-ma plate current, 2-ma screen current, and a cathode resistor of 1,000 ohms. Its cathode-bias voltage equals (*a*) 2 volts; (*b*) 3 volts; (*c*) 5 volts; (*d*) 7 volts.
7. An *RC*-coupled stage has 10-volt rms signal input without any fixed bias or cathode bias. The signal will develop grid-leak bias equal to (*a*) 0 volts; (*b*) 5 volts; (*c*) 3 volts; (*d*) 12.6 volts.
8. Three stages in cascade each have a gain of 10. The over-all gain equals (*a*) 30; (*b*) 100; (*c*) 300; (*d*) 1,000.
9. In a direct-coupled amplifier stage, (*a*) the control grid is more positive than the plate; (*b*) the control grid is more positive than the cathode; (*c*) a-c signal can be amplified as well as d-c voltage; (*d*) the cathode is more positive than the screen grid.
10. The plate voltage in an *RC*-coupled amplifier equals the full B+ value. Which of the following can be the trouble? (*a*) $R_L$ is open; (*b*) $R_f$ is open; (*c*) $C_f$ is shorted; (*d*) $R_k$ is open.

## QUESTIONS AND PROBLEMS

1. List four requirements of a vacuum-tube amplifier.
2. Draw the schematic diagram of a class A single-tuned r-f amplifier with cathode bias. (*a*) What makes the amplifier class A? (*b*) What makes it an r-f amplifier?
3. Define the following classes of operation: $A_1$, $A_2$, $AB_1$, $AB_2$, and C.
4. Draw the schematic diagram of a triode *RC*-coupled audio amplifier $V_1$ driving a pentode transformer-coupled power-output stage $V_2$. Both have cathode bias. Show an $R_fC_f$ decoupling filter in the plate circuit of $V_1$. No values are necessary.
5. (*a*) For the circuit in question 4, if the voltage gain is 50 for the first stage and 10 for the second stage, how much is the over-all gain? (*b*) With 0.2-volt grid-input signal to the first stage, how much is its plate signal output? How much at the plate of the second stage?
6. List three methods of bias and describe one characteristic of each.
7. List three methods of coupling and describe one characteristic of each.
8. (*a*) Give two features of a direct-coupled amplifier. (*b*) Give two features of a cathode-coupled stage.
9. (*a*) Give two troubles in components that can cause zero plate voltage. (*b*) Give one component trouble that can cause zero screen-grid voltage.
10. Referring to Fig. 1·27, calculate the resistance of $R_k$ for 8-volt cathode bias.
11. Define the following symbols: $E_{bb}$, $E_b$, $e_b$, $e_p$, $I_b$, $I_k$, $E_{c_1}$, $E_{c_2}$, $I_{c_2}$, $e_g$.
12. If $I_k$ is 22 ma and $I_b$ 17 ma, how much is $I_{c_2}$, assuming $I_{c_1}$ is zero?
13. How much $R_k$ is needed for 8-volt cathode bias with 16-ma $I_k$?
14. Calculate the value of $C_k$ needed to bypass a 500-ohm $R_k$ at 100 cps. (*Hint:* Make $X_{C_k}$ at 100 cps equal to $R_k/10$ or 50 ohms).
15. How much is $E_b$ with 250-volt $E_{bb}$, 12-ma $I_b$, and 10 K for $R_L$?
16. How much is $E_{c_2}$ with 6-ma $I_{c_2}$ and 30 K for $R_s$, connected to a supply of 250 volts?
17. A d-c amplifier has the following voltages measured with respect to chassis ground: cathode at +105 volts, grid at +95 volts, and plate at +300 volts. (*a*) How much is the grid-cathode bias? (*b*) How much is the plate-cathode voltage?
18. Referring to the load line in Fig. 1·27, change the $Q$ point to −6 volts but use the same grid drive of ±6 volts. Determine: $I_{b_o}$; minimum and maximum d-c values of $i_b$; peak-to-peak a-c value of $i_p$; minimum and maximum d-c values of $e_b$; peak-to-peak a-c value of $e_p$.

# Chapter 2 Transistor amplifiers

This unit describes the different types of amplifier circuits in which transistors are used. In comparison with vacuum tubes, where signal voltage is the main feature, in transistors the current is the important characteristic. Also, transistor circuits are generally miniaturized because of the small size of transistors. Finally, practical procedures are explained for servicing transistor circuits, particularly the precautions to keep in mind when making resistance checks with an ohmmeter. The topics are as follows:



## *2·1 Transistor characteristics*

The unique advantages of transistors are their small size, high efficiency compared with vacuum tubes, and their ability to function with extremely

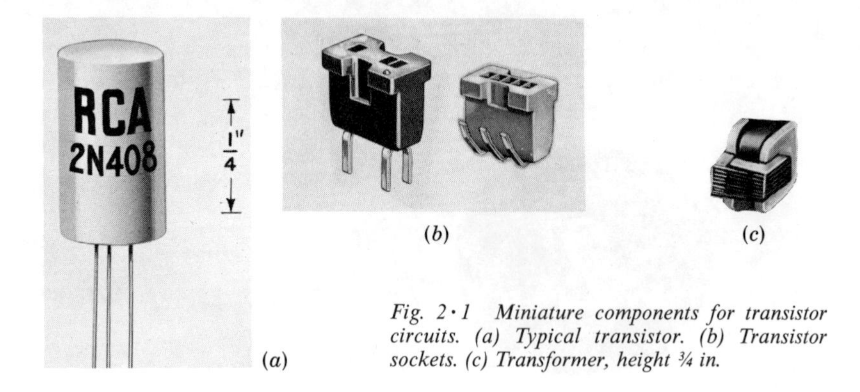


*Fig. 2 · 1 Miniature components for transistor circuits. (a) Typical transistor. (b) Transistor sockets. (c) Transformer, height ¾ in.*

small applied voltages. Physically, the body of a typical low-power transistor possesses an area of 3⁄16 inch square, while the leads, which do not have to be employed in their entirety, extend outward for a distance of 1½ inches or so (Fig. 2 · 1). With such small size, it would obviously be wasteful of space to use transistors with normal-size components, and in recognition of this, extensive miniaturization has taken place in every component employed with transistor circuits. Also assisting this trend has been the widespread employment of printed circuits, which not only facilitate the manufacture of transistor devices, but also enable the final product to possess an extremely compact and sturdy form. With transistors, miniaturization has achieved levels that would be quite impossible with vacuum tubes.

Transistors consist of combinations of P-type and N-type germanium (or silicon). N-type possesses an excess of negative charges, electrons; P-type has a deficiency of electrons. The positive charge for each missing electron is called a *hole*. A PNP transistor has a P-type emitter, an N-type base, and a P-type collector. An NPN unit, on the other hand, has an N-type emitter, a P-type base, and an N-type collector. Whenever either of these two units is used in a circuit, the following rules must be followed in applying the d-c operating voltages.

1. The emitter is always biased (with respect to the base) so that current will flow through the emitter-base circuit. In other words, the emitter is biased in the *forward,* or *low-resistance,* direction.
2. The collector is always biased (with respect to the base) so that no current will flow between collector and base. That is, the collector is biased in the *reverse,* or *high-resistance,* direction.

Because there are two kinds of transistors, PNP and NPN, and because they possess different internal structures, it is not surprising to find that the external voltages are applied in opposite polarities. In spite of this difference, however, both types of transistors perform the function of amplification in essentially the same fashion.

With this understanding of how transistors are biased generally, let us

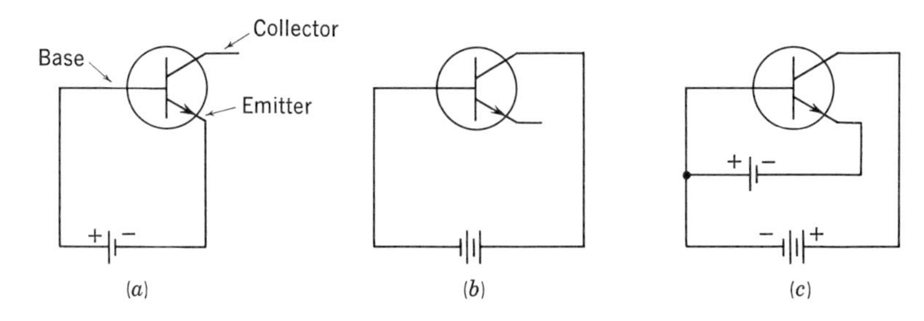


*Fig. 2 • 2 Biasing NPN transistor amplifier. (a) Forward bias on emitter to base. (b) Reverse bias on collector to base. (c) Total biasing.*

examine this procedure in somewhat greater detail. We start with NPN transistors and then extend the discussion to PNP units.

## 2 • 2 *Transistor biasing*

To achieve forward biasing in the base-emitter circuit of an NPN transistor, we connect a battery between these elements so that the negative terminal goes to the emitter (that is, the N element) and the positive terminal goes to the base (that is, the P terminal). In this way, the negative electric force of the battery repels the free electrons in the N-type emitter and drives them toward the base. At the same time, the positive electric force of the battery drives the positive holes of the base toward the emitter (see Fig. 2 • 2*a*).

In the collector circuit, the opposite situation exists. Here, the positive terminal of the battery connects to the N-type collector, attracting electrons away from the base-collector junction. By the same token, the negative terminal of the battery connects to the base and attracts holes away from the same junction. The collector thus is reverse-biased (see Fig. 2 • 2*b*). Both bias voltages are shown together in Fig. 2 • 2*c*. Now let us see what happens within the transistor under these conditions.

Since the emitter-base sections are biased in the forward direction, current will flow across their junction. Every time an electron from the emitter section crosses the junction and combines with a hole in the base section, an electron leaves the negative terminal of the battery and enters the emitter crystal. Since the battery cannot continue to supply electrons from the negative terminal without receiving an equivalent number at the positive terminal, for each electron leaving the negative terminal, the positive side receives an electron from the base section. This loss of electrons in the base creates holes, which then travel to the emitter for eventual combination with an electron.

If the center base section were made quite thick, then practically the entire current flow would occur in the manner just described and be entirely confined between emitter and base. There would be little current between base and collector because of the reverse biasing existing here.

The base strip is made exceedingly thin, however, and therefore transistor amplifying action can be achieved. For, with the base thin, electrons leaving the emitter pass right through the base section and into the collector region, where they see a positive attractive force that impels them on. Thus they travel through the collector section and around the external circuit back to the emitter again, completing their path of travel.

At this point the reader may wonder why the emitter current flows through the collector when it was specifically stated that the collector was biased in the reverse, or high-resistance, direction. If we disregard the base for a moment and simply consider the path from the emitter to the collector internally and from the collector to the emitter externally, we see that the two bias batteries are connected *series-aiding.* That is, in Fig. 2·2*c*, we go from the collector to the positive terminal of the bottom battery, then from the negative terminal of this battery to the positive terminal of the top battery, and from the negative terminal of this battery to the emitter. Thus any emitter electrons that can pass through the base region, without combining with the holes present here, will find the attracting force of the collector battery urging them on through the collector section. The reverse biasing between collector and base does not affect the *emitter* electrons that pass through the base and reach the collector.

With the base strip made very thin, the number of combinations between emitter electrons and base holes will be quite small, probably no more than 2 per cent of the total number of electrons leaving the emitter. The remaining 98 per cent of the electrons will reach the collector strip and travel through it. Thus, while the number of electrons leaving the emitter is a function solely of the emitter-base voltage, the element which receives most of this current is the collector. The analogy here to vacuum-tube behavior is very marked. In a tube, the number of electrons leaving the cathode (that is, the emitter) is governed by the grid-to-cathode voltage. It is the plate (that is, collector), however, which receives practically all these electrons. In a tube, the amount of current flowing is regulated by varying the grid-to-cathode voltage. In a transistor, the emitter-collector current is varied by changing the emitter-base voltage.

Note, too, that because the base current is very small, a change in emitter bias will have a far greater effect on emitter-collector current than it will on base current. This is also desirable, since it is the current flowing through the collector that reaches the output circuit. (By the same token, it is the current flowing through the plate circuit in a tube that is important.)

## 2·3 *Transistor gain*

We achieve a voltage gain in the transistor because the input, or emitter-to-base resistance, is low (as a result of the forward biasing between the two elements), whereas a high load resistance can be used because the collector-to-base resistance is high (owing to the reverse bias here). A typical value for the emitter-to-base resistance is about 100 ohms, and a typical value for the load resistor is 10,000 ohms. The current that reaches the collector is 98 per cent of the current leaving the emitter. If, now, we multiply the current gain (0.98) by the resistance ratio, 10,000/100, we shall obtain the voltage gain of the collector circuit over the emitter circuit. Numerically, this is

$$\text{Voltage gain} = \text{current gain} \times \text{resistance ratio} = 0.98\,\frac{10{,}000}{100} = 98$$

Thus we see that while the current gain here is actually a loss, it is more than made up by the extent to which the collector resistance exceeds the emitter resistance. Furthermore, this overwhelming differential in resistance will also provide a power gain. In other words, with a small amount of power in the input, or emitter-to-base, circuit, we can control a much larger amount of power in the output, or collector-to-base, circuit. Both of these characteristics are important; without them the transistor would have only limited application in electronics.

The voltage gain indicated above is that which would be obtained if the transistor operated into a very-high-impedance circuit. Actually, one of the problems which is encountered in cascaded transistor amplifiers is that of matching the relatively high output impedance of a prior stage with the low input impedance of the following stage. This point will be discussed in greater detail in Sec. 2·11.

## *2·4 Potential hills*

Because a complete understanding of what happens within the transistor is so vital to future circuit application, we can consider still another approach based on potential hills. This method (Fig. 2·3) reveals the

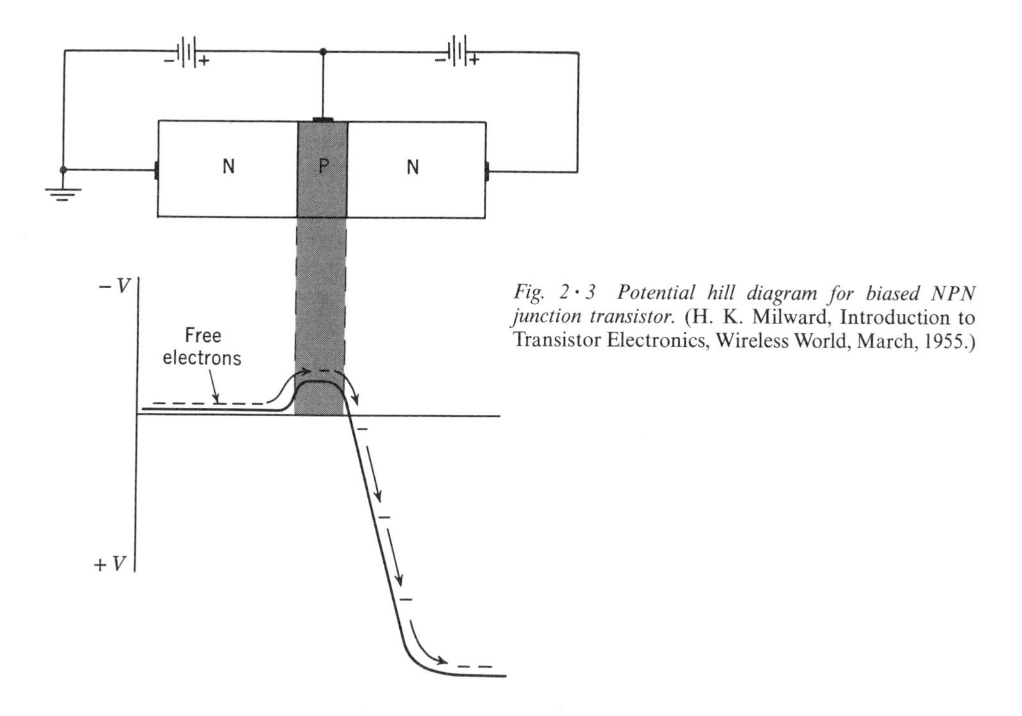


*Fig. 2·3 Potential hill diagram for biased NPN junction transistor.* (H. K. Milward, Introduction to Transistor Electronics, Wireless World, March, 1955.)

effects of emitter, base, and collector voltages and presents a simplified visual picture of junction-transistor operation.

When the emitter is biased in the forward direction and the collector in the reverse direction, electrons leaving the N-type emitter will see a small potential hill in front of them. This hill corresponds to a barrier voltage at the NP junction. Because of the energy supplied by the applied emf, however, the electrons will be able to surmount this hill. Once atop the hill, the "ground" levels off, and the electrons move through the P layer of the base quite readily. When they reach the junction between the base and the N-type collector, the electrons come under the influence of the positive battery potential, and they surge forward strongly. In the voltage diagram, this attraction is represented as a downward slope, which electrons (like human beings) find simple to traverse.

If the forward bias on the emitter is reduced, we are, in effect, raising the height of the base potential hill. Electrons leaving the emitter will find the higher hill more difficult to climb, and only those electrons possessing the greatest amount of energy will be able to reach its summit and, from there, move to the collector ahead. Current will consequently be reduced.

Similarly, increasing the forward bias on the emitter will reduce the height of the hill, enabling more emitter electrons to enter the base region. Thus the biasing voltages used in a transistor have a very important effect on its operation. Another significant controlling factor is the width of the base, as demonstrated by the next two illustrations.

Two three-dimensional representatives of this potential diagram are

*Fig. 2 · 4 Three-dimensional illustration of potential levels in biased NPN germanium transistor. The base here is shown wider than normal.* (Wireless World.)

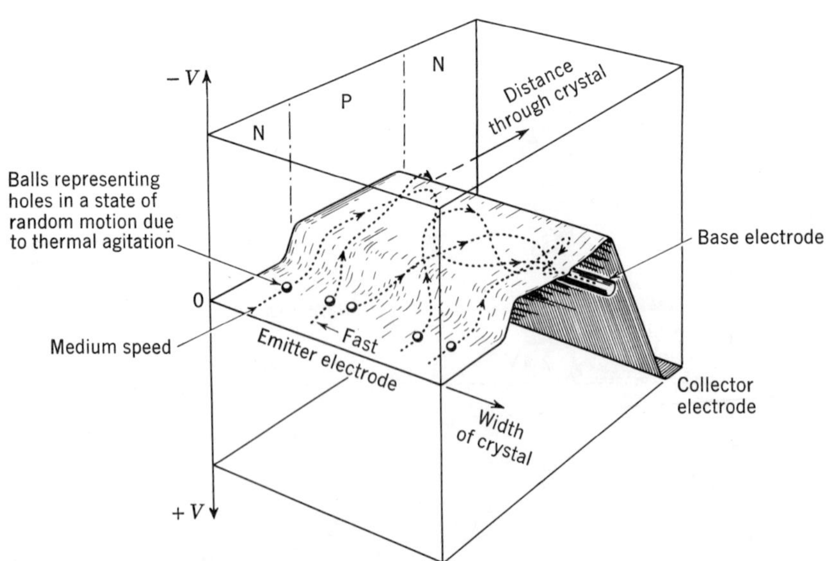

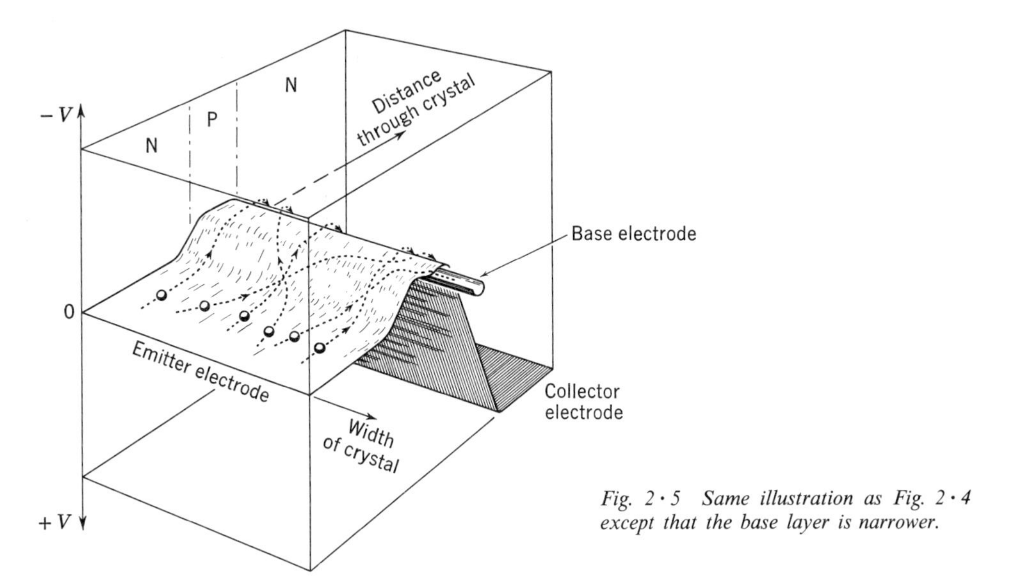


*Fig. 2·5 Same illustration as Fig. 2·4 except that the base layer is narrower.*

given in Figs. 2·4 and 2·5. The difference between the two drawings lies in the width of the base section. If the base section is wide, the tendency for emitter electrons (represented here by balls) to combine with holes at the base electrode is much greater than when the base section is narrow. In a physical model of these illustrations, the potential surfaces through the transistor are formed by a rubber membrane supported at several points. The holes in the base section are represented by a slight dip, or "valley," at the center of the base membrane. The wider the base section, the more difficult it is for the balls to roll through the base valley and over the edge of the precipice into the collector region without being trapped by the base dip.

On the other hand, if the base region is made quite narrow, any balls having enough energy to surmount the initial rise of the base hill will possess enough energy to reach the far edge of the base and fall down into the collector.

Thus, the width of the base section has a very direct bearing on transistor gain, both voltage and power. For if the percentage of current reaching the collector decreases to very small values, it will reduce the voltage gain in the same proportion. Power, being proportional to the square of the current, will be adversely affected to an even greater extent.

From the foregoing discussion we can formulate two rules concerning this and *all* transistors.

1. The emitter is biased in the forward, or low-resistance, direction.
2. The collector is biased in the reverse, or high-resistance, direction.

There are occasions, as with vacuum tubes, when it is desirable to bias

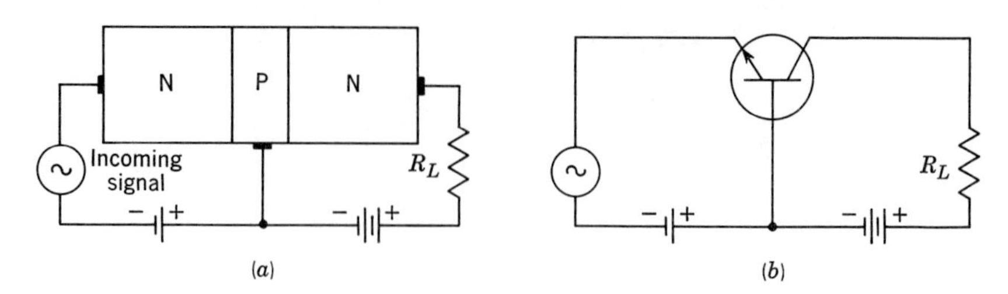


*Fig. 2·6 NPN transistor amplifier. (a) Circuit arrangement. (b) Schematic diagram.*

the transistor to cutoff. In the case of a vacuum tube, this is done by increasing the negative bias on the grid with respect to the cathode. For a transistor, cutoff is achieved by bringing the emitter-to-base bias to zero or even inserting a small amount of reverse-biasing voltage. In the majority of applications, however, the first statement is true.

That these rules are true can be seen if we consider their alternatives. If the emitter is biased in the reverse direction, it will not permit any electrons to reach the base region. And a reverse-biased emitter, with a reverse-biased collector, will produce a transistor in which current never passes.

If the emitter and collector are both forward-biased, then the general tendency will be for the emitter electrons to flow between emitter and base and for the collector electrons to flow between collector and base. In essence, we shall have two junction diodes possessing a common base. If the collector forward voltage is larger than the emitter voltage, some of the collector electrons will flow back to the collector via the emitter. But in any event, the desired amplification will not be obtained, and the purpose of the transistor will be defeated.

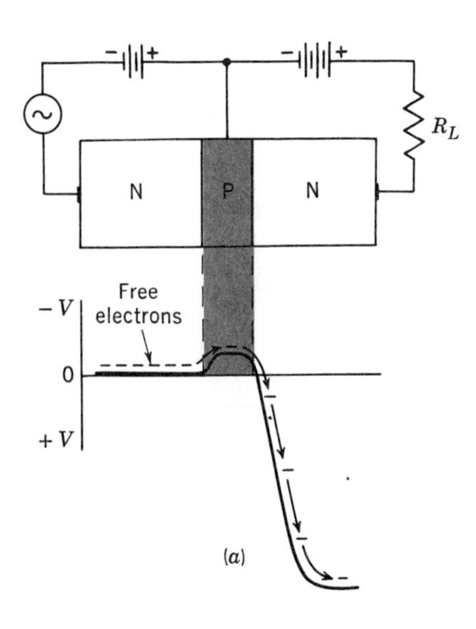


*Fig. 2·7 Transistor amplifier operation demonstrated by potential hills. (a) Signal voltage zero. (b) Signal voltage aids emitter-base forward bias. (c) Signal voltage opposes emitter-base forward bias.*

At this point, a note of caution regarding the application of any voltage to transistors. As we shall see later, the emitter bias voltage is quite small, on the order of 1 volt or less. The collector reverse voltage is generally much higher. If we should mistakenly connect the collector battery in the forward direction, the excessive current flowing through this section would develop enough heat to destroy the junctions and render the transistor worthless. Hence always be certain the collector voltage polarity is correct *before* connections are made.

## 2·5 *Transistor amplifier circuit*

We are now ready to connect this NPN transistor into an actual amplifier circuit with the signal input at one end and the load resistor at the other (see Fig. 2·6). The incoming signal is applied in series with the emitter-to-base bias, and the load resistor is inserted in series with the collector battery. When the signal voltage is zero, the number of electrons leaving the emitter and entering the base region is determined solely by the forward bias on the emitter. This situation can be represented by the potential distribution diagram shown in Fig. 2·7*a*. When the signal goes negative, it adds to the forward bias, further reducing the height of the base hill and causing more electrons to flow through the transistor (Fig. 2·7*b*). During the next half-cycle the signal goes positive, reducing the forward bias of the emitter and thereby reducing the number of electrons leaving the emitter to enter the base and collector. This condition is shown in Fig. 2·7*c*, where the height of the base hill has been increased.

At the other end of the transistor, these current fluctuations produce corresponding voltage variations across $R_L$, the load resistor. When the input signal is negative and the current increases, the bottom end of $R_L$ (in Fig. 2·7) becomes more negative. By the same reasoning, when the

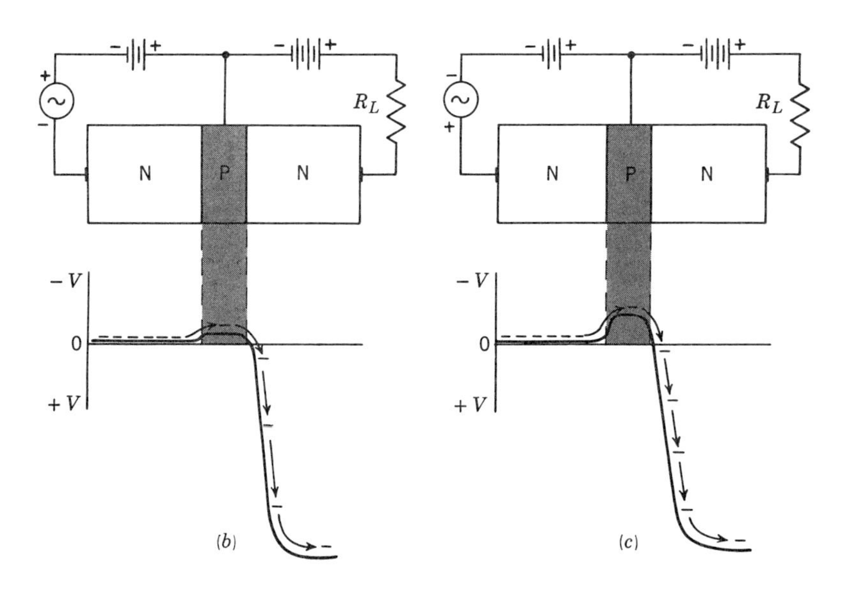

signal goes positive, current decreases and the bottom end of $R_L$ becomes relatively more positive.

Thus, through this transistor, amplification is achieved without the normal 180° phase shift we are accustomed to in vacuum tubes. This situation is not always true of transistors, and presently we will come across circuits where signal phase reversal does occur.

Another point to note here is that with this transistor, an increase in signal polarity (that is, positive) causes the transistor current to decrease in direct contrast to normal vacuum-tube amplifiers, when the signal is applied to the grid. We know also, however, that when the signal is applied to the cathode of a vacuum tube, a positive increase is equivalent to a more negative grid and the plate current decreases.

## 2·6 *PNP transistors*

In the formation of the initial transistor from a PN junction, we added a second N section to evolve an NPN transistor. We can approach the same problem by adding another P section to produce a PNP transistor (see Fig. 2·8). The emitter and collector sections are formed now of P-type germanium, while the base section consists of N-type germanium. Since this arrangement is actually the reverse of the NPN transistor—so far as material structure is concerned—we should expect differences in the mode of operation and in the polarity of the voltages applied to the emitter and collector. In spite of these differences, however, the emitter is still biased in the forward direction and the collector is biased in the reverse direction.

A typical bias setup with a PNP transistor is shown in Fig. 2·9. The positive side of the battery connects to the emitter, while the negative terminal of the battery goes to the base. The collector battery is attached in the reverse manner, with the negative terminal connecting to the collector and the positive terminal going to the base.

Holes are the current carriers in the P-type emitter and collector; in the N-type base section, electrons are the principal carriers. With the emitter-bias battery connected as shown, the positive field of the battery repels the positive holes toward the base. At the same time, the negative battery terminal at the base drives the base electrons toward the emitter. When a hole and an electron combine, another electron from the emitter section enters the positive battery terminal. This action creates a hole, which then

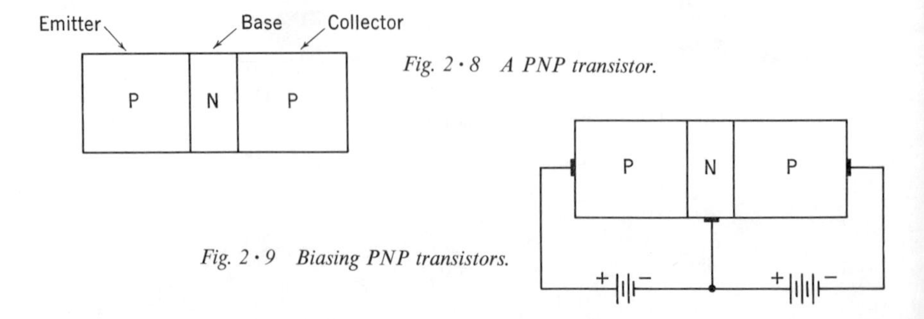


*Fig. 2·8 A PNP transistor.*

*Fig. 2·9 Biasing PNP transistors.*

starts traveling toward the base. At the same instant, too, that the first hole and electron combine, another electron leaves the negative battery terminal and enters the base. In this way, current flows through the base-emitter circuit.

In the PNP transistor, the holes are carriers in the emitter section, and when they cross the junction into the base region, a number of them combine with the base electrons. Well over 90 per cent of the holes, however, do not combine with base electrons; rather, they pass through the base region and continue on to the collector. Here they meet a negative attractive force and move toward the collector terminal. When the terminal is reached, an electron from the battery combines with a hole and effectively neutralizes it. At the same instant, an electron leaves the emitter region and starts on its way around the outer circuit to the collector battery.

Note, then, that although holes are the current carriers in P-type germanium, current conduction through the connecting wires of the external circuit is carried out entirely by electrons. This fits in with the current conduction that we are familiar with.

The incoming signal and the load resistor occupy the same positions in a PNP transistor amplifier that they do in an NPN transistor amplifier (see Fig. 2 · 10). Only the polarity of the biasing voltage is reversed.

## 2 · 7 *Comparison with vacuum tubes*

Transistors are designed to perform the same functions as vacuum tubes, and it is only natural to want to compare the two electrically to see where they differ and where they are similar. As a first step, let us consider these two components in the light of their internal operation. In a transistor, current flow through the various semiconductor sections is initiated by the flow of electrons or holes from the emitter section. In a vacuum tube, this initiation starts at the cathode. Thus we could say that the emitter in a transistor is equivalent to the cathode in a vacuum tube. The word *emitter,* of course, is a clue to the function of this element.

Current flow is received by the collector in the transistor and the plate in the vacuum tube. Hence these two elements can be considered to be equivalent in their functions. This comparison still leaves the grid in the vacuum tube and the base in the transistor. The equivalence of these ele-

*Fig. 2 · 10 PNP transistor amplifier. (a) Circuit arrangement. (b) Schematic diagram.*

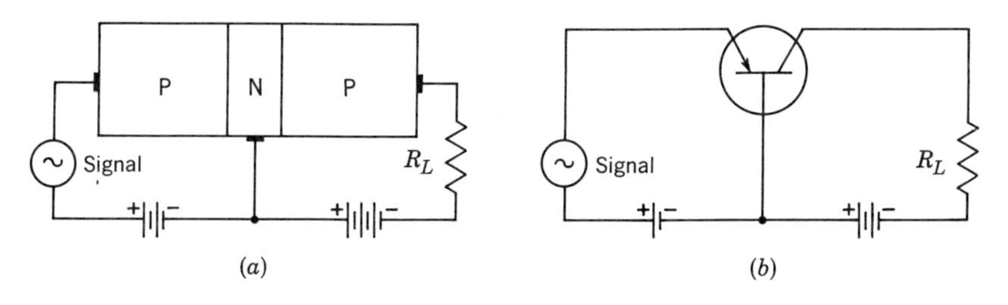

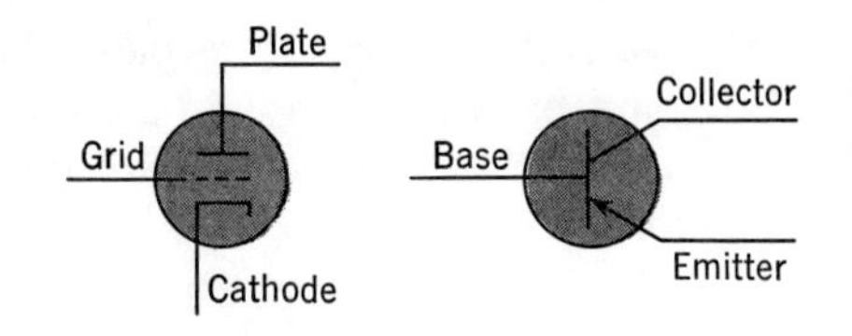


*Fig. 2·11 Comparable elements in triode vacuum tube and transistor: grid and base, cathode and emitter, plate and collector.*

ments is seen in the fact that current from emitter (or cathode) to collector (or plate) must flow through the base (or grid) structures. Current flow in both devices is governed by the potential difference between emitter or cathode and base or grid. Figure 2·11 illustrates this analogy between transistor and vacuum tube.

The next step is to consider both devices in terms of the d-c voltages which are applied to their elements. In a vacuum tube, the grid is almost always biased negatively with respect to the cathode. This bias makes the grid impedance very high, except at high frequencies where r-f losses enter the picture. The plate, on the other hand, is always given a potential which is positive with respect to the cathode. The purpose of the plate is to attract the electrons emitted by the cathode, and since electrons possess a negative charge, a positive potential is needed to attract them.

In the transistor, conditions are somewhat different, although we wish to accomplish the same purpose. To initiate a flow of current, there must first be a current between emitter and base, and the bias battery must be connected to produce it. This necessity is what determines the polarity connections of the bias battery. If the emitter is formed by P-type germanium, the base will contain N-type germanium, and current flow will occur between these sections when the positive battery terminal connects to the P-type emitter and the negative battery terminal to the base. We have spoken of this as *forward biasing,* and under these conditions the impedance of the emitter circuit is low. Here, then, is a marked departure from conventional amplifier practice as we know it now. In every transistor, the emitter-to-base circuit is *always* biased in the forward direction.

When we employ N-type germanium for the emitter and P-type germanium for the base, we must reverse the battery connections if we are to obtain the desired current flow through the emitter. Thus, the guiding thought in the emitter circuit is current flow, and we alter the battery conditions to suit the type of germanium being used in order to achieve this objective. Here is a radical departure from anything we have known in vacuum-tube practice, and it points up something which we have hinted at throughout the preceding discussion. That is, transistors are current-operated devices, while vacuum tubes are voltage-operated. For example, $\alpha$ is the symbol representing the ratio of $\Delta i_c/\Delta i_e$, where $\Delta i_c$ is the change in

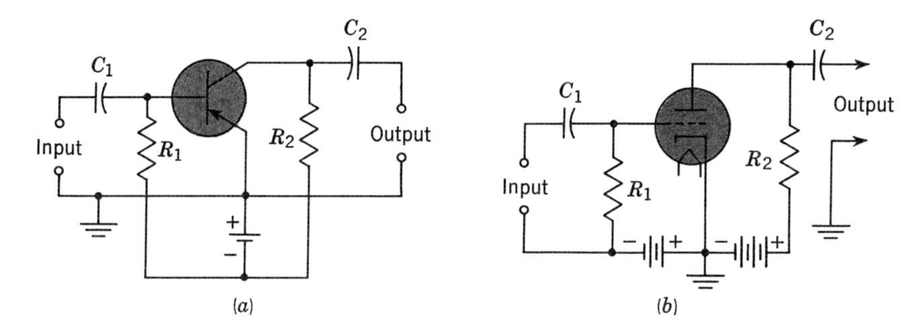


*Fig. 2·12 Comparison of common-emitter transistor circuit with triode vacuum-tube amplifier. (a) Transistor circuit. (b) Vacuum-tube circuit.*

collector current for a change in emitter current $\Delta i_e$. The counterpart of this symbol in the vacuum tube is $\mu$, the ratio of a voltage change in the plate circuit produced by a voltage change in the grid circuit. Again we see the emphasis on voltage in a tube as against current in the transistor.

## *2·8 Common-emitter amplifiers*

The most widely employed transistor circuit is one in which the input signal is applied between the base and the emitter; the output signal is taken from between the collector and emitter. The emitter, then, is common to both the input and output circuits, and so the amplifier is referred to as a *common-emitter* amplifier.

**Component functions.** Let us examine Fig. 2·12*a* closely to see what the various components do. The d-c power source is a small battery connected with its positive terminal to ground and its negative terminal to both base and collector. Because of this arrangement, the base is negative with respect to the emitter. Since the base-emitter circuit is to be biased in the forward, or low-resistance, direction, the base must consist of an N-type semiconductor and the emitter of a P-type semiconductor. Therefore we have a PNP transistor, as can be verified by noting further that the collector is biased negatively.

$R_1$ in Fig. 2·12*a* limits the current through the base-emitter circuit to the desired value. This value is governed by the operating point of the transistor. It can, to a large degree, be compared to choosing the grid bias for a vacuum-tube amplifier.

$R_2$ serves a double purpose. First, it serves to bring the negative battery voltage to the collector. Second, it acts as the load resistor for the transistor. The signal developed across $R_2$ is the voltage for the next stage.

$C_1$ and $C_2$ are coupling capacitors, bringing the signal into the base and removing it from the collector. They are also d-c blocking capacitors, keeping the d-c biasing voltages within the desired paths. In *RC*-coupled stages, these capacitors would keep the collector voltage of one stage from the input circuit of the following stage. When transformer coupling is used, the input capacitor (such as $C_1$) would prevent the low d-c resistance of the transformer winding from short-circuiting the bias voltage.

**Stabilized-emitter amplifier.** A form of common-emitter amplifier that is frequently seen is shown in Fig. 2 · 13. The chief difference between this circuit and that of Fig. 2 · 12*a* is the 5,000-ohm resistor $R_3$ and filter-bypass capacitor $C_3$, which have been inserted in the emitter lead. $R_3$ serves to stabilize the circuit by compensating for differences between transistors and by reducing the effects caused by temperature drift. Capacitor $C_3$ is shunted across the resistor to prevent degeneration and a reduction in gain. The added stability provided by degeneration may be desired in some instances, and then $C_3$ is omitted.

Temperature rise is a problem in transistors because as the temperature goes up it tends to cause electrons to break away from their bonds within the germanium crystal. This produces not only more electrons, but also more holes, which in turn will increase the amount of current flowing through the transistor for the same applied voltages. The greater currents will produce even more heat, which results in freeing more electrons and holes. This condition causes more current flow, and hence more heat, until the entire transistor becomes so hot that it is permanently damaged. This action is called *thermal runaway* in the transistor.

The insertion of a series resistor in the emitter leg is designed to prevent thermal runaway. As illustrated by the circuit of Fig. 2 · 13, most of the collector current will flow through the emitter resistor $R_3$. The voltage drop produced across $R_3$ serves to make the emitter negative with respect to ground. Note, however, that the base is also negative with respect to

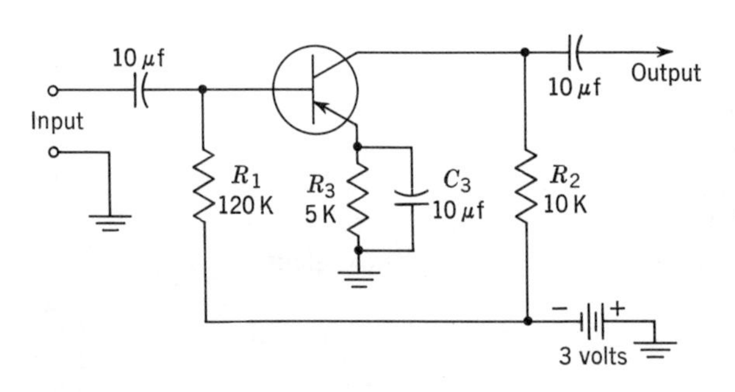


*Fig. 2 · 13 Common-emitter amplifier circuit with stabilizing resistor $R_3$. Its bypass capacitor $C_3$ prevents degeneration.*

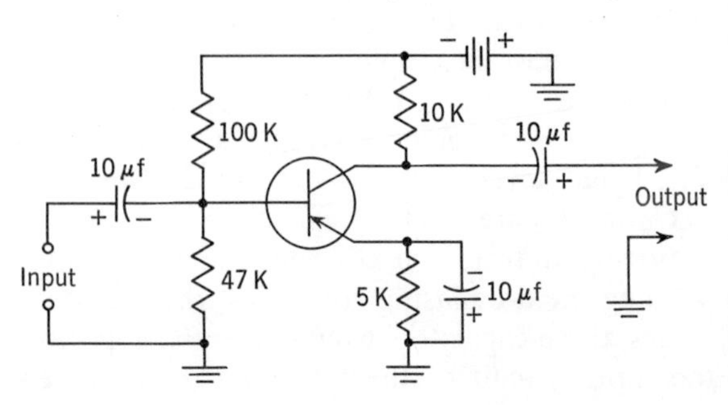


*Fig. 2 · 14 A more stable arrangement of the circuit in Fig. 2 · 13.*

ground, and hence the base-emitter voltage will be the difference between the battery voltage drop across $R_1$ and the smaller voltage drop across $R_3$. Now let us say that the collector current rises because of a temperature rise. This will cause the voltage drop across $R_3$ to increase, making the overall base-to-emitter voltage less negative than it was before. This voltage is actually working against the forward biasing voltage of the base-emitter circuit, resulting in *less* emitter current. Hence we are counteracting the rise in $i_c$ by decreasing $i_b$ and $i_e$. In this way, we achieve stabilization of our amplifier circuit.

A variation of this stabilization circuit is shown in Fig. 2·14. Here the base is connected to a voltage divider. This arrangement provides greater stabilization, but the additional resistor does absorb more power from the battery. In that sense, this circuit is less efficient.

**Signal phase reversal.** An interesting feature of the common-emitter form of connection is that a phase reversal occurs as the signal passes through the stage. In this respect it is similar to its vacuum-tube prototype, the grounded-cathode amplifier.

The reason for the reversal can be understood by considering the amplifier shown in Fig. 2·12*a*. The base-emitter circuit is biased in the forward direction, with the positive side of the bias battery connecting to the emitter and the negative side of the battery to the base. (In this way, the positive battery terminal repels the excess holes in the P-type emitter toward the PN junction, while the negative battery potential drives the electrons in the base to the same junction.) If we now apply a signal to the base, here is what will happen.

When the signal is positive, it will tend to reduce the bias potential applied between emitter and base. This means that the holes in the emitter and the electrons in the base will have less compulsion to overcome the inherent separating force at the junction and less current will flow. This, in turn, will reduce the collector current, providing less voltage drop across the load resistor. As a result, potential at the top end of $R_2$ will become more negative.

During the negative half-cycle of the signal, the total voltage in the emitter-base circuit will rise. This rise will increase the flow of current through the emitter, the collector, and $R_2$. The increased voltage drop across $R_2$ will make the top end of this resistor less negative or more positive. Thus, in common-emitter amplifiers, the output signal is 180° out of phase with the input signal.

## *2·9 Common-base amplifiers*

A second type of transistor amplifier is the common-base amplifier (see Fig. 2·15). The input signal is applied to the emitter, and the output signal is obtained at the collector. Thus the base is common to both the input and the output circuits. This is the type of circuit used in Figs. 2·6, 2·7, 2·9, and 2·10. When a signal is passed through this type of amplifier, it will be found to possess the same phase at the output that it did at the

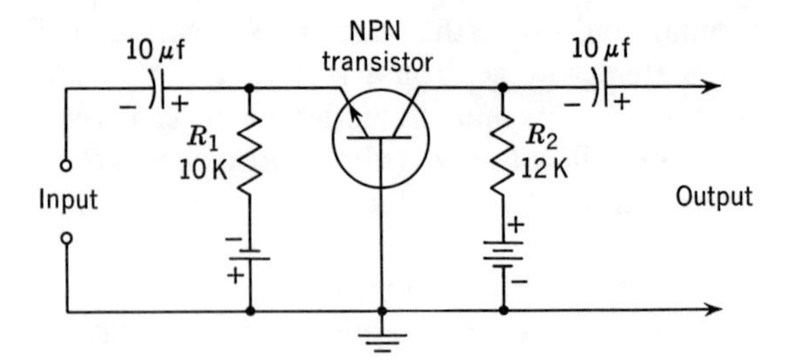


*Fig. 2·15 Common-base transistor amplifier circuit.*

input. To demonstrate this situation, the amplifier of Fig. 2·15 has been drawn using an NPN transistor, and the battery polarities have been chosen accordingly. Assume, now, that the incoming signal is positive at this instant. This positive voltage will counteract some of the normal negative bias between emitter and base and serve to reduce the current flowing through the transistor. This, in turn, will reduce the voltage drop across $R_2$, making the collector potential more positive. Thus, a positive-going input signal produces a positive-going output signal.

During the negative half-cycle of the input signal, the emitter will be driven more negative than it normally is with respect to the base. This will increase the flow of electrons from emitter to collector and cause the negative voltage drop across $R_2$ to increase. This will drive the collector more negative. Again we see that the polarity of the output signal is similar to that of the input signal.

The input impedance of a common-base junction transistor amplifier is low, with values of 100 to 300 ohms typical. The output impedance of this connection is quite high, being on the order of 200,000 ohms.

## 2·10 *Common-collector amplifiers*

The final transistor amplifier arrangement is the *common-collector* circuit. This is shown in Fig. 2·16*a*. The input signal is applied between base and ground, and the output signal is obtained from the emitter. The collector is effectively at a-c ground potential because $C_1$ presents a very-low-impedance path to ground for all a-c signals. Thus we can say that the input signal is effectively being applied between base and collector. By the same reasoning, the output is taken from a load resistor between emitter and ground or, what is the same thing, between emitter and collector. That is why we can call this arrangement a common-collector circuit.

This circuit has an input impedance on the order of several thousand ohms and an output impedance of less than 100 ohms. The voltage gain of a common-collector amplifier cannot exceed 1; in the circuit of Fig. 2·16*a*, it is about 0.8.

It will be seen that the common-collector amplifier is very similar to the vacuum-tube cathode follower. Both have high input impedance and low output impedance. Furthermore, in both there is no phase reversal of the signal as it passes through the stage.

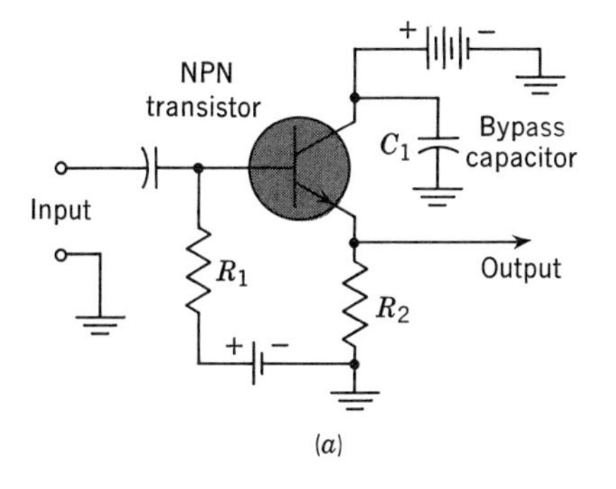


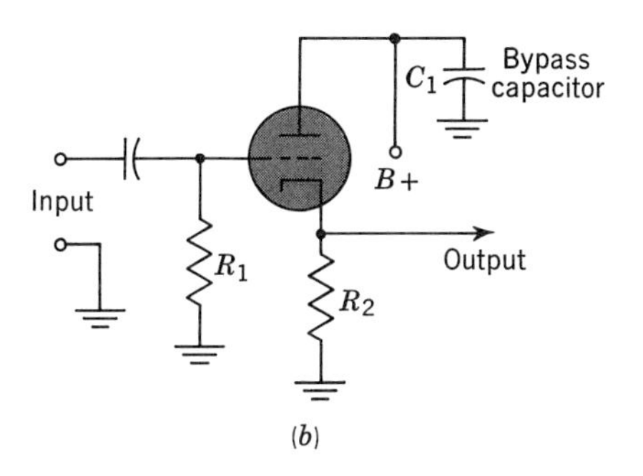


*Fig. 2·16 Comparison of common-collector transistor circuit with triode cathode follower. (a) Transistor circuit. (b) Vacuum-tube circuit.*

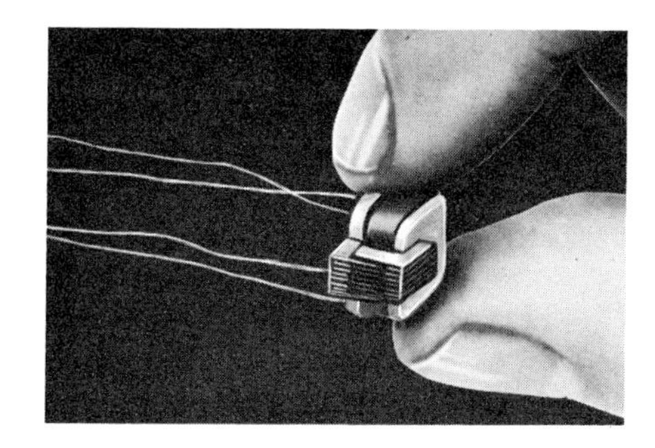

*Fig. 2·17 Miniature transformer for transistor circuits.*

Note that no matter how a transistor is connected, the emitter is biased in the forward direction, while the collector is biased in the reverse or high-resistance direction.

## 2·11 Cascaded amplifiers

Transistor amplifiers, like vacuum-tube amplifiers, are seldom used singly. Rather, it is more common to find them in groups, with two, three, or more stages following each other in order, that is, in cascade. When vacuum-tube amplifiers are used, it is a relatively simple matter to string them one after the other, because a conventional vacuum-tube amplifier has a much higher input impedance than output impedance. Hence, when we connect the input of one stage to the output of the preceding stage, we do not ordinarily affect the prior stage.

Consider, however, a transistor amplifier, say one designed with a common emitter. The input impedance is on the order of 1,000 ohms. The output impedance is more likely to be around 10,000 ohms. Obviously, a direct connection between these two stages would result in a significant loss in gain because of the mismatch. If we accept this reduced gain, it will be necessary to use more stages in order to obtain a desired amplification. Another solution is to insert a step-down transformer which will match the higher output impedance of one stage to the lower input impedance of the following stage. This solution has been used, and special miniature transformers (Fig. 2·17) have been devised for this purpose. Transformers, however, do not ordinarily possess the same flat frequency response that can be obtained from *RC* networks. Also, transformers are more costly, and hence it is often more desirable, from an economic standpoint, to add an extra amplifier stage and use *RC* coupling than revert to transformer coupling. Both methods are used, however, and typical amplifiers of both types will be examined.

**Transformer-coupled amplifiers.** A two-stage transformer-coupled common-emitter amplifier is shown in Fig. 2·18. The interstage transformers have a-c primary impedances of 10,000 ohms each and a-c secondary impedances of 1,000 ohms each. Capacitors $C_1$ and $C_2$ are 10 $\mu$f in value, and resistors $R_1$ and $R_2$ are both 150,000 ohms. The two resistors are needed to establish the proper forward bias for the base-emitter circuits, and the two capacitors are inserted to prevent grounding of the base bias through the low d-c resistance of the transformer secondary windings. In addition, $R_3$ and $R_4$ serve to stabilize the two transistors against temperature variations. $C_3$ and $C_4$ prevent degeneration of the signal.

**RC-coupled stages.** An *RC*-coupled common-emitter amplifier that will provide approximately the same amount of over-all power gain is shown in Fig. 2·19. Note that three stages are required because of the mismatch between the output of one stage and the input of the following stage.

*Fig. 2·18 A two-stage transformer-coupled transistor amplifier using the common-emitter circuit.*

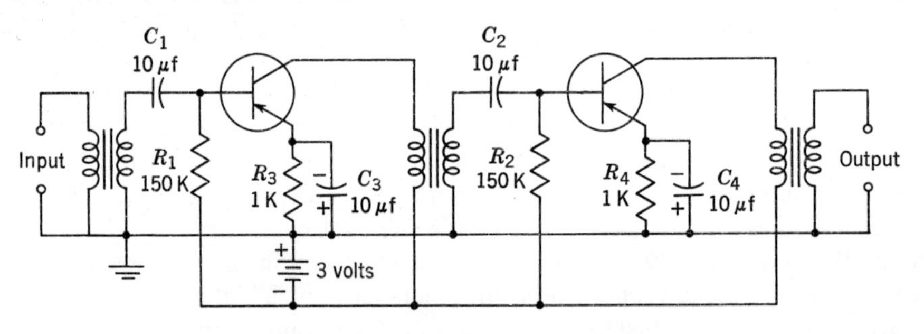


*Fig. 2·19 A three-stage RC-coupled transistor amplifier that will provide approximately the same over-all power gain as the circuit in Fig. 2·18.*

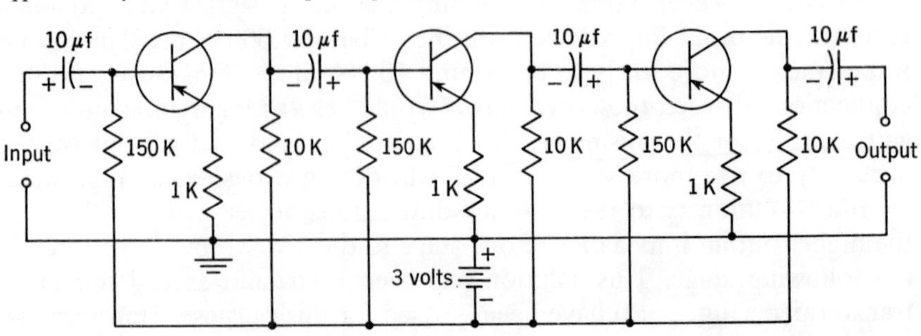

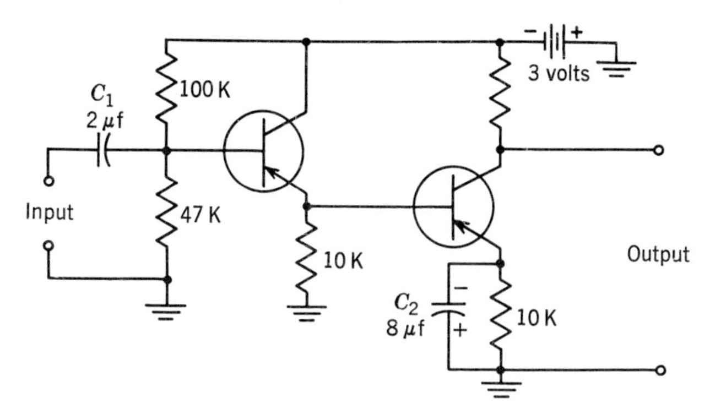


*Fig. 2·20 A two-stage amplifier with high input impedance and d-c stabilization.*

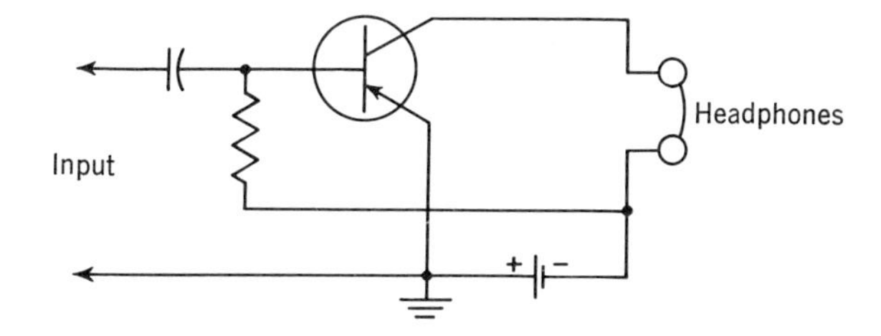


*Fig. 2·21 A direct-coupled transistor amplifier.*

A two-stage amplifier with high input impedance and stabilization against temperature changes is shown in Fig. 2·20. The higher input impedance is achieved through the use of a common-collector stage. The signal it develops is then directly coupled to a common-emitter amplifier. Insertion of 10,000-ohm resistors in the emitter leads of both transistors provides amplifier stabilization against temperature changes. The first 10,000-ohm resistor cannot, of course, be bypassed, since the signal is obtained from this point. In the second stage, however, an 8-μf bypass capacitor is employed.

It should be noted that electrolytic capacitors are used for *RC* coupling, in addition to bypassing. Large values of *C* are necessary because of the relatively low *R* in transistor circuits. Furthermore, the leakage current of electrolytic capacitors is permissible because of the low series resistance.

## *2·12 Direct-coupled amplifiers*

In this circuit, a d-c path exists from the output of one stage to the input of the next stage. In its simplest form, a direct-coupled stage would appear as shown in Fig. 2·21. Here, the output device, a pair of headphones, is directly connected to the collector element of the amplifier stage. In order to employ the phones in this manner, their impedance should match the amplifier output, their operation should not be affected by the collector current flowing through them, and their d-c resistance should not be too high, or the resulting voltage drop will reduce the collector voltage to too low a value. In place of phones, we might use a relay, a meter, or any one of a number of devices.

Another d-c amplifier is illustrated in Fig. 2·20. Here, a direct path exists between the emitter of the common-collector stage and the base of

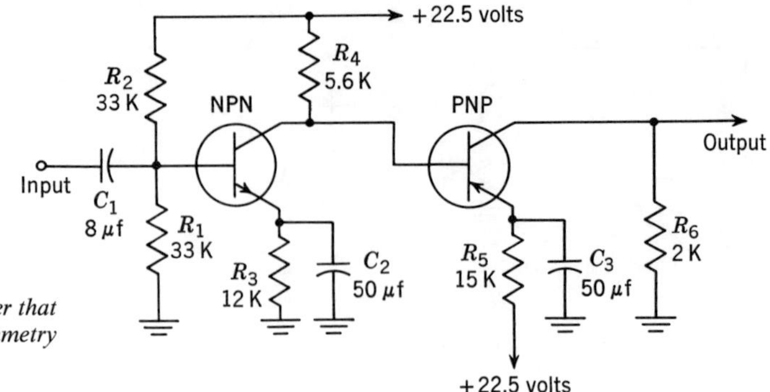


*Fig. 2·22 A direct-coupled amplifier that makes use of the complementary symmetry of NPN and PNP transistors.*

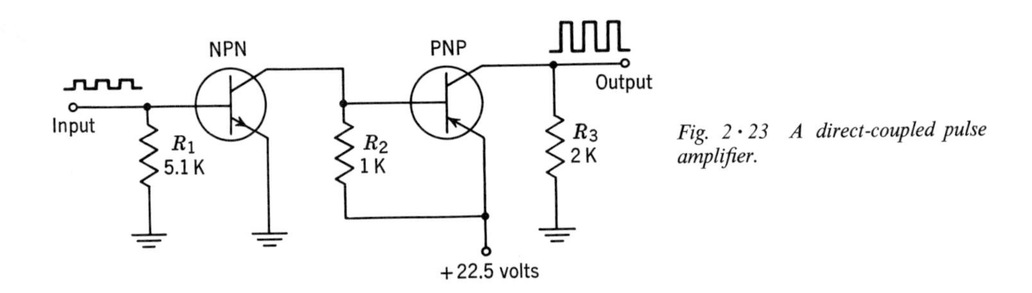


*Fig. 2·23 A direct-coupled pulse amplifier.*

the following common emitter. Any decrease in low-frequency response in this circuit would be entirely due to the input capacitor $C_1$ and the second emitter-bypass capacitor $C_2$.

**D-c amplifier using NPN and PNP transistors.** Another type of direct-coupled transistor amplifier takes advantage of the fact that there are two basic kinds of junction transistors: NPN and PNP units. Each is the symmetrical counterpart of the other, and the polarity of an input signal necessary to increase conduction in a PNP transistor is the opposite of that necessary to increase conduction in an NPN transistor.

A direct-coupled amplifier that makes use of this symmetry is shown in Fig. 2·22. The first transistor is an NPN unit; the second, a PNP type. The first stage is set up so that the collector current, flowing through its load resistor $R_4$, develops just enough negative voltage here to make the base of the PNP transistor negative with respect to its emitter. This establishes the proper conditions in the emitter-base circuit of the PNP unit to bias it in the forward direction. Thus, by the proper choice of resistor values and battery potential, both stages will operate as class A amplifiers.

The application of a signal to the base of the NPN stage will result in an amplified signal appearing across $R_6$. For example, when the signal at the base of the NPN transistor goes positive, an amplified negative voltage will appear across $R_4$ as the collector end of $R_4$ goes negative with respect to the battery end. This increasing negative voltage will provide an even greater forward bias for the base-emitter circuit of the PNP transistor, causing an increased flow of current through this unit. Electrons will flow up through $R_6$, making the top end positive with respect to the bottom end.

Thus, the positive signal applied to the input of this amplifier appears with the same polarity, but in amplified form, at the output. Note the simplicity of this arrangement, which requires no coupling capacitors and only one battery supply; both 22½-volt potentials shown would come from one source.

**D-c pulse amplifier.** Another two-stage amplifier designed along somewhat similar lines is shown in Fig. 2·23. This system has for its sole purpose the amplification of pulses, and as such, its mode of operation is modified accordingly. For example, if you examine the base-to-emitter circuits of both stages, you will note that no forward bias is employed. This means that the transistor is biased close to cutoff. The second stage, however, is not so close to cutoff as the first stage, because the small collector current that flows from the first transistor passes through $R_2$ and develops a small forward biasing voltage, which shifts the operating point of the second transistor away from cutoff.

In the circuit of Fig. 2·23, a positive pulse of 0.25-volt input to the first stage is amplified to a 20-volt peak at the output of the second stage. Since conduction is required only when the pulses are applied, cutoff biasing is used.

## 2·13 *Power transistors*

Just as we have voltage and power vacuum-tube amplifiers, so do we have voltage and power transistor amplifiers. An important consideration in the design of a power transistor is the ability to handle safely the power which is dissipated at the collector. To help remove this heat, power transistors are built with radiating fins (Fig. 2·24) and in metal cases (Fig. 2·25).

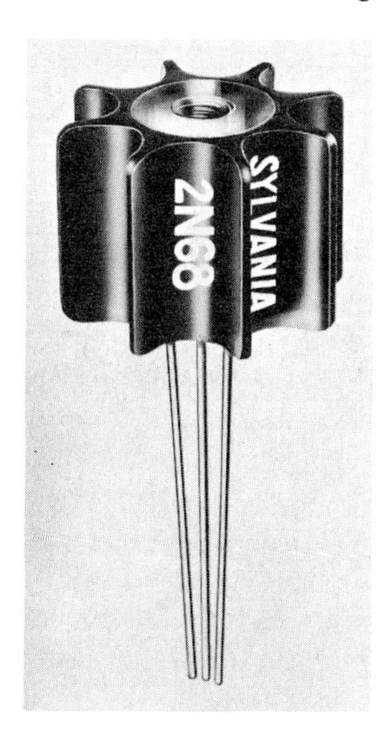


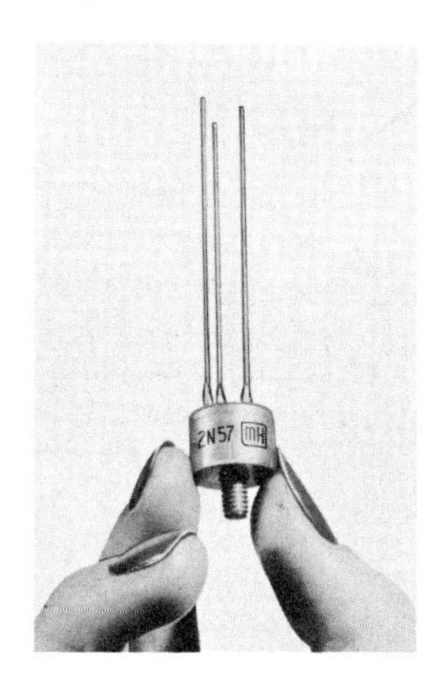


*Fig. 2·25 A power transistor housed in a metallic container.* (Minneapolis-Honeywell Regulator Co.)

*Fig. 2·24 A power transistor with radiating fins to help remove the heat.* (Sylvania Electric Products, Inc.)

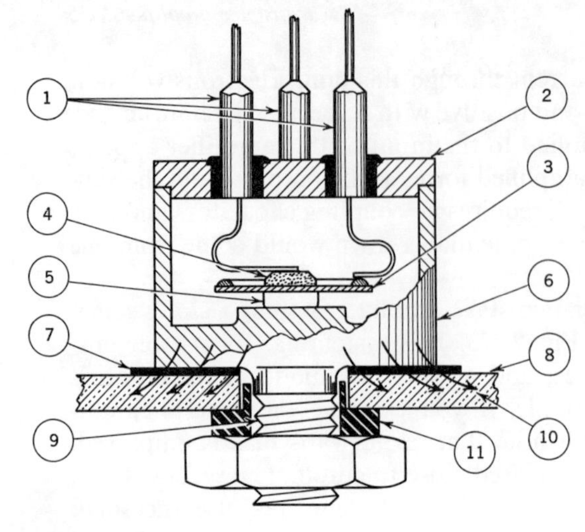


*Fig. 2 · 26 Cross-sectional view of the power transistor shown in Fig. 2 · 25. (1) Lead wires for base, collector, and emitter. (2) Metal top with insulators for base and emitter leads, hermetically sealed. (3) Base (germanium crystal wafer with nickel reinforcing ring). (4) Emitter (indium metal). (5) Collector (indium metal). (6) Metal case for crystal assembly. (7) Mica insulator. (8) Chassis, which usually serves as a heat sink. (9) Mounting stud, integral part of metal case. (10) Arrows indicating path of direct heat flow. (11) Plastic insulator bushing.*

*Fig. 2 · 27 A 10-watt power transistor. (RCA)*

Internal heat is transferred to the external metal case by metallic conduction with very little drop in temperature. The metal conducting surface may be soldered to any of the three transistor elements; in the power transistor of Fig. 2 · 25, the outer metal cup is soldered to the collector.

A cross-sectional view of this power transistor is shown in Fig. 2 · 26. Note how the collector (item 5) is set flush against an inner surface of the metal housing. If the design of the circuit permits, the transistor is mounted flush with the chassis, assuring the maximum dissipation of heat. In this case, items 7 and 11 of Fig. 2 · 26 would not be employed. If the collector must be electrically insulated from the chassis, however, there would be a small mica washer (item 7) on one side of the chassis and a nylon bushing (item 11) to insulate the stud and nut from the other side of the chassis.

Figure 2 · 27 shows another power transistor. Power transistors have also been developed with removable fin structures. This arrangement offers the advantage of fitting the fin structure to the particular application of the transistor. A small external structure would be employed for relatively low power outputs, and a larger fin structure would be used for higher power applications.

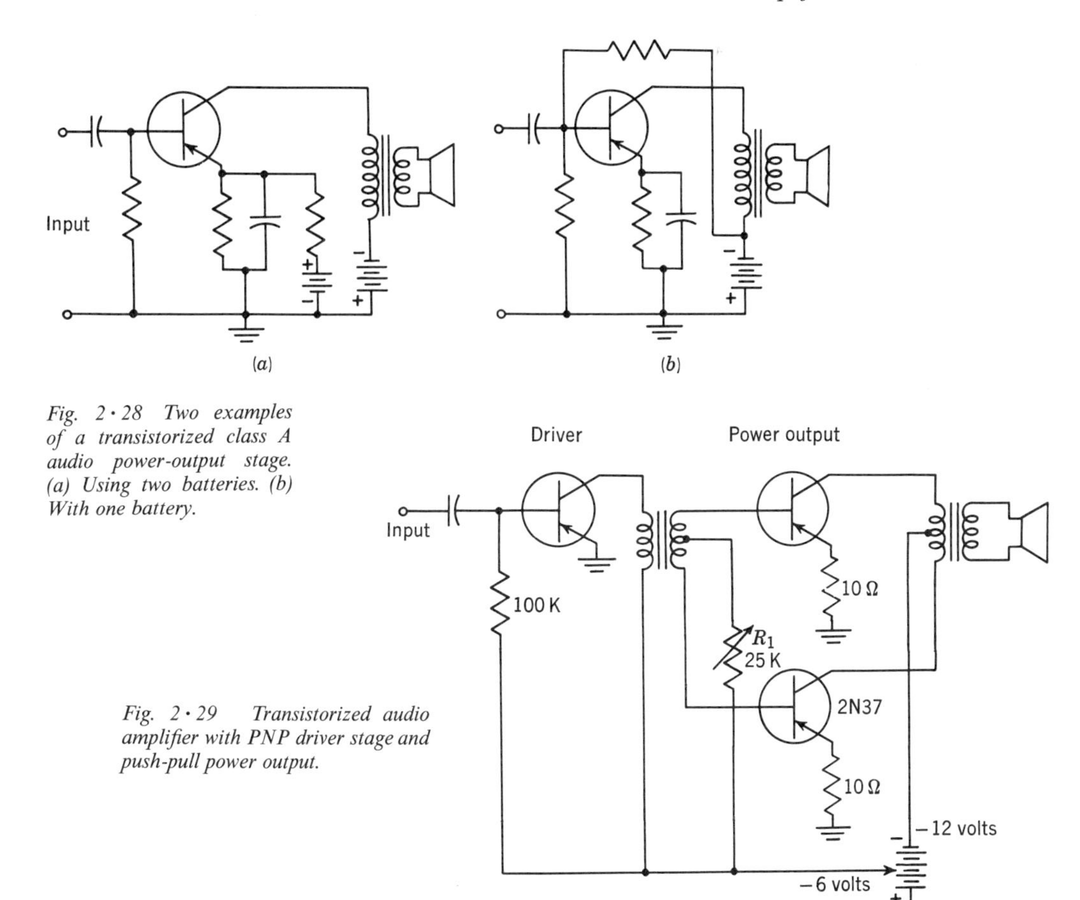


*Fig. 2·28 Two examples of a transistorized class A audio power-output stage. (a) Using two batteries. (b) With one battery.*

*Fig. 2·29 Transistorized audio amplifier with PNP driver stage and push-pull power output.*

## 2·14 Class A power amplifiers

Circuit arrangements of single-ended power amplifiers do not differ to any marked degree from corresponding voltage amplifiers. Figure 2·28 illustrates two class A power amplifiers designed to drive the loudspeaker of an audio amplifier. In one instance, two batteries are employed; in the other, a single battery. The output transformer would be designed to match the impedance of the collector in the primary and the loudspeaker in the secondary. The amount of power that may be obtained from this arrangement will be governed by the size of the battery and the permissible dissipation in the transistor itself. As in vacuum-tube practice, a single-ended audio amplifier can be operated only class A.

**Class A push-pull amplifiers.**[1] Power amplifiers can also be operated in push-pull. A typical illustration of an audio amplifier using a single driver stage and a class A push-pull output stage is shown in Fig. 2·29. All

[1] The details of push-pull circuits are explained in Sec. 3·6.

transistors are operated with grounded emitters; transformer coupling is employed between the driver and output stages and between the output amplifiers and the loudspeaker. The variable resistance $R_1$ is adjusted for a total collector current of 8 ma.

Push-pull amplifier operation results in the cancellation of second harmonics within the stage. For the same amount of distortion, a class A push-pull amplifier can be driven harder, providing greater output. It also means that we can obtain more than twice the output with push-pull operation than we can get using two similar transistors as single-ended amplifiers.

## 2·15 Class B power amplifiers

In class A push-pull operation, the average current that flows remains steady, whether or not a signal is being applied to the stage. More efficient operation can be achieved with class B operation, where each transistor is biased to cutoff. When no signal is applied, practically no current flows and no power is being dissipated.

The circuit of a class B push-pull amplifier is shown in Fig. 2·30. Three power transistors are employed; the first one serves as a class A driver amplifier and the remaining two serve as a class B output stage. Efficiency of the class B stage is close to 75 per cent. This is achieved because, with no signal, the total class B collector current is extremely low, since the stage is biased near cutoff. In a class A amplifier, the efficiency is perhaps half this amount or less.

Input signals are applied to the base of the first audio amplifier. A 2,200-ohm bypassed resistor in the emitter circuit of this driver stage provides thermal stabilization only; it does not introduce signal degeneration. However, just above it is an unbypassed 120-ohm resistor, and this does provide signal degeneration.

The output of the driver stage is transformer-coupled to the class B amplifier not only for proper impedance matching but also to provide two signals 180° out of phase with each other (as required by the class B

*Fig. 2·30 A two-stage audio amplifier. The push-pull output stage is operated class B.*

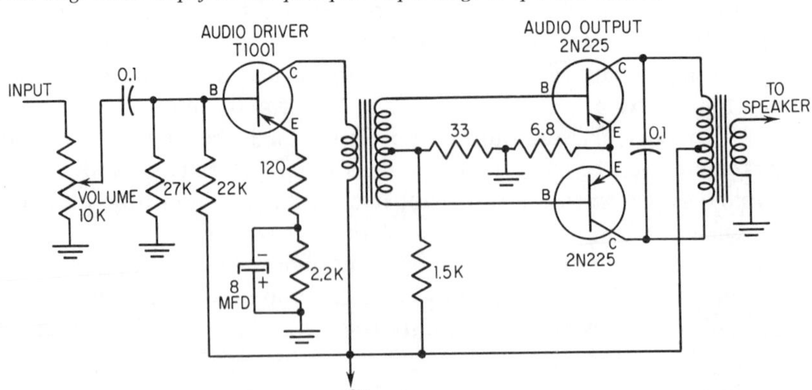

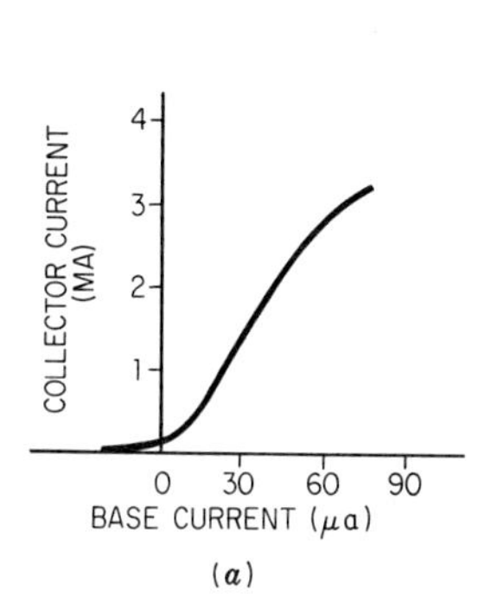


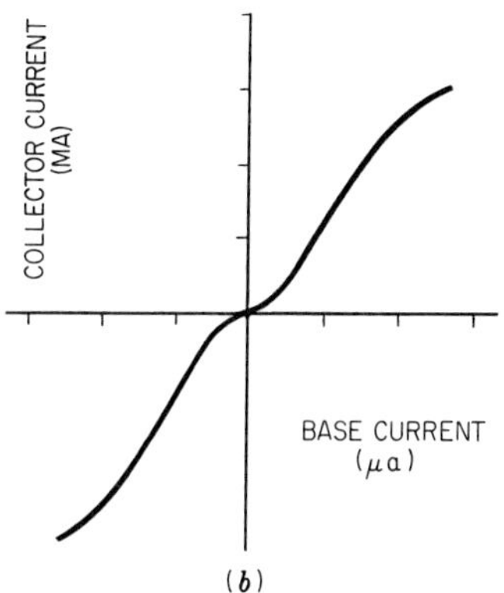


*Fig. 2·31 The mechanism of crossover distortion. (a) A single transistor. (b) Two transistors connected in push-pull. (c) Distorting effects caused by the jog in the transfer characteristic curve.*

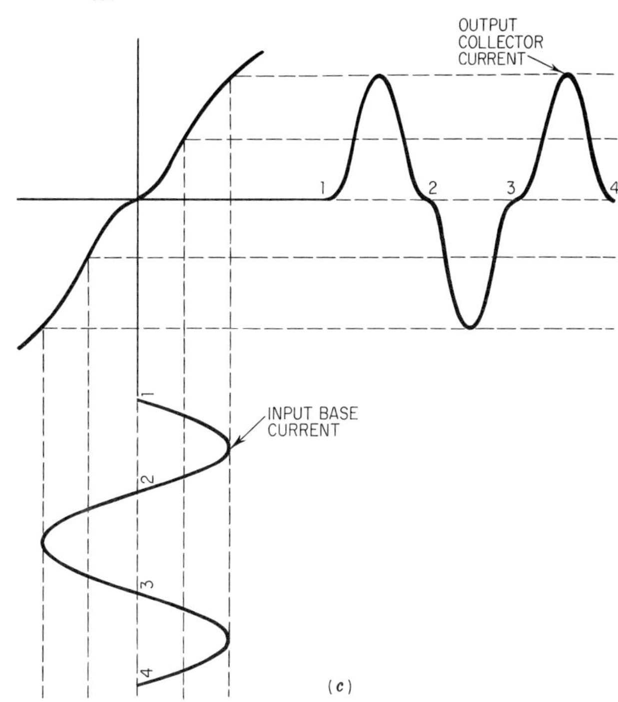


amplifier). A 6.8-ohm resistor in the emitter circuit of this output stage introduces a small amount of signal degeneration to improve the stability of the circuit. The base circuit contains a 33-ohm resistor, across which a small voltage from the negative 6-volt supply is developed. The purpose of this voltage is to reduce crossover distortion. (This will be explained presently.) The collector elements of the two output transistors connect to opposite ends of the output transformer, and the d-c voltage is brought in at the center tap. The 0.1-μf capacitor across the primary of the output transformer removes the highs from the output signal for a more mellow output tone. The speaker is an extremely small one which, because of its dimensions, naturally tends to emphasize the higher frequencies. This is counteracted somewhat by the 0.1-μf capacitor.

Class B audio amplifiers are favored in many transistor receivers not only for their greater power and reduced distortion but also because their current drain is practically zero when no signal is being received. If two class A output amplifiers were connected in push-pull, an average current would always flow and impose a constant drain on the battery. Since these are power transistors, their current requirements are fairly high and a significant amount of power would be dissipated.

Now let us examine the reason for the 33-ohm resistor in the base circuit of the output amplifier. Its purpose is not so much to provide base-emitter bias as it is to minimize a condition known as crossover distortion. When transistors are connected back to back in push-pull arrangements and the bias is zero, there is a region near cutoff where their respective

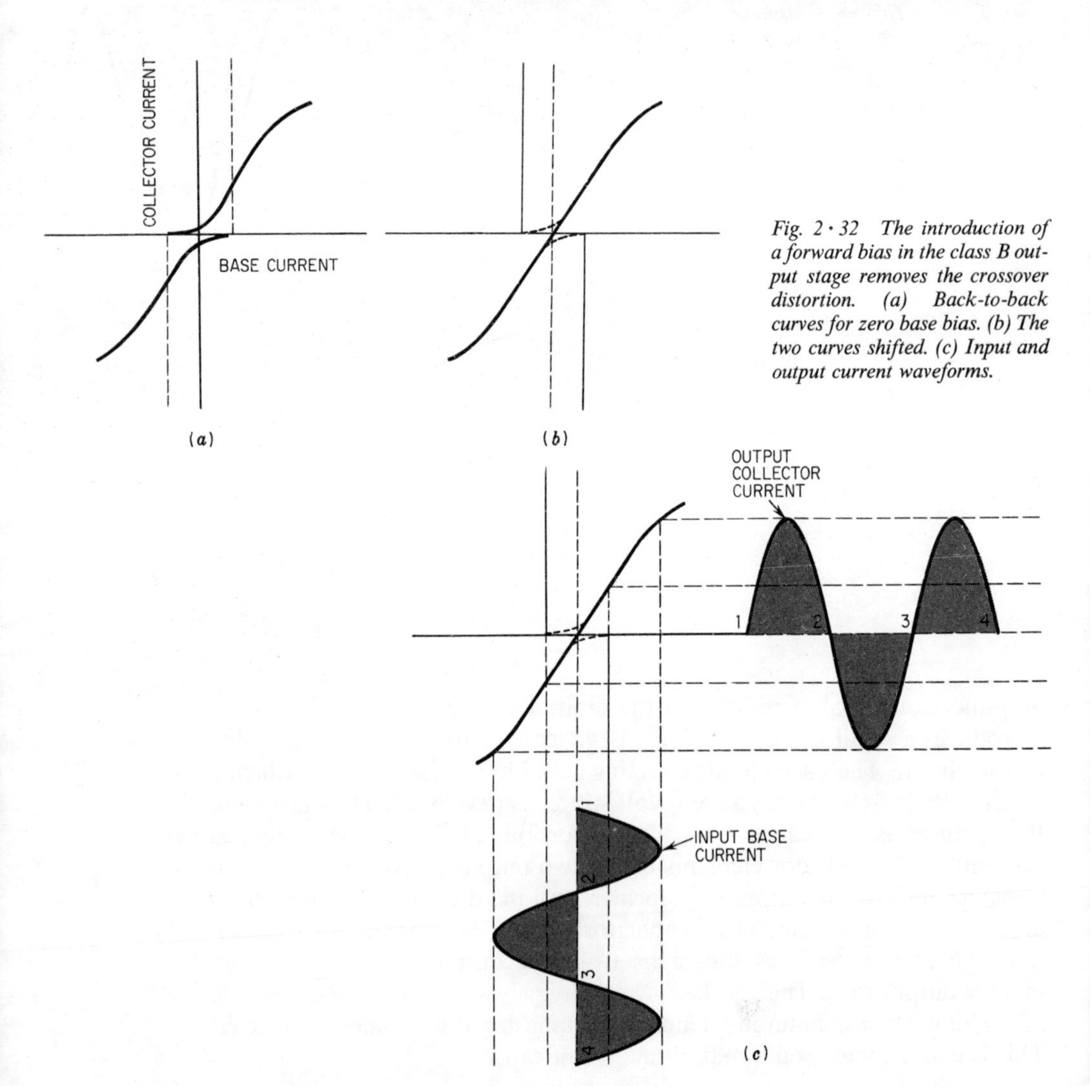


*Fig. 2·32 The introduction of a forward bias in the class B output stage removes the crossover distortion. (a) Back-to-back curves for zero base bias. (b) The two curves shifted. (c) Input and output current waveforms.*

characteristic curves tend to become nonlinear (Figs. 2·31*a* and *b*). Note the jog in both curves near the origin. If we now introduce a sinusoidal base current into the input circuit (Fig. 2·31*c*), we shall obtain the distorted collector current indicated to the right of the characteristic curve. This distortion becomes more severe as the signal level decreases.

To prevent crossover distortion, a small amount of forward bias is introduced between the base and emitter of each transistor. In Fig. 2·32*a*, the transfer characteristics of the transistors are shown back to back for zero base bias. These curves are not combined. The dashed lines indicate the base-current values when forward bias is applied to provide the overall dynamic operating curve of the amplifier. With this forward bias, the two curves must be shifted until the dashed lines are aligned with each other (Fig. 2·32*b*). Note that now the jog at the center of the curve has disappeared. If we apply a sine-wave signal, the undistorted output shown in Fig. 2·32*c* is obtained. The 33-ohm resistor in Fig. 2·30 provides this forward bias for the output stage. The output of the audio system in Fig. 2·30 is approximately 150 mw to a 3-in. speaker.

A second audio system is shown in Fig. 2·33. There is one small circuit variation in Fig. 2·33 that the reader should become familiar with. This occurs in the class B output stage, where two PNP transistors are employed. The voltage from the battery is applied to the emitter and base from the +12-volt line. The collectors, however, are returned to ground (negative side of the battery in this circuit). In a vacuum-tube amplifier, this would be equivalent to placing a large negative voltage on the cathode and returning the plate to ground. Since it is the relative potential between the two elements that produces current flow through the device, it makes little difference whether the cathode is made negative or the plate is made positive. The same type of reasoning applies to transistors. Note, however, that because a large positive voltage is applied to the emitter, the same

*Fig. 2·33 A typical audio amplifier system.*

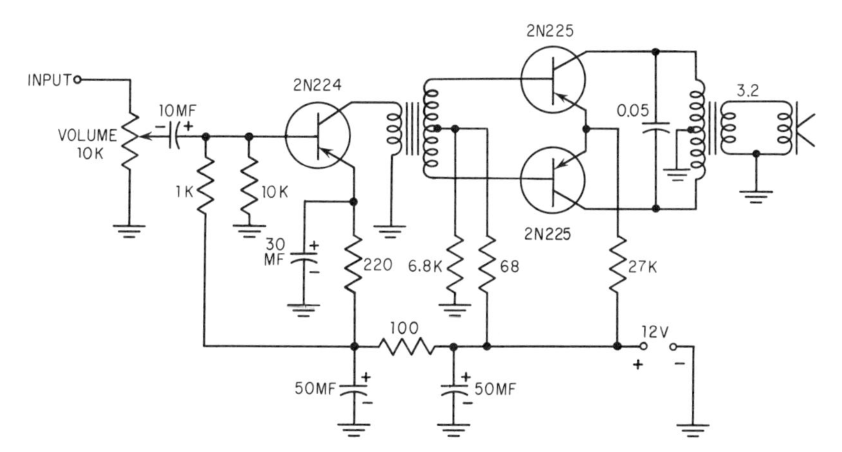

+12-volt line must also be connected directly to the base circuit. If the base were similarly grounded and the large positive voltage were applied to the emitter, excessive current would flow through the base-emitter circuit and destroy the transistors.

A similar voltage arrangement is employed in the driver stage preceding the class B output amplifier.

## *2·16 The tunnel diode*

This is a new semiconductor device capable of operation in the 1- to 10-kilomegacycle[1] range. Although used in the conventional applications of amplifying, oscillating, and switching, its principle of operation is entirely different from the transistor or the vacuum tube. The name "tunnel diode" has been adopted because the physical mechanism by which the device functions is caused by a complex quantum-mechanical tunneling process. In actual practice, the tunnel diode is basically a very heavily doped PN junction, and thus, as one might expect, it possesses many of the properties of a conventional diode.

The tunnel diode has two outstanding properties: (1) extremely high frequency response and (2) very low power consumption. For example, where a vacuum tube may operate at 100 Mc and consume 1 watt of power and a conventional transistor may operate at 100 Mc and consume 10 mw of power, a tunnel diode can operate at 500 Mc and consume 1 mw of power. The tunnel diode is not without its problems, however. Instability and unwanted signal feed-through plague the device, and these difficulties are not easily resolvable. The instability arises because a tunnel diode is a negative-resistance device and negative resistance is always difficult to control. Feed-through arises because the tunnel diode is a two-terminal device and its input and output terminals are the same. Compare this with the vacuum tube and transistor, where the third lead offers isolation between input and output.

The discussion to follow will present the operation, physical characteristics, and applications of the tunnel diode.

**Operation of the tunnel diode.** The operation of the tunnel diode is dependent upon a quantum-mechanical principle called tunneling. To appreciate its significance, let us review semiconductor-diode operation. Each diode contains two sections, a P section and an N section. The P section possesses an excess of holes, while the N section possesses an excess of electrons. Between the two sections there is a depletion layer containing very few free electrons or holes. That is, there are no mobile charges in this region. This depletion region amounts to a potential barrier, and to pass current through the diode, sufficient external voltage must be applied, with the proper polarity, to overcome the potential barrier. The proper polarity, of course, is connection of the negative battery terminal to the N section and the positive battery terminal to the P section.

In the absence of any external voltage or in the absence of sufficient

[1] Note that kilomegacycle or $10^9$ cps is the same as gigacycle (Gc).

voltage, very little current will pass across the junction. However, some carriers on both sides of the junction will attain enough thermal energy to surmount the barrier presented by the depletion layer and reach the other side. This, however, occurs only to those relatively few electrons and holes capable of attaining a sufficiently high energy state.

Classical physical theory states that unless a carrier possesses enough energy to overcome a potential barrier, it will never cross or surmount that barrier. The more recent quantum mechanics, however, contradicts classical physics in this instance and states that a carrier can reach the other side of a potential barrier even though it does not have enough energy to surmount the barrier. The carrier does this by tunneling through the potential hill. In fact, quantum mechanics can predict the probability of this occurrence. The probability of a carrier's tunneling through the potential hill is dependent on how far the particle must tunnel, i.e., the thickness of the potential hill. The probability of tunneling is nil unless the barrier is extremely narrow. This is proved by the fact that tunneling does not occur in a conventional PN junction.

To make a tunnel diode, the PN junction must be very heavily doped with impurities. The large number of impurities produces a very narrow depletion layer. With this thin layer, electrons can tunnel their way from the N region to the P region. This gives rise to an additional current in the diode at very small forward bias which disappears when the bias is increased. It is this additional current that produces the negative resistance in a tunnel diode. It has also been found that the electrons travel through the depletion layer at tremendously high velocities. This enables the tunnel diode to operate at far higher frequencies than conventional transistors. The theoretical frequency limit is in the neighborhood of 1 million Mc.

Figure 2 · 34 shows the voltage vs. current curve of a conventional PN junction diode. The forward voltage required to cause current flow is

*Fig. 2 · 34 The voltage vs. current characteristic curve of a conventional diode.*

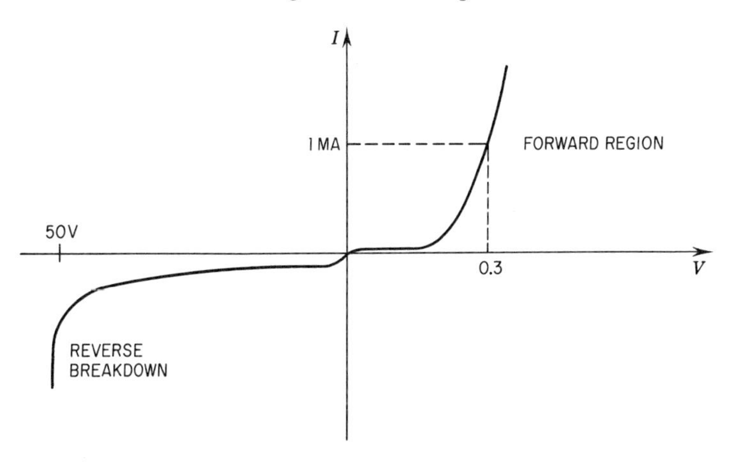

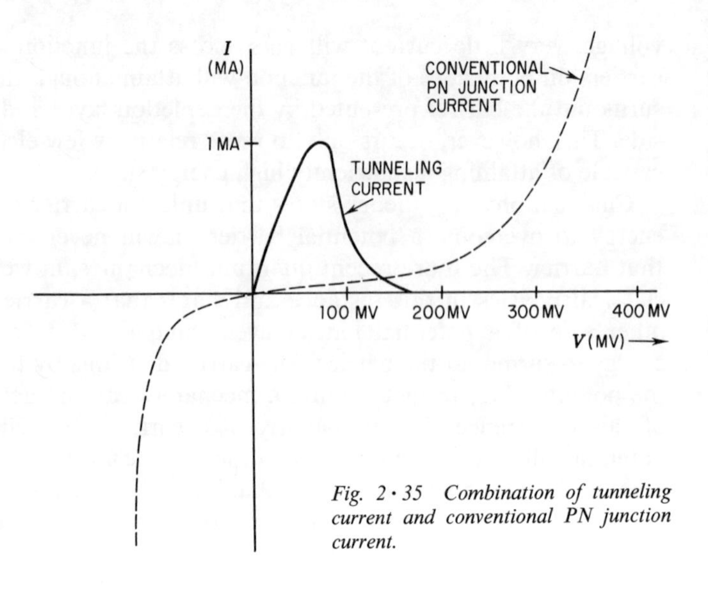


*Fig. 2 · 35 Combination of tunneling current and conventional PN junction current.*

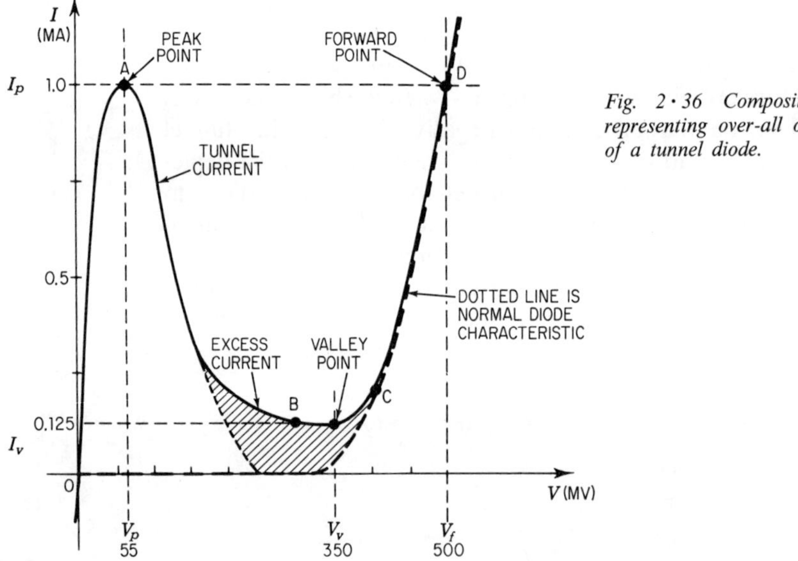


*Fig. 2 · 36 Composite characteristic representing over-all operational curve of a tunnel diode.*

approximately 0.3 volt for germanium diodes and 0.8 volt for silicon diodes. The reverse breakdown usually occurs between 20 and 200 volts. If we superimpose on this characteristic curve the tunneling current which occurs at a forward bias of about 50 to 100 mv, we obtain the characteristic of Fig. 2 · 35. The composite characteristic is shown in Fig. 2 · 36. Note that the tunneling current appears at very small voltages. As we continue to raise the applied voltage, the carriers receive enough energy to surmount the potential barrier and cross the junction in the conventional manner.

In the composite curve of Fig. 2·36, the dotted line in the valley region shows the curve one would expect to get if the two curves of Fig. 2·35 were superimposed. The solid line shows the actual curve which is obtained. The excess current which occurs in the valley region is as yet unexplained. The important characteristic to note is that the region between points *A* and *B* represents a negative resistance; i.e., as the voltage is increased, the current decreases (just the reverse of the behavior of an ordinary resistor).

Now let us see how the tunnel diode is employed in an oscillator and in an amplifier.

**Oscillators.** To achieve oscillation with a tunnel diode, it must be set up so that the negative resistance it provides is greater than the positive resistance of the resonant components in the circuit. A typical tunnel diode has a negative resistance of approximately −100 ohms when it is biased at the center of its negative-resistance region. To bias the diode correctly, approximately 125 mv is required. At this voltage, the current drawn by the diode is approximately 0.5 ma.

There are several ways to set up this bias circuit. We can connect the tunnel diode in series with a resistor and a small battery. If we assume a battery voltage of 3 volts, then a series resistor of 6,000 ohms would be required to limit the current flow to 0.5 ma. But the 6,000 ohms would completely overshadow the −100 ohms of the tunnel diode and prevent us from properly utilizing this negative resistance.

This difficulty can be circumvented by the resistive arrangement shown in Fig. 2·37. The necessary bias voltage is here developed across a 20-ohm resistor, a value considerably less than the −100 ohms of the tunnel diode.

We can now connect a resonant circuit in series with the tunnel diode, and if this circuit possesses a resistance less than 80 ohms, then the tunnel diode will completely balance out this resistive loss of the circuit and enable oscillations to take place. A suitable circuit capable of oscillating into the megacycle range is shown in Fig. 2·38.

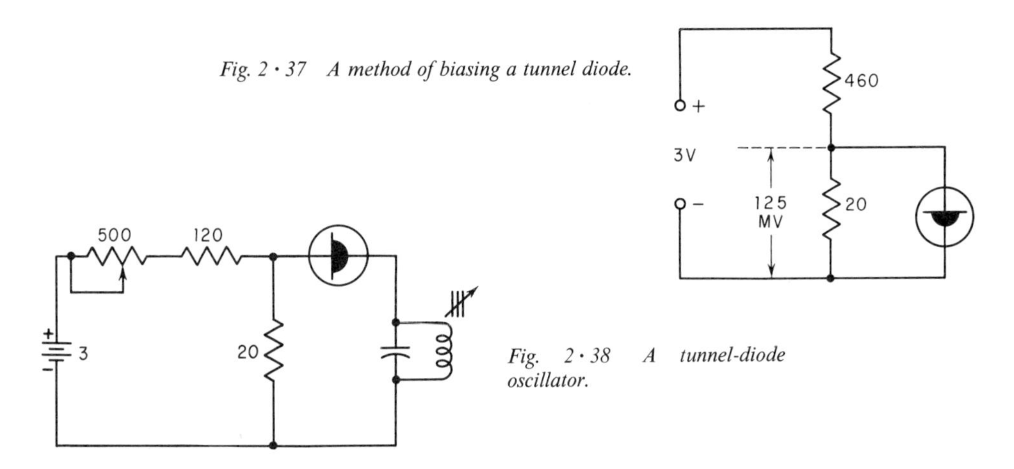


*Fig. 2·37 A method of biasing a tunnel diode.*

*Fig. 2·38 A tunnel-diode oscillator.*

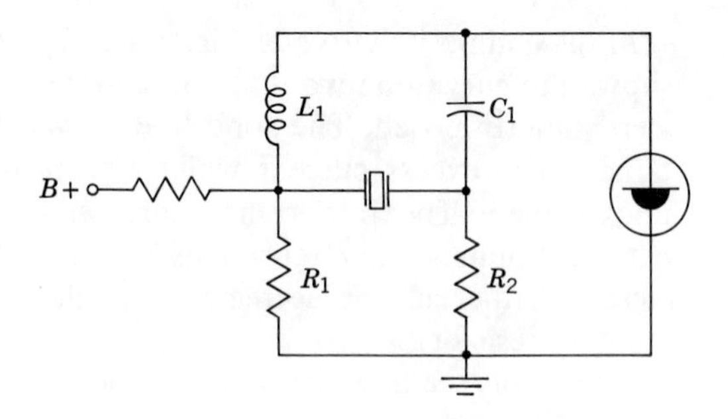


*Fig. 2 · 39 A tunnel-diode crystal-controlled oscillator.* (From GE "Tunnel Diode" Manual.)

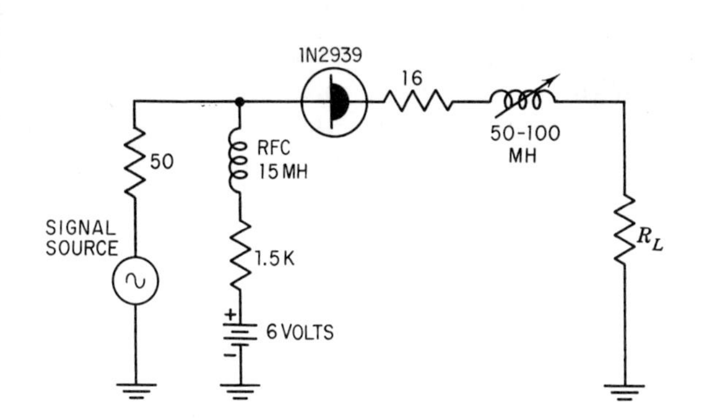


*Fig. 2 · 40 A tunnel-diode amplifier.*

A crystal-controlled oscillator using a tunnel diode is shown in Fig. 2 · 39. $R_1$ and $R_2$ are selected to each have about twice the value they should have to enable the negative resistance of the tunnel diode to dominate the circuit. A crystal is placed between these resistors. At all frequencies other than the resonance, the crystal impedance is high and the circuit is unable to function. At the resonant frequency, however, the crystal becomes a short circuit and $R_1$ is placed in parallel with $R_2$, thus reducing their total resistance value to half their individual values. This new value permits the circuit to oscillate freely at a frequency accurately governed by the crystal.

**Amplifiers.** If we increase the impedance of the external circuit connected across a tunnel diode until it equals the negative resistance of the tunnel diode, amplification rather than oscillation is obtained. This is done in the 100-Mc amplifier shown in Fig. 2 · 40. The IN2939 tunnel diode has a negative resistance which is just counterbalanced by the circuit positive resistance. The latter, in this instance, is equal to 50 ohms from the signal source, 50 ohms from the load, 2 ohms from the internal lead resistance of the IN2939, and 16 ohms from the series resistor. In this particular configuration, the 16-ohm resistor was added simply to achieve this counterbalancing.

While the actual mathematical justification for the above-indicated condition is quite complex, some inkling of how it is arrived at may be seen from an examination of the simple circuit shown in Fig. 2 · 41. Assume that the tunnel diode is biased to the center of its negative-resis-

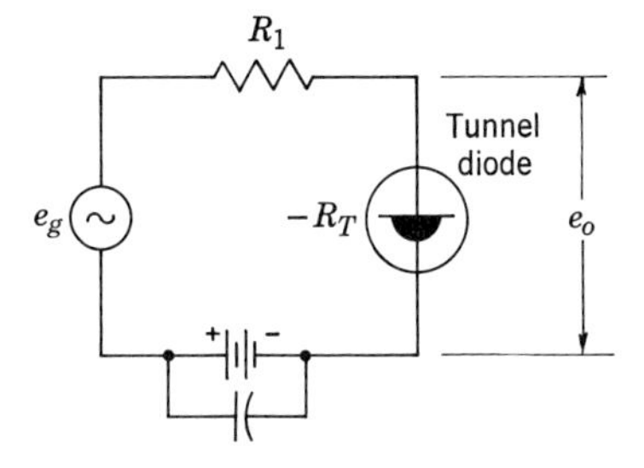


*Fig. 2·41 Simple circuit to demonstrate ability of tunnel diode to amplify.* (From Radio-Electronics, July, 1960.)

tance range and that $e_g$ represents an incoming signal and $e_o$ is the output signal. Then the voltage across the tunnel diode, with negative resistance $-R_T$, is

$$e_o = \frac{e_g(-R_T)}{R_1 + (-R_T)}$$

When $R_1$, the circuit positive resistance, exactly equals $-R_T$ in value, $e_o$ becomes infinite. In an actual circuit, of course, this does not happen, but the largest output is obtained when $R_1 = R_T$. At this point, the ratio of $e_o$ to $e_g$, or the voltage gain, is greatest.

## 2·17 *Servicing transistor circuits*

The servicing and repair of transistor circuits will not present too many problems provided a systematic and logical approach is taken. This includes first checking the applied voltages, particularly when they are supplied by a battery, to make certain that the voltages are at the proper value. Furthermore, if the voltages at the transistor elements are also correct, then the transistor itself may be tested. If this unit is found to be good, any coupling or bypass capacitors that are present in the circuit should be checked.

**Applied voltages.** The applied voltages should not vary more than 10 per cent from their specified values. If they are within this range, they may be presumed to be good. If the voltage reading is off by 20 per cent or more, the stage output may be weak or distorted, although it should not be dead. Since transistor characteristics are linear for very low voltages and currents, chances are that distortion will not occur until the battery voltage drops more than 20 per cent. There is, however, no set rule regarding this drop, and it is best to try a new battery when the voltage of the existing battery has decreased by that amount. If the voltage is supplied by an a-c power supply and its output is low, some defect in the supply is indicated unless a short circuit is present in the transistor amplifier, placing a large drain on the power supply. The latter effect can be determined by disconnecting the power supply and measuring its output voltage. If the voltage returns to normal, check the amplifier for a short circuit. If the power-supply voltage is still low after being disconnected from the amplifier, look for a defect in the supply.

**Transistor precautions.** Transistors may be checked either by testing them in a transistor tester or by substituting another transistor for the suspected unit. If the original transistor is plugged into a socket, removal of the unit for testing is readily accomplished. If the transistor is soldered into the circuit, however, it must first be unsoldered, and here special precautions are required. For example, the soldering-iron rating must not be higher than 35 or 40 watts. Further, to provide the transistor with the maximum protection while it is being soldered or unsoldered, it is good practice to grasp the terminal lead tightly with long-nose pliers positioned between the transistor body and the lead end. With this arrangement, any heat traveling along the wire will be shunted away from the transistor housing. It is desirable to retain the pliers on the wire for a short time after the iron has been removed to make certain that all heat has been dissipated. It is also good practice to provide such a heat shield when other wires are being soldered to any terminal lugs to which a transistor lead is attached.

Whenever voltages are to be applied to a transistor circuit, make certain the various transistors are firmly in place. Never insert or remove a transistor when voltages are present. This precaution will prevent the appearance of surge currents which, if they are powerful enough, can permanently damage a transistor. Always disconnect the applied voltage first. The sensitiveness of a transistor to surge currents should be borne in mind when a voltmeter is being used to check voltages at various points in a transistor circuit. Because of the closeness with which components are

*Fig. 2 · 42 A typical transistor housing and lead arrangement. Note how protruding tab identifies emitter lead. Base and collector leads then follow in rotation.*

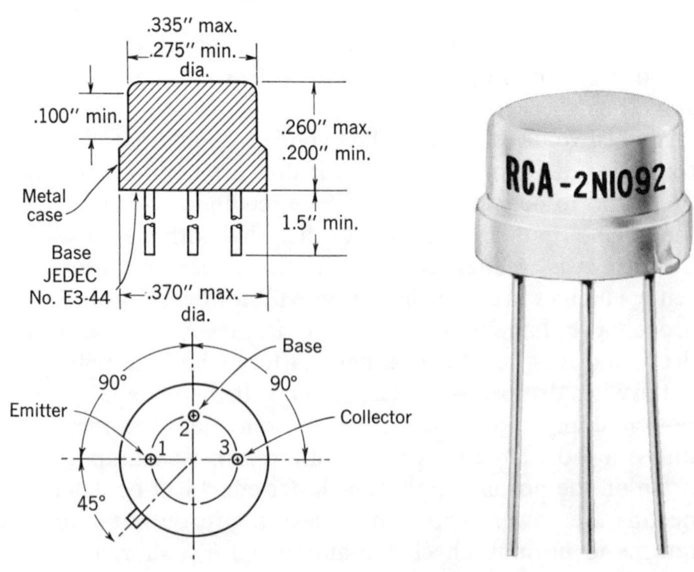

placed, if care is not taken it is easy for the probe tip accidentally to touch two closely spaced terminals. This simple slip may cause the battery or power supply to burn out or may result in a current surge through the transistor. Be especially careful not to let the probe make simultaneous contact with the collector and emitter electrodes. Extra emphasis is placed on this precaution because of the ease with which the mistake can be made. In vacuum-tube circuits, similar slips may occasionally cause a component to burn out, although they rarely affect tubes. In a transistor circuit, the transistor is usually the weakest link, and it becomes the victim.

**Identifying electrodes.** Whenever any measurements are to be made on a transistor circuit, it is important for you to be able to identify the emitter, base, and collector leads extending from the shell of the transistor. A number of manufacturers construct their transistors so that the three element leads are arranged in the manner shown in Fig. 2 • 42. Other methods of positioning transistor leads are shown in Fig. 2 • 43. Sometimes, a colored dot or stripe on the side of the case identifies the closest wire as a specific connection, say the emitter or base or collector. The remaining elements then follow in order.

A fourth wire, if used with a three-element transistor, connects to an internal shield, and this lead generally is grounded.

If there is ever any doubt about which lead represents a certain element in a transistor that is wired into a circuit, either consult the characteristic sheet published by the manufacturer on that transistor, or else trace the circuit leading to each transistor terminal until the proper identification has been made.

**Resistance measurements with the ohmmeter.** For all resistance measurements, the power normally supplied to the transistor circuit would be turned off. Suppose that you wish to check the value of resistance $R_1$ in the circuit of Fig. 2 • 44. You apply the two ohmmeter leads as shown by

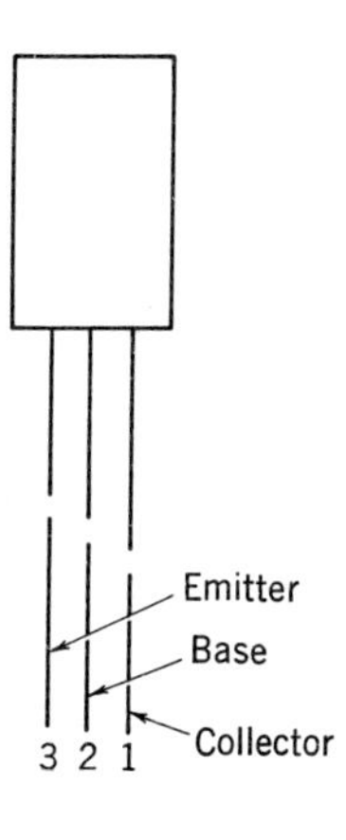


*Fig. 2 • 43 Additional method for transistor leads. From left to right: emitter, base and wide spacing for collector.*

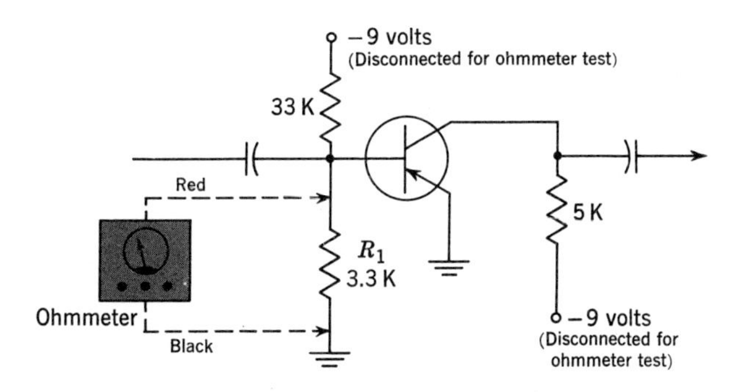


*Fig. 2 • 44 Checking a resistor with ohmmeter in transistor circuit.*

the dotted lines. Now it is important to note that in transistor circuits what you will read on the ohmmeter scale depends on how the leads are connected. You may read either the correct 3,300-ohm value of $R_1$ or practically zero ohms, even though $R_1$ is in good condition.

If the internal battery of the ohmmeter is so connected in its circuit that the red lead is positive and the black lead is negative, the ohmmeter will record the true value of $R_1$. Then the ohmmeter leads will apply a positive voltage to the base of the circuit transistor and a negative voltage to ground or to the emitter. Since the emitter connects directly to ground, the voltage will reverse-bias the base-emitter elements and prevent the transistor input circuit from placing a low resistance across $R_1$. As a result, the ohmmeter will read the true value of $R_1$.

Note, however, what happens when the ohmmeter leads apply a negative voltage to the top of $R_1$ at the base and a positive voltage to ground at the emitter. Now the base-emitter junction is forward-biased and its impedance is very low. In essence, then, we are shunting the very low resistance of the transistor across $R_1$. This reading will lead you to believe $R_1$ is defective, when in fact it is not. It may also cause the transistor to burn out because the low base-emitter resistance will cause a high current to flow.

Hence it is quite important to know not only the voltage polarity present at each ohmmeter lead, but also the value of this voltage. This value can be determined by connecting the ohmmeter leads to a d-c voltmeter. If resistance measurements are made in a transistor circuit and the transistor is easily removable, it should be removed. If the transistors are soldered into the circuit and not easily removed, extreme care should be taken in performing resistance measurements.

Be careful when capacitors are being checked, because here again a voltage must be applied to perform the test. For such tests, it is recommended that one lead of the capacitor be unsoldered from the transistor circuit and then the test performed.

## SUMMARY

1. The emitter is biased in the forward, or low-resistance, direction.
2. The collector is biased in the reverse, or high-resistance, direction.
3. Although the current reaching the collector is less than the current that started out from the emitter, power gain is achieved in a transistor because the collector resistance is so much greater than the emitter resistance.
4. In a common-emitter amplifier, the input signal is applied to the base, and the output signal is obtained from the load resistor in the collector circuit.
5. In a common-base amplifier, the input signal is applied to the emitter, and the output signal is obtained from a load resistor in the collector circuit.
6. In a common-collector amplifier, the input signal is applied to the base, and the output signal appears across a resistor in the emitter circuit.
7. A phase shift of 180° occurs in signals passing from input to output in common-emitter amplifiers. No signal phase shift takes place in common-base and common-collector amplifiers.

8. Signal gain is greatest in common-emitter amplifiers; gain is somewhat less in common-base amplifiers and less than 1 in common-collector amplifiers.
9. Transformer coupling between stages produces the best impedance match and hence greatest gain. *RC* coupling is more economical and possesses a wider bandwidth.
10. In direct-coupled amplifiers, there is a d-c path from the output of one amplifier to the input of the following amplifier. D-c amplifiers frequently use one type of transistor, such as an NPN, in one stage, followed by the opposite type of transistor, a PNP, in the next stage.
11. Power transistors are built with radiating fins or in metal housings to dissipate the heat which develops in these units. In power transistors, it is common practice to connect the outer casing to the collector as one way to effect heat removal.
12. Audio power amplifiers are designed for class A or class B operation. For class A, the output may use a single transistor or two in push-pull. For class B operation, a push-pull output must be employed.
13. The tunnel diode is a semiconductor device capable of operating as high as 10 kilomegacycles or more.
14. The tunnel diode is a negative-resistance device that achieves its operation by partially or totally canceling out the positive resistance existing in a circuit in which it is placed.
15. When servicing transistor circuits, first check the power supply or battery. If these are good, check the voltages at the transistor elements.
16. Transistors, to be checked, must be removed from the circuit. If this requires unsoldering, be careful not to let too much heat reach the transistors.
17. Always turn off the applied d-c voltages when transistors are being inserted into or removed from a circuit.
18. When using an ohmmeter to check resistance values in a transistor circuit, be careful how the ohmmeter voltage is applied. If due care is not taken, the transistor may be damaged or an incorrect reading may be obtained, or both.

## SELF-EXAMINATION

Here's a chance to find out how well you have learned the material in this chapter. Work the exercises; then check your answers against the Key at the back of this book. These exercises are for your self-testing only.

1. A PNP transistor has the following internal arrangement: (*a*) P-type emitter, P-type base, and N-type collector; (*b*) N-type emitter, P-type base, and P-type collector; (*c*) P-type emitter, N-type base, and P-type collector; (*d*) None of the above.
2. A potential hill in a transistor is increased in height when (*a*) the emitter is biased in the forward direction; (*b*) the emitter is reverse-biased; (*c*) the collector is biased in the forward direction; (*d*) all voltages are removed.
3. Transistor amplifiers resemble tube amplifiers because they (*a*) shift signal phase by 180°, no matter how the amplifier is wired; (*b*) amplify signals exactly in the same manner; (*c*) use a bias in the input circuit to set the amplifier to a specific operating point; (*d*) use d-c voltages of the same polarity in the input and output circuits.
4. In the circuit of Fig. 2·18, stabilization against changes in collector current due to temperature variations is the principal function of (*a*) $R_1$ and $R_3$; (*b*) $C_1$, $R_2$, and $C_2$; (*c*) $C_4$ and $C_3$; (*d*) $R_3$ and $R_4$.
5. Circuits in this chapter which are *all* common-base amplifiers are (*a*) Figs. 2·6, 2·7, 2·10, and 2·15; (*b*) Figs. 2·10, 2·12, 2·16, and 2·19; (*c*) Figs. 2·7, 2·13, and 2·20; (*d*) Figs. 2·14, 2·15, 2·21, and 2·23.
6. Which one of the following statements is true regarding direct-coupled amplifiers? (*a*) Direct-coupled amplifiers always have voltage gains less than 20; (*b*) Direct-coupled amplifiers always use one type of transistor (such as an NPN) followed by the other type (such as a PNP); (*c*) Pulse amplifiers are always direct-coupled; (*d*) A d-c path must exist between direct-coupled stages.

7. A transistor power amplifier (*a*) must be preceded by a class B driver stage; (*b*) utilizes push-pull arrangements exclusively; (*c*) alternates class A and class B stages; (*d*) possesses essentially the same form as any other transistor amplifier.
8. When a transistor circuit does not function properly, first (*a*) change all transistors; (*b*) check the applied d-c voltages; (*c*) test all resistors and then all capacitors; (*d*) apply a soldering iron to all wiring connections to uncover any poor connection.
9. The circuit of Fig. 2·18 consists of (*a*) a common-collector amplifier followed by a common-emitter stage; (*b*) two common-emitter amplifiers; (*c*) a common-base stage and a common-emitter stage; (*d*) two common-collector stages.
10. The transistors described in this chapter possess three leads for connections to the emitter, base, and collector. In most commercial units, these leads can be identified because (*a*) each lead is marked with a different color; (*b*) the emitter lead is between the collector and base wires, and closest to the base lead; (*c*) the base lead is between the collector and emitter wires; (*d*) the base lead is marked red.

## QUESTIONS AND PROBLEMS

1. Draw the circuit of a simple common-emitter amplifier using an NPN transistor. What electrode receives the a-c input signal? What electrode supplies output signal?
2. Show, by means of a numerical example, how the voltage gain of a transistor amplifier is greater than one even though the current gain is less than one.
3. Compare the manner in which a tube is cut off to the same action in a transistor.
4. Give two more ways in which tubes and transistors differ.
5. Explain how a resistor in the emitter leg of a common-emitter amplifier tends to stabilize the circuit against the effects of temperature changes.
6. What vacuum-tube circuit does the common-collector amplifier resemble most? Describe why briefly.
7. Prove that a signal passing through a common-emitter amplifier receives a 180° phase shift.
8. What do we mean by complementary symmetry in transistors? Draw the diagram of a circuit which utilizes this symmetry.
9. In what two ways do power transistors differ from other types of transistors? Give one application for power transistors.
10. Give two precautions to observe when servicing transistor circuits.
11. In what ways does a tunnel diode differ from a conventional three-element transistor?
12. How does a tunnel diode help develop oscillations in a tunnel-diode oscillator?
13. Why can a tunnel diode operate at extremely high frequencies?
14. What is the advantage of class B operation for power amplifiers, compared with class A?

# Chapter 3 Audio circuits

Audio equipment and audio amplifiers are explained in this unit. Besides being used for phonographs, tape recorders, and public address (PA) systems, audio circuits have additional applications. All radio and television receivers have an audio section for reproducing the desired voice and music information. Also, transmitters use audio circuits in the process of impressing the desired signal on the transmitted carrier wave. Finally, audio circuits include basic types of amplifiers. For example, the *RC*-coupled audio amplifier and the push-pull amplifier are fundamental circuits that have many uses. The topics are:



## *3·1 Sound waves and audio frequencies*

Sound is a wave motion of varying pressure in a material substance such as a solid, liquid, or gas. The pressure variations result from the mechanical vibrations of a source producing the sound waves. For instance, when the string on a violin is plucked, the vibrating string produces a wave motion of varying air pressure traveling outward from the source. The sound waves can be detected by the ear, producing the sensation of hearing. In this case, the sound waves are in air. There cannot be any sound waves in a vacuum.

A point of maximum air pressure in the sound wave is called a *compression;* a point of minimum air pressure is a *rarefaction.* The points of compression and rarefaction move outward from the source, propagating the sound waves in all directions. The velocity of sound is approximately 1,130 ft per sec in dry air at a temperature of 20°C. In metals and liquids, the velocity is much greater.

**Loudness.** This is the listener's sensation corresponding to the amplitude of the sound waves. Greater amplitude means a louder sound.

**Pitch or tone.** The frequency of the sound wave determines its pitch or tone as heard by the ear. The range of audible frequencies is approximately 16 to 16,000 cps. High frequencies correspond to high-pitched, or *treble,* tones. Low frequencies are low-pitched, or *bass,* tones. The wide range of frequencies for different sounds can be seen from the illustration in Fig. 3 · 1. For instance, all 88 piano keys have a frequency range of 30 to 4,100 cps, approximately; the high-pitched piccolo produces the frequency of 4,608 cps; the low-pitched bass viol has the frequency range of 40 to 240 cps; a man's speaking voice produces sound with a frequency of 128 cps; the frequency is 256 cps for the average woman's speaking voice, although a high-pitched soprano can produce 1,170 cps. The question then arises as to the source of the higher frequencies up to 16,000 cps. This question is answered by the presence of overtones or harmonic frequencies produced by the sound source.

*Fig. 3 · 1 Spectrum of audio frequencies* (Electronics.)

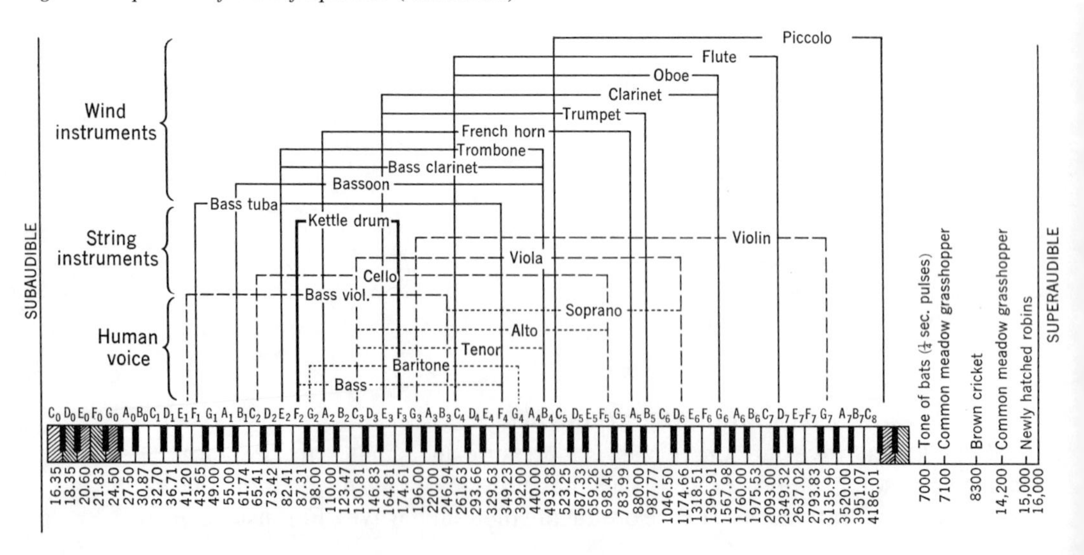

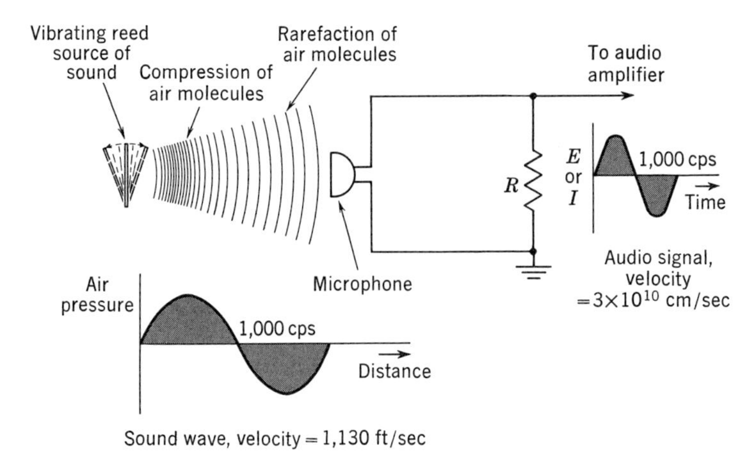


*Fig. 3·2 A 1,000-cps sound wave and its corresponding a-f variation in voltage.*

**Harmonic frequencies.** Consider a vibrating violin string as a source of sound. It may be producing sound waves with a frequency of 500 cps because the entire string length is vibrating at this frequency. This is called its *fundamental frequency* because it is the lowest frequency this string can produce. Since the string is not perfectly rigid, however, sections are also vibrating to develop additional frequencies. The frequency increases for shorter lengths of a vibrating string. Therefore the vibrating sections produce frequencies higher than the fundamental. Furthermore, these higher frequencies are exact multiples of the fundamental. One-half the string length vibrates at twice the fundamental frequency; similarly, one-third the string produces the third harmonic. In this way we could consider the string as producing an unlimited number of multiple harmonic frequencies. However, the amplitude of each harmonic decreases with higher numbers, so that generally it is not necessary to consider more than 10 to 20 harmonics.

In summary, then, a harmonic frequency is an exact multiple of the fundamental frequency. Even multiples are even harmonics; odd multiples are odd harmonics. In our example of the violin string vibrating in its entire length at 500 cps, this is its fundamental frequency. Sometimes this may be called the *first harmonic*. The second harmonic is double the fundamental, or 1,000 cps here; the third harmonic is 1,500 cps; the fourth harmonic is 2,000 cps; and so on. Considering frequencies up to the twentieth harmonic, then, this vibrating string is producing overtones with harmonic frequencies up to 10,000 cps. These higher harmonic frequencies must be included in the sound of the violin; otherwise, it will not have its characteristic *timbre*, or quality. It is the harmonics that make one source of sound different from another source producing the same fundamental frequency.

**Audio signal.** Sound waves are mechanical variations in air pressure which cannot be amplified electrically. The sound waves can be converted to electrical variations, however, as shown by the microphone in Fig. 3·2.

The microphone has a diaphragm that is moved physically by the variations of air pressure in the sound wave. The motion of the diaphragm then produces corresponding variations in voltage or current, resulting in an electric audio signal. The frequency of the audio signal corresponds to the frequency of the sound waves, including the harmonic frequency components. The greater the intensity of the sound, the greater the signal produced by the microphone. Still, there is little electrical output from a microphone even from sound waves of high intensity. Therefore the audio signal is coupled to an audio amplifier.

It is important to note that some audio circuits may not be able to operate over the full a-f range of 16 to 16,000 cps. Telephones use the restricted frequency range of 250 to 2,750 cps, and the speech is intelligible. For music and more natural speech, though, an audio range of 50 to 8,000 cps provides much better quality. Phonograph records generally do not have frequencies higher than 15,000 cps. In conclusion, the frequency span of 50 to 15,000 cps can be considered as a full range of frequencies for audio signals. This is the range used for high-fidelity audio modulation in the commercial FM broadcast band.

## 3·2 *Audio equipment*

Microphones and loudspeakers are two examples of *electromechanical transducers,* which are able to convert electrical energy into mechanical energy, and vice versa. A loudspeaker changes audio signal current to sound waves of varying air pressure by means of its vibrating cone. On the other hand, a microphone converts sound waves of varying air pressure into electric audio signals. Phonograph pickup cartridges fall into the same category. The phonograph pickup converts the vibrating motion of its needle or stylus into an audio signal output.

**Loudspeakers.** Typical construction of a loudspeaker is illustrated in Fig. 3·3. This is called an *electrodynamic* speaker as it has a lightweight voice coil that can easily move in and out. Attached to the voice coil is the cone, usually constructed of pressed paper or cloth. The voice coil has about 20 turns of fine enameled wire on a form of 1-in. diameter,

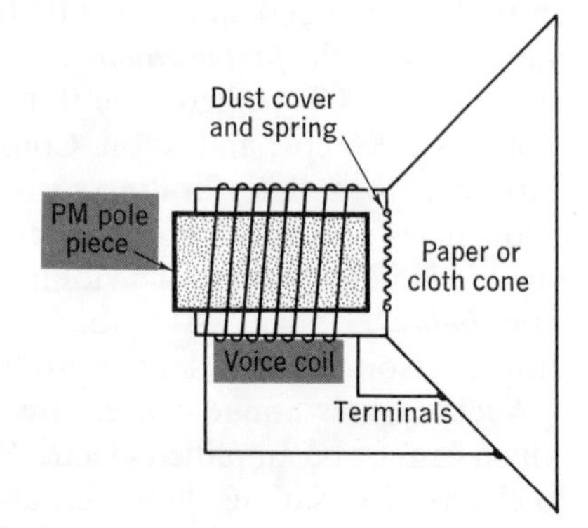


*Fig. 3·3 Construction of an electrodynamic loudspeaker.*

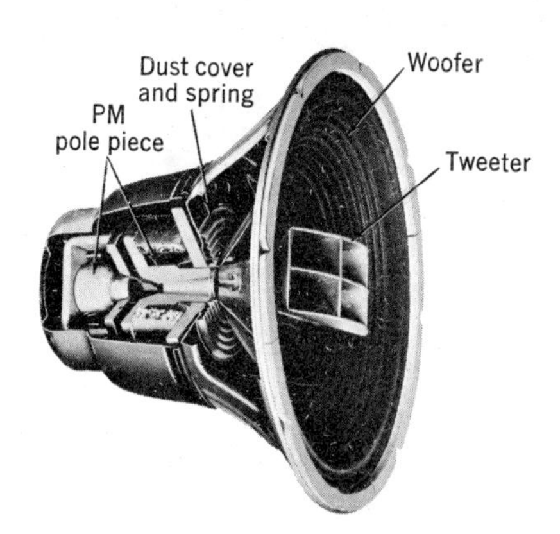


*Fig. 3·4 Coaxial loudspeaker, combining woofer and tweeter in one unit.* (Jensen Mfg. Co.)

approximately. It is positioned in the air gap of a fixed magnetic field, fitting over the center magnet slug. A permanent magnet provides the steady magnetic field of a PM speaker; in an electromagnetic (EM) speaker, a field coil supplies the fixed magnetic flux.

When audio signal current flows through the voice coil, the varying flux set up then reacts with the flux of the steady field, resulting in motor action. For the horizontal coil in Fig. 3·3, the motion is left and right. The voice coil is alternately attracted and repelled, moving in and out in accordance with the signal current. Its attached cone vibrates like a piston, compressing and rarefying the air to produce sound waves. Exact centering of the voice coil enables it to move across the air gap without rubbing against the field magnet. The two leads for the voice coil are connected by flexible braided wire to stationary terminals on the speaker frame. PM speakers require only these two connections because the permanent magnet supplies the field flux. In an EM speaker, there are additional leads to supply direct current for the field winding. Almost all modern speakers are PM speakers, using a magnet of 2 oz to 5 lb, or more. The better the speaker, the larger the magnet.

The speaker cone is usually circular to provide uniform radiation of the sound waves, although oval speakers are also used. Diameters may range from 1 to 15 in. for circular cones, the most common sizes being 4, 8, and 12 in. Large speakers are desirable for reproducing low audio frequencies. A speaker designed primarily for low frequencies is a *woofer;* a small speaker for high frequencies is a *tweeter.* The *coaxial* speaker in Fig. 3·4 combines a woofer and a tweeter in one unit.

The electrical characteristics of the speaker are given by the manufacturer's rating of impedance and power for the voice coil. The impedance is generally at 400 cps. Typical values are 3.2, 4, 8, or 16 ohms for the step-down transformer used with vacuum tubes and 10, 20, or 40 ohms for direct coupling in transistor circuits. The voice-coil impedance is important for matching to the impedance of the audio-output stage. If the speaker impedance is not specified, an approximate value can be obtained by measuring the voice-coil resistance with an ohmmeter and multiplying by 1.25.

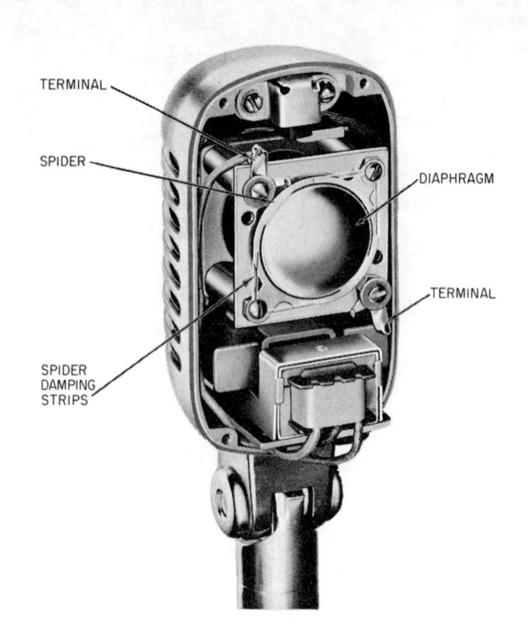


*Fig. 3·5 Cutaway view of moving-coil dynamic microphone.* (Shure Brothers Inc.)

The power rating states the maximum a-c power that can be supplied to the voice coil. Typical ratings are less than 1 watt to more than 50 watts, with 10 watts a typical value. With an average efficiency of 10 per cent for the loudspeaker, 10 watts of audio-input power produces 1 watt of sound power by the speaker cone. This is sufficient, however, to fill a large living room with very loud sound.

Earphones or headsets are for low-power applications. They operate on the principle of a vibrating metal diaphragm attracted or repelled by a small electromagnet in the earpiece, although crystal earphones are also used. The impedance is generally several hundred to several thousand ohms.

**Types of microphones.** A dynamic microphone uses the same principle as the dynamic loudspeaker but in reverse (Fig. 3·5). In fact, if you talk into a small loudspeaker, audio signal output can be taken from the voice-coil terminals. This is done in "intercom" systems, where you talk or listen with the same unit. For better frequency response and directional characteristics, though, there are many specialized types, including the ribbon and velocity microphones. These are EM microphones featuring high quality, a low impedance of several hundred ohms, usually at 1,000 cps, and very low output.

The crystal or ceramic type is a general-purpose microphone with high impedance and the relatively high output of 1 mv, approximately. Additional types are the capacitor microphone and the carbon microphone. These need a d-c supply for operation, however. The carbon microphone is used for telephones. Any type can be adapted to the form of a throat microphone or lapel microphone. An important component needed with low-impedance microphones is a microphone transformer that steps up the audio signal voltage coupled to the first audio amplifier. A microphone, audio amplifier, and loudspeaker form a PA system.

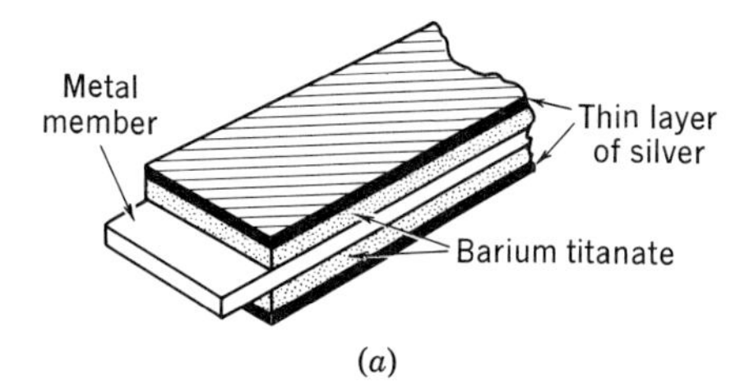


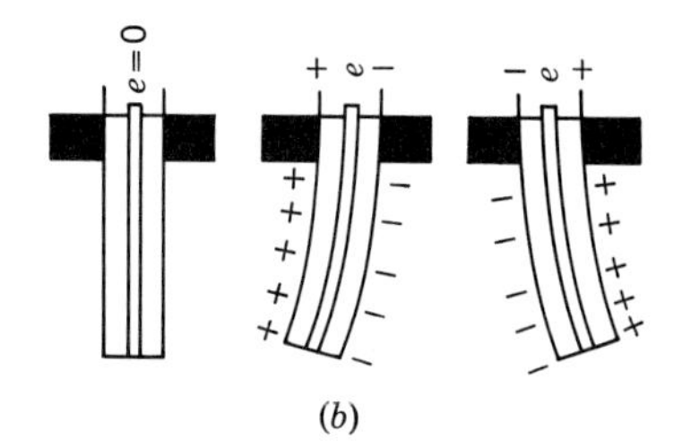


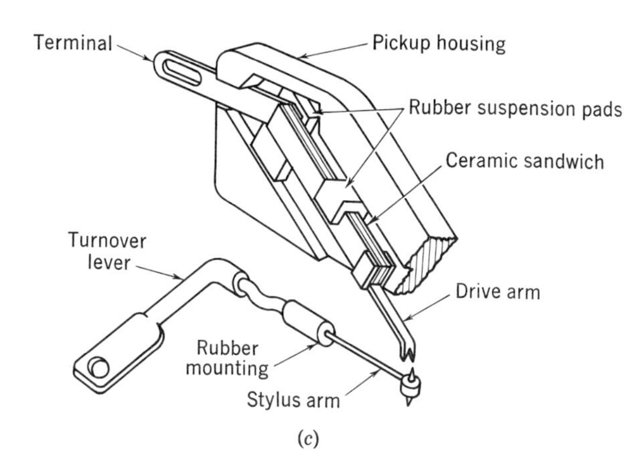


*Fig. 3 · 6 Ceramic phonograph cartridge. (a) Construction of ceramic transducer. (b) Piezoelectric effect. (c) Typical turnover cartridge for fine-groove and standard-groove records. Length is approximately 1 in.* (Sonotone Corp.)

**Types of phonograph pickups.** The most popular phonograph pickups are the crystal and ceramic type. In construction, the crystal cartridge is generally made of Rochelle salts, a sodium potassium tartrate compound; the ceramic material is usually barium titanate. Both types operate on the basis of the *piezoelectric effect,* whereby twisting or bending of the crystal or ceramic element produces audio-output voltage. This is illustrated in Fig. 3 · 6. As the needle or stylus in the cartridge tracks the grooves of the phonograph record, the physical vibrations are converted to audio-output signal. Ceramic and crystal pickups have similar electrical characteristics, but the ceramic unit is less sensitive to heat and humidity. Both have high impedance and produce a high output of about 1 volt.

Electromagnetic types of phonograph pickup convert the motion of the stylus into induced current. These are usually high-quality pickups but with low impedance and low output of about 10 mv.

**Phonograph records.** Records have narrow grooves with variations that correspond to the audio information. The old shellac records made for a speed of 78 revolutions per minute are becoming obsolete; in their place now are fine-groove vinyl records operating at the speed of 33⅓ or 45 rpm. These have about 250 grooves per inch. The combination of slower speed and finer grooves allows more playing time.

Two methods of cutting audio variations into the record grooves are vertical cut and horizontal or lateral cut. The lateral cut is used in records for sale to the public, while vertical cut is used for commercial recordings because it allows more grooves per inch. For stereo records, two separate audio signals are cut into both sides of one groove, at opposite 45° angles. As a result, vertical and lateral variations provide separate left and right audio signals, with a stereo cartridge. The stylus point required for fine-groove records is 0.001 in. or 1 mil, with 0.7 mil used for stereo records.

**Magnetic-tape recording.** In this application a thin plastic tape with a coating of very fine magnetic particles passes across a narrow air gap on the tape head (Fig. 3 · 7). Then the tape becomes magnetized by induction from the magnetic field of the current in the recording winding. The recording current consists of the desired audio signal superimposed on an a-c bias current with a frequency of about 200 kc. The a-c bias current is necessary to demagnetize the tape to remove any previous recorded information, for minimum distortion. On playback, the moving magnetized tape induces audio signal current in the playback winding. When it is desired to erase the audio signal from the tape, the r-f bias current is used by itself. The tape speed may be 1¾, 3½, 7, or 15 in. per sec.

## 3 · 3 *Audio preamplifier circuit*

A *preamplifier* circuit is a special voltage amplifier designed to deal with very low amplitude signals, such as the output of a microphone. A pre-

*Fig. 3 · 7 Record-playback head for magnetic tape recorder; height is ½ in.* (Shure Brothers, Inc.)

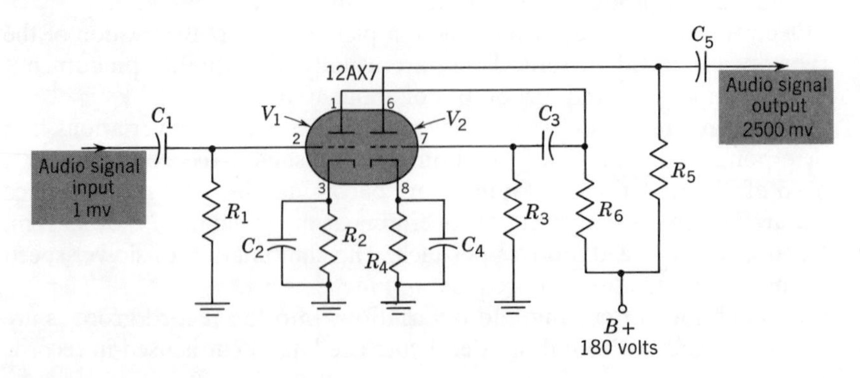


*Fig. 3 · 8 Audio preamplifier circuit for low-level audio signals.*

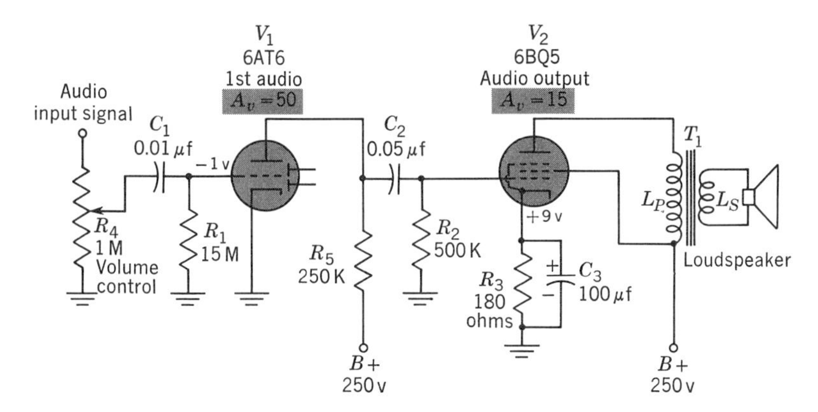


*Fig. 3 · 9 Typical audio amplifier circuit with power-output stage to drive loudspeaker.*

amplifier is also required for the low-level output of a magnetic phonograph pickup or the audio-output signal from magnetic tape. Essentially, the preamplifier uses the *RC*-coupled audio circuit, as shown in Fig. 3 · 8. Special attention is required, however, to reduce hum, microphonics, and tube noise because of the low signal level. A suitable tube is a high-$\mu$ triode with minimum cathode-heater leakage. The tube socket is shock-mounted on a spring to reduce vibrations. In addition, shielded input and output leads serve to minimize hum.

The dual triode 12AX7 in Fig. 3 · 8 provides two *RC*-coupled stages in cascade. With a gain of 50 for each stage, the total gain of the preamplifier is 2,500. Therefore, the output voltage from $V_2$ is 2,500 mv, or 2.5 volts, with 1 mv, or 0.001 volt, fed into $V_1$. Both stages operate class A, with cathode bias. In the first stage, $R_1C_1$ is the grid-input coupling circuit, while $R_2C_2$ provides cathode bias. The plate-load resistor for $V_1$ is $R_6$. Note that the high side of $R_6$ is coupled to the grid of $V_2$ by $R_3C_3$. In this stage, $R_4C_4$ provides cathode bias. The plate-load resistor for $V_2$ is $R_5$. Finally, the amplified signal output is coupled to the following audio amplifier circuit by $C_5$.

## *3 · 4 Typical audio amplifier circuits*

Figure 3 · 9 is the schematic diagram of a two-stage audio amplifier with the power-output stage driving a loudspeaker. The audio-input signal voltage required is 0.1 volt, or higher. This may be obtained from the preamplifier in Fig. 3 · 8, or directly from a radio or a crystal phonograph pickup which provides enough audio signal output. With a single 6BQ5 for the class A power-output stage, maximum power output is 5.7 watts.

**Circuit operation.** The high-$\mu$ triode section of the 6AT6 in Fig. 3 · 9 is the first audio amplifier; its duodiode section is not employed here. The amount of audio-input voltage is adjusted by the volume-control potentiometer $R_4$, while $R_1C_1$ forms the grid-coupling circuit. $R_1$ is 15 M to provide a contact bias of 1 volt for $V_1$. The plate-load resistor is $R_5$. With

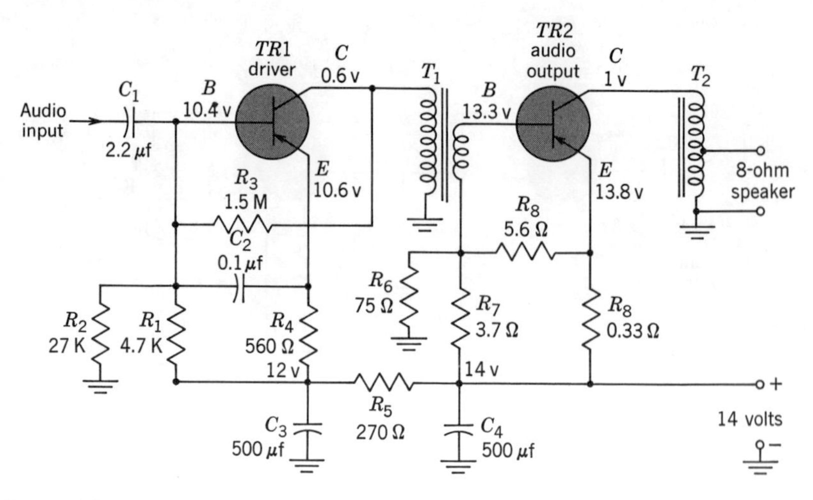


*Fig. 3·10 Audio amplifier using PNP transistors in the common-emitter circuit. Input of 0.1 volt at 400 cps produces 1 watt audio output.*

a gain of 50 for $V_1$, and $R_4$ set for an audio-input level of 0.1 volt, the audio signal output from $V_1$ coupled to $V_2$ equals 5 volts.

In the power-output stage, $R_2C_2$ is the grid-coupling circuit for $V_2$. Cathode bias of 9 volts is developed by $R_3$. Note the relatively large capacitance of 100 μf for cathode bypass $C_3$. This high value is necessary to have $X_c$ one-tenth, or less, of the 180-ohm $R_3$ for effective bypassing at low audio frequencies. However, $C_3$ can be a low-voltage electrolytic with a 25-volt breakdown rating, enabling it to be physically small in size. With a gain of 15, there is a 75-volt audio signal in the plate circuit of $V_2$ for a 5-volt signal from $V_1$. Plate-current variations in the primary of the audio-output transformer induce audio signal in the secondary, which is connected to the voice coil of the loudspeaker.

**Transistorized audio amplifier.** In Fig. 3·10, two PNP transistors are used in the common-emitter circuit as first and second audio amplifiers. The first stage TR1 has *RC* coupling for the input signal, but an interstage audio transformer $T_1$ is used for the output to drive the power stage TR2. With 0.1 volt audio signal input, the audio output is 1 watt. The audio output transformer $T_2$ is tapped as an autotransformer to match the 8-ohm loudspeaker.

The circuit in Fig. 3·10 is the audio section of an automobile radio with a 12 to 14-volt d-c supply. Note the inverted method of supplying d-c voltage for the collector-emitter reverse bias. The collector returns to chassis ground, while the supply voltage is connected to the emitter. As an example, for TR1 the emitter at 10.6 volts is 10 volts more positive than the collector at 0.6 volt. Therefore, the collector-emitter voltage equals 10 volts, negative at the collector side. Similarly, the reverse bias on TR2 is 12.8 volts, negative at the collector with respect to the emitter.

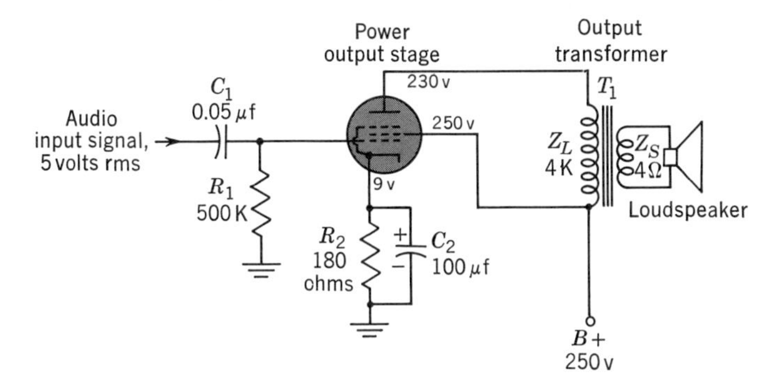


*Fig. 3 · 11 Pentode audio power-output stage.*

For forward bias, the base is slightly less positive than the emitter. The potential difference between emitter and base for forward bias equals 0.2 volt on TR1 and 0.5 volt on TR2.

## *3 · 5 The audio-output stage*

Figure 3 · 11 reveals more details of the audio-output stage with its power-output transformer. Note that one end of the primary winding connects directly to the plate, while the opposite end returns to B+. The screen grid is also connected to the latter point so that it can have full B+ voltage for maximum power output. The plate voltage is slightly lower because of the *IR* drop across the d-c resistance of the primary. This operation is possible because the beam-power tube has minimum screen current. The value of 4,000 ohms for $Z_L$ in the primary is its a-c impedance to the audio signal. This is the plate load for the amplifier. In the secondary, $Z_s$ is 4 ohms to match the loudspeaker 4-ohm voice coil. In terms of power rating, the output transformer and loudspeaker each should be rated no less than the maximum power output of the tube. A pentode, tetrode, or triode can be used in the audio-output stage.

**Impedance matching.** The output transformer matches the low impedance of the speaker's voice coil to the high resistance of the output tube. If the speaker were connected directly in the plate circuit, there would be practically no audio output because the tube would have no gain with a 4-ohm plate load. What value $Z_L$ should be for maximum gain with minimum distortion is specified in the tube manual. Typical values range from 2,000 to 5,000 ohms for a single tube. The voice coil of a dynamic loudspeaker, on the other hand, requires a low-impedance source. Both requirements are satisfied by the output transformer. The primary has a large number of turns for a high impedance, while the secondary has a few turns as a low-impedance source to drive the loudspeaker. Since the transformer has an iron core with practically unity coupling, the power delivered to the secondary equals the power supplied by the primary. Thus, the primary winding can provide the required $Z_L$ with relatively high voltage and low current, while the secondary delivers the output power at much lower voltage but at high current for the low-impedance voice coil. The ratio of turns required for the impedance match equals

$$\frac{T_p}{T_s} = \sqrt{\frac{Z_p}{Z_s}} \qquad (3\cdot1)$$

For instance, to match a 4,000-ohm plate load ($Z_p$) to a 4-ohm speaker ($Z_s$),

$$\frac{T_p}{T_s} = \sqrt{\frac{4{,}000}{4}} = \sqrt{1{,}000} = 31.3$$

That is, the primary must have 31.3 turns for each turn in the secondary.

**Parallel tubes.** For class A operation, two tubes can be connected in parallel to double the power output. As shown in Fig. 3·12, identical tubes are used, and like electrodes are tied together with short leads. As a result, the required plate-load impedance is one-half the value for one tube, the amplitude distortion is the same as for one tube, the required amount of grid-driving voltage does not change, but the total plate current is doubled. Parallel operation of audio-output tubes, however, does raise the possibility of r-f parasitic oscillations that would distort the output. This is the reason for the *parasitic suppressor* resistors $R_3$ and $R_4$ in the control-grid and screen-grid circuits in Fig. 3·12. Note that parallel tubes have parallel tube capacitances, increasing the capacitance for internal feedback between plate and grid.

A better way to use two tubes for more power output is the push-pull circuit, to be described next.

## 3·6 *Push-pull amplifiers*

Figure 3·13 shows how two tubes are connected in push-pull to provide double the power output of a single-tube amplifier. Triodes are illustrated here, but tetrodes or pentodes can also be used with their screen grids connected directly to B+. The most important part of the circuit is the push-pull output transformer in the plate circuit. This is center-tapped for the return to B+. Each plate circuit then uses one-half of the primary winding. Although the plate currents are opposite, they add in producing induced voltage in the secondary for the loudspeaker. This occurs because the grid-signal voltages for the two tubes are also opposite in polarity. Here the push-pull grid drive is provided by a center-tapped input transformer to provide equal grid-signal voltages of opposite phase for the two tubes. Cathode bias is developed by $R_1C_1$, although fixed bias can also be used. Since the cathode return current for both tubes flows through $R_1$, its voltage serves as the bias for both $V_1$ and $V_2$. For class A operation, $C_1$ may be omitted without any degeneration because the equal and opposite plate-current variations cancel in the common-cathode resistance.

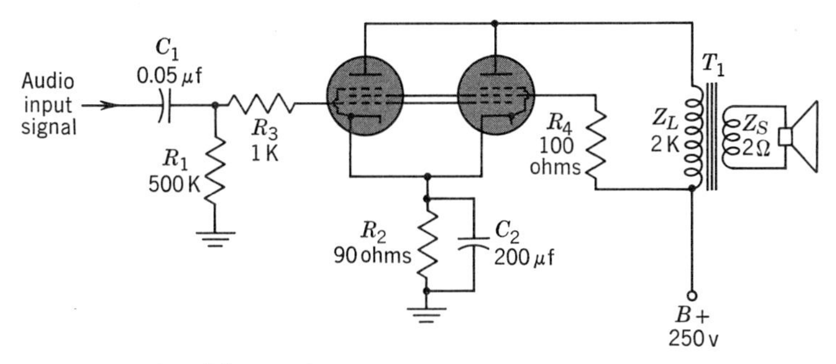


*Fig. 3·12 Parallel operation of power-output tubes.*

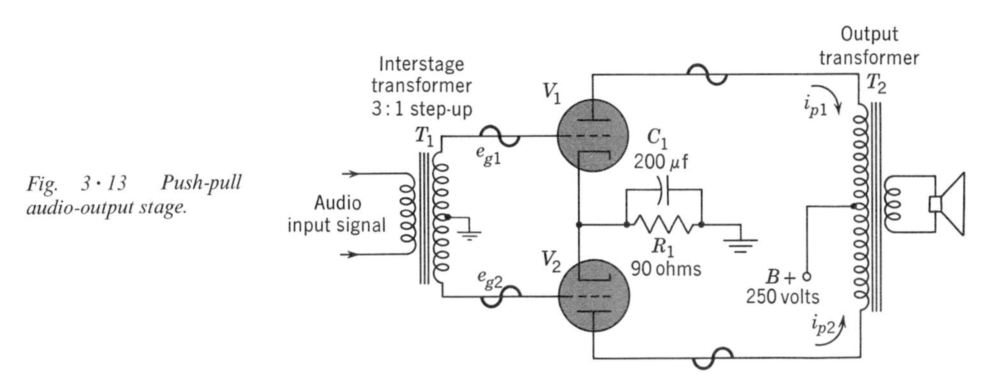


*Fig. 3·13 Push-pull audio-output stage.*

Class B or AB operation can be used for more power output and efficiency, but the distortion increases slightly. Such operation close to the grid-cutoff voltage is possible only with push-pull audio amplifiers because each tube supplies opposite half-cycles of output signal. It should be noted that the push-pull circuit can also be used for r-f amplifiers.

**Push-pull operation.** For class A operation, the equal and opposite plate-current variations of $i_{p_1}$ and $i_{p_2}$ are shown in Fig. 3·13. Note that $i_{p_1}$ flows down through the top half of $T_2$, returning to B+ through the center tap, while $i_{p_2}$ flows up through the other half of $T_2$. In addition, these two signal currents are varying in opposite phase because of the opposite polarities of the grid-driving signal. To see both effects on the output signal, assume that $i_{p_1}$ flowing down in the primary has a counterclockwise magnet field. Then $i_{p_2}$ has a clockwise field because it flows up with the same direction of winding. Both fields induce voltage in the secondary with opposite polarity for the same variations. The fact is, though, that the push-pull variations of signal are opposite. When $i_{p_1}$ is increasing because of positive grid drive, $i_{p_2}$ is decreasing because of negative grid drive. On the next half-cycle, the variations are reversed. Remember that a counterclockwise field expanding with increased current has the same flux variations as a clockwise field collapsing with less current. Both will induce the same voltage. As a result, push-pull operation enables the

*Table 3 · 1 Push-pull operation for circuit in Fig. 3 · 13*

| *Current* | *Direction* | *Field* | *Variation* | *Induced voltage* |
|---|---|---|---|---|
| $i_{p_1}$ | Down | Counterclockwise | Increase | + |
| | | | Decrease | − |
| $i_{p_2}$ | Up | Clockwise | Decrease | + |
| | | | Increase | − |

opposite plate currents to add in produced secondary voltage in the output transformer. These points are summarized in Table 3 · 1.

**Push-pull advantages.** The push-pull output circuit cancels all components of plate current that have the same phase in the two tubes. This really means everything except the push-pull input signal. The components canceled include hum from the B supply, hum due to cathode-heater leakage in the output tubes, and the average plate current in both tubes. Because the average plate current cancels in the output transformer, there is an important reduction of average magnetic flux. This permits more signal current to flow without saturating the iron core than would be possible with a single-ended transformer. Another advantage is less nonlinear amplitude distortion. When one tube is operating near zero grid voltage, the other tube is near cutoff. These two parts of the grid-transfer characteristic have opposite curvature. As a result, the amplitude distortion that results with maximum grid drive is reduced. This type of distortion, which causes unsymmetrical variations in plate current, is called second harmonic distortion (harmonic distortion is described in Sec. 3 · 9). Therefore, the push-pull arrangement eliminates second harmonics. This refers, however, only to the distortion generated by the output tube.

A disadvantage of the push-pull circuit is the need for supplying twice the grid-voltage input, compared with a single-ended stage, since each

*Fig. 3 · 14 Triode phase splitter with equal load resistances in plate and cathode circuits.*

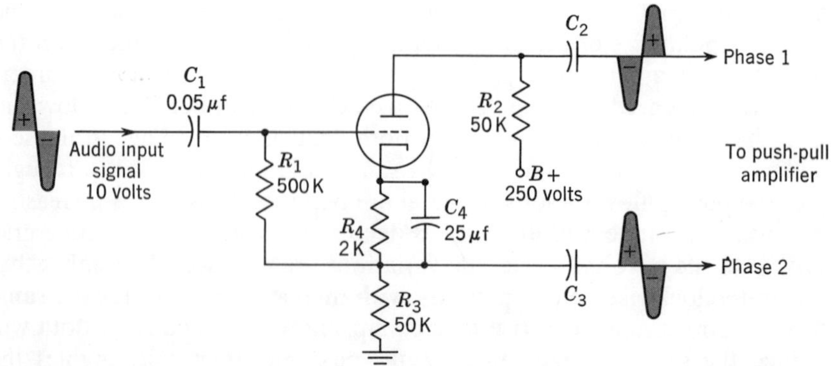

tube receives one-half the signal-input voltage. Also, two tubes in push-pull have twice the plate resistance of one tube. Another factor to consider is that all the advantages of push-pull are based on the assumption of equal and opposite plate currents. For best results some push-pull circuits use a pair of matched tubes. There may also be controls to adjust the plate currents for perfect balance.

**Push-pull output transformer.** The audio-output transformer for a push-pull circuit has a center-tapped primary. The lead to the tap is coded red for B+, while the two plate leads for the opposite tubes are usually blue or brown. The secondary connections are the same as for a single-ended transformer. The primary impedance of the push-pull transformer is figured as the total impedance across the entire winding, or plate-to-plate. The tube manual gives the load impedance required plate-to-plate, and this is usually double the value needed for one tube.

## *3·7 Phase-splitter circuits*

Although push-pull grid-input voltage can be supplied by a center-tapped interstage transformer, this job is usually done with a phase splitter or phase-inverter stage. Typical circuits are shown in Figs. 3·14 and 3·15. The phase splitter provides equal amplitudes and opposite polarities of output voltage to drive the grids of the push-pull power-output stage. Although an extra stage is required, the phase splitter is more economical and has better frequency response. Low-$\mu$ triodes are generally used.

**Cathode-follower phase splitter.** The phase splitter in Fig. 3·14 uses just one triode. Notice that in addition to the 50,000-ohm plate load $R_2$ there is another 50,000-ohm cathode-load resistor $R_3$. In a triode, the cathode current returning through $R_3$ is the same as the plate current. Leaving $R_3$ unbypassed therefore enables it to develop signal voltage. Since $R_3$ and $R_2$

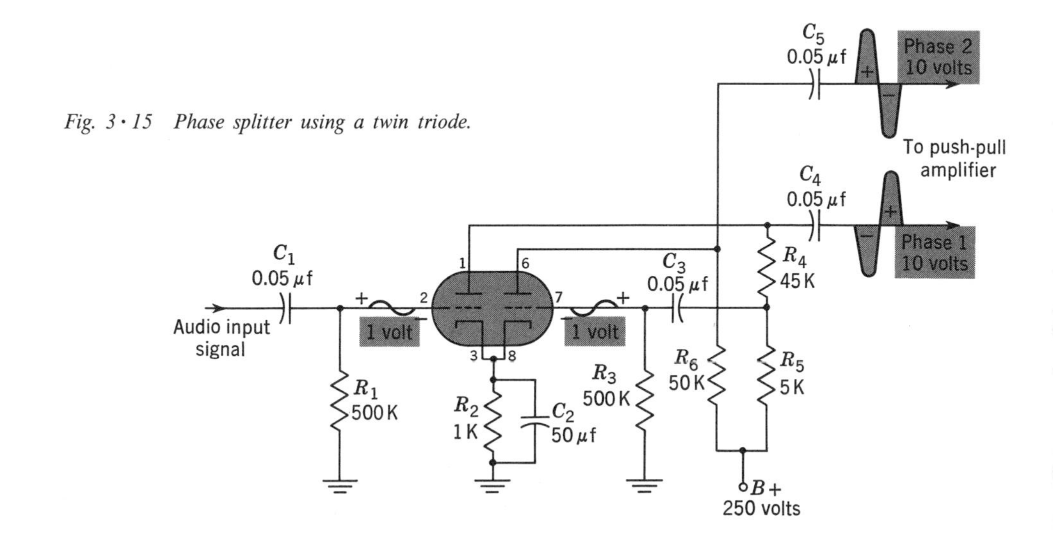


*Fig. 3·15 Phase splitter using a twin triode.*

are equal, both develop the same amount of output signal. The polarity of the signal across $R_3$, however, is opposite from the plate signal. The plate voltage is inverted from the polarity of the grid signal, but the cathode voltage follows the grid voltage with the same polarity.

The gain of the stage in each output channel is less than one because of degeneration introduced by the unbypassed cathode resistance $R_3$. Cathode bias is provided by $R_4C_4$. Grid resistor $R_1$ returns to the junction of $R_3$ and $R_4$, so that only the voltage across $R_4$ is connected between grid and cathode for bias. The output signals are taken from both the plate and cathode to drive the next stage. The plate signal output from $R_2$ is coupled to one control grid in the push-pull amplifier, while the cathode signal output from $R_3$ goes to the opposite control grid.

**Twin-triode phase inverter.** The phase splitter in Fig. 3 · 15 uses two triodes, but this circuit does provide more gain. Both triode sections are *RC*-coupled class A amplifiers with a gain of 10. Common-cathode bias is provided by $R_2C_2$. The 1-volt audio-input signal coupled to grid pin 2 is amplified to provide 10 volts of audio signal at plate pin 1. This is the output-signal voltage labeled phase 1. In addition, one-tenth of this plate signal is coupled back to grid pin 7 to be amplified by the second triode section.

Note that the plate-load resistance for plate pin 1 is 50,000 ohms, divided into two parts by $R_4$ and $R_5$. The 5,000 ohms of $R_5$ is one-tenth the total of 50,000 ohms. Therefore the voltage across $R_5$ is 1 volt. This is coupled by $C_3$ and $R_3$ to grid pin 7. The second triode section also has a gain of 10 with the 50,000-ohm plate load $R_6$. Thus, the audio-output voltage at plate pin 6 is also 10 volts. This is labeled phase 2. There are two opposite phases in the output signals because one was amplified by a single stage while the other was amplified by two stages. Both output voltages are equal in amplitude, however, because the signal amplified twice was divided down by a factor equal to the gain of each stage.

## *3 · 8 Tone-control circuits*

The response of an audio amplifier to different frequencies in the audio range determines the tone of the reproduced sound. Therefore, the tone can be controlled by varying the frequency response. Treble tone means greater response for the high audio frequencies, above 3,000 cps, approximately. Bass means more response for the low audio frequencies, from about 300 cps down. In many cases bass tone is obtained simply by reducing the gain for high frequencies, thereby emphasizing the relative response for bass frequencies. Typical circuits are shown in Fig. 3 · 16.

In Fig. 3 · 16*a*, $C_1$ is simply a bypass capacitor across the plate-load impedance $Z_L$. The range of frequencies bypassed depends on the value of $C_1$. A capacitance of 0.02 $\mu$f, approximately, is often used for $C_1$ to minimize the thin, tinny sound of a small speaker trying to reproduce high-frequency components outside its useful range.

The *RC* network across the primary of the output transformer in

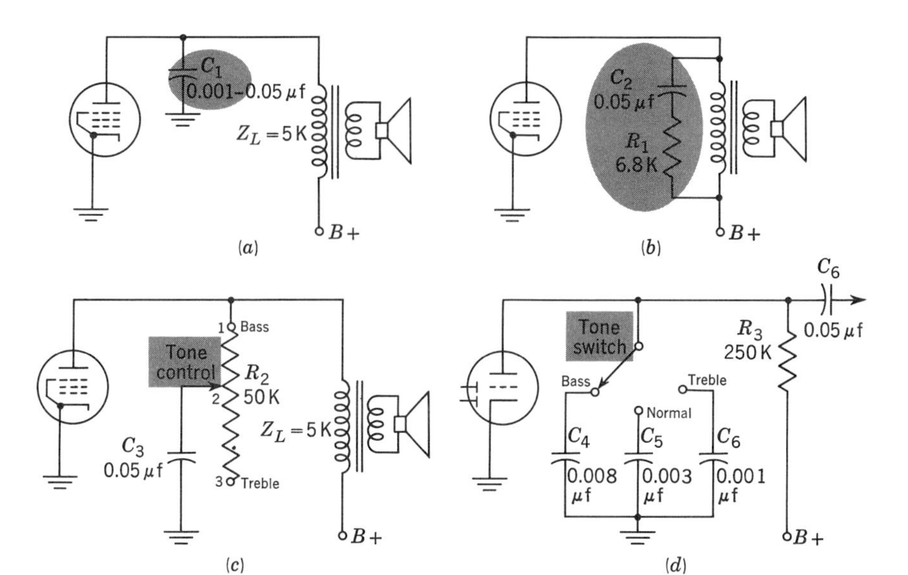


*Fig. 3 · 16 Tone-control circuits. (a) Fixed bypass capacitor in plate of output tube. (b) RC filter across primary of output transformer. (c) Variable RC filter. (d) Switch for different bypass capacitors.*

Fig. 3 · 16*b* has a dual function. Not only is $C_2$ a tone-control bypass, but its series resistance $R_1$ helps to maintain a uniform impedance across the output circuit. The nominal impedance of a loudspeaker applies at 400 cps, but it can be much higher for treble frequencies. With a beam-power tube or pentode in the output stage, it is especially important to have a constant-load impedance for minimum distortion. The values shown for $R_1$ and $C_2$ are typical.

**Variable tone control.** Figure 3 · 16*c* illustrates what is probably the most common tone-control circuit. Here $R_2$ is a variable resistance in series with $C_3$. When the variable arm of the tone control is moved closer to terminal 1, the series resistance is reduced, and $C_3$ bypasses more of the higher frequencies to provide the equivalent of improved bass response. At the opposite setting of the control, with the variable arm at terminal 3, $C_3$ cannot provide much bypassing because of the series resistance in $R_2$. This is the treble position. In some circuits there may be a resistance of several thousand ohms in the ground return for $C_3$, so that the volume will not be reduced too much in the bass position.

**Tone-control switch.** The circuit in Fig. 3 · 16*d* corresponds to Fig. 3 · 16*a*, but three capacitors are used with a switch to provide different degrees of high-frequency bypassing. In the bass position, $C_4$ has the largest capacitance for maximum bypassing of high frequencies; in the treble position, $C_6$ has the least capacitance.

**Tone-compensated volume control.** An interesting characteristic of hearing is the fact that when the volume is low, more bass response

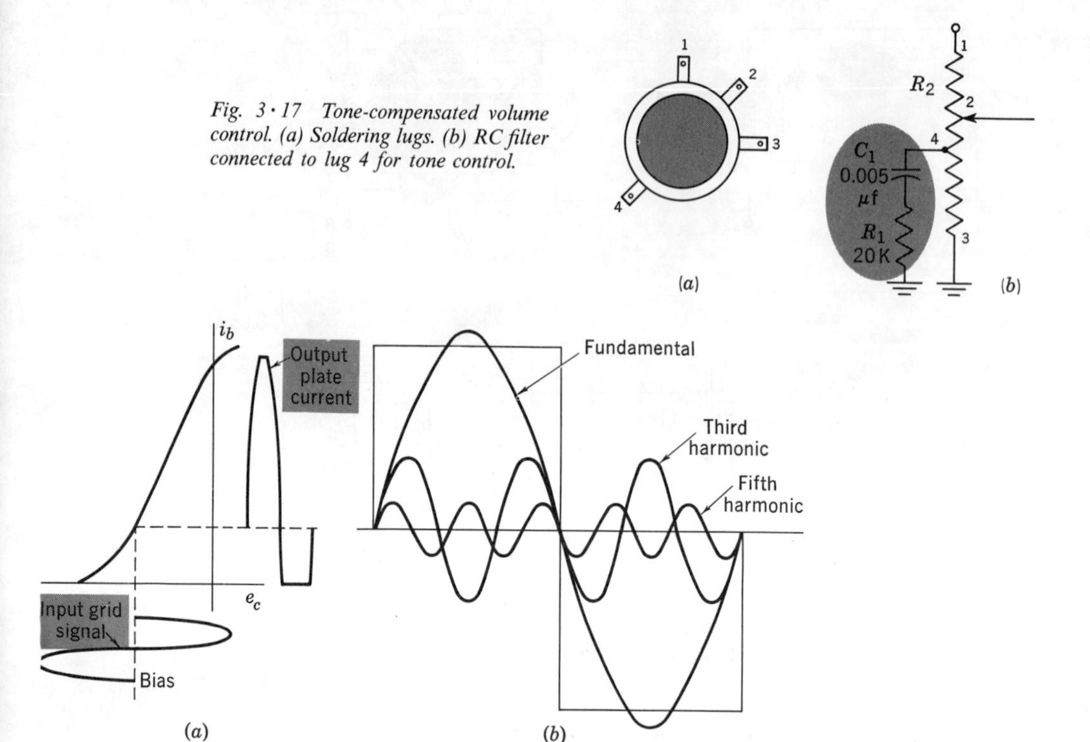


*Fig. 3·17 Tone-compensated volume control. (a) Soldering lugs. (b) RC filter connected to lug 4 for tone control.*

*Fig. 3·18 Nonlinear amplitude distortion and corresponding harmonic distortion, caused by too much input signal. (a) Sine-wave input flattened to produce square-wave output. (b) Harmonic components of square wave.*

is needed to have the reproduction sound like the original. For this reason, the volume control in Fig. 3·17 has an extra tap at lug 4 for a bass-boost circuit. Filter $R_1C_1$ is most effective when the variable arm is at a low-volume setting at the tap or closer to lug 3. For high-volume settings, the resistance between lugs 1 and 4 is in series with $R_1C_1$ and reduces their effect. Some controls have two taps for a greater range of tone compensation. A volume control that has provision for increasing the bass response in this way is sometimes called a *loudness* control.

## 3·9 Types of distortion

When the waveform of the output signal is not exactly the same as that of the input signal, distortion has been introduced. In amplifiers, it is important to minimize nonlinear amplitude distortion. This occurs when the relative amplitudes of the output-signal waveshape are not the same as they are in the input signal. Amplitude distortion of an audio signal generally makes the output sound either hoarse or garbled. A less obvious distortion is uneven frequency response; that is, the amplifier does not provide the same gain for the different frequency components in the signal.

The output signal may have excessive treble or bass tone, or insufficient frequency range.

**Amplitude distortion.** Operation over the nonlinear portion of the operating characteristic of a tube is the cause of amplitude distortion. This factor is especially important in the output stage. An extreme case is shown in Fig. 3 • 18*a*, where excessive input signal causes the tube to operate near saturation and beyond cutoff. Grid variations at the positive peak of the signal are lost because the plate current cannot increase above saturation. Negative grid drive beyond cutoff is also not reproduced. As a result, the signal variations of plate current have the square-wave form shown instead of a sine-wave shape.

The cause of the amplitude distortion here is excessive signal input, or *overload.* Similar distortion can be caused by a weak amplifier tube, low B+, too much or too little bias, current saturation in the output transformer, or an excessively high plate-load resistance $R_L$. Higher values of $R_L$ may produce more gain, but a smaller input signal can then cause overload distortion.

**Harmonic distortion.** Amplitude distortion is equivalent to harmonic distortion because changing the relative amplitudes is the same as adding harmonic frequency components not present in the input signal. This idea is illustrated in Fig. 3 • 18. The unsymmetrical square wave in *a* is equivalent to the square wave in *b*, plus an added d-c component. The increase in average plate current need not be considered except that it indicates amplitude distortion is present. Note that the square wave in *b* is composed of a fundamental sine wave at the same frequency as the distorted square wave, plus odd-harmonic frequency components. These harmonics were not present in the sine-wave input signal. Therefore the nonlinear amplification has introduced harmonic distortion in the form of new harmonic components.

Nonlinear amplification and its resultant harmonic distortion is produced mainly in the audio-output stage. Typical values are 5 per cent or less for the total harmonic distortion at full power output. Using less drive for less output reduces the distortion appreciably.

**Intermodulation distortion.** An additional effect of amplitude distortion is the fact that the harmonics introduced in the amplifier can combine with each other to produce new frequency components that are not harmonics of the fundamental. The appearance of these unrelated frequencies is called *intermodulation distortion.* These new frequencies are probably most responsible for the unpleasant rough sound of amplitude distortion because they are not harmonically related to the original frequency.

**Frequency distortion.** This type of distortion arises when the gain of the amplifier varies with frequency. Voice and music possess many frequency components corresponding to the different fundamental and harmonic frequencies produced by the source of sound. All these frequency components identify the sound. They may not all have the same relative amplitudes, but the amplifier should provide the same gain for all frequencies to maintain the character of the sound. Generally, the problem is amplifying the bass and treble frequencies. In the audio-output stage, the output transformer and loudspeaker are most important for a wide frequency range.

In the *RC*-coupled stage of Fig. 3·19*a*, the high-frequency response depends on a minimum value for stray capacitance $C_t$ and a relatively low value of $R_L$. Under these conditions, $R_L$ is not bypassed as easily by the decreasing reactance of $C_t$ at the higher audio frequencies. A typical value for $C_t$ is about 40 pf. Low-frequency response depends mainly on having large values for $C_c$ and $R_g$. In addition, low-frequency boost can be obtained by using $R_fC_f$ plate-decoupling filter shown.

The frequency response curve in Fig. 3·19*b* illustrates the relative gain at different frequencies for any *RC*-coupled amplifier. The middle range of frequencies has flat response because $R_L$ provides the same gain for different frequencies. For high frequencies, the gain decreases to 70.7 per cent of the mid-frequency response at the frequency $F_H = 1/(2\pi R_L C_t)$ as noted on the curve. With the values in Fig. 3·19*a*, amplifier gain is down to 70.7 per cent response at 80,000 cps. The response is flat up to the frequency that is one-tenth of $F_H$, which equals 8,000 cps here. At the low end, the gain is down to 70.7 per cent response at the frequency $F_L$, equal to $1/(2\pi R_g C_c)$, which is 16 cps here. The response is flat down to the frequency $10F_L$, or 160 cps. The drop-off at $F_L$ is due only to the $R_gC_c$ coupling

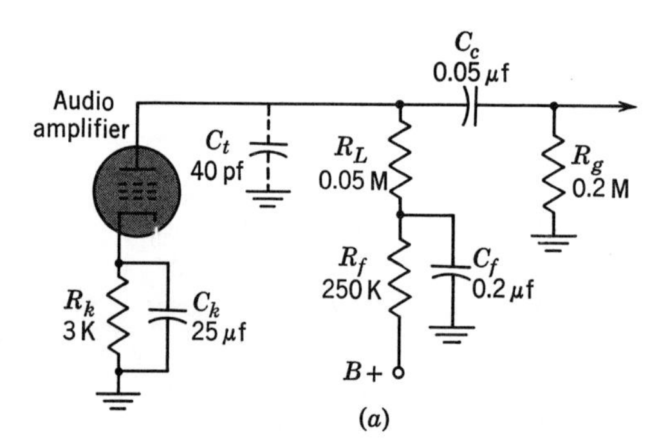


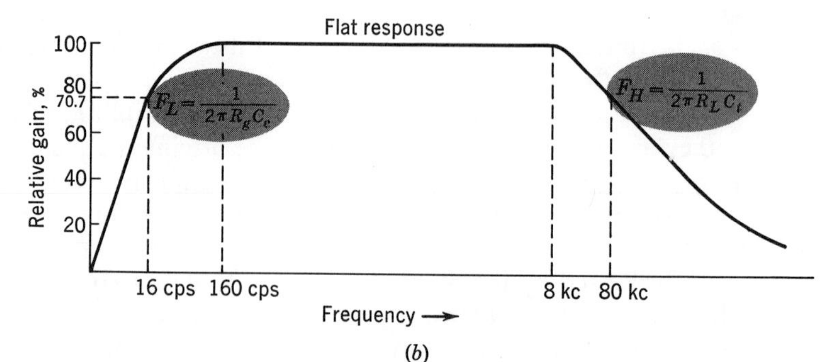


*Fig. 3·19 Sources of frequency distortion in RC-coupled amplifier. (a) Circuit with components. (b) Frequency response curve (effect of $R_fC_f$ filter not included).*

circuit, which does not include the effect of $C_f$. With the low-frequency boost provided by $R_fC_f$, the gain at $F_L$ can be made greater than the mid-frequency response.

## *3·10 Negative feedback*

A feedback circuit couples some of the output signal back to the input. When the feedback voltage is in phase with the input signal, positive feedback, or *regeneration,* occurs. In audio amplifiers, positive feedback is usually undesirable, even though it increases the gain. Regeneration can make the amplifier operate as an oscillator, causing an audio howl. This is sometimes troublesome with acoustic feedback between a microphone and loudspeaker.

Feedback voltage that has the opposite phase from the input signal is called *negative feedback, inverse feedback,* or *degeneration.* Negative feedback reduces the gain because the feedback voltage cancels part of the input signal. Negative feedback has the advantage, however, of reducing distortion. It is usually not too difficult to obtain as much gain as is needed, but distortion is a chronic problem. Therefore negative feedback is often employed to improve the quality of audio amplifiers. Negative feedback is particularly valuable for reducing the distortion of a pentode output stage driving the loudspeaker.

The reason negative feedback can reduce the distortion resides in the fact that any distortion in the output circuit is coupled back to the input circuit in opposite phase. This cancels some of the distortion, as well as part of the input signal. All forms of amplitude distortion and frequency distortion are reduced by negative feedback. The distortion is reduced in about the same proportion as the loss of gain.

**Unbypassed cathode-bias resistor.** A simple form of inverse feedback uses a cathode-bias resistor without any bypass capacitor, as in Fig. 3·20. The average value of plate current through $R_k$ determines the average cathode voltage, and this serves as cathode bias. Without the bypass, however, the cathode voltage also has variations corresponding to the signal. The cathode variations have the same phase as the input signal, since the cathode voltage follows the grid voltage. When the grid drive is more posi-

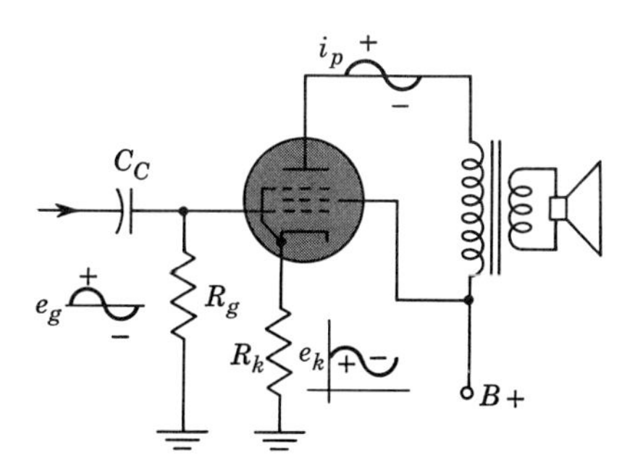


*Fig. 3·20 Inverse feedback with unbypassed cathode-bias resistor.*

tive, for instance, more plate current flows to make the cathode more positive with respect to the grounded side of $R_k$. This causes the cathode-signal voltage to cancel part of the input signal, as can be seen from the following example.

If the instantaneous value of grid-to-ground input signal increases by +8 volts, and this makes the cathode voltage rise by +2 volts, the net increase in grid-cathode voltage is only +6 volts. The effect of the varying cathode voltage, then, is to cancel a fraction of the signal input between control grid and cathode. As a result, both the gain and the distortion are reduced.

To calculate the reduced gain for pentodes,

$$A' = A \frac{1}{1 + g_m R_k} \tag{3·2}$$

where $A$ = normal gain
$A'$ = reduced gain with cathode degeneration
$R_k$ = cathode resistance, ohms
$g_m$ = transconductance of the pentode, mhos

For example, if the gain is 18 without degeneration, an unbypassed 200-ohm cathode resistor with a pentode having a $g_m$ of 4,000 $\mu$mhos will reduce the gain to 10, approximately.

**Plate-voltage feedback.** The feedback circuit in Fig. 3·21 couples a small part of the plate-voltage signal output back to its own grid-input circuit. This is negative feedback because the amplified plate-signal voltage has a polarity opposite to that of the grid-input signal. $C_1$ is a d-c blocking capacitor that couples only the a-c output signal to the voltage divider $R_1R_2$. The ratio of $R_2$ to the total of $R_1$ and $R_2$ determines what fraction of the plate signal will be fed back, since the voltage across $R_2$ is applied between grid and cathode. For the values shown, $R_2$ has one-tenth the total resistance of the divider. Therefore one-tenth the plate signal output is coupled back to the grid circuit. The capacitance of $C_1$ can be chosen large enough to provide equal feedback for all audio frequencies. A smaller capacitance, however, will provide more feedback for higher fre-

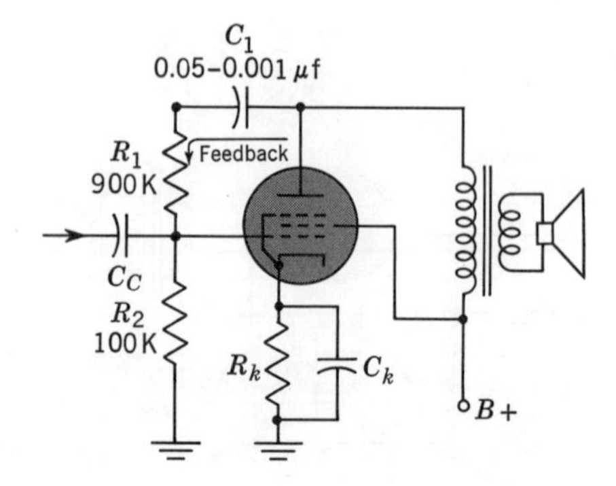


*Fig. 3·21 Negative feedback with plate signal coupled back to control grid.*

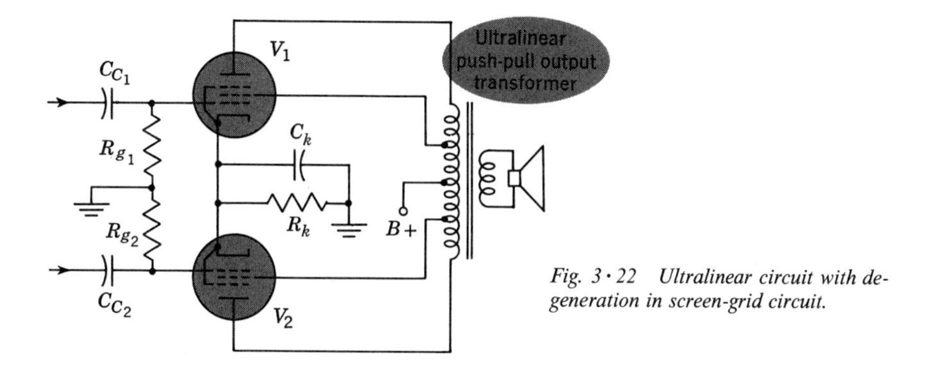


*Fig. 3·22 Ultralinear circuit with degeneration in screen-grid circuit.*

quencies and less feedback for lower frequencies. Thus, the gain is reduced more for the high audio frequencies, providing bass boost in the audio output.

The voltage feedback factor $\beta$ is defined as

$$\beta = \frac{\text{feedback voltage}}{\text{output voltage}} \tag{3·3}$$

Typical values of $\beta$ are 0.1 to 0.2. In this example of 10 per cent feedback, $\beta$ equals 0.1. The reduced gain with feedback can be calculated from the normal gain as follows:

$$A' = A\frac{1}{1 + A\beta} \tag{3·4}$$

where $A$ is normal gain; $A'$ is reduced gain; $B$ is feedback factor. For the example of 10 per cent feedback, or $\beta$ equal to 0.1, a normal gain of 10 is reduced to 5, approximately.

**Screen-voltage feedback.** The circuit in Fig. 3·22 uses an output transformer that has taps for the screen grid. This applies part of the plate-signal voltage to the screen grid. Usually the screen tap is 20 per cent of the plate winding. With the screen-grid voltage varying in opposite polarity to the input signal, degeneration is introduced. This arrangement is called as *ultralinear* circuit because the feedback improves the linearity of the pentode operating characteristic, making it function more like a triode. The ultralinear transformer is a convenient way of introducing degeneration into a push-pull output circuit, since balanced negative feedback is provided for both tubes.

**Feedback loop from loudspeaker.** Another circuit for providing negative feedback is shown in Fig. 3·23. The feedback voltage is taken from the secondary of the output transformer and coupled back to the cathode of $V_1$. In this arrangement, the negative feedback can cancel some of the distortion produced by the output transformer. Also, by including two ampli-

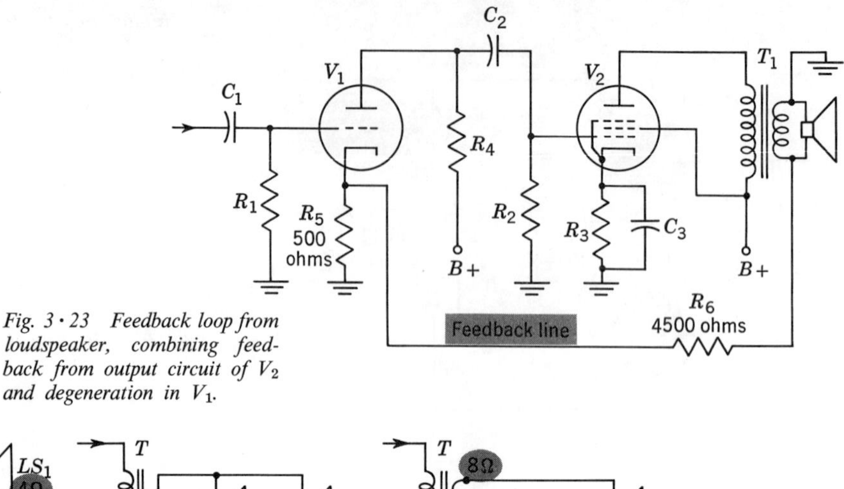


*Fig. 3·23 Feedback loop from loudspeaker, combining feedback from output circuit of $V_2$ and degeneration in $V_1$.*

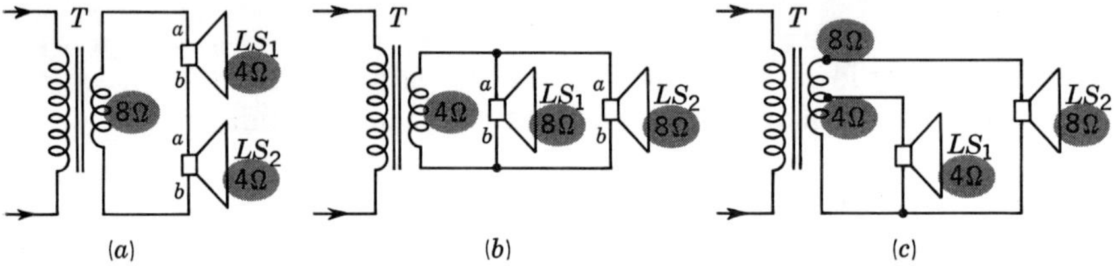


*Fig. 3·24 Multiple speakers across secondary of output transformer. (a) Series connections. (b) Parallel connections. (c) Different impedances across taps of secondary.*

fier stages in the feedback loop, we can make the degeneration more effective. Note that one side of the secondary winding on $T_1$ must be grounded to supply feedback voltage with respect to chassis ground. Which side is grounded depends on the windings. If the amplifier howls because of positive feedback, simply reverse the secondary connections.

## *3·11 Multiple speakers*

Several ways of connecting two speakers to one amplifier are shown in Fig. 3·24. In *a*, loudspeakers of equal impedance are connected in series across the secondary of the output transformer. The total impedance then is the sum of the individual values. In *b*, the total impedance of the two 8-ohm loudspeakers in parallel across the secondary is one-half of either value, or 4 ohms. In *c*, two loudspeakers with different impedances are shown connected to appropriate taps. This matches each speaker to the secondary for maximum volume. The load in the secondary is equivalent to the two speakers in parallel.

If speakers of unequal impedances are connected in series with each other, the largest impedance will develop the highest voltage and produce the most power. If they are connected in parallel, the lowest impedance will have the highest current and produce the most power. Keep in mind, though, that no matter how many speakers you connect, there cannot be more output than the power supplied by the output stage.

**Speaker phasing.** When multiple speakers are mounted in the same baffle, their connections must be phased to make the voice coils move the same way or some of the sound will cancel. To phase the speakers, find

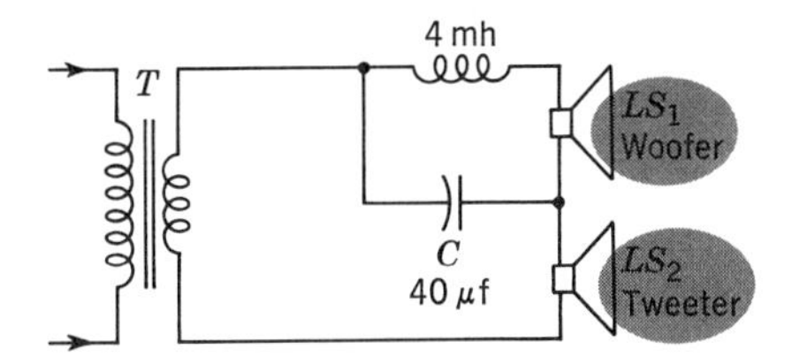


*Fig. 3·25 Frequency divider network for woofer and tweeter loudspeakers.*

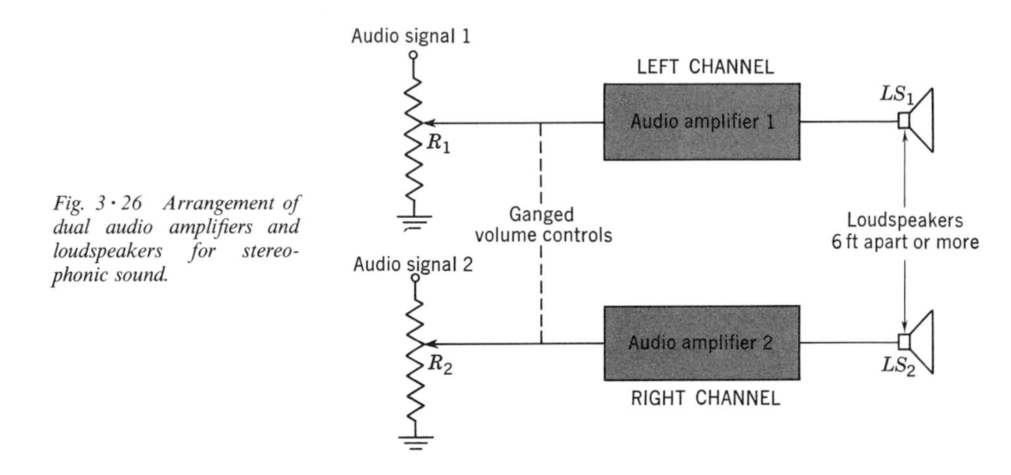


*Fig. 3·26 Arrangement of dual audio amplifiers and loudspeakers for stereophonic sound.*

out which terminal of the voice coil has the same effect on each loudspeaker. For instance, if the negative side of a dry cell makes the cone move outward, we can call this terminal *a*; the opposite end is terminal *b*. For series connections, the *a* and *b* terminals are connected in successive order as shown in Fig. 3·24*a*. For parallel connections all *a* terminals are tied together and all *b* terminals also. It is not usually necessary to phase the connections to loudspeakers in different locations.

**Frequency divider.** When woofer and tweeter loudspeakers are used, the frequencies for each are supplied by a *frequency divider network*. This is a filter to provide high frequencies for the tweeter and low frequencies for the woofer. One example is illustrated in Fig. 3·25. Both speakers are in series across the secondary, but the 4-mh choke prevents high-frequency current from flowing in the woofer $LS_1$. These frequencies are coupled by $C$ to the tweeter $LS_2$. At the crossover frequency, where $X_L$ equals $X_C$, both loudspeakers receive the same amount of input signal. The crossover frequency is generally in the range of 400 to 1,200 cps.

## *3·12 Stereophonic sound*

The most logical application of dual loudspeakers is to have each reproduce sound individually for the left and right ears, as we would hear the live sound. Such an arrangement for reproducing sound from the left and right directions in space is called *stereophonic* sound. However, two separate audio signals are required. Each provides the slight difference in sound that the left and right ears perceive when we listen in person. This gives a sense of direction to different components of the sound. The corresponding idea is to have left and right channels in the sound reproduction, as illustrated in Fig. 3·26. Note that separate and identical amplifiers are used for the audio signals in the left and right channels.

The two individual audio signals can be supplied from stereo records, from separate tracks on stereo tape recorders, or by multiplex transmission in the commercial FM broadcast band. The details of this FM multiplexing are described in Chap. 12 on Receiver Circuits.

## *3·13 The decibel unit*

The ear is more sensitive to a change in sound intensity at low-volume than at high-volume levels. For instance, a 1-watt increase of power output from 2 to 3 watts is readily recognized, but the same change from 20 to 21 watts is hardly noticeable. This means that the listener's impression of an increase or decrease in loudness depends on the logarithm of the ratio of the two powers. The logarithm of the power ratio is a unit called the *bel:*

$$\text{Number of bels} = \log \frac{P_2}{P_1}$$

or, the number of bels equals the logarithm of the power ratio. (Logarithms are explained in Appendix C.) The logarithm is to the base 10. The more commonly used unit is the *decibel,* however, abbreviated db, which equals one-tenth of a bel. One db equals a 26 per cent increase in power of 12 per cent increase in voltage, approximately. This amount of change is just perceptible to the ear.

**Decibel calculations.** The formula for calculating decibels is

$$\text{db} = 10 \times \log \frac{P_2}{P_1} \qquad (3\cdot5)$$

To use the formula, you can follow this procedure:

1. Find the ratio of the two powers, being sure to use the same units for both. For example, suppose we are comparing 1-mw input and 1-watt output. This ratio equals 1,000. Let $P_2$ in the numerator always be the larger power. Then the ratio will not be a fraction less than one, thereby eliminating the inconvenience of working with negative logarithms.
2. Find the logarithm of the power ratio. In this example the log of 1,000 equals 3.
3. Multiply the logarithm by 10 to find the number of decibels. The answer is 30 db.

In this example the answer is +30 db because the power increased. If the power were to decrease from 1 watt to 1 mw the answer would be −30 db. Always make the ratio of the two powers equal unity or more, so that positive logarithms can be used. Then decide whether the answer is positive or negative on the basis of an increase or decrease in power. The decibel power ratio is independent of the impedances that develop the two powers.

**Voltage ratios.** Since power is $E^2/R$, the decibel formula can be expressed in terms of voltages as

$$\text{db} = 20 \times \log \frac{E_2}{E_1} \qquad (3 \cdot 6)$$

The two voltages $E_1$ and $E_2$ must be across the same value of impedance, however. If not, multiply Eq. 3·6 by the correction factor $\sqrt{Z_1/Z_2}$.

For the example of an increase from 1 mv to 1 volt, the decibel voltage gain equals

$$\text{db} = 20 \log \frac{E_2}{E_1} = 20 \times \log \frac{1{,}000}{1} = 20 \times 3 = 60$$

The factor of 20 in the decibel voltage formula stems from the fact that power is proportional to the square of voltage. In logarithms, squaring a number corresponds to doubling its logarithm. Therefore, a voltage ratio always equals twice the decibels of the same numerical power ratio.

**Common decibel values.** Decibels are frequently used even for r-f applications, because decibel units simplify many power or voltage ratios and provide a basis of comparison related to its effect in audio reproduction. Certain decibel values are worth memorizing because they are used often. As listed in Table 3·2, double the power is 3 db; double the voltage is 6 db. Similarly, one-half power is $-3$ db and one-half voltage is $-6$ db. Note that zero decibel does not mean zero power or voltage but simply the fact that equal powers or voltages are compared. Then the ratio equals one and the log of one is zero.

If the common decibel values are memorized, many decibel calculations can be done quickly. The technique involves dividing the ratio into factors for which you know the decibel values and then adding the decibel values for all the factors. For instance, 200 is the same as $2 \times 100$. Therefore the decibel value for a power ratio of 200 equals 3 db for the factor of 2 and 20 db for the factor of 100, or a total of 23 db. This procedure is possible because adding the logarithms of numbers corresponds to multiplying the

*Table 3·2 Common decibel values*

| *Decibels* | *Power ratio* | *Voltage ratio* |
|---|---|---|
| 0 | 1 | 1 |
| 1 | 1.26 | 1.12 |
| 2 | 1.58 | 1.26 |
| 3 | 2 | 1.4 |
| −3 | ½ | 0.707 |
| 6 | 4 | 2 |
| −6 | ¼ | ½ |
| 10 | 10 | $\sqrt{10}$ |
| 20 | 100 | 10 |

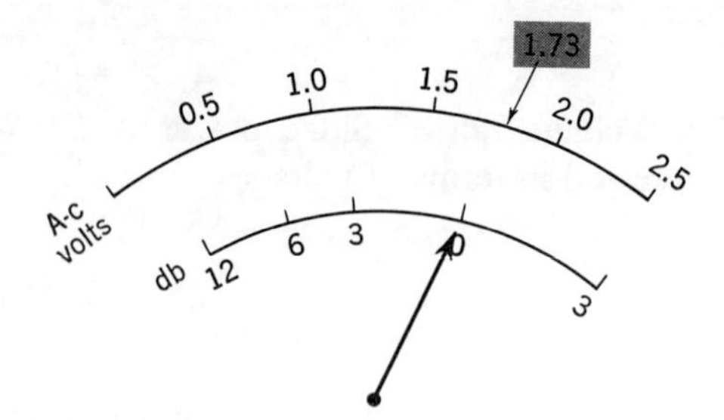


*Fig. 3 · 27 Typical decibel scale on a-c voltmeter. Zero db equals 1.73 volts across 500 ohms.*

numbers. For this reason, decibel gain or loss values for cascaded stages are added or subtracted. As an example, two stages in cascade each with a decibel voltage gain of 23 provide a total gain of 46 db.

**Decibel reference levels.** In order to make the decibel units more useful, standard reference levels are used for comparison. The most common reference is 6 mw. Then any other power level can be compared with this reference by the formula

$$\text{db} = 10 \log \frac{P}{0.006 \text{ watt}} \qquad (3 \cdot 7)$$

For voltage comparisons, the impedance of 500 ohms is specified. Then the 6-mw reference corresponds to 1.73 volts. With this reference, 0 db equals 6 mw, or 1.73 volts across 500 ohms. Negative decibel values mean that the power or voltage is less than the reference.

Since the decibel values correspond to different voltages across a fixed impedance, an a-c voltmeter can have a decibel scale for a specific reference. A typical decibel scale is shown in Fig. 3 · 27. Note that 1.73 volts corresponds to 0 db. Double this, or 3.46 volts, is +6 db; one-half, or 0.865 volt, is −6 db. These readings apply to a-c voltages measured across 500 ohms.

In telephone service the decibel reference generally used is 1 mw in 600 ohms, but this is usually indicated as *dbm*. Then

$$\text{dbm} = 10 \times \log \frac{P}{0.001 \text{ watt}} \qquad (3 \cdot 8)$$

For example, a power of 2 mw equals 3 dbm with the 1 mw reference; 100 mw equals 20 dbm; 1 mw is 0 dbm; ½ mw equals −3 dbm.

In radio-broadcast service, the reference of 1 mw in 600 ohms is used for defining the volume unit, or VU. Its main application is for a-c voltmeters calibrated in VU. The VU meter has standard characteristics to indicate relative volume levels for the complex voltage waveforms in typical voice and music programs.

## SUMMARY

1. Sound is a wave motion of varying air pressures. The velocity in air is approximately 1,130 ft per sec. The amplitude of the sound waves determines the loudness; the frequency determines the pitch or tone of the sound.
2. Harmonic frequencies are exact multiples of the fundamental frequency.

3. Common types of electromechanical transducers for audio circuits are microphones, phonograph pickups, loudspeakers, and headsets. A PA system consists of a microphone feeding an audio amplifier which drives a loudspeaker.
4. An audio preamplifier circuit consists of *RC*-coupled stages to amplify a low-level audio signal.
5. A typical audio amplifier consists of one or more *RC*-coupled stages to drive a power-output stage, transformer-coupled to the loudspeaker.
6. The output transformer has a voltage step-down turns ratio to match the low impedance of the loudspeaker's voice coil to the high impedance of the output tube's plate circuit. The turns ratio equals the square root of the impedance ratio.
7. The audio power-output stage can use one tube, two tubes in parallel, or two tubes in push-pull. Push-pull operation can be class A, B, or AB, but a single-ended audio amplifier must operate class A.
8. The push-pull circuit uses a center-tapped output transformer in the plate circuit, with equal and opposite grid-driving voltages. Push-pull operation provides maximum power output because more grid drive can be used with the least distortion.
9. A phase-inverter or phase-splitter circuit is a stage that provides equal and opposite output voltages to drive a push-pull amplifier.
10. Tone-control circuits generally adjust the gain for the high audio frequencies. Bypassing the high frequencies corresponds to bass response; more gain for the high frequencies provides treble response.
11. The main types of distortion are amplitude distortion and frequency distortion. Nonlinear amplification results in amplitude distortion. Frequency distortion means nonuniform gain for all the frequencies present in the signal.
12. Negative feedback, inverse feedback, or degeneration means feeding part of the output signal back to the input circuit, out of phase with the input signal. Negative feedback decreases the gain of the stage but reduces the distortion. Two common forms of negative feedback are an unbypassed cathode-bias resistor and a circuit to couple part of the output-signal voltage back to the grid-cathode circuit.
13. Multiple speakers can be connected in series or parallel, as shown in Fig. 3·24. When the speakers are in one enclosure, they must be connected to make the cones move in phase with each other.
14. In stereophonic sound, two individual audio systems are used to reproduce sound from the left and right directions, as your ears would hear the live sound.
15. The decibel is a logarithmic unit for a power or voltage ratio:

$$\text{db} = 10 \log \frac{P_2}{P_1} \qquad \text{for power ratios}$$

$$\text{db} = 20 \log \frac{E_2}{E_1} \qquad \text{for voltage ratios (across the same } R \text{ or } Z\text{)}$$

Common references are 6 mw in 500 ohms for db and 1 mw in 600 ohms for dbm.

## SELF-EXAMINATION

Here's a chance to find out how well you have learned the material in this chapter. Work the exercises; then check your answers against the Key at the back of this book. These exercises are for your self-testing only.

1. The difference between sound waves and electric audio signal is that (*a*) sound waves have lower frequencies; (*b*) sound waves have lower velocity; (*c*) audio signal always has lower amplitude; (*d*) audio signal always has lower frequencies.
2. The third harmonic of a 200-cps fundamental frequency is (*a*) 200 cps; (*b*) 300 cps; (*c*) 600 cps; (*d*) 800 cps.
3. The low impedance of a dynamic loudspeaker is a result of (*a*) a large field magnet on a PM speaker; (*b*) the few turns of the voice coil; (*c*) a large paper cone with big loudspeakers; (*d*) a small baffle for enclosing the loudspeaker.

4. A single-ended output transformer matches a 16-ohm loudspeaker to a 1,600-ohm plate load. The required turns ratio is (*a*) 1:10; (*b*) 1:16; (*c*) 1:1,000; (*d*) 1:1,600.
5. Insufficient capacitance for the coupling capacitor in an audio amplifier can cause (*a*) amplitude distortion; (*b*) high-frequency attenuation; (*c*) intermodulation distortion; (*d*) low-frequency attenuation.
6. Opposite polarities of audio signal voltage can be obtained from the (*a*) control grid and cathode in one stage; (*b*) plate circuits of two stages in cascade; (*c*) plate and screen grid in one stage; (*d*) plate and B+ connections in one stage.
7. A push-pull amplifier circuit requires (*a*) equal input voltages of the same phase; (*b*) an output transformer with a center tap in the secondary; (*c*) an output transformer with a center tap in the primary; (*d*) opposite polarities of screen-grid voltage.
8. In tone-control circuits (*a*) treble response means maximum high-frequency response; (*b*) treble response means maximum response for the middle range of audio frequencies; (*c*) bass response means emphasizing the high audio frequencies; (*d*) bass response means bypassing the low audio frequencies.
9. Two 8-ohm speakers in series have a combined impedance of (*a*) 4 ohms; (*b*) 8 ohms; (*c*) 16 ohms; (*d*) 64 ohms.
10. A gain of 33 db means a power gain of (*a*) 13; (*b*) 20; (*c*) 40; (*d*) 2,000.

## QUESTIONS AND PROBLEMS

1. Describe briefly three types of electromechanical transducers for audio circuits.
2. Define the following: PM loudspeaker, woofer, tweeter, ceramic phonograph cartridge, audio preamplifier.
3. List the fundamental frequency and frequencies up to the fifth harmonic for a 1,000-cps signal.
4. (*a*) Give one advantage and one disadvantage of using negative feedback; (*b*) How can a negative feedback circuit be used to provide bass boost?
5. Explain briefly the operation of the variable tone-control circuit in Fig. 3·16*c*.
6. Explain the difference between frequency distortion and harmonic distortion.
7. Calculate the decibel gain or loss for the following: (*a*) 2 mw in and 200 mw out; (*b*) 5 mv in and 500 mv out; (*c*) 10 volts in and 5 volts out; (*d*) 8 watts in and 4 watts out; (*e*) 10 volts in and 7.07 volts out.
8. What is the difference in the reference for db and dbm?
9. Draw the schematic diagram of a triode cathode-follower phase splitter driving a push-pull output stage coupled to one loudspeaker. Use two pentodes for the push-pull stage. Values of the components are not required. Tone control and negative feedback need not be shown.
10. Give two methods of providing negative feedback in an audio amplifier.
11. Give one function for each of the following components in the audio amplifier circuit in Fig. 3·9: $C_1$, $R_5$, $R_2$, $R_3$, $T_1$.
12. Give the voltages with polarity for collector reverse bias and base forward bias on TR1 and TR2 in Fig. 3·10.
13. In Fig. 3·9: (*a*) Calculate the total cathode current through $R_3$. (*b*) If $I_{c_2}$ is 6 ma, how much is $I_b$? (*c*) If the resistance of $L_p$ is 200 ohms, how much is $E_b$?
14. In Fig. 3·9: (*a*) How much is the over-all gain $A_v$ for the two stages? (*b*) Convert this voltage ratio to decibels.
15. Calculate the turns ratio needed for an audio output transformer to match a 16-ohm speaker to a 5,600-ohm plate load.
16. Referring to Fig. 3·19, calculate the frequencies $F_L$ and $F_H$ with the values of 20 pf for $C_t$, 5 K for $R_L$, 0.5 M for $R_g$, and 0.5 $\mu$f for $C_c$.
17. Describe the effect on the audio output from the loudspeaker, for the following troubles in the audio amplifier in Fig. 3·9: $L_p$ open, $R_3$ open, $C_3$ open, $R_5$ open, and $C_1$ open.
18. Describe the effect on the audio output from the loudspeaker for the following troubles in the audio amplifier in Fig. 3·10: $C_1$ open, $R_5$ open, $C_3$ shorted, and $T_2$ open.

# Review of chapters 1 to 3

## SUMMARY

1. Voltage gain is the ratio of output-signal voltage to input-signal voltage. A voltage amplifier has a high-impedance plate load for maximum voltage gain; a power amplifier provides maximum signal current through a low-impedance plate load.
2. The control-grid bias is a steady negative d-c voltage to set the operating point for signal variations in the amplifier. The main methods are (1) fixed bias, (2) cathode bias or self-bias, (3) grid-leak bias or signal bias, and (4) contact bias.
3. The amount of negative grid voltage that cuts off plate current is the grid-cutoff voltage.
4. A class A amplifier allows plate current for the complete signal cycle of 360°; class B, 180°; class C, about 120°, or much less than 180°.
5. Amplifiers are usually connected in cascade with the plate signal output of one stage coupled to the control-grid input of the next stage. The over-all voltage gain equals the product of the individual values.
6. The main types of coupling between stages are (1) *RC*, (2) impedance, (3) transformer, and (4) direct. R-f amplifiers generally have tuned coupling circuits. Direct-coupled stages amplify d-c voltage as well as a-c signal.
7. The cathode-follower or cathode-coupled stage has the load impedance in the cathode circuit, instead of the plate circuit.
8. In transistors, the emitter always has forward bias for low resistance to the base; the collector has reverse bias and high resistance. These requirements apply to PNP and NPN transistors.
9. The three basic types of transistor amplifier circuits are common-emitter, common-base, and common-collector. Gain is greatest for common-emitter circuits. The common-collector circuit corresponds to a cathode-follower stage.
10. Transformer coupling between transistor stages produces the best impedance match and the most gain. *RC* coupling is more economical, however, and has more uniform frequency response for audio amplifiers.
11. Precautions to keep in mind for transistor circuits are: (1) check the battery supply and transistor voltages; (2) it is useful to meter the transistor currents; (3) turn off applied voltage when inserting or removing transistor; (4) be careful not to apply excessive heat to transistor; (5) ohmmeter measurements made with transistor out of circuit eliminate problems of the ohmmeter's internal battery affecting the transistor bias.
12. Sound is a physical wave motion of varying air pressures. The velocity of sound waves in air is approximately 1,130 ft per sec. Amplitude determines loudness; frequency determines pitch or tone.
13. Sound waves in the audible range can be converted to audio-frequency electric signals by a microphone or phonograph pickup, as examples of transducers. A loudspeaker and headset convert audio signals into sound waves.
14. For either sound waves or audio signals, harmonics are exact multiples of the fundamental signal frequency.
15. A typical audio amplifier consists of one or more *RC*-coupled stages driving a power-output stage, transformer-coupled to the loudspeaker.
16. The audio-output transformer has a voltage step-down ratio to match the low-impedance voice coil to the high-impedance plate circuit.
17. Audio amplifiers operate class A, except that push-pull operation can be class B or class AB.
18. A push-pull circuit uses two tubes with a center-tapped output transformer. Equal and

opposite input signals are needed to drive the grid circuit. Push-pull operation provides maximum power output with minimum distortion.

19. A phase inverter or phase splitter provides equal and opposite output voltages to drive a push-pull amplifier.
20. Tone-control circuits vary the frequency response of the amplifier. Less response for the high audio frequencies provides bass; more high-frequency gain means treble response.
21. Amplitude distortion results from nonlinear amplification of the signal voltage; frequency distortion means the gain is not the same for different frequencies in the signal.
22. Coupling some of the output signal back to the input circuit out of phase with the input signal is called negative feedback, inverse feedback, or degeneration. Negative feedback reduces gain but also minimizes distortion.
23. The decibel is a logarithmic unit for power or voltage ratios. Double the power equals 3 db; double the voltage is 6 db; −3 db is a voltage ratio of 0.707, or 70.7 per cent.

## REFERENCES (*Additional references at back of book.*)

### *Books*

Crowhurst, N. H., *Basic Audio Course,* John F. Rider, Publisher, Inc.
Everitt, W. L., *Fundamentals of Radio and Electronics,* Prentice-Hall, Inc.
Ghirardi, A. A., and J. R. Johnson, *Radio and Television Receiver Circuitry and Operation,* Rinehart & Company, Inc.
Hellman, C. I., *Elements of Radio,* 3d ed., D. Van Nostrand Company, Inc.
Henney, K., and G. A. Richardson, *Principles of Radio,* 6th ed., John Wiley & Sons, Inc.
Hickey, H. V., and W. M. Villines, *Elements of Electronics,* 2d ed., McGraw-Hill Book Company.
Kiver, Milton S., *Transistors,* 3d ed., McGraw-Hill Book Company.
Marcus, A., and W. Marcus, *Elements of Radio,* 3d ed., Prentice-Hall, Inc.
Sheingold, A., *Fundamentals of Radio Communication,* D. Van Nostrand Company, Inc.
Slurzberg, M., and W. Osterheld, *Essentials of Radio,* 2d ed., McGraw-Hill Book Company.
Van Valkenburgh, Nooger, and Neville, Inc., *Basic Electronics,* John F. Rider, Publisher, Inc.
Watson, H. M., H. E. Welch, and G. S. Eby: *Understanding Radio,* 3d ed., McGraw-Hill Book Company.

## REVIEW SELF-EXAMINATION

Here's another chance to check your progress. Work the exercises just as you did those at the end of each chapter and check your answers.

Which phrases in the right-hand column belong to the phrases in the left-hand column? Answer by placing the correct letter from the right column next to the corresponding phrase at the left. For example, the answer to 1 is (*c*) because "less negative grid bias" at the right matches "increased plate current" at the left. Each numbered phrase has one and only one corresponding answer.

1. Increased plate current ______
2. High voltage gain ______
3. Cathode bias ______
4. Grid-leak bias ______
5. Minimum distortion ______
6. Maximum efficiency ______
7. Plate current for 180° ______
8. Cathode follower ______
9. D-c amplifier ______
10. Forward bias ______
11. Reverse bias ______
12. R-f amplifiers ______

(*a*) Gain less than 1
(*b*) Emitter
(*c*) Less negative grid bias
(*d*) Collector
(*e*) Impedance match
(*f*) Has no coupling capacitor
(*g*) Tuned coupling circuit
(*h*) Multiply gain
(*i*) Input signal to emitter
(*j*) Proportional to grid drive
(*k*) $I_k R_k$
(*l*) High value of $R_L$

13. Cascaded stages ______
14. Transformer coupling ______
15. Common-emitter circuit ______
16. Common-base circuit ______
17. 60 cps ______
18. 14,000 cps ______
19. Phase inverter ______
20. Bass response ______
21. Amplitude distortion ______
22. Zero plate voltage ______
23. Zero screen-grid voltage ______
24. Degeneration ______
25. Voice coil impedance ______
26. −3 db ______

(*m*) Class A operation
(*n*) Class B operation
(*o*) Class C operation
(*p*) Input signal to base
(*q*) Low audio frequency
(*r*) Incorrect bias
(*s*) Open plate circuit
(*t*) Open screen-grid dropping resistor
(*u*) Open cathode-bypass capacitor
(*v*) 3 to 16 ohms
(*w*) One-half power
(*x*) Bypass high audio frequencies
(*y*) Drives push-pull amplifier
(*z*) Second harmonic of 7,000 cps

# Chapter 4 Radio-frequency circuits

This unit explains the operation of amplifiers that are tuned to a specific frequency. Both single-tuned and double-tuned stages are described in detail, including the effect of mutual coupling in a tuned transformer. In addition to the analysis of the resonant response of tuned circuits, typical r-f circuits are described. The topics are:



## *4·1 Functions of the r-f amplifier*

The block diagram of a typical r-f amplifier is shown in Fig. 4·1. Coming into the amplifier are three bands of signals, each 10 kc wide. One is centered around 990 kc and extends from 985 to 995 kc. The second band has a center frequency of 1,000 kc and extends from 995 to 1,005 kc. The third band extends from 1,005 to 1,015 kc with a center frequency of 1,010 kc. All three bands have signal amplitudes of 1 mv. Assume now that the amplifier is tuned to a resonant frequency of 1,000 kc and that it amplifies signals between 995 and 1,005 kc with a gain of 100. For signals below 995 kc or above 1,005 kc, the gain will be much lower than 100.

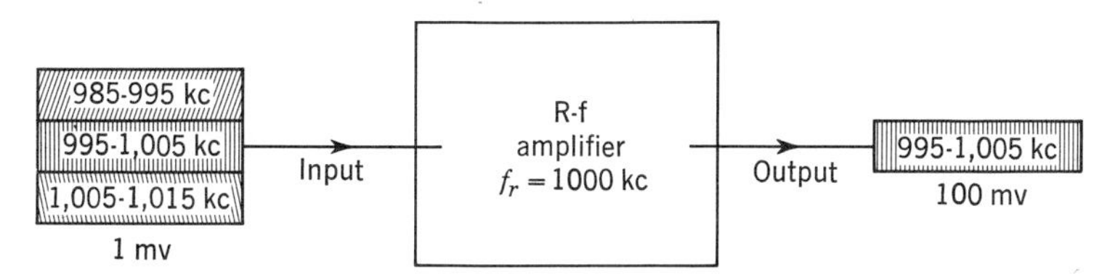


*Fig. 4·1 Function of a tuned r-f amplifier.*

As shown in Fig. 4·1, only the band of signals between 995 and 1,005 kc passes through with amplitudes of 100 mv. The other bands are said to be *attenuated,* because they have much lower amplitudes. This type of amplifier, which has the ability to select certain signals from a large group of signals, is a *tuned* amplifier. Amplifiers for r-f signals are generally tuned because the *LC* values needed for resonance are practical. Figure 4·2 shows some typical *LC* components for tuned circuits.

**Selectivity.** This can be defined as the ability of a tuned amplifier to provide maximum gain for signals within a definite band of frequencies. For all other signal frequencies, the amplifier provides very little gain. For example, the amplifier in Fig. 4·1 selects the band of signals extending from 995 to 1,005 kc by amplifying signal voltage at these frequencies. There is very little output for frequencies from 985 to 995 kc and from 1,005 to 1,015 kc. The selectivity results from using a tuned circuit in the amplifier.

**Gain.** Another important requirement of tuned r-f amplifiers is the need to provide enough gain or amplification of the very weak signals coming into the amplifier. In radio and television receivers, for instance, the signal induced in the antenna may have an amplitude of only several microvolts. Before the signal can be detected, it must have an amplitude of at least several volts. The tuned amplifier stages must therefore provide an over-all gain of about 1,000,000. The number of stages necessary to accomplish this much over-all gain will depend on how much gain each individual r-f amplifier stage can contribute. Remember that the total gain of cascaded stages equals the product of the individual gain values. For example,

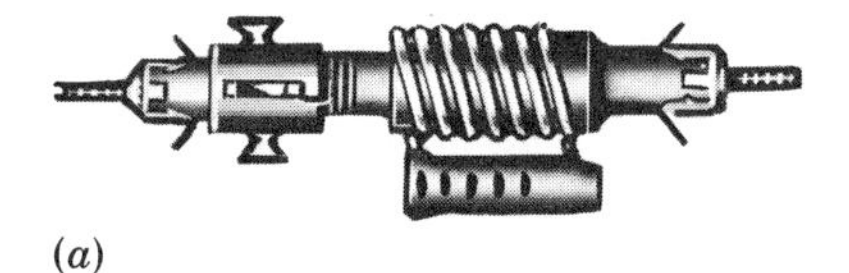

(*a*)

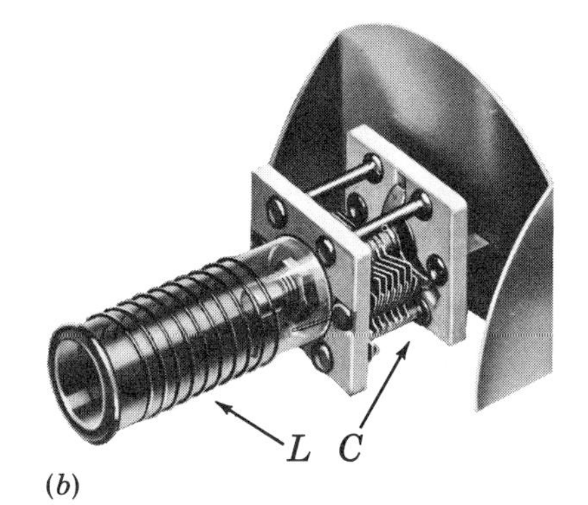


(*b*)

*Fig. 4·2 R-f tuned circuits. (a) Single-tuned circuit resonant at 27 Mc, with adjustable slug. (b) Wavemeter as example of single-tuned circuit.* (James Millen Mfg. Co., Inc.)

if each r-f amplifier stage in a radio receiver can provide an average gain of 100, only three r-f stages will be required to provide an over-all r-f gain of 1,000,000. If, however, each r-f amplifier stage in a television receiver can provide an average gain of only 10, it will take *six* r-f stages to provide a total r-f gain of 1,000,000. It will be shown later that the gain produced at the center frequency of an r-f amplifier stage depends on the type of tuned circuit associated with the stage, and on the width of the band of signals that must be passed by the stage. In general, the greater the bandwidth is, the smaller the gain.

**R-f amplifier tubes.** Pentodes are preferred for r-f amplifiers because of their low interelectrode capacitance between the control grid and plate. Also, high transconductance ($g_m$) is desirable. Miniature tubes are preferable for higher frequencies because of their lower interelectrode capacitance. Similarly, when transistors are used for r-f amplifiers, low capacitance with a high value for the $\alpha$ cutoff frequency is important.

**Fidelity.** Another requirement of r-f amplifiers is faithful reproduction of the signal coming into the amplifier, or fidelity. The amplifier must not distort the incoming signal in any way. Fidelity in r-f amplifiers depends largely on the proper operation of the amplifier itself as well as proper design of the tuned circuits. In terms of the tuned circuit, it must have the bandwidth required for the signal frequencies to be amplified.

**Noise.** The last important requirement of r-f amplifiers is the ability to eliminate or suppress interfering signals which are usually grouped together as "noise." Some of the noise is actually generated within the vacuum tube used in the circuit. This is called *tube noise*. In applications of r-f amplifiers for very weak signals, it is important to select tubes that are relatively free from tube noise.

The relative amount of noise present in an amplifier is usually expressed by the *signal-to-noise ratio* of the amplifier. For example, if the r-f signal output of an amplifier is 500 $\mu$v and the noise voltage is 50 $\mu$v, the signal-to-noise ratio equals 10. There are no units because it is a ratio. In decibel units a voltage ratio of 10:1 equals 20 db.

## 4·2 *Typical r-f amplifier stage*

An r-f amplifier consists of an amplifier tube associated with a tuned plate load. The combination actually used in any particular circuit depends on the function. For example, in a typical r-f amplifier stage employed in a radio receiver, a band of signals 10 kc wide centered around 1,000 kc may be amplified. This can be accomplished with the combination of vacuum tube and tuned circuit shown in Fig. 4·3. The input signal $e_1$ is applied between control grid and cathode in series with a bias voltage $E_{cc_1}$. The plate load is a tuned circuit consisting of a coil in parallel with a variable tuning capacitor $C$. The inductance of the coil is $L$ and $R_S$ is its a-c series resistance. The parallel-resonant circuit is connected in series with the plate-supply voltage $E_{bb}$.

The screen grid of the pentode amplifier has its d-c voltage supplied from B+ through the dropping resistor $R_1$ with $C_1$ the screen-bypass capacitor. In the plate circuit, $C_2$ bypasses the r-f signal around the B supply. In this way the low-potential side of the tuned circuit is returned to chassis ground and cathode for r-f signal. At the plate, the output r-f signal is coupled to the next stage by coupling capacitor $C_c$.

Since the cathode is connected to chassis, this arrangement is called a *grounded-cathode* circuit. Even if the stage has cathode bias, the cathode-bypass capacitor returns the cathode to ground. Essentially, then, the cathode is grounded. The a-c input signal is coupled to the control grid, while the output signal is taken from the plate circuit.

**Equivalent circuits.** Before analyzing the operation of the tuned amplifier in Fig. 4·3, we can simplify it into an equivalent circuit. With this equivalent circuit, we can see more clearly how the gain of the amplifier depends on the plate-load impedance.

First, as shown in Fig. 4·4*a*, all the d-c supply and bias-voltage sources are omitted. This is done because only the *a-c* operation of the circuit is being analyzed. Since capacitors $C_1$, $C_2$, and $C_c$ usually have values between 0.01 and 0.1 $\mu$f, and their reactance at frequencies to which the circuit is normally tuned (550 to 1,650 kc) is only several ohms, they can be completely ignored in the equivalent a-c diagram. Therefore the equivalent a-c circuit in Fig. 4·4*a* contains the tube, the input signal $e_1$, and the

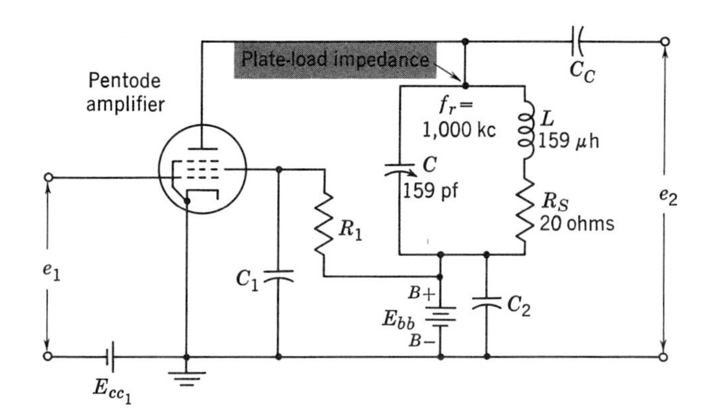


*Fig. 4·3 Typical r-f amplifier stage. Grounded-cathode circuit with single-tuned plate load.*

*Fig. 4·4 (a) Simplified a-c circuit of Fig. 4·3, omitting d-c supply voltages and bypass capacitors. (b) Equivalent a-c plate circuit.*

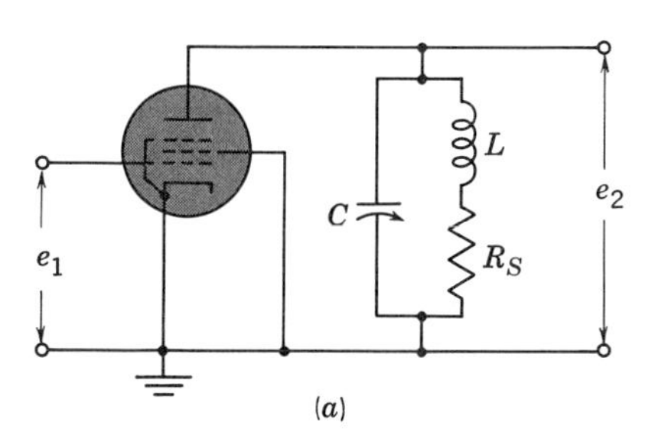


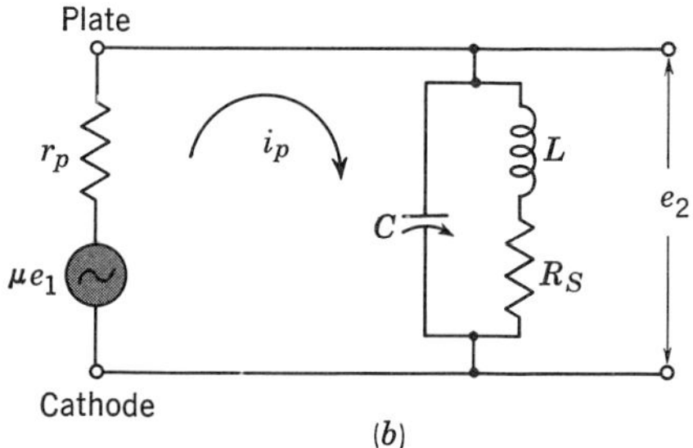

parallel-tuned circuit. The output-signal voltage $e_2$ appears across the tuned plate load.

The second step in reducing the original circuit is shown in Fig. 4·4*b*. This is the equivalent a-c plate circuit. Here the vacuum tube is replaced by an equivalent signal source $\mu e_1$ in series with the internal plate resistance of the tube, $r_p$, between plate and cathode. The circuit is completed by returning from plate to cathode through the plate load, which is the parallel-resonant circuit with $L$ and $C$. The equivalent signal source is made $\mu e_1$ because $\mu$ is the amplification factor of the tube. Then $\mu e_1$ represents the maximum possible amplification that may be produced by the tube. The equivalent a-c circuit is completed by representing the circuit current by the symbol $i_p$. This is the plate signal current.

**Gain of the amplifier.** To see how the gain depends on the plate load, the impedance of the tuned circuit at resonance will be represented temporarily by the symbol $Z$. This makes the circuit a series combination of $r_p$ in series with $Z$. The applied voltage is the amplified signal $\mu e_1$. Applying Ohm's law for a-c series circuits, the signal plate current $i_p$ is obtained as follows:

$$i_p = \frac{\text{applied voltage}}{\text{total opposition}} = \frac{\mu e_1}{r_p + Z}$$

The output voltage $e_2$ equals this value of $i_p$ multiplied by the plate-load impedance $Z$. Therefore

$$e_2 = i_p \times Z$$

Substituting for $i_p$,

$$e_2 = \frac{\mu e_1}{r_p + Z} \times Z$$

Finally, the gain of the stage represented by the symbol $A_v$ equals the output voltage $e_2$ divided by the input voltage $e_1$. Therefore

$$A_v = \frac{e_2}{e_1} = \frac{\mu e_1 Z}{e_1(r_p + Z)}$$

Canceling $e_1$,

$$A_v = \mu \frac{Z}{r_p + Z} \tag{4·1}$$

This formula shows that the voltage gain equals the $\mu$ of the tube times a factor that is the ratio of the external $Z$ to the sum of $Z$ and the internal $r_p$.

In a pentode, $r_p$ is very high, typically 1 M or more. The impedance of the resonant plate load is generally about 50,000 ohms or less. Since $r_p$ is much greater than $Z$, the sum of the two can be represented just by $r_p$, approximately. The denominator in formula (4·1) then is $r_p$. Grouping the $\mu$ and $r_p$ therefore gives

$$A_v = \frac{\mu}{r_p} \times Z$$

Furthermore, the ratio $\mu/r_p$ in a vacuum tube is equal to its $g_m$. Then

$$A_v = g_m Z \qquad (4 \cdot 2)$$

This formula states that the gain equals the $g_m$ in mhos multiplied by the plate-load impedance in ohms. For example, with an impedance $Z$ of 50,000 ohms and a $g_m$ of 2,000 $\mu$mhos, the gain is

$$A = 2{,}000 \times 10^{-6} \times 50{,}000 = 2 \times 50 = 100$$

Formula (4·2) applies only to the case of an amplifier where the external $Z$ is much smaller than the internal $r_p$ of the tube. Since a tuned pentode amplifier fits these requirements, its gain can be considered as $g_m Z$.

The $g_m$ for a vacuum tube is essentially constant if the tube is operated as a class A stage. However, $Z$ varies with frequency for a parallel-resonant circuit. Therefore the gain of a tuned amplifier will vary with frequency in the same way as the impedance $Z$ of the tuned circuit.

## *4·3 Response of single-tuned circuit*

The operation of the tuned circuit in Fig. 4·3 will now be considered at a typical signal frequency of 1,000 kc. The tuned plate-load impedance is an application of parallel resonance. At the resonant frequency, $X_L$ equals $X_C$. The $Q$ of the resonant circuit equals $X_L/R_S$. At the resonant frequency, the impedance is maximum, equal to $Q \times X_L$. The following typical values will be used for the various letter symbols previously referred to: $e_1 = 1$ volt, $\mu = 2{,}000$, $r_p = 1$ M, $g_m = 2{,}000$ $\mu$mhos, $L = 159$ $\mu$h (microhenrys), and $R_S = 20$ ohms. To tune the load to a center resonant frequency of 1,000 kc, the tuning capacitance $C$ can be calculated by the parallel-resonance formula. The required value of $C$ is 159 pf for resonance with the 159-$\mu$h inductance at the frequency of 1,000 kc which is 1 Mc or $10^6$ cps.

The reactance of inductance $L$ and capacitance $C$ can now be calculated as follows:

$$X_L = 2\pi f_r L = 2 \times 3.14 \times 10^6 \times 159 \times 10^{-6} = 1{,}000 \text{ ohms}$$

and $$X_C = \frac{1}{2\pi f_r C} = \frac{1}{2 \times 3.14 \times 10^6 \times 159 \times 10^{-12}} = 1{,}000 \text{ ohms}$$

Since the $Q$ of the parallel-tuned circuit is defined as $X_L/R_S$ at the resonant frequency, this circuit has a $Q$ equal to 1,000/20. This equals a $Q$ of 50 at 1,000 kc. Also, the impedance of the parallel circuit at the resonant frequency is an equivalent pure resistance numerically equal to $Q$ multiplied by $X_L$. Therefore this parallel-resonant circuit has a maximum impedance of $50 \times 1{,}000$, or 50,000 ohms. The stage will have a maximum gain of 100 when it is tuned to 1,000 kc and the signal coming into the stage has a frequency of 1,000 kc. For frequencies other than 1,000 kc, the impedance is less than 50,000 ohms, and the gain is less than 100. The variation in gain for different frequencies above and below the resonant frequency $f_r$ can be seen from the curves in Fig. 4·5.

**Relative gain or selectivity.** Sometimes, instead of considering the actual gain, it is more convenient to examine the *relative gain,* or *selectivity,* of the circuit. Selectivity is defined as the ratio of the actual gain at any frequency to the *maximum* gain of the stage. If the selectivity of a stage is represented by the symbol $S$, the expression for selectivity can be written as follows:

$$S = \frac{A_{\text{off res}}}{A_{\text{max}}} \qquad (4 \cdot 3)$$

Since the maximum gain in a single-tuned stage occurs at resonance, its selectivity at $f_r$ equals one. For the stage in Fig. 4·3, the gain is 89.2 at a frequency of 995 kc compared with the gain of 100 at 1,000 kc. Therefore the selectivity of the stage at 995 kc is 89.2/100, or 0.892. The selectivity of the circuit at other frequencies can be calculated in a similar manner, with the results shown in the curves in Fig. 4·5.

**Selectivity curve.** When the selectivity values are plotted against frequency, as in Fig. 4·5, the resulting graph is a selectivity curve. This curve actually illustrates the relative gain of the amplifier for frequencies around resonance. Three curves are shown for different values of $Q$, but they all illustrate the characteristics of a single-tuned amplifier. Note that the maximum selectivity is 1.000 and that it occurs at the resonant frequency. At any frequency off resonance, the selectivity is always *less* than 1.000 in a single-tuned circuit, as shown in the figure. Another interesting fact about the selectivity curve is its symmetry about the resonant frequency for frequencies fairly close to resonance, in this case about 30 kc below and above resonance. This means that the same selectivity occurs at a frequency a few kilocycles above resonance as at a frequency the same few kilocycles below resonance.

**Bandwidth.** The curves in Fig. 4·5 show that the selectivity of a single-tuned circuit starts falling off as soon as the frequency is changed from the resonant frequency of the tuned circuit. This means that the circuit never responds as well to any frequency off resonance as it does to the resonant frequency. As the graph shows, however, the circuit responds *better* to frequencies closer to resonance than it does to frequencies farther off resonance. For all types of tuned r-f circuits, the arbitrary dividing point is a

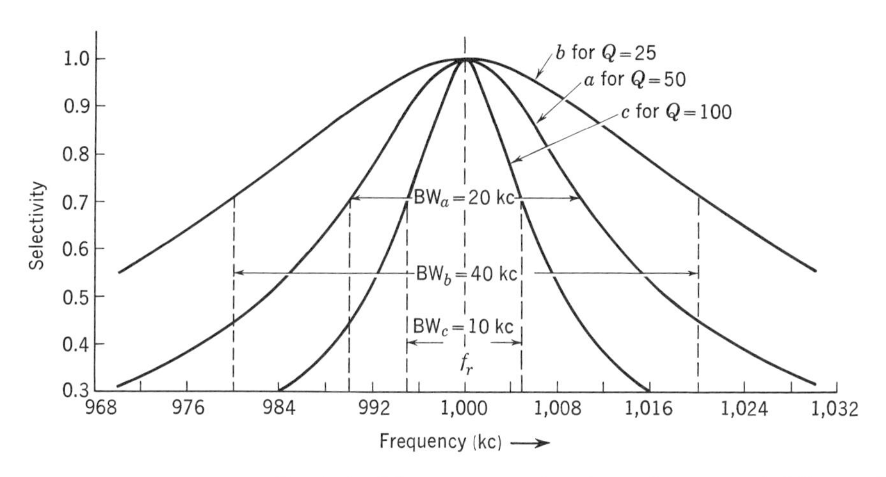


*Fig. 4·5 Selectivity curves of single-tuned circuits with different values of Q.*

selectivity of 0.707. The *difference* between the frequencies at which the selectivity is 0.707 is called the *bandwidth* of the stage, and the end points of the bandwidth are called *band limits* or *half-power frequencies*. In the circuit just described, the band limits are 990 and 1,010 kc, and the bandwidth of the circuit is 1,010 − 990, or 20 kc.

The bandwidth of a single-tuned stage can be determined without going through the calculations and plotting of a selectivity curve. It can be calculated from the formula

$$\text{BW} = \frac{f_r}{Q} \tag{4·4}$$

where $f_r$ is the resonant frequency of the circuit, and $Q$ is the ratio of $X_L$ at resonance to $R_S$. The bandwidth will be in the same units as $f_r$. Thus, for the circuit just described the bandwidth is 1,000 kc/50, or 20 kc. This is the case shown for the selectivity curve *a* in Fig. 4·5.

**Effect of *Q* on circuit response.** As the $Q$ of the circuit increases, the bandwidth of the circuit decreases, and vice versa. For a given circuit inductance and capacitance at a given frequency, the $Q$ of the circuit can be varied by varying the series coil resistance $R_S$.

For example, suppose everything in the circuit is kept constant at the values previously given, but $R_S$ is made 40 ohms instead of 20 ohms. The circuit $Q$ is now 1,000/40, or 25. Therefore the bandwidth is now 1,000/25, or 40 kc. The band limits are now 980 and 1,020 kc, or twice as far away from 1,000 kc as they were before. Furthermore, the circuit now has a maximum impedance at 1,000 kc of 25 × 1,000, or only 25,000 ohms. Therefore its maximum gain at 1,000 kc is now only $2{,}000 \times 10^{-6} \times 25{,}000$, or only 50.

The reduced gain with lower $Q$ does not show in the selectivity curves in Fig. 4·5, because only relative gain is indicated. However, curve *b* for the $Q$ of 25 falls off more slowly than curve *a*. This means a broader response for the circuit with lower $Q$. As a final comparison, curve $c$ shows the narrower bandwidth of 10 kc for a higher $Q$ of 100 if $R_S$ is reduced to 10 ohms. This circuit with the highest $Q$ would allow the most gain for the resonant frequency.

**Varying C for circuit tuning.** Suppose the tuning capacitor in Fig. 4·3 is varied, so that the circuit is resonant at 1,500 kc instead of 1,000 kc, with $L$ still equal to 159 $\mu$h. The required tuning capacitance can be calculated by the resonance formula to be 70.6 pf. This makes $X_C$ and $X_L$ equal to 1,500 ohms at resonance. If $R_S$ is 20 ohms, the $Q$ of the circuit at resonance is now 1,500/20, or 75. However, the circuit bandwidth is still only 20 kc, since $f_r/Q = 1{,}500/75 = 20$ kc. In this case, the band limits or half-power frequencies are approximately 1,490 and 1,510 kc, or 10 kc below and above 1,500 kc, respectively.

The maximum impedance of the circuit, at 1,500 kc, is $75 \times 1{,}500$, or 112,500 ohms for $Z$. This makes the maximum gain, equal to $G_mZ$, $2{,}000 \times 10^{-6} \times 112{,}500$, or 225. Note that this is 2.25 times the maximum gain obtained when the circuit was tuned to 1,000 kc. Thus, although the circuit bandwidth remains the same when the circuit is tuned to 1,500 kc instead of 1,000 kc, the maximum circuit impedance and the maximum stage gain increase by a factor of 2.25.

Suppose the circuit is tuned to 600 kc by varying the tuning capacitor to 442 pf. This makes $X_C$ and $X_L$ equal to 600 ohms at resonance. With $R_S$ still 20 ohms as before, the $Q$ of the circuit at resonance (600 kc) becomes 600/20, or 30. The circuit bandwidth is therefore 600/30, or 20 kc —the same as it was at 1,000 kc and at 1,500 kc. The band limits are now approximately 590 and 610 kc.

However, the maximum circuit impedance at 600 kc equal to $Q \times X_L$ is now $30 \times 600$, or only 18,000 ohms. Therefore the maximum stage gain becomes $2{,}000 \times 10^{-6} \times 18{,}000$, or 36. In this case, although the bandwidth again remains constant at 20 kc, the maximum circuit impedance and the maximum stage gain are reduced by the factor 0.36 from their values at 1,000 kc.

This is the general behavior of a single-tuned circuit which is tuned to different resonant frequencies by varying the circuit capacitance:

1. Bandwidth remains constant.
2. $Q$ varies directly with the resonant frequency.
3. Maximum circuit impedance and stage gain vary directly with the *square* of the resonant frequency.

**Varying L for circuit tuning.** Suppose the circuit of Fig. 4·3 had a variable inductance, and it was varied to make the circuit resonant at 1,500 kc, with $C$ remaining constant at 159 pf. The required inductance value is found by the resonance formula to be 70.6 $\mu$h. The values of both $X_L$ and $X_C$ are now only 667 ohms. If $R_S$ is again assumed to be 20 ohms, the circuit $Q$ at 1,500 kc is now 667/20, or about 33.3. The circuit bandwidth is therefore 1,500/33.3, or 45 kc instead of 20 kc as it was before, an increase of 2.25 times. This change of bandwidth in the tuned circuit is a disadvantage where the signal being amplified has the same bandwidth no matter what its center frequency happens to be, as for different stations in the commercial AM broadcast band. To receive these signals properly, the tuned circuits in a receiver must also provide the same bandwidth regardless of the center frequency to which it is being tuned.

The maximum impedance at 1,500 kc in this case is $33.3 \times 667$, or about 22,200 ohms; the maximum stage gain at 1,500 kc then becomes $2{,}000 \times 10^{-6} \times 22{,}200$, or about 44.4—both lower than at 1,000 kc by the factor 0.444.

Now suppose the circuit is tuned to 600 kc by varying the inductance to 442 $\mu$h. This makes $X_L$ and $X_C$ equal to about 1,667 ohms, and the circuit $Q$ equal to 1,667/20, or about 83.3. The circuit bandwidth is now 600/83.3, or only 7.2 kc. The plate impedance $Z$, equal to $Q \times X_L$, is $83.3 \times 1{,}667$, or about 138,800 ohms. Also, the maximum stage gain is $2{,}000 \times 10^{-6} \times 138{,}800$, or about 277.6—both greater than at 1,000 kc by the factor 2.776. However, the high gain is obtained at the expense of a very narrow bandwidth.

The behavior of a single-tuned circuit which is tuned to different resonant frequencies by varying the inductance can be summarized as follows:

1. $Q$ varies *inversely* with the resonant frequency.
2. Maximum circuit impedance and stage gain vary inversely with the square of the frequency.
3. Circuit bandwidth varies directly with the square of the resonant frequency.

## 4·4 *Shunt damping resistor*

Figure 4·6 shows a single-tuned stage, as before, but the tuned circuit has a shunt damping resistor $R_D$ in parallel with $L$ and $C$. The shunt resistor is often used to broaden the circuit response, using the same components in the $LC$ circuit. Now the shunt resistor $R_D$ determines the $Q$ of the resonant circuit, instead of the coil's series resistance. $R_D$ is generally

*Fig. 4·6 Single-tuned r-f amplifier with shunt damping resistor.*

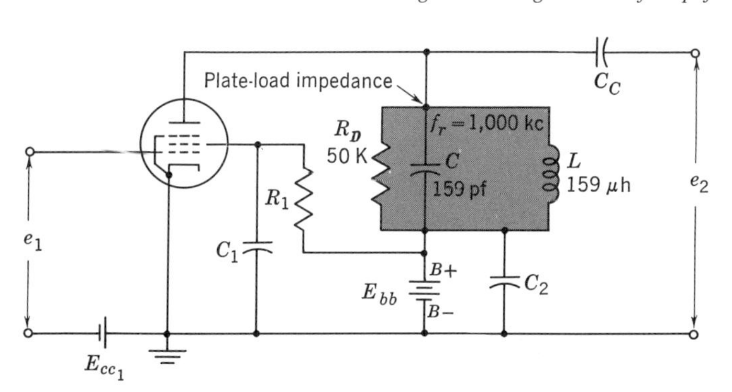

a carbon resistor of 10,000 to 50,000 ohms. Connected as a parallel branch, its resistive branch current reduces the $Q$ of the $LC$ circuit. In this case the $Q$ is equal to the ratio of the shunt resistance $R_D$ to either $X_L$ or $X_C$ at resonance. For example, if $X_L$ and $X_C$ each equal 1,000 ohms at resonance, and $R_D$ is 50,000 ohms, then

$$Q = \frac{R_D}{X_L \text{ or } X_C} = \frac{50{,}000}{1{,}000} = 50 \tag{4·5}$$

With a $Q$ of 50 because of the shunt damping resistance, the $Q$ is lower and the bandwidth increased. The term *damping* refers to the action of $R_D$ in lowering the $Q$ of the resonant circuit.

Note that the formula for $Q$ with the shunt resistance $R_D$ is the reciprocal of the $Q$ formula for resistance in series with $X_L$. Furthermore, the lower the resistance of the shunt $R_D$, the lower the $Q$ of the parallel-resonant circuit becomes because the lower shunt resistance draws more resistive current in its branch.

## 4·5 Single-tuned transformer-coupled stage

This circuit uses a single-tuned circuit in the secondary of the r-f coupling transformer $T_1$ (see Fig. 4·7). The secondary inductance $L_2$ and the variable-tuning capacitor $C$ form the resonant circuit. The a-c signal input is provided by transformer coupling from the primary winding $L_1$. The mutual inductive coupling is indicated by $L_M$. $T_1$ is an air-core transformer as shown by the typical r-f transformer in Fig. 4·8.

As in the single-tuned stage previously described, the circuit in Fig. 4·7 uses a pentode vacuum tube. However, instead of a parallel-tuned plate load and the coupling capacitor $C_c$ of Fig. 4·3, the circuit here uses the

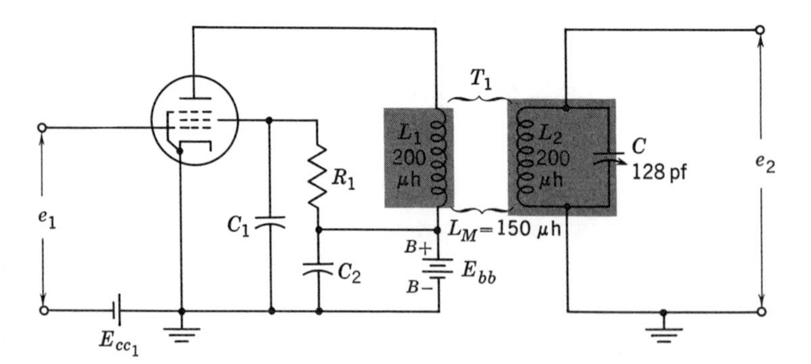


*Fig. 4·7 Single-tuned transformer-coupled r-f amplifier stage.*

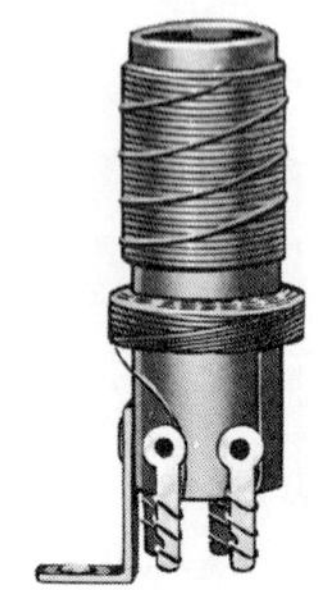

*Fig. 4·8 Typical r-f transformer for AM broadcast band.* (J.W. Miller Co.)

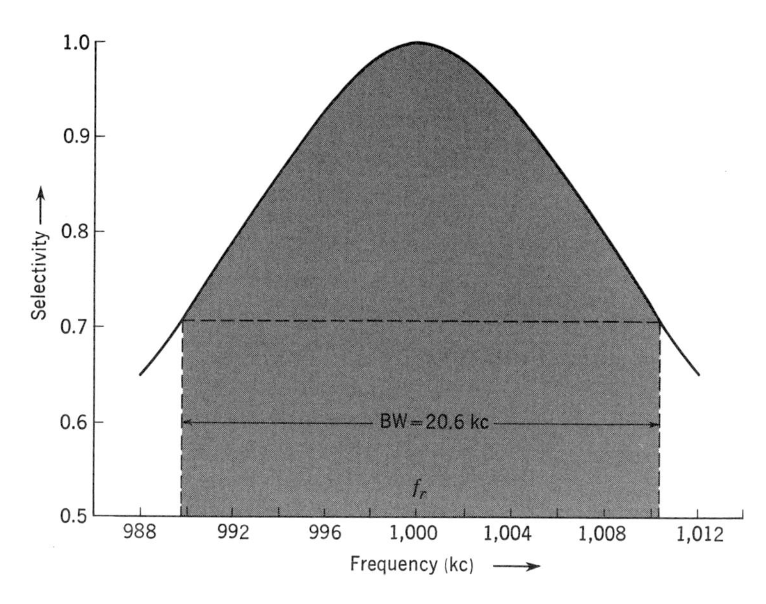


*Fig. 4·9 Selectivity curve of tuned-secondary transformer-coupled circuit.*

r-f transformer $T_1$. The plate of the vacuum tube returns to B+ through the untuned primary inductance $L_1$. As a result, the signal in the primary is coupled by transformer action into the secondary inductance $L_2$. The output voltage $e_2$ is applied to the grid of the next stage. Maximum signal voltage is produced at the resonant frequency of the $LC$ circuit in the secondary.

In terms of its resonance effect, the single-tuned circuit in the secondary has the same response as the single-tuned circuits described before. However, the mutual inductance $L_M$ provides another factor that affects the $Q$ and bandwidth. For this reason, the transformer-coupled single-tuned circuit is generally used for the tuned r-f stage in radio and television receivers. Figure 4·9 shows a typical response curve for the single-tuned transformer-coupled stage. The curve is plotted for the values given in the circuit in Fig. 4·7.

## *4·6 Double-tuned transformer-coupled stage*

Another type of circuit arrangement, the double-tuned transformer-coupled circuit, is shown in Fig. 4·10. This circuit is used commonly for fixed-tuned stages, where it is unnecessary to vary the tuning of different radio frequencies. A very important example is the intermediate-frequency (i-f) amplifier in radio receivers. Typical frequencies for i-f amplifiers are 455 kc in AM radios and 10.7 Mc in FM radios. Again an r-f pentode is used. The supply voltage and bypass capacitor requirements are the same as before. The difference between this circuit and the one shown in Fig. 4·7, however, is that here the transformer has a tuned primary as well as a tuned secondary. A typical double-tuned transformer for 455 kc is

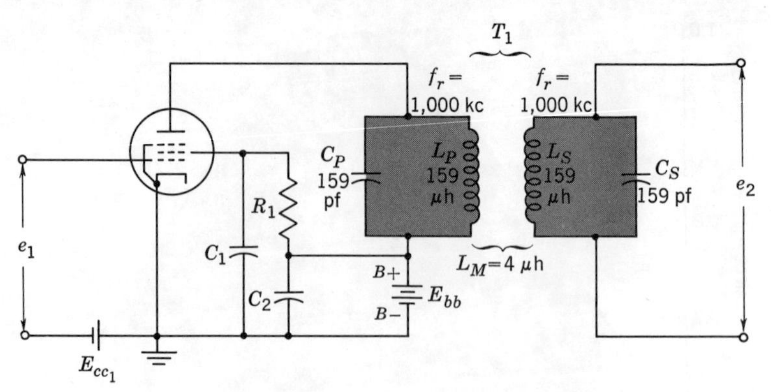


*Fig. 4·10 Double-tuned transformer-coupled r-f amplifier.*

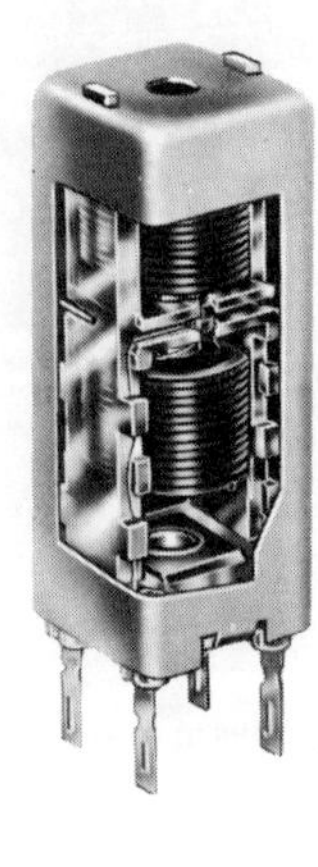

*Fig. 4·11 Double-tuned transformer for 455 kc, with adjustable slugs top and bottom.* (J.W. Miller Co.)

shown in Fig. 4·11. Generally, each *LC* circuit is resonant at the same frequency. The use of two tuned circuits, however, enables the stage to have greater selectivity and bandwidth.

**Critical coupling.** Double-tuned circuits are usually fixed-tuned because the coupling must be fairly close to a critical value to obtain more bandwidth and selectivity than a single-tuned stage. In a tuned r-f circuit with variable tuning, it is difficult to provide the required coupling at different resonant frequencies.

In order to use specific values for the coupling we can review some of the characteristics of an r-f transformer. First, the turns ratio concept used with iron-core transformers does not apply because the r-f transformer cannot have an iron core. The effect of the primary on the secondary is given by the formula

$$L_M = k\sqrt{L_p L_s} \qquad (4\cdot6)$$

where $L_M$ is the mutual inductance between primary and secondary, in the same units as $L_p$ and $L_s$. The factor $k$ is the coupling coefficient. Transposing, we can have the formula

$$k = \frac{L_M}{\sqrt{L_p L_s}} \qquad (4\cdot7)$$

The coupling coefficient $k$ is a numerical factor without any units because it is a ratio of two inductance values. The greater the mutual inductive coupling $L_M$ between $L_p$ and $L_s$, the higher is the value of $k$.

In Fig. 4·12, the response curve *a* shows the selectivity of the double-tuned circuit in Fig. 4·10 for the case of a coupling coefficient $k$ equal to 0.025. This results from the fact that the mutual inductance $L_M$ equals 4 μh. Then

$$k = \frac{L_M}{\sqrt{L_p L_s}} = \frac{4}{\sqrt{159 \times 159}} = \frac{4}{159} = 0.025$$

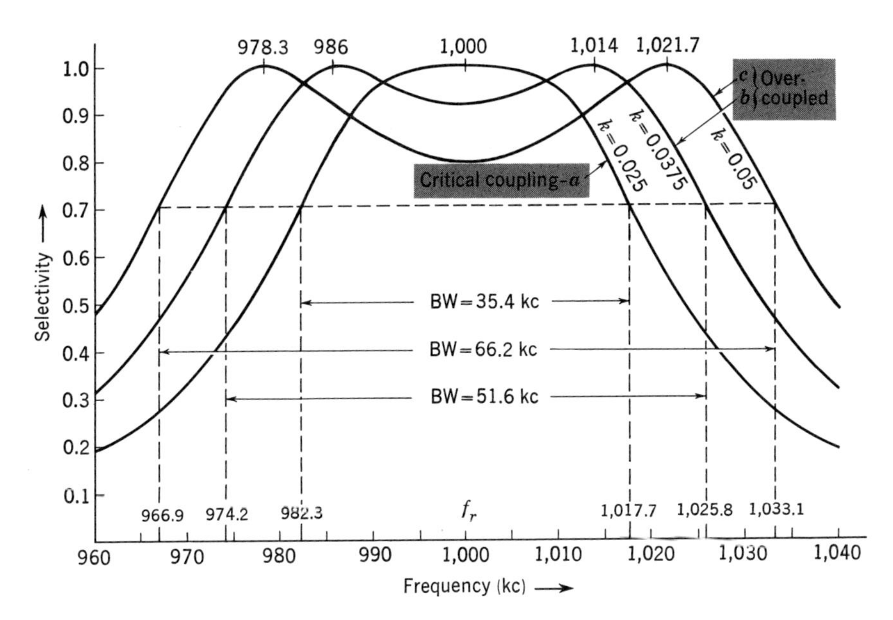


*Fig. 4 · 12 Selectivity vs. frequency for double-tuned transformer-coupled stage in Fig. 4 · 10. The Q is 40 for each tuned circuit.*

Furthermore, the $Q$ of the tuned circuit in Fig. 4 · 10 is taken as 40, for a typical value. This is based on 25 ohms of series resistance $R_s$ in each tuned circuit. With a $Q$ of 40 and the coupling coefficient of 0.025, the product of these two factors equals one. In other words, this value of $k$ equals the reciprocal of $Q$. Or

$$k = \frac{1}{Q} \qquad \text{(for critical coupling)} \qquad (4 \cdot 8)$$

This value of $k$ equal to $1/Q$ is called *critical coupling*. With critical coupling, the response of the double-tuned circuit has its maximum selectivity and bandwidth, while providing uniform response for frequencies at and near resonance. This is shown by curve $a$ in Fig. 4 · 12. If less than critical coupling is used, the bandwidth and selectivity are reduced. This can be called *loose coupling*. Then the advantage of the double-tuned coupling is lost, since the response approaches that of a single-tuned circuit as the coupling is made looser.

**Tight coupling.** A coupling factor greater than the critical value is called *tight coupling*. Or the primary and secondary are said to be overcoupled. Two examples are shown by curves $b$ and $c$ in Fig. 4 · 12. As the coupling is made tighter, the bandwidth increases, at the expense of less gain for the resonant frequency $f_r$. Tight coupling is often used with double-tuned circuits to obtain the increased bandwidth. In many applications, the response can be considered satisfactory as long as the response at $f_r$ does not dip below 70 per cent of the maximum response.

**Transistorized i-f amplifier.** Figure 4·13 shows a typical double-tuned transistor amplifier for the intermediate frequency of 455 kc. $T_1$ is the input and $T_2$ is the output transformer. Both have the primary and secondary tuned to this resonant frequency. The adjustable slugs are used for tuning or *aligning* the transformers to 455 kc. The tap on the primary winding provides an impedance match from 15,000 ohms for the collector circuit to approximately 500 ohms in the base circuit. An NPN transistor is used in a common-emitter circuit. $R_3$ with its bypass $C_3$ provide emitter bias, similar to cathode bias in a tube. The divider with $R_1$ and $R_2$ determines the d-c voltage for the base. $C_1$ bypasses $R_1$ for the base input signal. Finally, the collector voltage is essentially 9 volts, as the d-c resistance of the windings in an r-f or i-f transformer is close to zero ohms.

## 4·7 *Stagger-tuned amplifiers*

This type of amplifier contains two or more single-tuned stages in cascade, each tuned to a different resonant frequency. If the resonant frequencies and circuit constants of the individual stages are carefully selected, a stagger-tuned amplifier can provide excellent bandwidth and selectivity.

When tuned stages in cascade are resonant at the same frequency, this method is called *synchronous tuning*. Then the over-all bandwidth is reduced, as described in Sec. 10·4. In amplifiers where more bandwidth is needed with single-tuned stages, staggered tuning is preferable.

**Staggered pair.** The simplest kind of stagger-tuned system, consisting of two stages, is called a *staggered pair* (see Fig. 4·14). It consists of two single-tuned stages, referred to in the figure as Stage 1 and Stage 2, similar to the single-tuned stage shown in Fig. 4·3. Cathode bias is used here, however, provided by $R_k$ and $C_k$. The screen grid is supplied with positive d-c voltage from the B supply, while the plate returns to B+ through the parallel-tuned plate load. Instead of the small series resistance $R_S$ of Fig.

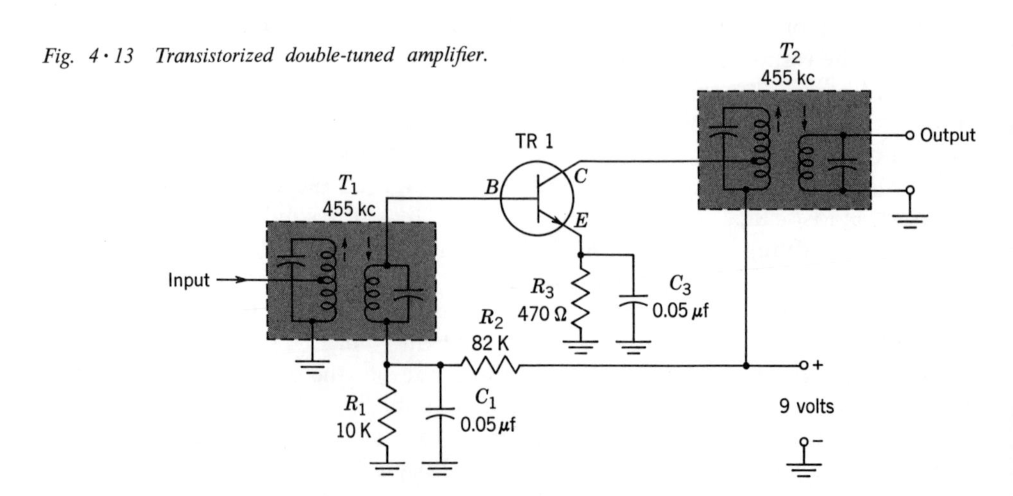


*Fig. 4·13 Transistorized double-tuned amplifier.*

4 · 3, however, each stage here has its own *shunt* damping resistor, $R_1$ and $R_2$, respectively. Since the reactance of the coupling capacitors $C_c$ is negligibly small, the shunt resistors can be assumed to be effectively in parallel with their respective tuned circuits. Therefore Stage 1 can be said to include $L_1$, $C_1$, and $R_1$ in parallel as its plate load, and Stage 2 to include $L_2$, $C_2$, and $R_2$ in parallel as its plate load.

In order to investigate the behavior of the staggered pair, typical values will be assumed. Assume that each stage uses a pentode, with $\mu = 2{,}000$, $r_p = 1$ M, and $g_m = 2{,}000$ $\mu$mhos. In Stage 1, $C_1 = 159$ pf, $L_1 = 161$ $\mu$h, and $R_1 = 71{,}100$ ohms. The values of $C_1$ and $L_1$ cause Stage 1 to be resonant at 993 kc. Therefore $X_{C_1}$ and $X_{L_1}$ equal 1,007 ohms at 993 kc. The $Q$ of such a circuit is defined as the ratio of the shunt resistance to either $X_C$ or $X_L$ at resonance. In this case, therefore, the $Q$ of Stage 1 is 71,100/1,007, or 70.7.

In Stage 2, assume that $C_2 = 159$ pf, $L_2 = 157$ $\mu$h, and $R_2 = 70{,}300$ ohms. The values of $C_2$ and $L_2$ cause Stage 2 to be resonant at 1,007 kc. Therefore $X_{C_2}$ and $X_{L_2}$ equal 993 ohms at 1,007 kc. The $Q$ of Stage 2 is therefore 70,300/993, or 70.7 also.

The gain at resonance in Stage 1, which is also the maximum for this stage by itself is $g_m \times R_1$, or $2{,}000 \times 10^{-6} \times 71{,}100$, which equals 142.2. Similarly, the gain at resonance in Stage 2 is $g_m \times R_2$, or $2{,}000 \times 10^{-6} \times 70{,}300$, or 140.6. It should be noted that maximum gain occurs at 993 kc in Stage 1, and at 1,007 kc in Stage 2.

The values of gain for the staggered pair are shown by the graphs in Fig. 4 · 15. Note that each stage provides a gain of 100 at 1,000 kc. The gain is the same for both because 1,000 kc is 7 kc above the resonant frequency of Stage 1 and 7 kc below the resonant frequency of Stage 2. The over-all gain at 1,000 kc is therefore 100 × 100, or 10,000. This is the maximum over-all gain of the staggered pair. The effective center fre-

*Fig. 4 · 14 Circuit diagram for a staggered pair of single-tuned stages.*

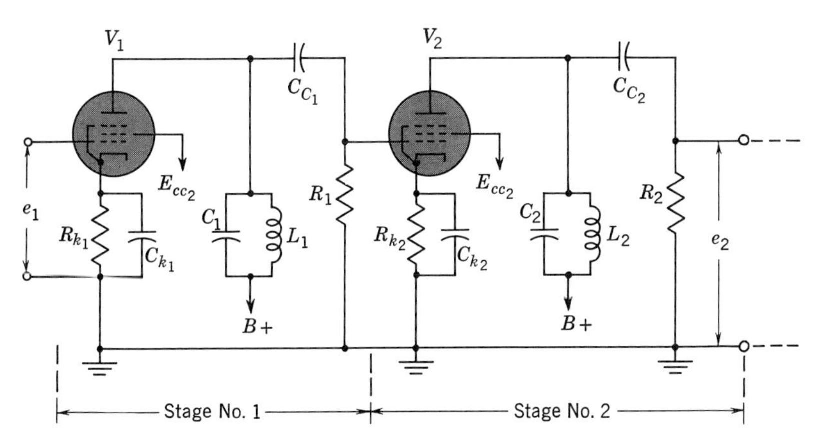

quency of the staggered pair can be said to be 1,000 kc, or the average of the two individual center frequencies.

Note that the curve of over-all gain vs. frequency is approximately symmetrical about 1,000 kc, as was also true for a single-tuned stage. The over-all gain remains fairly constant, however, or is *flatter,* for a greater range of frequencies on either side of 1,000 kc, compared with a single-tuned stage. For example, the over-all gain at 995 and 1,005 kc, 5 kc on either side of 1,000 kc, is seen to be approximately 9,750, or only 2.5 per cent below the maximum gain. In the single-tuned stage of Fig. 4·3, however, the gain at 995 and 1,005 kc had fallen off by more than 10 per cent from the maximum value. Another desirable feature of the response of the staggered pair is the more rapid falling off of the response curve after the limits of the relatively flat portion of the curve are reached, in this case, approximately 995 and 1,005 kc. This means sharper skirts at

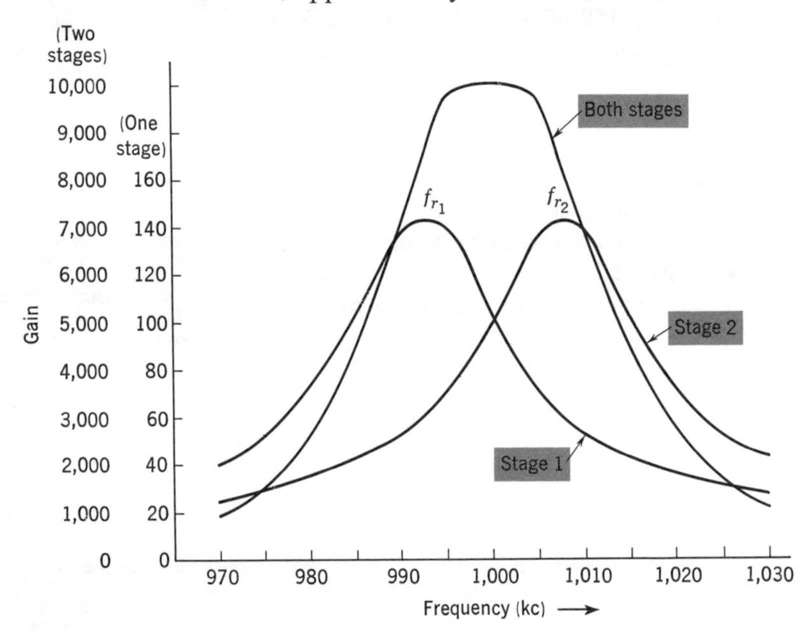


*Fig. 4·15 Gain vs. frequency for the staggered pair in Fig. 4·14.*

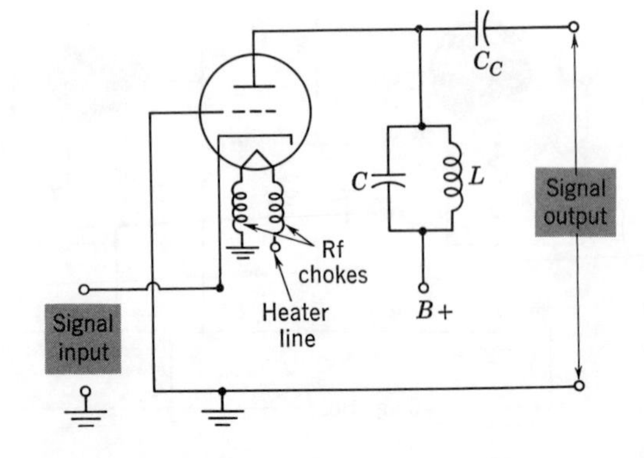


*Fig. 4·16 Ground-grid amplifier.*

the edges of the over-all response curve, corresponding to sharper selectivity.

If the band limits for a staggered pair are again defined as the frequencies at which the over-all gain has fallen off to 0.707 of its maximum value, it can be seen from Fig. 4·15 that the band limits are 990 and 1,010 kc, giving a bandwidth of 20 kc. It is actually greater than the bandwidth of each individual stage, which is about 14 kc for Stage 1, and 14.2 kc for Stage 2. This increase in bandwidth of a staggered pair over the bandwidths of the individual stages is another important feature. It should be noted, though, that stagger-tuned stages are generally used only for fixed-tuned amplifiers, where the correct resonant frequencies can be maintained.

## 4·8 *Grounded-grid amplifier*

In each of the r-f amplifiers considered so far, the vacuum-tube arrangement was assumed to be the conventional, or grounded-cathode, circuit. Another type of circuit is the *grounded-grid* amplifier shown in Fig. 4·16. The circuit consists of a triode, with the input signal $e_{in}$ applied between cathode and grid. However, the grid returns to ground instead of having the cathode grounded. Therefore the input signal is coupled to the cathode. The cathode then has signal voltage with respect to chassis ground. The output voltage $e_{out}$ is taken across the plate-load impedance, as usual, to be coupled to the next stage through $C_c$. The plate-load impedance here is a single-tuned $LC$ circuit.

R-f chokes are necessary in the heater line for a grounded-grid stage. Since the cathode is not grounded, r-f signal voltage at the cathode could be coupled into the heater line by cathode-heater leakage inside the tube. This may enable the r-f signal to reach other r-f stages and cause undesired oscillations. R-f heater chokes are made of a few turns of heavy wire than can conduct the heater current. Their reactance is high for the r-f signal but negligibly small at 60 cps.

The gain of a grounded-grid stage, if the possible internal resistance of the signal source is neglected, will be approximately the same as a grounded-cathode stage. However, the grounded grid acts as an effective shield between plate and cathode. The grounded-grid amplifier is able to use triode tubes at high frequencies without danger of oscillations, which is one of the serious limitations of the grounded-cathode circuit. Since triodes produce less tube noise than pentodes, a grounded-grid triode can be used in preference to pentodes for r-f amplifiers where minimum noise is important.

## 4·9 *Cascode amplifier*

An application of the grounded-grid circuit is the *cascode* r-f amplifier, shown in schematic diagram form in Fig. 4·17. A double triode is used. The circuit consists of a triode grounded-cathode stage coupled to a triode grounded-grid stage. The load on Stage 1 consists of a single-tuned coil $L_1$ which resonates with the stray shunt capacitance $C_{t_1}$. $L_1$ is varied to tune

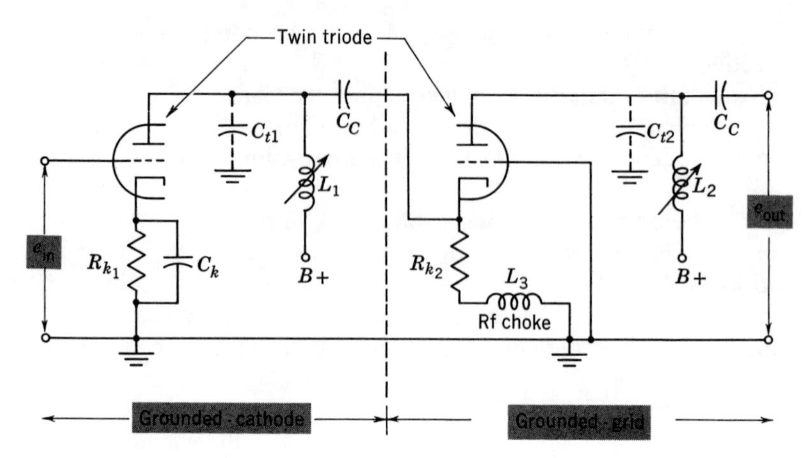


*Fig. 4·17 Cascode r-f amplifier.*

the resonant circuit. Its output is coupled through capacitor $C_c$ to the cathode of the grounded-grid section. Then the output of the second stage appears across the resonant plate circuit tuned by $L_2$. Both triode sections form one cascode r-f amplifier. Finally, the amplifier r-f output signal is coupled by $C_c$ to the next amplifier. $R_{k_1}$ and $R_{k_2}$ are cathode-bias resistors, $R_{k_1}$ being bypassed by capacitor $C_k$. However, $R_{k_2}$ is not bypassed, because the cathode has r-f signal voltage. The r-f choke $L_3$ provides a high impedance to the signal coupled from Stage 1.

The gain provided by a cascode r-f amplifier is only a little greater than would be provided by either a grounded-cathode or a grounded-grid stage alone, but the selectivity is much better. To obtain the same selectivity from either a grounded-cathode stage or a grounded-grid stage alone, a pentode would be required. Since triodes have less tube noise, the cascode amplifier using triodes has the advantage of being relatively noise-free, while providing all the advantages usually associated with pentodes. In addition, the tendency in a triode r-f amplifier to produce oscillations is reduced in the cascode circuit because the grounded-grid portion of the amplifier loads downs the grounded-cathode portion, so that it operates as a low-gain stage. As a matter of fact, with a typical value of gain at resonance for a cascode amplifier equal to 20, the grounded-cathode portion contributes about 1.25 and the grounded-grid portion about 16.

## 4·10 Wavetraps

A resonant circuit that is used to reject an undesired frequency is called a *wavetrap,* or *trap circuit*. Its action depends on the resonance effect in an *LC* circuit tuned to the rejection frequency. Three examples of trap circuits are illustrated in Fig. 4·18. In *a,* the parallel-resonant trap $L_2C_2$ is connected in series with the lead to the grid in the next amplifier. Since the trap is tuned to the undesired frequency, parallel resonance provides a

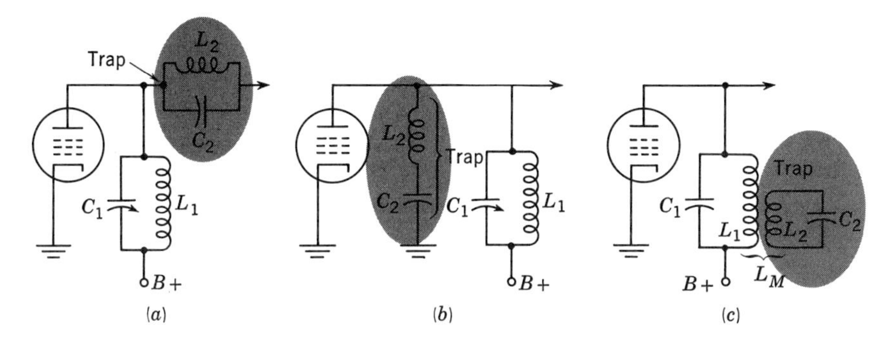


*Fig. 4·18 Wavetrap circuits. (a) Parallel-resonant circuit in series with load. (b) Series-resonant circuit in shunt with load. (c) Absorption trap inductively coupled to plate load.*

high impedance at the rejection frequency. With the trap in series, its high impedance results in maximum voltage across the trap and minimum voltage coupled to the grid-cathode circuit of the next stage. Then there is little voltage at the rejection frequency in the next stage, because most of the undesired voltage is across the $L_2C_2$ trap.

The same results can be obtained with the series-resonant trap in *b*. Here the series-resonant trap is in shunt with the plate load. At the undesired frequency to which $L_2C_2$ is tuned, series resonance provides a very low impedance. There is practically a short circuit across the plate load at the rejection frequency, resulting in very little gain at the trap frequency.

In *c*, the trap is inductively coupled to the plate load. At resonance, maximum current flows in the $L_2C_2$ trap circuit at the frequency to be rejected, and maximum power is absorbed from the plate-load circuit. For the undesired frequency, therefore, the effective plate-load impedance is reduced by the absorption trap, and the stage has very little gain at the trap frequency.

In all cases, the trap is tuned to the undesired frequency to reject voltage at this frequency. Trap circuits are useful in rejecting r-f interference of a specific frequency. If the interfering frequency is not too close to the desired signal frequency, the trap can be tuned to reject the undesired signal.

## *4·11 Wide-band amplifiers*

All the amplifier circuits considered so far are examples of r-f amplifiers because they amplify a band of radio frequencies. Some may have a narrow bandwidth, others a wide bandwidth, depending on the $Q$ of the tuned circuits. However, the term *wide-band amplifier* refers to a circuit that can amplify audio frequencies as well as radio frequencies. The difference between a wide-band amplifier and an r-f amplifier, both with the same amount of bandwidth, is illustrated by the response curves in Fig. 4·19. In *a*, the response curve is for an r-f amplifier tuned to 42 Mc. Its bandwidth is 4 Mc, meaning it can amplify the band of frequencies from 40 to

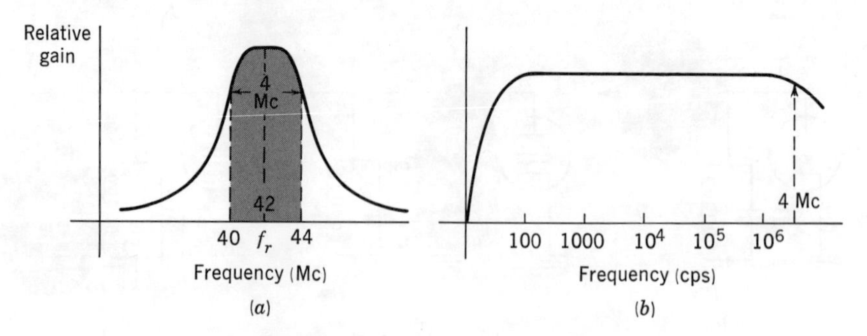


*Fig. 4·19 Comparison of response curves for tuned r-f amplifier and wide-band amplifier. (a) Tuned response centered at 42 Mc. (b) Wide-band response from low audio frequencies up to 4 Mc.*

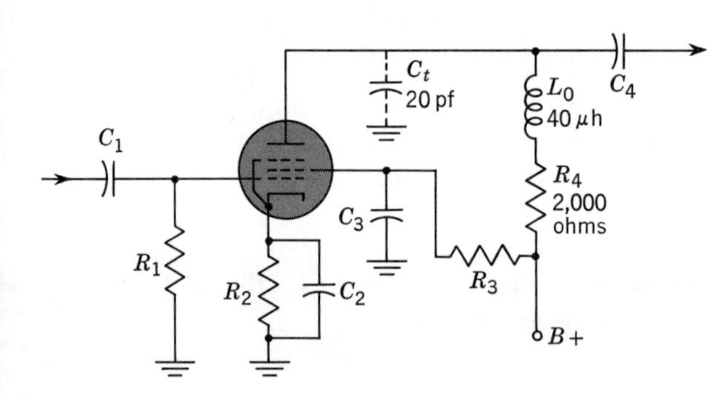


*Fig. 4·20 Wide-band amplifier circuit, consisting of RC amplifier with high-frequency peaking coil $L_o$.*

44 Mc. In *b*, the response for the wide-band amplifier is entirely different, although it also has a bandwidth of 4 Mc. Here the bandwidth includes audio frequencies, from below 100 cps, and radio frequencies up to 4 Mc. This is really an example of amplifying a wide range of frequencies because the ratio of the highest to the lowest frequency is very great.

There are many applications where such a wide-band amplifier is necessary. Two examples where both audio and radio frequencies must be amplified are found in the amplifier circuits for oscilloscopes and the video amplifier circuits in television receivers.

The wide-band amplifier circuit is basically an *RC*-coupled audio amplifier, modified to allow amplification of higher frequencies also. A typical circuit is shown in Fig. 4·20. $R_4$ is the plate-load resistor, while $C_4$ is the coupling capacitor for the *RC*-coupled amplifier. Note, however, the low resistance of $R_4$, which is 2,000 ohms here, compared with a typical value of 0.5 M for an audio amplifier. The low value of plate-load resistance in the wide-band amplifier is necessary to extend the high-frequency response. Reducing the size of $R_4$ minimizes the effect of the stray shunt capacitance $C_t$ as a parallel reactance that tends to bypass higher frequencies. In addition, the small r-f coil $L_o$ is inserted to resonate with $C_t$ neutralizing its effect for radio frequencies. $L_o$ is called a *peaking coil* because it boosts the gain for high frequencies. Other circuits can be used with one or more peaking coils. Also the bandwidth may be more or less than 4 Mc. Nevertheless, the two main features of all wide-band amplifiers are the use of a relatively low value of plate-load resistance and the addition of peaking coils to boost the high-frequency response of the *RC*-coupled amplifier.

## SUMMARY

1. R-f amplifiers have a tuned plate load, so that only a band of frequencies at and near resonance can be amplified.
2. Selectivity is the ratio of gain for frequencies off resonance to the gain at the resonant frequency.
3. The signal-to-noise ratio is often important for r-f amplifiers when a very low signal level in the order of microvolts must be amplified. This is equal to the ratio of signal voltage to noise voltage.
4. Pentodes are generally used for r-f amplifiers because of their low interelectrode capacitances. Where signal-to-noise ratio is important, though, triodes may be used, because they have lower tube noise.
5. R-f amplifiers used for tuning to different resonant frequencies generally use the single-tuned transformer-coupled circuit with the secondary tuned to the desired frequency.
6. Fixed-tuned amplifiers are generally double-tuned. In this type of stage, both the primary and secondary of the coupling transformer are resonant at the desired frequency. Staggered single-tuned stages can also be used.
7. With a double-tuned circuit, increasing the coupling $L_M$ increases the bandwidth. When $k$ equals $1/Q$, the transformer has critical coupling. This provides the greatest bandwidth without double peaks.
8. The grounded-grid amplifier has the control grid grounded, while the input signal is coupled to the cathode. The output signal is taken from the plate circuit. This arrangement is often used with triodes for r-f amplifiers, since the grounded grid shields the output circuit from the input circuit.
9. A stagger-tuned amplifier uses two or more single-tuned stages resonant at slightly different frequencies. Staggering the resonant frequencies increases the over-all bandwidth. This arrangement is used for fixed-tuned amplifiers, providing about the same bandwidth and selectivity as in double-tuned stages.
10. The cascode amplifier consists of a double triode where the grounded-cathode section drives the cathode of a grounded-grid section. The over-all gain is about equal to a pentode stage but with the low signal-to-noise ratio of a triode amplifier.
11. A wavetrap is a resonant *LC* circuit tuned to reject an undesired frequency. See Fig. 4·18.
12. A wide-band amplifier amplifies audio and radio frequencies. The circuit is essentially an *RC*-coupled audio amplifier with a low plate-load resistance and peaking coils to boost the high-frequency response. See Fig. 4·20.

## SELF-EXAMINATION

Here's a chance to find out how well you have learned the material in this chapter. Work the exercises; then check your answers against the Key at the back of this book. These exercises are for your self-testing only.

1. The gain of a single-tuned r-f amplifier is maximum at the resonant frequency of the *LC* circuit because of (*a*) series resonance; (*b*) parallel resonance; (*c*) low $Q$; (*d*) interelectrode capacitance.
2. The $Q$ of a single-tuned *LC* circuit is lower if its (*a*) shunt resistance is increased; (*b*) shunt resistance is decreased; (*c*) series resistance is decreased; (*d*) capacitance is decreased.

3. In a grounded-grid amplifier circuit, the (*a*) output signal is coupled to the control grid; (*b*) output signal is taken from the cathode; (*c*) cathode-heater capacitance is negligible; (*d*) input signal is coupled to the cathode.
4. The response curve of a double-tuned circuit has two peaks when the transformer has (*a*) loose coupling; (*b*) critical coupling; (*c*) tight coupling; (*d*) $Q$ equal to 1.
5. Grounded-cathode r-f amplifiers generally use pentodes because they have low (*a*) tube noise; (*b*) grid-plate capacitance; (*c*) plate resistance; (*d*) amplification factor.
6. If the $Q$ of a single-tuned stage is doubled, its bandwidth is then (*a*) doubled; (*b*) halved; (*c*) the same; (*d*) four times as great.
7. Which of the following resonant circuits has the greatest bandwidth? (*a*) $f_r$ is 50 Mc; $Q$ is 50. (*b*) $f_r$ is 455 kc; $Q$ is 100. (*c*) $f_r$ is 1 Mc; $Q$ is 100. (*d*) $f_r$ is 1 Mc; $Q$ is 10.
8. In order to amplify the band of frequencies from 10 cps to 2 Mc, the best type of amplifier circuit to use is (*a*) double-tuned circuit with tight coupling; (*b*) single-tuned circuit with a small damping resistor; (*c*) staggered pair centered at 1 Mc; (*d*) wide-band amplifier.
9. Referring to the circuit in Fig. 4·18*a*, if the desired frequency is 25.75 Mc and the undesired frequency is 21.25 Mc, then (*a*) $L_2C_2$ is tuned to 25.75 Mc; (*b*) $L_1C_1$ is tuned to 21.25 Mc; (*c*) $L_2C_2$ is tuned to 21.25 Mc; (*d*) $L_1C_1$ is tuned to 51.5 Mc.
10. $C$ is usually varied instead of $L$ in tuning an r-f amplifier because the bandwidth (*a*) increases with the resonant frequency; (*b*) decreases with the resonant frequency; (*c*) remains constant; (*d*) varies with $Q$.

## QUESTIONS AND PROBLEMS

1. Describe briefly three requirements of r-f amplifiers.
2. Draw the schematic diagram of a single-tuned capacitively coupled amplifier stage, using a pentode with cathode bias.
3. Draw the schematic diagram of a double-tuned stage, using a pentode with cathode bias.
4. A single-tuned stage is resonant at 10 Mc. At this frequency $X_L$ and $X_C$ each equal 1,200 ohms. The series resistance is negligible, but the shunt damping resistance is 24,000 ohms. (*a*) How much is the $Q$? (*b*) How much is the bandwidth? (*c*) How much is $L$, in microhenrys?
5. Draw three selectivity curves on one graph, showing the effect on bandwidth when a double-tuned circuit is (*a*) undercoupled, (*b*) critically coupled, (*c*) overcoupled.
6. Give two features of the grounded-grid amplifier and its operation, compared with a grounded-cathode circuit.
7. Name two types of wavetraps and indicate how each is connected in an amplifier circuit.
8. (*a*) What are the two main sections in a cascode amplifier circuit? (*b*) What features makes the cascode circuit useful as an amplifier for weak r-f signals?
9. In a staggered pair, one amplifier is resonant at 10 Mc and the other at 12 Mc, each with a bandwidth of 1 Mc. Show the selectivity curves for each stage and the over-all response of the staggered pair, all on one graph.
10. (*a*) What is meant by a wide-band amplifier? (*b*) Give one way in which this circuit differs from a tuned r-f amplifier circuit. (*c*) Give one way in which this circuit differs from an audio amplifier circuit.
11. Referring to Fig. 4·10, give the function of $T_1$, $R_1$, and $C_1$.
12. Referring to Fig. 4·13 give the function of $T_1$, $T_2$, $R_1$, $R_2$, and $R_3$.
13. For the transistor in Fig. 4·13: (*a*) If the emitter voltage $V_E$ is 0.9 volt, calculate the average d-c emitter current $I_E$. (*b*) Calculate the amount of base voltage provided by $R_1$ in the voltage divider with $R_2$.
14. Referring to Fig. 4·20, give the function of $R_1$, $C_1$, $R_2$, $C_2$, $R_3$, $C_3$, $R_4$, and $L_o$.
15. If the wavetrap in Fig. 4·18*a* is tuned to reject 4.5 Mc, what value of $L_2$ will be necessary with 40 pf for $C_2$?

# Chapter 5 Oscillators

An oscillator is a circuit using either electron tubes or transistors to generate a-c output. Its frequency may be quite low for audio oscillators, or quite high for r-f and microwave oscillators. There are many applications. An oscillator is a very important part of every transmitter and superheterodyne receiver. In addition, oscillators are required for signal generators, frequency meters, and other test instruments. This unit describes typical oscillator circuits. The topics are as follows:

5·1 Oscillator Requirements
5·2 Oscillator Operation
5·3 How a Tuned Circuit Oscillates
5·4 Tickler-coil Oscillator
5·5 Hartley Oscillator
5·6 Colpitts Oscillator
5·7 Electron-coupled Oscillator
5·8 Tuned-grid Tuned-plate Oscillator
5·9 Crystal Oscillator
5·10 Resonant Lines
5·11 Cavity Resonators and Waveguides
5·12 Magnetrons
5·13 Klystrons
5·14 Traveling-wave Oscillators
5·15 Masers and Lasers

## *5·1 Oscillator requirements*

An oscillator circuit is one which delivers an a-c output voltage, usually at a definite frequency and with a specified waveform. It does this without receiving any external signals; in other words, it is a self-generating circuit.

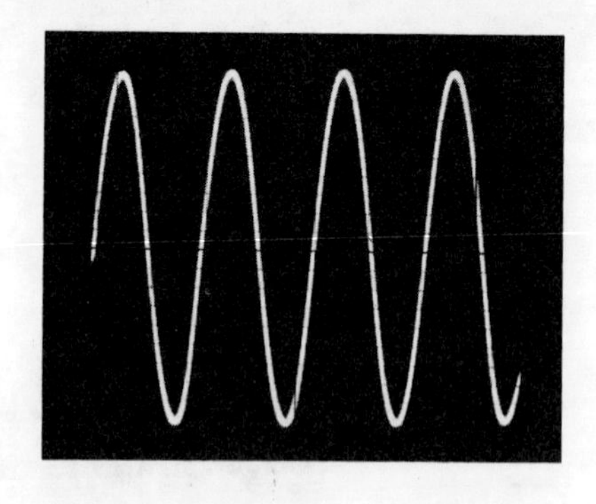

*Fig. 5 · 1 Oscilloscope photograph of sine-wave output of an r-f oscillator. Four cycles are shown.*

*Fig. 5 · 2 Two additional waveforms that can be generated by oscillators. (a) Pulses. (b) Sawtooth waves.*

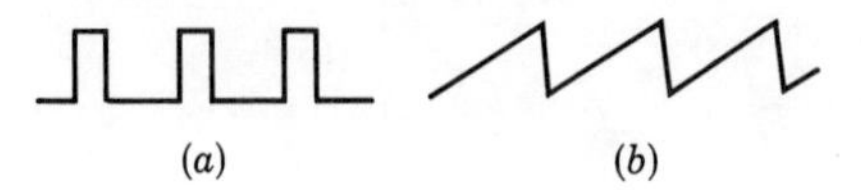


However, the oscillator circuit requires d-c power to sustain the oscillations.

The output voltage which an oscillator develops, in its most common form, is a sine wave (Fig. 5 · 1). Oscillators can also produce pulses or sawtooth waves (Fig. 5 · 2). Usually, sine waves are generated by one of the types of r-f feedback oscillators described here. For nonsinusoidal waveforms, the oscillators explained in Chap. 14 on Pulse Circuits are generally used.

Before we examine an oscillator to see how it functions, let us first note some of the requirements expected of these circuits.

1. An important characteristic of an oscillator is its frequency. When an oscillator is set to function at a certain frequency, it should be capable of remaining at that value within narrow limits for as long as the oscillator is in use. In other words, the oscillator should be drift-free.
2. The amplitude of the signal produced by the oscillator should remain at whatever value it is set.
3. The oscillator should produce an output voltage having negligible distortion. The appearance of distortion means that frequencies other than the one desired also appear in the output, together with the desired frequency.

## *5 · 2 Oscillator operation*

Basically, an oscillator is an amplifier which derives its input signal from its own output. That is, a small portion of the output signal is fed back to the input circuit in such a way that it maintains the system in operation. This basic principle of oscillator operation is illustrated in Fig. 5 · 3.

The amplifier shown in Fig. 5 · 3*b* has a gain of 10. It produces an output of 10 volts when 1 volt is applied to its input terminals from some external signal source through the switch in position 1. Note that the input and output voltages have the same polarity. That is, they are in phase. To obtain input and output voltages which are in phase from the box labeled "amplifier" in Fig. 5 · 3, it is necessary to have either two conventional stages of amplification or else use some other method to provide the same phase.

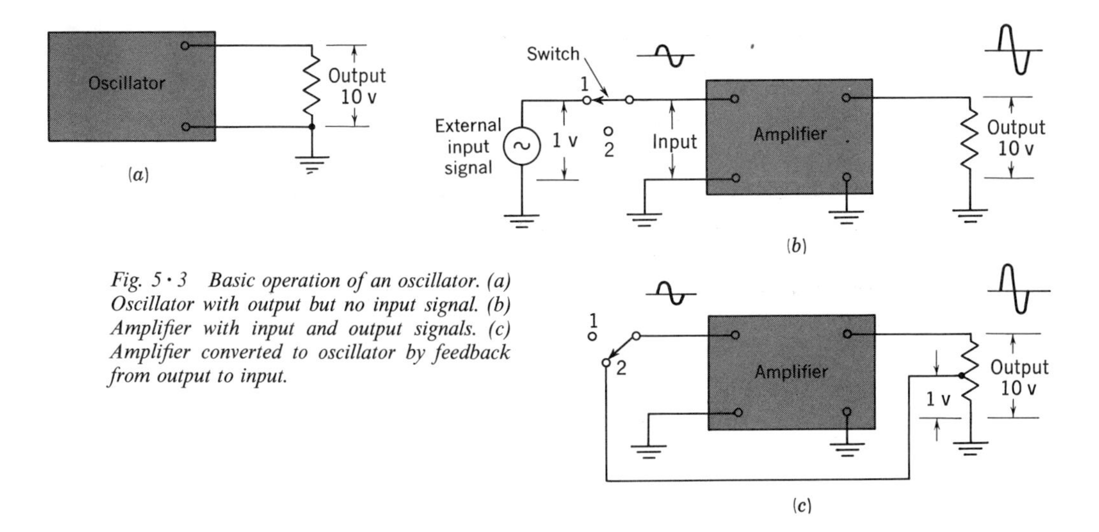


*Fig. 5 · 3 Basic operation of an oscillator. (a) Oscillator with output but no input signal. (b) Amplifier with input and output signals. (c) Amplifier converted to oscillator by feedback from output to input.*

In Fig. 5 · 3*c*, switch contact 2 has been connected to the load resistor. Thus, when the amplifier input switch is thrown from position 1 to position 2, the external input voltage, in *b*, is replaced by its exact duplicate—obtained, however, from a tap on the output load resistor of the amplifier. The new input voltage is identical to the original external input signal in both amplitude and phase. Therefore the signal obtained from the external source is no longer needed, and the amplifier continues to produce an output voltage so long as the feedback path from the output to input is not disturbed. Under these conditions the amplifier is said to *oscillate.*

In practical oscillators, the oscillations are started by variations in plate current as the tube heats up. They build up because of the feedback path and the amplifying action of the circuit. It is not necessary to use an external generator and a switch. These elements were merely used to illustrate the principle of oscillations.

Thus, to produce oscillations in a circuit, two conditions must be met. First, it is necessary that there be feedback from output to input circuits in such a way as to add to, or reinforce, the voltage on the grid. This is called *positive,* or *regenerative,* feedback. Second, it is necessary that the feedback be sufficient to transfer enough voltage back to the input circuit to keep the output at the desired level.

The feedback may be accomplished by inductive, capacitive, or resistive coupling. In general, the frequency of the oscillations produced in a circuit depends upon the values of the inductance and capacitance in the circuit. Thus, by using the proper coils and capacitors, it is possible to generate oscillations from the very low audio frequencies to the very high radio frequencies.

Note that the vacuum tube itself does not oscillate; the oscillations actually take place in a tuned circuit. The vacuum tube functions as an electric valve which automatically controls the release of energy into this circuit to maintain oscillations.

## 5·3 How a tuned circuit oscillates

We have just indicated that the oscillations take place in a tuned circuit. Now let us examine such tuned circuits to see how this action takes place. A coil *L*, a capacitor *C*, a battery, and a double-throw switch *S* are connected into a circuit, shown in Fig. 5·4*a*. In Fig. 5·4*b*, the switch has been moved to contact 1, putting the capacitor across the battery. Immediately, electrons will start flowing and charge the capacitor as indicated. After the capacitor has been charged, the switch is moved to point 2 (Fig. 5·4*c*). The capacitor now starts to discharge, creating a field around the coil. The arrows indicate the direction of current.

After a brief interval, the capacitor will discharge completely, and the potential across the two plates will drop to zero. The flow of electrons tends to cease, causing the magnetic field to start collapsing. In collapsing, the field induces a voltage across the coil which aids the continued flow of electrons to the upper capacitor plate, since a magnetic field acts to prevent any change in the flow of electrons. This causes additional electrons to leave the bottom plate and accumulate on the upper plate, making

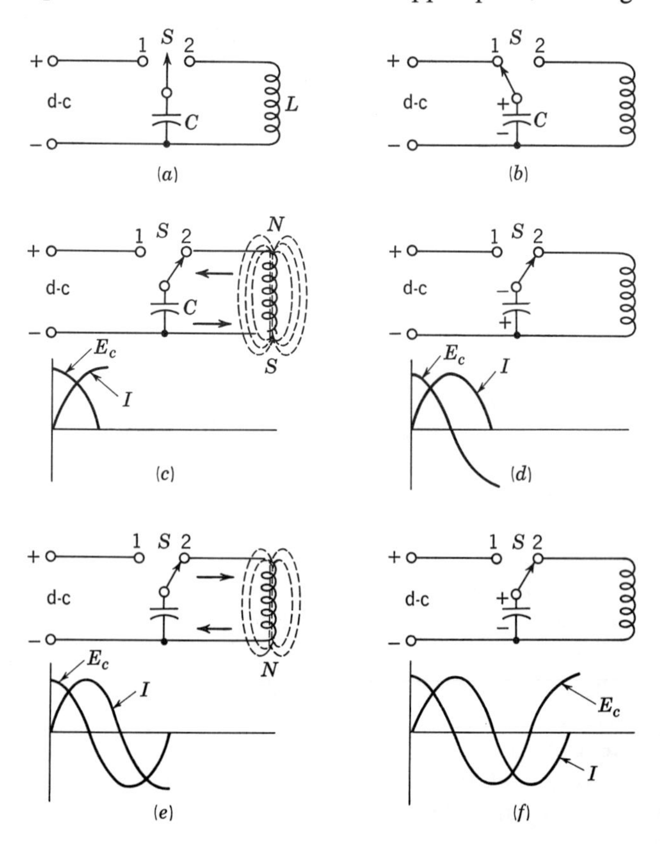


*Fig. 5·4 How a tuned circuit generates an alternating voltage. See text for explanation of each step in the oscillatory cycle.*

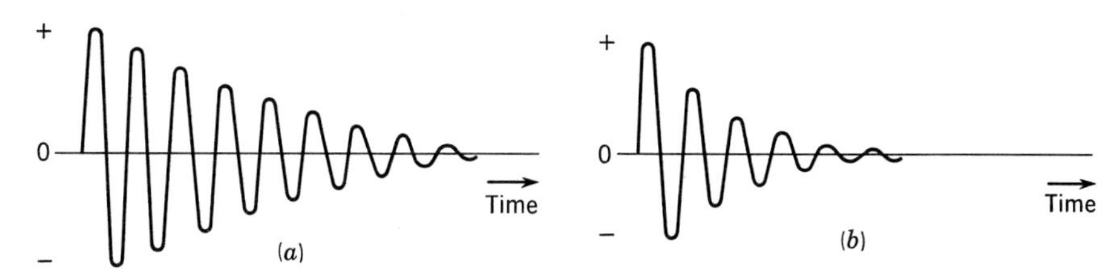


*Fig. 5·5 Oscillations in a tuned circuit will die out unless the lost energy is replaced periodically. (a) Slow decay with low resistance. (b) Faster decay because of higher resistance.*

it negative with respect to the bottom plate. When the field around the coil has collapsed completely, the flow of electrons to the upper plate stops (see Fig. 5·4*d*).

Once more the capacitor starts to discharge, but this time in the opposite direction (see Fig. 5·4*e*). The electrons, in traveling through the coil, create another expanding magnetic field. When the capacitor has discharged, the magnetic field starts to collapse. The collapsing field aids the flow of electrons to the bottom plate, again making it negative with respect to the top plate. In this way, a current oscillates back and forth around this tuned circuit, generating the alternating voltage illustrated by the sine waves shown in Fig. 5·4*c* to *f*.

The oscillating, or alternating, current has a certain frequency which is determined by the length of time required for the charging and discharging of the capacitor through the inductance. The larger the values of $L$ and $C$, the longer the required time and the lower the frequency.

The action just outlined would continue indefinitely if the wires through which the electrons traveled possessed no resistance and if no energy were lost from the magnetic field developed around the coil. Wires do possess resistance, however, which absorbs energy, and magnetic fields do not return to the circuit all the energy they get from it. As a result of these two losses, the oscillations in a circuit die away, as shown in Fig. 5·5. If the circuit resistance is low, the decrease in amplitude of the oscillators will be gradual, as in Fig. 5·5*a*. If the circuit resistance is high, however, the oscillations will die down fairly quickly, as in Fig. 5·5*b*. Since one of the major requirements of an oscillator is to produce a constant output voltage, some means must be developed to prevent either of these conditions from taking place.

In the circuit of Fig. 5·4, the power lost in the circuit could be replenished by quickly swinging the switch from position 2 to position 1, letting the battery recharge the capacitor to its full value, and then flipping the switch back again to position 2. While this procedure would work to an extent, it would certainly be inconvenient. A more efficient approach is to substitute a vacuum tube for the switch. This is done in the circuit of Fig. 5·6.

The vacuum tube, a battery, and the resonant circuit are all placed in series with each other. Since the tube is an inanimate object, some action or some event must occur which will tell it when to conduct. At these

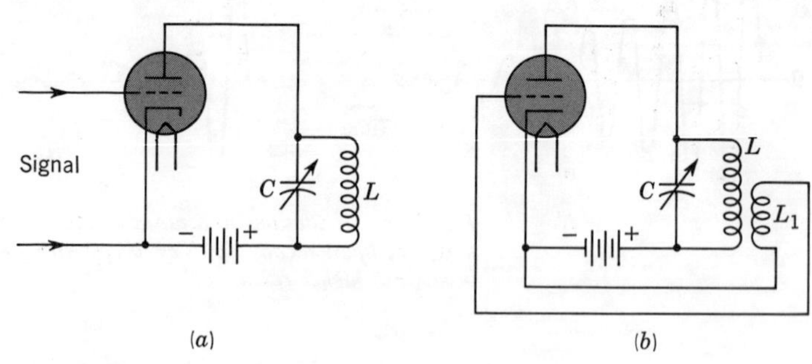


*Fig. 5 · 6 How a vacuum tube maintains steady oscillations in a tuned circuit. (a) Tuned circuit as plate load. (b) Feedback added from plate to grid.*

instants, the tube will pour energy into the tuned circuit, and the oscillations there will be able to continue at their normal amplitude.

For this purpose of control, a small coil $L_1$ is positioned close to the resonant circuit coil $L$ (see Fig. 5 · 6*b*). When oscillations are present in the tuned circuit, a voltage will be induced in $L_1$, and this voltage will be applied to the control grid of the vacuum tube. If coil $L_1$ has been wound properly, its voltage will periodically drive the grid of the tube positive, and during these instants the tube will conduct current, recharging $C$ in the tuned circuit.

If coil $L_1$ is wound in the opposite manner, oscillations will not take place because the tube will not conduct at the proper instants. This is most important and should be observed carefully whenever an oscillator is wired together. The frequency of the circuit will be determined by the values of $L$ and $C$.

It has been previously mentioned that an oscillator is somewhat like an amplifier, in which a part of the amplified output is fed back from the plate circuit to the grid circuit. In Fig. 5 · 6*b*, the feedback is accomplished inductively with $L_1$, but other forms of coupling may be used, as we will see in a subsequent discussion. In all instances, it is important that the voltage fed back possess the proper phase (that is, polarity) and magnitude.

In Fig. 5 · 6*b*, coil $L_1$ is connected directly to the control grid of the vacuum tube. No other bias is shown. In practical oscillator circuits, there is usually a high negative bias on the grid which permits plate current to flow only during a small part of each cycle, when the a-c grid voltage is near its positive peak. Also, the grid of the tube is permitted to draw current. This is how the grid-leak bias is produced. The energy for this grid current must be supplied by the oscillating current in the tuned circuit.

**Frequency of oscillations.** The frequency at which oscillations take place in a vacuum-tube oscillator is determined by the resonant frequency of the tuned circuit. The approximate frequency of oscillations may be determined by the relationship

$$f_r = \frac{1}{2\pi\sqrt{LC}}$$

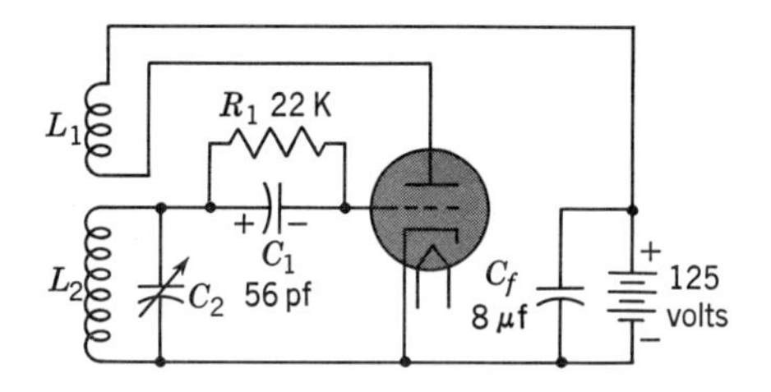


*Fig. 5·7 A tickler-coil feedback oscillator circuit.*

where $f_r$ = frequency, cps
$L$ = inductance of tuned circuit coil, henrys
$C$ = capacitance of tuning-circuit capacitor, farads

From this equation, we see that decreasing either the capacitance or the inductance raises the value of the frequency. If the inductance or the capacitance is increased in value, the frequency will decrease.

*Example.* Determine the frequency of oscillations in the circuit of Fig. 5·6*b* when $L$ is 16 μh, and $C$ is 100 pf. Converting to proper units for the formula

$$16\ \mu h = 16 \times 10^{-6}\ \text{henry}$$
$$100\ \text{pf} = 100 \times 10^{-12}\ \text{farad}$$

Then

$$f_r = \frac{1}{2\pi\sqrt{LC}} = \frac{1}{2\pi\sqrt{16 \times 10^{-6} \times 100 \times 10^{-12}}} = \frac{1}{2\pi\sqrt{1{,}600 \times 10^{-18}}}$$

$$= \frac{1}{6.28 \times 40 \times 10^{-9}} = \frac{1{,}000}{6.28 \times 40 \times 10^{-6}} = 3.98 \times 10^{6} = 3.98\ \text{Mc, approximately}$$

## 5·4 *Tickler-coil oscillator*

The tickler-coil oscillator employs the arrangement shown previously in Fig. 5·6*b* with the slight change that the tuning circuit is in the grid circuit (see Fig. 5·7). The other coil, which provides the feedback and is known as the *tickler coil,* is in the plate circuit.

The frequency of oscillation is the resonant frequency of the tuned circuit $L_2C_2$. The grid current, flowing through the resistor $R_1$, known also as a *grid leak,* provides the proper negative grid bias. Capacitor $C_1$ bypasses the r-f currents around $R_1$ to keep the bias constant. The values of $R_1$ and $C_1$ are selected so that the grid is biased negatively to a considerable extent with respect to the cathode. Practically all oscillators use grid-leak bias because it provides a more stable arrangement, as the bias depends on the amount of feedback.

It may be of interest to examine one complete cycle of operation of this oscillator in order to gain a better insight into its operation.

As a first step, the filament of the tube is permitted to heat up. B+ voltage is applied to the plate now, and since the grid voltage is initially zero, plate current will start to flow. This flow of plate current through coil $L_1$ will induce a voltage in $L_2$. The coils are so wound that the grid end of $L_2$ now becomes positive with respect to the cathode end. Under these conditions, plate current through the tube increases; at the same time, grid current flows in the grid circuit, charging $C_1$ to the polarity shown. Note, however, that even though the charge on $C_1$ is such as to make the grid negative with respect to its cathode, still the positive induced voltage across $L_2$ is strong enough to overcome this negative $C_1$ voltage and keep tube current flowing.

The increase in plate current just mentioned induces an even greater positive voltage in $L_2$, which, in turn, further raises the plate current, and so on until the plate current has reached the saturation limit of the tube. When this is reached, the current through $L_1$ can no longer increase. Therefore $L_1$ is no longer able to induce a voltage in $L_2$.

With the induced voltage gone, the negative voltage across $C_1$ causes the plate current to drop. Less current makes the magnetic field collapse. Then this reduced current in $L_1$ induces a voltage in $L_2$ which is opposite in polarity to the previous induced voltage. That is, the grid end of $L_2$ now becomes negative with respect to the cathode end (previously it was positive). Thus the grid is now subjected to an even greater negative voltage, which further reduces the plate current. The end result of this negative voltage is to drive the tube rapidly into cutoff.

The tube will remain cut off until $C_1$ has had a chance to discharge enough of its charge through $R$ to bring the grid voltage back above cutoff. Once this point is reached, plate current starts flowing again, and the entire sequence repeats itself.

The first oscillation is weak. Like a playground swing, the oscillator does not reach full swing until after several pushes are applied. As long as the tube continues to supply more energy than is lost in resistance, however, the oscillations will become stronger and stronger until the tube is driven well into saturation. Actually, from the instant the oscillator starts to work, the amplification of the circuit becomes less and less until the ratio of output to input is equal to one. That is,

$$\frac{\text{Output signal}}{\text{Input signal}} = 1$$

The whole circuit might even be considered as becoming less and less efficient. Amplification for the first cycle might be 10, gradually dropping for each succeeding cycle, until some place along the line the amplification is equal to one.

If the decrease in amplification stops at one, everything is fine. The amplitude and frequency of the oscillator will be constant. But if the gain drops below one, the oscillator will eventually run down.

The build-up of the grid-leak bias follows the build-up of the oscillations. On the first cycle, the grid is only slightly positive. The small number of electrons picked up by the grid charge the grid capacitor $C_1$. When the cycle reverses itself, the charge on $C_1$ decreases by discharge current through $R_1$.

On the next oscillation, the grid again becomes positive and picks up additional electrons. These add to the ones already in the grid circuit. When the cycle reverses itself, the extra electrons make the grid swing still more negative.

The process of picking up and adding to the electrons already in the circuit continues at a reduced rate, until the number of electrons picked up by the grid equals the number lost by discharge through the grid resistor. When this point is reached, the circuit amplification is one, and the oscillator is stable.

**Measuring the grid-leak bias.** The negative grid voltage is an indicator of an oscillator's "state of health." If the voltage value is normal, then the oscillator is working fine. If it is too low, it means that the oscillations are weak. And if it is completely absent, the circuit is not functioning at all. Here, then, is a good point to check when you suspect that an oscillator is not performing as it should. Measure the amount of negative grid-leak bias with a d-c voltmeter, preferably with a VTVM for minimum loading effect.

**Transistorized oscillator circuit.** Figure 5•8 corresponds to Fig. 5•7 but an NPN transistor is used instead of the triode vacuum tube. The values here are for operation at 1 to 2 Mc, as determined by the $L_2C_2$ resonant circuit. The tap on $L_2$ matches the high impedance of the parallel-resonant circuit to the relatively low impedance of the emitter. Note that $L_2C_2$ in the emitter circuit corresponds to having the tuned circuit in the cathode circuit for a tube. $L_1$ is the tickler coil for feedback from collector to emitter. $C_1$ with $R_1$ couples the tuned circuit to the emitter and provides signal bias similar to grid-leak bias. The $R_2R_3$ voltage divider supplies the required d-c voltage for the base. Since the base is grounded by $C_3$ for a-c signal, this is essentially a grounded-base circuit.

All the r-f feedback oscillator circuits described in this chapter can use either a tube or a transistor. The main features of a transistorized circuit usually are lower d-c supply voltages, lower impedances, and larger bypass or coupling capacitances for any given frequency.

## 5•5 *Hartley oscillator*

The circuit of the tickler coil may be rearranged slightly in the manner shown in Fig. 5•9. In spite of this change, operation is still very much the

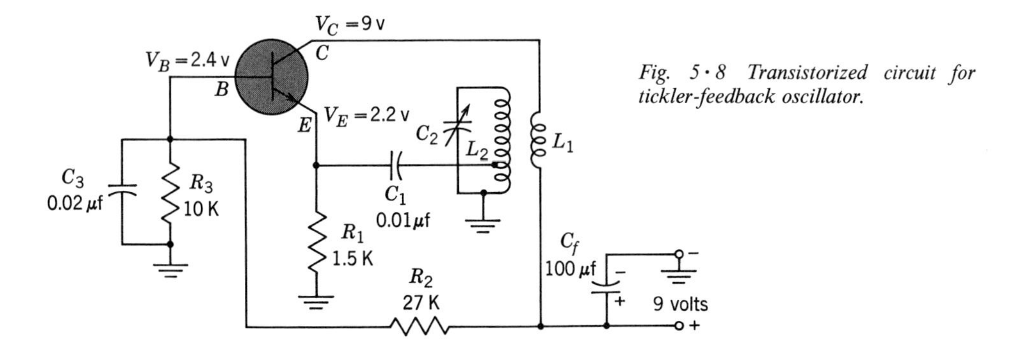


*Fig. 5•8 Transistorized circuit for tickler-feedback oscillator.*

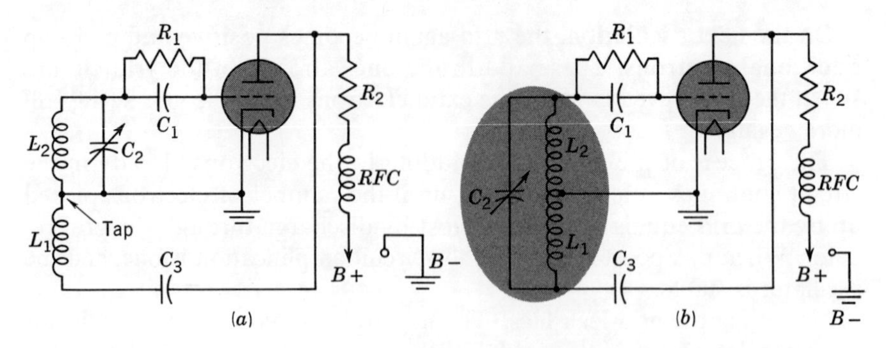


*Fig. 5·9 Hartley oscillator, showing how the Hartley is derived from the tickler-coil oscillator in a to the Hartley circuit in b.*

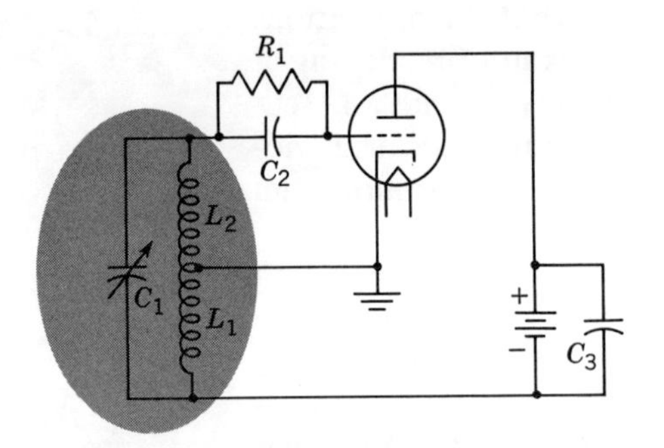


*Fig. 5·10 A series-fed Hartley oscillator.*

same, with feedback of energy from the plate to the grid circuit by means of inductive coupling between $L_1$ and $L_2$.

Instead of two coils, the Hartley has one coil with a tap. The resonant frequency is determined by the values of $L_2$, $C_2$, and $L_1$. $C_1$ and $R_1$ provide the operating grid-leak bias. $C_3$ is primarily employed to prevent the d-c plate voltage from shorting to ground at the tap between $L_1$ and $L_2$. The plate-load resistor is marked $R_2$, and *RFC* is an r-f choke designed to keep the r-f currents out of the d-c power supply.

**Series and parallel feed.** The method of supplying d-c plate voltage in Fig. 5·9 is called *parallel feed* because $R_2$ with its r-f choke is actually in parallel with $L_1$. This can be seen from the fact that both $L_1$ and $R_2$ connect directly to the tube plate. At the same time, the far end of $L_1$ goes to ground, while $R_2$ and *RFC* connect to ground by way of the battery or power supply.

It is also possible to supply the d-c voltage to the tube plate by way of $L_1$, and this arrangement is known as *series feed* (see Fig. 5·10). Parallel feed is perhaps more advantageous because capacitor $C_3$ prevents the d-c voltage from reaching the tuning coil or the tuning capacitor. Since high voltages are quite common in such circuits, use of the parallel feed prevents accidental contact with this voltage by persons tuning the circuit.

**Coupling the oscillator output.** After oscillations have been developed in an oscillator, some means must be provided to couple some of this energy to other circuits where it can be used. This transfer of energy can

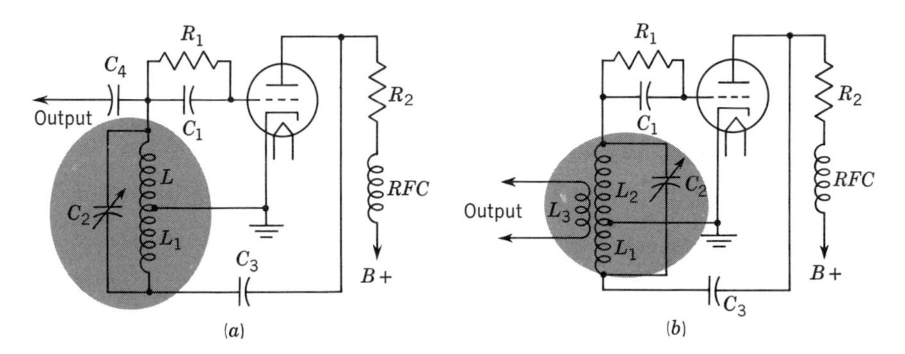


*Fig. 5 · 11 Two methods of coupling energy out of an oscillator. (a) Capacitive coupling by $C_4$. (b) Inductive coupling by $L_3$.*

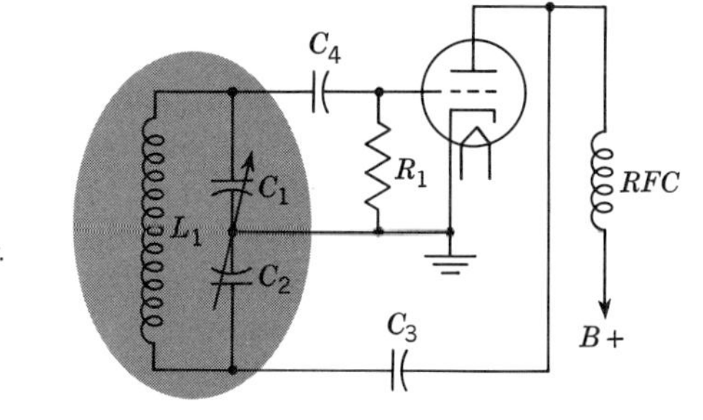


*Fig. 5 · 12 A Colpitts oscillator.*

be achieved capacitively, as in Fig. 5 · 11*a*, or inductively, as in *b*. In *a*, $C_4$ is connected to one end of the tuning coil. The other end of this capacitor would then connect to whatever circuit is to receive this signal. In *b*, $L_3$ is positioned close to $L_1$ and $L_2$, and energy is taken from the tuning circuit by inductive coupling. Loose coupling is generally necessary to prevent the load from affecting the frequency of the oscillator.

## 5 · 6 Colpitts oscillator

The Colpitts oscillator, shown in Fig. 5 · 12 is essentially the same circuit as the Hartley oscillator, except that a pair of capacitors, $C_1$ and $C_2$, are used in place of the tap on the inductance. Feedback of energy from the plate to the grid circuit now occurs capacitively, by means of the voltage divider effect between $C_1$ and $C_2$.

Tuning of the oscillator can be achieved either by varying the inductance of $L_1$ or by using two variable capacitors and ganging their shafts together. This is the method indicated in Fig. 5 · 12. Grid-leak bias is developed by $C_4$ and $R_1$. Note that this bias differs slightly in form from that shown in the preceding circuits. This approach is required here, however, because whatever negative voltage is developed across $C_4$ and $R_1$ must be applied between control grid and cathode of the tube. With $C_1$ and $C_2$ blocking any d-c path between grid and cathode, $R_1$ must be connected as shown.

The resonant frequency of a Colpitts oscillator is still governed by the formula $f_r = 1/(2\pi\sqrt{LC})$. Note, however, that $C$ represents the two tuning-circuit capacitances connected in series. For example, if $C_1$ and $C_2$ are each 300 pf, then the combined tuning capacitance equals 150 pf.

## 5·7 Electron-coupled oscillator

The electron-coupled oscillator is a composite circuit combining an oscillator and an amplifier in one network. The Hartley, Colpitts, or any other type of oscillator can be used. As indicated in Fig. 5·13, the cathode, control grid, and screen grid form a Hartley oscillator circuit with the screen grid acting as the plate. Feedback takes place between the screen-grid circuit and the control-grid circuit. The frequency of oscillations is controlled by the values of $L_1$ and $C_1$, and this section of the tube functions as an independent unit even if the plate voltage is removed. The screen-grid voltage, however, must still be present.

In the plate circuit there is still another parallel-resonant circuit, $L_2$ and $C_4$. The electrons that reach the plate must pass the control and screen grids; hence the current will arrive at the plate with the frequency of oscillations determined by $L_1$ and $C_1$. Since the electrons arrive in pulses, $L_2$ and $C_4$ will be pulsed into oscillations and kept going as long as the circuit is in operation. Note, however, that the only coupling between the grid- and plate-tuned circuits is through the electron beam, which is the reason for the name of *electron-coupled oscillator.*

The oscillations which develop in the plate tuning circuit are more powerful than the oscillations in the grid circuit. Furthermore, since the electron beam is the coupling medium between the oscillator and $L_2C_4$, and since the screen grid serves as a shield between the circuits, variations

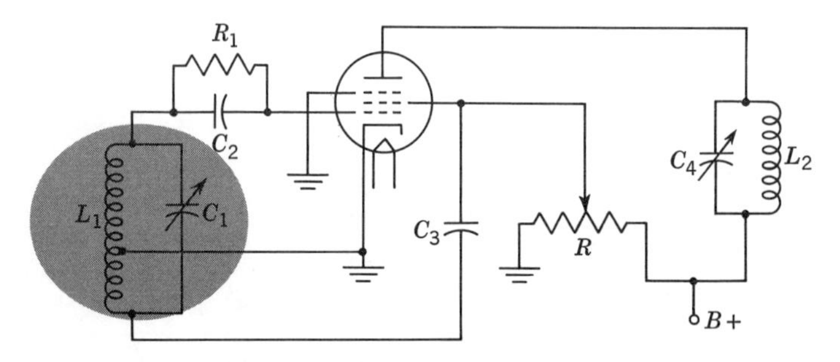


*Fig. 5·13 An electron-coupled oscillator.*

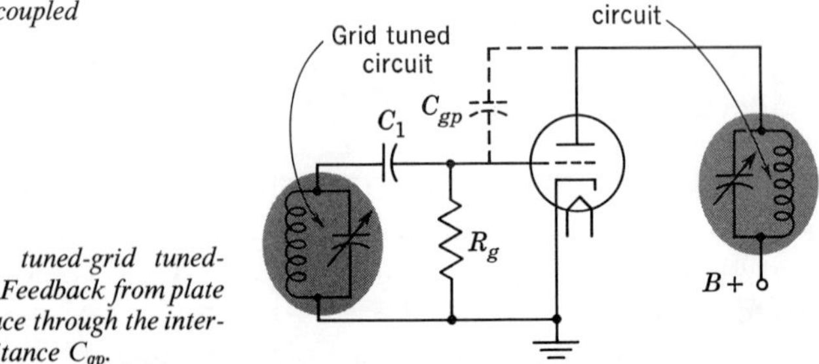


*Fig. 5·14 A tuned-grid tuned-plate oscillator. Feedback from plate to grid takes place through the interelectrode capacitance $C_{gp}$.*

in the plate circuit will have little effect on the oscillator. This makes for a more stable arrangement. The output from this network is obtained from the plate circuit, and variations in the load, which in the oscillators previously described can cause frequency changes, here produce little effect. This, too, is beneficial.

An interesting feature of electron-coupled oscillators is that when $L_2C_4$ is tuned to a frequency which is a multiple of the natural frequency of the oscillator, the circuit provides frequency multiplication. As an example of a frequency doubler, if $L_1C_1$ is tuned to the fundamental frequency of 10 Mc, the $L_2C_4$ plate circuit can be tuned to the second harmonic for output at 20 Mc.

## *5 · 8 Tuned-grid tuned-plate oscillator*

In the Colpitts oscillator, capacitive feedback provided the needed energy that kept the circuit in operation. Capacitive feedback is employed also in the tuned-grid tuned-plate (TGTP) oscillator, shown in Fig. 5 · 14. Here, however, the capacitor providing this feedback path is the capacitance existing between plate and grid of the vacuum tube. In Fig. 5 · 14, this capacitance, labeled $C_{gp}$, is shown in dotted lines between these two elements.

Two resonant circuits are employed in this oscillator, one in the grid circuit and the other in the plate circuit. Each circuit has its own *LC* components, but the tuned circuits are coupled through the grid-to-plate capacitance of the tube. Frequency of operation depends on each tuned circuit, therefore. If it is desired to operate this oscillator over a range of frequencies, both tuning circuits must be provided with variable components (either a variable capacitor or a variable inductance) so that both tuning circuits can have their frequency changed.

It may prove helpful, in understanding the operation of the TGTP oscillator, to redraw the circuit to the form shown in Fig. 5 · 15. Note the close resemblance to the Hartley oscillator, particularly the way in which the cathode connects between the two tuning circuits. (Remember that there is no mutual coupling between the plate and grid coils here, although such coupling does exist in the Hartley.)

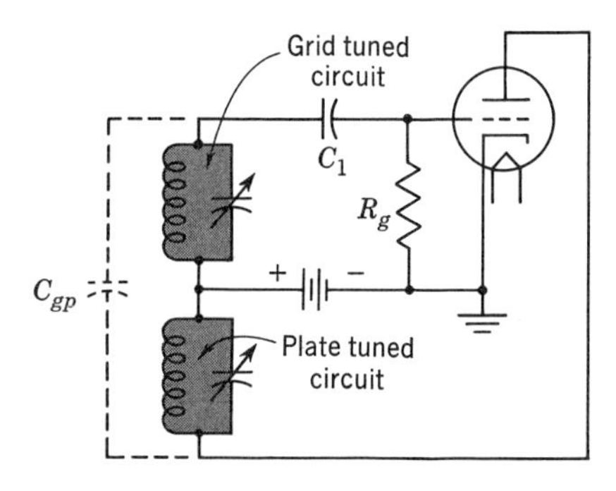


*Fig. 5 · 15 The tuned-grid tuned-plate oscillator circuit rearranged to show close resemblance to the Hartley oscillator.*

To adjust the oscillator to resonance, it is necessary to tune the plate circuit to a slightly lower frequency than the grid circuit. This is necessary in order to have the two tuning circuits resonate with the interelectrode capacitance $C_{gp}$.

## 5·9 Crystal oscillators

Certain crystalline substances, such as quartz, Rochelle salts, and tourmaline, exhibit an unusual electrical feature. If you apply a mechanical pressure to the crystal, it will generate an a-c voltage. By the same token, if a voltage is applied across the crystal, it will undergo a change in its physical shape, resulting in mechanical vibrations. This relationship between mechanical and electrical effects is known as the *piezoelectric* effect.

The frequency of the a-c voltage developed by the crystals is practically constant, and it is this characteristic that makes these units so valuable for use in r-f oscillators.

Of the three crystalline substances mentioned above, quartz is the most suitable for crystal oscillators. The usual shape of quartz crystals, as shown in Fig. 5·16, is a six-sided figure with a point at one end. An oscillator crystal is described according to the way it is cut from the raw crystal. Six axes are present in the crystal, three axes being formed by the three sets of corners labeled $X$, $X'$, and $X''$. The other three axes are perpendicular to the sides of the crystal and are labeled $Y$, $Y'$, and $Y''$.

In order for the crystal to exhibit a piezoelectric effect, the cut must be made along one of the six axes, parallel to the vertical length of the crystal. If a cut is made along the $X$ axis, the crystal is described as an $X$

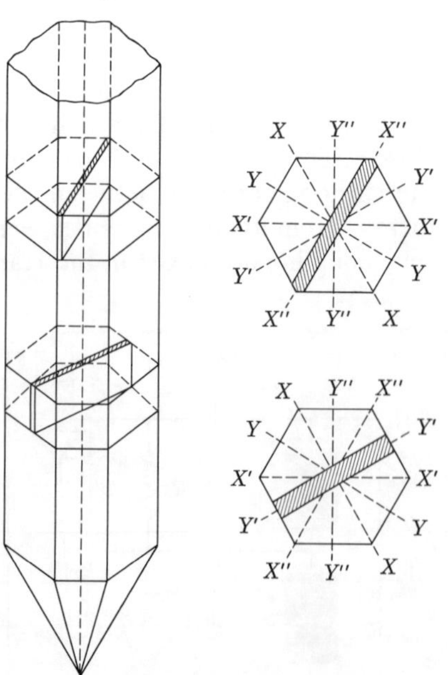


*Fig. 5·16 A quartz crystal and its cuttings.*

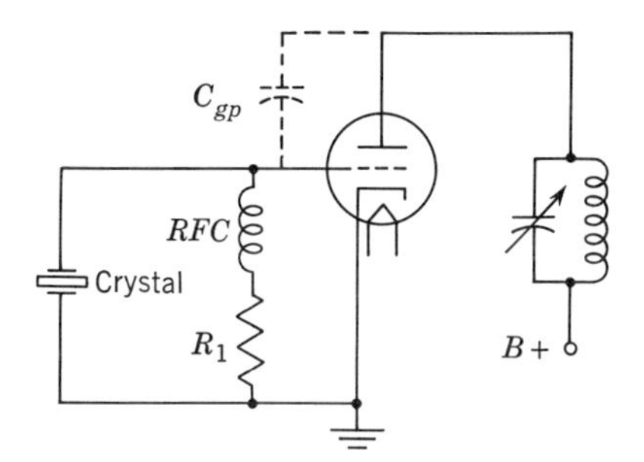


*Fig. 5 · 17 A crystal oscillator circuit. A triode is used here but a pentode can be used as well.*

crystal. A cut that is made halfway between the *X* axis and the *Y* axis is an *X-Y* cut. Additional patterns of cutting are being used in order to gain the most efficient operation and provide greater stability in the frequency of vibration.

**Factors affecting crystal frequency.** Before we discuss crystal oscillator circuits, it might be helpful to note the two factors which govern the frequency of oscillation of crystals. One is the crystal thickness. The thicker the crystal, the lower the frequency of oscillation. As higher and higher frequencies are desired, thinner and thinner crystals must be employed. Eventually the crystal becomes so thin that it is much too fragile to use. Generally, crystals are not employed above 10 Mc; if higher frequencies are needed, they are obtained by using a low-frequency crystal and doubling or tripling this frequency.

Another governing factor of crystal frequency is its temperature. When the temperature of a crystal changes, so does its frequency. Thus, to achieve the greatest stability, crystals should be kept in thermostatically controlled ovens where the temperature can be carefully controlled. This is done in commercial broadcast stations, where the law requires a high degree of accuracy in maintaining the broadcast signal on an assigned frequency. In less critical applications, the crystal is simply kept in a special holder designed for this purpose without any attempt being made to maintain a constant temperature. Oscillator stability is still excellent and superior in many respects to conventional oscillators.

**Oscillator circuits.** Crystals have been employed in a number of circuits, but the two to be described here are the most popular. The first circuit (Fig. 5 · 17) is similar to the TGTP oscillator, with the crystal replacing the tuned circuit previously used in the grid circuit. Feedback of energy still takes place through the plate-to-grid capacitance of the tube. The oscillations occur at the resonant frequency of the crystal. The plate circuit, as before, is tuned just slightly below this frequency.

Grid-leak bias is developed by the grid current flowing through $R_1$. A special grid-leak capacitor is not required, the crystal holder itself serving this purpose.

If too much voltage is fed back from the plate to the grid, the crystal will vibrate so violently that it may crack. This is a possibility often diffi-

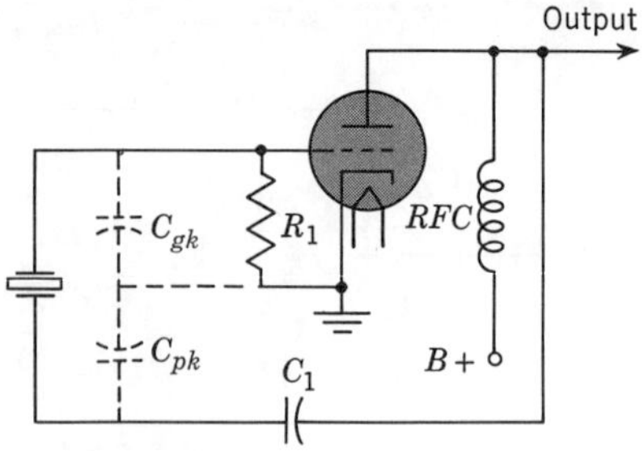

*Fig. 5·18 A Pierce crystal oscillator circuit.*

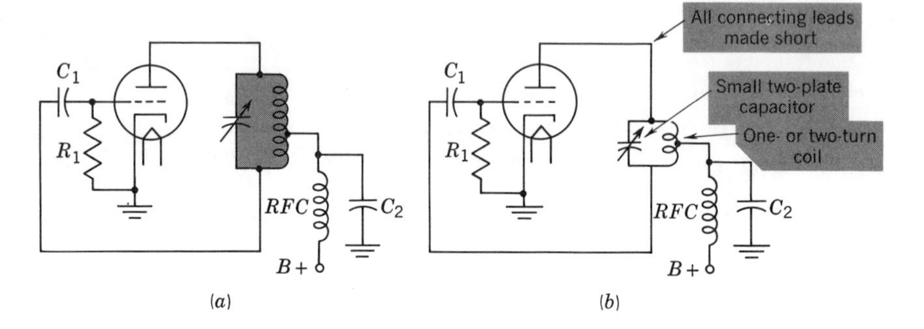


*Fig. 5·19 Transition in oscillator design from low frequencies to high frequencies. (a) Low-frequency circuit. (b) Very high-frequency circuit.*

cult to avoid with a triode tube because of the relatively high plate-to-grid capacitance. For this reason, pentodes are often favored for this circuit. The grid-to-plate interelectrode capacitance of a pentode is considerably smaller, reducing the feedback voltage. If the internal capacitance of the tube is too small, a small external capacitance can readily be added.

The second crystal oscillator is the Pierce oscillator circuit shown in Fig. 5·18. The crystal is connected directly from grid to plate by capacitor $C_1$. The circuit is the crystal modification of the Colpitts oscillator, with the tuned circuit replaced by the crystal and the voltage division achieved through the plate-to-cathode capacitance ($C_{pk}$) and grid-to-cathode capacitance ($C_{gk}$) of the tube. These small capacitances are represented by the dotted lines. The amount of feedback depends on the grid-to-cathode capacitance. This capacitance is not critical, and, ordinarily, it is not necessary to change the capacitor when changing crystals for different frequencies.

Capacitor $C_1$ keeps the d-c plate voltage off the crystal and provides an r-f feedback path back to the crystal. Resistor $R_1$ is the grid-leak resistance.

## 5·10 Resonant lines

Up to this point we have been discussing oscillators that can operate quite efficiently at frequencies up to 100 Mc, approximately, with conventional coils and capacitors. When we attempt to raise the frequency beyond this point, however, we run into some difficulty. In order to have the resonant frequency of a tuned circuit increase, either $L$ or $C$, or both, must be decreased. This means using fewer turns for the coil and fewer plates for the tuning capacitor. But there is a limit to this process. At the end nothing

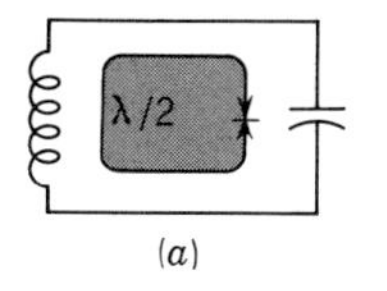


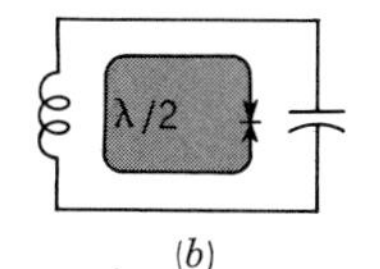


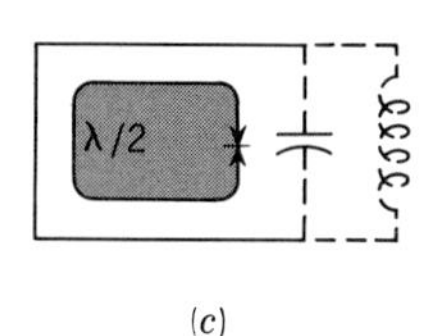


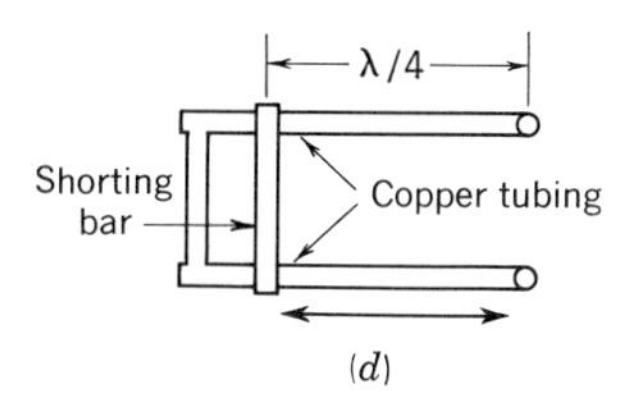


*Fig. 5·20 The development of a high-frequency tuning circuit as L and C are reduced. (a) Basic circuit. (b) Reduced inductance and capacitance, increased frequency. (c) Inductance and capacitance further reduced. (d) Final tank circuit.*

would remain of the capacitor except two very small plates, and of the inductance a turn or so of wire. In Fig. 5·19, this process has been applied to the Hartley oscillator. Illustration *a* shows the usual form of this oscillator at the low frequencies; *b* shows how a higher frequency (or shorter wavelength) is obtained by utilizing a smaller capacitance and a coil of few turns.

When even higher frequencies are desired, we must further reduce both $L$ and $C$, as illustrated in Fig. 5·20, where $\lambda/2$ corresponds to one-half wave[1] at the resonant frequency. Eventually, the capacitance of the tuning capacitor and the inductance of the coil will become less than the distributed capacitance and inductance of the leads. After all these reductions, the circuit will look like Fig. 5·20*c*—nothing but a partial loop of wire. To increase both $L$ and $C$, you lengthen the loop. To decrease $L$ and $C$, you shorten the loop.

The actual tuning circuit you will use is shown in Fig. 5·20*d*. The circuit is tuned by moving a shorting bar so that it either increases or decreases the length of the tubing, and with this, the frequency of the oscillator. The electrical length of each rod is adjusted by the shorting bar to be one-quarter wavelength. The two legs together will have an electrical length of one-half wavelength.

The final tuning circuit shown in Fig. 5·20*d* is known as a *Lecher line*, or, more popularly, as a *transmission-line tuner*.

An actual high-frequency oscillator circuit using this type of tuner is shown in Fig. 5·21*a*. The parallel transmission lines are one-quarter wavelength long at the frequency that it is desired to operate the oscillator. B+ voltage is fed to the plate of the tube through the transmission line. A blocking capacitor inserted between the grid terminal and the line prevents B+ from reaching the grid of the tube. The grid-leak bias is developed across $R_1$.

[1] For definitions of half-wave and quarter-wave lengths, see Chap. 9 on Antennas and Transmission Lines.

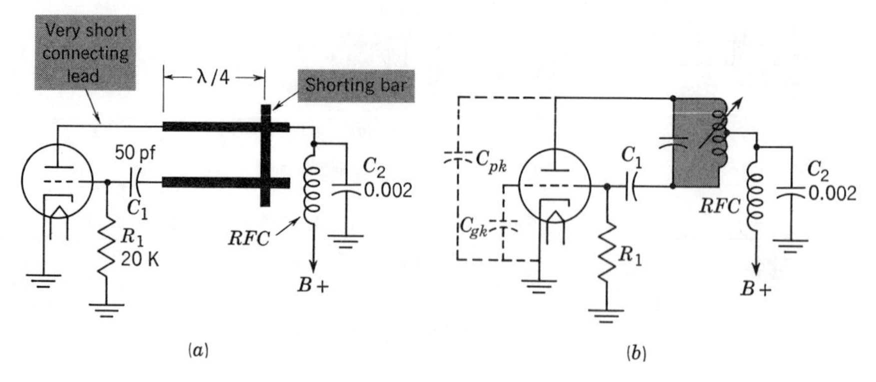


*Fig. 5·21 Ultraudion oscillator circuit. (a) High-frequency oscillator using a transmission-line tuning circuit. (b) The low-frequency equivalent circuit of this oscillator.*

The low-frequency equivalent circuit of this oscillator is shown in Fig. 5·21*b*. This is known as an *ultraudion oscillator* circuit. It is actually a modified Colpitts, in which use is made of the internal plate-to-cathode and grid-to-cathode capacitances of the tube for the feedback of energy. Note that $C_{pk}$ and $C_{gk}$ form a voltage divider across the tuned circuit.

## 5·11 Cavity resonators and waveguides

Although resonant lines can be useful for the VHF band of 100 to 300 Mc, at the higher frequencies in the UHF band of 300 to 3,000 Mc resonant cavities are used for microwave oscillators. A cavity resonator is a closed metal chamber that corresponds to a tuned circuit with very high $Q$. The resonant frequency of the cavity depends on its size. UHF cavity resonators can be compared with acoustic resonators for sound. As an example, the "boomy" sound in a room with smooth reflecting walls results from standing waves that reinforce the sound at particular frequencies. Similarly, a cavity resonator can magnify the effect of its resonant frequency, by means of standing waves for the electric and magnetic fields in the cavity.

In order to illustrate the resonance characteristics, Fig. 5·22 shows how a cavity can be developed by the addition of quarter-wave transmission lines, shorted at the end. Several lines in parallel are shown in *a*. If we continue to add more and more lines in parallel, the final result is a metal container as in *b*. This area enclosed by the metal conductors with just a small opening for input signal is a cavity resonator. The shape can be cylindrical, as shown, spherical, square, or rectangular. For any shape, though, the cavity is resonant for a particular frequency. In general, the resonant frequency has the wavelength that allows a quarter wave to approximate the radius of the cavity. Or, the cavity resonator must be one-half wavelength or any exact multiple of a half wave, at the operating frequency.

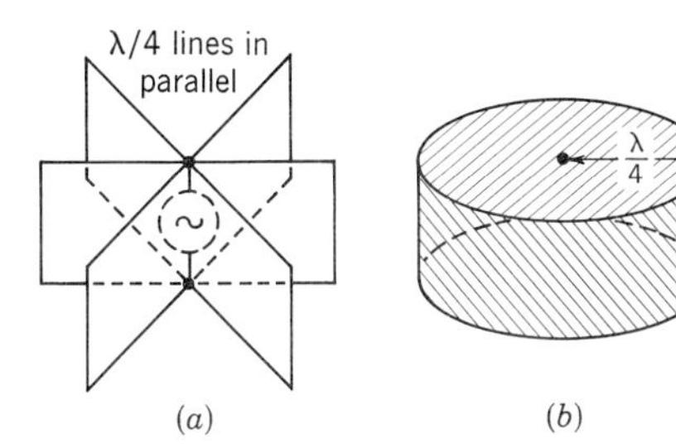


*Fig. 5·22 Development of cavity resonator from resonant lines. (a) Quarter-wave lines in parallel. (b) Resonant cavity.*

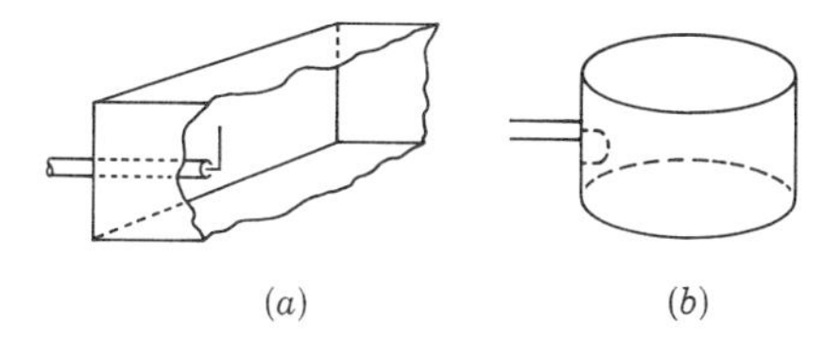


*Fig. 5·23 Methods of coupling to cavity resonators and waveguides. (a) Capacitive coupling to electric field. (b) Inductive coupling to magnetic field.*

Energy can be injected or taken out by coupling to the electric and magnetic fields in the cavity resonator. This is often done with a coaxial transmission line, using the two methods in Fig. 5·23. The probe in *a* is just an extension of the inner conductor of the coaxial cable, placed parallel to the electric lines of force in the field. This arrangement corresponds to capacitive coupling. In *b*, the loop of conductor encloses magnetic lines of force, which corresponds to inductive coupling. Tuning of a cavity resonator is accomplished by changing its size, either physically or electrically, to determine its resonant frequency.

Another device for particular use in the UHF band is the metal enclosure or waveguide shown in Fig. 5·24. A waveguide has the function of a transmission line but waveguides are much more efficient above 1,000 Mc. The metal tube forming a waveguide should not be considered as a conductor in the same sense as electrical wiring. Actually, the waveguide is made of metal in order to form a boundary for the electromagnetic field inside. The energy of the electric and magnetic fields moves from the input to output end of the waveguide by reflections from the inner walls. Skin effect on the inside surface of the metal enclosure prevents any

*Fig. 5·24 Rectangular waveguides.* (De Mornay Bonardi.)

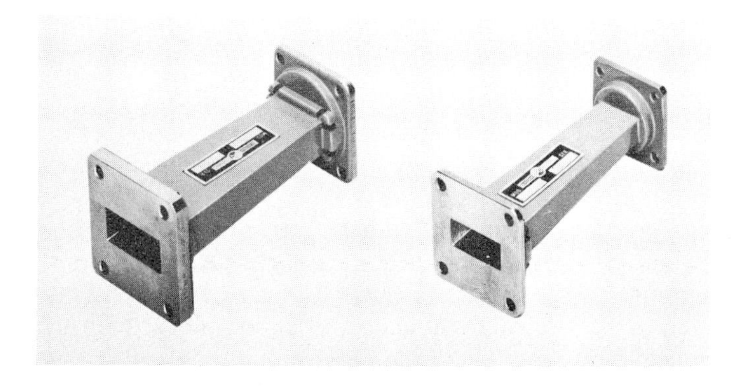

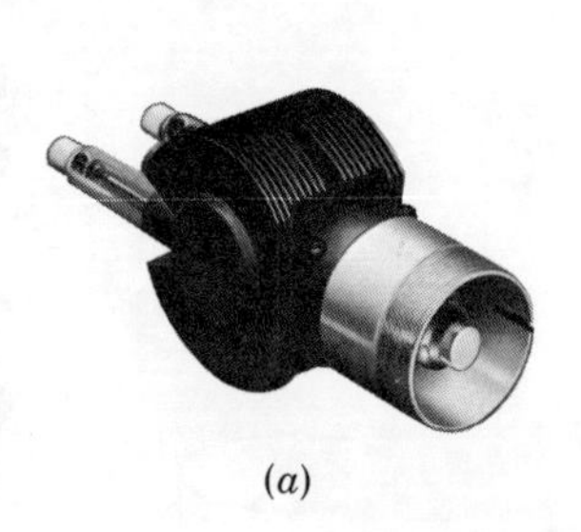
(a)

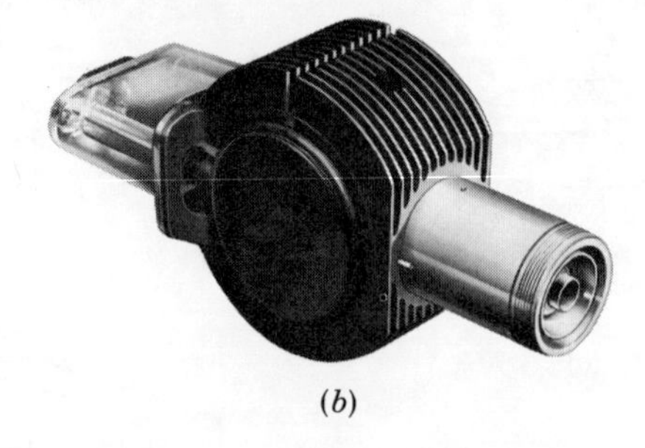
(b)

*Fig. 5·25 Magnetron oscillators. (a) For 900 to 970 Mc. (b) For 8,500 to 9,600 Mc.*

appreciable radiation outside the waveguide. The inside surface is usually coated with gold, silver, or aluminum for minimum losses, plated over brass.

In summary, then, a waveguide can be considered as a guide for propagation of magnetic and electric fields within an enclosed conductor. *TM propagation* refers to transverse-magnetic, which means that the magnetic field is transverse or perpendicular to the direction of motion through the waveguide. Similarly *TE propagation* refers to transverse-electric, when the electric field is in the transverse direction.

## 5·12 Magnetrons

One application of cavity resonators is in magnetron oscillators. The external appearance of a magnetron is shown in Fig. 5·25. Note the fins designed to prevent the unit from overheating during operation. Another feature of the magnetron is permanent magnets. As we shall see, an important requirement for the proper operation of a magnetron is a fairly strong magnetic field. The field is a fixed field (that is, it does not vary), and it may be obtained from permanent magnets that are part of the magnetron structure. In larger magnetrons, an electromagnet is employed.

Internally, a magnetron appears as shown in Fig. 5·26. At the center is a filament which, when heated by current flowing through it, emits electrons. Surrounding the filament is a circular anode block in which a number of cavities are machined. These cavities are hollow chambers, and it is possible to develop magnetic and electric fields inside each chamber. For the magnetron, these chambers or cavities serve as resonant tuning circuits because it has been found that magnetic and electric fields of certain frequencies can be developed for a given cavity size more strongly than fields of other frequencies. (This, of course, is similar to conventional coil and capacitance resonant circuits.) There are a number of such cavities, and they all combine to permit a considerable amount of energy to develop. This energy is then coupled out of the magnetron by means of a single loop of wire which is positioned in one cavity.

So much for the physical structure of a magnetron; now let us see how it develops its oscillations electrically.

**Magnetron operation.** If the filament is heated by a direct current and a voltage is connected between the filament and anode block, with the block positive, any electrons emitted by the filament will be drawn to the

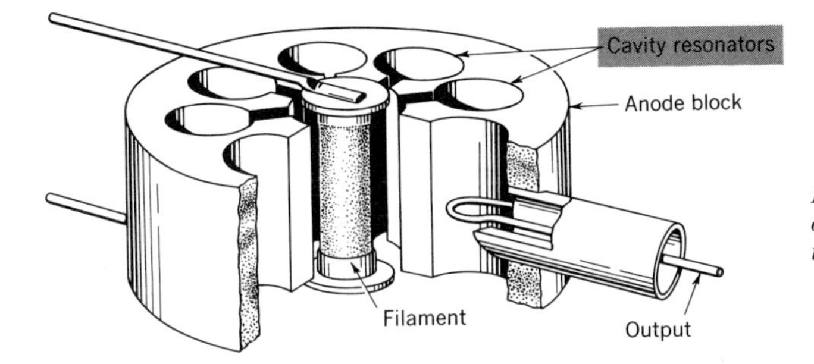


*Fig. 5·26 A cutaway view of the interior of a magnetron.*

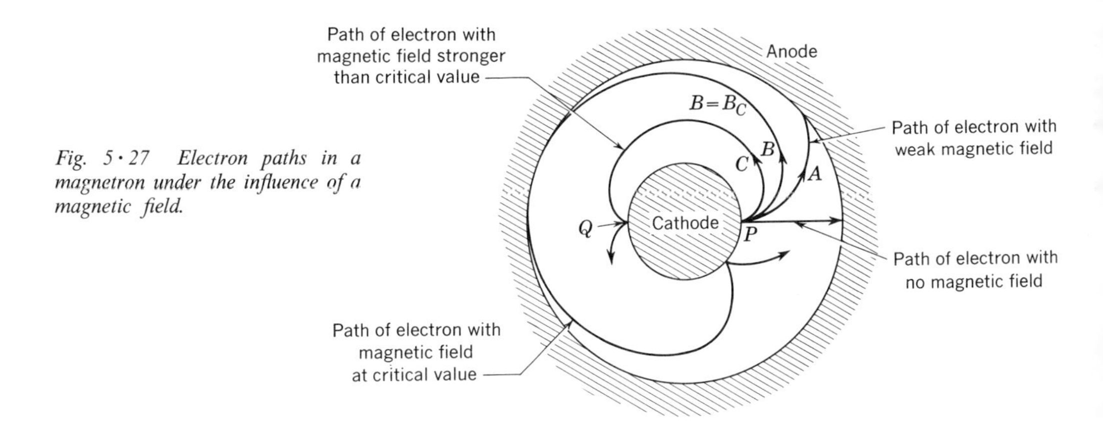


*Fig. 5·27 Electron paths in a magnetron under the influence of a magnetic field.*

anode. With nothing else brought into the magnetron system, this is all that would happen, and no oscillations would take place.

To the magnetron, let us add a steady magnetic field whose lines of force are parallel to the cathode. This is an axial field. What will the path of an electron be under these circumstances? To obtain the answer, it is first necessary to recall that an electron in motion is itself surrounded by a magnetic field. When the electron is more or less stationary near the cathode, the magnetic field is weak; but as the electron moves, the field builds up.

When we add an external magnetic field to the magnetron, it will interact with the magnetic field set up by the electron. Thus, the moment an electron leaves the cathode and starts moving toward the anode, it will find itself being acted on by the added magnetic field. The force arising from this interaction will be at right angles to both the added magnetic field and to the line of motion of the electron, forcing the electron to move in more or less of a curved or spiral path (Fig. 5·27). If the magnetic field is weak, the electron path will be only slightly curved, as shown by path $A$. The electron will still reach the anode, however, and give up whatever energy it gained in going from cathode to anode.

As the magnetic field strength is gradually increased, the electron path will tend to curve more and more until, at a certain value of magnetic field, the electron will just barely graze the anode wall and then be brought back to the cathode (see curve *B* of Fig. 5·27). The field strength at which this occurs is called the *critical magnetic-field strength.*

If the magnetic-field strenght is increased beyond this critical value, the electron will never reach the anode (see curve *C* of Fig. 5·27). It will start out for the anode from point *P*, but the force produced by the magnetic field will be so great that the electron will be quickly brought back to the cathode, perhaps at point *Q*. It reaches point *Q* with little or no velocity, and it no sooner comes to rest than the positive anode voltage once more attracts it. Again the electron starts for the plate, again it travels through its circular path, ending up now at some new point on the cathode. The same sequence continues to recur, with the electron hopping its way around the cathode cylinder and never reaching the plate.

Thus, for a given potential difference between anode and cathode, the distance an electron travels from the cathode depends upon the strength of the magnetic field present between the two electrodes. We can achieve a similar variation in electron motion by altering the positive anode voltage. An increased voltage will cause the electron to move in a larger circle; a lowered voltage will reduce the circle.

In the magnetron, the problem is to adjust the magnetic-field strength and anode potential so that electrons will curve properly to obtain the desired oscillation and efficiency of operation.

**Magnetron-oscillator operation.** In the multicavity magnetron shown in Fig. 5·26, oscillations are established in the various cavity resonators by the interaction of electrons rotating in the space between the cathode and anode and the r-f fields set up in each of the cavity resonators. The actual velocity and curvature of the electrons are determined by the value of d-c voltage used and the strength of the magnetic field, and this velocity may be set at any desired value. Maximum output of any frequency is obtained when the d-c voltage and the strength of the magnetic field are adjusted so that the electrons transfer energy to the cavity resonators at the proper time, thereby building up large oscillations in the cavity resonators. At other values of voltage and magnetic-field strength, the motion of the electrons will not be such as to aid and reinforce any oscillations set up in the cavity resonators. Then the output energy will be either zero or considerably smaller than the maximum obtainable under the proper operating conditions.

To operate the magnetron, a positive d-c voltage is connected between the cathode and anode block, and an axial magnetic field is provided by an external magnet. Electrons emitted by the cathode are attracted to the highly positive anode block. The moment the electrons start traveling toward the anode, however, they interact with the axial magnetic field, and are forced to travel in the circular path previously described. These electrons, traveling toward the anode, shock-excite the cavity resonators in

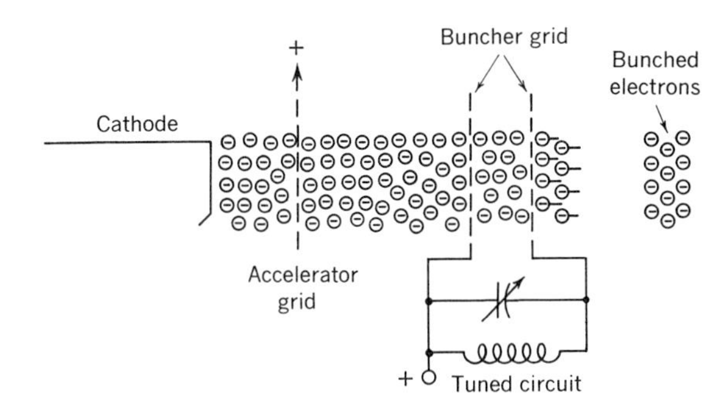


*Fig. 5·28 Electron gun and buncher grids of a klystron tube.*

the same manner that a conventional oscillator is shock-excited into oscillations when power is suddenly applied, and electrons flow through the tube. As a result of this shock excitation, the cavity resonators will oscillate at their resonant frequencies.

**Magnetron tuning.** Some method of tuning a magnetron is desirable, and this is accomplished, as in conventional circuits, by changing the circuit inductance or the circuit capacitance. To vary the resonator inductance, a pin, known as a *tuning pin,* is inserted in each of the resonator cavities by some form of mechanical assembly. These tuning pins are composed of some nonmagnetic substance, and as they move deeper into the cavity, they reduce the volume available for the magnetic flux, thereby reducing the inductance and causing the frequency to increase. Tuning ranges as great as $\pm 7$ per cent of the average frequency have been accomplished in this manner.

Frequency tuning by means of capacitance variation can be achieved by a ring, shaped like a cooky cutter, moved in or out of grooves cut in the anode block. The deeper the ring moves into the groove, the higher the magnetron frequency. In the design of magnetrons, it has been found that it is easier to vary the inductance for the high frequencies and the capacitance at the low frequencies.

## *5·13 Klystrons*

Another approach to UHF oscillations is employed by the klystron tube. This tube operates on a principle of velocity modulation or change in the speed of electrons passing through it. By means of this change in electron speed, the tube produces bunches of electrons separated by spaces in which there are few electrons.

Since klystron tubes depend on bunches of electrons, let us see how these are produced. The first step is to produce a stream of electrons, all traveling at the same speed. The electrons are emitted by a heated cathode and accelerated forward by a positive grid (see Fig. 5·28). Most of the electrons miss the grid wires and pass through the grid to form a beam of electrons, all moving forward at the same speed.

The beam of electrons is then passed through a pair of closely spaced grids, called *buncher grids,* each of which is connected to one side of a tuned circuit. The grids are very close together, so that the electrons spend only a relatively short time between them. The tuned circuit and the grids are at the same d-c potential as the accelerator grid. In addition, a high-frequency a-c voltage is present across the tuned circuit. This a-c voltage causes the velocity of the electrons leaving the buncher grids to differ, depending on the time at which they pass through the grids.

The manner in which the buncher produces groups of electrons can be understood by considering the motion of individual electrons. If an electron reaches the grids when the r-f voltage between them is zero, it will pass on unaffected. However, any electron reaching the buncher grids when the r-f voltage is going positive will receive a greater acceleration than if the r-f voltage were not present.

Similarly, when the r-f voltage is going negative, each oncoming electron will be subjected to less positive voltage than usual, and its velocity will be reduced. In this way, the electrons will tend to have different velocities, dependent upon the time in the r-f cycle at which they pass the two grids.

At this point, the electron beam may be said to be *velocity-modulated* because we have altered the velocity of the electrons as they passed through the buncher grids. This particular fact is, of itself, not significant. If we let the electrons travel long enough, however, those electrons which were accelerated in going through the buncher grids will eventually catch up with those electrons that passed through shortly before this and were slowed down. The result will be bunches, or clusters, of electrons.

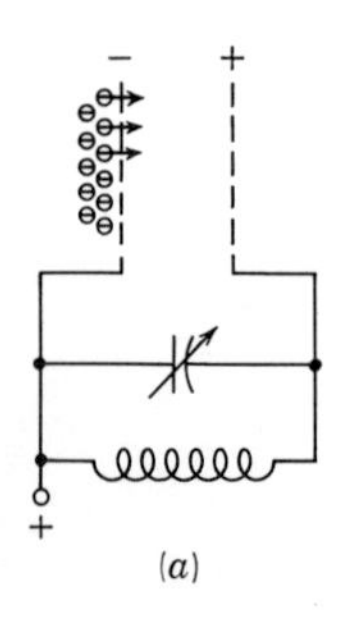


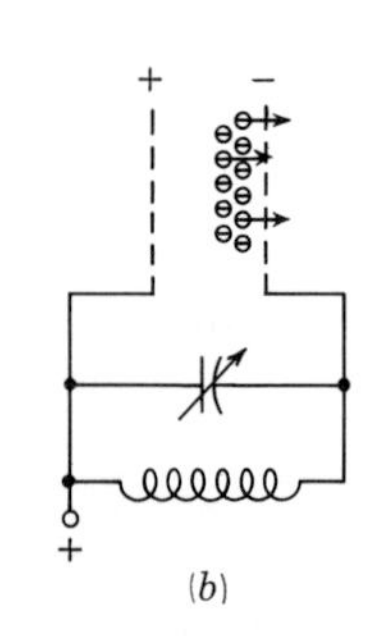


*Fig. 5 · 29 The change of catcher grid polarity with oscillations in the tuned circuit. (a) One polarity. (b) Reversed polarity.*

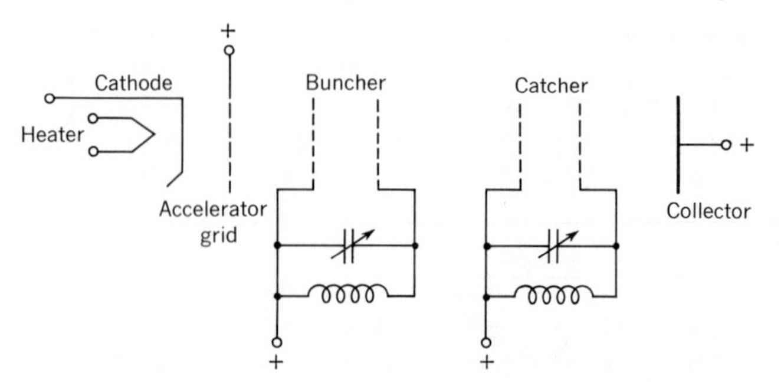


*Fig. 5 · 30 Schematic showing the elements of a klystron tube.*

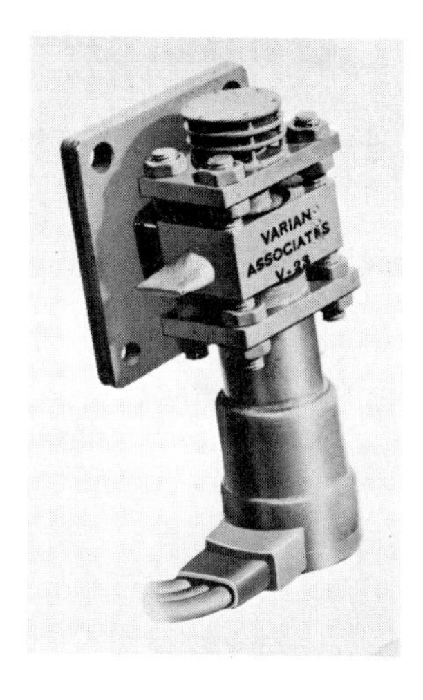


*Fig. 5·31 The physical appearance of one type of klystron.*

*Fig. 5·32 A klystron tube with cavity resonators.*

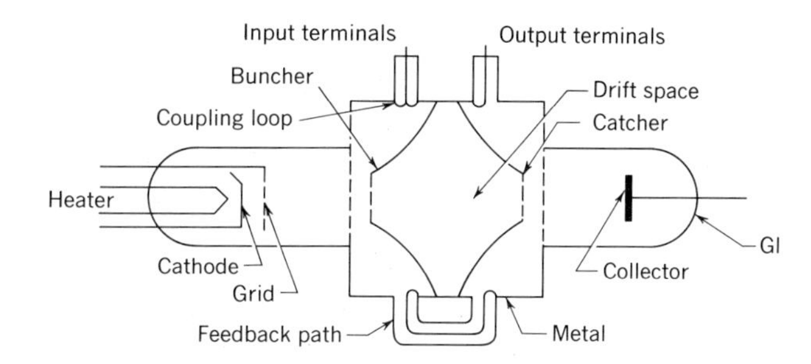


These bunches are allowed to pass through a similar second set of grids called *catcher grids,* coupled to another oscillating circuit. If the relative grid potentials are as shown in Fig. 5·29, when each bunch of electrons reaches the first grid of this set, the field is such that it slows down the electrons and thus absorbs energy from them. By the time the bunch of electrons reaches the second grid of the set, the relative grid potentials are reversed, as shown in Fig. 5·29, because it takes the group of electrons approximately one-half cycle to go from one grid to the other. Therefore the second catcher grid also slows down the electrons, absorbing more energy from them.

By this time, the bunches of electrons have been greatly reduced in speed. Since they have now served their purpose, they are collected by a plate and transferred back to the cathode to complete the circuit.

A klystron tube is shown schematically in Fig. 5·30. Physically, the tube appears in one form as shown in Fig. 5·31. The buncher and catcher tuner circuits may be small coils and capacitors, but for the very high frequencies at which these tubes operate, cavity resonators are ordinarily used. Energy may be coupled into or out of these tuned circuits by using small one-turn loops properly positioned in the resonator.

If the output from the catcher is fed back into the buncher, and if the proper phase and energy relations are maintained between buncher and catcher, the tube operates as an oscillator, and in order to do so successfully, the tube must amplify. Such amplification occurs because the electrons pass the buncher grids in a continuous stream, whereas they go by the catcher grids in bunches.

The klystron may be used as an amplifier, oscillator, or mixer. Cavity resonators are generally used for the tuned circuits, as shown in Fig. 5·32. These cavities are so small at microwave frequencies that they may be sealed inside the tube envelope. In this case, the cavity is tuned by varying the spacing of the cavity grids. Thus, a slight flexing of the tube varies the effective capacitance of the tuned cavity circuit.

In another type of construction, the grid connections are brought out through the envelope of the tube, and an external cavity is used, clamped around the tube. In such a system, the cavity is tuned by changing its effective inductance. This can be done, for example, by screwing plugs into the surface of the cavity.

**Reflex klystron oscillators.** Oscillations in a two-resonator klystron are maintained by feeding back energy from the output to the input. It is possible, however, to obtain oscillations by the use of a single-cavity resonator. Such oscillators, known as *reflex oscillators,* have been developed and are used extensively.

## *5 · 14 Traveling-wave oscillators*

The traveling-wave oscillator takes still another approach to the problem of generating microwave frequencies. It is designed to overcome the rather low efficiency of the klystron, a characteristic which is due to the short contact between the electron beam and the r-f field of the cavity resonator. If the beam and the energy were permitted to remain in contact longer than the time afforded them by the narrow resonator gaps, the efficiency of the energy transfer would be higher. Unfortunately, the transit time of the electrons when crossing the resonator gap must be kept small, and this prevents extended interaction between the beam and the field.

A typical traveling-wave tube is shown in Fig. 5 · 33*a*. This tube consists of a tightly wound helix of wire mounted in a glass tube and supported by four ceramic rods. The length of this helix may be 6 in., or more. The ends of the helix are connected to short, straight conducting stubs parallel to the axis of the tube and close to the glass wall. The stubs pick up or transfer electromagnetic (EM) energy to two waveguide sections through which the tube projects (see Fig. 5 · 33*b*). The tube is enclosed in a circular case to shield it from stray electromagnetic fields.

If energy is fed into the input waveguide, it will travel along the helix and appear at the output waveguide. Since the wire is wound in the form of a spiral, however, and the waves follow the twisting wires, the actual forward velocity of the wave is only about one-tenth that of light.

If the tube contained only the helix, energy traveling along the coil from end to end would experience a considerable loss because of the resistance of the wires. The tube, however, in addition to the helix, has an electron gun at one end and a collector plate at the other. When the gun is emitting electrons, EM waves traveling along the helix receive a gain rather than a loss.

A beam of electrons formed by the gun is shot through the tube, passing through the center of the helix winding and impinging on the collector plate. The helix winding is given a relatively high positive potential (approximately 1,600 volts), drawing a beam current of about 10 ma. If the current is properly adjusted through an external focusing coil, which completely surrounds the tube, the major portion of the beam reaches the collector plate. At the acceleration voltage of 1,600 volts, the electrons travel axially

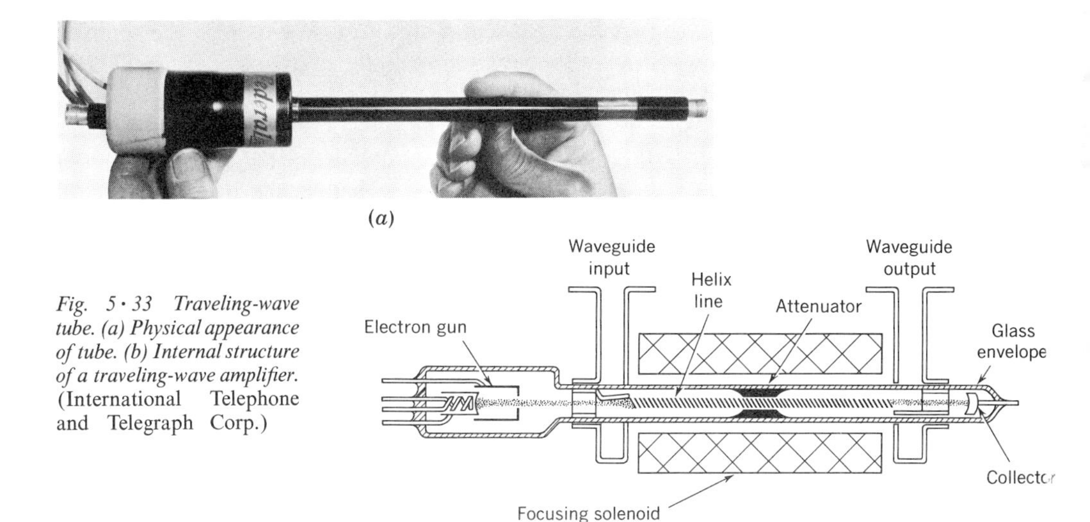


*Fig. 5·33 Traveling-wave tube. (a) Physical appearance of tube. (b) Internal structure of a traveling-wave amplifier.* (International Telephone and Telegraph Corp.)

through the tube with a velocity which is only slightly greater than the forward velocity of the EM wave along the helix.

The amplification of the tube depends upon the interaction of the EM field which is set up along the axis of the tube and the electrons traveling through the coil. The EM field carried by the helix contains a magnetic and an electric component. Both these components extend axially along the tube. Since the electrons are traveling parallel to the magnetic field, there will be no interaction between them. There will, however, be an interaction between the electric component of the EM wave and the beam electrons.

The EM energy of the wave enters through the input waveguide and travels along the helix. An electron leaving the electron gun and entering this field will do so at one of two times: when the electric field of the wave is positive at the gun or when it is negative. If the field is positive, the electron will receive a greater acceleration than it would have received from the d-c voltage alone. If the field is negative at this instant, the electron will not receive as much acceleration.

Now, the electron velocity is adjusted so that it is slightly greater than the forward wave velocity. This is an important point to remember. If an electron emerges from the electron gun at a time when the electric field of the wave is positive, it will travel even faster and move fairly rapidly out of the positive electric field and into the preceding negative field. Here it will gradually slow down with the result that it will remain in this negative or decelerating field for a longer period of time than it remained in the positive electric field. It will, therefore, give up more energy to the wave than it absorbed from it.

Electrons which emerge from the electron gun when the electric field of the wave is negative will receive less acceleration than they would have in the absence of the field. The field will slow them down and force them to remain for a comparatively longer time in this negative electric field. The result, again, is a transfer of energy from the electrons to the wave.

Since the electrons are traveling faster than the wave, we can visualize the wave as standing still and the electrons slowly moving past it. Wherever the electric field of the wave is positive, the electrons will move forward faster than at those points where the electric field is negative. Consequently, at any one instant of time, more electrons are being slowed down than are being speeded up. Therefore more energy is being received by the wave than it is giving. The amplitude of the wave will increase as it travels through the tube. At the end of the helix, the energy is radiated into the waveguide and conducted away from the tube.

The gain achievable in the traveling-wave tube depends on the relative velocity between the wave and the electron beam. Beam velocity is a function of the d-c accelerating voltage. In order that reasonable voltages may be used, the helix is wound so that the forward velocity of the EM wave is reduced to a value about one-tenth that of light. At this velocity, an acceleration voltage of 1,600 volts forces the electrons to move slightly faster than the wave. Experimentally it was discovered that the tube gain rises as the beam velocity increases until this velocity slightly exceeds that of the wave. Beyond this, the gain decreases again.

Gain is also governed by the length of the helix. The longer the tube, the greater the contact between the electron beam and the wave, and the greater the possibility of energy transfer from the beam to the wave. If the tube is made too long, however, difficulties in beam focusing and beam spreading arise, both of which result in a reduction in gain. The tube shown in Fig. 5·33 was designed for a mid-frequency of 3,600 Mc. It has good gain and a bandwidth of 800 Mc, which is extraordinarily large.

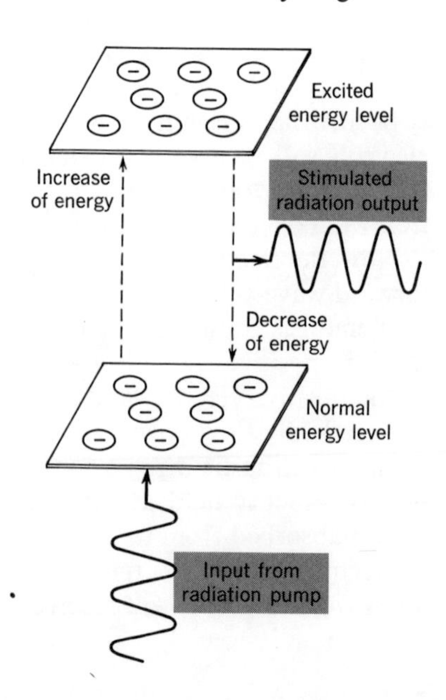


*Fig. 5·34 How radiation pump changes energy level of electrons in atom to stimulate emission of light.*

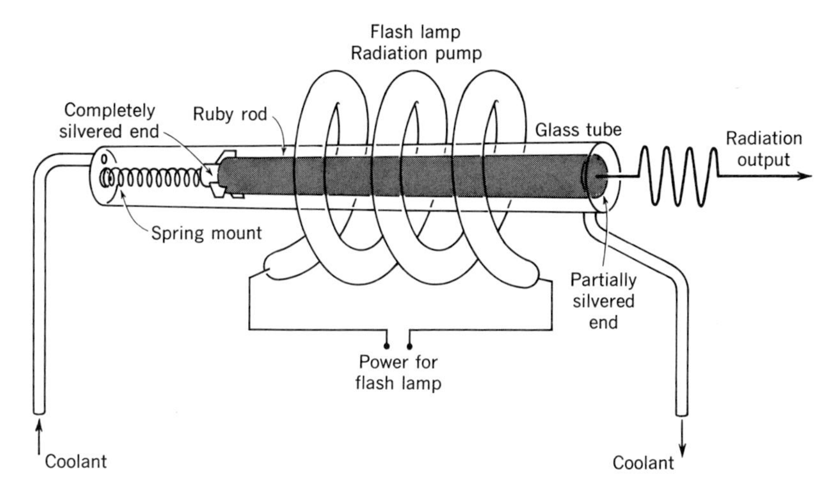


*Fig. 5·35 Construction of a ruby maser.*

## 5·15 Masers and Lasers

The term maser is an abbreviation for microwave amplification by stimulated emission of radiation. Similarly, a laser is a device for light amplification by stimulated emission. The emitted radiation is at microwave frequencies for masers, while lasers produce output of light, usually in the infrared or visible red spectrum. For both lasers and masers, the stimulated emission refers to the fact that electrons in the atoms are forced to change their energy levels. The resulting radiation when electrons drop to a lower energy level can be produced at the desired frequency or wavelength. The stimulation is produced by applying input from a source called a *radiation pump*. As an example, a laser made from a ruby crystal for infrared output can use green light for the radiation pump, to produce output with a frequency of $300 \times 10^{12}$ cps or a wavelength of $1 \times 10^{-4}$ cm.

Masers and lasers can be used as either amplifiers or oscillators, for extremely high frequencies starting with microwaves at the low end up to visible light. Originally, masers were developed for efficient amplification of microwaves, with the advantage of eliminating tube noise. Then the same effect was extended to the visible spectrum. The output of a laser is called *coherent light*, because it includes an extremely narrow band of wavelengths and is propagated in a very intense beam without spreading out too much.

The basis of laser action is the fact that all light results from electrons in the shells or energy levels in the atoms. When energy is absorbed by the atom its electrons can be raised to a higher energy level. Because this level is not the stable condition for the electrons, they then drop back to a lower energy level. As the electrons drop to a lower level, energy is released in the form of electromagnetic radiation. This cycle of gain and loss in energy is illustrated in Fig. 5·34. The frequency or wavelength of the emitted radiation depends on the difference between the two energy levels for the electrons. Furthermore, this factor depends on the material used for stimulated emission. A gas mixture of helium and neon can be used for masers, a solid ruby crystal for a laser with infrared output, and semiconductor materials such as gallium arsenide are being developed for output of visible light, to give just a few of the possibilities. In many cases the material to be stimulated is cooled to very low temperatures by cryogenic equipment, so that less pumping energy is required.

The construction of a ruby maser is illustrated in Fig. 5·35. The radiation pump here is a flash lamp, which is similar to a neon bulb. This radiation supplies energy to the solid ruby rod to stimulate its emission. The radiated output is transmitted through the partially silvered end at the right end of the ruby crystal. The other end is completely silvered, so that the radiation can build up its intensity by repeated reflections between the two ends. Liquid nitrogen is used for the coolant. The output of the maser can be taken through a waveguide, for microwave radiation. For light, the output is a visible beam.

## SUMMARY

1. An oscillator is basically an amplifier which derives its input signal from its own output.
2. In practical oscillators, the oscillations are started by variations in plate current as the tube heats up. They build up because of the feedback path and the amplifying action of the circuit.
3. A tuned circuit develops oscillations by the interchange of energy between the tuning capacitor and the inductance. To maintain the oscillations steady, a vacuum tube periodically replenishes the energy lost in the tuned circuit. The frequency of oscillations in a tuned circuit is given by $f_r = 1/(2\pi\sqrt{LC})$.
4. Practically all oscillators use grid-leak bias, produced by the feedback voltage.
5. In the tickler-coil oscillator, energy is fed back from the plate to the grid circuit inductively by a small tickler coil.
6. The Hartley oscillator uses a single coil and a tap to achieve the same results as the two separate coils in the tickler-coil arrangement.
7. The electron-coupled oscillator combines an oscillator and an amplifier in one network. The coupling between the oscillator and the amplifier sections is achieved through the electron beam in the tube.
8. The Colpitts oscillator uses capacitors to couple the energy from the plate to the grid.
9. The interelectrode capacitance between the grid and plate provides the feedback path in a tuned-grid tuned-plate oscillator.
10. Certain crystalline substances exhibit the piezoelectric effect. If an electric voltage is applied across the crystal, it will vibrate mechanically. Conversely, if the crystal is subjected to mechanical pressure, it will generate an a-c voltage.
11. Crystal oscillators possess a very high degree of frequency stability. This may be even further enhanced by keeping the crystal in a temperature-controlled oven.
12. In VHF oscillators, the conventional tuning coil and capacitor are replaced by sections of transmission lines.
13. A magnetron is a UHF oscillator which consists of a central cathode and a circular anode block that surrounds the cathode. A series of cavity resonators are machined into the anode block. Oscillations are produced by the interaction of the emitted electrons and a magnetic field.
14. A cavity resonator may be tuned inductively by inserting a small pin into the resonator. It may be tuned capacitively by moving a ring in or out of grooves cut into the anode block.

15. Klystron oscillators operate on the principle of velocity modulation or changing the speed of electrons passing through it.
16. Internally, a klystron consists of a pair of buncher grids and a pair of catcher grids, together with a collector plate. The buncher grids modulate the velocity of the electrons passing through them. This causes the electrons to form into bunches. Then the bunches of electrons transfer energy to the cavity resonator connected to the catcher grids. In a reflex klystron oscillator, one set of grids serves as buncher and catcher.
17. A traveling-wave tube consists of a tightly wound helix mounted in a glass tube. Electrons from an electron gun at one end travel through the helix and are collected by a collector element at the other end of the tube. Energy traveling along the helix interacts with the electron beam, absorbing energy from the beam.
18. Gain of a traveling-wave tube is governed by the length of the helix and the relative velocity of the energy wave and the electron beam.
19. Lasers and masers provide amplication by stimulated emission of radiation, resulting from changes in the energy level of electrons in the atom. The stimulation is a result of electromagnetic energy input from a radiation pump.

## SELF-EXAMINATION

Here's a chance to find out how well you have learned the material in this chapter. Work the exercises; then check your answers against the Key at the back of this book. These exercises are for your self-testing only.

1. A 100-pf capacitor is combined with a 100-mh coil in a tuning circuit. The resonant frequency is approximately (*a*) 480 cps; (*b*) 4.8 Mc; (*c*) 24 Mc; (*d*) 48 kc.
2. The purpose of the grid-leak capacitor and resistor in an oscillator is to (*a*) maintain the frequency constant; (*b*) provide a more stable output voltage; (*c*) take the place of the d-c power supply; (*d*) permit the frequency to be varied.
3. For maximum efficiency, oscillators can operate (*a*) class A; (*b*) class AB; (*c*) class B; (*d*) class C.
4. Energy is usually taken from an oscillator circuit by (*a*) inserting a resistor in the grid circuit; (*b*) coupling a small coil close to the tube; (*c*) capacitive or inductive coupling from the tuned circuit; (*d*) connecting a capacitor across the power supply.
5. Oscillators are generally tuned by varying the (*a*) capacitance or inductance of the tuned circuit; (*b*) grid voltage; (*c*) plate voltage; (*d*) filament voltage.
6. The oscillator circuit that uses a tapped coil in the tuned circuit is (*a*) Colpitts; (*b*) Hartley; (*c*) ultraudion; (*d*) Pierce.
7. In high-frequency circuits, tuning is frequently done by varying the (*a*) shape of a coil; (*b*) length of a transmission line; (*c*) relative position of two transmission lines; (*d*) interelectrode capacitance of the tube.
8. Which of the following is *not* found in a magnetron oscillator? (*a*) A magnetic field; (*b*) cavity resonators; (*c*) an anode; (*d*) a grid.
9. In a reflex klystron oscillator, (*a*) there is no feedback of energy; (*b*) there is one tuning circuit; (*c*) some electrons are captured by the collector; (*d*) electron bunching does not occur.
10. In a traveling-wave tube, (*a*) electrons travel at the speed of light; (*b*) the helix receives the input signal through a waveguide; (*c*) the helix is part of the electron-gun structure; (*d*) beam defocusing is used.

## QUESTIONS AND PROBLEMS

1. Give the function of each of the following parts in an oscillator: (*a*) *LC* circuit; (*b*) amplifier tube; (*c*) B+ supply voltage.
2. List four types of circuits for an r-f feedback oscillator, stating how the feedback is obtained for each.
3. Draw the circuit diagram of a Hartley oscillator, using a triode with series feed for the plate-supply voltage.

4. What is meant by an electron-coupled oscillator? Give one advantage of this circuit.
5. (*a*) What is the main advantage of a crystal oscillator circuit? (*b*) Name two types of crystal-controlled oscillator circuits.
6. (*a*) As the frequency of an oscillator is increased to the VHF and UHF ranges, is more or less $L$ and $C$ needed? (*b*) What determines the minimum $L$ and $C$ the circuit can have?
7. Describe briefly the internal construction of a magnetron.
8. What is the purpose of the buncher and catcher grids in the klystron?
9. Give two ways to alter the resonant frequency of a cavity resonator.
10. Describe briefly the internal construction of a traveling-wave tube.
11. What is the difference between a maser and a laser?
12. What is the function of the radiation pump for a maser or laser?
13. List the frequencies (in cps) in the VHF and UHF bands.
14. List the wavelengths (in cm) for the visible light spectrum.
15. In an r-f feedback oscillator with grid-leak bias, why does measuring the d-c bias indicate the intensity of the r-f oscillations?
16. How much $L$ (in microhenrys) is necessary with 40-pf $C$ for an oscillator frequency of 5 Mc?
17. Calculate one wavelength (in cm) at 300 Mc, assuming the velocity of propagation is the same as the speed of light.
18. Why is it impractical to use cavity resonators and waveguides for a frequency such as 5 Mc?

# Chapter 6 Power supplies

The amplifier and oscillator circuits in the previous chapters need a power supply for operation. With transistors, the d-c bias voltages are generally supplied by a battery, or a low-voltage power supply of 9 to 45 volts. For electron tubes, power is needed to heat the filament, in addition to the d-c voltages required for plate and screen. The heater voltage can be a-c or d-c, but the B+ of 250 to 350 volts for plate and screen must be a steady d-c voltage. These requirements are described in the following topics:

6·1 Functions of the Power Supply
6·2 The Power Transformer
6·3 Rectifiers
6·4 Half-wave Rectifiers
6·5 Full-wave Rectifiers
6·6 Bridge Rectifier Circuits
6·7 Voltage-doubler Circuits
6·8 Filters
6·9 Voltage Dividers
6·10 Transformerless Power Supply
6·11 Heater Circuits
6·12 Typical A-c Power Supply
6·13 Voltage Regulators
6·14 Vibrators
6·15 Dynamotors
6·16 Power-supply Troubles

## *6·1 Functions of the power supply*

The power used to heat the filament or heater in electron tubes is an *A supply*. D-c high voltage for the plates and screens is obtained from the

*Fig. 6·1 Power-supply chassis. Width of chassis is 7 in.* (Heath Company.)

*Fig. 6·2 Functions of the components in a power supply.*

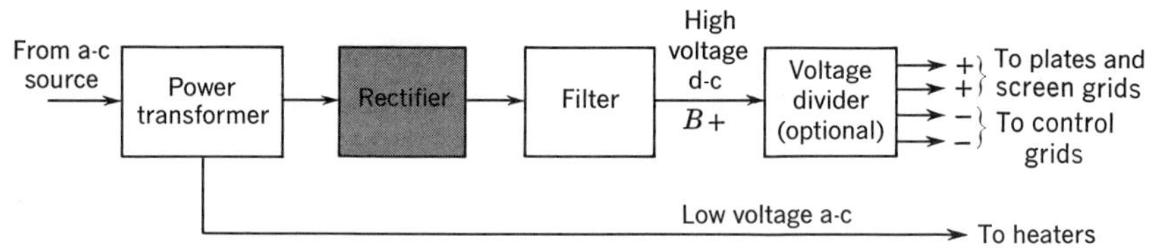


*B supply*. Control-grid bias is furnished by the *C supply*. This nomenclature dates from the days when only batteries supplied all power for electronic equipment. These were called *A*, *B*, and *C* batteries. Now that one power supply takes the place of these batteries, it must deliver the same three types of power they did. A typical power supply is shown in Fig. 6·1.

In terms of the B supply, the following requirements must be met. The proper value of B+ voltage must be supplied, at the required amount of load current. Also the ripple voltage, which is the a-c component in the d-c output, must be reduced for minimum hum. Furthermore, the d-c voltage output of the power supply should be constant regardless of varying load current. This is known as *voltage regulation* of the power supply.

To do all these jobs a typical power supply consists of the components shown in Fig. 6·2. The a-c power-line voltage is connected to the primary winding of the transformer. The power transformer must step up or step down the available a-c voltage to the required value. Then the secondary windings can supply low voltage to the heaters of the various tubes and high voltage to the rectifiers. The rectifier changes alternating to direct current, but this d-c output is pulsating, or fluctuating. This fluctuating ripple will produce hum in the amplifiers and must be removed. The job of removing the pulsations is reserved for the filter.

In addition, a voltage divider can be used to make available different values of B or C voltages as required by the various plates, screen grids, and control grids of the tubes being supplied. However, the voltage divider is optional.

## 6 · 2 *The power transformer*

One purpose of the power transformer (Fig. 6 · 3) is to step up the a-c line voltage which is applied to the primary winding, to make it possible for the power supply to deliver the proper value of B+ voltage. The transformer also steps down the line voltage to the values needed to operate the heaters of tubes.

Power transformers are rated according to the voltage they can develop across the high-voltage secondary (or in some cases at the output of the filter), while delivering full rated current. A typical rating may be 400 volts at 90 ma. Transformers with a center-tapped high-voltage secondary winding are rated by the voltage at each end with respect to the center tap—350–0–350 at 100 ma. Low-voltage secondary windings, which deliver power to the heaters, are rated by the voltage and current they can deliver. A typical example is 6.3 volts at 2 amp.

**Types of power transformers.** Figure 6 · 4 shows some of the variations which may be found in power transformers. The unit in *a* has a tapped

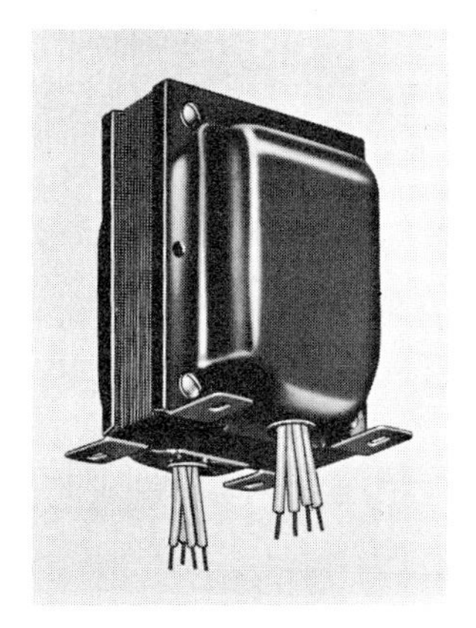

*Fig. 6 · 3 Typical power transformer. Height is 5 in.* (Chicago Standard Transformer Corp.)

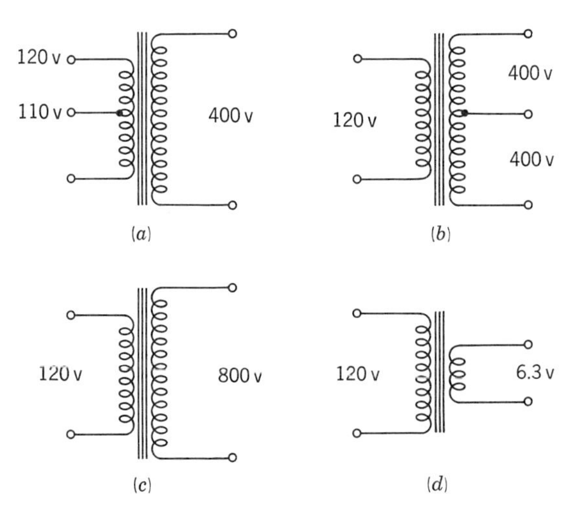


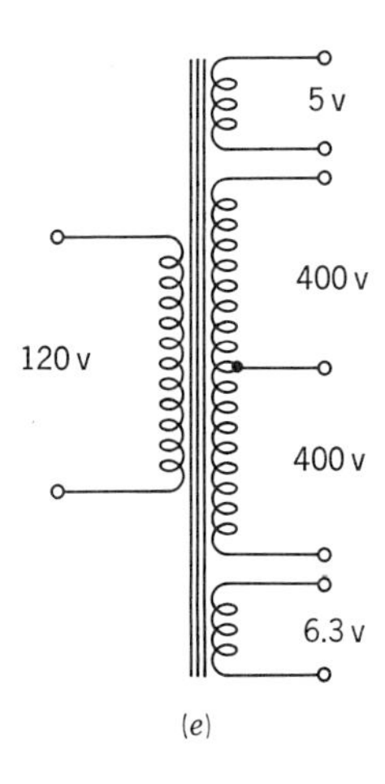


*Fig. 6 · 4 Types of power-transformer windings. (a) Tapped primary. (b) Center-tapped secondary. (c) Single high-voltage secondary. (d) Low-voltage secondary. (e) Multiple secondary windings.*

primary winding, enabling the transformer to operate on either 110- or 120-volt input. The same type of winding can be used in transformers for 120- or 240-volt input. Figure 6 · 4*b* shows a transformer having a center-tapped high-voltage secondary winding for a full-wave rectifier.

In systems delivering very high voltage, the transformer is frequently used only for this function (Fig. 6 · 4*c*). For the opposite extreme, windings on a step-down filament transformer are shown in *d*. In B supplies as used in radio receivers, however, one transformer delivers the high voltage to the rectifier as well as the heater voltages for the amplifier tubes (Fig. 6 · 4*e*). A separate winding is necessary for the rectifier heater when it has the B+ output voltage on it.

A 1:1 transformer may be used where the line voltage is sufficient to meet the circuit requirements but where the electronic equipment must not be grounded through the power line. This type of transformer is called an *isolation transformer*.

**Transformer windings and leads.** The leads from a transformer may be identified in several ways:

1. By checking the diagram on the transformer or accompanying it.
2. By standard color codes as shown in Appendix F.
3. By making resistance and voltage checks. This may be done by first checking with an ohmmeter to find those leads which show continuity. These windings usually have two leads or, in the case of a center-tapped winding, three leads. Then the resistance of the windings can be measured with an ohmmeter. Typical values are:

Primary winding—5 to 20 ohms
High-voltage secondary winding—200 to 400 ohms
Heater windings—approximately 1 ohm

Connect the primary winding to a source of line voltage and measure the voltages on the heater windings to discover the 5-volt winding for the rectifier and the 6.3-volt winding for the amplifiers.

One lead may not be color-coded or show continuity to any other lead. This is used for the Faraday shield, designed to reduce capacitive coupling between the primary and secondary windings. Simply tie this lead to the chassis or common ground.

## *6 · 3 Rectifiers*

In diode tubes, current can flow only from cathode to plate. If the applied voltage makes the plate negative with respect to the cathode, no current will flow. In this way the tube acts as an electronic switching device. During the half-cycle when the plate is positive, the current is permitted to flow through the load. During the next half-cycle, when the plate is negative, the tube creates the effect of an open circuit. Thus, the rectifier allows current to flow in only one direction in the output circuit.

Rectifier tubes may be of the high-vacuum or gaseous types (Fig. 6 · 5). High-vacuum tubes are used where the power supply is required to deliver

*Fig. 6 • 5 Rectifier tubes. (a) Dual-diode full-wave vacuum-tube rectifier. (b) Gas rectifier. Top cap is the plate.* (RCA.)

currents of less than 300 ma, approximately. The gaseous-type tube is generally used where the circuit must deliver higher amounts of current.

Current flowing through a rectifier will produce a voltage drop, thus reducing the voltage available to the load. Vacuum-tube rectifiers have a fairly high internal resistance, which varies for specific rectifier tubes, ranging from 1,500 to 6,000 ohms. The resulting internal voltage drop may be 15 to 60 volts. Furthermore, the voltage drop increases with the amount of load current, which results in poor voltage regulation.

In gas-filled tubes, ionization of the gas causes the space charge to be reduced, allowing a low value of internal resistance. In addition, the internal resistance will vary inversely with the current, thereby causing the internal voltage to remain constant and produce good voltage regulation. For example, the type 83 mercury-vapor rectifier has a constant voltage drop of 15 volts.

**Selenium rectifiers.** This type of rectifier consists of an aluminum plate which has one surface coated with selenium. The chemical properties of these materials permit current to flow easily from the aluminum to the selenium, but offer a high resistance in the reverse direction. Therefore more current flows in the forward direction. An important advantage of the selenium rectifier is that no heater is required.

The current rating of the selenium rectifier is determined by the surface area of the plate. The voltage rating is approximately 30 to 50 volts. In order to raise the voltage rating, many plates are added in series; to increase the current capacity, plates are added in parallel. A typical selenium rectifier is shown in Fig. 6 • 6*a*.

**Silicon rectifiers.** Out of the study of semiconductors has come the silicon diode, an extremely efficient and compact rectifier, providing a constant-output voltage. Figure 6 • 6*b* shows typical silicon rectifiers. As with all semiconductors, no heater is required. Silicon diodes are used more then selenium rectifiers because of the higher current ratings and small size. A typical low-cost silicon rectifier only ½ in. long is rated at

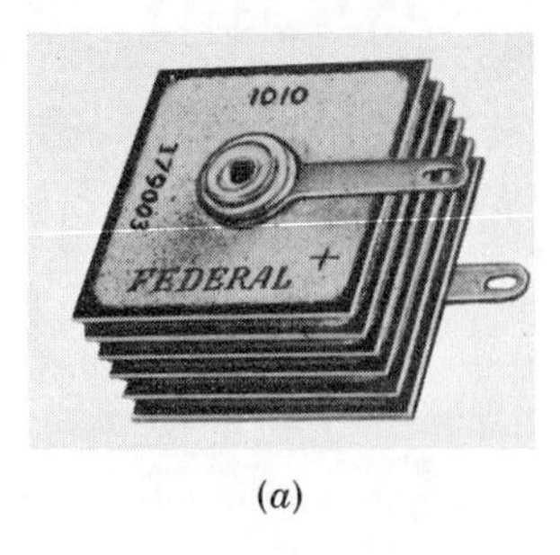


(a)

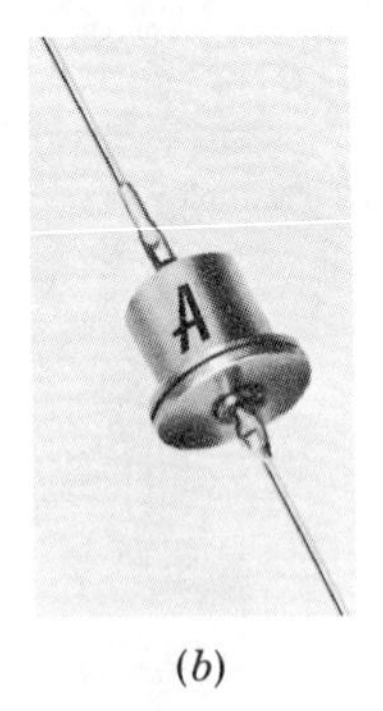


(b)

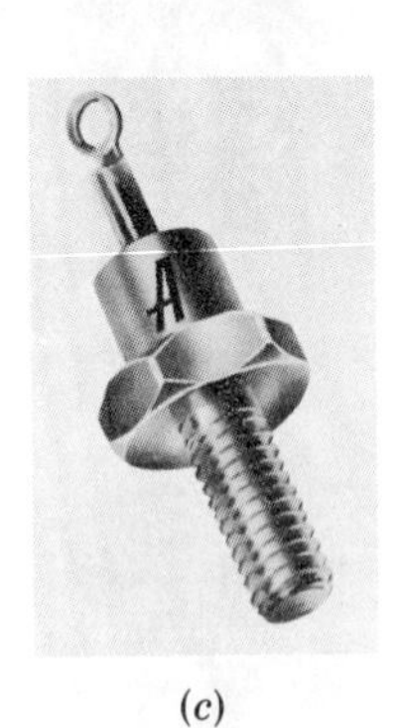


(c)

*Fig. 6·6 Diode rectifiers for power supplies. (a) Selenium metallic rectifier. (b) and (c) Silicon semiconductor diodes.*

750 ma d-c load current with a peak inverse voltage rating of 450 volts. This rating gives the maximum amount of voltage the rectifier can withstand in the nonconducting direction without internal arcing. The actual peak inverse voltage applied to a rectifier is about double the peak value of its a-c input.

The symbol for semiconductor diodes is shown in Fig. 6·7*b*, compared with a vacuum tube in *a*. The arrow symbol, which applies to selenium, silicon, and all solid-state devices, shows current can flow in only one direction. However, this indicates hole current, which is opposite from electron flow. Note that the bar in the symbol corresponds to the cathode in a tube, while the arrowhead indicates the anode or plate. Positive voltage applied to this terminal allows the diode to conduct.

## 6·4 *Half-wave rectifiers*

The function of the rectifier circuit is to convert the a-c input to d-c output. Either a vacuum tube can be used for the rectifier, as in Fig. 6·8*a*, or a semiconductor diode as in *b*. In both *a* and *b*, two cycles of alternating voltage are shown applied to the primary of the transformer. By the action of the transformer, this voltage is increased across the secondary and is applied between plate and cathode of the rectifier. The alternating voltage drives the plate positive for one-half of the cycle and negative for the other half-cycle. However, current will flow only when the plate is positive. Therefore, current can flow for only one-half of each cycle, producing a pulsating d-c voltage across the load resistor as shown.

Operation of the half-wave rectifier circuit in Fig. 6·8 can be summarized as follows:

1. The positive half-cycle of the a-c input voltage makes the diode plate positive.
2. Plate current can flow in only one direction, from cathode to plate inside the tube, through the secondary of the power transformer, and back at cathode through the load resistance $R_L$. This path is indicated by the arrows for $I$ in the figure.

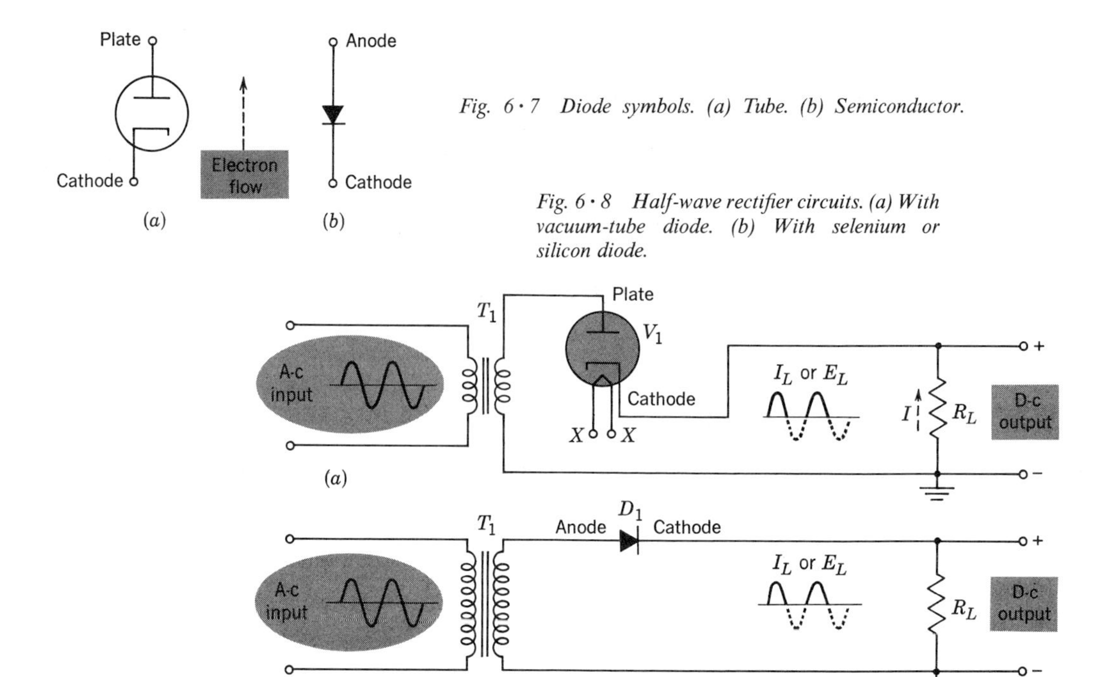


Fig. 6·7 *Diode symbols. (a) Tube. (b) Semiconductor.*

Fig. 6·8 *Half-wave rectifier circuits. (a) With vacuum-tube diode. (b) With selenium or silicon diode.*

3. The resultant $IR$ drop produces a d-c voltage across $R_L$ for the load.
4. Assuming 60-cycle a-c input, the fluctuations in the d-c output have the same frequency of 60 cps.

Notice that the a-c input voltage is applied to the plate circuit in Fig. 6·8*a*, while $R_L$ is connected in the cathode-to-ground circuit. The same idea applies to *b*, but the arrow in the symbol for a semiconductor rectifier shows the direction of current opposite from electron flow. As a result, the d-c output voltage across $R_L$ is positive at the cathode side, with respect to chassis ground. Finally, the heater terminals marked XX in *a* connect to a heater winding (not shown) on the power transformer. The semiconductor diode in *b* does not need any heater voltage.

**Inverted power supply.** This means a rectifier circuit to provide negative d-c output voltage, instead of positive voltage. If we take the circuits in Fig. 6·8 as examples, they can be changed to negative d-c output by reversing the diode connections. In other words, connect the a-c input to the cathode and connect the ungrounded side of $R_L$ to the plate or anode. Then the d-c output voltage across $R_L$ will be negative with respect to chassis ground.

## 6·5 *Full-wave rectifiers*

The full-wave rectifier in Fig. 6·9 consists of two half-wave rectifiers which operate during opposite halves of the input cycle. This circuit can utilize two individual diodes, or the two rectifying elements may be contained in one unit. Note that the high-voltage secondary winding is center-tapped. This provides two equal voltages for operation of the two half-wave diodes. Since the voltage at one end of the secondary winding is opposite in polarity to the voltage at the other end, one diode will always be conducting. In this way both halves of the input cycle are utilized. Because of this, the full-wave rectifier can supply twice the load current of a half-wave rectifier.

Operation of the full-wave rectifier circuit in Fig. 6·9 can be summarized as follows:

1. The center tap on the secondary of the power transformer returns through the load $R_L$ to the common cathode for both diodes.
2. When the diode plate $V_1$ is driven positive by a-c input voltage, the current flows through the diode and through the top half of the secondary to the center tap and returns to the cathode through $R_L$.
3. On the next half-cycle of a-c input voltage, $V_2$ is driven positive. Current now flows through this diode and through the bottom half of the secondary to the center tap and returns to its cathode. Note that this current $I_2$ is in the same direction as $I_1$ through $R_L$.
4. Thus both diodes produce d-c output voltage across $R_L$ in the same polarity, with the cathode side positive.
5. Assuming 60-cycle a-c input voltage, the fluctuations in the d-c output have the frequency of 120 cps because both half-cycles of the input produce an output voltage.

The full-wave rectifier circuit is generally used where appreciable direct

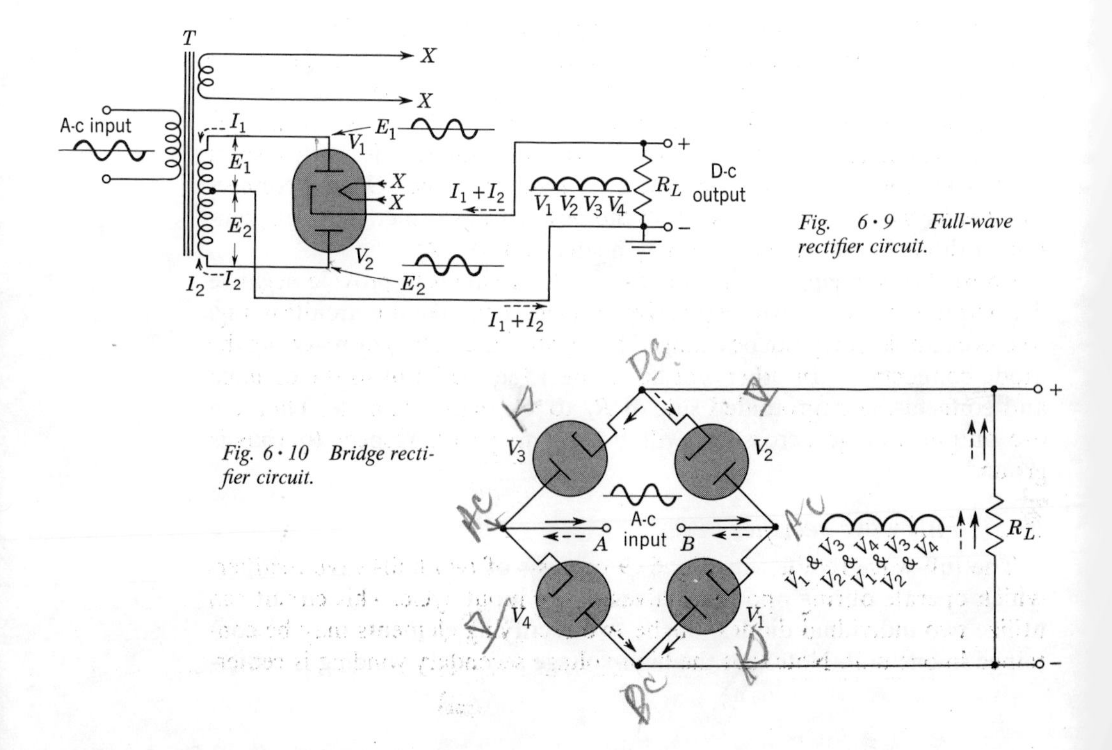


Fig. 6·9 Full-wave rectifier circuit.

Fig. 6·10 Bridge rectifier circuit.

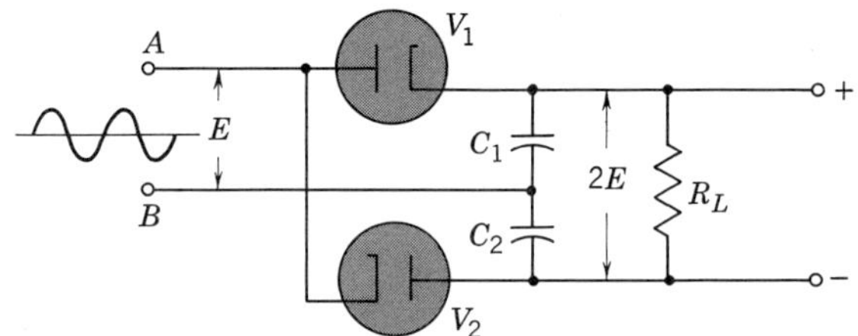


*Fig. 6 · 11 Full-wave voltage-doubler circuit.*

load current is required. Also, better filtering is possible because of the higher-ripple frequency. A center-tapped high-voltage transformer is needed, however, to provide equal and opposite a-c input voltages.

## 6 · 6 Bridge rectifier circuits

To utilize the advantages of the full-wave rectifier, but eliminate the need for a center-tapped high-voltage transformer, a bridge rectifier circuit can be used (Fig. 6 · 10). The circuit operates as follows: When point $A$ of the a-c input is positive, it causes the plate of $V_3$ to become positive. At the same time, point $B$, being negative, applies a negative voltage to the cathode of $V_1$. Current flows through $V_1$ to the load, through $V_3$, and back to the source, as shown by the solid arrows. Current cannot flow through $V_2$ because its plate is negative; also, the cathode of $V_4$ is positive so that it will not conduct either.

During the next half-cycle, the polarity of the a-c input voltage reverses, and point $A$ becomes negative. This causes a negative voltage to be applied to the cathode of $V_4$. At the same time, point $B$ is positive, making the plate of $V_2$ positive. Current will now flow as shown by the broken-line arrows through $V_4$ to the load, and then through $V_2$ back to the a-c input power line. Since current flows through the load in the same direction during both halves of the input cycle, full-wave rectification is achieved.

The bridge rectifier uses the full secondary voltage just as the half-wave circuit did. Both half-cycles, however, produce d-c output, as in a full-wave rectifier. The bridge circuit requires four rectifiers, though.

## 6 · 7 Voltage-doubler circuits

Voltage doublers are rectifier circuits arranged to produce an output voltage which is twice the voltage applied. Figure 6 · 11 shows the circuit of a full-wave voltage doubler. Its operation is as follows: During the half-cycle when point $A$ of the a-c input voltage is positive, current will flow through diode $V_1$, charging capacitor $C_1$. During the next half-cycle, point $B$ is positive, and the current through $V_2$ will charge $C_2$. Since the output voltage is taken across the two capacitors in series, it will equal twice the input voltage.

For good regulation, the values of $C_1$ and $C_2$ should be very large. The larger these capacitors are, however, the higher the charging current. The

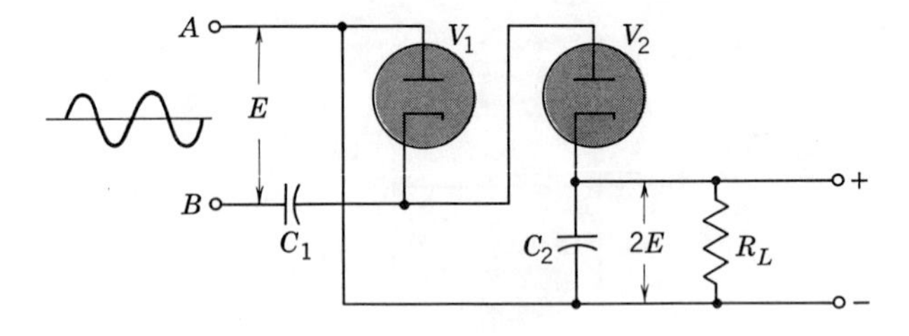


*Fig. 6 · 12 Half-wave or cascade voltage-doubler circuit.*

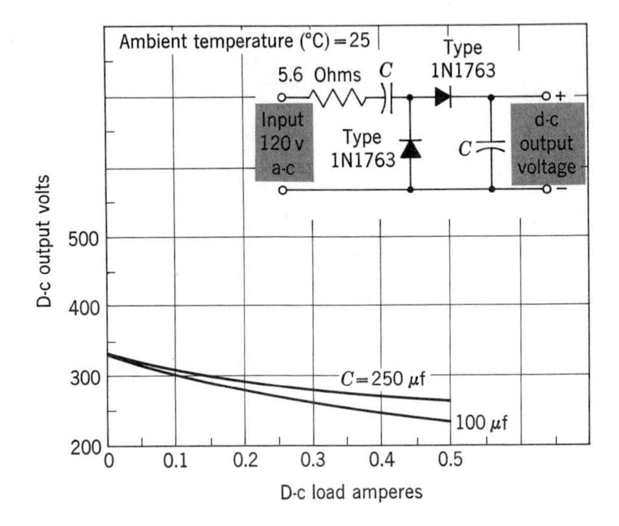


*Fig. 6 · 13 Half-wave voltage-doubler circuit with silicon diodes.* (RCA).

current in the circuit must never exceed the peak plate current of the rectifier tubes. Typical values for these capacitors are 40 to 100 μf.

Figure 6 · 12 shows a cascade, or half-wave, voltage-doubler circuit. The operation of this circuit is as follows: During the half-cycle when point $A$ is positive, diode $V_1$ conducts, charging capacitor $C_1$. During the next half-cycle, point $B$ is positive, and diode $V_2$ conducts, charging $C_2$. When $V_2$ conducts, the charge on $C_1$ is equal to the peak value of the applied voltage. Therefore the input voltage for $V_2$ is the sum of the applied voltage and the voltage across $C_1$. This causes the output voltage across $C_2$ to equal twice the peak value of the applied voltage.

Voltage-doubler circuits are used to supply high voltage with a small value of load current. The cascade circuit is more popular because the negative side of the d-c output circuit is common to one side of the a-c input. Then both these points can be returned to the common-chassis ground. A typical application of the voltage doubler is in a transformerless B supply, where about 240 volts d-c output can be obtained from 120 volts a-c input. A typical circuit using silicon diodes is shown in Fig. 6 · 13. Note the current-limiting resistor of 5.6 ohms to protect the semiconductor diodes against damage from peak surge currents.

## 6·8 Filters

The voltage obtained from the rectifier is undirectional, but it varies in amplitude. The fluctuations above and below the average d-c value represent a ripple voltage. Thus the pulsating direct current may be considered as a constant-amplitude direct current with an a-c ripple superimposed on it. This ripple causes hum. It is the function of the power-supply filter to remove this a-c component or reduce it to a minimum.

Filtering is accomplished by connecting shunt capacitors and series inductors or resistors in the output circuit of the rectifier. The choke coil, which is connected in series with the rectifier, has the function of smoothing the current. The capacitors are used to smooth the voltage.

The filter capacitor charges through the low rectifier internal resistance. This enables the capacitor to reach full charge very rapidly. The discharge path through the relatively high resistance of the load has a long *RC* time-constant circuit. This causes the capacitor to discharge slowly. The slow discharge prevents the capacitor voltage from decreasing to any great extent, and the voltage is kept constant.

The effectiveness of a filter is measured by the percentage of ripple in its output:

$$\% E_{\text{ripple}} = \frac{E_{\text{ripple}}}{E_{dc}} \times 100 \qquad (6 \cdot 1)$$

*Example.* The output voltage of a power-supply unit is 300 volts, and the rms value of the ripple voltage is 0.6 volt. What is the per cent of ripple voltage?

$$\% E_{\text{ripple}} = \frac{E_{\text{ripple}}}{E_{dc}} \times 100 = \frac{0.6}{300} \times 100 = 0.2 \text{ per cent}$$

The application of the power supply determines the maximum allowable ripple, which is generally 1 per cent or less.

*Fig. 6·14 Power-supply filters with capacitor input. (a) L type. (b) π type. (c) With filter resistor instead of choke. (d) Two π sections.*

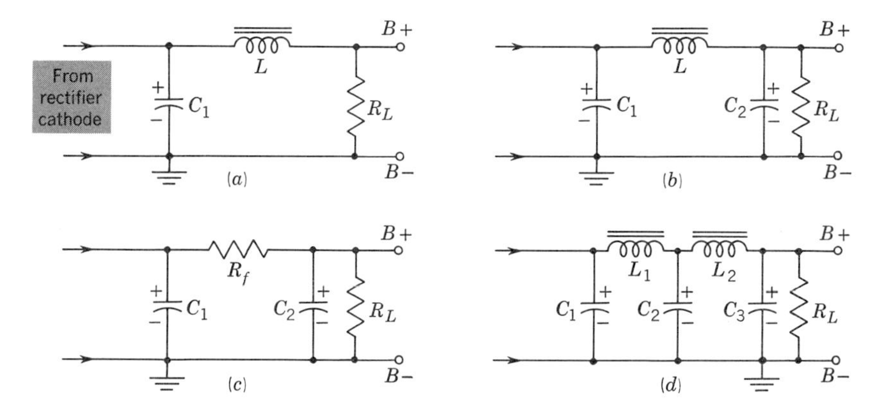

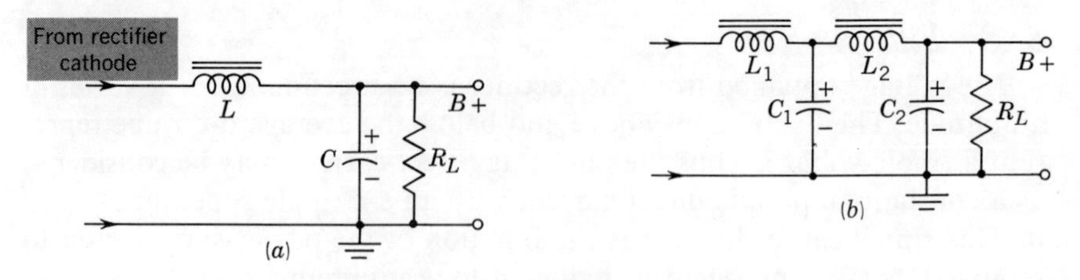


*Fig. 6 · 15 Power-supply filters with choke input. (a) L type. (b) Two L sections.*

Power-supply filters are classified as either capacitor-input (Fig. 6 · 14) or choke-input (Fig. 6 · 15). Notice that $C_1$ in the capacitor-input filter can charge to the peak value of the a-c input voltage because this path has no series choke. As a result, more d-c output voltage can be obtained than with a choke-input filter.

Referring to Fig. 6 · 14, an L-type capacitor-input filter is shown in *a*. The $\pi$-type filter in *b* is most common, though, with both an input filter capacitor $C_1$ at the cathode of the rectifier and output filter capacitor $C_2$ across the load. In many circuits, a filter resistor is used instead of the choke, as in *c*. The filter resistor is cheaper and saves space. For extremely good filtering with little ripple, two $\pi$ sections can be used, as in *d*. In all cases, the filter capacitors are electrolytics, connected in the polarity indicated, with the capacitance values needed for good filtering and a voltage rating equal to or greater than the d-c output voltage to prevent breakdown in the electrolytic.

Choke-input filters are shown in Fig. 6 · 15. Here the rectifier current must flow through an inductance in its return path to cathode. Therefore, the filter capacitor cannot charge to the peak value of the a-c input voltage, and the d-c output voltage is lower than with a capacitor-input filter. However, the choke-input filter has better voltage regulation with varying load current. This is defined as follows:

$$\%\text{ voltage regulation} = \frac{E_{\text{no load}} - E_{\text{full load}}}{E_{\text{full load}}} \times 100 \qquad (6 \cdot 2)$$

*Example.* The d-c output voltage drops from 300 volts with no load to 240 volts at full load.

$$\%\text{ voltage regulation} = \frac{300 - 240}{240} \times 100 = \frac{60}{240} = 25 \text{ per cent}$$

If the full-load voltage equals the no-load voltage, this fraction will equal zero. Therefore the lower the per cent figure, the better is the voltage regulation.

Typical values for filter capacitors are 4 to 100 $\mu$f. Figure 6 · 16 shows a typical dual-section filter for the input and output capacitors. Usually both sections have one common negative; here this is the metal can. Relatively

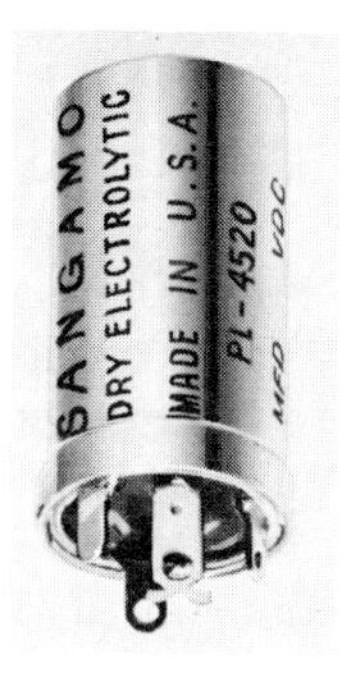


*Fig. 6 · 16 Dual-section filter capacitor can. Ratings are 40–40 μf and 150 volts. Height is 2 in.* (Sangamo Elec. Co.)

*Fig. 6 · 17 Power-supply filter choke. Channel frame mounting; width of unit 3 in.; inductance 8 henrys; resistance 250 ohms.* (United Transformer Corporation.)

large filter capacitances are needed for half-wave rectifiers and voltage-doubler circuits. Also, larger values of direct load current, corresponding to a lower load resistance, require more capacitance for adequate filtering. The voltage rating of the filter capacitor is often 150 to 450 volts. For the filter choke, typical values are 1 to 10 henrys (Fig. 6 · 17).

## *6 · 9 Voltage dividers*

The voltage divider is a resistor, or series of resistors, connected across the output of the power supply and tapped at a number of points along its length to provide a selection of output voltages (see Fig. 6 · 18). Its function is to deliver power to a load which requires various voltages and currents. Generally, the load consists of amplifier tubes connected to the power supply for plate voltage, screen-grid voltage, and control-grid voltage where fixed bias is used.

The total resistance of the voltage divider connected across the power-

*Fig. 6 · 18 Voltage divider. Its total resistance is a bleeder across the power supply. The taps provide different values of B+ and C− output voltages. (a) Circuit. (b) Photograph of 20-K wirewound resistor, 4½ in. long, 50-watt rating.* (International Resistance Co.)

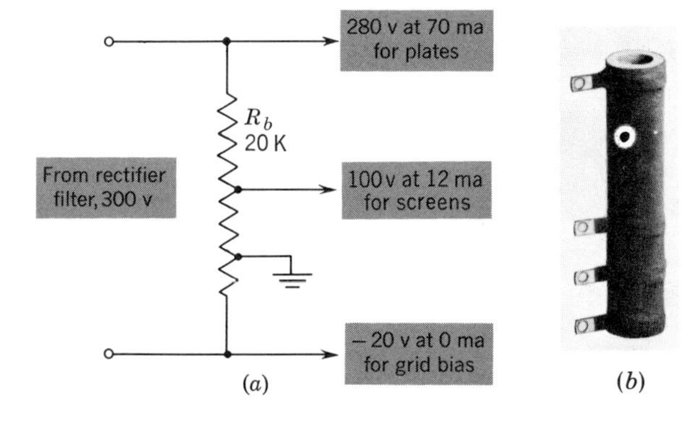

supply output acts as a *bleeder resistor*. The bleeder appears to the power supply as a fixed load. No matter what other equipment is connected to the power supply, the bleeder draws a constant current from the filter so long as the rectifier is operating. The current drawn by the bleeder is generally 10 to 20 per cent of the rectifier output. In this way, the bleeder tends to stabilize the output voltage, providing good voltage regulation. The bleeder resistor is also a path through which the filter capacitors can discharge when the power supply is turned off, eliminating the hazard of shock.

In the voltage divider of Fig. 6 · 18, note that the voltage from the power supply equals the total of the maximum positive and negative voltages on the divider. Here the total is 300 volts to provide +280 and −20 volts. All similar load voltages are supplied in parallel from one tap. The tubes requiring 280 volts for plate voltage connect to this tap on the divider. The load current of 70 ma at this tap is the sum of the individual values of plate current. Similarly, all tubes requiring 100 volts for the screen grid connect to this tap. The load current here of 12 ma is the sum of the individual screen currents. The tap at −20 volts is for grid bias. Its load current is 0 ma because we assume no grid current. For the case of a total bleeder resistance of 20,000 ohms across the 300-volt supply, the fixed bleeder current equals 15 ma. The total load current through the rectifier, then, equals 70 + 12 + 15 for a sum of 97 ma.

Since bleeder resistors generally dissipate appreciable power, good ventilation is important. Typical ratings are 20 to 100 watts. When the bleeder resistor is used as a voltage divider, different amounts of load current flow in each section, and the power rating must be high enough for the section that dissipates the most power. If one section opens, it may not be necessary to replace the whole divider; the open section can be bridged with an individual resistor having the required resistance and power rating.

## 6 · 10 *Transformerless power supply*

The voltage from an a-c power source may be supplied directly to the rectifier, as shown in Fig. 6 · 19. This eliminates the bulk and cost of a power

*Fig. 6 · 19 Transformerless power supply with half-wave rectifier for B+ and series heaters across the power line.*

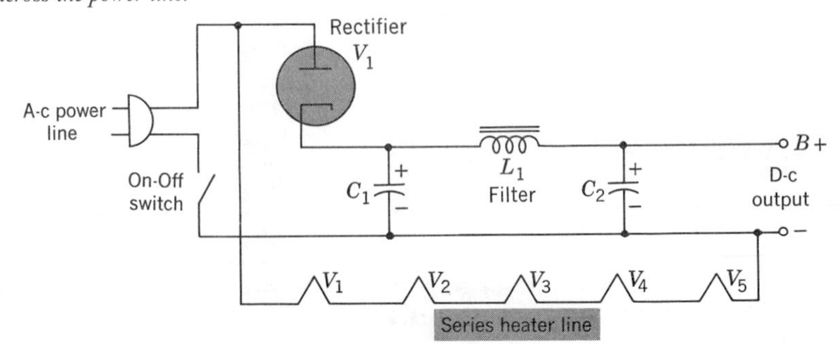

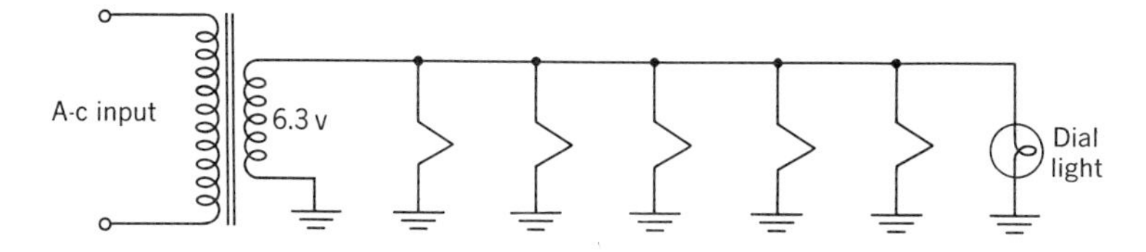


*Fig. 6·20 Parallel heaters across 6.3-volt transformer winding.*

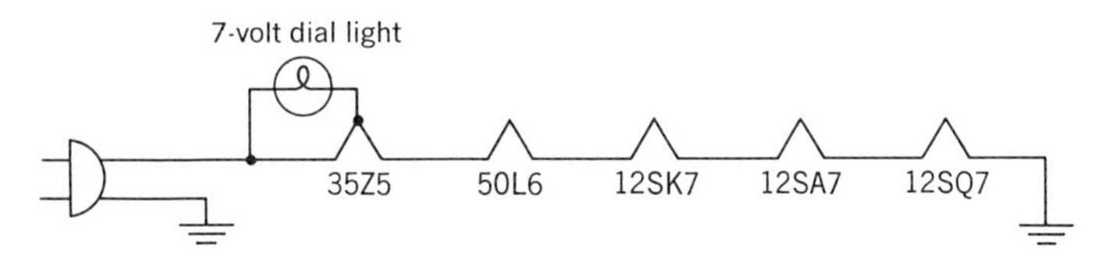


*Fig. 6·21 Series heaters across 120-volt power line.*

transformer. The rectified output voltage, however, will be approximately equal to the a-c voltage input to the rectifier. For higher output, a voltage-doubler circuit can be used.

Since a transformer is not used, the power supply can be operated directly from either an a-c or a d-c power line. With a-c input, the rectifier will change the alternating current to pulsating direct current and send it on to the filter. When direct current is applied, the rectifier simply acts as a resistance in series with the filter. However, the power plug must be inserted properly to have the positive line connected to the plate of the rectifier tube. For either a-c or d-c input, the heaters are connected in series across the power line.

This type of power supply is commonly used where small or economical power supplies are required. To reduce the size further, the filter choke may be replaced by a resistor.

## *6·11 Heater circuits*

**Parallel connections.** In a-c power supplies, the heaters of all tubes are supplied by the power transformer. Low-voltage secondary windings of the transformer are used to deliver the proper voltage and current for the heaters. Tubes are chosen with heaters requiring the same amount of voltage, often 6.3 volts. Then all the 6.3-volt heaters are connected in parallel across the low-voltage winding (see Fig. 6·20). The total current in the heater winding equals the sum of all the individual heater currents.

**Series connections.** This arrangement is used in a-c/d-c power supplies because there is no power transformer. For the series connections, tubes are selected with heaters rated for the same operating current. The total voltage across the heater line equals the sum of all the heater voltages (see Fig. 6·21). It is important to note that with a series heater string, when one heater opens, no current can flow in the entire string.

The series heater circuit must be arranged so that capacitive coupling

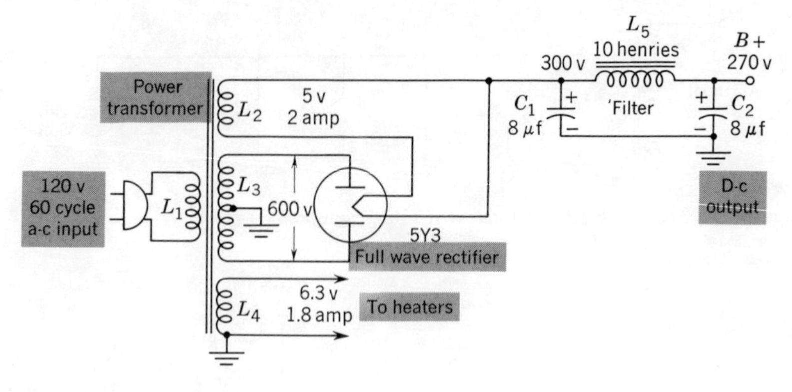


*Fig. 6·22 Circuit of typical a-c power supply.*

between the heater and cathode will not induce 60-cycle voltage into the signal circuits. Otherwise, the receiver will have 60-cycle hum in the audio output. The tubes most sensitive are those with small input signals. To prevent hum, the first audio amplifier (12SQ7) is placed closest to ground, followed by the converter tube and the other tubes, as shown in Fig. 6·21.

## 6·12 *Typical a-c power supply*

All the components in a complete B+ supply using a power transformer are illustrated in Fig. 6·22. This circuit operates only with a-c input because of the power transformer. Note the center-tapped high-voltage secondary $L_3$ for the full-wave rectifier, the secondary $L_2$ to supply filament voltage for the rectifier, and the heater winding $L_4$ for the remaining tubes. With this arrangement, the heaters are connected in parallel across $L_4$. A $\pi$-type capacitor-input filter is used for the B+ output voltage.

The unfiltered high-voltage output at the input capacitor $C_1$ is 300 volts, but the filtered output for B+ equals 270 volts; the remaining 30 volts is the *IR* drop across the filter choke. This is a d-c voltage drop across the d-c resistance of the filter choke. In many cases, an audio power-output stage may use the unfiltered B+ voltage because of its higher value. Furthermore, hum due to ripple in the B+ is less of a problem in the last audio amplifier, where the audio signal level is high. In any case, the load current for the rectifier equals the total of all the individual plate and screen-grid currents for the amplifiers connected to the B+ output of the power supply.

A separate filament winding is used for the rectifier because the high-voltage B+ is present here. Therefore, this winding cannot be connected to chassis ground; otherwise, the B+ voltage would be shorted. The 6.3-volt winding for the amplifier tubes is grounded at one end to reduce the amount of wiring for the parallel connections. However, it may be center-tapped to chassis ground for minimum hum.

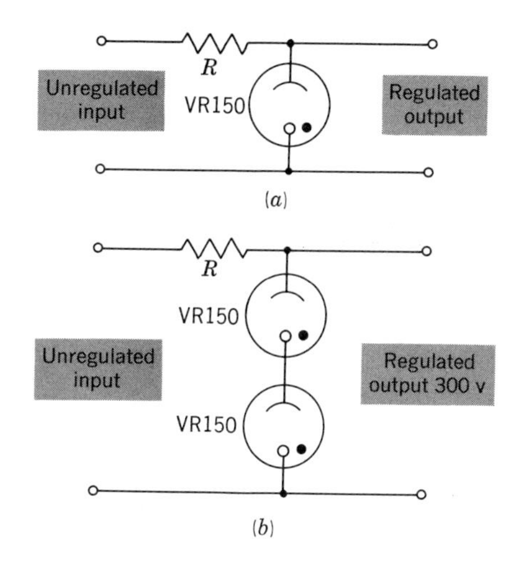


*Fig. 6·23 Glow-tube voltage-regulator circuits. (a) Single tube for 150 volts. (b) Two tubes in series for 300 volts.*

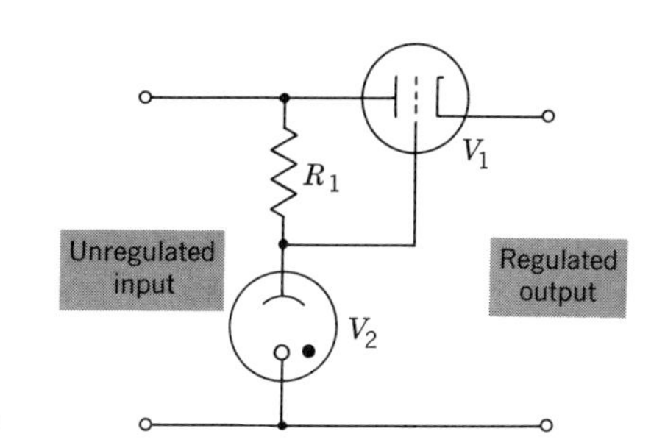


*Fig. 6·24 Voltage-regulator circuit.*

## 6·13 *Voltage regulators*

A voltage regulator is a device connected in the output of a power supply to keep the output voltage constant regardless of variations in load current. The basic action of voltage regulation is that of an automatic resistance which changes with variations in current, in order to maintain a constant voltage drop.

**Glow-tube regulator.** One of the most commonly used voltage regulators is the glow tube. This is a two-element cold-cathode tube, filled with a combination of helium and argon gases. The VR 75, the VR 105, and the VR 150 are examples of such tubes (VR indicates a voltage-regulator tube). Each is designed to provide regulation at the specified value of voltage. Figure 6·23 shows two circuit arrangements with the VR 150. When the rated voltage is applied across the tube, ionization takes place. The degree of ionization is controlled by the amount of current passing through the tube. With more current, more ionization results, reducing the internal resistance. Therefore, since current and resistance vary inversely, the voltage drop across the tube remains fairly constant over the operating range.

Two important requirements for operation of glow tubes are that the starting voltage must be slightly higher than the operating voltage to make the tube ionize, and a series resistance of several hundred ohms is necessary to limit the current through the tube. It should be noted that semiconductors, generally called *zener diodes* or *voltage-reference diodes,* can be used instead of the glow tubes.

**Voltage-regulator circuit.** Figure 6·24 shows an improved voltage regulator using a glow tube $V_2$ and vacuum-tube triode $V_1$. Resistor $R_1$ and tube $V_2$ form a voltage divider to keep a constant voltage on the grid of $V_1$. This bias on $V_1$ is the difference between the voltage across the glow tube $V_2$ and the voltage across the load in the output. Furthermore, the bias controls the amount of plate current in $V_1$ and therefore its effective resistance. The internal plate resistance of $V_1$ is in series with the output. An increase in the load current would tend to decrease the output voltage. However, the increased current through $V_1$ lowers its series resistance. Thus it has a lower internal voltage drop, allowing the regulated output voltage to stay the same. When the load current decreases, $V_1$ conducts less. The internal resistance rises, increasing the voltage drop across the tube, and the output voltage remains constant.

**Voltage-regulating transformer.** When current through a coil rises, the magnetic flux also increases until the core becomes saturated. Then the flux maintains a constant strength. The saturation effect reduces the reactance of the coil. A voltage-regulation method utilizing the effect of saturation is shown in Fig. 6 · 25. Windings $L_1$ and $L_2$ provide high voltage for the power supply; $L_3$ with $L_4$ and $L_5$ with $L_6$ are the voltage-regulating windings.

Primary windings $L_3$ and $L_5$ are in series opposition, so that the energy they couple into secondary windings $L_4$ and $L_6$ cancel. This prevents the voltage-regulating windings from introducing any a-c voltage into the power-supply output. Secondary winding $L_2$ is used with a full-wave rectifier. The rectifier current flows through the secondary windings $L_4$ and $L_6$. Since $L_4$ and $L_6$ are in series with the load, an increase in load current increases the current in these windings. This increased current saturates the core of the transformer, lowering the inductive reactance of the primary windings. Since the primary windings $L_3$ and $L_5$ are in series, lowering their reactance increases the voltage across $L_1$. This induces more voltage across $L_2$ for the power supply. As a result, the increased a-c input voltage compensates for the increased load current, allowing constant B+ output voltage from the power supply.

## 6 · 14 Vibrators

A vibrator converts direct current from a low-voltage source, such as a storage battery, into alternating current which can be stepped up and rectified for high-voltage d-c output. The vibrating mechanism is essentially

Fig. 6 · 25 Circuit of voltage-regulating transformer.

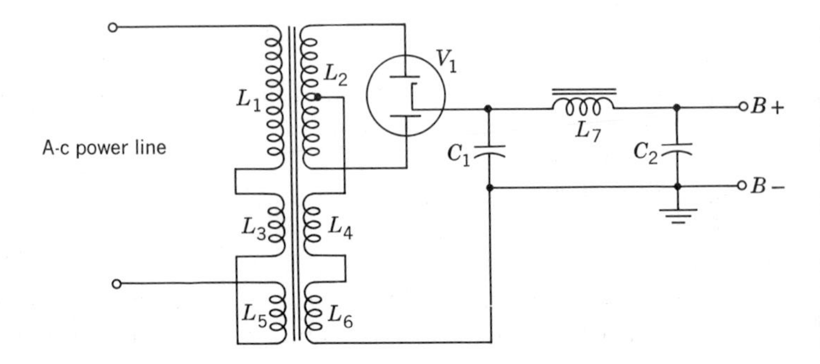

(a)

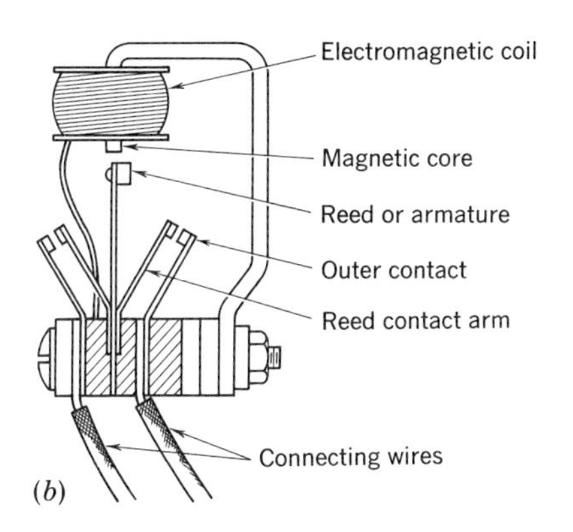


(b)

*Fig. 6·26 Typical vibrator. (a) Photograph. (b) Internal construction.*

a high-speed reversing switch. It automatically opens and closes sets of contacts, as an electromagnet, when d-c power is applied. The vibrator type of power supply was commonly used to supply B+ voltage in automobile radios. However, transistorized receivers are generally used now, and these can operate directly from the 12-volt car battery.

The vibrator consists of these parts: a heavy frame, an EM coil and core, a flexible reed or armature, and one or more contacts mounted on each side of the armature (see Fig. 6·26). In construction, the coil is mounted at one end of the frame, and the armature is rigidly fastened to the opposite end. The armature contacts and the outer contacts are mounted on arms at either side of the armature. The whole vibrator assembly is sealed in a can lined with sponge rubber, which creates a shielded shock mount. Leads from the contacts are brought out as metal prongs on the base of the can (see Fig. 6·26). The vibrator unit is plugged into a socket like a tube, affording the convenience of quick replacement of defective units.

In a power supply, the vibrator is connected between the primary winding of the power transformer and the battery supplying the power. When the vibrator is actuated, battery current through the primary winding is interrupted at the vibrator frequency, producing sharp pulses of current through the transformer primary. These current pulses flow in alternating directions through the transformer primary because of the action of the vibrator contacts, which interrupt and reverse the current flow. The a-c component of this pulsating direct current induces a voltage in the transformer secondary. Because of the high step-up ratio, the induced voltage is many times greater than the primary voltage. This high secondary voltage is then rectified and filtered to reduce ripple in the rectified d-c output. The frequency of the vibrator pulses is generally about 250 cps.

The basic types of vibrators include two variations:

1. The *synchronous type,* which converts direct to alternating current and

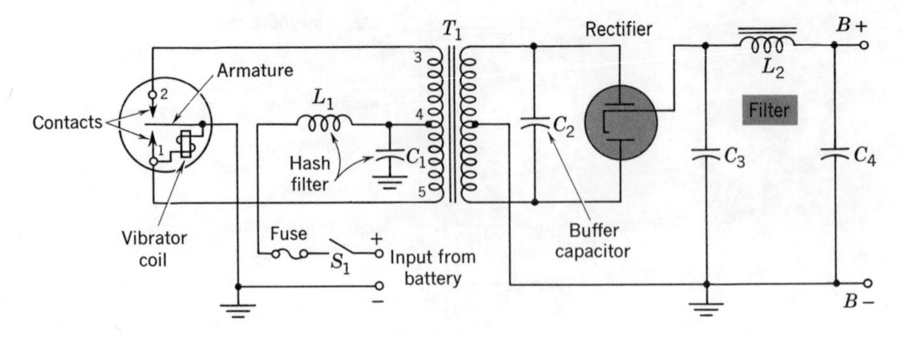


*Fig. 6 · 27 Power-supply circuit using nonsynchronous vibrator.*

also changes the high-voltage a-c output of the transformer back to direct current, eliminating the need for a separate rectifier.

2. The *nonsynchronous* type, which requires a rectifier.

**Nonsynchronous vibrator.** The circuit of a nonsynchronous-vibrator power supply is shown in Fig. 6 · 27. The buffer capacitor $C_2$ across the secondary winding has a critical value of capacitance. Its function is to absorb the surges of current that occur when the primary circuit is broken. The magnetic field collapses very rapidly, inducing high voltages in the secondary winding. If $C_2$ is not used, excessive sparking develops at the vibrator contacts, shortening the life of the vibrator. Typical values of buffer capacitors range from 0.0005 to 0.03 $\mu$f, with a breakdown rating of 1,000 to 2,000 volts. When a defective vibrator is replaced, a new buffer capacitor should also be used.

*Fig. 6 · 28 Motor and generator windings of a dynamotor.*

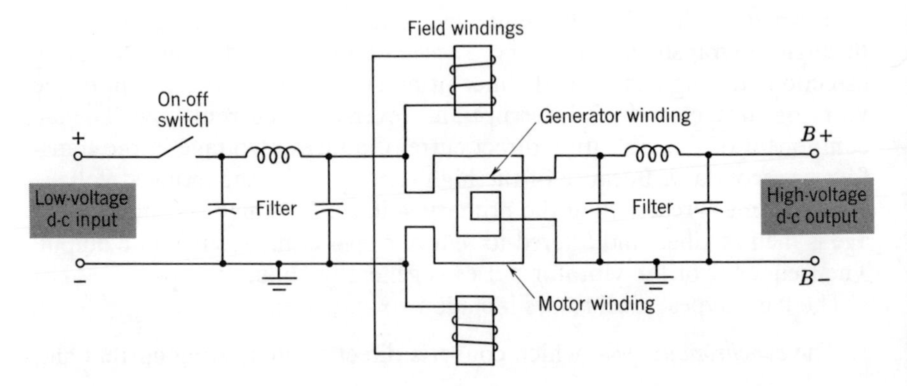

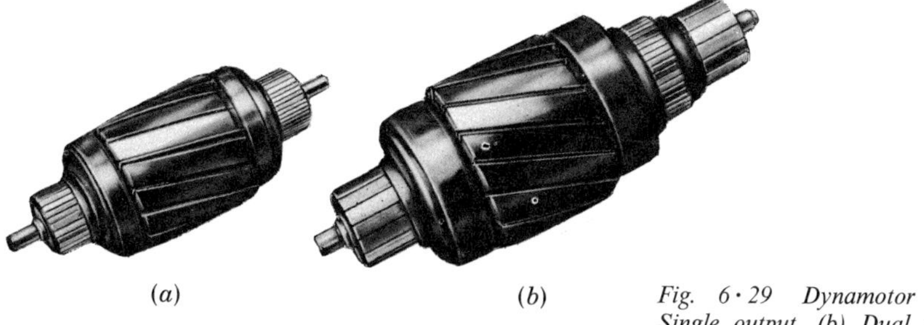


*Fig. 6 • 29 Dynamotor armatures. (a) Single output. (b) Dual output.* (Bendix Aviation Corp.)

## *6 • 15 Dynamotors*

A dynamotor is a motor-generator unit designed to convert low-voltage direct current to high-voltage direct current. The motor and generator use the same field windings and share the same armature, but they use separate commutators. See Figs. 6 • 28 and 6 • 29.

Figure 6 • 28 shows the wiring diagram of a typical dynamotor. The d-c input is applied to the field and motor windings in parallel. This input is applied to the armature by the commutator at one end. The output is taken from the generator windings through another commutator at the opposite end of the armature. Because the motor and generator use the same field windings, the ratio of input voltage to output voltage depends upon the ratio of turns in the motor winding to the turns in the generator winding. Changing the number of turns in the motor winding will change the speed of rotation as well as the output voltage. Changing the number of turns in the generator winding will change only the output voltage. Dynamotors generally operate on a battery input voltage of 6 to 32 volts. They deliver as much as 1,000 volts or more d-c output.

For efficient operation of a dynamotor, the input leads must be short and heavy. Also, the unit should be spring-mounted to eliminate mechanical vibrations. To minimize r-f interference, the hash filter capacitors are grounded at a common point on the frame. These are usually mica capacitors possessing the required capacitance and voltage rating.

## *6 • 16 Power-supply troubles*

The power supply is a common source of trouble because the components operate with high voltage, high current, or both. Some common troubles are:

**Defective rectifier.** No B+ voltage output, although a-c input voltage is normal. No audio output in radio receiver.

**Weak rectifier.** Low B+ voltage. Weak and distorted audio output.

**Open input filter capacitor.** Low B+ voltage and hum in sound.

**Shorted input filter capacitor.** Excessive current in the rectifier which can burn it out or burn open the power transformer. In a vacuum-tube rectifier, the plates will glow red hot. If there is a fuse in the primary, it will blow.

**Open output capacitor.** Hum and possibly motorboating in the sound, with B+ voltage normal or only slightly lower.

**Shorted output capacitor.** Excessive current in rectifier and filter choke.

**Open filter choke.** No B+ output.

**Open heater in series string.** Set is completely dead; no heater current in any tube.

## SUMMARY

1. In an a-c power supply, the main components are the power transformer, the rectifier, and the filter. The transformer steps up the a-c voltage to the rectifier and provides low-voltage for filaments; the rectifier converts the high voltage a-c to fluctuating d-c voltage; the filter removes the fluctuations to provide steady d-c voltage output.
2. Types of rectifiers include vacuum tubes, gas tubes, selenium disk rectifiers, and silicon semiconductor rectifiers. Tubes require heater voltage, while the others do not; semiconductor rectifiers have a lower internal voltage drop.
3. A half-wave rectifier circuit utilizes just one-half of the a-c input cycle. The ripple frequency in the d-c output is 60 cps. An inverted rectifier can provide negative d-c voltage.
4. A full-wave rectifier circuit utilizes both halves of the a-c input cycle. Its ripple frequency is 120 cps. The full-wave circuit can supply more load current than the half-wave rectifier.
5. A bridge rectifier uses four diodes to provide full-wave rectification of the a-c input voltage without the need for a center-tapped transformer.
6. A voltage doubler uses two diodes to provide d-c output voltage equal to twice the a-c input voltage.
7. A transformerless power supply can operate with either d-c or a-c input. With d-c input, it must have the correct polarity to make the rectifier plate positive; with a-c input, the half-wave rectifier provides d-c output. In either case, all the heaters are in a series string across the power-line input voltage.
8. A capacitor-input filter has a filter capacitor connected directly at the rectifier cathode. A choke-input filter has a choke in series with all the rectifier current. The capacitor-input filter provides higher d-c output voltage with less ripple; the choke-input filter has better voltage regulation.
9. Voltage regulation means that the d-c output of the supply remains constant with varying amounts of load current. Formula (6 · 2) defines the voltage-regulation factor in per cent. The lower the percentage, the better is the voltage regulation.
10. A vibrator converts low-voltage direct current to fluctuating current that can be stepped up by a power transformer to provide high-voltage alternating current in the secondary. A nonsynchronous vibrator needs a rectifier to provide high-voltage d-c output; the synchronous vibrator has an extra set of contacts functioning as a mechanical rectifier to supply d-c output.
11. A dynamotor is a motor-generator operated from a low-voltage d-c source to produce high-voltage d-c output. For mobile transmitters and receivers, the dynamotor can provide the required B+ voltage from a low-voltage battery.
12. Troubles in the power supply can cause no B+ voltage output with normal a-c input, low B+ voltage, and hum due to insufficient filtering. With series heaters, an open in one means that none of the tubes can light.

## SELF-EXAMINATION

Here's a chance to find out how well you have learned the material in this chapter. Work the exercises; then check your answers against the Key at the back of this book. These exercises are for your self-testing only.

1. A typical value of filter capacitor for 60-cycle ripple is (*a*) 20 pf; (*b*) 1,000 pf; (*c*) 30 μf; (*d*) 10 farads.
2. In a series heater string, if one heater is open, the applied voltage is across the (*a*) tube with the lowest heater voltage rating; (*b*) tube with the highest heater voltage rating; (*c*) tube with the highest current rating; (*d*) open heater.
3. Which of the following advantages does a full-wave rectifier have over a half-wave rectifier? (*a*) The output voltage is lower with more ripple. (*b*) The ripple frequency is lower. (*c*) Each diode can cool off during half of each input cycle. (*d*) The tube will conduct during both halves of the input cycle.
4. What is the frequency of the ripple voltage at the output of a full-wave rectifier operating from a 60-cycle supply? (*a*) 30 cycles; (*b*) 60 cycles; (*c*) 120 cycles; (*d*) 240 cycles.
5. A typical inductance for a filter choke in a 60-cps supply is (*a*) 100 μh; (*b*) 2 mh; (*c*) 30 mh; (*d*) 8 henrys.
6. In comparison with an *RC* power-supply filter, a filter with a choke provides (*a*) more filtering action, and larger d-c voltage drop; (*b*) less filtering action, and larger d-c voltage drop; (*c*) less filtering action, and smaller d-c voltage drop; (*d*) more filtering action, and smaller d-c voltage drop.
7. With 300 volts rms applied to each plate in a full-wave rectifier, the peak positive applied voltage for each diode equals (*a*) 300 volts; (*b*) 420 volts; (*c*) 600 volts; (*d*) 1,000 volts.
8. In a full-wave rectifier, the most negative point in the circuit is (*a*) either plate; (*b*) either cathode; (*c*) chassis ground; (*d*) the center tap on the high-voltage secondary.
9. One advantage of using a choke-input *LC* filter for a power supply is that the (*a*) d-c output voltage of the filter is higher; (*b*) direct load current is increased; (*c*) a-c input voltage is increased; (*d*) circuit has good voltage regulation.
10. What is the regulation of a filter that has a no-load output voltage of 300 volts and a full-load output voltage of 280 volts? (*a*) 4.2 per cent; (*b*) 7.1 per cent; (*c*) 6.7 per cent; (*d*) 9.6 per cent.

## QUESTIONS AND PROBLEMS

1. Draw a diagram of a full-wave power supply including a $\pi$-type filter.
2. Name the three main parts of an a-c power supply. Briefly explain the function of each.
3. Draw a diagram of a half-wave power supply using a silicon diode, including a $\pi$-type filter with a resistor instead of a choke.
4. Trace the operation of a cascade voltage-doubler circuit through one complete cycle.
5. Give one advantage and one disadvantage of a capacitor-input filter compared with a choke-input filter.
6. Give two functions of the voltage divider in a power supply.
7. Compare the dynamotor and the vibrator types of B supply, in terms of (*a*) function and (*b*) operation.
8. Define the following terms: peak inverse voltage, voltage regulation, bleeder current, per cent ripple, and rectifier d-c load current.
9. Give one advantage and one disadvantage of a full-wave bridge rectifier circuit.
10. Give one trouble in the power supply that can cause hum.

# Review of chapters  to 

## SUMMARY

1. R-f amplifiers are tuned stages, amplifying just the band of frequencies around the resonant frequency of the *LC* circuit. Pentodes are generally used because of the small grid-plate capacitance and high plate resistance compared with triodes.
2. A single-tuned amplifier uses a parallel-resonant *LC* circuit as the plate-load impedance. Gain is maximum, or peaked, at the resonant frequency. Cascaded stages where each is tuned to a different frequency form a stagger-tuned amplifier.
3. The $Q$ of the resonant circuit equals $X_L/R_s$ for series resistance or $R_D/X_L$ with a shunt damping resistor.
4. A double-tuned amplifier uses two tuned circuits with mutual coupling $L_M$. Increasing $L_M$ increases the bandwidth. When $L_M$ equals $1/Q$ the transformer is critically coupled, for the greatest bandwidth without double peaks; less coupling is loose coupling; more is tight coupling or overcoupling.
5. When triodes are used because of their lower noise, the r-f amplifier often uses the grounded-grid circuit. Input signal is coupled to the cathode because the grid is grounded. Output signal is taken from the plate.
6. A cascode amplifier consists of a grounded cathode section driving a grounded grid section.
7. A wide-band amplifier, or video amplifier, amplifies audio and radio frequencies. The circuit in an *RC*-coupled audio amplifier with a low value of $R_L$, plus peaking coils for r-f gain.
8. An r-f feedback oscillator is basically an r-f amplifier with positive feedback from the output circuit to its own input. The tuned circuit provides oscillations at its resonant frequency. The amplifier enables part of the output to be fed back to the input.
9. The Hartley oscillator circuit uses a tapped coil for inductive feedback.
10. The Colpitts oscillator circuit uses a capacitive voltage divider for feedback.
11. The tuned-grid tuned-plate oscillator circuit uses feedback through the grid-plate capacitance of the tube.
12. Crystal-controlled oscillator circuits have the advantage of excellent frequency stability.
13. Any of the oscillators can be used in an electron-coupled circuit, which combines in one stage the oscillator and an amplifier for oscillator output.
14. Oscillators generally use grid-leak bias, which is produced by the grid feedback. Measuring this negative bias with a d-c voltmeter is a convenient method of checking operation. Typical bias values are $-5$ to $-50$ volts. Zero grid bias means no feedback, which indicates no oscillations.
15. In high-frequency oscillators, the conventional *LC* circuit can be replaced by a quarter-wave section of transmission line, or by resonant cavities.
16. A magnetron is a high-frequency oscillator consisting of a central cathode and a circular anode block around the cathode. A series of cavity resonators are machined into the anode block. In addition, an external magnet produces a field of magnetic lines parallel to the cathode. Oscillations are produced by interaction of the emitted electrons and the magnetic field.
17. A klystron oscillator operates on the principle of velocity modulation, changing the speed of electrons passing through the tube. A klystron tube internally contains a pair of

buncher grids, a pair of catcher grids, and a collector plate. The buncher grids modulate the velocity of the electrons. Then the bunches of electrons transfer energy to the cavity resonator connected to the catcher grids. In a reflex klystron, one set of grids acts as buncher and catcher.

18. A traveling-wave tube contains a tightly wound helix. Electrons from the electron gun at one end travel through the helix to be collected at the other end. Energy traveling along the helix interacts with the electron beam, absorbing energy from the beam. In a backward-wave oscillator, a backward-moving wave is employed to interact with the electron beam.
19. In a power supply, the rectifier changes a-c input to d-c output. Vacuum tubes, gas tubes, or semiconductor diodes can be used for the rectifier. The filter smooths the fluctuations to provide steady d-c output. A low-pass filter is needed, using series $L$ or $R$, with shunt capacitance.
20. A half-wave rectifier uses just one-half of the a-c input cycle. The ripple frequency is 60 cps in the d-c output (60-cycle a-c input). A full-wave rectifier uses both halves; the ripple frequency then is 120 cps. More load current can be supplied by a full-wave power supply.
21. A voltage doubler uses two diodes to provide d-c output voltage twice the a-c input voltage.
22. A transformerless power supply can operate with either a-c or d-c input. The heaters are in series.
23. A capacitor-input filter has a shunt filter capacitor at the rectifier cathode. A choke-input filter has a choke in series with all the rectifier current. The capacitor-input filter provides more d-c output voltage with less ripple; the choke-input filter has better voltage regulation.
24. Voltage regulation means that the d-c output voltage does not go down when the load current increases.
25. Common troubles in power supplies are: (*a*) Excessive ripple in the B+ voltage because of poor filtering, causing hum; (*b*) zero B+ voltage, resulting in no output from all the amplifiers connected to the power supply; (*c*) open heater in a series string, which means that none of the tubes can light.

## REFERENCES (*Additional references at back of book.*)

### *Books*

Hickey, H. V., and W. M. Villines, *Elements of Electronics,* 2d ed., McGraw-Hill Book Company.
Kaufman, M., *Radio Operator's Q and A Manual,* John F. Rider, Publisher, Inc.
Marcus, A., and W. Marcus, *Elements of Radio,* 3d ed., Prentice-Hall, Inc.
Sheingold, A., *Fundamentals of Radio Communication,* D. Van Nostrand Company, Inc.
Shrader, R. L., *Electronic Communication,* McGraw-Hill Book Company.
Slurzburg, M., and W. Osterheld, *Essentials of Radio,* 2d ed., McGraw-Hill Book Company.

## REVIEW SELF-EXAMINATION

Here's another chance to check your progress. Work the exercises just as you did those at the end of each chapter and check your answers.

Answer true or false

1. R-f amplifiers often use resonant $LC$ circuits.
2. Pentodes have more grid-plate capacitance than triodes.
3. The screen-grid voltage for a pentode r-f amplifier should be a steady d-c voltage.
4. Cathode bias is a suitable method for class A amplifiers.
5. Parallel resonance can provide a high value of plate-load impedance at the resonant frequency of the $LC$ circuit.
6. An overcoupled double-tuned stage can provide more bandwidth than a single-tuned stage.

7. Reducing the value of shunt damping resistance lowers $Q$ and increases bandwidth.
8. With 1,000 ohms $X_L$ and 10,000 ohms shunt damping resistance, $Q$ equals 0.1.
9. An r-f amplifier using the cascode circuit has a good signal-to-noise ratio.
10. In a grounded grid amplifier, the input signal is applied to the plate.
11. A video amplifier can be used for the band of frequencies from 30 cps to 4 Mc.
12. An r-f feedback oscillator oscillates at the resonant frequency of the tuned circuit.
13. An electron-coupled oscillator has the advantage of isolating the oscillator circuit from the load.
14. The Hartley and Colpitts oscillator circuits each use one tuned circuit common to plate and grid.
15. Crystal-controlled oscillators are seldom used because the oscillator frequency drifts.
16. An r-f voltmeter is necessary to measure grid-leak bias on a 10-Mc oscillator.
17. A magnetron needs a magnetic field to react with electrons emitted from the cathode.
18. In a reflex klystron one set of grids acts as both buncher and catcher.
19. A backward-wave oscillator contains a helix.
20. A power transformer changes alternating current to direct current.
21. In a directly heated rectifier tube, B+ voltage is on the filament.
22. Gas-tube rectifiers can supply more load current than vacuum-tube rectifiers.
23. Less filtering is needed for the d-c output of a full-wave rectifier, compared to a half-wave rectifier.
24. With a transformerless power supply, the heaters are in series.
25. A voltage doubler circuit uses two diode rectifiers.
26. The internal voltage drop of a silicon rectifier is much less than a vacuum-tube rectifier.
27. Semiconductor rectifiers generally need a current-limiting resistor.
28. With an inverted rectifier, the a-c input is applied to cathode for negative d-c output at the plate.
29. 100 volts d-c output with 1-volt ripple equals 1 per cent ripple.
30. With 160 volts output at no load and 120 volts at full load, the voltage regulation is 40 per cent.

# Chapter 7 Modulation and transmitters

This unit describes how the r-f output of an oscillator circuit can be transmitted. Photographs of typical transmitters are shown in Figs. 7·1 and 7·2. The transmitting equipment generally includes amplifiers for the oscillator output and provision for modulating the r-f wave. This means varying either its amplitude or its frequency in step with a lower-frequency voltage that corresponds to the desired intelligence. The r-f wave then is the carrier for the modulation information, which may be audio intelligence for sound, video information for television pictures, or any other type of information to be transmitted. The topics are as follows:

7·1 Transmitter Requirements
7·2 Principles of Modulation
7·3 Percentage of Modulation
7·4 Sidebands
7·5 Plate Modulation
7·6 Control-grid Modulation
7·7 High- and Low-level Modulation
7·8 AM Transmitters
7·9 Single-Sideband Transmission
7·10 Frequency Modulation
7·11 FCC Allocations

## *7·1 Transmitter requirements*

An oscillator is a generator of r-f energy. Its output may be controlled easily so that intelligence can be transmitted. However, the relatively low power available, the instability of the output frequency and amplitude, the high-frequency limitations of certain oscillator circuits, as well as other reasons, require that most modern transmitters consist of power amplifiers,

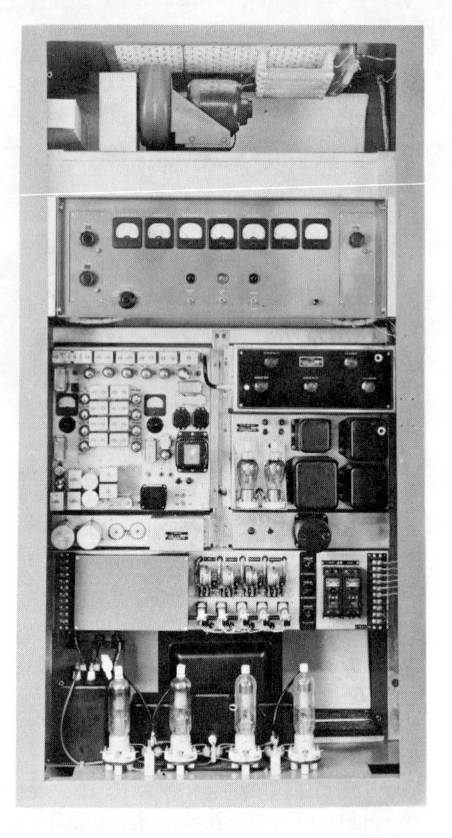

*Fig. 7·2 A 1,000-watt FM transmitter for television sound signal.*

*Fig. 7·1 A 100-watt transmitter for c-w and radiotelephone operation in the amateur bands.* (Heath Company.)

power supplies, and some method of controlling the r-f energy, in addition to the oscillator. The radio transmitter is therefore a power source with a means of converting the d-c power into r-f energy and a means of controlling it.

Many devices have been employed since the beginning of radio to generate r-f energy. Among these are the high-frequency alternator, the Poulsen arc, and the oscillatory spark discharge. Today, however, the vacuum tube is used for this purpose. Transistors are also beginning to be used also in this type of equipment, although essentially only in the low-power range.

**Types of waves.** Four basic types of radio waves have been used: damped waves, continuous waves, interrupted continuous waves, and modulated waves. Damped waves may be understood by considering the simple tuning fork. When the fork is struck, its first vibration has the greatest amplitude because it has the greatest energy. As it vibrates, it loses energy at each vibration, and therefore each successive vibration has less amplitude than the preceding one. Figure 7·3*a* illustrates this type of wave. Similar waves are set up by a spark transmitter. However, since damped waves are undesirable for many reasons, they are not employed today for radio communications.

Consider the wave set up by the bowing of a violin string. If the bowing puts just enough energy into the string to keep it vibrating at a constant amplitude, each wave that leaves the string will have the same amplitude.

This is a continuous wave (c-w) and is illustrated in Fig. 7·3*b*. These undamped waves are produced by a vacuum-tube transmitter, without modulation. If a continuous wave is interrupted at regular intervals without loss of amplitude, the wave is known as an interrupted continuous wave (i-c-w), as shown in Fig. 7·3*c*.

If, in the case of the violin string, the energy transferred to the string by the bow is varied, the amplitude of the vibration and the waves also vary. This results in a modulated continuous wave (m-c-w), as shown in Fig. 7·4. In transmitting speech or music by radio, it is customary to use the audio signal to vary the r-f wave to produce the modulated output. The frequency of the wave being modulated is known as the *carrier frequency* and the wave itself as the *carrier wave*. Modulation is thus a process of combining audio-signal energy with r-f energy. Three examples of modulation are:

1. *Radiotelegraph.* In the radiotelegraph transmitter, a telegraph key is used as a switch to turn the r-f carrier wave on and off according to some prearranged system, such as the Morse code (see Appendix H). The interrupted carrier may be either c-w or m-c-w.

2. *Radiotelephone.* The carrier wave of a radiotelephone transmitter is modulated according to an a-f signal, such as voice or music.

3. *Facsimile and television.* The carrier waves from facsimile or television

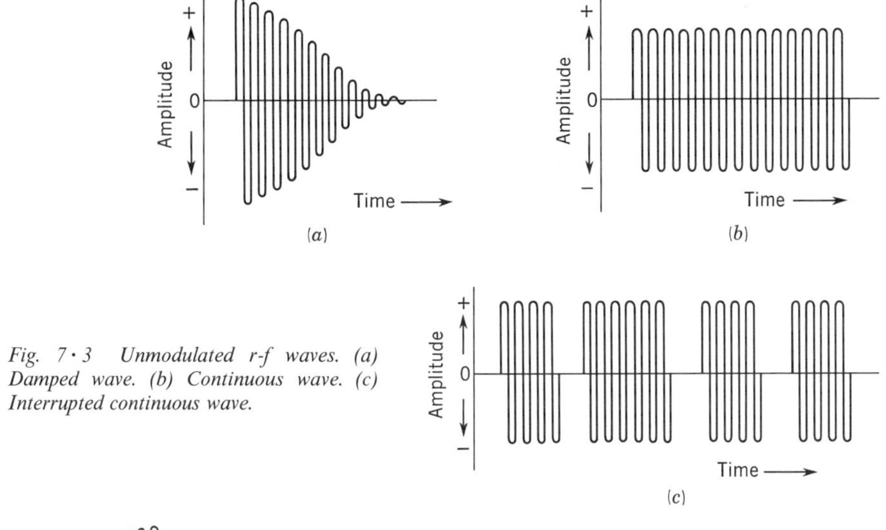


*Fig. 7·3 Unmodulated r-f waves. (a) Damped wave. (b) Continuous wave. (c) Interrupted continuous wave.*

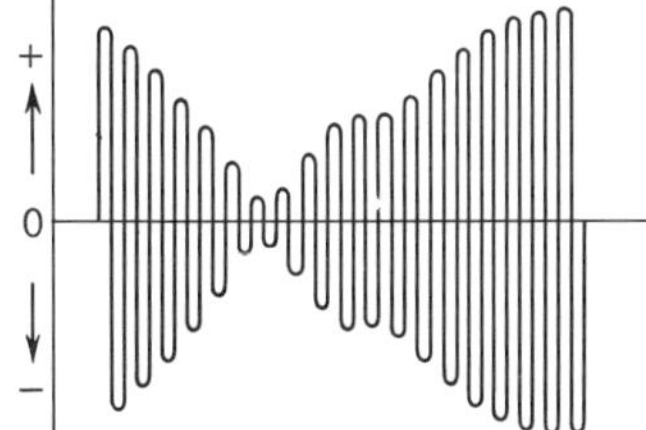


*Fig. 7·4 An amplitude-modulated r-f wave.*

transmitters are modulated by impulses that vary according to the amount of light reflected from the image to be transmitted. For facsimile, still pictures are transmitted; television shows motion pictures.

**The need for radio communication.** As is well known, speech is conveyed from the speaker to the listener by sound waves. These waves are propagated through the medium of the air. Two persons may converse as long as they are close enough to each other to hear intelligible transmission of the sound waves. If the distance between them is increased, a point will be reached at which communication is impossible by this means. Communication may be reinstated by using cupped hands or a megaphone to channel more sound energy in the direction of the listener. As the separation of speaker and listener is further increased, though, a point is reached at which communication again ceases to be effective. Now communications can be resumed by means of a public-address system to increase the sound energy to an intelligible level, but this, too, has a limited range. Wired communication by telephone is of course the next choice, but it is an expensive method and requires the use of wires. Radio communication is the most practical answer to the problem of long-distance communications because it is a wireless system.

Any electric circuit carrying an alternating current will radiate a certain amount of electrical energy in the form of electromagnetic waves. This radiation is normally very small and is of little or no consequence in communications work. The amount of energy radiated from a circuit is determined by its length.

Thus a 60-cps a-c power line would have to be almost 1,550 miles long before efficient radiation would occur. On the other hand, a wire 20 ft long would radiate a considerable amount of 3,000-kc energy. It is apparent that the size of the efficient radiator or antenna is inversely proportional to the wavelength. High-frequency waves can be radiated satisfactorily by a small antenna, while low-frequency waves require a large radiating system.

Our two communicants may, of course, attempt to connect the output of their microphone to an a-f amplifier and its output, in turn, directly to an antenna. There are a number of reasons why this would be unsatisfactory. Since the antenna should be of the order of a half wavelength long, it would be necessary for the high audio frequency of 15,000 cps to have an antenna 6.2 miles long. For the lowest audio frequency to be transmitted, say 50 cps, the antenna length should be 1,860 miles long. Inasmuch as both these lengths are not practical, we see that it is not possible to communicate efficiently by this method at these frequencies. Even if it were possible to erect an antenna of proper length, it would be correct for only one frequency. Thus, a communications system utilizing a-f radiation is not feasible.

If, however, an r-f wave is employed, we obtain efficient radiation. If different radio frequencies are used for different pairs of communicants, the system will be selective. It is necessary, therefore, to obtain the audio signals and translate them to an equal bandwidth at a much higher frequency in order to radiate the intelligence. The process of modulation accomplishes this.

## *7·2 Principles of modulation*

Modulation is a process of combining high- and low-frequency currents. There are three waves to consider:

1. *Carrier wave.* The carrier or high-frequency wave is produced by a high-frequency oscillator. This wave is necessary to transmit radio energy over long distances.

2. *Audio wave.* The low-frequency energy used to modulate the carrier wave is assumed here to be in the audible range from 16 to 16,000 cps. However, the same idea applies to all types of modulating information.

3. *Modulated wave.* A modulated wave is produced when the audio wave is combined with the carrier wave.

Modulation may be defined as the process of varying the r-f signal to be transmitted in accordance with some intelligence. During modulation, the simple r-f wave is converted into a complex wave by the speech or music from the audio amplifiers.

**Frequency modulation and amplitude modulation.** There are two common ways of modulating a radio carrier: frequency or amplitude modulation. Frequency modulation (FM) is a method by which the frequency of the carrier wave is made to change in accordance with an audio voltage. Phase modulation is sometimes treated as a third type, but it can be considered as a method of frequency modulation. During amplitude modulation (AM), the carrier frequency remains constant, but its amplitude is made to change in accordance with the audio signal. Thus, when the audio signal is impressed on the modulated stage, the carrier wave becomes larger or smaller as the audio wave becomes stronger or weaker. Two applications of amplitude modulation are the radio signal in the standard radio broadcast band and the picture signal for television broadcasting; two examples of frequency modulation are the radio signal in the FM broadcast band and the sound signal for television broadcasting. These methods apply to broadcast stations regulated by the Federal Communications Commission (FCC) in the United States.

**An AM system.** The block diagram of a typical AM radiotelephone transmitter is shown in Fig. 7·5. Note that the final r-f power amplifier has two input signals. One is the r-f signal from the r-f section, while the other is the audio modulating voltage from the audio section of the transmitter.

The output of the transmitter can be varied by varying the voltage on one of the electrodes of the final r-f power-amplifier tube. For instance, if the plate voltage on the final amplifier were to be varied at an audio frequency, the output of the r-f amplifier, and hence of the transmitter, would be varied at the same rate. This is plate modulation, which is the most popular method of amplitude modulation.

In order to vary the plate voltage of the final r-f amplifier at an audio

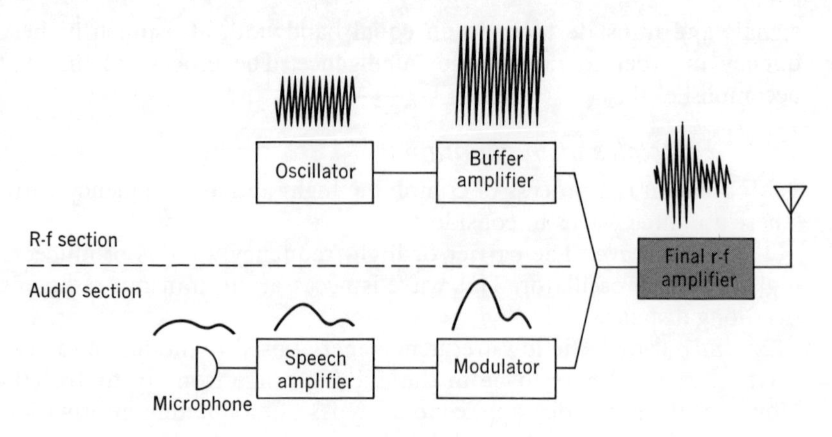


*Fig. 7 · 5 Block diagram of an amplitude-modulated transmitter.*

frequency, it is necessary, first of all, to produce an audio voltage. This is done with a microphone. The output of a microphone is, however, very small (usually much less than 1 volt), while the plate voltage of the r-f amplifier is quite high. The addition of a small audio voltage to a high plate voltage would result in hardly any variation of the plate voltage. It is necessary, therefore, to amplify the output of the microphone before it is applied to the plate of the power amplifier. This amplification is usually accomplished in at least two stages. The output of the microphone is fed to the grid of a class A voltage amplifier, merely to step up the voltage. This first voltage amplifier is called the *speech amplifier*. The voltage output of the speech amplifier is used to drive the grid of an audio-power amplifier. This second amplifier is called the *modulator*.

The manner in which the a-f signal is amplified and then applied to the r-f carrier is shown graphically by the waveforms in Fig. 7 · 5. The r-f oscillations without modulation are known as the *r-f carrier*. With audio variations in r-f amplitude produced by the modulator, the r-f output becomes a modulated r-f carrier.

## *7 · 3 Percentage of modulation*

The effect of a modulated wave in terms of audio output from a receiver is proportioned to the degree or per cent of modulation. The degree of modulation is expressed by the percentage of maximum amplitude deviation from the unmodulated value of the r-f carrier. This value will depend upon the ratio of a-f to d-c voltage. An example is illustrated in Fig. 7 · 6*a*. If the d-c plate voltage to the r-f amplifier is 100 volts and the a-f voltage is 50 volts, the two voltages will add, when they are acting in the same direction, to provide 150 volts. They will subtract, when they are acting in opposite directions, to give 50 volts. The plate voltage on the r-f amplifier will then vary between 50 and 150 volts. Since the variation of

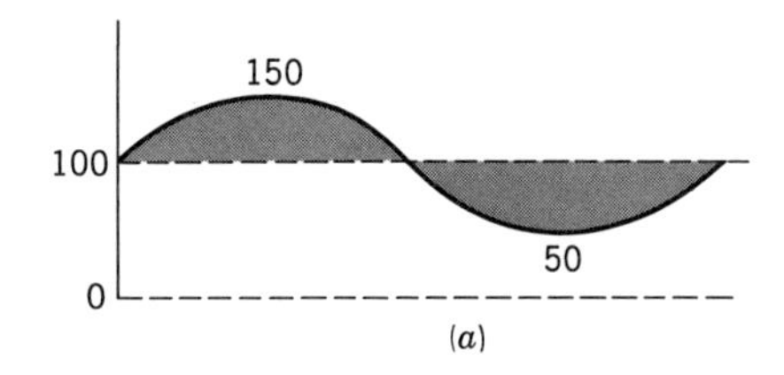


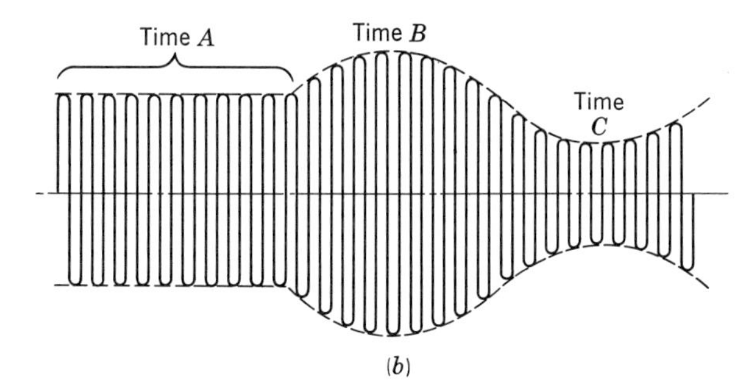


*Fig. 7 · 6 Illustration of 50 per cent modulation. (a) Instantaneous plate voltage on modulated r-f amplifier. (b) Modulated r-f output.*

50 volts on either side of the d-c voltage is one-half of the d-c voltage of 100 volts, the transmitter is said to be modulated 50 per cent.

This same result may be shown in terms of the r-f output of the transmitter, as in Fig. 7 · 6*b*. The amplitude of the carrier, which is the r-f wave produced without any audio voltage on the plate of the r-f amplifier, is shown during time *A* of Fig. 7 · 6*b*. Notice that this carrier is of constant amplitude. As soon as an a-f voltage is applied when the modulator is in operation, the plate voltage, and hence the r-f output, begins to vary. During time *B* in Fig. 7 · 6*b* the r-f wave has reached an amplitude 50 per cent greater than during period *A*. Notice now that when the plate voltage decreases, the r-f output decreases. During time *C* the r-f wave has reached an amplitude 50 per cent less than the unmodulated wave at *A*. Thus the percentage or degree of modulation may also be defined as the percentage of variation of the modulated wave compared with the unmodulated wave.

If the d-c voltage is 100 volts and the audio voltage is also 100 volts, the instantaneous plate voltage will vary between zero and 200 volts, as shown in Fig. 7 · 7. Whenever the instantaneous plate voltage varies between zero and twice its modulated value, there is 100 per cent modulation. Note that the modulated r-f carrier wave now varies between twice its unmodulated level and zero.

It is important that the amplitude be varied as much as possible because the output in a radio receiver varies with the amplitude variations of the received signal. This is why a station comparatively low in power, if fully modulated, can produce a stronger signal at a given point than a much higher powered, but under-modulated, transmitter located the same distance from the receiver. There is, however, a limit to the permissible percentage of modulation. This limit is 100 per cent.

To understand more clearly this limitation of 100 per cent modulation, assume that a given transmitter is actually modulated 150 per cent. With

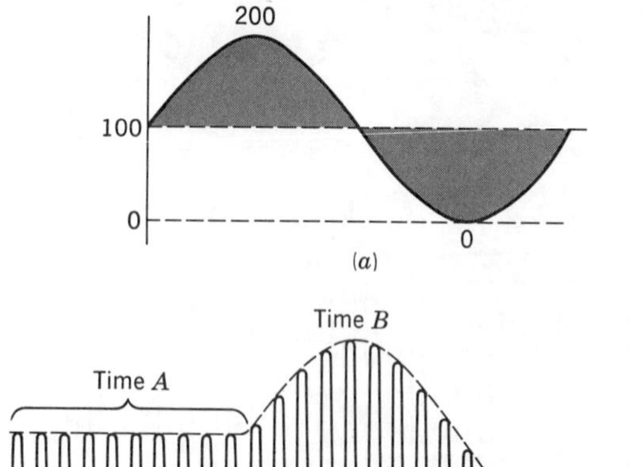


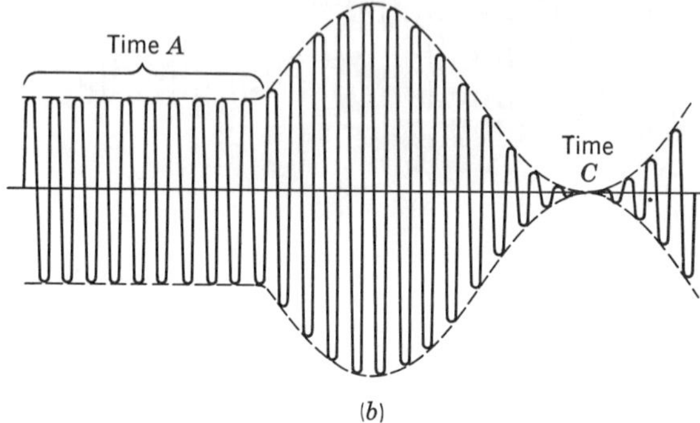


*Fig. 7·7 Illustration of 100 per cent modulation. (a) Instantaneous plate voltage on r-f amplifier. (b) Modulated r-f output.*

a d-c voltage of 100 volts, this would require an audio voltage of 150 volts. As shown in Fig. 7·8*a*, the two would add to give 250 volts; then the plate voltage would swing back through zero, down to minus 50 volts, and back to zero. During the swing from zero to 250 volts and back to zero, plate current would flow. But during the swing from zero to −50 volts and back again to zero, no plate current could flow in the modulated r-f amplifier. During this period, the transmitter would effectively be shut off. This condition produces an overmodulated r-f wave, as shown in Fig. 7·8*b*. We see, then, that overmodulation distorts the signal.

## *7·4 Sidebands*

Assume that an a-f signal whose frequency is 5,000 cycles (5 kc) is modulating an r-f carrier of 1,000 kc, as shown in Fig. 7·9. Notice that during the process of modulation the modulated wave consists of the three components labeled *L*, *C*, and *U*. Each is a radio frequency. One equals the carrier frequency plus the audio frequency (1,000 kc plus 5 kc, or 1,005 kc); another equals the carrier frequency minus the audio frequency (1,000 kc minus 5 kc, or 995 kc). The former is called an *upper sideband frequency* and the latter a *lower sideband frequency*. The third component is the constant-amplitude carrier wave that exists without any modulation.

In other words, the resultant amplitude-modulated wave with audio modulation represents the sum of the carrier and its sideband frequencies. In the case of the 5,000-cycle modulation, there are r-f sideband frequen-

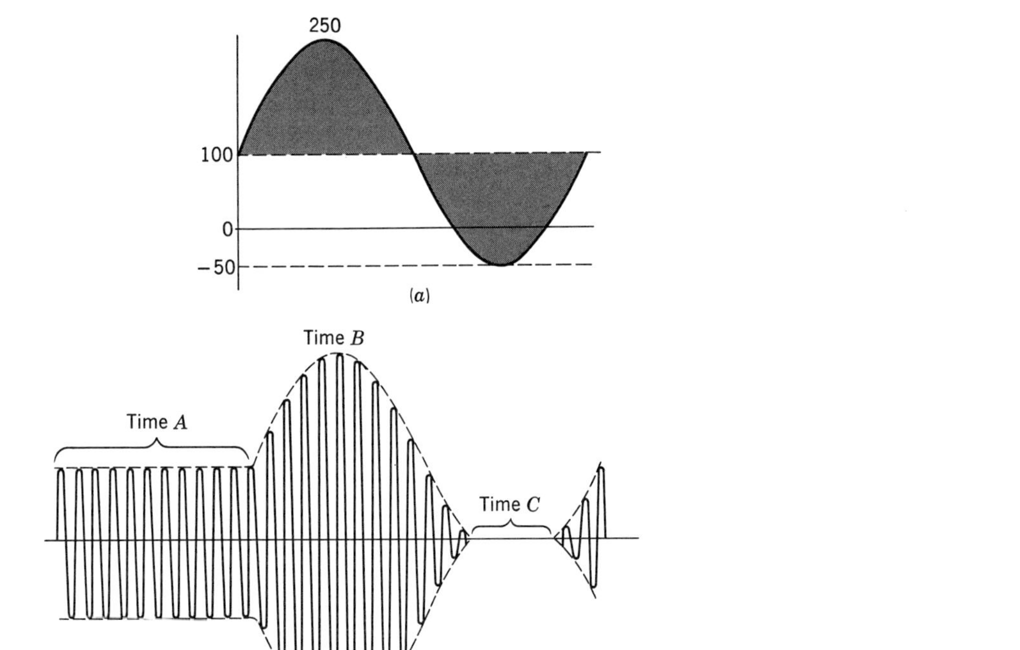


*Fig. 7 · 8 Overmodulation. (a) Instantaneous plate voltage. (b) Modulated r-f output.*

cies 5 kc above and 5 kc below that of the carrier frequency, or a bandwidth of 10 kc for the sidebands. The intelligence in a modulated wave is due entirely to the sidebands and their relation to the carrier.

In actual speech, many audio frequencies are used to modulate the carrier wave. There will be a pair of r-f sideband frequencies (one upper and one lower) for each audio frequency, and there will be an entire band, or group, of frequencies resulting from speech modulation. The form of such a carrier is shown in Fig. 7 · 10. Note that the outline of the amplitude variations is called the *modulation envelope* of the r-f carrier. Do not con-

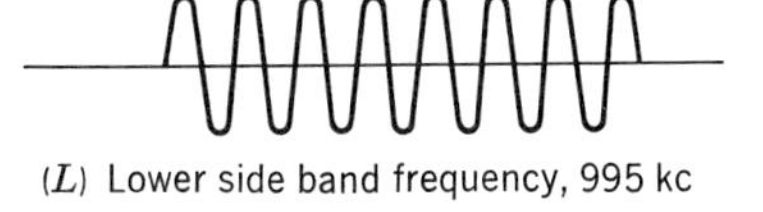

(*L*) Lower side band frequency, 995 kc

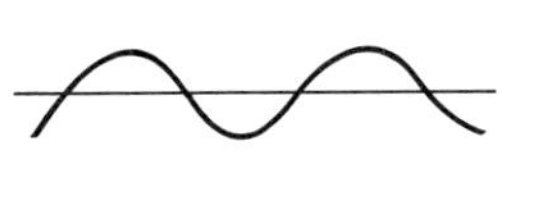

Audio frequency input, 5 kc

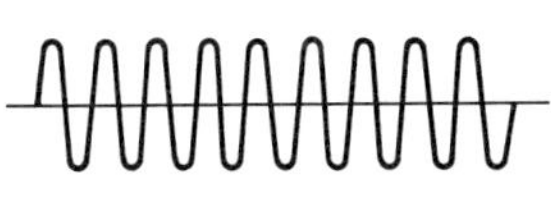

(*C*) Carrier frequency, 1,000 kc

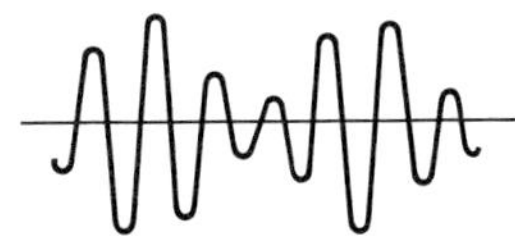

(*X*) Resultant modulated wave

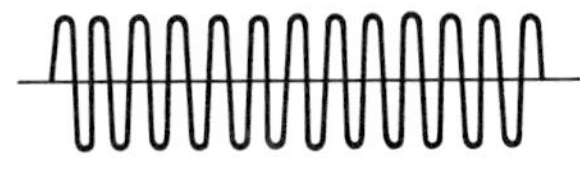

(*U*) Upper side band frequency, 1,005 kc

*Fig. 7 · 9 The relationship of the carrier frequency and its side frequencies produced by modulation.*

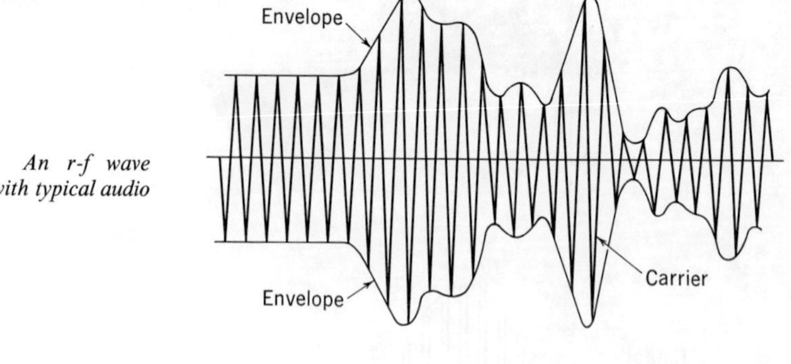


*Fig. 7·10 An r-f wave modulated with typical audio signal.*

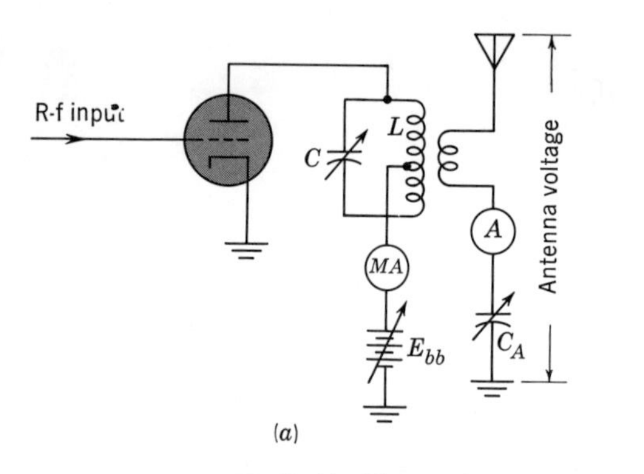


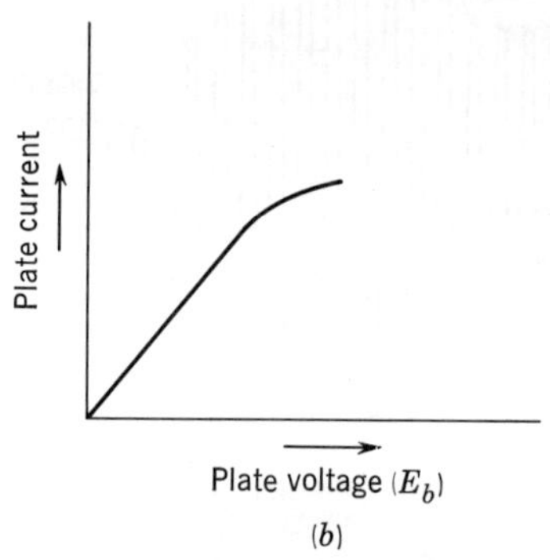


*Fig. 7·11 Plate modulation. (a) Circuit for varying plate voltage on the r-f amplifier. (b) Graph of plate current vs. plate voltage.*

fuse the envelope with the sidebands. The envelope is a duplicate of the audio-modulating voltage. The sidebands include radio frequencies. The r-f upper and lower sidebands each have a bandwidth great enough to include the entire range of audio frequencies.

The process of amplitude modulation normally produces both the upper and lower sidebands. When both sidebands are transmitted with the carrier, this is *double-sideband transmission*. It is possible, however, to transmit just one sideband, either the upper or lower, since they both have the same modulation information. The undesired frequencies can be removed by a sideband filter circuit. When only one sideband is used, it is called *single-sideband transmission*.

## *7·5 Plate modulation*

In this AM system, the audio-modulation voltage is applied directly to the plate circuit of the r-f amplifier. Plate modulation is one of the most common methods of modulation. It has good linearity, is easily adjusted, and employs a relatively small r-f amplifier tube. The modulator, however, must supply a relatively large amount of audio power to modulate the output wave.

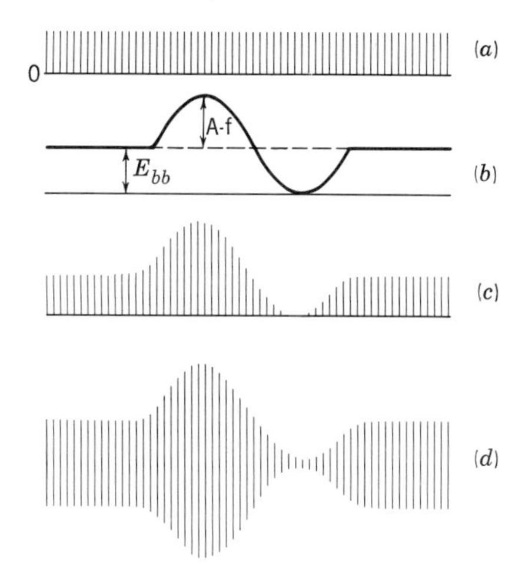


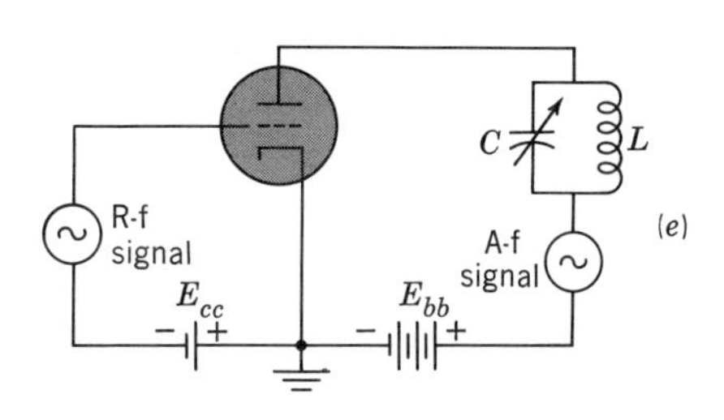


*Fig. 7·12 Plate modulation with audio signal. (a) Plate current with no modulation. (b) Plate voltage with audio modulation. (c) Resulting plate current with audio modulation. (d) Modulated r-f voltage developed across LC plate circuit. (e) Plate-modulated amplifier circuit.*

**Plate-modulation circuit.** The idea of plate modulation is illustrated by the circuit in Fig. 7·11. The plate-supply voltage $E_{bb}$ is adjustable from zero volts to some maximum value. When $E_{bb}$ is zero, the plate current is zero. Then no r-f voltage appears across the tuning circuit, and the antenna current is zero. As $E_{bb}$ is increased, plate current and antenna current are increased. For example, when the plate voltage is doubled, the plate current and antenna current also double. The graph in Fig. 7·11*b* shows how the amount of plate current depends on the plate voltage. If a continuous sine-wave variation of plate voltage is introduced, the plate current and r-f antenna current will follow this sine wave. Therefore if an audio-modulator stage is connected so that it varies the plate voltage according to variations in a voice wave, corresponding variations in plate and antenna currents are produced. The plate modulation is made possible because a class C amplifier can be adjusted for a linear relationship between the plate-supply voltage $E_{bb}$ and the r-f output current.

The operation of a plate-modulated r-f amplifier is further illustrated in Fig. 7·12 for 100 per cent modulation by a sine-wave audio voltage. The a-f modulation voltage should have a peak value equal to $E_{bb}$ for 100 per cent modulation. The waveforms appear as shown in *a*, *b*, *c*, and *d*.

1. The sharp pulses of plate current in *a* are obtained in the class C amplifier with no modulation. Each plate-current pulse is of approximately 120° duration and occurs at the frequency to which the r-f amplifier is tuned.
2. With 100 per cent modulation, the plate voltage varies from twice normal on positive modulation peaks to zero on negative modulation peaks. These variations in plate voltage are shown in Fig. 7·12*b*.
3. The r-f plate current pulses follow the waveform for the applied plate voltage, as shown in Fig. 7·12*c*.

4. The final waveform in *d* shows how the flywheel effect of the tuned tank circuit carries over the portion of each r-f cycle in which no plate current flows, and the remainder of the modulated wave is filled in. The result is the amplitude-modulated r-f wave. The same waveform appears in the antenna circuit, transformer-coupled to the tuned circuit.

**The Heising modulation system.** Plate modulation in which both the audio-modulator stage and the modulated r-f amplifier stage obtain their plate voltage from a common source through a modulation choke is called the *Heising,* or *constant-current,* modulation system. The essential components of such a system are shown in Fig. 7 · 13. $V_1$ is the modulated r-f amplifier tube; $V_2$ is the modulator tube; $L_2$ is the modulation choke. $R_2$, in conjunction with $C_5$, drops the d-c plate voltage of $V_1$ to a value which is equal to the peak value of the a-f voltage appearing across $L_2$. This condition is required in order to achieve 100 per cent modulation. At the same time, $R_2$ is shunted by $C_5$ so that the audio-modulating voltages can pass around $R_2$ without any loss. $C_6$ is an r-f bypass capacitor.

Combination grid-leak bias and battery bias provide three to four times cutoff bias voltage in the modulated r-f amplifier tube $V_1$. In the grid circuit, $C_1$ is an r-f bypass capacitor, and $L_1$ is an r-f choke. $C_2$ is a neutralizing capacitor, used to prevent oscillations in the amplifier. Tuning capacitor $C_3$ and the primary winding of the r-f transformer $T_2$ form a parallel-resonant tuned circuit. $C_4$ tunes the antenna circuit to resonance. The audio-input transformer $T_1$ couples an a-f voltage into the grid circuit of the modulator tube. $E_{bb}$ is the common plate-voltage supply for both $V_1$ and $V_2$.

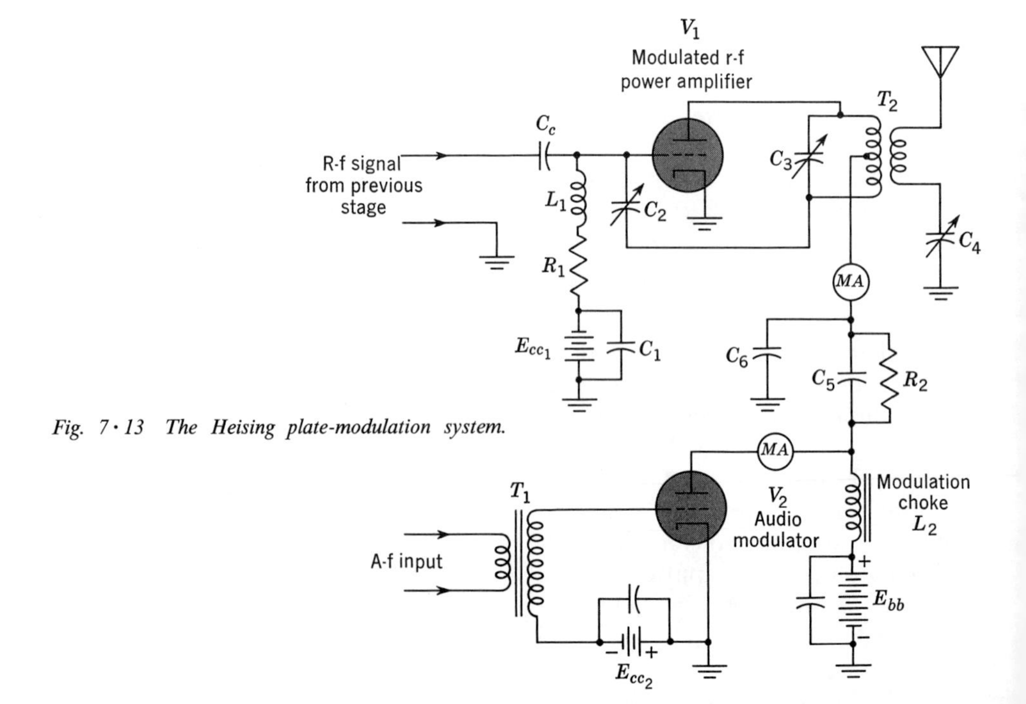


Fig. 7 · 13 The Heising plate-modulation system.

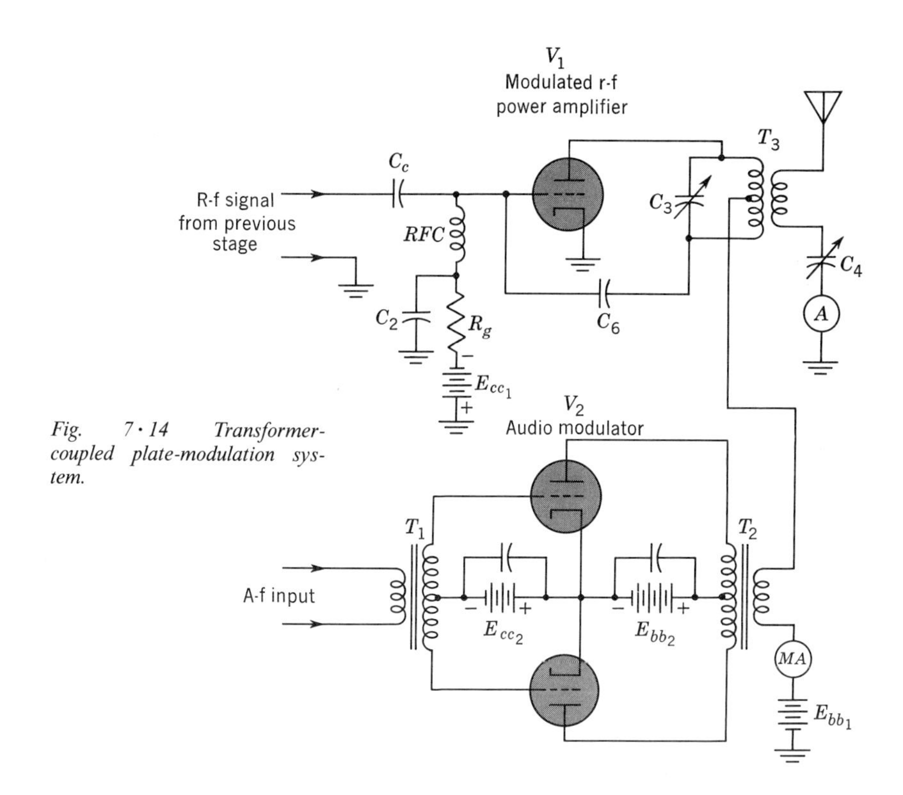


*Fig. 7·14 Transformer-coupled plate-modulation system.*

The r-f signal from a previous stage (oscillator, buffer, or multiplier) is amplified by $V_1$. This signal is sufficient to drive the tube into the plate-current saturation region even on peaks of modulation. $V_1$ delivers a signal to the tuning circuit and causes an r-f voltage to develop across the tank. A corresponding voltage is induced in the antenna circuit because of the inductive coupling between the two windings of $T_2$.

When an a-f voltage is applied to the input of the modulator $V_2$, it will be amplified by the tube, and an a-f output voltage will develop across $L_2$. When this voltage is negative at the plate end of the choke, it opposes the d-c voltage supplied by $E_{bb}$. This causes a decrease in the plate voltage of $V_1$. During 100 per cent modulation, the peak a-f voltage across $L_2$ is equal and opposite to the normal plate-to-cathode voltage of $V_1$, and produces zero plate voltage for $V_1$. The r-f output is then zero. This condition exists only at the instant the modulating voltage reaches its negative peak.

Positive a-f voltage across $L_2$ adds to the d-c plate voltage $E_b$, thereby increasing the plate voltage of $V_1$. At the positive peak of modulation, the plate voltage of $V_1$ is twice $E_{bb}$. The plate current of $V_1$ and the r-f antenna voltage increase in like measure, actually causing the antenna power to rise to four times the average carrier power without modulation. This is so because if we double both antenna current and antenna voltage, their product (which is power) is $2I$ times $2E$, or $4EI$.

When the a-f voltage is constantly supplied to the grid of the modulator tube $V_2$, the modulation voltage developed across the modulation choke

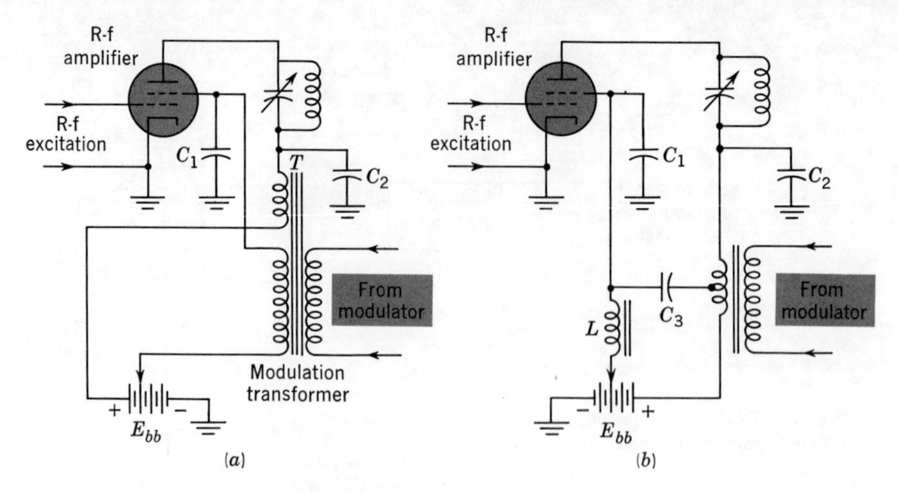


*Fig. 7 · 15 Modulating the screen-grid voltage in a pentode, in addition to the plate modulation. (a) Separate winding in modulation transformer for screen-grid voltage. (b) Capacitive coupling of audio voltage to screen grid.*

$L_2$ acts as an a-c generator in series with the d-c plate supply $E_{bb}$. Since the plate voltage to the r-f amplifier varies according to an a-f modulating signal, the plate current and the r-f tank voltage vary accordingly. The r-f antenna voltage varies as shown in Fig. 7 · 12*d*, and thus the transmitter produces a modulated wave in the antenna circuit.

**Transformer-coupled modulation system.** A modulation transformer used in place of the choke produces better results. The transformer per-

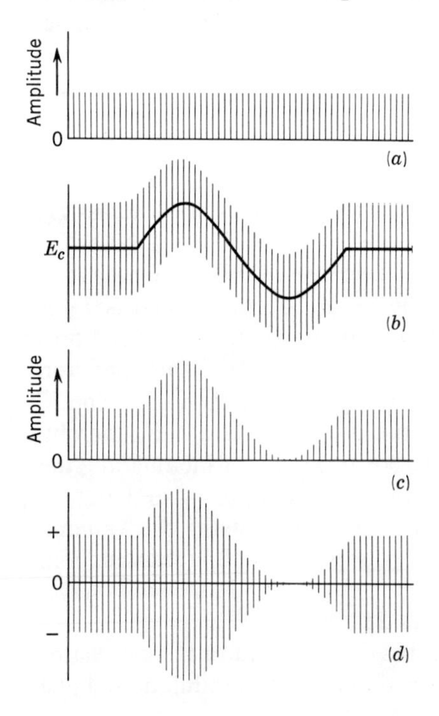


*Fig. 7 · 16 Control-grid modulation. (a) R-f plate current with no modulation. (b) Combined a-f and r-f voltage in grid circuit. (c) Resulting plate current with modulation. (d) Modulated r-f output. (e) Voltages in control-grid modulation circuit.*

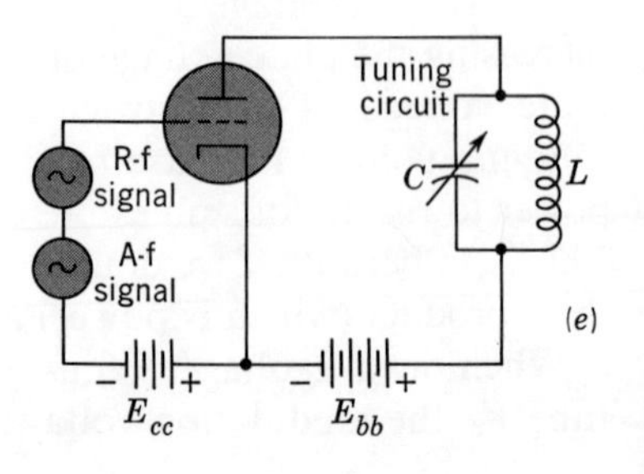

mits separate d-c plate-supply voltages of the modulator and modulated power amplifier. Figure 7·14 illustrates this type of modulation system. $T_2$ is the modulation transformer. The a-f signal output from the modulator is transferred inductively from the primary to the secondary of this modulation transformer; hence, the a-f signal in the secondary is placed in series with the d-c plate supply of the modulated r-f amplifier stage. The peak a-f modulation voltage, appearing across the secondary of the modulation transformer, must equal the r-f amplifier plate-supply voltage $E_{bb}$ for 100 per cent modulation. Except for the improved method of obtaining voltage relationships between the modulator and the modulated r-f power amplifier, the transformer-coupled system and the Heising system of modulation operate in the same way.

Since the plate current in tetrodes and pentodes is relatively independent of plate voltage, the screen grid in such tubes must also be modulated to obtain 100 per cent modulation. The percentage of voltage change in both the plate and screen-grid circuits should be made approximately the same. In Fig. 7·15*a*, the modulation transformer $T$ has a separate winding to provide audio voltage for the screen grid. $C_1$ is an r-f bypass capacitor. In *b*, $C_3$ is an audio coupling capacitor to provide modulating voltage for the screen grid. $L$ is an audio choke.

## 7·6 Control-grid modulation

A class C amplifier can be control-grid-modulated if the r-f signal and grid-bias voltages are properly adjusted. The audio voltage is injected in series with the grid-bias voltage. The main advantage of the grid modulation system is that only a relatively small amount of audio power is re-

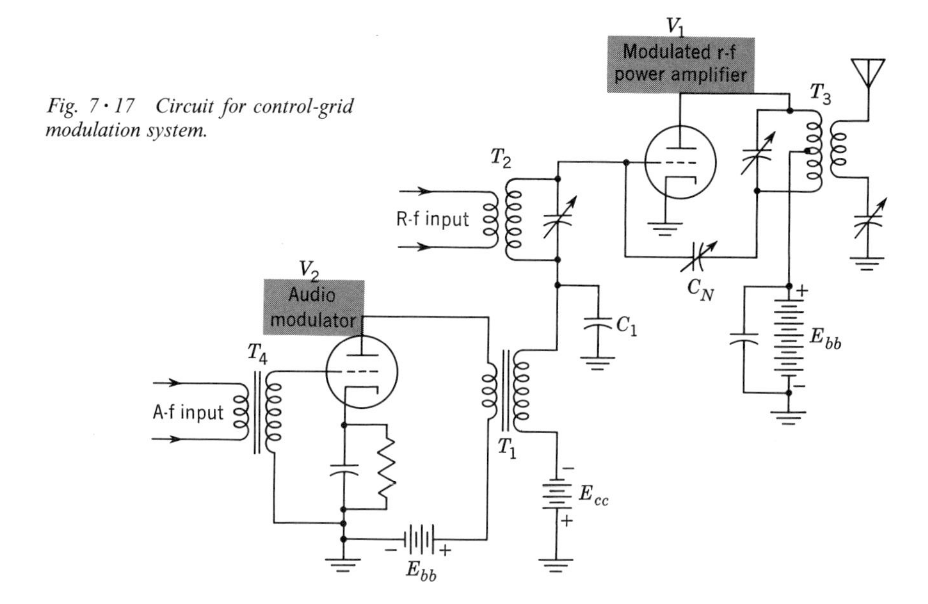


*Fig. 7·17 Circuit for control-grid modulation system.*

quired to modulate the r-f stage completely. The efficiency of the modulated r-f amplifier, however, is reduced considerably, because the r-f amplifier requires a reduction in excitation voltage for modulated operation.

Voltage and current relationships of the grid-modulated amplifier are shown in Fig. 7 · 16. An actual circuit is shown in Fig. 7 · 17. Note that the d-c plate voltage $E_{bb}$ for the modulated amplifier $V_1$ is constant. However, the a-f modulation voltage, which appears across the secondary of the modulation transformer $T_1$, is connected in series with the d-c grid bias $E_{cc}$. As a result, the operating point of the class C amplifier $V_1$ is varied by the a-f modulating voltage. The resulting variations in plate current of $V_1$ are the same as those of a plate-modulated amplifier (see Fig. 7 · 16).

**Suppressor-grid modulation.** Suppressor-grid modulation is similar in operation to control-grid modulation. In the suppressor-grid modulation system, the r-f output of a pentode class C amplifier varies at an a-f rate because an a-f modulation voltage is superimposed on the suppressor-grid bias. Linearity is poor in this system, but since the modulation circuit is removed from the r-f circuit, the r-f amplifier is less critical in its operation and adjustment than in grid modulation. As another method, modulation may also be accomplished in the cathode circuit. The *cathode modulation* is, in reality, a combination of both grid and plate modulation.

## *7 · 7 High- and low-level modulation*

Methods of modulation can be classified as either high-level or low-level, according to the amount of audio power required. When we speak of high-level modulation, we mean that the plate circuit of the last r-f stage, or the stage that feeds the antenna, is modulated. This type of modulation has one distinct advantage. All stages preceding the final or modulated stage are straight r-f amplifiers of the carrier wave. They may be of the high-gain type and do not have to operate as linear amplifiers. In general, we always assume that plate modulation is employed when we refer to high-level modulation. However, grid modulation may be employed in the final amplifier, and in many instances it is referred to as high-level modulation. In all cases, high-level modulation needs more audio power than low-level modulation.

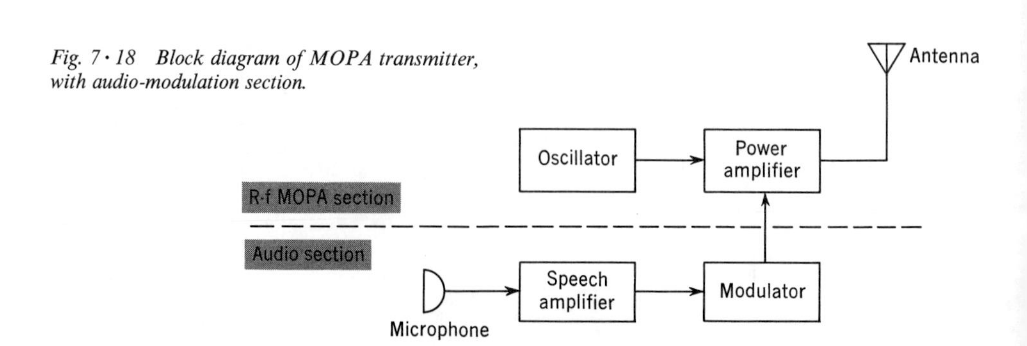


*Fig. 7 · 18 Block diagram of MOPA transmitter, with audio-modulation section.*

When the low-level method of modulation is employed, all modulation takes place in one of the low-power r-f stages and not in the final stage. All amplifiers following the low-level modulated stage must be linear amplifiers capable of amplifying the modulated carrier. Either grid or plate modulation may be employed for low-level modulation.

## *7·8 AM transmitters*

A radio transmitter is a device which produces r-f energy that is controlled by the information or intelligence to be transmitted. Nature has many simple transmitters that convey information to us. One example is a flash of lightning that a radio picks up in the form of static. When a huge electric spark jumps from cloud to cloud, or from earth to cloud, electromagnetic waves of r-f energy are produced. These include many different frequencies rather than one definite frequency and therefore are received as static, or *noise*. An electric shaver, a sparking electric motor, an automobile ignition system, and a leaky high-voltage power line all transmit static. Basically, these are all sources of r-f energy and may be considered as high-frequency generators. This energy is not, however, controlled by any intelligence or information. In transmitters, the r-f output can be controlled by modulation.

**Master oscillator power amplifier.** As shown in Fig. 7 · 18, a combination of an oscillator and a stage of power amplification forms a simple transmitter called a *master oscillator power amplifier* (MOPA). The oscillator may be variable or crystal-controlled. The power amplifier can be keyed or modulated to control the high-frequency energy by the intelligence to be transmitted. The keying turns the output on and off by closing and opening a circuit that controls the plate current. It would be best to key or modulate such a stage in the plate circuit in order to reduce any possible loading of the oscillator. The keying can also be done in the cathode circuit. The modulated output of the power amplifier is coupled to the antenna where the r-f energy is converted to radiation in space. A disadvantage of the MOPA circuit, however, is that keying or modulating the power amplifier varies the load on the oscillator, which can change its frequency.

**Buffer amplifiers.** In most transmitters, more than one stage of power amplification is used to obtain the required output. A first amplifier stage is then utilized to isolate the oscillator from any effects produced by the other tubes: voltage changes, modulation, and so on. This is known as a *buffer amplifier*. The buffer thus serves two functions: it amplifies the weak signals of the oscillator, and it isolates the oscillator from the rest of the transmitter. Sometimes more than one buffer stage is used.

**Frequency multipliers.** It is often necessary to operate the transmitter oscillator at a lower frequency than is desired for transmission, particularly a crystal-controlled oscillator. For this purpose, frequency multipliers are used. Figure 7 · 19 shows a block diagram of a transmitter using two frequency multipliers for output at four times the oscillator frequency.

Frequency multiplying—doubling, tripling, and so on—is a process of

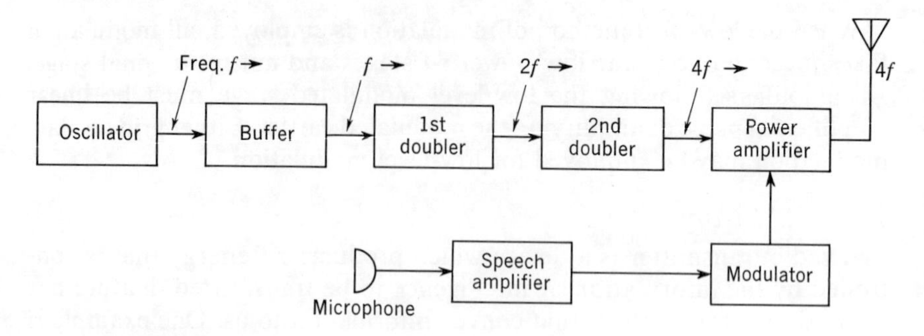


*Fig. 7·19 Block diagram of transmitter with two frequency doublers for output at four times the oscillator fundamental frequency. Note buffer stage between the oscillator and the first doubler.*

tuning an r-f amplifier plate-tuned circuit to a multiple frequency, or harmonic, of the fundamental oscillator frequency. You recall that harmonics are frequencies which have a ratio of 2:1, 3:1, 4:1, or more, to the fundamental frequency, or first harmonic. If the crystal has a first harmonic of 5 Mc, its second harmonic will be 10 Mc, the third 15 Mc, and so on. For example, suppose you have a crystal that has been ground to a frequency of 5 Mc, but you desire a frequency of 30 Mc. You can tune the plate tank of the first multiplier to a frequency of $5 \times 3$, or 15 Mc. Then tune the plate tank of the second multiplier stage to the second harmonic of 15 Mc, or $15 \times 2 = 30$ Mc. This is your required frequency.

When frequency multiplication is required, the signal is deliberately distorted to form strong harmonics, and the desired harmonic frequency is selected with a resonant circuit tuned to the desired harmonic. Since the output of a class C amplifier is greatly distorted, frequency multipliers are generally operated in this manner. In fact, the tubes of some frequency multipliers are biased far more negatively than an ordinary class C amplifier, in order to introduce the greatest possible distortion. The higher the grid bias, however, the greater the grid excitation, or drive, required. The plate-tank circuit is tuned to the harmonic desired. The flywheel effect of the plate-tank circuit will make up the remaining portion of the sine wave of the harmonic-frequency peaks furnished by the vacuum tube.

Three important conditions must prevail in order for an efficient frequency multiplier to be obtained: high grid drive or excitation, high grid bias, and a plate-tank circuit tuned to the desired harmonic. If the second harmonic is selected, the circuit is called a *frequency doubler;* if the third harmonic is used, the circuit is called a *frequency tripler,* and so on.

**Power amplifiers.** The power amplifier stage or stages supply the energy needed to carry the signal through hundreds or thousands of miles of space to the receiver. This stage may use a single tube, or two tubes in parallel, or push-pull, or parallel-push-pull. The circuit used will depend upon the type of tubes and the amount of power needed. The power-amplifier stage normally operates class C, unless the transmitter is modulated in an early stage. Then all stages following the modulated stage operate class B.

Biasing systems for transmitters may use either fixed bias from a power supply, grid-leak bias, or cathode self-bias. With either of the first two systems, however, some sort of protective system is needed to keep the tube from destroying itself by excessive plate current in case the bias supply fails. You will find that this protective system consists of cathode bias in combination with the normal bias. The cathode bias is considered safety bias. Some transmitters use overload relays to protect the power-amplifier tube. The overload opens the relay and prevents excessive plate current from damaging the tube.

**Tuning.** It is important that all radio transmitters be properly tuned to insure efficient operation on the assigned frequency. Plate-current meters are used to indicate proper adjustment of the r-f stages. All stages, with the exception of the oscillator, are always adjusted or tuned for minimum plate current. This is done by watching the d-c milliammeter in the plate circuit while tuning the plate-tank circuit to resonance. A sharp dip in d-c plate current indicates maximum impedance of the tank circuit when it is set to parallel resonance at the desired frequency.

If a stage is not tuned to resonance, direct current to the plate will be high. Therefore high plate dissipation, power loss, and low a-c output will result. It should be noted that when a stage is loaded by another stage or by an antenna, the plate current of the stage in question must be rechecked for circuit resonance (minimum plate current) after loading.

If grid-current meters are available in the transmitter, the grid-input stage is tuned so that maximum grid current is drawn. This shows maximum signal drive. If no grid-current meter is available, grid-circuit resonance is indicated by a sharp increase of plate current in the previous stage, which shows it is being loaded.

In many cases the direct current to the plate is indicated by a meter in the cathode-return circuit, instead of the plate meter. The cathode meter is an arrangement for greater safety, since there is no high voltage. Tuning is done the same way as with a plate meter because the cathode meter indicates plate current. The specific tuning procedure for the r-f power amplifier in a transmitter is explained in detail in Sec. 8 · 8 of the next chapter.

## *7 · 9 Single-sideband transmission*

Brief mention was made at the start of this chapter of the fact that since both sidebands of a modulated signal possess identical signal information, it is not really necessary to send both sidebands. Further, since the carrier itself possesses no information concerning the intelligence to be sent, it too can be dispensed with. The result is a signal which not only occupies less spectrum space but concentrates the available power where it will do the most good, in terms of the intelligence to be transmitted. It is for these reasons that increasing use is being made of single-sideband communications.

When an r-f oscillator is modulated by an audio signal, two sidebands are automatically produced. This is shown in Fig. 7·20*a*. Assuming that the audio signal has a frequency of 1 kc, we find that the upper sideband will have a frequency 1 kc above the carrier. By the same token, the lower sideband has a frequency that is 1 kc below the carrier.

Furthermore, at 100 per cent modulation, each sideband will have 25 per cent of the power contained in the carrier. If the carrier has an output power of 100 watts, then each sideband will have a power output of 25 watts.

Now, suppose we suppress or remove the carrier. See Fig. 7·20*b*. This enables us to raise the power in each sideband to 50 watts without exceeding the plate-dissipation limits of the amplifier tube in the stage where these signals are present. (Before the removal of the carrier this could not be done because the carrier signal produced power dissipation in this tube.)

If, finally, the carrier and one of the sidebands are removed, and only the single lower sideband remains (Fig. 7·20*c*), then it can be raised in power to 100 watts. This will produce an even more powerful signal in the receiver and the demodulated sound output will be considerably stronger than under any of the two previous conditions.

And, of course, it is quite obvious that a single sideband, by itself, will occupy less spectrum space than either of the other situations shown in Fig. 7·20. Thus, single-sideband operation permits us to place more different frequency signals in a given spectrum space, a feature which is highly desirable in today's crowded communications channels.

The single-sideband system is not without its disadvantages. To appreciate what these are, it is necessary to take another look at the AM signal of Fig. 7·20. If we could obtain a highly selective receiver that could tune only to the upper sideband of 601 kc, and not at all to the other two signals (i.e., the carrier and the lower sideband), we would hear *nothing* at the output of the receiver. That is, no 1-kc tone would be developed.

The same result would be obtained if we selected only the 599-kc signal, i.e., the lower sideband.

But, if we permitted *both* the carrier and one sideband to be received,

*Fig. 7·20 The power distribution in 100 per cent amplitude-modulated signals. (a) Carrier with both sidebands. (b) Two sidebands but carrier suppressed. (c) Single sideband; suppressed carrier.*

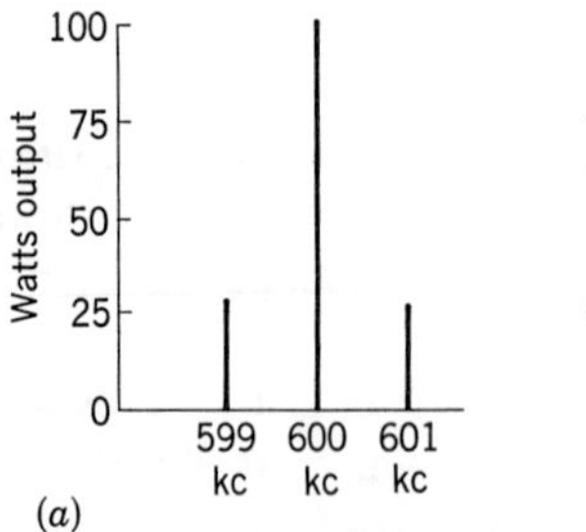


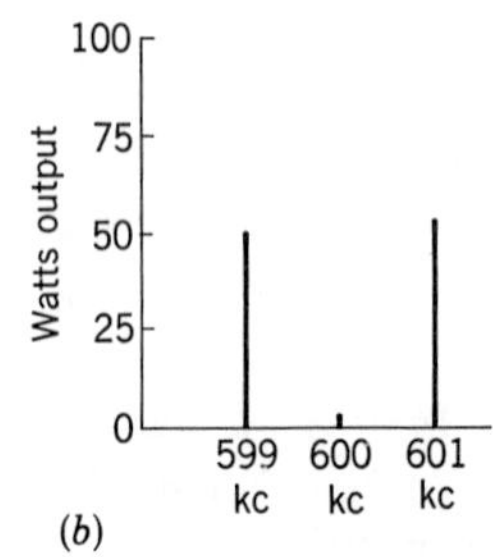


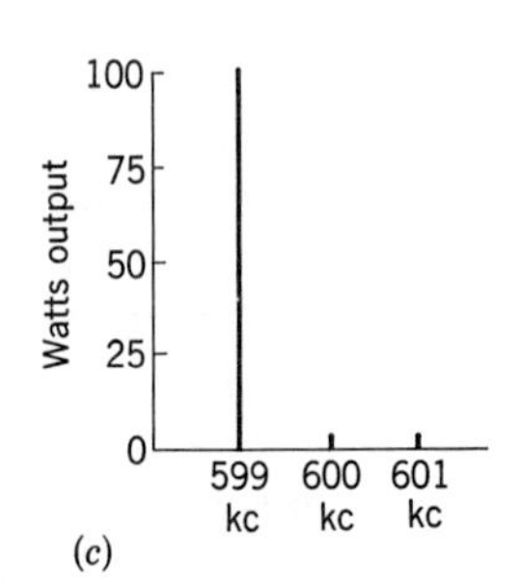

then the 1-kc tone would be heard at the receiver speaker. In short, we require the beating or mixing of the carrier and one sideband (either one) to give us the audio signal we want.

From this we see that when only one sideband is sent and received, then at the receiver we must develop an artificial carrier to mix with the sideband that is picked up. Furthermore, the frequency of this reinserted carrier must be kept at precisely the same frequency as the original carrier; othewise the final audio output will *not* be the same as the original audio modulation developed at the transmitter.

For this reason, and several others, a single-sideband receiver is much more critical in operation than a conventional receiver where the carrier and its sidebands are picked up. Still, the requirements of a good single-sideband (SSB) receiver can be met fairly economically and this system of communications is widely used.

**SSB modulators.** A number of methods exist to produce a single-sideband signal, but the circuit most generally used to suppress the carrier (the first step) is the balanced modulator. A typical circuit is shown in Fig. 7·21.

The control grids of both tubes in the modulator are connected together and both receive the r-f carrier signal from the previous oscillator. The plate of each tube, however, is connected to opposite ends of the primary winding of the output transformer $T_1$. This means that the plate currents of each tube flow through the output transformer in the opposite directions, with the net effect that their magnetic fields cancel each other and *no* output carrier signal is developed across the secondary winding.

Thus, since each modulator tube receives the same input, and produces no net resultant carrier output, the carrier never appears in the output circuit.

*Fig. 7·21 A balanced modulator designed to produce sideband output only, without the carrier.*

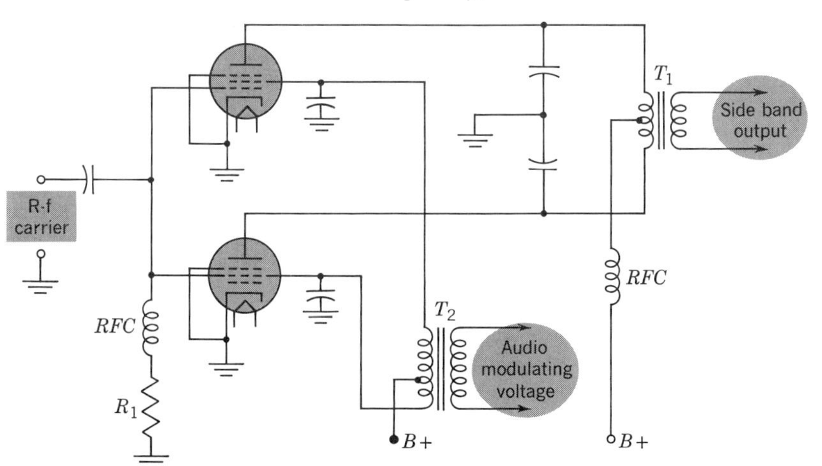

Audio voltages are applied to the modulator tubes through another transformer $T_2$, which brings these voltages to the screen grids of the tubes. The audio voltages on the screen grids are 180° out of phase with each other. Thus, when the plate current of one tube is made to increase by the audio modulation voltage, the plate current of the other tube is decreased. This unbalance will cause r-f voltages to appear across the secondary of $T_1$ and these r-f voltages, varying at an audio rate, will be the upper and lower sidebands. There will be no carrier present, however, since the modulator is balanced for the carrier. It is not balanced for the sidebands and these do develop.

In essence, then, we have two sidebands. Since we require only one, a subsequent filter is used to remove the undesired sideband, resulting in a single-sideband signal. It is this signal that is transmitted.

## 7·10 Frequency modulation

Noise in the output of a radio receiving set may be defined as any sound or disturbance which was not originally present at the microphone of the radio transmitter and which interferes with understanding the message coming over the air. Noise may come from many sources, such as automobile ignition systems, lightning and magnetic storms, diathermy machines, atmospherics, or interfering radio stations. These disturbances are like radio signals in character, appear at all radio frequencies, and affect the amplitude of the r-f signal by distorting the wave. This is one of the great disadvantages of AM waves, since both natural and man-made noise disturbances (static) combine with the incoming r-f wave at the receiving antenna. The combination results in an r-f wave which varies in amplitude according to the static impulses as well as the original (audio) modulating signal. Both the modulating and static impulses therefore will be heard in the loudspeaker of the receiver. To eliminate this fault, some method of modulation is required in which the character of the desired modulation

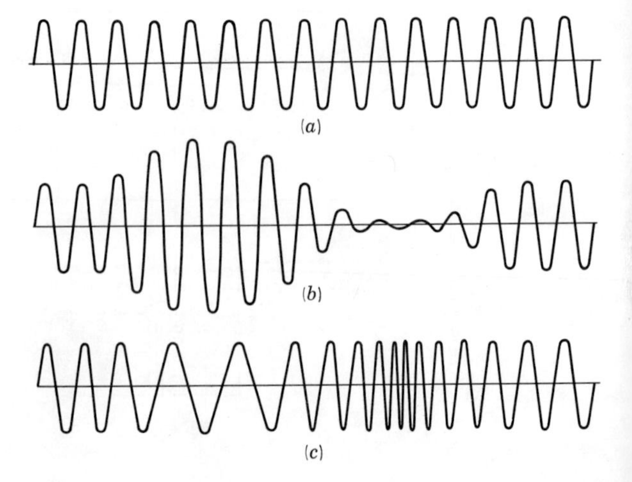


*Fig. 7·22 Comparison of amplitude and frequency modulation. (a) Unmodulated r-f carrier wave. (b) AM wave. (c) FM wave.*

is different from the amplitude variations caused by static impulses. This modulation method is known as frequency modulation.

The frequency of a carrier wave is equal to the number of cycles per second. This frequency, known as the carrier frequency, can be varied or changed a slight amount on either side of its average or assigned value by means of the a-f modulating signal. These frequency changes can be detected by FM receivers designed to respond to the frequency-modulated r-f waves. The changes in frequency of the transmitter take place within certain specified limits in accordance with the voice or speech to be transmitted. The amplitude of the r-f carrier remains constant, with or without modulation. A radio receiver which is sensitive only to variations in frequency of the incoming carrier and which discriminates to a large extent against variations in amplitude is used to receive FM signals. Since static crashes, man-made interference, and other disturbances cause a much larger effective change in the amplitude of an incoming carrier than in its frequency, this system of communication gives very-high-quality reception with an almost complete absence of noise.

The essential difference between frequency modulation and amplitude modulation is shown in Fig. 7 • 22, where *a* represents an unmodulated r-f carrier, *b* shows the result of amplitude-modulating the carrier, and *c* shows the result of frequency-modulating the carrier. In *b*, during the modulation period, the amplitude rises and falls in accordance with an impressed a-f signal. In *c*, during the modulation period, the frequency increases and decreases in accordance with the audio signal, but the amplitude remains constant.

**Principles of frequency modulation.** The simplest form of frequency modulator is that of a capacitor microphone shunting a tuned oscillatory circuit, as shown in Fig. 7 • 23*a*. This simple circuit will help to explain the fundamental principles of all FM transmitters.

The circuit shown in Fig. 7 • 23*a* is a Hartley oscillator with a capacitor microphone *M* connected across the oscillator-tuning capacitor *C*. Electrically, this microphone is nothing more than two plates of a capacitor, one of which is the diaphragm. Sound waves, striking the microphone, compress and release the diaphragm, thus causing the capacitance to vary, since the capacitance of any capacitor depends, in part, upon the distance between the two plates. It will be recalled that the frequency of an oscillator may be varied by a change in either the inductance or capacitance of its tuned circuit. In this case, a variation in the capacitance of the microphone causes the resonant frequency of the oscillator tank circuit to shift alternately to frequencies above and below the original, or resting, frequency. This shifting of frequencies takes place whenever the diaphragm of the microphone moves.

The positive half-cycle *A* of the sound wave in Fig. 7 • 23*b* strikes the diaphragm *D* of the microphone. The microphone is shown in enlarged form to the left in the figure. When the sound wave strikes the diaphragm, it moves inward from the position of rest to position *A*. Since the distance

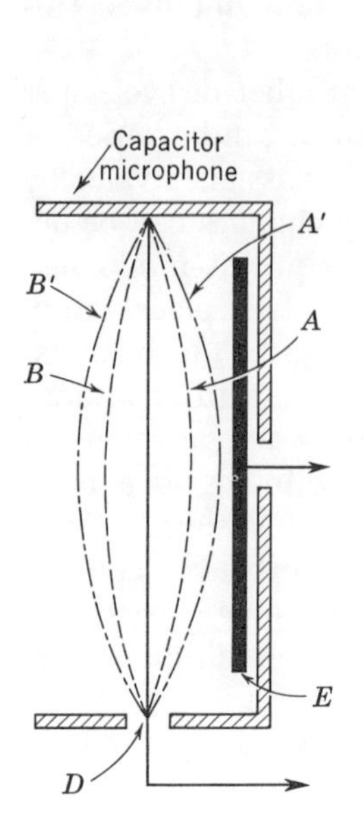


*Fig. 7 · 23 An arrangement for producing frequency modulation. (a) Circuit where capacitor microphone varies r-f oscillator frequency. (b) Sound waves into microphone. (c) FM output of oscillator.*

between the movable diaphragm $D$ and the fixed plate $E$ has been decreased, the capacitance has been correspondingly increased. Then the oscillator frequency decreases from the resting frequency, as shown during time $A$ in Fig. 7 · 23*c*. At the end of the first half-cycle, the diaphragm returns to its position of rest, and the frequency of the oscillator is again at the resting value. During the negative half-cycle $B$ of the audio wave, the diaphragm moves to position $B$, increasing the distance between the plates. Now the resultant decrease in capacitance increases the oscillator frequency, as shown during time $B$ in Fig. 7 · 23*c*. At the end of the alternation, the diaphragm $D$ returns to its position of rest, and the oscillator resumes its resting frequency instantaneously before the action is repeated for the next audio cycle.

The frequency or pitch of the audio signal applied to the microphone determines the number of times per second that the diaphragm vibrates between the two extreme positions. This equals the number of times per second that the oscillator frequency varies across its resting frequency between its high and low values.

Another important detail to notice at this point is that if the audio signal strength (amplitude) is increased, as shown by the dotted line in Fig. 7 · 23*b*, the movement of diaphragm $D$ will be over a greater distance, or from $A'$

to $B'$, and will result in a greater change in capacitance and a greater change in frequency. Thus the amplitude of the modulating signal determines the change in frequency on both sides of the resting frequency.

The amount of change in frequency either side of the resting frequency is known as *deviation*. In the commercial FM broadcast band, the maximum deviation allowed for any channel is set at 75 kc. This means that the strongest audio signal that can be used for modulating a transmitter is limited to that value which will cause a maximum deviation of 75 kc on either side of the resting frequency. This makes available a total of 150 kc, known as the carrier *swing*, over which the frequency of any one station may vary. A band of 25 kc is also provided for separation purposes between channels. This 25-kc band is called the *guard band*. Thus the channel allotted to each station consists of two deviation ranges of 75 kc each, plus the guard band on either side, or a total of 200 kc.

With any of the systems of obtaining frequency modulation used at the present time, the amount of deviation obtained at the point of modulation is small compared with that required for successful transmission of FM signals. In order to increase the amount of initial deviation to a suitable value, a system of frequency multiplication is used. If two frequencies, such as 6.00 and 6.025 Mc, having a difference of 25 kc, are applied to the input of a broadly tuned tripler, the output frequencies will be 18 and 18.075 Mc, respectively. There is now a difference of 75-kc between these two frequencies, or three times the original frequency difference. The varying frequencies produced at the modulating point are therefore applied to a series of multipliers, and the amount of initial frequency change, or deviation, obtained is multiplied to a suitable value before the signal is applied to the power amplifier and then to the transmitting antenna.

In amplitude modulation the amplitude of the carrier varies between zero and twice its normal value for 100 per cent modulation. There is also a corresponding change in power; consequently, additional power must be supplied to the carrier during modulation peaks. Hence the tubes cannot be operated at maximum efficiency at all times. In frequency modulation, however, so-called 100 per cent modulation has a different meaning. As shown in Fig. 7 · 23*c*, the amplitude of the signal remains constant regardless of modulation, since the modulating signal varies only the frequency of the oscillator. Therefore the tubes may be operated at their maximum efficiency at all times. This ability represents one of the important advantages of frequency modulation.

Modulation of 100 per cent in frequency modulation indicates a variation of the carrier by the amount of the full permissible deviation. Examples of three different amounts of deviation are shown in Fig. 7 · 24. The vertical line *RO* represents the resting frequency, which will be assumed to be 20 Mc. If the oscillator producing this frequency is modulated with a weak audio signal of 500 cycles, the oscillator frequency will deviate from the resting frequency to, for example, 19.99 Mc, then back across the resting frequency to 20.01 Mc, and back to the resting frequency. This example is

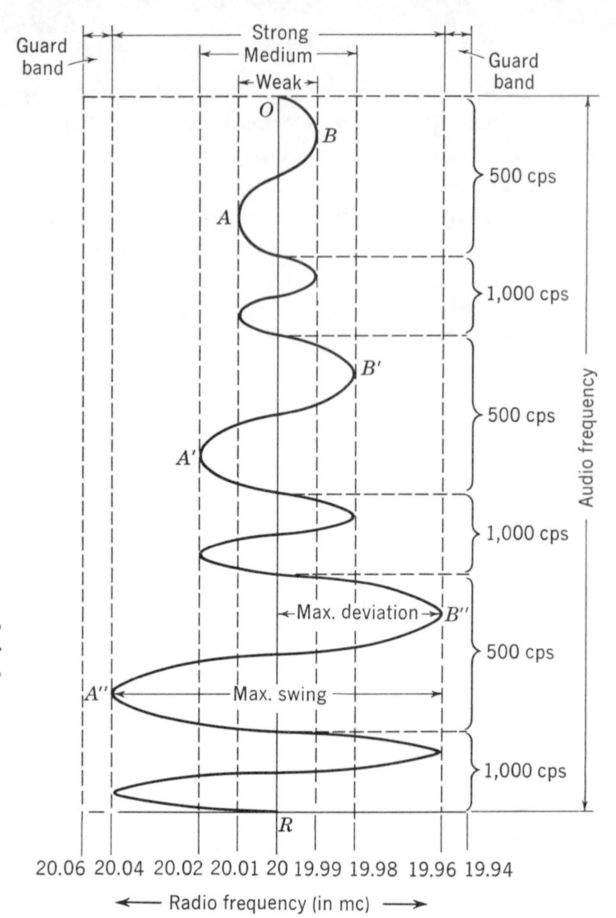


*Fig. 7·24 Graphic representation to show how the audio amplitude varies the amount of FM deviation, while the audio frequency is the rate of deviation.*

illustrated at the top of the figure. The rate of this deviation is 500 times per second. If the frequency or pitch of the modulating signal is changed to 1,000 cps (the amplitude remaining the same), the swing between $A$ and $B$ will occur at a rate of 1,000 times per second. Increasing the amplitude of the modulating audio signal to a medium value will increase the deviation to $A'$ on one side of $RO$, and to $B'$ on the other. The rates of frequency change will still be 500 to 1,000 times per second, respectively. Increasing the modulating signal amplitude still more will cause the maximum carrier to swing between points $A''$ and $B''$, giving the maximum allowable deviation. The rate of change again is the frequency of the modulating signal.

**Methods of frequency modulation.** An FM transmitter must fulfill two important requirements: the frequency deviation must be symmetrical about a fixed frequency, and the deviation must be directly proportional to the amplitude of the modulation. Several methods of frequency modulation fulfill these requirements. The arrangement described in Fig. 7·24, known as a *mechanical modulator,* is the simplest system of frequency modulation, but is seldom used. The two important types of frequency modulation used in FM radio equipment are known as the *reactance-tube modulating system* and the *Armstrong phase-modulating system.* The main difference between these two systems is that the reactance-tube circuit modulates the r-f wave at its source, which is the oscillator, while in phase modulation the r-f wave is modulated in some stage following the oscillator. Also, a crystal-controlled oscillator is generally used in the Armstrong transmitter, while the reactance tube must be used with a variable-frequency oscillator. The results of each of these systems are the same; that is, the FM wave created by either system can be received by the same receiver.

The reactance-tube system of frequency modulation is shown in the block diagram of Fig. 7·25. The oscillator is self-excited, usually operating in a Hartley circuit. Another tube, called the *reactance tube,* is connected in parallel with the tank circuit of the oscillator stage. By means of a suitable circuit, this reactance tube can be made to act as either a capacitive or an inductive reactance. This reactance of the tube is varied in accordance with the audio-modulating frequency. The frequency of the oscillator is changed because of the changing reactance connected across its tank circuit. Thus an FM signal appears in the output of the oscillator stage. The frequency-modulated carrier then passes through a frequency doubler in order to increase both the carrier frequency and the deviation. A power amplifier feeds the final signal into a suitable antenna.

## 7·11 FCC allocations

Every radio transmitter requires a certain amount of the radio spectrum or bandwidth for its operation. Designated channels or frequencies have been assigned to all countries of the world for radio broadcasting and other radio communications facilities. In many cases, channels are shared by more than one station, since there are insufficient channels to allocate exclusive use to all stations. Low-powered stations can serve a small community and need provide signals only for distances of, say, up to 100 miles, so that other stations in distant cities may use the same frequency without danger of interference. On the other hand, an international broadcast station, transmitting between countries and around the world, requires exclusive use of its channel. Each country in the world regulates the

*Fig. 7·25 Block diagram of FM transmitter with reactance tube to produce frequency modulation.*

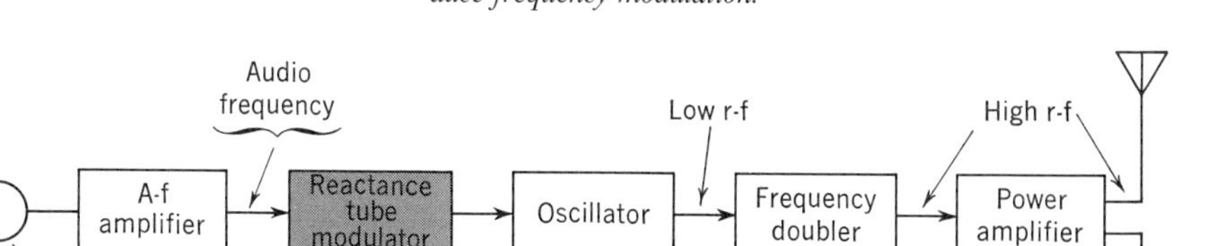

channels assigned to it by an international agreement. The FCC allocates and regulates the channels available to the United States so that different services will not interfere with each other. (See Appendix B.)

The standard AM broadcast band extends from 535 to 1,605 kc. An AM transmitter is assigned a definite carrier frequency within this band. The FCC permits only a 20-cycle plus or minus variation from the assigned frequency. This high degree of accuracy in assigned frequency is necessary to prevent interference with other stations. Broadcast-band stations are spaced 10 kc apart, and any deviation from the assigned operating frequency may produce sideband frequencies in the channel of adjacent stations. In the United States, the standard broadcast stations are divided into three categories, depending on the nature of the service they are to render:

1. Clear-channel stations for servicing very large areas. These stations usually operate with 50-kw power output and are granted exclusive use of the specified channel.
2. Regional stations for covering areas consisting of large combined urban and rural districts. Maximum power output for these stations is 5 kw. These station assignments are duplicated around the country. Particular care is taken so that by selection on a geographic basis interference between stations is negligible. Where adjacent channels are separated by 10 kc, the highest audio modulation frequency that can be used is 5,000 cps to keep the sideband frequencies within the assigned channel.
3. Local stations for very limited coverage, operating on a maximum of 250 watts. These assignments are also duplicated across the country.

Standard broadcast stations are sometimes licensed to operate on higher power during the daytime than at night. This reduces the interference caused by increased sky-wave transmission through the atmosphere after sundown.

The commercial FM broadcasting band has been allotted the 88–108-Mc portion of the spectrum. The total bandwidth for each assigned channel is 200 kc. This includes a 25-kc "guard band" on either side of the maximum frequency deviation of 75 kc to prevent adjacent channel interference during overmodulation. The highest audio-modulating frequency is 15 kc for FM broadcasting.

Commercial television broadcasting stations operate in the following

*Table 7·1 Short-wave bands*

| MINOR BANDS, Mc | MAJOR BANDS, Mc |
|---|---|
| 3.200—3.400 | 9.500— 9.775 |
| 4.750—5.060 | 11.700—11.975 |
| | 15.100—15.450 |
| 5.950—6.200 | 17.700—17.900 |
| 7.100—7.300 | 21.450—21.750 |
| | 25.600—26.100 |

*Table 7 · 2 Amateur radio bands, Mc*

| | |
|---|---|
| 1.800— 2.000 | 420.000— 450.000 |
| 3.500— 4.000 | 1,215.000— 1,300.000 |
| 7.000— 7.300 | 2,300.000— 2,450.000 |
| 14.000— 14.350 | 3,300.000— 3,500.000 |
| 21.000— 21.450 | 5,650.000— 5,925.000 |
| 28.000— 29.700 | 10,000.000—10,500.000 |
| 50.000— 54.000 | 21,000.000—22,000.000 |
| 144.000—148.000 | Above 30,000.000 |
| 220.000—225.000 | |

frequency bands: 54–72, 76–88, and 174–216 Mc for the VHF channels and 470–890 Mc for the ultrahigh-frequency (UHF) channels. The prescribed television channel width is 6 Mc. This includes the modulated picture-carrier signal and sound signal. The picture signal is amplitude modulated, but the sound is an FM signal, with a maximum frequency deviation of 25 kc.

Short-wave, or international broadcasting, stations transmit in the region between 3 and 30 Mc, in bands allocated for this purpose. There are six major and four minor international short-wave bands, as listed in Table 7 · 1.

Amateur or "ham" broadcasting bands are also designated by international agreement. The assignments are made at approximately harmonic intervals over the entire frequency spectrum. The amateur bands are listed in Table 7 · 2.

## SUMMARY

1. A transmitter is a device to produce r-f energy that is controlled by the intelligence to be transmitted.
2. Four types of radio waves are damped waves, continuous waves (c-w), interrupted continuous waves (i-c-w), and modulated waves (m-c-w).
3. Modulation is a process of combining signal voltage with a higher-frequency r-f carrier wave.
4. Modulation may be accomplished by keying in a radiotelegraph transmitter, by voice or music in a radiotelephone transmitter, or by light-dependent impulses in a facsimile or television transmitter.
5. All electric circuits carrying alternating current radiate certain amounts of energy in the form of electromagnetic waves. Efficient radiation will be obtained if the current has a sufficiently high frequency.
6. As a result of amplitude modulation, the modulated wave consists of the carrier, the upper sideband, and the lower sideband. Each sideband differs in frequency from the carrier by the amount of the audio frequency.
7. The intelligence of the modulated wave is entirely within the sidebands.
8. Percentage of amplitude modulation is determined by the ratio of the audio to the unmodulated carrier amplitude.
9. Amplitude modulation may be accomplished in the plate, grid, screen, suppressor, or cathode circuits of an r-f power amplifier.
10. High-level modulation usually means modulating in the plate circuit of the final r-f power-amplifier stage. Low-level modulation is the modulation of any r-f stage before the final amplifier.

11. The MOPA is a simple transmitter consisting of a master oscillator and a power amplifier. It is used in portable equipment and where low power output is satisfactory.
12. The main function of a buffer amplifier is to isolate the oscillator from the rest of the transmitter in order to reduce the loading effect.
13. Frequency multipliers are harmonic generators operated class C in order to produce the maximum harmonic output. Doublers and triplers are most common.
14. Single-sideband transmission uses only one sideband of the modulated signal, for reduced bandwidth and less power.
15. Frequency modulation (FM) is less susceptible to noise interference. In this respect it is superior to amplitude modulation (AM).
16. To achieve frequency modulation, we use the audio modulating voltage to vary the frequency of an r-f carrier. The audio frequency determines the rate at which the r-f carrier shifts back and forth. The amplitude of the audio-modulating voltage determines how far from its central or resting frequency the FM carrier shifts. The amount of shift from center is the frequency deviation.
17. In commercial practice, the maximum deviation followed for any channel is set at $\pm 75$ kc, a total swing of 150 kc. In addition, two guard bands of 25 kc each are placed at each end of the band.
18. Two methods of producing frequency modulation are employed commercially. One is the reactance-tube system; the other is the Armstrong phase-modulation system.

## SELF-EXAMINATION

Here's a chance to find out how well you have learned the material in this chapter. Work the exercises; then check your answers against the Key at the back of this book. These exercises are for your self-testing only.

1. When human voice and music are transmitted, the type of communication employed is known as (*a*) radiotelegraphy; (*b*) audio frequency; (*c*) wired radio; (*d*) radiotelephony.
2. A radio transmitter is essentially a device for (*a*) producing carrier frequencies; (*b*) producing r-f energy that is controlled by some intelligence; (*c*) producing an audio voltage; (*d*) modulating a carrier with r-f energy.
3. Which of the following is commonly found in a broadcast transmitter? (*a*) Class C audio amplifiers; (*b*) crystal-controlled oscillator; (*c*) class A final r-f power amplifier; (*d*) tuned modulator.
4. In a plate-modulated class C amplifier, the (*a*) carrier power is supplied by the modulator; (*b*) sideband power is always equal to the carrier power; (*c*) sideband power is obtained from the modulator; (*d*) carrier power is supplied by the speech amplifier.
5. In order to obtain a frequency of six times the oscillator frequency, we use (*a*) three doublers; (*b*) two triplers; (*c*) one tripler and one doubler; (*d*) two doublers.
6. Amplifiers following the modulated stage must be (*a*) harmonic devices; (*b*) linear devices; (*c*) class C-operated amplifiers; (*d*) nonlinear devices.
7. Buffer amplifiers are primarily (*a*) audio amplifiers; (*b*) frequency multipliers; (*c*) degenerative amplifiers; (*d*) r-f isolating stages.
8. In broadcast transmitters, the oscillator is *not* modulated, because (*a*) it would not produce sidebands; (*b*) it is a low-power stage; (*c*) its frequency might change; (*d*) it is a nonlinear device.
9. If 1-volt audio signal produces 10-kc frequency deviation, 2 volts will produce a deviation of (*a*) 2 kc; (*b*) 10 kc; (*c*) 20 kc; (*d*) 75 kc.
10. If a 500-cycle 3-volt audio signal produces a frequency deviation of 30 kc, the rate of the frequency swings is (*a*) 500 cps; (*b*) 3 kc; (*c*) 75 kc; (*d*) 500 kc.

## QUESTIONS AND PROBLEMS

1. In a MOPA transmitter for radiotelephone operation, name the two stages in the r-f section and the two stages in the audio section.

2. Give two methods of modulating the final r-f power amplifier, describing briefly how each method varies the amplitude of the r-f carrier.
3. Define percentage of amplitude modulation. Give one disadvantage of overmodulation. Why is too little modulation undesirable?
4. In the Heising plate-modulation circuit in Fig. 7·13, give the function of $L_2$, $R_1$, $C_5$, and $C_6$. What would be a typical value for $C_5$ if $R_2$ were 2,000 ohms?
5. In the transformer-coupled plate-modulation circuit in Fig. 7·14, give the function of $C_2$, $C_3$, $C_4$, $T_1$, and $T_2$.
6. Give two functions of a frequency multiplier stage in an FM transmitter.
7. What are three differences between an AM signal and an FM signal?
8. What is meant by a reactance tube? How is it used in an FM transmitter?
9. List the frequencies for the following radio services: (*a*) AM standard radio broadcast band; (*b*) FM commercial broadcast band; (*c*) Amateur band that is the second harmonic of the 1.8–2.0-Mc band.
10. A 1-Mc carrier is amplitude-modulated by a-f voltage with a frequency of 5,000 cps. (*a*) What are the resultant sideband frequencies, in Mc? (*b*) What bandwidth is required for this AM signal, in kc?

# Chapter 8 Transmitter circuits

This unit describes how an r-f oscillator can be combined with r-f power amplifiers in transmitter circuits. There is also a description of circuits to key the carrier wave on and off for transmitting dots and dashes in the Morse code (Appendix H). Also, the practical procedures in tuning a transmitter are explained in detail. The topics are as follows:

8·1 Types of Emission
8·2 Power Oscillator as a Transmitter
8·3 R-f Stages in Transmitters
8·4 R-f Power-amplifier Circuits
8·5 Neutralizing Circuits
8·6 Frequency Multipliers
8·7 Interstage Coupling and Driving Power
8·8 Power-amplifier Tuning
8·9 Keying Methods
8·10 Parasitics
8·11 Large Power Tubes
8·12 Sources of Power for Transmitters

## *8·1 Types of emission*

The transmitter circuits for a particular job depend on the amount of output power required, the carrier frequency, and the type of emission. The emission refers to the modulation characteristics of the signal. Without any modulation, the transmission is a continuous wave, or just the carrier wave, which is called *c-w* emission. When the unmodulated carrier wave is interrupted for dots and dashes, this method is *i-c-w* emission. When the carrier wave is modulated by an audio tone, generally 1,000 cps, the emission is *m-c-w*. Both i-c-w and m-c-w are used for radiotelegraph

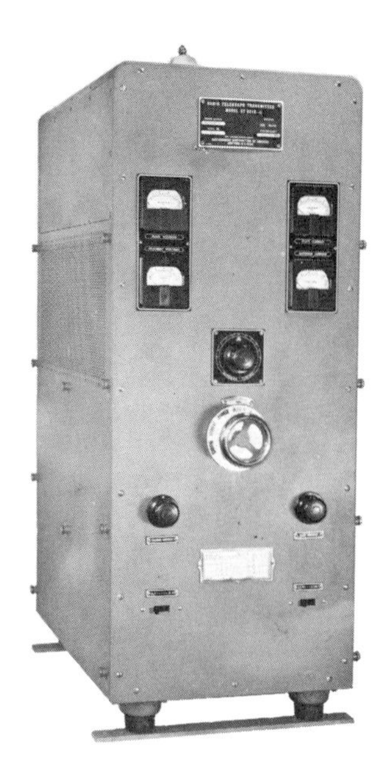

*Fig. 8 · 1 Marine transmitter. Frequency range 375 to 500 kc; power output 200 watts; for c-w or m-c-w transmission.* (Radiomarine Corp. of America.)

transmission of dots and dashes. With i-c-w, the modulation is inserted at the receiver, by means of a beat-frequency oscillator (BFO) to produce an audible tone. Radiotelegraph transmission without the audio tone is also called c-w emission because there is no tone modulation. Figure 8 · 1 shows a marine radiotelegraph transmitter, for either c-w or m-c-w emission. In radiotelephone transmission, the carrier is modulated by the audio signal corresponding to the voice or music information.

These types of emission are classified by the FCC, as listed in Table 8 · 1. The letter A is used to indicate amplitude modulation, while F is for frequency modulation. Also, telegraphy without modulation is classified by the number 1, with modulation by number 2, while number 3 is used for telephony.

The standard frequency sources indicated for AO emission are transmitted by the National Bureau of Standards radio station WWV at Beltsville, Md. In addition to its other services, WWV transmits 2.5 Mc, 10 Mc, 15 Mc, 20 Mc, and 25 Mc as frequency standards. It may be of interest to note that broadcast stations in the United States east of the Mississippi River have

*Table 8 · 1 Types of emission for transmitters*

| TYPE | CHARACTERISTICS |
|---|---|
| A0, continuous waves | No signaling; used for standard frequency source |
| A1, c-w telegraphy | For Morse code; no audio-modulated tone |
| A2, m-c-w telegraphy | For Morse code with audio-modulated tone |
| A3, radiotelephone | Includes commercial AM radio broadcasting |
| F3, radiotelephone | Includes commercial FM radio broadcasting |
| A4, facsimile broadcasting | Double sidebands transmitted with carrier |
| A4A, facsimile broadcasting | Single sideband with reduced carrier |
| A5C, television picture | Vestigial sidebands indicated by *C* |

call letters starting with W, while K is for stations in the west. There are a few exceptions for stations that probably started before this convention was adopted.

## *8·2 Power oscillator as a transmitter*

Originally, a power oscillator was coupled directly to an antenna and used for the transmission of intelligence by telegraphic code. At that time, transmission was at low radio frequencies, and the effects of antenna variations on the operating frequency were relatively unimportant. Even today, where the utmost in simplicity and reliability is desired, as in shipboard emergency service, such transmitters are still employed. In fact, some frequency instability may even be desirable for attracting attention to a vessel in distress.

A typical emergency transmitter, where the power oscillator is coupled directly to an antenna, is shown schematically in Fig. 8·2. Note that alternating voltage with a frequency of 350 cps is used directly on the plates of the tubes, supplied by the power transformer $T$. This produces a completely modulated wave having sidebands spaced 350 cps from the carrier, resulting in a fairly wide band in the frequency spectrum. This enables the transmitted signal to be heard more readily by receivers within range which are not tuned exactly to the carrier frequency. A Colpitts circuit is used for the oscillator, with the r-f output inductively coupled to the antenna by $L_3$ and $L_4$. The tubes $V_1$ and $V_2$ are in parallel for the oscillator signal, with both grids tied together and both plates coupled to one side of the tank circuit.

The sidebands are produced by alternating voltage on the plates from the 350-cycle a-c power line. As a result, the r-f output from the oscillator is

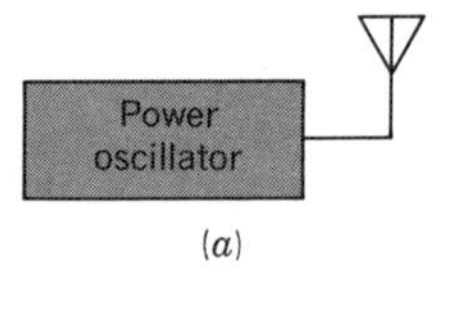


Fig. 8·2 Power oscillator as transmitter. (a) Block diagram. (b) Schematic diagram of emergency shipboard transmitter.

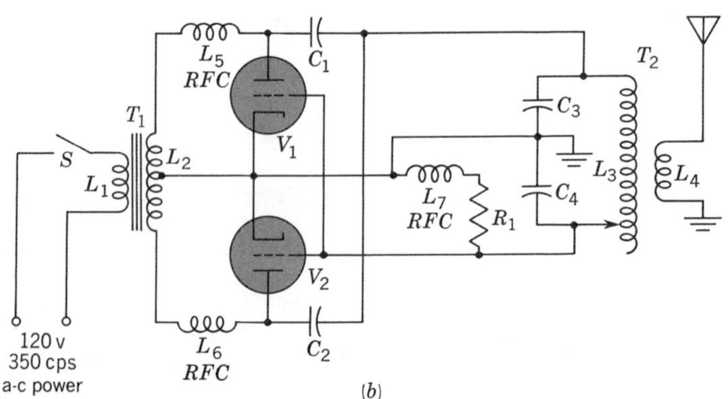

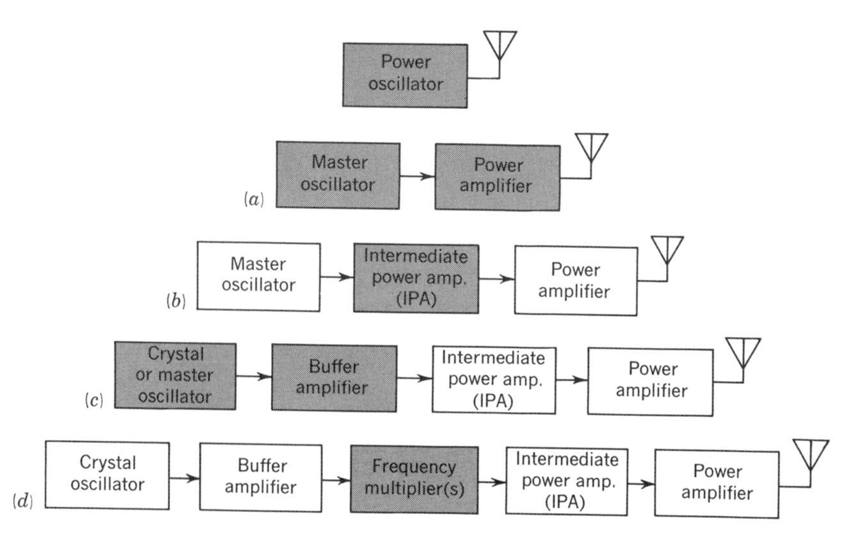


*Fig. 8·3 General arrangements for transmitter stages. (a) MOPA. (b) With IPA stage. (c) With buffer and IPA stages. (d) With buffer, IPA, and frequency-multiplier stages.*

modulated with 350-cps audio variations in the plate-supply voltage. Thus, the transmitter does not need the additional complications of a B supply and a modulator.

## *8·3 R-f stages in transmitters*

Where greater frequency stability is required, a low-power oscillator is used, and one or more r-f amplifiers raise the signal to the desired power level. Also, the power supply normally provides filtered B+ for the plate and screen elements, so that the transmitted signal will not have any spurious modulation.

Figure 8·3 shows several transmitter arrangements with one or more r-f stages following the oscillator. In the MOPA transmitter in *a*, the master oscillator drives the final power-amplifier stage, which is coupled to the antenna. The r-f amplifier thus separates the oscillator from the varying load of the antenna. When the oscillator can be tuned to different frequencies, it is known as a *variable-frequency oscillator,* or VFO.

For transmitters operating on higher frequencies, where frequency-stability requirements become even more important, the circuits usually include additional r-f amplifier stages. Also, more power output may require an intermediate power amplifier (IPA) stage to drive the final stage. This is shown in *b*. The buffer stage in *c* is an r-f amplifier which isolates the oscillator from the IPA stage. A crystal oscillator is shown for improved frequency stability. For VHF operation of the transmitter, usually one or more frequency-multiplier stages provide output at a harmonic of the crystal oscillator frequency. This is illustrated in *d*.

## *8·4 R-f power-amplifier circuits*

The term *r-f amplifier,* as applied to transmitters, generally means a power amplifier. An r-f power amplifier may be class A, B, or C, depending on the particular requirements. Class C operation is most common, however. A class C amplifier has grid bias considerably greater than cutoff. As a result, plate current flows for less than 180° during each cycle of the operating frequency. The amplifier has a tuned load impedance that is resonant to the operating frequency, in order to produce sine-wave output. Class C operation is characterized by high efficiency. It is used to generate large amounts of r-f power at a single frequency.

**General considerations.** Since a class C amplifier operates with a grid bias in the order of 1½ to two times the cutoff value, the plate current flows in extremely short pulses and by no means even approaches a reproduction of the grid signal. Consequently, the harmonic distortion present in the plate current is extreme. It is the function of the tuned output circuit, or tank circuit, to filter out most of the harmonics present in the plate-current pulses in order to re-create a sinusoidal voltage across the load. Also, the tank circuit has the function of supplying the negative half-cycles of alternating plate current through its flywheel, or oscillatory action.

The efficiency of the class C amplifier is high because plate current is permitted to flow only during the period when the instantaneous plate voltage is low. Under such conditions, the average voltage drop across the tube during plate-current flow is low, and a minimum amount of power is dissipated at the plate. More of the input power is therefore available for useful output power. Well-adjusted amplifiers can provide plate efficiencies in the order of 70 to 80 per cent.

To achieve the flow of plate current in short pulses and to obtain a reasonable amount of average plate current, it is necessary to drive the grid voltage well into the positive region. As a result, rectified grid current flows, and a certain amount of power is dissipated in the grid resistor. This power must be supplied from whatever source drives this stage. The input grid signal to the amplifier is its *grid drive,* or *excitation voltage.*

**Typical circuits.** Several arrangements for a class C r-f power amplifier are illustrated in Figs. 8·4 to 8·7. The circuits differ by the method

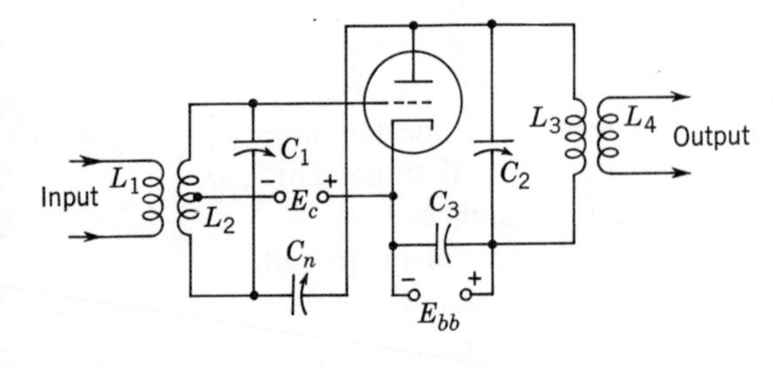


*Fig. 8·4 Class C amplifier circuit with inductive coupling, grid neutralization, fixed bias, and series feed for B+ ($E_{bb}$).*

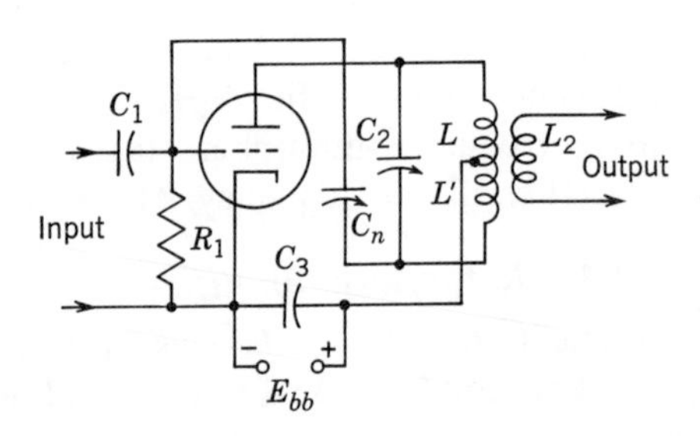


*Fig. 8·5 Class C amplifier circuit with capacitive-input and inductive-output coupling, plate neutralization, grid-leak bias, and series feed for B+ ($E_{bb}$).*

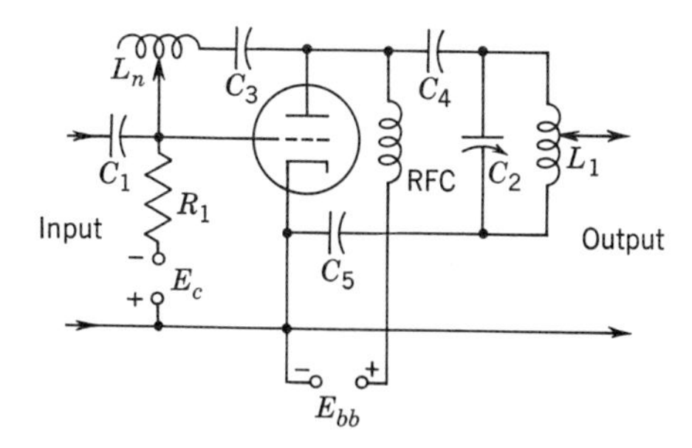


*Fig. 8·6 Class C amplifier circuit with capacitive-input and direct-output coupling, inductive neutralization, grid-leak bias and fixed bias, and shunt feed for $B+(E_{bb})$.*

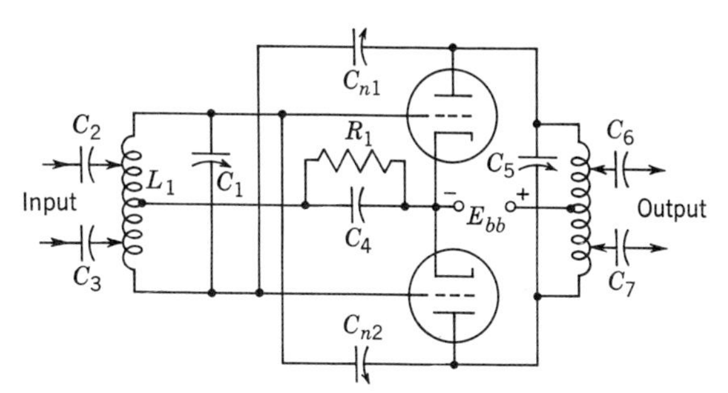


*Fig. 8·7 Class C amplifier circuit with capacitive coupling, cross neutralization, grid-leak bias, and series feed for $B+(E_{bb})$.*

of obtaining grid bias, by the input and output coupling, and by the manner of feeding d-c voltage to the plate of the tube. Also, different neutralizing circuits are used. These serve to cancel feedback, which can make the amplifier oscillate.

In the amplifier shown in Fig. 8·4, the input signal is inductively coupled to the grid of the tube. The power developed by the amplifier is also inductively coupled to the load. Bias for the amplifier is obtained from a fixed source, such as a generator, rectifier-filter system, or battery. Since the d-c plate current of the tube must flow through the tuned circuit, the amplifier is termed *series-fed.* With this particular arrangement, grid neutralization is used, although other methods are possible.

Figure 8·5 illustrates another amplifier. In this circuit, excitation is coupled to the grid by capacitive coupling. Grid-leak bias is provided by $R_1C_1$. The output circuit, however, uses inductive coupling for transfer to the load. Here, plate neutralization is used to prevent oscillation. For this reason, the output-tuned circuit is center-tapped, like the grid circuit in Fig. 8·4.

The circuit shown in Fig. 8·6 differs in several respects from the two previous circuits. The major difference is an adjustable coil $L_n$ for neutralization. Also, it should be noted that the tuned output circuit carries no d-c plate current. An r-f choke is used for plate-supply voltage and to conduct d-c plate current. Since this coil is in parallel with the tuned output circuit, the arrangement is said to be *shunt-fed,* in terms of the B supply. Bias for this amplifier is obtained from a fixed source, plus the grid-leak bias from $R_1C_1$. This type of bias supply is called *combination* bias. Although the grid obtains its excitation by capacitive coupling, the load is coupled to the output circuit by a direct connection to coil $L_1$.

A push-pull amplifier is shown in Fig. 8·7. In this circuit, both input and output circuits are tuned. Each uses capacitive coupling for the excitation and for the load. It will be noted that the circuit is completely symmetrical even to the neutralizing arrangement (*cross neutralization*). The amplifier uses grid-leak bias only and is therefore dependent upon a constant excitation signal. Should excitation fail, the plate current will become excessive unless additional bias is used. Since the tuned output circuit carries d-c plate voltage and current, this is also a series-fed amplifier.

## *8·5 Neutralizing circuits*

An amplifier will oscillate if enough energy of the correct frequency and phase is fed back from the output circuit to the control grid of the tube. The energy transfer can be accomplished through the grid-plate capacitance of the tube or by mutual coupling external to the tube. By suitable shielding of input and output circuits of the amplifier, however, the external coupling can be reduced to prevent oscillation. Neutralization may still be necessary, though, because of internal feedback through the tube.

At moderate frequencies, tetrodes, pentodes, and beam-power tubes usually will not have enough internal feedback to cause oscillation. With triode tubes, however, the grid-plate capacitance is large enough to require definite steps to prevent oscillation. This process, though fundamentally a stabilization process, is known as *neutralization*. Neutralization arrangements generally use some method to introduce a voltage on the grid that is equal and opposite to the voltage fed back from plate to grid by the grid-plate capacitance $C_{gp}$. These two voltages therefore cancel to prevent oscillations and allow the stage to operate as an amplifier.

**Grid neutralization.** The neutralizing arrangement shown in Fig. 8·4 is the *grid,* or *Rice,* system of neutralization. In this circuit, voltage fed back to the grid end of the input coil through $C_{gp}$ is canceled by equal and opposite voltage fed to the opposite end of $L_2$ through the neutralizing capacitor $C_n$.

The neutralizing action occurs in $L_2$. The top or grid end of $L_2$ has applied between it and the center tap the voltage fed back from the plate to the grid through $C_{gp}$. This voltage develops a certain current flow through the top half of $L_2$. At the same time, the same feedback voltage is applied, through $C_n$, to the bottom half of $L_2$. The current produced by this voltage flows in a direction opposite to the current in the top half of $L_2$ and the fields of the two currents effectively cancel, thereby eliminating the effect of the feedback.

**Plate neutralization.** Figure 8·5 illustrates the neutralizing circuit called the *plate, Hazeltine,* or *neutrodyne* system. With this arrangement, any voltage fed back to the grid by $C_{gp}$ is canceled by equal and opposite voltage fed from the plate circuit to the same point through the neutralizing capacitor. Here the plate coil is tapped; in grid neutralization the grid coil is tapped.

**Inductive neutralization.** The neutralization illustrated in Fig. 8·6 is commonly used in broadcast transmitters and in other transmitters operating continuously on one frequency. The neutralizing inductor $L_n$ is adjusted

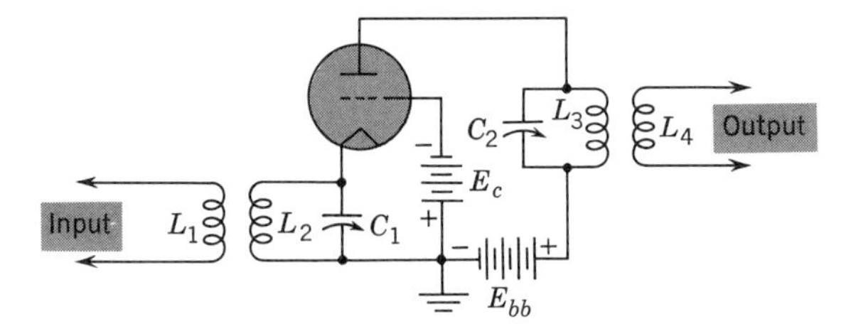


*Fig. 8·8 Grounded-grid amplifier circuit.*

for parallel resonance at the operating frequency with the grid-plate capacitance $C_{gp}$. The combination offers a high impedance to any energy being transferred from the plate circuit to the grid circuit. Since this impedance is maximum at but one frequency, a particular adjustment is effective at that frequency only.

**Cross neutralization.** In Fig. 8·7 is shown the conventional manner of neutralizing a push-pull amplifier. Note that each plate is cross-coupled to the opposite grid by the neutralizing capacitors.

**Grounded-grid neutralization.** At very high frequencies, the inductances of the various connecting leads, including those of the tube itself, make it almost impossible to secure effective neutralization. Then the grounded-grid circuit is used (see Fig. 8·8). In this circuit, the control grid acts as a shield between plate and cathode, effectively preventing energy transfer between the input and output circuits. This system of neutralization has the advantage of being effective over a wide range of frequencies and requiring no adjustment.

**Neutralization of tetrodes and pentodes.** A screen-grid tube has a very low value of $C_{gp}$ but high power amplification or high power sensitivity. In many circuits it may not be necessary to neutralize a screen-grid tube because of the low $C_{gp}$. Relatively small amounts of r-f energy fed back through the grid-plate capacitance can often be sufficient to cause oscillation, however, especially at the higher frequencies where the reactance of $C_{gp}$ becomes small. In such cases, it may be necessary to use a neutralization circuit to prevent oscillation, as with triodes.

## *8·6 Frequency multipliers*

If, in the previous single-tube amplifiers, the plate-tank circuit is tuned to twice the excitation frequency, the stage will deliver output power at the second harmonic of the input frequency. Similarly, when the tank is resonated to other exact multiples of the excitation frequency, the amplifier will deliver power at those harmonics. There is no power output at lower frequencies. Such an r-f amplifier is a frequency multiplier. Multipliers are commonly used in high-frequency transmitters so that the crystal oscillator can operate at a low and, generally, more stable frequency.

The performance of a frequency multiplier is similar to that of a straight

*Table 8 · 2 Plate conduction angle for frequency multipliers*

| *Harmonic* | *$\theta_p$, degrees of fundamental frequency* | *Relative power output, per cent* |
|---|---|---|
| 1 (fundamental) | 120–160 | 100 |
| 2 | 90–120 | 65 |
| 3 | 80–120 | 40 |
| 4 | 70– 90 | 33 |
| 5 | 60– 72 | 25 |

class C amplifier, although the portion of the cycle during which plate current flows is more important. The plate conduction angle $\theta_p$ states what part of the input cycle drives the grid voltage past cutoff to allow plate-current flow. Assuming sufficient grid drive, plate-current flow depends on the amount of bias. More negative bias results in a shorter plate-current flow.

The output at a harmonic of the exciting frequency is extremely sensitive to $\theta_p$. Table 8 · 2 lists the range of plate conduction angles for various harmonics to give optimum performance of the frequency multiplier. It also lists the approximate percentage of power output that can be expected compared with the same amplifier operating as a straight class C amplifier. Because of the rapidly decreasing amounts of power output obtainable with increasing harmonic number, it is usually not practical to attempt frequency multiplication by more than a factor of 5. Doublers and triplers are most common.

To obtain more output power from a frequency multiplier, it is possible to employ two tubes instead of one. It should be borne in mind, however, that the ordinary push-pull amplifier discriminates against even-order harmonics. It can be used successfully, however, for odd-harmonic multiplication.

The so-called *push-push* amplifier, shown in Fig. 8 · 9, can be employed for even harmonics, such as the second, fourth, and so on. In this amplifier, the two grids are effectively in push-pull, but the two plates are in parallel. The two tubes have opposite polarities of input voltage. When one grid is driven positive, it produces a pulse of plate current. On the alternate half-cycle, the opposite grid is positive, producing another pulse of plate current. As a result, the number of pulses of plate current is doubled. A more symmetrical voltage is developed across the tank circuit, therefore, because the tank does not have to coast as much as it would with a single tube. For a doubler, the tank tuned to the second harmonic of the input will receive a plate-current pulse every cycle instead of every other cycle.

Because of the difference between input and output frequencies, it is usually not necessary to provide neutralization for frequency multipliers, even with triodes.

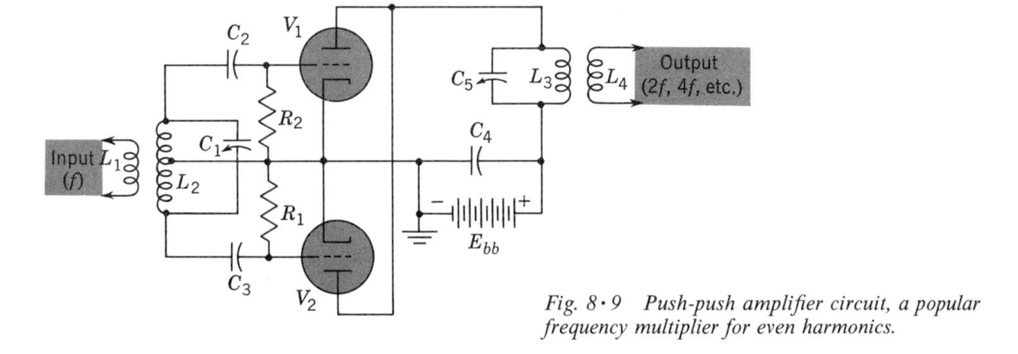


*Fig. 8 · 9 Push-push amplifier circuit, a popular frequency multiplier for even harmonics.*

## *8 · 7 Interstage coupling and driving power*

An important consideration in r-f amplifiers is the method of coupling the output of one stage to the input of the succeeding stage. Interstage coupling circuits must transfer energy efficiently at the operating frequency, discriminate against harmonics of the operating frequency, provide d-c isolation between stages if required, and be capable of adjusting the amount of energy to be transferred. There are three general types of coupling employed. These are inductive, indirect inductive, and capacitive coupling. It must be remembered that the applications shown are by no means the only ones, but illustrate common practice only.

Figure 8 · 10 shows two examples of capacitive coupling which differ only in the manner of varying the degree of coupling. At *a*, the amount of coupling is varied by adjusting coupling capacitor $C_3$. At *b*, $C_3$ is fixed and the coupling is varied by moving the adjustable tap on the tank inductor $L_1$ of the driving stage. The coupling capacitor should have a value which will provide a low reactance at the operating frequency. In addition it should be made specifically for radio frequencies (with low inductance) and should have a voltage rating high enough to withstand the maximum voltages developed between the plate circuit of the driver and the grid circuit of the driven stage. Generally mica capacitors are used.

The inductive coupling shown in Fig. 8 · 10*c* is a very efficient means of transferring energy. This method inherently discriminates against harmonics of the operating frequency and hence is particularly suitable for this purpose. The amount of energy transfer may be varied by physically moving the secondary of the transformer closer to or farther away from the primary, thereby changing the amount of coupling between $L_1$ and $L_2$. Or the coupling may be varied by increasing or decreasing the number of secondary turns by means of adjustable taps on the secondary. This method is often undesirable, however, since unused turns can cause arc-over or parasitic oscillations. The secondary may be tuned or untuned to the operating frequency, as desired. Tuning the secondary provides harmonic discrimination, but has the disadvantage of adding an additional operating control to the transmitter.

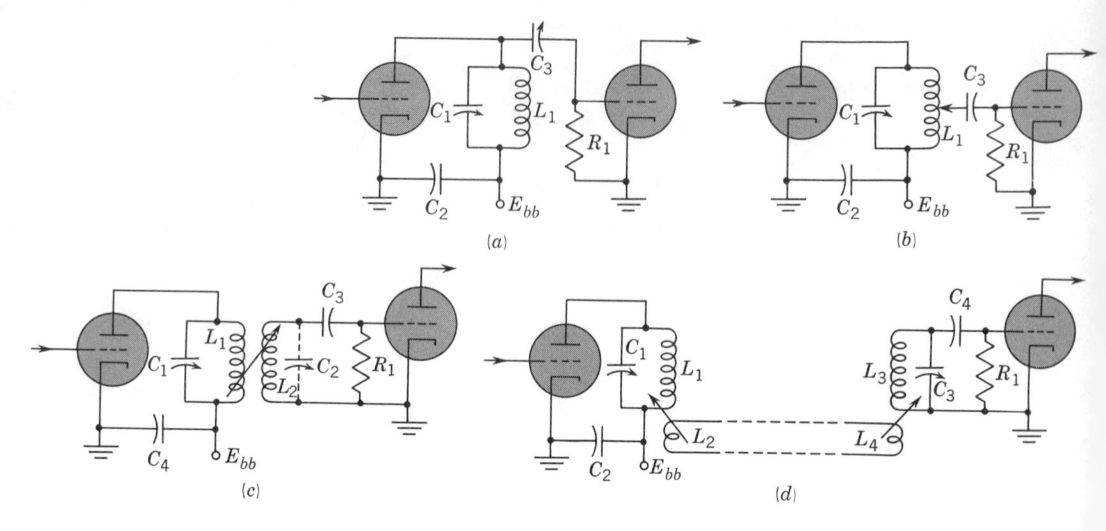


*Fig. 8 · 10 Methods of interstage coupling. (a) Capacitive. (b) From tap on coil $L_1$. (c) Inductive. (d) Link coupling.*

**Link coupling.** The coupling methods illustrated in Fig. 8 · 10*a*, *b*, and *c* involve high impedances on both sides of the coupling device. They therefore require that the driving stage and the driven stage be physically close together to avoid undue losses, radiation, and insulating problems. The method shown in Fig. 8 · 10*d* permits the two stages to be separated rather easily. This arrangement, called *link coupling*, is used quite extensively. It consists of a few turns coupled to the plate tank coil and feeding another coil of a similar number of turns coupled to the grid tank of the driven stage. Because of the turns ratios of the two transformers involved, the impedance of the connecting transmission line may be quite low. Although the coupling between the windings of each transformer is shown to be variable here, the usual practice is to vary the coupling at one transformer only.

The link windings should be coupled to their respective tank coils at points of minimum r-f voltage. In single-ended stages, as shown, this places the coupling at the ground end of the coils. In push-pull stages, the coupling is made at the midpoint of the tank coils. This reduces the amount of energy coupled to the coils by capacitive means, thereby reducing the transfer of harmonics to the output signal.

If desired, one side of the transmission line may be grounded, since no direct current is carried by it. Since the transmission-line impedance is low, insulation requirements are easily met. The line may take the form of a twisted pair, a ribbon line, a coaxial cable, or an open-wire line. Although link coupling is not as efficient as direct inductive coupling, its advantages are many. As with direct inductive coupling, link coupling inherently suppresses the transfer of harmonic frequencies.

**Driving power.** The amount of signal-power input required to drive a power r-f amplifier is usually given in a tube manual for transmitter tubes. This power, it must be realized, represents only that amount dissipated in the grid-bias circuit and in the internal grid-cathode circuit of the driven tube. In addition, driving power is also consumed by losses in the tubes, sockets, circuit components, wiring, and so on. The general rule up to a frequency of 30 Mc, approximately, is that twice the computed driving power for the driven tube is required from the driving stage to compensate for the additional losses. Above this frequency, from three to ten times the computed power is required, depending on the frequency and the circuit.

In addition to the above requirements, it is good practice to supply enough additional driving power to produce saturation in the driven tube. Saturation is realized when a substantial increase or decrease in driving power is not accompanied by an appreciable increase or decrease in power output of the driven stage. This provision is a safety factor to prevent loss of output power through deterioration of driving tubes, misadjustments of the driving stages, and so on. It is important, however, not to exceed the maximum grid-input ratings of the driven tube.

## *8·8 Power-amplifier tuning*

A typical power amplifier (PA), together with its source of driving power, is shown in Fig. 8·11, to be used as a reference here for the tuning procedure.

**Neutralizing.** Since the amplifier uses a triode, it will be necessary to adjust the neutralization before any other adjustments are made. When the amplifier $V_2$ is first neutralized, the plate-supply switch $S$ is opened. The plate circuit of the driver $V_1$ is then tuned to resonance by varying $C_1$. Also, the coupling is adjusted so that a reasonable amount of PA grid current is indicated on the grid ammeter labeled $I_c$. The driver may be resonated by

*Fig. 8·11 Adjustments and meters for tuning r-f power amplifier. See text for tuning procedure.*

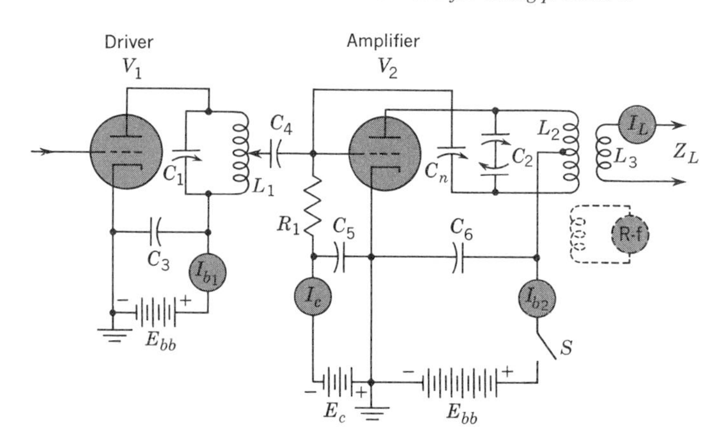

tuning its tank circuit for minimum driver plate current $I_{b_1}$ or for maximum amplifier grid current $I_c$.

If the amplifier is not neutralized, a small amount of r-f current will appear in the amplifier tank circuit $L_2C_2$, even without B+ voltage. The r-f signal in the tank is indicated by a small coil and r-f galvanometer, temporarily coupled to $L_2$. This is indicated by dotted lines in the figure.

The amplifier tank should now be tuned to resonance by varying $C_2$, still without B+. Resonance is indicated by a maximum reading on the r-f galvanometer. Then the neutralizing capacitor $C_n$ is adjusted for minimum r-f current, and the tank capacitor $C_2$ is again adjusted for maximum r-f tank current. This process is repeated over and over until the r-f tank current is either zero or at a minimum value as $C_2$ is tuned through the resonant point. After each adjustment of $C_n$, the driver tank should be retuned to resonance.

If you do not have an r-f indicator for tank current, the grid ammeter can be used as a sensitive indicator of neutralization. With no plate voltage on the amplifier, it will be found that the grid current shows a sharp variation as the plate tank is tuned through resonance. The neutralizing capacitor $C_n$ should be adjusted until the variation of grid current is zero or at a minimum as the plate tank is tuned through resonance.

At this point the power amplifier is neutralized enough to permit application of plate voltage by closing switch $S$.

**Tuning.** After neutralization is completed, the amplifier is ready to be tuned for maximum power output at the desired frequency. This adjustment is usually carried out by a trial-and-error method, using power output and d-c grid and plate current as indicators of correct operation. The following procedure can be used:

1. Adjust the PA input and output coupling so that they are at a minimum. We do not want too much load current or excessive drive until the stage is tuned.
2. Make sure the driver plate circuit is tuned to resonance. This can be done by varying $C_1$ for minimum plate current on the meter $I_{b_1}$.
3. When plate voltage is applied to the final amplifier, immediately tune its plate tank to resonance (minimum plate current on the meter $I_{b_2}$).
4. Retune the driver plate circuit ($V_1$) and the final plate current ($V_2$).
5. Note PA grid current, plate current, and power output. Normally they will all be lower than desired because of the loose input and output coupling. If so, proceed as indicated below.
6. Increase excitation from the driver by increasing the coupling until grid current of $V_2$ is as desired. Note the plate current in the final stage.
7. If this plate current is still too low, increase output coupling until plate current is the desired amount.
8. If grid current is too low as this point, or if it is not possible to increase the plate current sufficiently by increased output coupling, it will be necessary to increase excitation again from the driver.

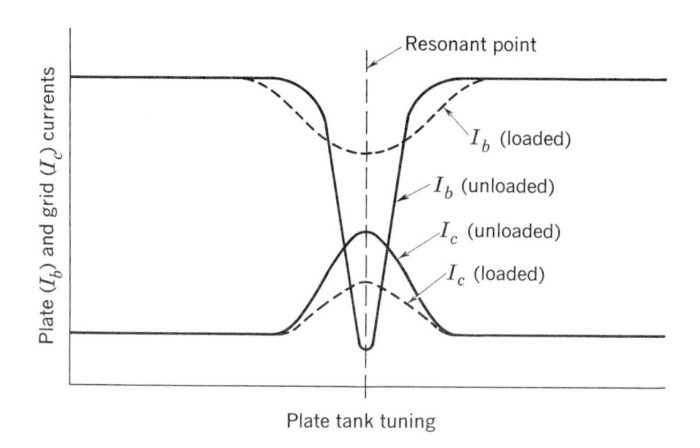


*Fig. 8·12 Variation of plate and grid currents with tuning of plate tank circuit.*

Steps 5 to 8 are repeated until the desired results are obtained, with the required amount of plate current and grid current in the final amplifier. It is important to make all these adjustments in small steps and to begin with low values of coupling.

Caution must be exercised throughout the adjustment process so that maximum ratings of the tube are not exceeded. After every change in loading or excitation, the tuning of both the driver tank and the amplifier tank should be checked to see that they are in resonance. Unless the loading is too great, resonance in the plate tank circuit is always indicated by minimum plate current on the d-c meter.

The effect of the load on the tuning of the final amplifier is shown by the curves in Fig. 8·12 for both plate current and grid current. Notice that the dip of $I_b$ to minimum is not so sharp at resonance when the plate circuit is loaded (tight coupling). Also, the grid current does not rise so sharply when the plate circuit is loaded.

After the amplifier has been adjusted, the neutralization adjustment should be checked. In the initial neutralization process, the effect of the amplifier grid-plate capacitance was largely eliminated. With the amplifier now delivering power, there is always the possibility of energy feedback outside the tube. For complete neutralization, the decrease and increase of plate current and the opposite increase and decrease of grid current should each by symmetrical around the resonant point as the tank circuit is tuned through resonance. If not, the neutralizing capacitor $C_n$ should be readjusted slightly.

## *8·9 Keying methods*

Telegraph transmitters convey information by turning the transmitter on or off in accordance with some prearranged code. This may be the Morse code or accepted machine and business codes in general use. Controlling a transmitter's output power to produce the characters of the code is accomplished by a keying circuit. Typical keys are shown in Fig. 8·13. For convenience, keying of a transmitter is generally done in a section of

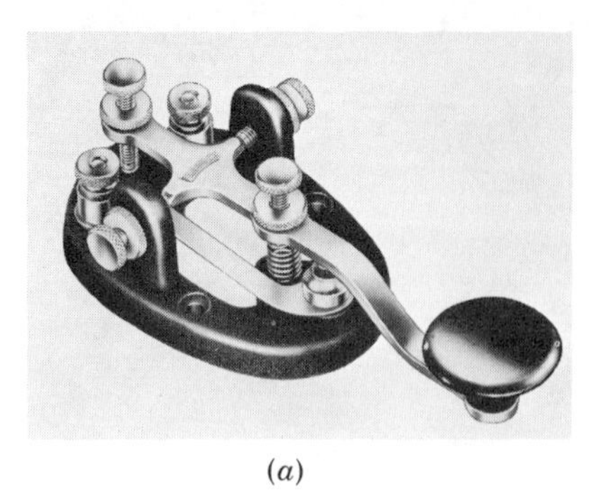

(a)

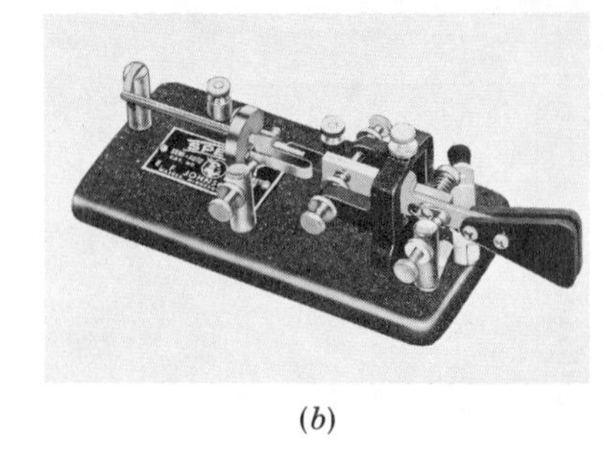

(b)

*Fig. 8 · 13 Telegraph keys. (a) Standard manual model. (b) Semi-automatic model.* (E. F. Johnson Co.)

the circuit where the power to be controlled is small. Here the current and voltage to be interrupted are small and the requirements of keys or relays are easily met. Figure 8 · 14 illustrates six methods of keying.

**Primary keying.** One of the simplest ways of keying a transmitter is shown by Fig. 8 · 14*a*, where the power supply to the transmitter tube plates is interrupted by means of a key in the primary of the plate power transformer. This method is very often used in the smaller transmitters because it has the advantage of simplicity. The speed of the code characters is limited, however, to that produced by hand keying.

**Plate-circuit keying.** This method of keying, illustrated in Fig. 8 · 14*b*, is a simple scheme, also used in many low-power transmitters. It is similar to primary keying except that the plate supply to the transmitter is interrupted after the power supply has been converted from alternating to direct current.

**Cathode keying.** Perhaps the most commonly used system of keying is the cathode-keying circuit shown in Fig. 8 · 14*c*. Here, with the key open as shown, the open cathode assumes the same positive potential as the plate. Since the grid of the tube is at ground or more negative than ground, plate current is effectively interrupted. Upon closing, however, the tube electrodes again assume their normal potential, plate current again flows, and amplification is resumed once more.

Since the key, when open, has full plate voltage across its contacts, good insulation plays an important part in this system. The current to be interrupted is relatively small, and generally no difficulty is encountered on this score.

**Grid-block keying.** Figure 8 · 14*d* shows a method of keying that applies enough additional negative bias to the grid when the key is open to prevent the grid drive from producing any plate current. In the diagram, $R_1$ provides excessive negative grid bias. When the key is closed, it shorts $R_1$, removing the blocking bias from the grid. The amplifier then operates normally with grid-leak bias provided by the input signal.

The source of the additional keying bias may be derived from batteries,

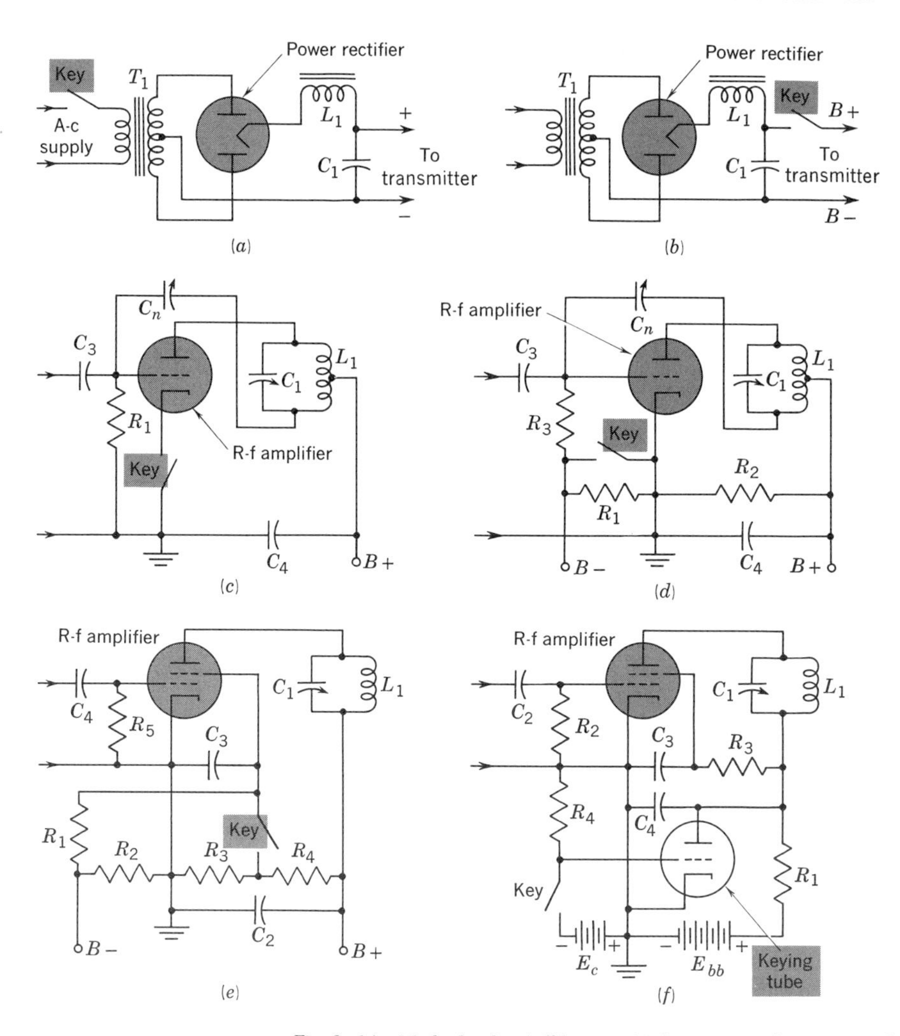


*Fig. 8·14 Methods of on/off keying. (a) In primary of power transformer. (b) In B+ supply line. (c) Cathode keying. (d) Blocked-grid keying. (e) Screen-grid keying. (f) Plate absorption keying.*

from a rectifier and filter, or from a voltage divider across the plate supply. The voltage-divider method is shown in Fig. 8·14*d*. With this type of keying, both the current and voltage to be controlled by the key may be made quite small, which is a great advantage.

**Screen keying.** It is often desired to utilize the screen grid to control the output of the keyed amplifier. One manner of accomplishing this is shown in Fig. 8·14*e*. With this particular arrangement, a negative voltage is applied to the screen when the key is open and effectively prevents any operation of the amplifier. When the key is closed, a positive voltage is applied to the screen and normal operation is resumed.

This method has the same general advantages as grid-block keying. It may also be applied to the suppressor grid of a pentode if such a tube is used.

**Absorption keying.** In Fig. 8 · 14*f*, an entirely different scheme is used. In the previous methods, keying was accomplished by direct interruption of power, by removing or applying a voltage, or by reversing the polarity of a voltage. However, the absorption method uses a separate keying tube to interrupt radiation from the transmitter.

With the key open as shown, the grid of the keying tube is at the same potential as the cathode. Then direct plate current produces a large voltage drop across resistor $R_1$. This voltage drop is in series with the supply for the plate and screen of the r-f amplifier. Therefore the net voltage at the plate and screen is reduced by the amount of the voltage drop, and the output of the amplifier decreases to a low value. The output is reduced enough to prevent excitation of the following stage. When the key is closed, a negative bias, sufficient to cut off the plate current, is applied by $E_c$ to the grid of the keyer tube. Without any plate current in the keyer tube, the voltage drop across $R_1$ is zero. This allows full B+ voltage for the plate and screen of the r-f amplifier. Then normal output is produced. The r-f output is enough to produce normal excitation of the following stage, and the transmitter again radiates.

Because of the relatively small values of voltage and current to be controlled by the key, this system can be used for extremely high keying speeds. If the key is replaced by an electronic device producing zero and negative voltages, keying speeds as high as 1,000 words, or more, a minute may be used.

In all the descriptions of keying systems, reference has been made to a key. This key may actually be replaced by a relay, or, in some instances, by a tube that alternately conducts and stops conducting in accordance with some control in its grid circuit. Only in very-low-power transmitters is the keying actually accomplished with a hand key as shown. When high voltages are considered, the reasons are obvious.

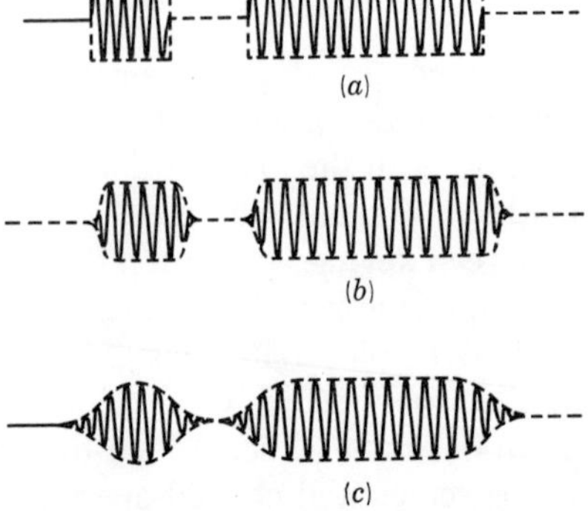

(a)

(b)

(c)

*Fig. 8 · 15 Envelopes for keyed continuous wave. (a) Wave with key clicks. (b) Wave with filtering—no clicks. (c) Excessive filtering—blurred and indistinct.*

**Key clicks.** While perfect reproduction of the original keying would seem desirable, this is not so. Figure 8 · 15*a* shows such a reproduction. An analysis of these waveforms would indicate that a great number of harmonics of the keying frequency are present because of the steep wave fronts and abrupt transitions from ON to OFF (also called *mark* to *space*). By the introduction of delay or filtering circuits in the controlled circuit, the wave fronts of the characters may be made to have a slight slope. This makes the changes from mark to space more gradual. In this way, the harmonics of the keying frequency, which appear as clicks or thumps in nearby receivers, can be reduced to a point where they are no longer objectionable. A satisfactory keying envelope is shown in Fig. 8 · 15*b*.

The usual manner of introducing delay or filtering is by inserting a small inductor in series with the current being controlled. To provide a sloping transition from mark to space, it is usual to have capacitance across the controlled circuit. If too much delay or too much filtering of the keying is attempted, the characters will become excessively rounded, as in Fig. 8 · 15*c*. This results in a radiated signal that is indistinct and often impossible to read.

**Frequency-shift keying.** Frequency-shift keying was developed shortly after World War II. The advantages are so great that in long-distance high-speed point-to-point telegraph service, it is used to the exclusion of the older on/off type of keying.

As the name implies, this type of keying does not turn the transmitter on and off, but allows it to run continuously. The intelligence to be transmitted is made to control the transmitter output frequency in such a way that it emits one frequency for mark and another frequency for space. At present, the usual spacing between the two frequencies is 400 to 1,800 cps. It is usual to have the mark frequency the higher of the two, but it is not mandatory. Most receivers designed for this keying system have provisions for receiving either mark or space as the upper frequency.

Of the many ways of generating the frequency-shift signal, Fig. 8 · 16 illustrates one system in common use. In this arrangement, the transmitter obtains its original frequency not from a conventional oscillator but from a mixer. This stage delivers to the transmitter a signal which is the sum of a signal from a crystal oscillator and a signal from a master oscillator. The master oscillator has a nominal frequency of 200 kc, but its actual frequency is controlled by a reactance modulator into which the keying signal is fed. By the action of the keying signal, which is either an on/off d-c voltage or a reversing d-c voltage, the reactance modulator either subtracts from or adds to the reactance of the master oscillator tank, thus producing a shifting frequency. The output of the mixer is then a signal whose frequency is shifted according to the keying signal. Generally, this frequency-shifted signal is generated in the 1,500- to 2,500-kc range. It is then amplified and multiplied by succeeding stages in the transmitter. The calibrating crystal oscillator, the test mixer, and the indicator are provided for the accurate setting of the center frequency of the master oscillator to 200 kc.

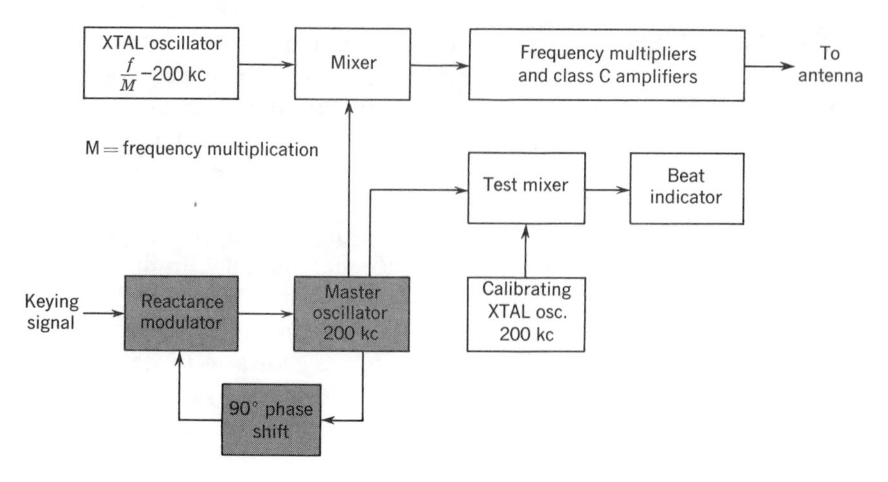


*Fig. 8·16 Block diagram of a frequency-shift telegraph transmitter.*

Class C amplifiers are used in the transmitter following the formation of the frequency-shifted signal because of their high efficiency. Distortion in these amplifiers is no problem, since the amplitude of the signal is constant and the frequency shift is a very small percentage of the center frequency.

## *8·10 Parasitics*

Parasitic oscillations are any oscillations developed in oscillators or amplifiers other than the desired operating frequency. These undesired oscillations not only give rise to spurious radiation and interferences but consume power that could otherwise be used for increased output at the desired frequency. In especially bad cases, the parasitic oscillations may cause overheating and breakdown of the components or tubes.

Parasitic oscillations are generally produced by resonances in the input and output circuits other than those designed for the operating frequency. These resonances are generally found in circuits formed by the inductances of connecting wires, combined with the stray capacitances to ground and to other parts of the physical circuit.

In the usual physical design of a power amplifier, using manufactured components, the low-potential ends of the input and output circuits are well grounded by short, heavy leads. However, the high potential ends are usually connected with relatively long leads to grid and plate. Usually connections to neutralizing capacitors are also relatively long. All such leads have a small amount of inductance, and their reactances at very high frequencies become substantial. If, by chance, such reactances combine with the reactances of stray capacitances at the same or nearly the same very high frequency in the input and output circuits of the stage, oscillations are likely to occur. When considering such possibilities, many forms of oscillator circuits may be designed. Figure 8·17 illustrates a

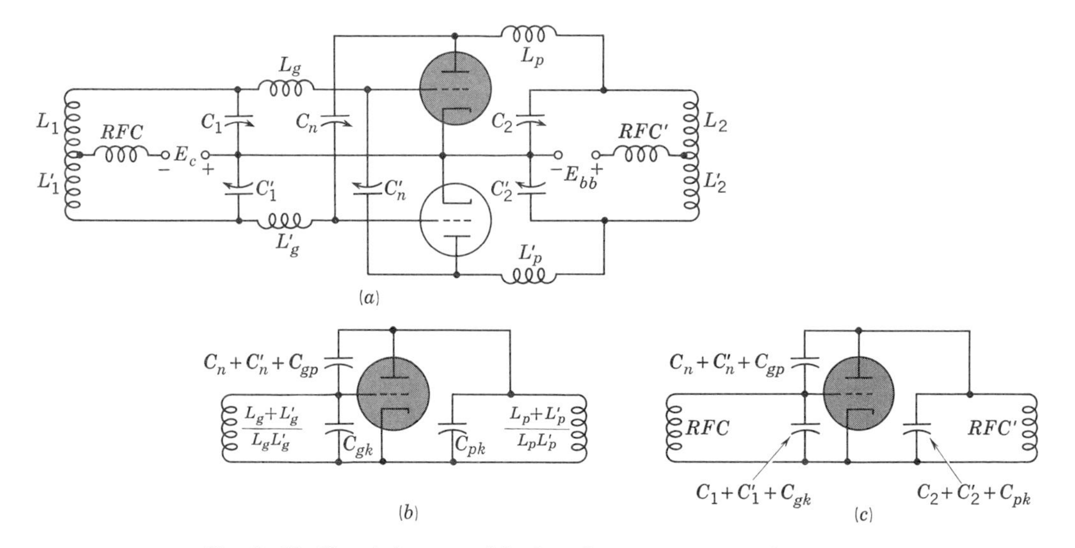


*Fig. 8·17 How inductance of leads and stray capacitance form parasitic circuits. $L_p$, $L'_p$, $L_g$, and $L'_g$ are lead inductances. (a) Actual circuit of push-pull r-f power amplifier. (b) High-frequency circuit equivalent to tuned-grid tuned-plate oscillator. (c) Equivalent circuit for low-frequency parasitic oscillations.*

common form of push-pull amplifier and the possibilities of both high- and low-frequency parasitic oscillations.

**Detection of parasitics.** The existence of parasitics may be detected in several ways. In general, parasitic oscillations produce signals which interfere with other services. If, however, the parasitics involve considerable power, they usually result in erratic operation of the stage involved or in the actual overheating or breakdown of components of the stage.

If parasitics are suspected, the usual procedure for investigation is: (1) remove excitation from the stages, (2) decrease the bias to a point where normal plate current is flowing, (3) and then explore the circuit with a neon bulb connected to a small coil. In this manner, points of high r-f potential at the parasitic frequency may be located. It is also very helpful to employ an absorption-type wavemeter to identify the frequency or frequencies of the parasitics.

**Elimination of parasitics.** Once the nature and cause of a particular parasitic is determined, the usual procedure is to alter the reactances involved by changing the lead dress. In general, short, heavy connecting leads are of utmost importance. The indiscriminate use of r-f chokes should be avoided, since these are particularly troublesome in causing parasitics. If chokes must be used, dissimilar types or values should be used in input or output circuits. A common preventive measure is to insert a noninductive resistor in parallel with a small inductance in either the grid or plate leads but not in both in the same stage. The resistor is often called a *parasitic suppressor,* or *stopper*. Simple circuits are much less likely to have parasitics than more complex arrangements.

## *8 · 11 Large power tubes*

In class C power amplifiers designed for output of a few watts, small vacuum tubes of the type commonly used in receiving equipment are employed. For greater power output, of course, larger tubes are required. The power that a particular tube can handle is determined by the plate voltage that can be applied safely, the amount of current that can be delivered by the cathode, and the amount of power that can be dissipated within the tube without dangerously overheating it.

**Air-cooled tubes.** In order to handle high anode voltages, it is necessary to provide sufficient insulation between electrodes. In air-cooled tubes the glass envelope provides this insulation in the same way as the envelope of a receiving tube. The heat dissipated within the tube is radiated to the envelope, which, in turn, conducts it to the surrounding air. Normally, air currents aid in carrying away heat, but often fans or blowers are used for cooling.

**Water-cooled tubes.** As the size of air-cooled tubes is increased in area to permit greater power handling, the glass envelope becomes too unwieldy and fragile. The largest plate dissipation capable of being handled in glass tubes is about 1,500 watts. At ordinary efficiencies, this tube can provide a-c power output of about 3 kw. To handle more power than this, it is necessary to cool the tube by water, in particular, the anode.

In water-cooled tubes, the anode is made part of the envelope, the remainder of the envelope being composed of glass (Fig. 8 · 18). The glass portion is also used to support the grid and filament. The anode is commonly made of copper and is surrounded externally by a jacket through which cooling water is circulated. Because of the high thermal conductivity of the copper and high thermal absorption of the water, a great deal of heat may be carried away from the anode in this manner.

**Cathodes.** In the smaller air-cooled tubes, oxide-coated filaments or cathodes are utilized, similar in all respects to receiving tubes. As the size of the tube increases, however, the greater plate voltages used result in

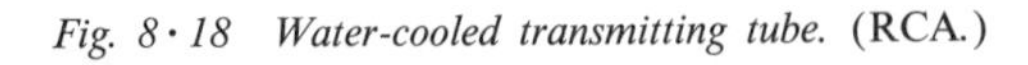

*Fig. 8 · 18 Water-cooled transmitting tube.* (RCA.)

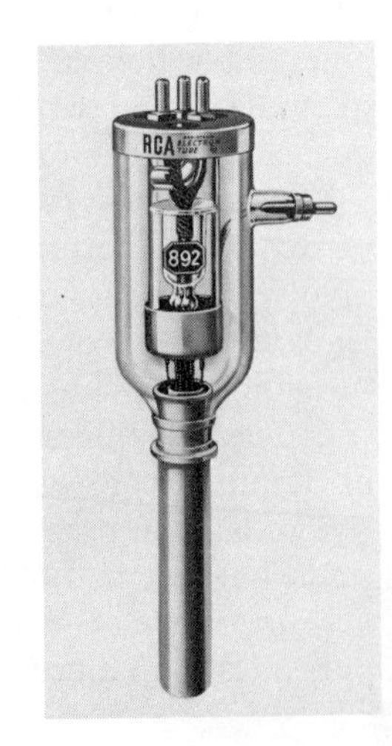

ion bombardment of the cathode. Thoriated tungsten filaments are required to withstand the ion bombardment. In high-power water-cooled tubes, the ion bombardment is generally so severe it is necessary to use filaments of pure tungsten.

**Other electrodes.** In air-cooled tubes, the anode is made of carbon, tantalum, or molybdenum to withstand the continual high temperature to which the anode is subjected. Tantalum has the advantage of producing less gas at these temperatures, but is the most expensive material. As stated before, copper is invariably used for the anode in the water-cooled types. The grids of air-cooled types generally use molybdenum, but tantalum and tungsten are sometimes used, depending upon the particular service for which the tube is designed. Because of its ability to withstand extremely high temperatures, tungsten is always used for the grids of water-cooled tubes.

In the largest sizes of air-cooled tubes and in all water-cooled tubes, only triodes are available.

## *8·12 Sources of power for transmitters*

**Filament power.** Because of the convenience and economy with which the relatively low voltages required for filament operation may be obtained from transformers, a-c voltage is generally used to heat the filaments of transmitter tubes. The voltage applied to thoriated-tungsten or oxide-coated filaments should not be permitted to vary more than ±5 per cent from the values specified in the tube data.

For tungsten filaments, the actual voltage to be used depends upon the filament emission desired. Nominal values are specified in the tube data, but the actual voltage used may be less. The voltage finally decided upon can be determined by experiment. The usual procedure is to adjust the amplifier to design conditions using the filament voltage specified in the data. After adjustment is completed, the filament voltage is reduced until a decrease in output is just detected. At this point, the filament voltage is noted. Since the regulation of the supply line is known, the filament voltage to be used in operation is the sum of the minimum filament voltage needed for full output plus the regulation of the supply line.

Filament power should always be applied to water-cooled tubes in steps to prevent magnetic effects from damaging the filaments because of the extremely high starting currents encountered. Generally, tungsten filaments have a hot-cold resistance ratio of about 14/1. If the use of relays to provide the filament voltage in steps is not desired, then high-reactance filament transformers may be used to limit the initial current.

*Of all electrode voltages, the filament voltage is the most critical and the major factor in determining tube life and correct operation.* For accurate adjustment of filament voltage, it is usual to provide a rheostat or potentiometer between the supply line and the primary of the filament transformer.

**Grid bias.** Grid bias may be provided by a separate power supply or a resistor in the grid or cathode circuit. Fixed bias is obtained from an independent battery, d-c generator, or rectifier-filter system. The fixed-bias supply must have good voltage regulation because the flow of grid current varies the load on the bias supply. Grid-leak bias is obtained by rectification of a portion of the input signal applied to the grid. Although this type of bias is the most economical and, to a large extent, self-adjusting, it does not provide protection against excessive plate current in the event of failure of the excitation. It is therefore used in combination with some other means to provide this protection.

Cathode bias is obtained by means of a resistor in series with the cathode or filament return to ground. Plate and grid currents develop a voltage drop in the resistor, thus making the cathode or filament positive with respect to the grid. This method cannot generate a bias greater than cutoff, since it depends upon plate current to produce the bias. It is, however, a very convenient means of protection to the tube if excitation fails or is removed. This feature may be achieved by making the resistor large enough that the voltage drop produced by a safe amount of plate current will be equal to the grid bias required to limit the plate current to a safe value. The cathode bias on a power amplifier is generally considered *safety bias.*

**Plate voltage.** Small mobile and portable transmitters obtain plate voltage from batteries or from motor-generator sets driven from batteries. This type of power, however, is uneconomical for the large transmitters. Shipboard transmitters commonly obtain anode power from d-c generators driven by a motor powered from the ship's mains. Some of the larger ships have a-c mains which are used to supply plate transformer and rectifier-filter systems for use in the transmitters.

Most fixed equipments normally obtain their plate-supply power from the secondary windings of high-voltage transformers connected to commercial power lines. For small transmitters, single-phase full-wave rectification is used, but for large units, three-phase rectification is almost invariably used.

## SUMMARY

1. Types of emission are classified in Table 8 · 1.
2. The r-f stages in a transmitter may include: (*a*) power oscillator as transmitter; frequency stability is poor because antenna is varying load on oscillator; (*b*) MOPA with master oscillator and power amplifier; (*c*) intermediate power amplifier (IPA) to drive the final PA; (*d*) buffer amplifier to isolate oscillator from power stage; (*e*) frequency-multiplier stage to provide harmonic of oscillator frequency. The multiplier has its plate-tank circuit tuned to the desired harmonic of the input frequency. Doublers and triplers are often used.
3. A master oscillator may be crystal-controlled for greater frequency stability or a variable-frequency oscillator (VFO) for greater convenience.
4. R-f power amplifiers in transmitters generally operate class C to provide maximum power output with the highest efficiency for one frequency. The plate-tank circuit provides sine-wave output although the plate current flows in pulses. The efficiency is high for class C operation because plate current flows for less than one-half the input cycle, and maximum current flows when the plate voltage is minimum.

5. In a series-fed stage, B+ is applied through the plate-tank circuit, resulting in direct plate current through the tuned circuit; with shunt feed, B+ is applied in parallel with the plate-tuned circuit.
6. An r-f amplifier can be capacitively coupled or inductively coupled to the succeeding stage. Inductive coupling helps prevent higher harmonic frequencies from being coupled. Link coupling is a form of inductive coupling.
7. A push-push amplifier uses two tubes with push-pull input signal of opposite phases for the two grids, but the two plates are in parallel for output signal. This type of amplifier is preferred for a frequency-doubler stage.
8. Neutralization is the process of feeding back part of the plate signal through a neutralizing capacitor $C_n$ to the grid circuit, in opposite phase from the feedback through $C_{gp}$ in the tube. $C_n$ is adjusted to prevent the amplifier from oscillating.
9. Parasitics are oscillations at frequencies other than the desired output, causing unwanted radiation and loss of power. Short, heavy connecting leads, proper lead dress, and stopper resistors are used to prevent parasitic oscillations.
10. An r-f power amplifier is tuned by adjusting the plate-tuned circuit either for minimum direct plate current or maximum grid drive on the following stage.
11. In keying a transmitter for radiotelegraph emission, the radiated output is turned on and off, or frequency-shift keying can be used. In the latter method, the transmitter is on continuously, but its frequency is shifted by 400 to 1,800 cps to have different frequencies for mark and space.
12. In large power tubes the filament voltage is critical. Usually a voltmeter and rheostat are provided to set the filament voltage at the best value for normal operation. Pure tungsten and thoriated tungsten filaments are generally used. For tubes rated above 1,500-watt plate dissipation, approximately, the anode is water-cooled.

## SELF-EXAMINATION

Here's a chance to find out how well you have learned the material in this chapter. Work the exercises, then check your answers against the Key at the back of this book. These exercises are for your self-testing only.

1. The frequency stability of a transmitter is best improved by using (*a*) a crystal oscillator; (*b*) a larger tube; (*c*) more grid drive; (*d*) a higher plate voltage.
2. The plate efficiency of a class C amplifier is high because (*a*) a resonant circuit is used as a load impedance; (*b*) the plate current flows for 180° intervals; (*c*) the grid is driven positive during part of the cycle; (*d*) the plate current flows when the instantaneous plate voltage is low.
3. An amplifier employing plate neutralization does *not* oscillate, because (*a*) energy transfer from the plate circuit through $C_{gp}$ is canceled by an equal and opposite voltage coupled to the grid by the neutralizing capacitor; (*b*) the circuit formed by $C_{gp}$, the tank inductance, and the neutralizing capacitor are tuned to the operating frequency; (*c*) currents flowing from grid to plate are canceled; (*d*) voltages appearing on the grid from the plate circuit are reinforced by similar voltages from the neutralizing capacitor.
4. A class C amplifier can be operated successfully as a frequency multiplier because the (*a*) grid current is rich in harmonics; (*b*) plate current is rich in harmonics; (*c*) plate current is a sine wave; (*d*) plate-tank circuit is tuned to the fundamental frequency.
5. Modulated c-w radiotelegraph transmission is what class of emission? (*a*) A0; (*b*) A1; (*c*) A2; (*d*) A3.
6. A class C amplifier, designed to operate at 10 Mc, is found by calculation to require 80-watt excitation power. The stage providing this excitation should be able to deliver (*a*) 80 watts; (*b*) 88 watts; (*c*) 160 watts; (*d*) 240 watts.
7. With plate voltage removed, the neutralizing capacitor of a power amplifier is adjusted for (*a*) zero or minimum grid current; (*b*) zero or minimum plate current; (*c*) minimum variation in grid current as the plate tank is tuned through resonance; (*d*) maximum variation in grid current as the plate tank is tuned through resonance.

8. In selecting a keying system for a transmitter, it is important to choose one that controls (*a*) as large an amount of power as possible; (*b*) as high a voltage as possible; (*c*) a small amount of voltage and current; (*d*) as high a frequency as possible.
9. Parasitic oscillations in a power amplifier are generally produced by (*a*) resonances in the input and output circuits other than those designed for the operating frequency; (*b*) weak tubes that develop a change in interelectrode capacitances; (*c*) too low an $L/C$ ratio in the tank circuit; (*d*) too much excitation, too much grid bias, or both.
10. In frequency-shift keying, the (*a*) oscillator output is keyed ON and OFF for mark and space; (*b*) oscillator frequency is doubled for mark compared to space; (*c*) transmitter cannot be used for radiotelegraphy; (*d*) transmitter frequency is changed about 1 kc between mark and space.

## QUESTIONS AND PROBLEMS

1. Give two reasons why a class C amplifier uses a resonant circuit as a load impedance.
2. What two factors determine the length of time a pulse of plate current flows in a class C amplifier?
3. How does a neutralizing circuit prevent oscillations? Name two types of neutralizing circuits.
4. Give two reasons why class C amplifiers are not generally used to multiply frequencies by more than a factor of four or five.
5. Define the following terms for stages in a transmitter: VFO, IPA, buffer, driver, final PA.
6. Why is it necessary to have more power to drive a class C amplifier than indicated by typical operating conditions listed for the tube?
7. A class C amplifier is apparently neutralized completely in the absence of plate voltage. When the plate voltage is applied, the amplifier breaks into oscillation. What is the probable cause of oscillation, and how may it be remedied?
8. Describe briefly the methods of cathode keying and blocked-grid keying for radiotelegraph transmission.
9. Why is it good practice to use as few chokes as possible in a class C amplifier?
10. Describe an economical biasing method that provides tube protection as well as some degree of self-adjustment for a class C amplifier that is to be driven with a constant-amplitude signal.

# Chapter 9 Antennas and transmission lines

The wireless part of radio is explained here, that is, what electromagnetic waves are, how they are produced, and why they can be radiated through space by an antenna to transmit the desired signal to the receiver. Different types of transmitting and receiving antennas are described, since both have essentially the same requirements but reverse functions. The transmission line is the connecting link to the antenna. In this unit, the topics are as follows:



## *9·1 Electromagnetic waves*

When a constant electric current flows through a conductor, a stationary magnetic field will exist in the space surrounding the conductor. The strength of this field varies with the distance from the conductor, being greater near the wire and weaker farther away. If the magnetic field changes in intensity or is moved, a voltage will be induced in an adjacent conductor when it is cut by the lines of force. This voltage, or emf, will produce an electric force or field in the conductor. If a complete circuit is available, current will flow.

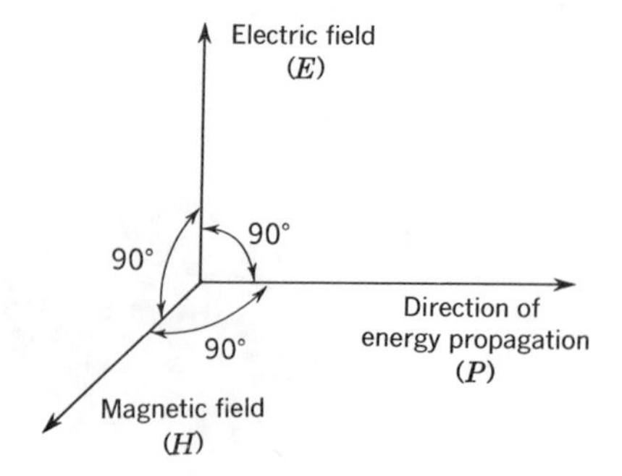


*Fig. 9 · 1 Components of an electromagnetic wave in space.*

It is not necessary, however, to have a physical conductor. An electric field will be established by the varying magnetic field between any two points in space. Thus a changing magnetic field is accompanied by an associated electric field; similarly, a changing electric field is accompanied by an associated magnetic field. The electric field (labeled $E$) and the magnetic field (labeled $H$) are at right angles to each other in space; furthermore, both $E$ and $H$ are at right angles to the direction in which the wave or energy travels. Both fields form what is known as the *electromagnetic wave*.

Figure 9 · 1 shows the direction of the respective fields as well as the direction of propagation ($P$). The energy in an electromagnetic wave is divided equally between the magnetic and the electric field components. Electromagnetic waves may be regarded as a disturbance of space. Light, radiant heat, and radio waves are propagated through space and are actually different types of electromagnetic waves. All electrowaves are propagated with the same velocity—the velocity of light. This is denoted by the letter $c$ and is equal to 300,000,000 meters per second, or 186,000 miles per second in space. The frequency $f$ of the wave is the same as the frequency of the originating current and is equal to the number of complete oscillations per second of this current. The distance in space occupied by one complete oscillation is the wavelength $\lambda$ (the Greek letter *lambda*).

In free space, the wavelength $\lambda$, the frequency $f$, and the velocity of propagation $c$ are related by the equation

$$\lambda = \frac{c}{f} = \frac{300{,}000{,}000}{f} \qquad (9 \cdot 1)$$

where $\lambda$ is in meters, $f$ is in cycles per second, and 300,000,000 is the velocity of light in meters per second.

Thus a radio station operating on a frequency of 1,000 kc would have a wavelength of

$$\lambda = \frac{300{,}000{,}000}{1{,}000{,}000} = 300 \text{ meters}$$

The strength, or intensity, of the wave is basically a measure of the emf produced by the electric field component and is usually expressed in terms of microvolts per meter. A conductor 1 meter long held at right angles to the magnetic component will have a voltage induced in it equal to the value of the field intensity. For example, an electromagnetic wave cutting a conductor 1 meter long induces a voltage of 8 $\mu$v. This corresponds to a field intensity of 8 $\mu$v per meter. The same wave, cutting a wire 3 meters long, will induce a voltage of 24 $\mu$v across the conductor; however, the field intensity is still 8 $\mu$v per meter.

## *9·2 Principles of radiation*

After an r-f signal has been generated by a transmitter, there must be some way to radiate the signal into space and also a way by which the signal energy can be picked up by a receiver. Antennas are used for this purpose. The antenna, or aerial, is just a wire conductor in space.

The transmitter antenna radiates the high-frequency signal and sets up disturbances in space in a manner similar to that of a pebble dropped in water. In order to appreciate how energy is radiated from an antenna, consider a simple r-f oscillator or generator whose output voltage is applied to a series-resonant circuit, as shown in Fig. 9·2. At a certain instant, the current in this circuit may be assumed to be in a direction such that the top plate of capacitor $C$ is charged positively and the bottom plate negatively. In other words, electrons have been conducted away from the top plate and accumulated on the bottom plate.

The charging action is shown in more detail in Fig. 9·3. Electrostatic (or electric) lines of force will be created first between the central portions of the capacitor plates shown in Fig. 9·3*a* and progress toward the outer ends as the charge spreads over the plate area. These lines of force will be in the direction indicated by the arrows in Fig. 9·3*b* during the time that the top plate is positive. Because the lines are all in the same direction, they react on each other in the same manner as like charges; that is, they tend to repel each other. As a result of this repulsion, some of the lines of force will ultimately extend beyond the space between the capacitor plates, as shown in Fig. 9·3*c*.

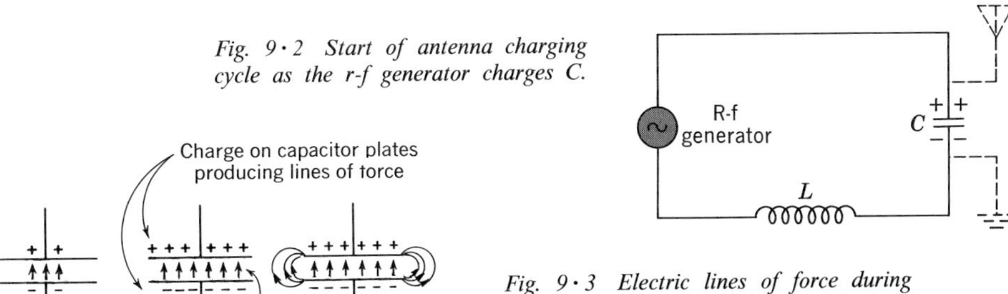


*Fig. 9·2 Start of antenna charging cycle as the r-f generator charges C.*

*Fig. 9·3 Electric lines of force during charging cycle of a capacitor. Note how the field spreads out from the center to the two ends.*

As the impressed emf passes through its positive maximum value and begins to decrease, the lines of force will start to collapse. The lines in the center portion of the capacitor plates collapse first, and the adjacent lines move in toward the center, where they collapse in turn. During the time that the voltage on the capacitor is passing through zero, there will be no lines of force between the two capacitor plates. As the voltage builds up in the opposite direction, lines of force will be created again between the capacitor plates, beginning at the middle and traveling toward the ends. These lines of force, of course, will be in a direction opposite to those which existed during the previous half-cycle. By opposite direction we mean they will be pointing upward instead of downward.

Lines of force might be regarded as having the equivalent of mechanical inertia. When they are moving rapidly in one direction and are suddenly required to reverse their direction, those on the extreme edge are unable to make this change with sufficient rapidity and are forced out of the system, as shown in Fig. 9 · 4. In this way, a certain amount of the energy stored in the capacitor is lost from the system or is radiated during each cycle.

Now, suppose that one plate of this capacitor is actually a wire extending out from the r-f generator and insulated from the ground while the other capacitor plate is connected to ground. As a matter of fact, we can actually have earth ground serving as the other capacitor plate. (As we shall see presently, it is not necessary that one capacitor plate be grounded. Both sections of an antenna can be kept away from ground and do just as effective a job of radiation.) Electric lines of force between the wire and ground will be detached from this system in the manner described. After the current has passed through 2½ cycles, the electric field in one direction about this antenna would be indicated by Fig. 9 · 5. The field has left the antenna and is traveling out in space on its own.

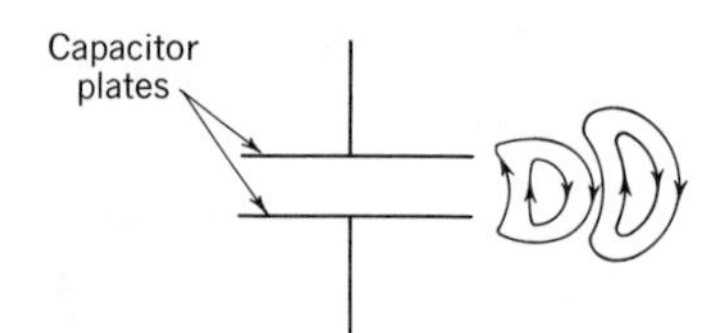


*Fig. 9 · 4 Electric lines of force radiating into space from the changing field of a charged capacitor.*

*Fig. 9 · 5 Electromagnetic wave after it has left the antenna. The variations of field intensity in space are indicated by the sine curve.*

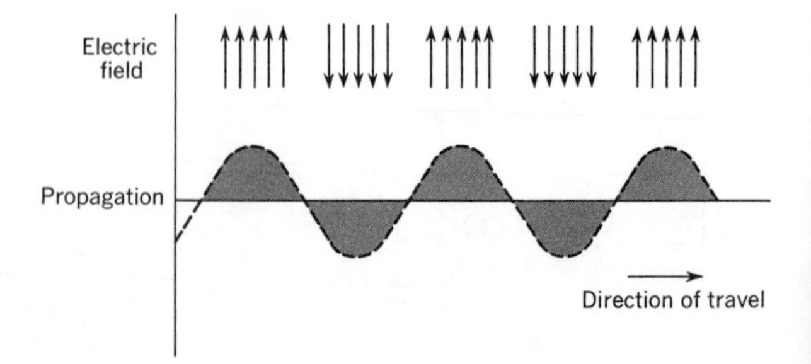

Observing this field at any instant along any straight line, as, for example, along the surface of the earth, it appears that there are alternate equidistant sections of upward and downward directions of the lines of force, so that if indicated graphically, as in Fig. 9·5, the intensity of the field at various points could be as represented by the sine curve. These lines continue to move outward. Therefore they constitute a moving electrostatic field and give rise to a corresponding magnetic field. The magnetic lines (which are not shown) are perpendicular to the electric lines and to the direction of travel. In this manner an electromagnetic wave is radiated from a wire through which a high-frequency current flows. The higher the frequency, the higher is the efficiency of radiation.

In a similar way, energy is radiated from all types of antennas. In all instances, the escape of electrostatic energy may be thought of as being due, first, to the repelling effect between like electrostatic lines of force and, second, to the inertia of such lines. Finally, the magnetic field may be considered as being produced by changes in the electrostatic field.

**Radiation and induction fields.** In addition to the energy radiated from an antenna, there is, of course, the energy contained in the usual electric and magnetic fields associated with any circuit through which current is flowing. To distinguish the radiated lines of force from the lines which are not detached from the circuit, the former is called the *radiation field* and the latter the *induction field*. It is the radiation field which is utilized in radio communication, although it is possible to signal over relatively short distances by means of the induction field. The total field about an antenna can be shown to consist of three components:

1. The electric field due to the charges on the antenna
2. The induction field consisting of its magnetic component and its corresponding electric component
3. The radiation field with its respective electric and magnetic components

Beyond a distance of a few wavelengths, the radiation field, which falls off in intensity less rapidly than the others, is the only one which needs to be considered. Undoubtedly every reader has had experience with induction and radiation fields, often without being aware of them. For example, when a house is not far from a busy highway, it will frequently be found that interference flashes will be present on a television receiver screen. This interference is produced by the electric ignition systems in the passing vehicles. The energy which reaches the television antenna is the radiation field of the ignition system. While the frequency of operation of this ignition system (including ignition coil, spark-plug firing, and so on) is fundamentally low, enough harmonics are developed which can have an effect on high-frequency television receivers. The same automotive electric system also develops an induction field which is quite strong when you are close to the car.

Another common example of radiation is the interference with radio reception caused by electric shavers or fluorescent lights. If you have a port-

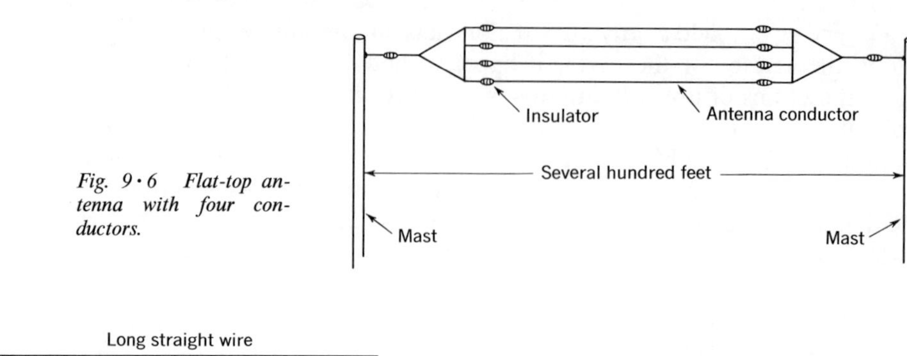


*Fig. 9·6 Flat-top antenna with four conductors.*

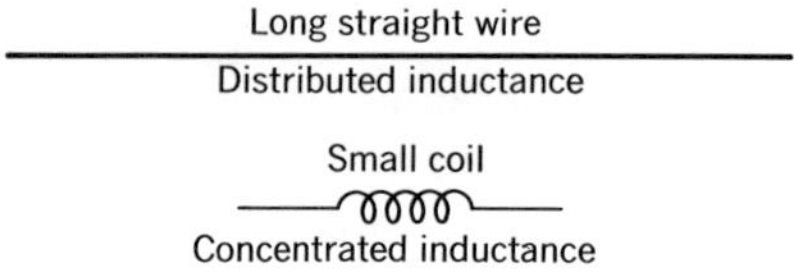

Long straight wire

Distributed inductance

Small coil

Concentrated inductance

*Fig. 9·7 Comparison of distributed inductance and concentrated or lumped inductance.*

able radio, you can easily demonstrate the fact that the amount of interference will decrease as the radio receiver is moved away from the source of radiation. Many other evidences of the existence of electromagnetic radiation are to be found in the sparking of motors, appliances, and electric circuits.

## *9·3 Antenna requirements*

In the very early days of radio, it was almost a necessity to erect antennas as high as possible in order to receive stations which were not always in the immediate vicinity. Receivers were crude, according to present-day standards, and the amplification they provided was so small that a large signal pickup by the antenna was of prime importance. Figure 9·6 shows an antenna structure commonly used with early radio receivers.

The purpose of any receiving antenna is to intercept passing electromagnetic waves so that a voltage is developed in the antenna wire (or wires). The magnitude of this voltage depends on many factors, such as the power used at the transmitter, the distance between transmitter and receiver, and which way the antenna is facing with respect to the oncoming wave. The antenna, at a given transmitting station, is called upon to radiate only a single frequency. At the receiver, however, we wish to tune in many stations at different times. Thus the transmitting antenna can be accurately designed to radiate its single frequency with maximum efficiency, while the usual receiving antenna must be able to accommodate stations over a range of frequencies.

**Inductance and capacitance of antenna.** While it might appear that an elevated antenna is simply a conductor suspended in space to intercept electromagnetic waves, it does have definite electrical characteristics. As a long straight wire, it possesses the properties of inductance, and a magnetic field is set up about its length as current flows through it. This inductance value is not large; a wire in coil form, as in Fig. 9·7, can be made to have

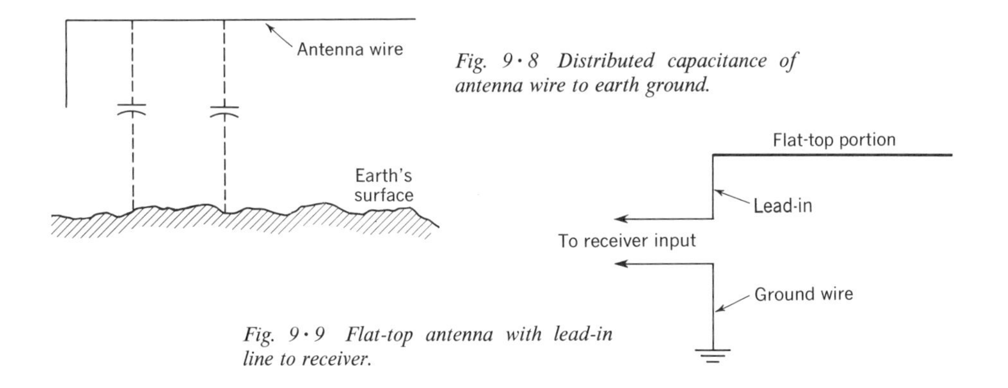


*Fig. 9·8 Distributed capacitance of antenna wire to earth ground.*

*Fig. 9·9 Flat-top antenna with lead-in line to receiver.*

many times the inductance of the straight wire. We cannot substitute an elevated coil of wire and expect it to act effectively as an antenna, however, for a coil would intercept only a relatively small amount of the electromagnetic wave because of its small physical size.

The elevated antenna must be well insulated from ground in order to act like a capacitor, consisting of two conducting surfaces separated by a dielectric. Figure 9·8 shows the antenna wire as one plate of such a capacitor, while the surface of the earth acts as the other plate. The air between the antenna and ground serves as the dielectric. The total capacitance value of such a system is very small and could be represented by a very small capacitor having a solid dielectric, such as mica. Elevating a tiny mica capacitor would not be a suitable substitution for the wire antenna for the same reason that the concentrated inductance would not serve the purpose—because of the small physical size. Both the inductance and capacitance of the antenna are said to be *distributed,* which means they extend for some distance in space rather than being confined. It is apparent that the elevated wire possesses resistance also. Every conductor has resistance, and the antenna is no exception.

Since the elevated wire possesses inductance and capacitance, it will have a resonant frequency. The formula for the frequency to which an antenna resonates is the same as for any other electric circuit with inductance and capacitance: $f_r = 1/(2\pi\sqrt{LC})$.

**Antenna structures.** Consider the simplest type of antenna, which may consist of a single elevated wire suspended between two structures but well insulated from them. The section of wire that connects the antenna to the receiver is called the *leadin.* Also, because a connection to the other plate of our antenna capacitor is needed, a ground wire from the receiver to the earth is provided, as in Fig. 9·9. To make a better connection with the earth, we generally terminate the ground wire in some metallic structure which penetrates the earth's surface, such as an iron stake or, more conveniently, a water pipe which runs underground. The ground system for a broadcast antenna may consist of many miles of buried wire. The transmission coverage of the transmitter depends upon keeping all resist-

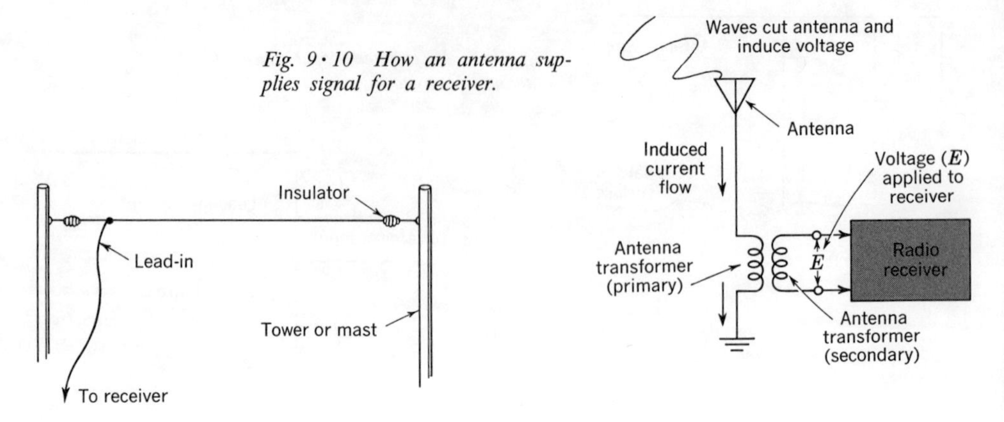


*Fig. 9 · 10 How an antenna supplies signal for a receiver.*

*Fig. 9 · 11 L-type antenna.*

ance in the antenna circuit at its lowest possible value, and an elaborate ground system helps to make good contact with the earth.

The electromagnetic field sent out by the transmitter travels equally well in all directions unless obstructions are placed in its path. In some cases, metal reflectors are used to focus the action of the waves so that they favor a given direction. Consider the case, however, where this does not take place. As the transmitted waves cut across the receiving antenna, a small emf is generated, which causes a current to flow in the antenna system. This current, flowing through the input of the receiver, sets up a varying magnetic field about the antenna transformer, which cuts the secondary winding and induces in it a voltage to be passed on to the balance of the receiver for amplification and detection (see Fig. 9 · 10). The energy of the receiver antenna is minute, and every possible precaution must be taken to conserve it. Note also that every transmitter in operation at a given instant is inducing voltages in the receiving antenna. The job of selecting and amplifying one desired signal and excluding all other signals is a function of the tuned circuits in the receiver.

Antennas may take any one of a number of shapes and sizes, depending largely on the type of service and frequency of operation. Radiation systems used in high-power long-wave stations are sometimes a mile or more long,

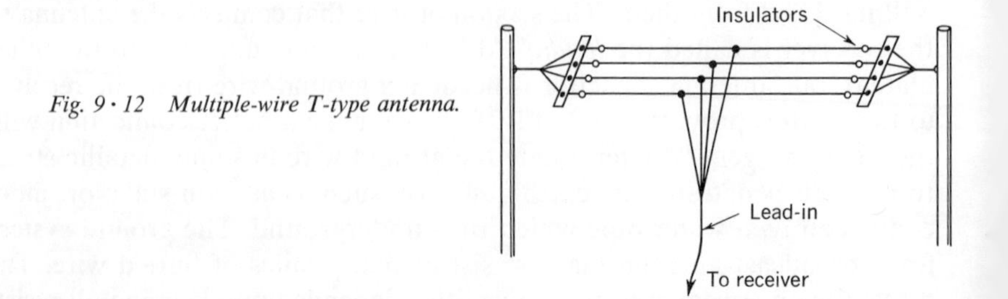


*Fig. 9 · 12 Multiple-wire T-type antenna.*

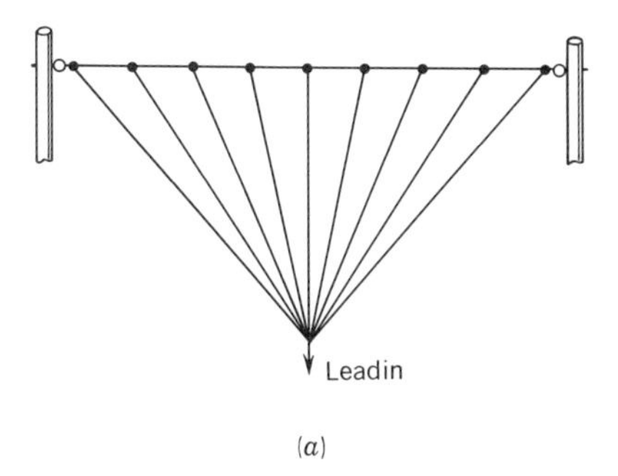


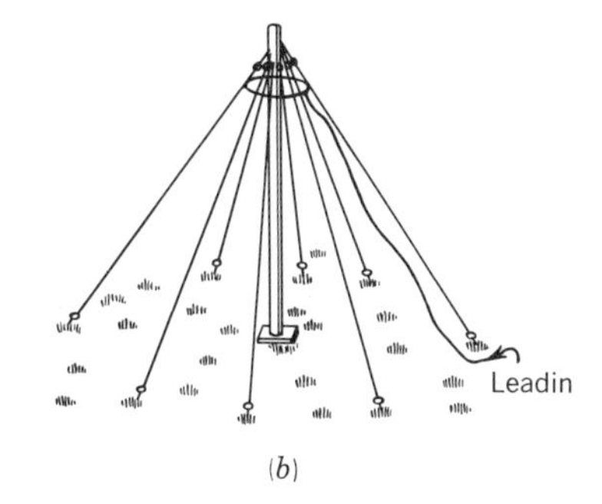


*Fig. 9 · 13 Two types of transmitting antennas. (a) Fan type. (b) Umbrella type.*

while the antenna for a short-wave station is much shorter. Television antennas are only a few feet long. The length of microwave antennas may be only a few inches. The larger the antenna, however, the more energy it will pick up.

There are various types of antennas used for transmission of radio signals. Figure 9 · 11 shows the single-wire flat-top type with its leadin being taken from one end of the horizontal wire. This is commonly referred to as an *L-type* antenna. Another simple antenna is the *T type,* in which the leadin comes from the flat-top center rather than from one end. Figure 9 · 12 illustrates a multiple-wire T antenna. This type usually consists of four to six wires in the flat-top section. The natural frequency varies with the area of the flat top. This type of antenna is quite common on ship radio stations. Figure 9 · 13*a* shows a *fan* type of antenna composed of a number of wires positioned in the shape of a fan. Figure 9 · 13*b* represents the general arrangement of the *umbrella* type of antenna. The wires are insulated from the mast and ground by insulators. All the wires join together at the top, and the leadin is taken from this point.

## 9 · 4 The dipole antenna

You will recall from Sec. 9 · 2 that electromagnetic radiation is considered to be the result of the mutual repulsion of electrostatic lines of force and their inertia. Thus, when electric lines move rapidly in one direction and then suddenly are required to reverse their direction, some lines are unable to make the change and are forced out of the immediate system (Fig. 9 · 4). The capacitor enclosing the field will lose, as radiation, some of the energy stored in the system. Since any two conductors separated by an insulator constitute a capacitor, consider the field between two parallel wires separated by an air dielectric (Fig. 9 · 14*a*). The electric lines of force will be at right angles to the wires, and when radiation occurs the lines at the ends will find it most convenient to leave the system as radiation. This is a very inefficient method of obtaining radiation. By bending back the wires, as in Fig. 9 · 14*b*, the lines of force are less confined by the system, and thus more of them are allowed to escape during the radiation process. As the wires are

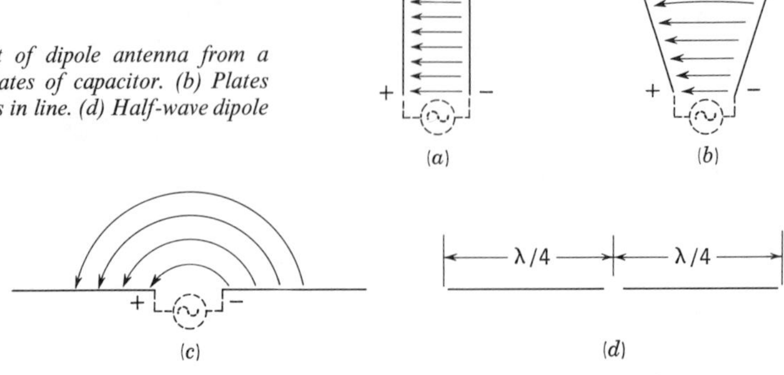


*Fig. 9 · 14 Development of dipole antenna from a capacitor. (a) Parallel plates of capacitor. (b) Plates spread out. (c) Conductors in line. (d) Half-wave dipole antenna.*

separated farther, more radiation is obtained. Figure 9 · 14*c* shows the condition for maximum radiation. Here the two wires are in line with each other so that the electric field is coupled to the greatest possible amount of space.

It should also be noted here that for maximum field strength the two wires are of opposite polarity. Then all the lines leaving the positive conductor can terminate on the negative conductor. The length of the wire will also determine the field strength. When each wire is made one-quarter wavelength long, for the operating frequency, the electric field will be maximum and efficient radiation will occur (Fig. 9 · 14*d*). This type of system, consisting of two quarter-wave conductors in a line, is called a *dipole.* Note, also, that the total dipole length is one-half wavelength. This is the simplest, or basic, type of resonant antenna.

There are a number of ways of arranging wires and structures to provide what is called an antenna system. Many complex configurations are used with our present-day communications apparatus, but the dipole is still used extensively. The dipole antenna is simply a wire whose length is one-half of the wavelength of the frequency being utilized. Thus, if a station is operating on a wavelength of 150 meters, a dipole for that wavelength would be 75 meters long.

In order to understand how the dipole functions, it is necessary to consider the voltage and current distribution along its length. When the antenna is excited by the transmitter, as in Fig. 9 · 15*a*, the electric impulses will proceed at a uniform speed in both directions to the open ends. When an open end is reached, the impulse or wave can go no farther, and therefore starts back again; the impulse is said to be *reflected.* The open end is the equivalent of an open circuit or a point of high impedance. The wave, thus, is reflected from the point of high impedance and travels back toward the input or starting point at the center, where it is again reflected. The wave in this continuing process of reflection will gradually dissipate its energy in the resistance of the wire.

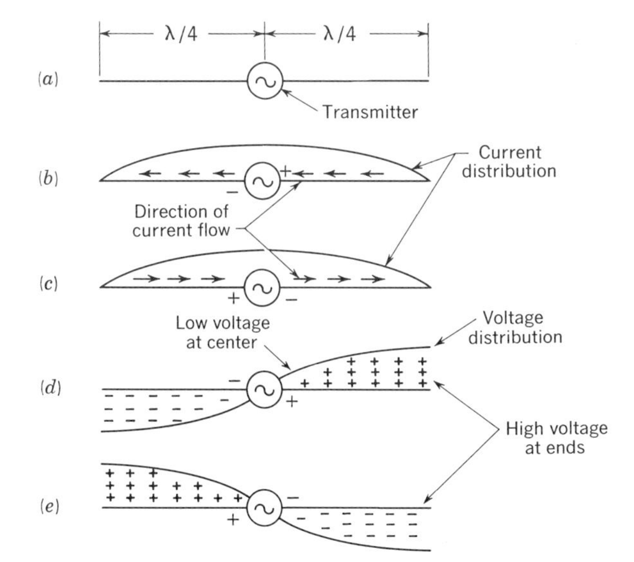


*Fig. 9·15 Current and voltage distribution on a dipole. (a) Half-wave antenna. (b) Current distribution for one half-cycle. (c) Current for opposite half-cycle. (d) Voltage distribution for one half-cycle. (e) Voltage distribution for opposite half-cycle.*

On the other hand, if, when the wave reaches the starting point, it is reinforced by the transmitter with a sufficient amount of energy to replace the losses due to the resistance, a regular back-and-forth movement, or oscillation, of energy will be established along the wire.

This can be compared to the swing of a pendulum. If the pendulum is struck at the proper instant, it will continue to swing indefinitely. On the other hand, if it is hit once, and no more, then the arc over which it swings will become progressively smaller, because of air resistance, and eventually the pendulum will come to rest.

**Current and voltage distribution.** All along the length of the antenna, the reflected impulses either add to or subtract from the original impulses traveling toward the open end of the antenna. The net result of the addition or subtraction at any point depends upon the distance from the open end and the frequency of the impulses fed to the antenna, that is, the frequency of the transmitted wave. At one open end of the antenna, the resultant voltage will be the sum of the outgoing and reflected pulses; at a distance from the open end equal to one-fourth the wavelength, the outgoing and reflected impulses will cancel each other, and the resultant potential or voltage will be zero (Fig. 9·15).

During one-half of a cycle, when the left side of the r-f generator is negative and the right side is positive (Fig. 9·15*b*), electrons will leave the negative side of the r-f generator and travel toward the left. At the same time, the positive polarity on the right side of the generator will attract electrons from the right-hand section of the antenna wire. On the next half-cycle, the generator will cause electron flow in the opposite direction, as shown in Fig. 9·15*c*.

Over a complete cycle, it will be found that the maximum amount of current is present at the center, where the r-f generator is located. Out toward the two ends, the current value will decrease until it drops to zero. This decrease is indicated in Fig. 9·15*b* by the curved line which is drawn from one end of the dipole to the other; it is a shorthand notation that says the current is maximum in the center, the line shown reaching its highest point here. From the center, the curved line gradually slopes down, reaching zero at both ends.

Since an antenna wire possesses impedance and there is current flow, the antenna will also have voltages present at various points along the wire length. This voltage distribution is shown in Fig. 9·15*d*. During one-half of the cycle, the left end of the dipole will be maximum negative. On the next half-cycle, the polarity of the voltage is reversed. The voltage increases toward the ends of the wire because the impedance increases the farther we are from the center. At the center of the wire, the impedance is low and so is the voltage. At the ends, the impedance is highest and, again, the voltage follows suit. It is important to know the current and voltage distribution along the antenna because this determines the best way to feed the antenna. Note that when the voltage is maximum, the current will be minimum; wherever the current is maximum, the voltage will be very low.

*Fig. 9·16 Current and voltage standing waves on a half-wave antenna.*

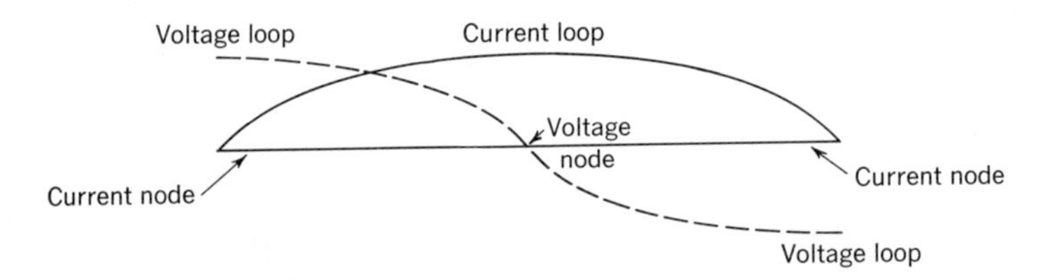


*Fig. 9·17 Directional response of a half-wave dipole. Maximum signal is radiated or received perpendicular to the wires in the broadside direction; minimum signal is off the ends.*

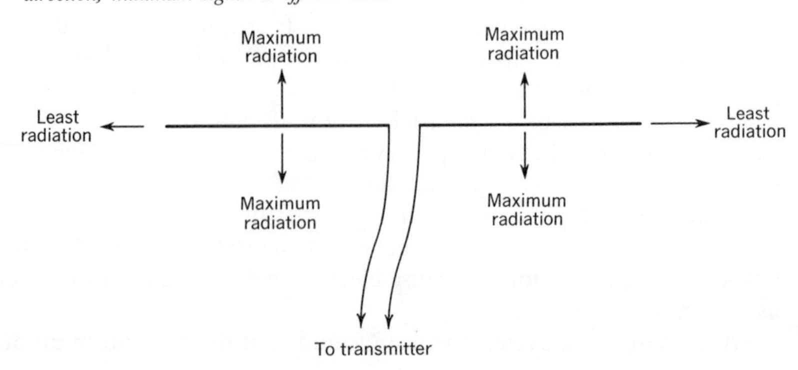

**Standing waves.** The variation of voltage and current along the length of the antenna as a result of the combination of the reflected and the incident energy is known as a *standing wave.* The points of minimum voltage or current are *nodes,* and the maximum points are *loops*. Figure 9 · 16 illustrates the standing waves along the half-wave antenna. Note that consecutive current nodes and consecutive voltage loops are one-half wavelength apart. Standing waves are also of great importance in connection with r-f transmission lines.

**Dipole antenna length.** The fundamental half-wavelength antenna, or dipole, when mounted high enough to be completely free of the influence of the earth, would have a physical length exactly equal to one-half wavelength. Since no practical antenna is completely free from the earth's influence, the actual physical length of the dipole is approximately 95 per cent of the electrical length. Thus a half-wave dipole for a station operating on 50 meters would be 95 per cent of 25 meters, or 23.75 meters, in length. The physical length of a half-wave antenna can be calculated for any given frequency from the following equation:

$$\text{Length (ft)} = \frac{492 \times 0.95}{\text{frequency (Mc)}} = \frac{467.4}{\text{frequency (Mc)}} \qquad (9 \cdot 2)$$

An antenna for 100 Mc, for example, would be 4.674 ft long for half-wave operation.

**Directional pattern.** As shown in Fig. 9 · 17, a dipole radiates maximum energy perpendicular to the axis of its wire. This is the *broadside* direction. The radiation for all other angles with the axis will depend upon the direction. As the angle decreases from 90° to zero, the radiation reduces from its maximum value to zero. A graph showing the relative radiation or field strength for different directions is called the *directional pattern* or, simply, the *pattern* of the antenna. Every antenna has a horizontal and a vertical pattern.

The horizontal pattern shows the signal radiated in different directions of the compass, that is, north, east, west, and south. The vertical pattern shows how the radiated signal will vary as you move upward, above the ground. For example, in Fig. 9 · 18*a*, the dipole antenna is held vertically. Let us suppose that it is transmitting a signal. Then, if we were to take a small receiver tuned to this frequency and walk completely around the antenna, we would receive the signal with equal strength at all points. To indicate this graphically, we draw the circle shown in Fig. 9 · 18*b*. This is the horizontal radiation pattern for vertical antennas.

The same antenna also has a vertical transmitting pattern. To see what this is, let us stand a short distance away from the antenna and, with our portable receiver, see how strong a signal we can pick up in various positions. If the vertical wire is up in the air and we stand directly below the wire, very little signal will be picked up. As we move vertically upward, the signal intensity increases until it reaches a maximum when we are

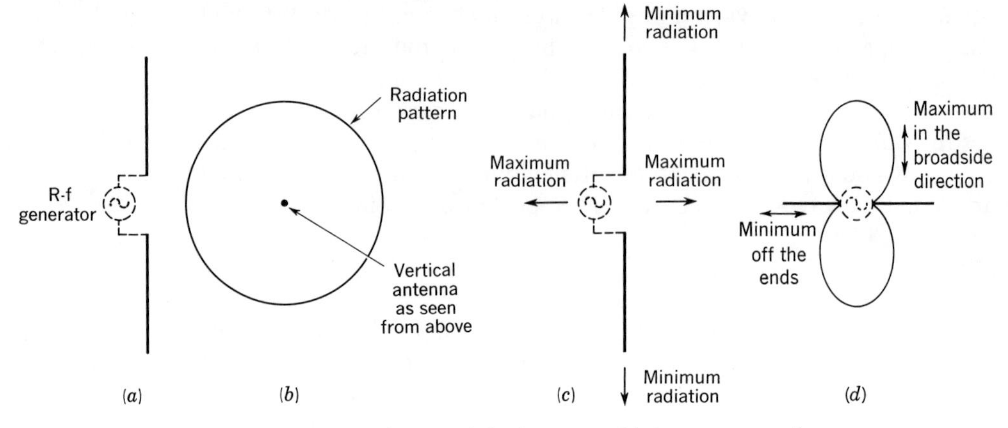


*Fig. 9 · 18 Directional response of a vertical dipole antenna. (a) Antenna mounted vertically. (b) Horizontal radiation pattern is uniform around the antenna. (c) Vertical radiation pattern. (d) Horizontal half-wave dipole, with its horizontal pattern.*

directly opposite the center of the wire. If we continue to move upward, the signal level decreases, becoming very small when we are directly above the antenna wire (see Fig. 9 · 18*c*).

Figure 9 · 18*d* shows the horizontal directivity pattern of a horizontal dipole. This type of half-wave antenna is generally used for television receivers and transmitters. The figure-eight pattern in polar coordinates shows the signal is maximum broadside or perpendicular to the line of the antenna; the signal is minimum off the ends. Combinations of the basic half-wave dipoles can provide a more directional response. These combinations form an *antenna array.*

It is interesting to note that while we have been discussing the radiation patterns of transmitting antennas, everything stated applies equally if the same antennas are used for reception. That is, they will receive signals best from those directions in which they transmit best. This is an important characteristic to keep in mind.

**Polarization.** An electromagnetic wave, as previously noted, consists of a magnetic field and an electric field at right angles to each other. Both fields are perpendicular to the direction of propagation. Radio waves are perpendicular to the direction of propagation. Radio waves are therefore transverse waves, since their fields are at right angles to the direction of energy travel. Radio waves, like light waves, may be polarized.

The polarization of a radio wave is defined to be the same as the position of its electric lines of force. Hence a vertical antenna radiates a vertical electric field because the electric lines of force are perpendicular to the ground, and the wave is said to be *vertically polarized.* A horizontal antenna radiates a *horizontally* polarized wave. Experience has revealed that the greatest signal is induced in the receiving antenna if it has the same polarization (is held in the same manner) as the transmitting antenna. As the distance between the receiving and transmitting antennas increases, however, the likelihood of a shift in polarization in the transmitted signal likewise increases.

Assume that a vertical radiator is being used at a transmitter, and a short dipole with an indicating meter at its center is held within a few wavelengths of the vertical antenna. It will be found that the greatest current will be induced in the dipole when it, too, is perpendicular to the ground. Hence the wave is still vertically polarized. Now, suppose the short dipole is taken a greater distance away and a radio receiver is substituted for the meter. The results obtained may now be different. A louder signal may be heard when the dipole is turned so as to be in a horizontal position as the result of a twisting of the wave polarization because of the conditions of the medium of propagation, such as the earth. This twisting effect is very noticeable on the longer wavelengths of the broadcast band but is not so much in evidence on the shorter wavelengths.

## *9·5 Hertz and Marconi antennas*

Simple antennas are usually classified in two fundamental groups; the Hertz antenna and the Marconi antenna. The operation of both types is based upon the theory of the basic dipole antenna. As we have seen, the dipole, or fundamental antenna, is a conductor in free space whose electrical length is one-half the wavelength of the operating frequency. The Hertz antenna is a half-wave antenna and hence is the minimum length required for basic or resonant operation. The antenna conductor has inductance, capacitance, and resistance distributed along its entire length. Therefore every length of wire will have a definite resonant frequency.

A resonant antenna has a current loop at the center feed point, while current nodes are at the ends. The voltage loops are at the ends, and the voltage node is at the center. This standing-wave condition produces efficient radiation.

**Radiation resistance.** The impedance along the length of the Hertz half-wave antenna or along any resonant antenna will vary, since the current and voltage vary with length. The impedance is maximum at the ends and minimum at the center for the Hertz antenna. The impedance of a half-wave antenna at its center is approximately 73 ohms. Very little of this impedance is due to the actual ohmic resistance of the wire. Almost all the 73 ohms is due to the *radiation resistance,* an equivalent resistance related to the power radiated by the antenna. The product of the radiation resistance and the square of the antenna current is equal to the radiated power. Note that the radiation resistance refers to the center point of the Hertz antenna.

**Harmonic antennas.** Harmonic frequency operation of antennas is often employed to obtain radiation characteristics different from those of the simple half-wave Hertz. Thus, if the frequency of the applied wave is doubled and the antenna length is unchanged, two standing waves will appear, as shown in Fig. 9·19*a*. The antenna is now said to be operating

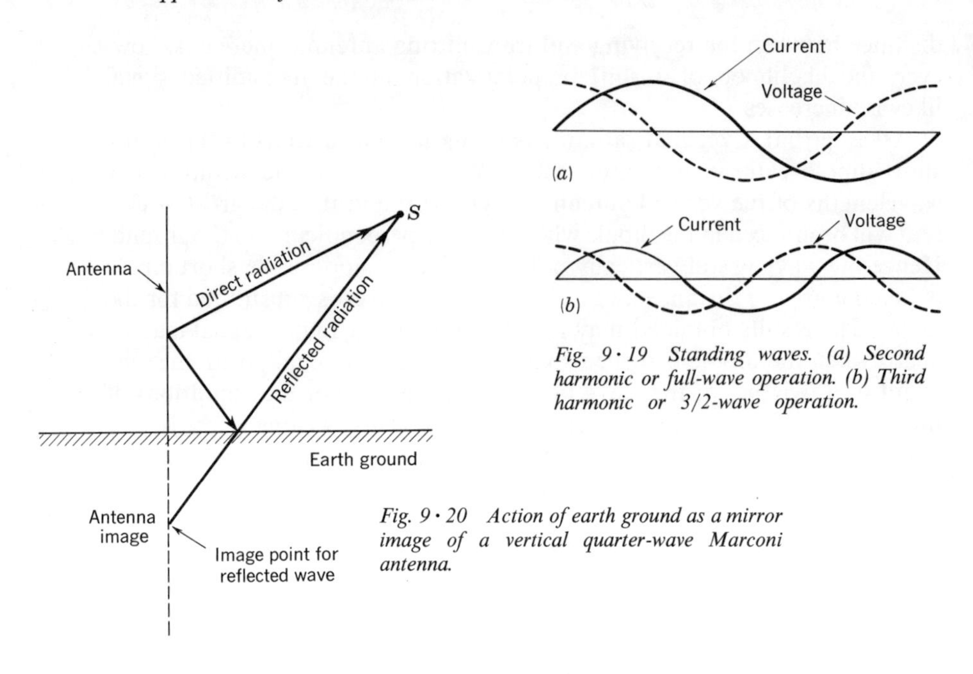


*Fig. 9·19 Standing waves. (a) Second harmonic or full-wave operation. (b) Third harmonic or 3/2-wave operation.*

*Fig. 9·20 Action of earth ground as a mirror image of a vertical quarter-wave Marconi antenna.*

at its *second harmonic*. An antenna can be resonant at harmonics of three, four, or more times the fundamental frequency (see Fig. 9·19*b* for the standing waves in a 3/2-wave antenna). The directional pattern and the impedance of the antenna change with harmonic operation. In general, the higher the harmonic, the greater the radiation off the ends of the antenna.

**Antenna doublet.** Hertz-type antennas are usually supported above the ground, and neither the antenna nor its feeder system depends upon the ground as part of the current-carrying or radiating circuit. The dipole structure is also referred to as a *doublet* antenna, independent of ground. It is used mostly at the higher frequencies in the VHF and UHF bands, since its basic half wavelength is generally too long for practical use at lower frequencies.

**Grounded antennas.** If an antenna is close to the ground or is connected to it, the ground itself becomes a part of the radiating system. A vertical conductor near ground or connected to it creates an electrostatic field which extends through the ground just as though the ground did not exist and the wire were extended into the space occupied by the ground. The earth acts as a conducting surface, or mirror, and reflects an image of the field above it, as illustrated in Fig. 9·20. A quarter-wave vertical antenna connected to the ground will have a quarter-wave image within the ground. Thus a quarter-wave grounded vertical radiator has an effective additional quarter wavelength added to it by the image. Then its voltage and current distribution becomes the same as that for a half-wave Hertz antenna in space (Fig. 9·21).

The grounded quarter-wave antenna is called the *basic Marconi* antenna.

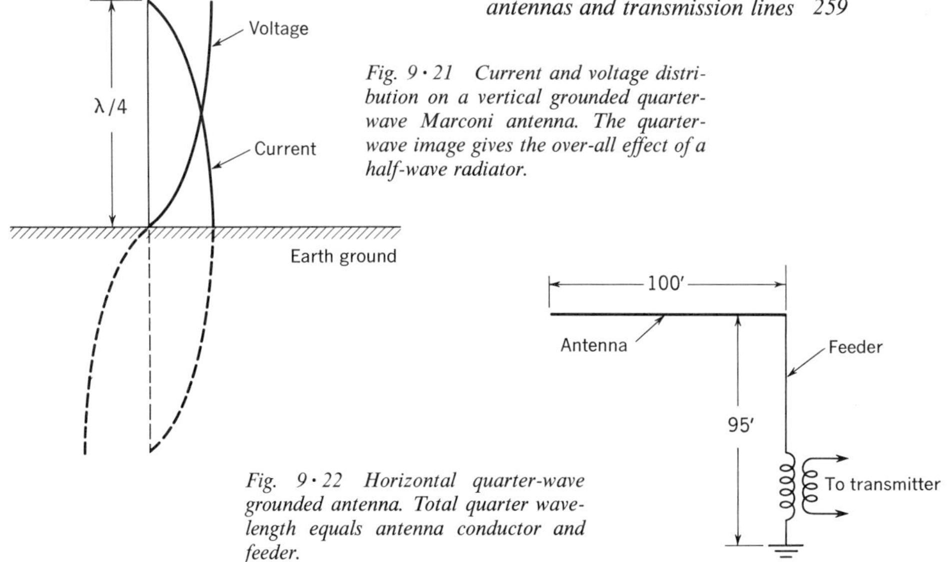


*Fig. 9 · 21 Current and voltage distribution on a vertical grounded quarter-wave Marconi antenna. The quarter-wave image gives the over-all effect of a half-wave radiator.*

*Fig. 9 · 22 Horizontal quarter-wave grounded antenna. Total quarter wave-length equals antenna conductor and feeder.*

The radiation pattern for the Marconi is the same as for the Hertz. The Marconi antenna radiates a vertically polarized wave. This wave is most useful when it is desired to communicate with ground waves, as in the radio broadcast band. Hence most broadcast stations use vertical antennas of the quarter-wave Marconi type. Another distinct advantage of the grounded antenna at the lower frequencies is that, for a given frequency, the Marconi antenna will be half the length of a Hertz.

If the Marconi is used, the ground plays a very important part in the operation, and for this reason a good ground is essential. An artificial ground system may be employed in cases where the actual ground conductivity is too low. A good ground system can be obtained by burying several hundred feet of copper wire from 6 to 12 in. deep directly under the antenna. If the soil is sandy or rocky, it is very difficult to get a good ground even though this system is used. When a good ground is not available, it is sometimes necessary to employ a *counterpoise.* A counterpoise consists of a number of wires placed radially and directly under the antenna. Precautions should be taken to insulate the counterpoise as well as the antenna.

Horizontal antennas may also be operated as grounded systems. In this case, the image is also horizontal. The total lengths of the radiating system must include the distance from the far end of the antenna to ground, as in Fig. 9 · 22. The length of a quarter-wave antenna is determined from the formula

$$\text{Length} = \frac{\text{wavelength}}{4.2} \tag{9·3}$$

The antenna length is in the same units as the wavelength. For example,

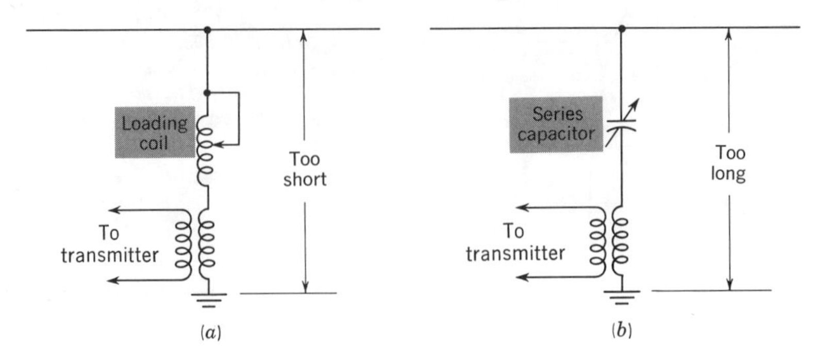


*Fig. 9·23 Loading an antenna to change its electrical length. (a) Inductive loading to lengthen an antenna that is too short. (b) Capacitive loading to shorten an antenna that is too long.*

what is the length of a quarter-wave antenna operating on a frequency of 1,200 kc? First, find λ:

$$\text{Wavelength} = \lambda = \frac{300{,}000{,}000 \text{ meters}}{1{,}200{,}000 \text{ cycles}} = 250 \text{ meters}$$

Then,

$$\text{Length} = L = \frac{250}{4.2} = 59.5 \text{ meters}$$

The electrical length of a Marconi antenna must always be a quarter wavelength, or odd multiples of a quarter wavelength, such as ¾ or 5⁄4 wavelength. This size will always place a voltage node at the grounded end of the antenna.

**Antenna loading.** If the antenna is not of the correct electrical length, it can, like any tuned circuit, be tuned to the correct frequency by adding either inductance or capacitance in series. This is known as *loading* the antenna.

If the antenna does not have the required inductance to bring its wavelength to the desired value or, in other words, if the antenna is not long enough to resonate at the desired frequency, additional inductance can be placed in the leadin, or feeder wire. Then the antenna is loaded for resonance at the desired wavelength. Since such a *loading coil* cannot radiate to any extent and since its resistance must be added to the resistance of the antenna system, loading creates additional losses. Figure 9·23*a* shows how the antenna is loaded by placing an inductance in series with the leadin.

If the antenna is too long to tune to the desired wavelength, its natural wavelength can be reduced by placing a capacitor in series with the leadin and ground. This reduces the effective capacitance of the antenna and, as a result, reduces its wavelength. Here again, loss of power in the capacitor must be considered. Figure 9·23*b* shows how the series capacitor is connected to reduce the length of the antenna.

## 9·6 Antenna types

An antenna that is used frequently is the inverted L. This type of antenna, illustrated in Fig. 9·24*a*, consists of a single wire running parallel to the ground and a leadin wire which connects the transmitter or receiver to the horizontal wire. The inverted L is basically a Marconi antenna with its total electrical length from ground to the open end one-quarter wavelength. The horizontal section is suspended from towers or poles that are insulated from the antenna by porcelain insulators. This type of antenna is easy to construct and has good performance characteristics. Care should be taken to insure that every portion of the antenna is well-insulated from ground except the grounded end.

The T antenna, shown in Fig. 9·24*b*, is another Marconi-type antenna. The leadin is connected to the center of the overhead antenna, and the total length from the grounded end to either open end is one-quarter wavelength. Both the T and inverted L antennas receive or transmit from the vertical and horizontal wires. Therefore, they can be used for both vertically and horizontally polarized waves.

A special type of antenna, known as a *loop* antenna, provides a highly directional pattern. The structure is simply that of a large rectangular coil, as shown in Fig. 9·25*a*. If the two vertical wires are separated by a dis-

*Fig. 9·24 Two common types of grounded Marconi antennas. (a) L antenna. (b) T antenna.*

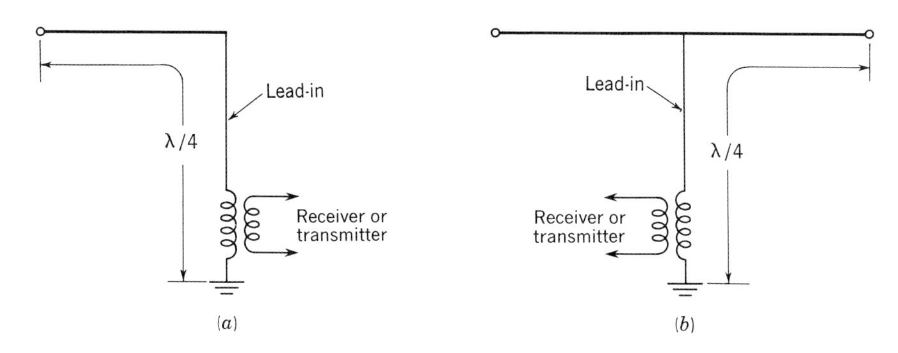


*Fig. 9·25 Directional characteristics of loop antenna. (a) Maximum signal off the ends. (b) Minimum signal from the broadside direction.*

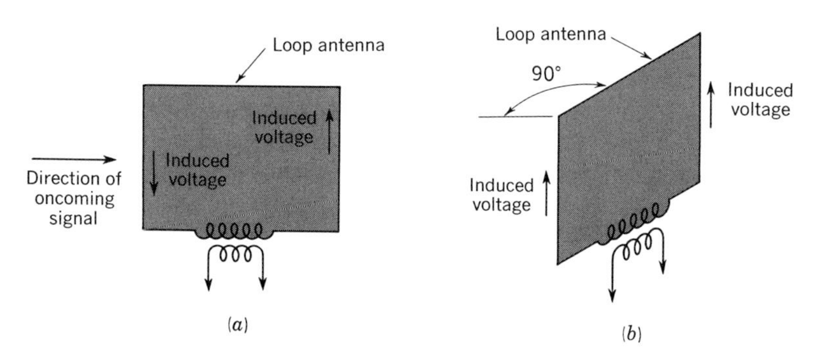

tance of one-half wavelength, a radiated wave passing the antenna will induce voltages of opposite polarity in them. If the vertical wires are then connected to the horizontal wires, the induced currents add and provide maximum signal because of the half-wave spacing. On the other hand, if the loop is rotated 90° (as in Fig. 9·25*b*), the antenna is broadside to the oncoming signal. The passing wave now cuts both vertical wires at the same time, and the induced voltages and currents will be in the same direction, bucking or canceling out each other. At angles between 0° and 90°, there will be a varying phase difference between the two induced voltages, and the resultant net signal will vary from a maximum of 0° to a minimum of 90°.

For practical size considerations, the distance between the vertical wires is greatly reduced from the one-half wavelength spacing. A large number of turns of wire are used in the loop in order to obtain a greater total induced voltage. Because of its highly directive characteristic, the loop antenna is used in direction-finding equipment (Chap. 17).

## 9·7 Directional arrays

In long-distance transmission, the signal is directed toward the receiver to increase the amount of signal picked up and enable it to override interference.

The directional patterns resulting from a number of individual antennas depend on both the spacing and the relative phases of the antenna fields. This makes possible many combinations. In Fig. 9·26, two vertical antennas are represented by dots *A* and *B*. Assume that they are positioned one-half wavelength apart and each receives the same signal at the same time. This means they are fed in phase with each other. If the field intensity is now measured at various points about the antennas, it will be distributed

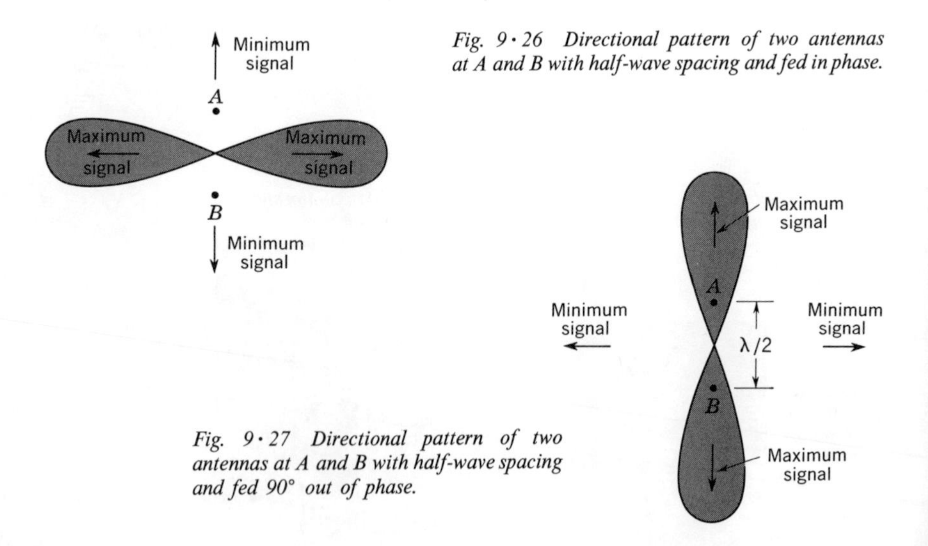


*Fig. 9·26 Directional pattern of two antennas at A and B with half-wave spacing and fed in phase.*

*Fig. 9·27 Directional pattern of two antennas at A and B with half-wave spacing and fed 90° out of phase.*

as shown. The signal strength is very low in a line with the radiating elements. It gradually increases to maximum in a line, at right angles to the axis, formed by $A$ and $B$.

If the antennas are spaced one-half wavelength apart and fed signals 90° out of phase, the pattern of Fig. 9·27 results. Note that now the entire radiation pattern is shifted 90° from what it is in Fig. 9·26. In a similar manner, by adding more antennas to the grouping and varying the phase of the currents fed to them, we can obtain almost any desired radiation pattern.

## *9·8 Radio-wave propagation*

The forward movement of the radio wave is called *propagation,* since it spreads outward from the source. The wave from a transmitting antenna travels by two paths toward the receiving antenna. In AM radio broadcasting, the primary coverage area is determined by the *ground wave,* or that part of the energy that travels close to the surface of the earth. For long-distance reception and transmission, the *sky wave* is utilized. When making use of the sky wave, we depend on reflection from one or more of the ionized layers which exist at great heights in the atmosphere. The phase relations of the ground wave and the sky wave determine the strength of the received signal at any point. The waves radiated from an antenna are thus divided into ground waves and sky waves (see Fig. 9·28).

The division of the radiated signal into ground waves and sky waves depends on the frequency of operation and the type of antenna used for the transmission. Generally speaking, low-frequency signals are transmitted predominantly by means of the ground wave. On the other hand, high-frequency signals are sent from point to point either by means of the sky wave or by direct (line-of-sight) waves.

The absorption of the waves by the ground increases with frequency. Short waves (that is, high frequencies) are accordingly subject to an especially strong absorption in the ground. The range of the ground radiation therefore decreases with decreasing wavelength.

**The ionosphere.** Investigation has shown that the sky wave travels upward at various angles. If the medium did not change sharply, the wave

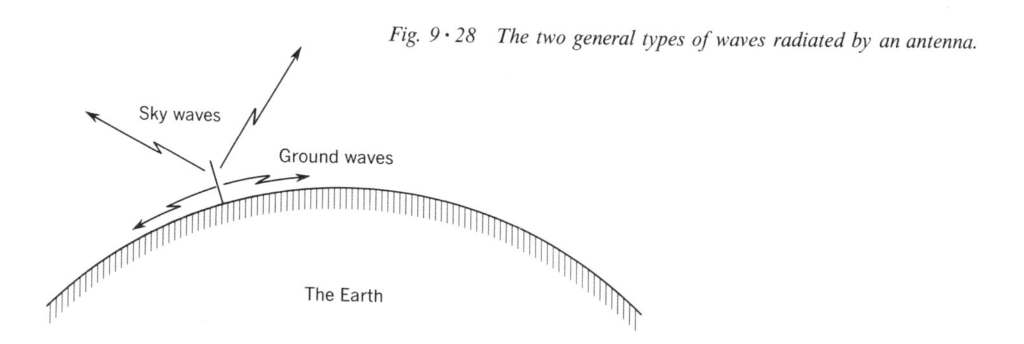


*Fig. 9·28 The two general types of waves radiated by an antenna.*

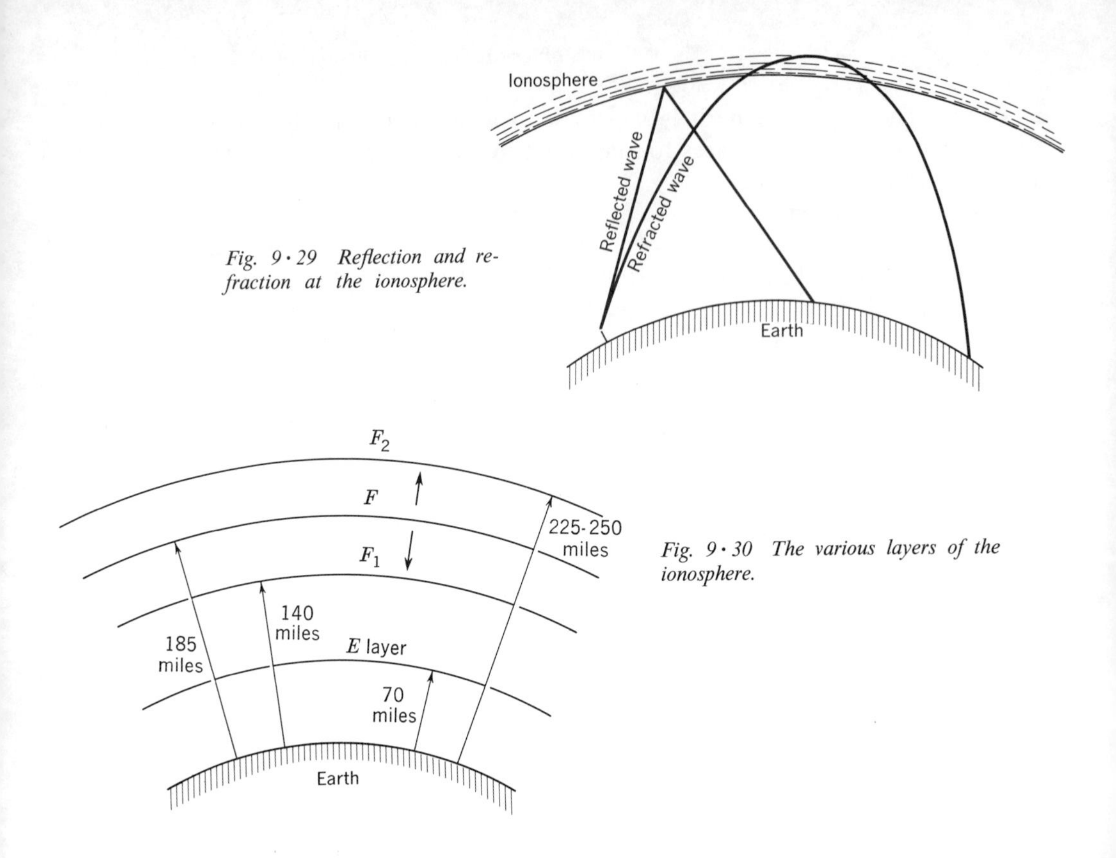


*Fig. 9·29 Reflection and refraction at the ionosphere.*

*Fig. 9·30 The various layers of the ionosphere.*

would continue until lost somewhere in space. The sharp change in medium is due mostly to the ionization of gas molecules and atoms that are found in the upper regions of the earth's atmosphere. Ionization consists of the separation of a gas molecule into a positive ion and one or more electrons. It can be brought about by ultraviolet radiations from the sun.

These ionized layers make possible long-distance radio communication by sky waves in the range of wavelengths extending from 10 to 160 meters because of refraction of the radio waves. In reflection, a wave hits an obstacle and bounces back; in refraction, the wave enters the new medium and is bent so much that it finally returns to earth (Fig. 9·29). Since it is the process of refraction that is responsible for the usefulness of the ionized layers in our stratosphere, it is advisable to investigate more closely the various effects of these ions on radio waves.

The layers of ions, or the *ionosphere* as they are collectively called, exist at distances from 70 to 250 miles above the surface of the earth. Another name sometimes applied to this region is *Kennelly-Heaviside* region. Analysis has disclosed that, although ionized gas molecules are present throughout the entire ionosphere, there are distinct layers. The lowest layer, called the *E layer,* is usually found about 70 miles above the earth. This height has been shown to be practically constant. The formation of the ions in this layer is dependent partly on the amount of ultraviolet rays received from the sun; hence, the density of the *E* layer varies from hour to hour. The

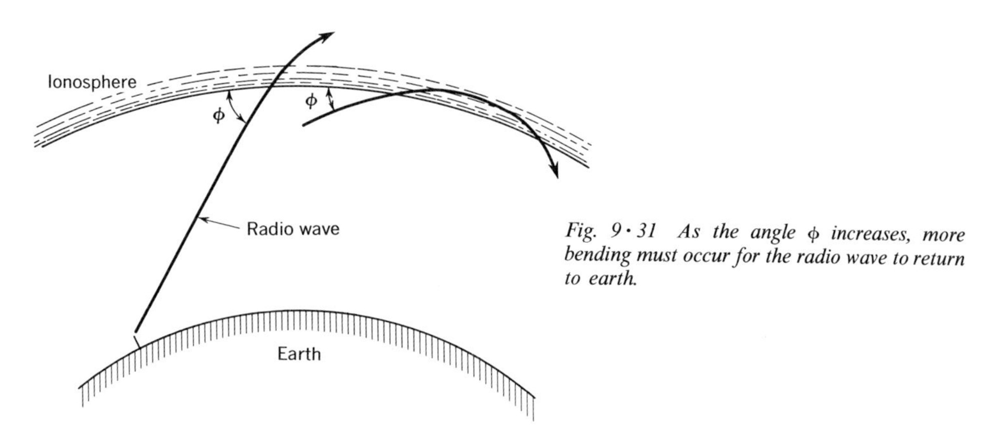


*Fig. 9 · 31 As the angle ϕ increases, more bending must occur for the radio wave to return to earth.*

greatest density occurs about noon, and the least value occurs during the dark hours of the morning.

At higher levels, the next region of maximum ion concentration occurs at a distance of about 185 miles above the earth. This region has been labeled the *F layer,* but it exists as a single layer only at night. During the day, it breaks into two separate groups of ions, called the $F_1$ and $F_2$ layers. The $F_1$ layer is usually found 140 miles above the ground, while the $F_2$ layer occurs at about 225 to 250 miles. At night these two regions combine again to form the single $F$ ionized layer. The three are shown in Fig. 9 · 30, together with the lower $E$ layer. The $F_2$ layer has the greatest degree of ionization and electron density. It is this layer that refracts the radio waves of higher frequency, the waves that other layers cannot turn back. If a wave suffers no refraction, or at least not enough bending, upon entering this final ($F_2$) layer, it continues on into outer space until all its energy is dissipated.

**Wave bending.** A radio wave, as it travels up through the lower atmosphere, suffers only a slight bending because of the decreasing air density. The bending is not great enough in itself to force the wave back to earth. The ionosphere, on the other hand, does have sufficient refracting power, and it is this power which makes long-distance communications possible.

Whether a radio wave will be bent back depends upon three conditions: the angle at which the wave enters the ionosphere, the frequency of the wave, and the electron density of the layer. The importance of the angle may best be seen by reference to Fig. 9 · 31. The greater the angle $\phi$, the more the wave has to be bent, or refracted, in order for it to return to earth. Waves entering at very small angles need only slight bending to have their direction changed enough to cause them to return to earth again. The frequency of the radio wave causes an alteration in the vibration of the free electrons in the ionosphere, and so it, too, is directly related to the degree of refraction suffered by the radio wave. Experimental observations have shown that the strong refracting power of the ionosphere is directly due to these free electrons. Anything affecting their motion will likewise affect the refractive power of the ionosphere. Finally, the density of the charged particles in the layers themselves will, in the final analysis, determine just how much the wave is bent. As the density increases, so will the amount of refraction increase for any one wave. Although these conditions have been discussed separately, they influence each other, as will be shown.

**Critical frequencies of ionosphere.** Radio waves may be sent upward at almost any desired angle, but the wave that will require the greatest amount of bending is one that is sent vertically upward or, what is the same thing, at right angles to the earth's surface. The frequency of the wave that can be bent back after entering a certain layer at this vertical angle is known as the *critical frequency* for that particular layer. If a wave of higher frequency tries to enter this ionosphere layer at right angles, it will usually not be refracted enough to return to earth and will continue upward. This wave of higher frequency may, however, enter the ionosphere at a smaller angle, say 50°, and have a better chance of being returned to earth. Of course, if the frequency is very high, not even a small angle of incidence will help. In this case the wave will continue into space.

At various times of the day and during the year, the density of a given layer changes, whereupon its critical frequency changes. As the density increases, the critical frequency will rise in value. As the density decreases, the critical frequency becomes lower. The critical frequency may thus be looked upon as an index of the ability of the ionosphere to return a wave to earth. All lower frequencies will be refracted to earth no matter what the angle used to transmit them upward, whereas higher frequencies may be returned only if the proper angle is used. The lowest possible angles achievable practically are about 4° to 6° with the horizon. If a wave cannot be returned at this angle, sky-wave transmission cannot be used for this particular wavelength.

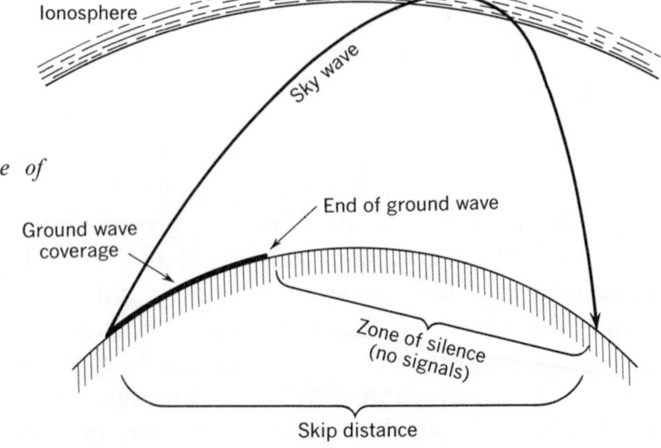


*Fig. 9·32 Skip distance and zone of silence in radio propagation.*

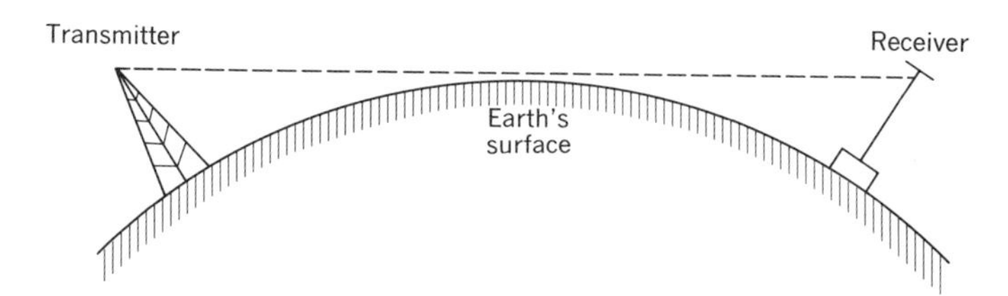


*Fig. 9 · 33 Line-of-sight transmission.*

**Skip distance.** Now both ground and sky waves may be combined into one figure to form a composite picture (Fig. 9 · 32). Unless the sky waves come back to the surface of the earth just where the ground waves end, there will be a region in which signals from this particular antenna cannot possibly be received. The distance from the end of the ground wave to the place where the sky wave first comes back is referred to as the *zone of silence,* while the total distance from the transmitter to this sky wave is called the *skip distance.* The only remedy for this situation is to change the angle of the transmitting antenna until a point is found at which the sky waves will be brought down in this area.

Many times sky waves have come down at points well within the region of the ground wave itself. Then a given antenna receives both these signals at the same time. This phenomenon usually results in interference, since the paths followed by the ground and sky waves in reaching any one common point are almost always different in length and give rise to different phase conditions at the time of arrival.

Another possibility to consider is the fact that multiple refractions and reflections can take place before the wave reaches a certain point on the earth. In fact, signals have been known to hop their way around the globe and arrive at their starting point much later. It takes approximately 0.14 sec for a signal to go once around a major circle of earth. Each time one of these refractions (at the ionosphere) and reflections (at the surface of the earth) takes place, energy is lost by the traveling wave. After a few of these hops, the wave is much too low in intensity to be picked up. Usually a signal must have a strength of 10 $\mu$v per meter to be intelligible as far as entertainment is concerned. For code reception, however, values as low as 1 $\mu$v per meter will suffice. These are the lowest limits. For effective reception, much higher intensities are to be desired.

**Very-high-frequency waves.** For the VHF band of 30 to 300 Mc, there is little bending of the waves in the normal ionosphere layers. Moreover, the ground-wave range is also extremely limited because of high ground absorption at these frequencies. For the VHF range, then, r-f energy from the transmitting antenna is radiated in a direct path through the atmosphere to the receiving antenna. This is called *line-of-sight* transmission. The transmitting and receiving points should be sufficiently high to provide such a transmission path. In calculating VHF range, the curvature of the earth as well as the intervening terrain must be taken into account (see Fig. 9 · 33).

The height of the antennas determines how far apart they may be located and still receive the VHF signal. The VHF horizon distance can be calculated by a simple formula. This formula is strictly accurate only over water or when the intervening ground is almost level, but it serves as a useful guide in less ideal conditions. When the height in feet ($H$) of a transmitting antenna above ground level is known, the distance to the VHF horizon in miles can be found from the relation

$$\text{VHF horizon distance} = 1.41\sqrt{H}$$

The VHF horizon distance found by the above equation assumes that the distance receiving antenna is at ground level. When the distant receiving antenna is elevated, as is more often the case, the total VHF path is the sum of the two VHF horizon distances for each of the antennas, as computed by the above formula.

Owing to the optical nature of these very short waves, pronounced radio shadows are noticeable. Obstructions, such as mountains, forests, and tall buildings, cause the signals to be sharply attenuated or to disappear entirely when they come between transmitter and receiver. Such shadows are caused by the inability of the waves to bend around the obstructions. This effect becomes more pronounced with higher frequencies.

On low frequencies, static is, at one time or another, a source of interference. In the VHF spectrum, there is no static to contend with. Communications may be carried on even during severe electric storms. VHF signals are, however, subject to man-made interference, such as electrical noises from automobiles and electric trains. This type of interference may be eliminated or greatly reduced by proper shielding or by special noise filters.

**Forward-scatter propagation.** It has been found that useful radio signals at all frequencies can be received consistently at distances well beyond the

*Fig. 9·34 Illustration of forward-scatter propagation.* (Bell Telephone Laboratories, Inc.)

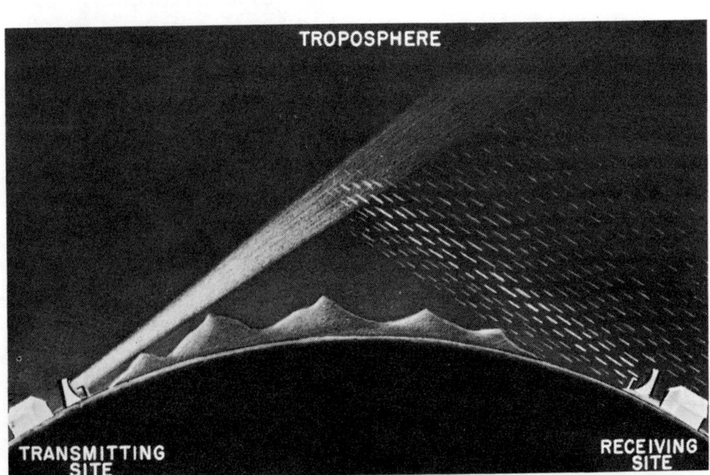

horizon. Effective communication therefore can be maintained at distances previously considered too short for good ionospheric transmission and too long for conventional VHF or UHF line-of-sight propagation. This technique, called *scatter* propagation, is illustrated in Fig. 9 · 34. It has opened up many radio transmission possibilities, particularly in the area of fixed point-to-point communications.

Beyond-the-horizon, extended-range, or forward-scatter propagation may be accomplished by either of two types of transmission. The first type, *ionospheric* scatter, is obtained by means of radio waves scattered from the *E* layer of the ionosphere. Communication at frequencies of 25 to approximately 65 Mc is possible for distances of over 1,200 miles by this method. This form of transmission is less subject to complete interruption than the normal high-frequency transmissions using sky waves reflected from the ionosphere. The received signal, though fluctuating, is always present, so that reliable communication can be obtained. Telegraph and teletype, as well as voice, transmission can be handled by this method.

The second type of scatter is *tropospheric* transmission. It is obtained by the scattering of electromagnetic waves in the troposphere, the lowest part of the earth's atmosphere (see Fig. 9 · 34). Troposphere scatter is obtained over a frequency range of approximately 100 to 10,000 Mc. This method of long-range communication is applicable to the transmission of almost all types of radio signals.

Scattering may be produced when an electromagnetic wave strikes a material that is too rough to provide regular mirror-like reflection. Rough sections of the earth's surface or waves of water will produce scattering. When a wave travels through a nonhomogeneous medium or strikes particles, such as rain drops, snow, sleet, hail, or fog, it may be scattered. The degree of scattering depends upon the frequency of the wave and the size, shape, refractive index, and distribution of the atmospheric particles. Scattered reflections are, therefore, random and diffused. Their polarization is, of course, random.

Large parabolic antennas (Fig. 9 · 35) are particularly well suited for scatter transmission and reception because of their ability to concentrate the signal energy they receive. These antennas may be as much as 120 ft in diameter.

## *9 · 9 Transmission lines*

The transfer of electrical energy to or from an antenna is usually accomplished by a transmission line, which consists of two conductors with uniform spacing. For example, the energy from a transmitter to its antenna, which may be located several hundred feet away, is transferred by means of a transmission line. Likewise, when a receiving antenna picks up a signal, it forwards this to the receiver by means of a transmission line. This is especially true with television receivers, where the antenna is frequently mounted on the roof.

All transmission lines have resistance, inductance, and capacitance

*Fig. 9·35 A 120-ft antenna designed for scatter propagation.* (D. S. Kennedy & Co.)

associated with them. The resistance is determined by the size and type of conductor. Since a capacitor consists of two conductors separated by a dielectric, the two wires of the line and their spacing determine its capacitance. This capacitance is not lumped in one small unit, but is spread out, or distributed, along the entire length of the line. The presence of inductance is explained by the fact that magnetic lines of force link the currents flowing in the conductors. The inductance is also distributed along the entire length of the line.

**Types of transmission lines.** The two most common types of transmission lines are the open-wire line and the coaxial line. The open-wire line consists of two parallel wires maintained at a fixed spacing by insulated supports, or spacers, as illustrated in Fig. 9·36*a*. The coaxial line in Fig. 9·36*b* has the inner conductor insulated from an outer conductor. Insulating

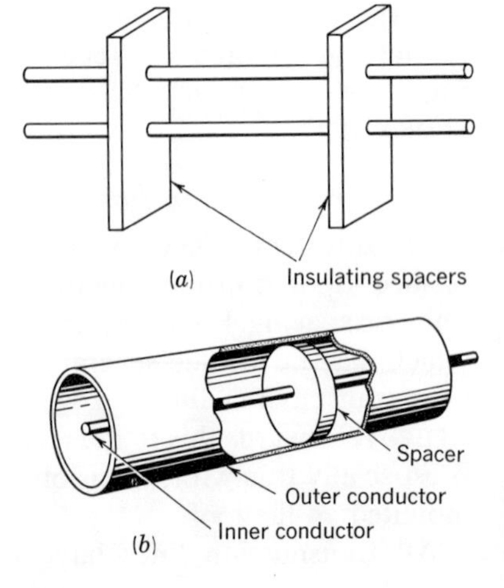


*Fig. 9·36 Two types of transmission line. (a) Open-wire line. (b) Coaxial line.*

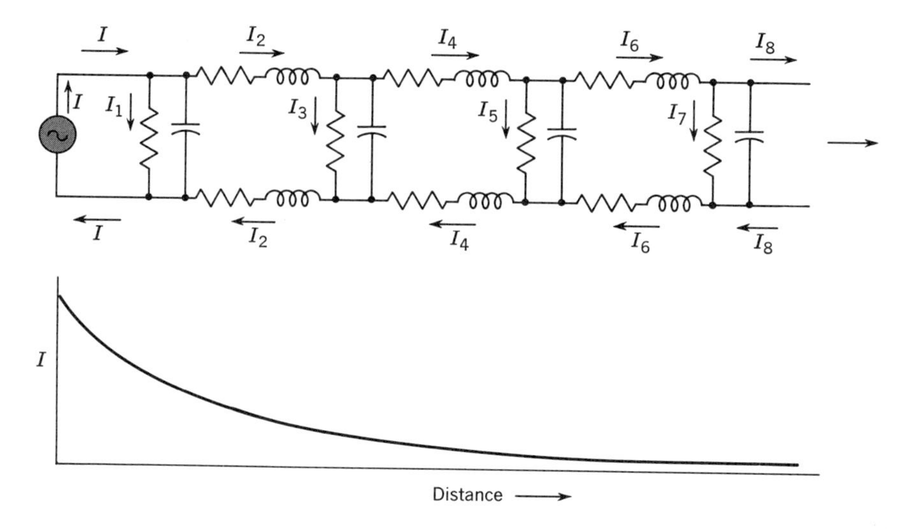


*Fig. 9·37 How the current in the line decreases at greater distances from the generator.*

spacers or beads placed at regular intervals separate the inner and the outer conductors. In some types of coaxial line, rubber or plastic insulation is employed between the two conductors. For flexibility, the outer conductor is often a metallic braid.

**Characteristic impedance.** To determine the behavior of transmission lines, consider an alternating voltage applied to one end of an extremely long transmission line (Fig. 9·37). Let the current that leaves the generator be labeled $I$. As this current proceeds down the line, part of it ($I_1$) is diverted through the shunt resistance and capacitance. The current remaining ($I_2$) is equal to $I$ minus the amount diverted ($I_1$). Farther on, the same process takes place again. It can be seen that as this continues, the resultant current will eventually be reduced to zero. In an actual cable, where the inductance, capacitance, and resistance are uniformly distributed along the wires, the decrease of current is gradual.

Investigating the voltage between the wires as one progresses along the line, the same sort of diminishing effect is found because of the various voltage losses which occur across the series resistances (shown in Fig. 9·38*a*). In Fig. 9·38*b*, it can be seen that the voltage across the line at any point will be less and less the farther it is from the start of the line.

If the transmission line is infinite in length, then whatever current is sent into it from the generator will eventually decrease to zero. The same effect as an infinite line may be achieved, however, by taking *any* length of line and terminating it in a special impedance known as its *characteristic impedance,* or *surge impedance.* This impedance is a characteristic of the line, determined by its resistance, capacitance, and inductance.

In a line terminated in an impedance equal to its characteristic impedance

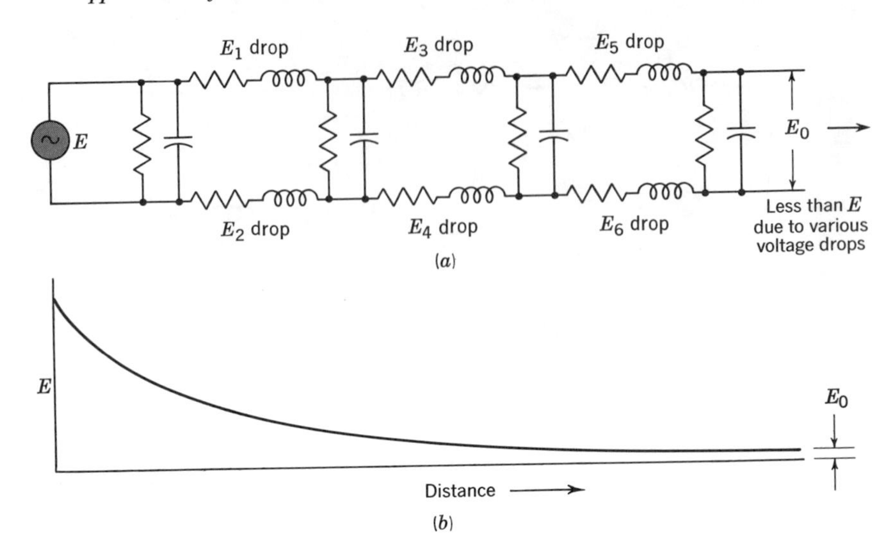


*Fig. 9·38 How the voltage across the line decreases at greater distances from the source.*

all power sent down the line will be fully absorbed by the load, just as it would be absorbed eventually in an infinite line. Figure 9·39 shows this process pictorially.

When power is to be transmitted by the line from one point to another, it is imperative that the load impedance be equal to the characteristic impedance of the line. Under this condition, the maximum amount of power will be transferred from the line to the load at the end of the line. Any mismatch results in a correspondingly lower amount of power reaching the receiving terminals. Lower transfer to the load on mismatch is due to the fact that energy of any type, whether it be light waves, heat rays, or radio waves, continues to flow smoothly only as long as the conditions of the transmission media remain constant. When the line is terminated in a load whose impedance differs from the characteristic impedance, we have an altered condition and the distribution of the arriving energy must change. Part of the energy is absorbed by the load, while part of it travels back. As far as the load is concerned, the energy reflected represents a loss because it is not received.

A line that is infinitely long absorbs all the energy introduced to it, without any reflection occurring. To obtain this condition in practice, using practical lengths of line, we need merely place an impedance across the end of the line which is equal in value to the characteristic impedance. So far as the energy traveling toward this end is concerned, it sees no change in conditions, and the transfer of power from the line to the load occurs smoothly, without any reflections. The load, in this case, will absorb the energy just as fast as it arrives. The line is properly terminated and behaves like an infinite line.

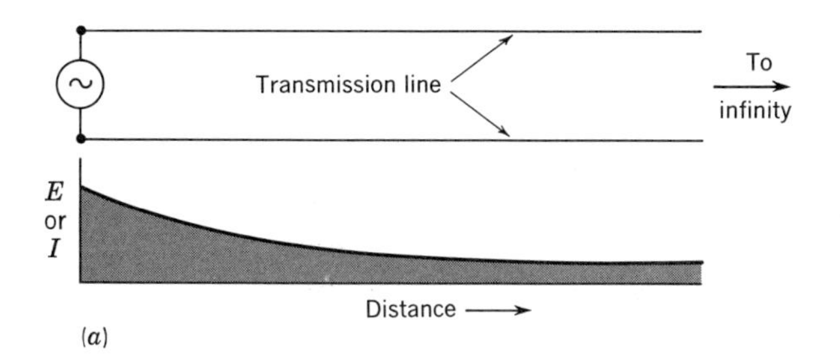


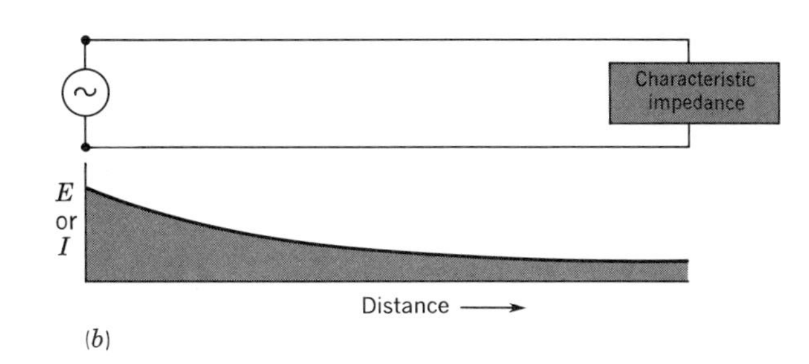


*Fig. 9·39 Characteristics of a continuous transmission line. (a) Infinitely long line. (b) Same results with line terminated in its characteristic impedance.*

Whenever a line is properly matched, it is said to be a *flat line*. With a mismatch, however, the reflection from the end of the line may either add to or subtract from the oncoming wave, resulting in a variation of voltage and current along the line. The greater the mismatch, the greater the variation from point to point. We may now say that the line is resonant because standing waves of voltage or current are present. In fact, a simple method of ascertaining whether a transmission line is properly terminated is to take a voltmeter and measure the voltage at several points along the line. If very little change occurs as the meter is moved up and down the line, we can assume that the line is flat, or properly terminated. If the values fluctuate widely, the line is mismatched.

## *9·10 Feeding and matching antennas*

The process of introducing energy into an antenna is known as *feeding* the antenna. The point at which the energy is introduced is called the *feed point,* and the transmission line used is called the *feed line* or *feeder*. Antennas are usually either voltage-fed or current-fed, depending upon whether the feed point is a point of maximum voltage or maximum current. Power may be supplied to an antenna by direct coupling to the transmitter, by a resonant line, or by a nonresonant flat line. All these methods may be employed with either Hertz or Marconi antennas. The method of coupling energy to the antenna and the type of feeder system used should not affect the standing waves on the antenna or the radiation pattern. In some cases, the antenna is fed at some point other than at voltage or current maximum in order to match the antenna to the impedance of the feeder.

**Direct coupling.** Direct coupling without a transmission line is feasible only when the antenna is close to the transmitter output stage. Either

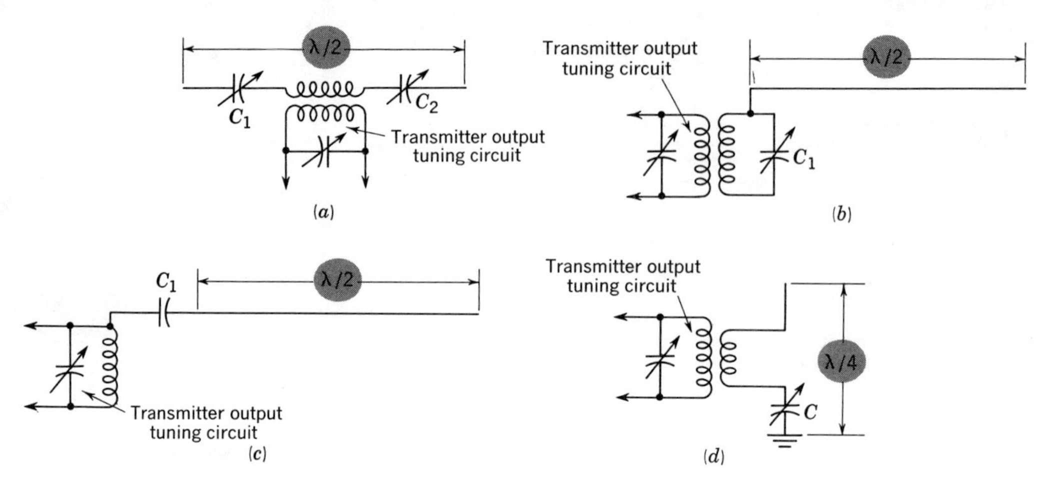


*Fig. 9·40 Methods of feeding antenna. (a) Current feed at center of dipole; series tuning of antenna. (b) Voltage feed at end of dipole; inductive coupling. (c) End or voltage feed again but with capacitive coupling. (d) Inductive coupling to Marconi antenna.*

inductive or capacitive coupling may be employed. Figure 9·40*a* illustrates current feed to a half-wave Hertz antenna. This is a low-impedance circuit and therefore requires a low-impedance coupling circuit. Capacitors $C_1$ and $C_2$ tune the inductance to series resonance. An inductively coupled end-fed Hertz antenna is shown in Fig. 9·40*b*. The end of the antenna is connected to a high-impedance point of the tuned secondary. The circuit is tuned by the variable capacitor $C_1$ in the coupling circuit.

The antenna may be directly connected to a point of high impedance on the transmitter output tuning circuit, as shown in Fig. 9·40*c*. $C_1$ is inserted in series with the antenna to block the d-c voltage of the transmitter tuning circuit from the antenna. Figure 9·40*d* indicates a quarter-wave Marconi antenna being effectively fed by a low-impedance coupling circuit.

**Resonant line feeders.** Resonant lines are often used to feed antenna systems because they are easily adjusted. They are usually connected to the antenna at a voltage or current loop. Open-wire lines are best suited for this application. The resonant feeder may be considered as an antenna folded back on itself. It is normally employed in lengths which are an exact multiple of a quarter wavelength. Figure 9·41 shows typical quarter-wave and half-wave resonant lines used as feeders. For end-fed antennas, if there is an odd number of quarter waves in the line, series tuning is required at the transmitter end. The series tuning of the *LC* circuit is shown in *a*. If the line is an even number of quarter waves, parallel tuning is used; the parallel tuning is shown in *b*. These tuning requirements are reversed with center-fed antennas (see *c* and *d*).

In Fig. 9·41, the dotted lines show the current distribution along the transmission line and along the antenna. The voltage distribution along these same points is not shown, but wherever the current is low, the voltage is high, and wherever the current is high, the voltage is low. Thus, in

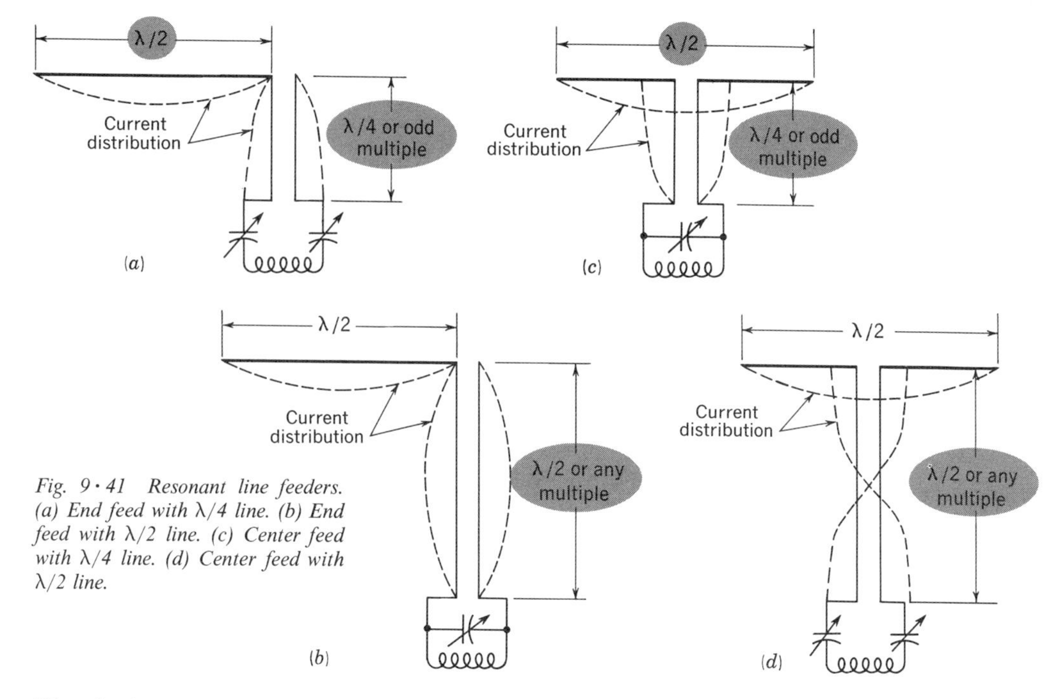


*Fig. 9·41 Resonant line feeders. (a) End feed with λ/4 line. (b) End feed with λ/2 line. (c) Center feed with λ/4 line. (d) Center feed with λ/2 line.*

Fig. 9·41*a*, the transmission line attaches to the end of the half-wave antenna where the current is low and the voltage would be high. To match this condition, the transmission line, at the point of connection, also has low current and high voltage. The same matching requirements apply to the other arrangements shown.

**Nonresonant feeders.** Since standing waves exist on resonant feeders, the high currents and voltages at the loop points give rise to considerable line loss. In installations where relatively long transmission lines are neces-

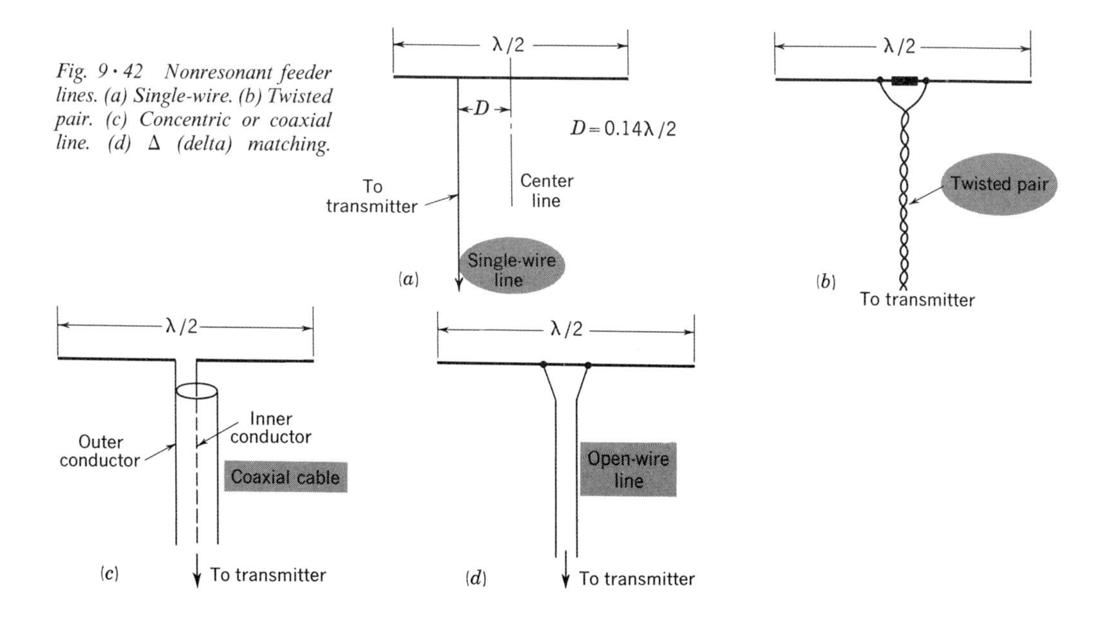


*Fig. 9·42 Nonresonant feeder lines. (a) Single-wire. (b) Twisted pair. (c) Concentric or coaxial line. (d) Δ (delta) matching.*

sary, nonresonant lines are used to eliminate this undesirable loss. A line terminated in its characteristic impedance will absorb the arriving energy and act, therefore, like an infinite line. No reflection occurs, and hence no standing waves appear. The line, being flat, has little radiation, and the only losses are due to the normal wire and insulation resistances. Tuning of the lines is not necessary, but it must be noted that the nonresonant line, when adjusted, will be matched only at the adjustment frequency.

The adjustment required is simply that of finding a proper impedance value along the antenna length which will match the characteristic impedance of the line and prevent reflection. In order to accomplish this matching, the antenna itself is made to resonate at the operating frequency, and the line is then connected to it at the proper spot to effect the match.

A single wire has a characteristic impedance of approximately 550 ohms. When it is used, as in Fig. 9·42*a*, to feed an antenna, it is connected to a point approximately 14 per cent of a half wavelength from the center of the half-wave antenna where the same impedance exists.

A pair of insulated wires twisted together can be constructed to have a characteristic impedance of approximately 73 ohms. The twisted pair can therefore be connected directly to the center of a half-wave Hertz antenna to achieve a good match (Fig. 9·42*b*). Similarly, Fig. 9·42*c* indicates the use of a coaxial or concentric line whose characteristic impedance is 72 ohms to center-feed a Hertz antenna. The outer conductor may be grounded if desired.

Most practical open-wire transmission lines have characteristic impedance values between 200 and 1,000 ohms. When these lines are to be used as antenna feeders, the delta matching system of Fig. 9·42*d* is commonly employed. In this method, the section of line next to the antenna is fanned out to the points on the antenna that match the impedance of the line.

## *9·11 UHF antennas and waveguides*

In the ultra-high-frequency region from 300 to 3,000 Mc, the same line-of-sight transmission features just discussed are obtained. Perhaps the only major differences that occur are, first, a decrease in the size of the antenna elements [see Eq. (9·2)] and, second, increasing attenuation caused by rain, fog, and the dust particles that are found in the air. The latter characteristic stems from the fact that with sufficient increase in frequency, the wavelength of the signals becomes short enough so that rain and dust particles represent significant obstacles to these passing signals.

With the reduction in antenna size, it becomes possible to increase the number of elements per array to achieve a higher gain. Several typical arrays of this kind are shown in Fig. 9·43. Note the fairly large number of dipole elements, all conveniently mounted on a relatively small structure. Note, too, the use of a screen behind the dipoles as a reflecting element in place of individual reflectors behind each dipole. A screen of this kind is far more effective in shielding the array from any signals attempting to approach the antenna from the opposite direction.

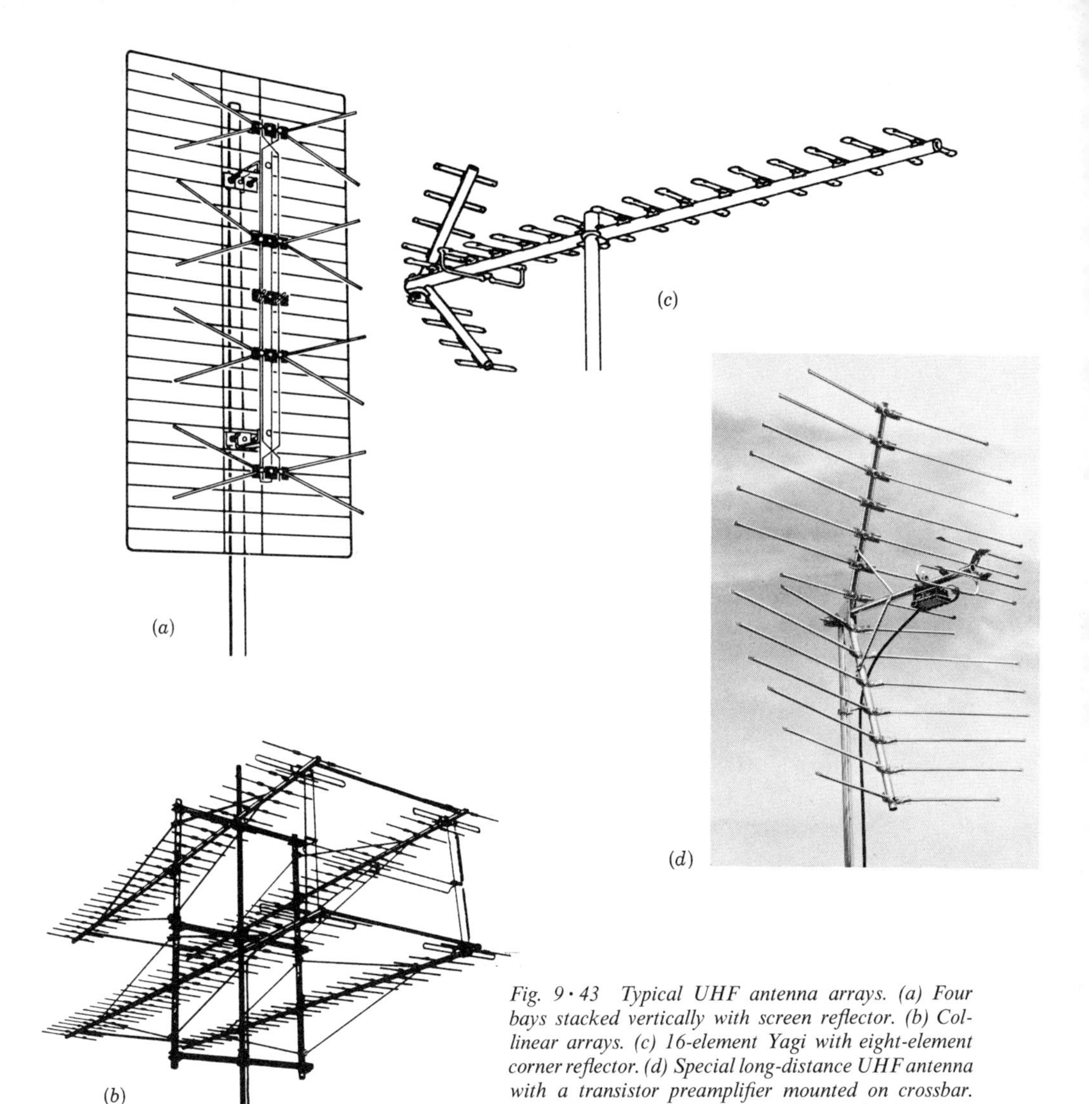


*Fig. 9·43 Typical UHF antenna arrays. (a) Four bays stacked vertically with screen reflector. (b) Collinear arrays. (c) 16-element Yagi with eight-element corner reflector. (d) Special long-distance UHF antenna with a transistor preamplifier mounted on crossbar.* (Winegard Co. *a* and *d*; The Finney Co. *b* and *c*.)

Another popular UHF antenna uses part or all of a parabolic surface as a reflector for high gain and sharp directivity. Parabolic reflectors are well known for their use in car headlights to provide a high concentration of light in one (i.e., the forward) direction. In similar fashion, parabolic reflectors can serve to receive and transmit electromagnetic waves. See Fig. 9·44.

The beam width of a parabolic reflector is closely related to the area across the mouth of the reflector. As this area becomes greater, in comparison to the signal wavelength, the antenna gain increases accordingly. Hence, to achieve a useful gain, this type of structure can only be employed at extremely high frequencies.

Instead of utilizing the entire parabolic reflector, it is possible to use

*Fig. 9·44 UHF antenna with parabolic reflector behind it.*

*Fig. 9·45 UHF antenna with a section of a parabolic reflector behind it.*

only a section, as shown in Fig. 9·45. The gain and beam sharpness of this modified structure are somewhat less than that obtainable with a complete reflector, but at the lower end of the UHF region a full parabolic reflector may still be too bulky physically to be economically (or structurally) feasible.

Fairly common, too, is the array shown in Fig. 9·46. This is called a *truncated paraboloid* and it can be energized by either a single dipole or a waveguide positioned as shown.

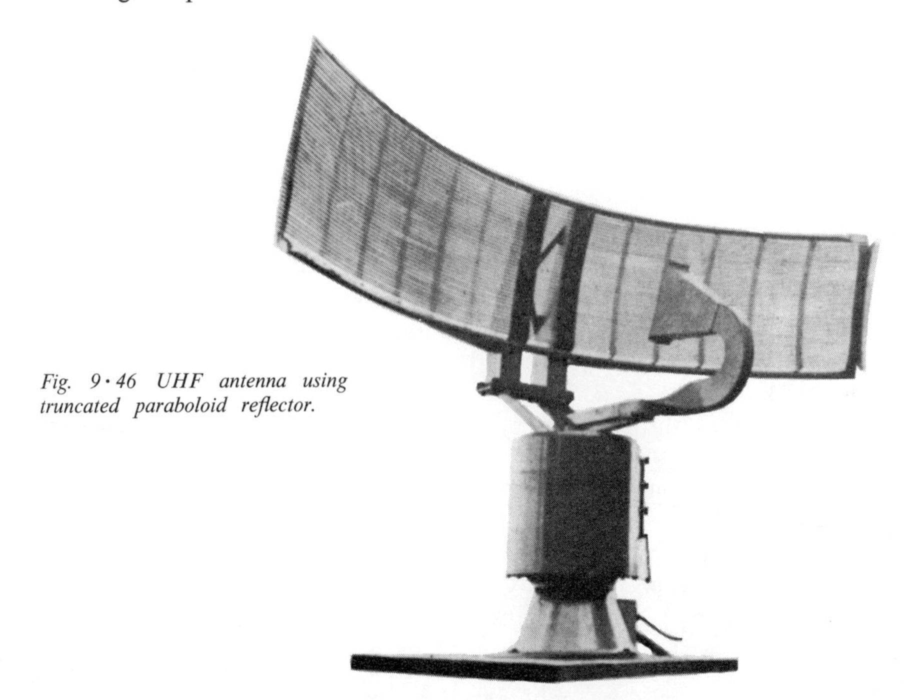

*Fig. 9·46 UHF antenna using truncated paraboloid reflector.*

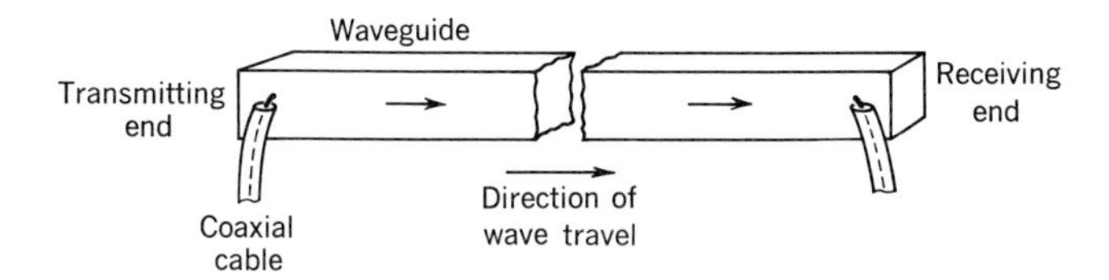


*Fig. 9·47 A rectangular waveguide being used to direct energy from a UHF generator to open space.*

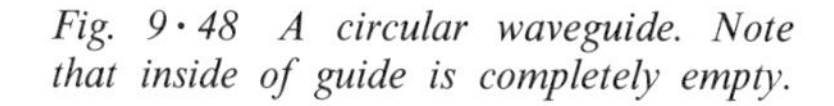

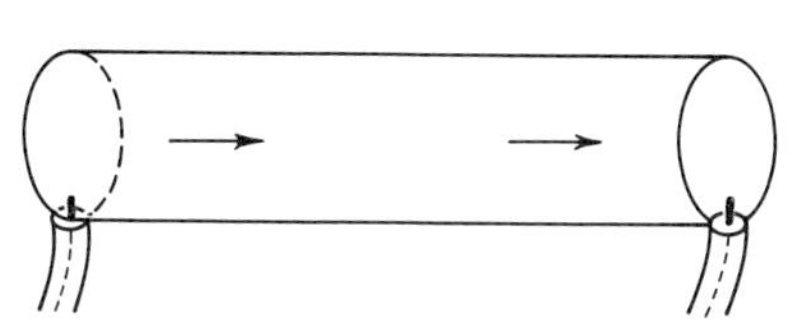

*Fig. 9·48 A circular waveguide. Note that inside of guide is completely empty.*

**Waveguides.** At frequencies of 1,000 Mc and beyond, it is quite common to transfer signals between nearby points (i.e., from antenna to receiver, or from transmitter to antenna) by hollow enclosures known as waveguides instead of conventional transmission lines. These hollow enclosures are constructed of metal and may be rectangular or circular (Figs. 9·47 and 9·48). The cross-sectional area of such waveguides is generally related to the frequency of the signals to be transmitted through them, becoming smaller as the signal frequency rises.

A signal can be introduced into a waveguide by inserting an antenna rod into the waveguide as shown in Fig. 9·47. The waves radiated by the rod will travel through the waveguide by bouncing back and forth across the guide. In this respect, the action is similar to that which is obtained at radio broadcast frequencies where the radiated signals are generally confined by the ground and the ionosphere. See Fig. 9·49. In a waveguide, we simply add two side panels so that the traveling energy is totally confined, whereas with a radio broadcast signal, the electromagnetic energy may travel horizontally in all directions.

At the other end of the waveguide, another rod is positioned to pick up the arriving signal. This energy is then taken by means of a coaxial cable to its terminating point, be it a radiating antenna, or a receiver.

Energy losses in waveguides are considerably lower than in coaxial or

*Fig. 9·49 The ionosphere and the earth act as a gigantic waveguide for low-frequency signals.*

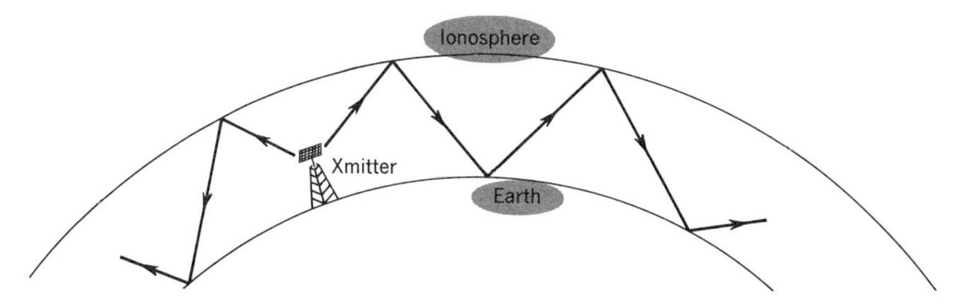

other transmission lines. Waveguides, however, become physically and economically feasible only at very high frequencies where their dimensions reach manageable proportions. Generally speaking, the minimum width of a guide must be no less than one-half wavelength of the signal to be transmitted through it. At 1,000 Mc, this results in a width of at least 6 in.; at 500 Mc, the width would be 12 in. A waveguide a foot wide is fairly bulky and, at lower frequencies, would be bulkier still.

## SUMMARY

1. An electromagnetic wave consists of an electric and a magnetic field. The two components are at right angles to each other and to the direction of energy propagation.
2. In free space, the wavelength of an electromagnetic wave is given by $\lambda = c/f$.
3. The intensity of a wave is a measure of the emf produced by the electric-field component and is expressed in microvolts per meter.
4. Antennas radiate the signal into space for transmission and extract signal energy from space for reception.
5. The total field about an antenna consists of the electric field due to the charges on the antenna, the induction field associated with the current, and the radiation field.
6. The dipole is the basic antenna, consisting of two quarter-wave conductors in a line. Maximum current appears at the center, with maximum voltage at the ends.
7. The polarization of a radio wave is defined as the plane of the electric-field component of the wave.
8. The Hertz antenna is an ungrounded half-wave antenna.
9. The Marconi antenna is a grounded quarter-wave antenna. It uses the ground as a reflecting surface to obtain the effect of equivalent half-wave operation.
10. A harmonic antenna operates at an exact multiple of its resonant frequency.
11. Adding series inductance to an antenna increases its effective length. Adding series capacitance to an antenna decreases the effective length.
12. More than one antenna used in combination constitute an antenna array. Arrays generally have better directional characteristics than a single antenna.
13. The ionosphere, or Kennelly-Heaviside region, consists of ionized layers of atmosphere.
14. Skip distance is the distance between the transmitter and the point at which the sky wave returns to the earth.
15. Fading is caused by two or more parts of a wave reaching a receiver by paths of different lengths. The difference in lengths causes the signals to arrive with different phases.
16. Line-of-sight waves travel in straight-line paths. They are very useful in the VHF and UHF bands and for microwaves.
17. Open-wire, twisted pair, and coaxial cable are common types of transmission line. At ultrahigh frequencies, these are replaced by waveguides.
18. The process of introducing energy into an antenna is known as feeding the antenna. Two general types of feeders are employed, resonant and nonresonant.

## SELF-EXAMINATION

Here's a chance to find out how well you have learned the material in this chapter. Work the exercises; then check your answers against the Key at the back of this book. These exercises are for your self-testing only.

1. The wavelength of an electromagnetic wave is equal to (*a*) velocity divided by frequency; (*b*) velocity multiplied by frequency; (*c*) frequency divided by velocity; (*d*) frequency multiplied by a constant.
2. Radio waves travel with a velocity of (*a*) 186,000 ft per sec; (*b*) 300,000 meters per sec; (*c*) 186,000 miles per sec; (*d*) 300,000 miles per sec.
3. A series of ionized layers of atmosphere used for short-wave communications is called

the (*a*) Hertzian layers; (*b*) Marconi-Sommerfield layers; (*c*) atmosphere; (*d*) Kennelly-Heaviside layers.

4. Line-of-sight transmission is used mainly for (*a*) VHF communication; (*b*) long-distance radiotelephony; (*c*) skip-distance propagation; (*d*) broadcast-band radio.
5. Characteristic impedance is (*a*) the surge impedance of the ionosphere; (*b*) the input impedance of an infinite line; (*c*) always maximum for open-wire lines; (*d*) the impedance of the reflected wave.
6. A simple one-half-wavelength antenna radiates the strongest field (*a*) at 45° to its axis; (*b*) parallel to its axis; (*c*) at right angles to its axis; (*d*) at 60° to its axis.
7. The current maximum in a Marconi antenna is found at (*a*) the base of the antenna; (*b*) the center of the antenna; (*c*) some point between the base and the center of the antenna; (*d*) the top of the antenna.
8. If an antenna is too short for the wavelength being used, the effective length can be increased by adding (*a*) capacitance in series; (*b*) inductance in series; (*c*) resistance in parallel (*d*) resistance in series.
9. Standing waves exist on an antenna because (*a*) the antenna has resistance losses; (*b*) radiation fields cannot induce current; (*c*) the antenna is not the proper length; (*d*) there is reflection from the open end.
10. The radiation resistance of a one-half wavelength antenna is approximately (*a*) 37 ohms; (*b*) 73 ohms; (*c*) 300 ohms; (*d*) 1 K.

## QUESTIONS AND PROBLEMS

1. Calculate the wavelength for radio waves at the following frequencies: 1 Mc, 10 Mc, 100 Mc, and 1,000 Mc.
2. Explain briefly why high frequencies are radiated more efficiently than low frequencies.
3. What is meant by distributed inductance? Distributed capacitance?
4. For a half-wave dipole antenna, mounted horizontally, (*a*) what is its directional response in the vertical plane? (*b*) how much is its radiation resistance at the center?
5. Calculate the length in feet of a half-wave dipole for (*a*) the center of the FM broadcast band, at 98 Mc; (*b*) the low end of the television broadcast channels, at 54 Mc.
6. What are the directional characteristics of a loop antenna?
7. Define the following terms in the propagation of radio waves: (*a*) ground wave; (*b*) ionosphere; (*c*) skip distance; (*d*) zone of silence; (*e*) line-of-sight transmission.
8. What is meant by the characteristic impedance of a transmission line? Give two advantages of terminating a line in its characteristic impedance.
9. What is meant by a harmonic antenna? At what frequency will a dipole antenna 4.67 ft long be operating at the third harmonic of its half-wave fundamental?
10. Illustrate by diagrams how to load an antenna to make it (*a*) electrically longer for a lower frequency, (*b*) electrically shorter for a higher frequency.

# Review of chapters 7 to 9

## SUMMARY

1. A transmitter produces r-f energy that is radiated by the antenna. Types of radio waves are damped waves, continuous waves (c-w), interrupted continuous waves (i-c-w), and modulated continuous waves (m-c-w).
2. Modulation is a process of making a signal voltage vary one characteristic of the higher-frequency r-f carrier wave. In amplitude modulation, the amplitude of the carrier is varied by the modulating signal. The envelope of the AM carrier is an outline of its amplitude variations. In frequency modulation, the frequency of the carrier is varied by the modulating signal. Frequency modulation is more immune to noise than amplitude modulation.
3. The AM wave consists of the carrier, upper sideband, and lower sideband. Each sideband differs from the carrier by the amount of the audio frequency.
4. Per cent modulation in AM is determined by the ratio of the audio to the unmodulated carrier amplitude.
5. High-level modulation usually means modulating the plate circuit of the final r-f power amplifier. Low-level modulation is in a preceding stage. Modulation can be accomplished in the plate, grid, screen, suppressor, or cathode circuit of an r-f amplifier.
6. Two methods of producing frequency modulation are the reactance-tube circuit and the Armstrong phase-modulating system.
7. In frequency modulation, the variation of the carrier from its center frequency is the amount of frequency deviation. The amount of modulating voltage amplitude determines the deviation. In the commercial FM broadcast band, maximum deviation is 75 kc, for 100 per cent modulation. For the FM sound in television broadcasting, 25 kc is maximum deviation.
8. In the r-f section, a MOPA transmitter includes a master oscillator, which drives the power amplifier coupled to the antenna. The master oscillator may be crystal-controlled for better frequency stability, or it may be a variable-frequency oscillator (VFO).
9. A buffer amplifier is an r-f stage to isolate the oscillator from the power amplifier.
10. An intermediate power amplifier (IPA) is a medium-power stage to drive the final amplifier.
11. R-f power amplifiers generally operate class C for maximum power output with greatest efficiency.
12. In a stage using series feed for plate voltage, B+ is applied to the plate coil; with shunt feed there is no B+ voltage in the plate-tank circuit.
13. Neutralization is a method of feedback to neutralize the tube's grid-plate capacitance so that the amplifier will not oscillate.
14. Parasitics are undesired oscillations in the amplifier, resulting from stray capacitance and inductance, especially with long connecting leads.
15. An r-f power amplifier is tuned by adjusting for minimum direct plate current in the stage or maximum grid drive on the following stage.
16. In radiotelegraph emission the transmitter output is turned on for dot or dash (mark), and off for the space between. Blocked-grid keying is one method of turning off the r-f output. In frequency-shift keying, however, the transmitter's carrier frequency is shifted by 400 to 1,800 cps to have different frequencies for mark and space.
17. In large power tubes, the filament voltage is critical for normal operation and tube life. Tubes rated above 1,500 watts plate dissipation are generally water-cooled.
18. The electromagnetic radio wave consists of an electric field and a magnetic field. Both com-

ponents are at right angles to each other and to the direction of energy propagation. The polarization is defined as the plane of the electric field.

19. In free space the wavelength of an electromagnetic wave is $\lambda = c/f$, where $c$ is the velocity of light. Long wavelengths correspond to low frequencies; short waves mean high frequencies.
20. The intensity or strength of an electromagnetic wave is indicated in microvolts per meter.
21. A transmitting antenna radiates r-f signal into space; a receiving antenna extracts r-f signal from a radio wave in space.
22. The dipole is the basic antenna, consisting of two quarter-wave poles insulated from each other. At the center, current is maximum and impedance minimum; voltage and impedance are maximum at the ends.
23. The Hertz antenna is an ungrounded half-wave dipole antenna.
24. The Marconi antenna is a grounded quarter-wave antenna. It uses ground as a reflecting surface for the equivalent of half-wave operation.
25. A harmonic antenna operates at an exact multiple of its resonant frequency.
26. Adding series inductance to load an antenna increases its electrical length.
27. The ionosphere, or Kennelly-Heaviside region, consists of ionized layers of atmosphere. Radio waves reflected from the ionosphere are sky waves.
28. Line-of-sight waves travel mainly in straight-line paths. The higher the frequency, the greater is this effect. Line-of-sight transmission applies to microwaves and the VHF and UHF bands.
29. A transmission line has two conductors with constant spacing to provide uniform characteristics in delivering signal to or from the antenna. Common types are open-wire line and coaxial cable.
30. The characteristic impedance of a transmission line is the impedance that would be present across the line if it were infinitely long. Wider spacing between conductors results in higher impedance. A line terminated in its characteristic impedance is nonresonant.

## REFERENCES (*Additional references at back of book.*)

### *Books*


Griffith, B. W., Jr., *Radio-Electronic Transmission Fundamentals,* McGraw-Hill Book Company.

Hellman, C. I., *Elements of Radio,* 3d ed., D. Van Nostrand Company, Inc.

Henney, K., and G. A. Richardson, *Principles of Radio,* 6th ed., John Wiley & Sons, Inc.

Hickey, H. V., and W. M. Villines, *Elements of Electronics,* 2d ed., McGraw-Hill Book Company.

Marcus, A., and W. Marcus, *Elements of Radio,* 3d ed., Prentice-Hall, Inc.

Sheingold, A., *Fundamentals of Radio Communication,* D. Van Nostrand Company, Inc.

Shrader, R. L., *Electronic Communication,* McGraw-Hill Book Company.

Slurzberg, M., and W. Osterheld, *Essentials of Radio Electronics,* 2d ed., McGraw-Hill Book Company.


## REVIEW SELF-EXAMINATION

Here's another chance to check your progress. Work the exercises just as you did those at the end of each chapter and check your answers.

Answer true or false

1. An A0 emission is unmodulated.
2. The carrier-wave frequency must be much higher than the frequency of the modulating voltage.
3. With audio modulation the envelope of the carrier is an audio frequency.
4. With audio modulation the sidebands of the carrier are radio frequencies.
5. With 1,000 cps audio-modulating a 1,000-kc carrier, the sideband frequencies are 99 kc and 101 kc.

6. Modulating the control-grid circuit of an IPA stage is high-level modulation.
7. With 100 per cent modulation, the peak of the modulated carrier wave is twice the unmodulated carrier amplitude.
8. Maximum deviation for the commercial FM broadcast band is 15 kc.
9. Radiotelegraph transmission with 1,000-cps audio tone modulating the carrier is an example of m-c-w.
10. The reactance tube circuit can be used for frequency modulation.
11. A crystal-controlled oscillator is called a VFO.
12. A MOPA transmitter usually has a buffer amplifier for isolation.
13. VHF transmitters can use frequency multipliers with a crystal oscillator.
14. An amplifier with shunt feed for B+ voltage can have an r-f choke in shunt with the plate-tank circuit.
15. In tuning the plate tank watch the d-c milliammeter in the plate circuit dip to minimum.
16. Maximum grid-leak bias on an r-f amplifier indicates maximum signal drive from the preceding stage.

# Chapter 10 Principles of receivers

This unit shows how r-f and a-f circuits with a detector can be combined to form a receiver that reproduces audio signal from the modulated radio wave. The block diagram in Fig. 10·1 illustrates how the receiver does this job. R-f tuning requirements, typical detector circuits, and volume control are explained in detail. Also included is alignment of the r-f tuned circuits for best performance. The topics are as follows:



## *10·1 The t-r-f amplifier*

The basic circuit of a typical tuned-radio-frequency (t-r-f) amplifier is shown in Fig. 10·2. The r-f signal input, from either the antenna or a previous r-f stage, is inductively coupled into the tuned circuit by means of r-f transformer $T_1$. Its secondary inductance $L_s$ is tuned to resonance at the desired radio frequency by varying the tuning capacitor $C_T$. At resonance, maximum voltage is developed across $C_T$ for signal currents at the resonant frequency, impressing maximum signal voltage across the grid-to-cathode circuit of the amplifier tube. In the amplifier output circuit, the signal variations in plate current flow through the primary of $T_2$. By inductive coupling through the transformer, then, the amplified signal is coupled to the next stage.

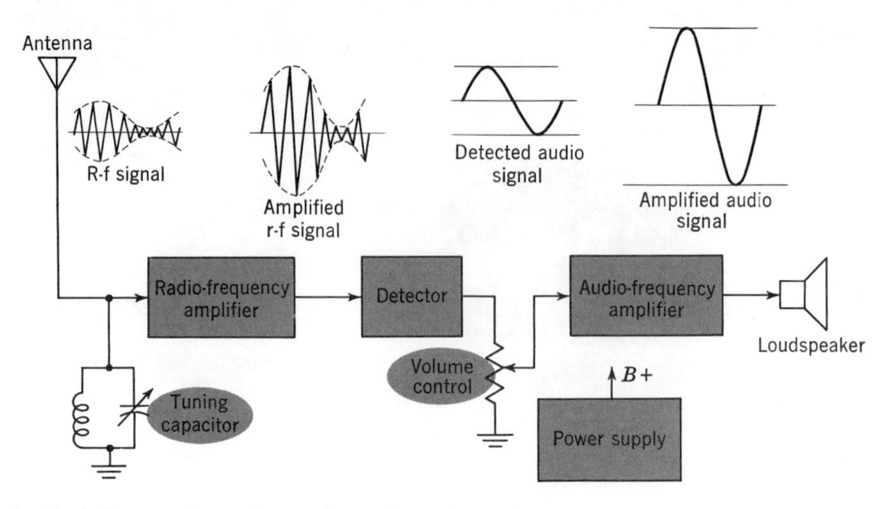


*Fig. 10 • 1 Block diagram of a receiver, with signal waveshapes for each section.*

**Circuit components.** $R_s$ is the screen voltage-dropping resistor to provide a typical value of about 100 volts at the screen grid from the 250-volt B supply. The screen-bypass capacitor $C_s$ keeps the screen-grid voltage constant, bypassing a-c signal variations around $R_s$. The typical cathode-bias value of 3 volts is produced by the cathode resistor $R_k$ with the cathode-bypass capacitor $C_k$.

R-f amplifier tubes are usually pentodes. At radio frequencies, triodes have the disadvantage of a large plate-to-grid capacitance that allows some of the plate signal to feed back to the grid circuit. If there is enough feedback the stage may oscillate, producing output at the oscillator frequency instead of amplifying the desired signal. The gain of a t-r-f stage is about 15.

**Transistorized r-f amplifier circuit.** Corresponding to Fig. 10 • 2, the transistorized r-f amplifier in Fig. 10 • 3 shows a common-emitter circuit. Note that negative supply voltage is used for the PNP transistor. The reverse bias on the collector is $-12$ volts, with $-0.2$ volt on the base for

*Fig. 10 • 2 Typical t-r-f amplifier. D-c voltages indicated with respect to chassis ground.*

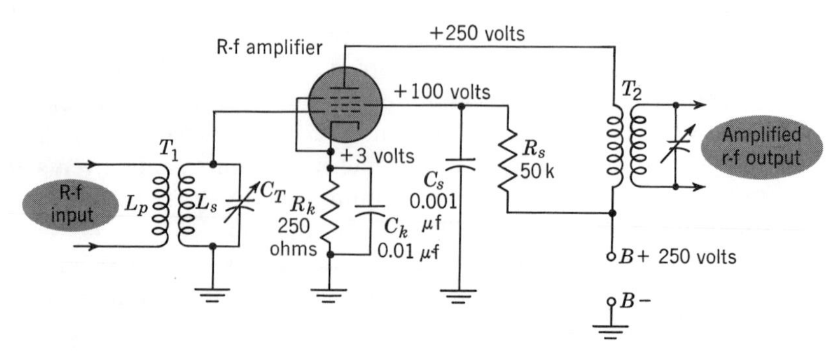

forward bias. Voltage division by $R_1$ and $R_2$ provides the required base voltage. $C_1$ bypasses $R_1$ for the r-f signal frequencies. The r-f transformers $T_1$ and $T_2$ are shown with voltage step-down from $L_p$ to $L_s$ for an impedance match to the low resistance of the base input circuit.

**R-f response curve.** The characteristics of a tuned amplifier are indicated by means of an r-f response curve, such as the one shown in Fig. 10·4. The variations of amplifier gain are plotted for different frequencies, showing the amount of signal-voltage output at the resonant frequency $f_r$, and for other frequencies close to resonance.

The ideal response, indicated by the dotted rectangle, illustrates the tuning characteristics desired for the receiver. This shows maximum output at resonance with a total bandwidth of 10 kc. The 10-kc range is for sideband frequencies of the modulated carrier wave to which the circuit is tuned. The straight side responses show that there is no gain outside the desired 10-kc band, completely rejecting all undesired signal frequencies. The flat-top characteristic allows all the sideband frequencies to receive as much amplification as the carrier at the resonant frequency.

The actual response of the t-r-f amplifier does not have these ideal characteristics, however. Actually, the response is exactly the same as the resonance curve of the amplifier tuned circuit. This is illustrated by the solid curve in Fig. 10·4. The bandwidth here is 10 kc measured between the two points where the gain is 71 per cent of maximum, or 3 db down from 100 per cent response. This is suitable even though it only approximates the ideal response.

*Sensitivity* is the ability of the receiver to produce a usable amount of audio output at the loudspeaker from weak r-f signals in the antenna circuit. The more tuned amplifier stages in the receiver, the greater is the sensitivity, since each amplifier contributes gain. Although the audio stages also amplify the selected signal, the receiver sensitivity is basically determined by the tuned amplifiers. These must furnish the detector and audio stages with useful signal that has much more amplitude than static, noise, and other interfering signals. In the response curve of Fig. 10·4, higher amplitude indicates greater sensitivity.

*Fig. 10·3 Transistorized r-f amplifier circuit.*

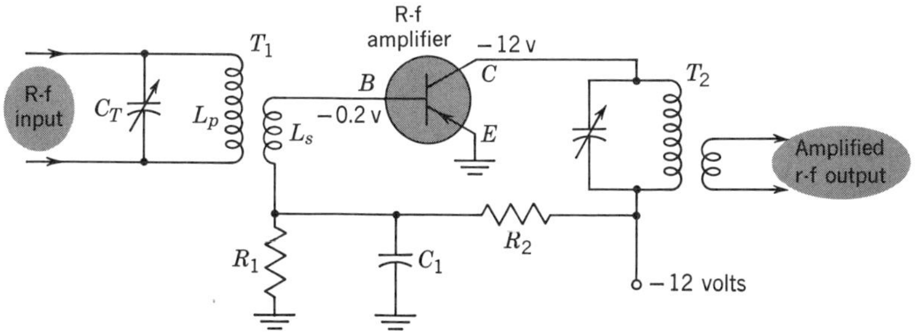

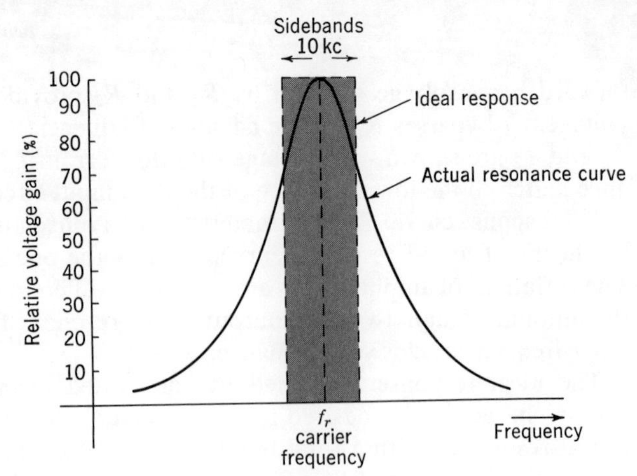


*Fig. 10·4 Response of t-r-f stage.*

The sensitivity of a receiver is often specified by the amount of r-f signal required at the antenna to produce 1 volt at the detector. As an example, 10-μv sensitivity for a high-gain receiver means this amount of antenna signal receives enough amplification to produce 1-volt output from the detector. This is a practical amount of output to drive the audio amplifier. Typical sensitivity values are 250 to 5,000 μv for average receivers for the AM radio broadcasting band.

*Selectivity* is the ability of the receiver to tune in the carrier-wave frequency of the one desired station, while rejecting undesired frequencies. This is determined entirely by the tuned stages in the receiver because the audio circuits cannot distinguish between signals from different stations. For any one tuned circuit, its $Q$, which determines the sharpness of resonance, limits the selectivity. The more tuned circuits in the receiver, the greater is the selectivity. Selectivity is indicated by the steepness of the side responses and the narrowness of the response curve. The sharper the response curve of a tuned circuit, the greater is its selectivity. For this reason, the graph of the r-f response in a tuned amplifier is its *selectivity curve.* The selectivity curve should not be too sharp, however, because the response curve must be wide enough to include the sidebands associated with the modulated carrier wave.

## *10·2 The r-f tuning circuit*

In the t-r-f amplifier the $LC$-tuned circuit selects the desired signal. The tube itself can function only as an amplifier because it does not discriminate between signal frequencies. This ability of the tuned circuit to select the desired frequencies is the result of resonance.

The tuning circuit in the receiver usually consists of a coil having a fixed value and a variable capacitor to accomplish the tuning, as illustrated by $L_sC_T$ in Fig. 10·2. The variable capacitor is generally of the air-dielectric type, with rotor plates that are moved in and out of mesh with fixed stator plates, providing a simple and stable mechanical device for tuning. Rotating

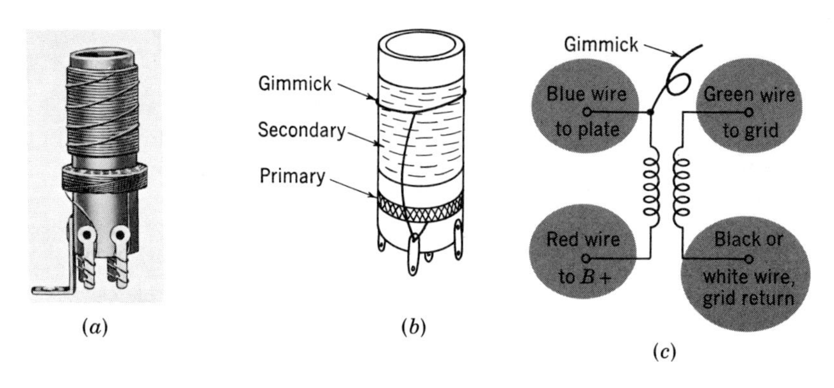


*Fig. 10 · 5 Broadcast-band r-f transformer. Note color code of wiring diagram for interstage transformer.*

the capacitor completely out of mesh provides minimum capacitance. This establishes the high-frequency limit of the band through which the circuit can be tuned for a fixed inductance value. For the AM radio broadcasting band, this value is 1,605 kc. When the rotor plates are completely in mesh, capacitance is maximum, and the circuit is tuned to the lowest frequency in the tuning range, generally 535 kc. For intermediate positions of the capacitor rotor plates, the tuned circuit is resonant to different frequencies between the lowest and the highest in the tuning range.

Some receivers, particularly automobile radios, use *permeability tuning* for the *LC* circuit. In this method, the capacitor is fixed and the inductance is varied by moving a powdered-iron slug mounted at the center of the coil.

**The r-f transformer.** Figure 10 · 5 shows a typical unit for a vacuum-tube circuit. The lower coil on the air-core transformer is the primary inductance while the upper winding is the secondary. The primary has several millihenrys of inductance to serve as an r-f choke. Note that the primary is not made to resonate at the desired frequency because of the possibility of oscillations in the r-f amplifier as the secondary circuit is tuned over its range.

The secondary inductance is about 250 $\mu$h to tune with a variable capacitor having a range of 20 to 365 pf. The d-c resistance values may be 40 ohms for the primary coil and 4 ohms for the secondary, approximately.

The r-f transformer $T_1$ in Fig. 10 · 2 is called an *antenna transformer* because it couples signal from the antenna into the first r-f amplifier. The transformer $T_2$ is an *interstage r-f transformer,* coupling signal from one stage to the next. For both functions the construction is essentially the same.

The r-f transformer generally has a single turn called a *gimmick loop* to provide some capacitive coupling between the primary and secondary. As illustrated in Fig. 10 · 5, the loop connects to the high potential side of the primary, and is wrapped around the secondary insulation. The gimmick provides capacitive coupling equal to a few picofarads in addition to the transformer action. The capacitive coupling is used to neutralize part of the inductive coupling in the transformer, which increases at the high-frequency end of the band. Most r-f transformers have the gimmick, but often it is not indicated on the schematic diagram.

**Resonant frequency of tuned secondary.** The $LC$-tuned circuit in the grid of the r-f amplifier is resonant at the frequency $f_r = 1/(2\pi\sqrt{LC})$.

*Example.* If resonance is at 1,000 kc at the center of the AM broadcasting band, when the tuning capacitor is set at 100 pf, the required inductance value is

$$L = \frac{1}{4\pi^2 f_r^2 C} = \frac{0.025}{1 \times 10^{12} \times 100 \times 10^{-12}} = \frac{0.025}{100} = 0.00025 \text{ henry} = 250 \ \mu\text{h}$$

With the fixed inductance value of 250 μh, the variable capacitor tunes the $LC$ circuit to the frequencies below 1,000 kc when $C$ is more than 100 pf, or above 1,000 kc when $C$ is less than 100 pf.

**The tuning range.** The range of frequencies through which the coil can tune with a variable capacitor depends on its maximum and minimum capacitance values. In a typical tuning circuit for the AM broadcasting band, the minimum capacitance, including stray capacitance, may be 40 pf while the maximum capacitance is 360 pf. The ratio of these two extreme values is 9 to 1. This is not the range of frequencies that can be tuned, though, because the resonant frequency of a tuned circuit decreases inversely as the square root of the capacitance. Therefore the ratio of maximum to minimum frequencies for the capacitance ratio is only $\sqrt{9}$, or 3 to 1. The actual frequencies are determined by the inductance value as well as by the capacitance, but the tuning range with this variable capacitor can only include a band where the highest frequency is three times the lowest frequency. For instance, this could cover the 540- to 1,620-kc range.

**Tuning capacitors.** As shown in Fig. 10·6, the variable air capacitor has a set of rotor plates that moves in and out of mesh with the stator plates without touching them. The amount of shaft rotation is about one-half

*Fig. 10·6 (a) Construction of a tuning capacitor. Note parallel trimmer capacitance. (b) Schematic symbol for each section.* (J. W. Miller Co.)

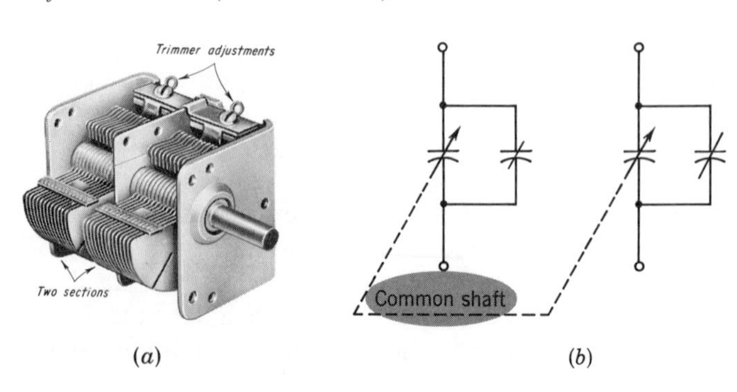

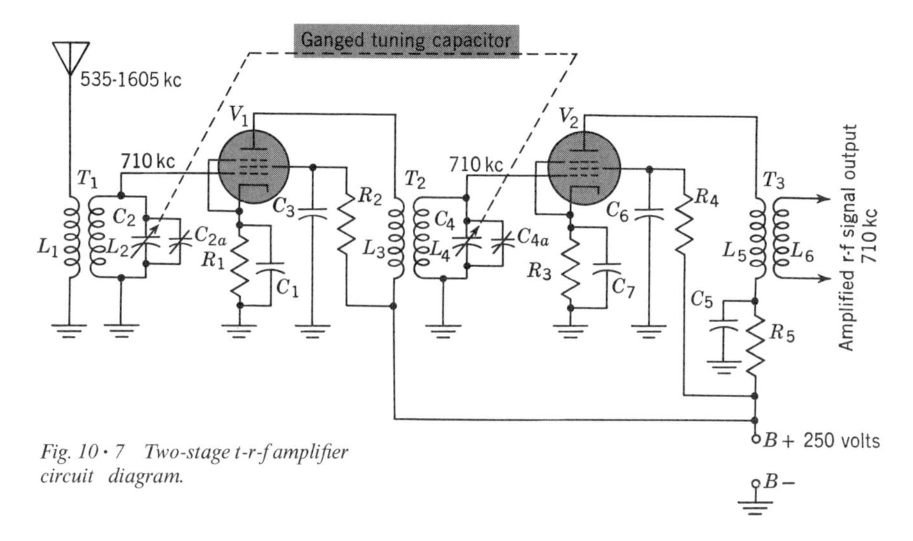


*Fig. 10·7 Two-stage t-r-f amplifier circuit diagram.*

revolution between full mesh for maximum capacitance and the out-of-mesh position for minimum capacitance. For the AM broadcast band the capacitance range is generally 20 to 353 pf with 17 plates.

When the tuning capacitor is connected into its circuit, the stator is connected to the high side of the circuit, that is, the point of high-signal potential. This is usually the control grid of the r-f amplifier, or the base in a transistor. The rotor is connected to chassis ground. The rotor is grounded instead of the stator, to minimize the detuning effect produced by hand capacitance when the capacitor's shaft is touched while tuning.

The rate at which the resonant frequency of the tuned circuit varies as the capacitor is rotated depends upon the construction of the capacitor. The main types are:

1. *Straight-line capacitance* (SLC). Each degree of shaft rotation provides an equal change in capacitance. This type has semicircular plates.
2. *Straight-line wavelength* (SLW). Each degree of shaft rotation provides an equal change in wavelength for the resonant frequency of the *LC* circuit.
3. *Straight-line frequency* (SLF). Each degree of shaft rotation provides an equal change in resonant frequency for the *LC* circuit. This type has plates offset from center.

## 10·3 Multiple t-r-f stages

To increase the r-f gain, two or more stages can be cascaded, as shown in Fig. 10·7. Both stages are similar to the t-r-f amplifier of Fig. 10·2. Tracing the circuit, we find signal from the antenna is coupled by the antenna transformer $T_1$ into the tuned circuit $L_2C_2$ of the first r-f amplifier $V_1$. The resonant circuit is tuned to select the desired signal frequencies, which provide maximum voltage output. In Fig. 10·7, the circuit is tuned to 710 kc. This signal voltage, connected between grid and cathode of the amplifier, produces plate-current variations through the plate-load impedance $L_3$ that correspond to the desired signal.

The plate-current variations through the primary of the r-f interstage transformer $T_2$ induce the desired signal voltage in the secondary, coupling the desired signal to the second r-f amplifier. The grid circuit of the second stage is also tuned to the incoming-signal frequency of 710 kc, since the tuned circuit $L_4C_4$ is resonant at the same frequency as $L_2C_2$. Maximum signal voltage is applied to the grid of the second stage at the resonant frequency, and the desired signal is amplified again. The second r-f amplifier has a plate decoupling filter $R_5C_5$. This serves to keep r-f signal currents out of the B supply to prevent feedback from the plate circuit of $V_2$ to the output circuit of $V_1$. The latter stage is also connected to the B supply.

**Ganged tuning.** The dotted lines joining tuning capacitors $C_2$ and $C_4$ in Fig. 10·7 indicate that they are varied together to tune both resonant circuits at the same time. This is called *ganged tuning.* It is accomplished by using a ganged tuning capacitor, which has two or more rotors on a common shaft, as illustrated in Fig. 10·6. When permeability tuning is used, instead of ganged tuning capacitors, the position of the tuning slug in each of the variable-inductance coils can be varied simultaneously.

**Alignment of tuned circuits.** An important problem with multiple-tuned circuits is setting each tuned circuit to exactly the same resonant frequency. If any resonant circuit were tuned to a frequency higher or lower than the others, much of the advantage gained with multiple-tuned stages would be lost. The inductances in the tuned circuits may all be very close to the same value, but at any position of the ganged tuning capacitor there is generally a slight difference in resonant frequency between individual circuits because of small differences in stray capacitance. Therefore it is necessary to provide some way of adjusting each variable capacitor to bring each tuning circuit to the same resonant frequency. The procedure of adjusting the circuits to put them all "in line" at the same resonant frequency is called *alignment.*

**Trimmer capacitance.** The alignment of t-r-f stages is usually done by means of small variable mica capacitors having a maximum capacitance of about 20 pf. One such capacitor is connected in parallel with each section of the main tuning capacitor, as illustrated by $C_{2a}$ and $C_{4a}$ in the schematic diagram of Fig. 10·7. Physically, the trimmer is often built as a part of the gang capacitor. Several thin sheets of mica are placed between the metal frame of the tuning capacitor and a small metal spring plate connected to the stator. As a result, the soldering lug at the top connects to both the stator of the main tuning capacitor and the high-potential side of the trimmer. The rotor of the main tuning capacitor and the low-potential side of the trimmer are part of the frame, grounded to the chassis. Tightening the small screw in the trimmer capacitor brings the top plate closer to the bottom plate to increase the capacitance. See the trimmer capacitors in Fig. 10·6.

The trimmer capacitance has its greatest effect at the high-frequency end of the tuning range, where the capacitance of the main tuning capacitor is minimum. A numerical example will show why this is so. Assume that the trimmer is at 10 pf when the main tuning capacitance is set at 30 pf. Of the total 40-pf capacitance in the tuned circuit, the trimmer provides 10/40, or 25 per cent. At the low-frequency end, however, with a tuning capacitance of 360 pf, a trimmer capacitance of 10 pf is only 10/360 of the total capacitance in the tuned circuit, or approximately 3 per cent.

**Slotted plates.** The two end plates of the rotor in each section of the gang generally have four or five radial slots so that the plates can be bent to adjust the capacitance value. These allow independent adjustments of the tuning capacitance at different points in the frequency range, particularly for the low end where the trimmer has little effect.

## *10·4 R-f selectivity*

Not only does a cascaded r-f amplifier increase the amplitude of the desired signal, but the over-all selectivity of multiple-tuned circuits is much sharper than the response of a single stage because each amplifier multiplies the output of the previous stage. The product of the responses for the individual tuned stages increases the selectivity to a marked degree, providing much greater rejection of undesired signals. This effect is illustrated by the relative response curves in Fig. 10·8.

Some numerical values will show why the selectivity is so much sharper with two tuned stages. Assume that the first r-f amplifier has an actual gain of 10 at its resonant frequency, 1,000 kc, corresponding to the 100 per cent point of maximum gain at $f_r$. In addition, let the value of the input signal be 1 mv, which is also the amplitude of the input voltage at the frequencies between 900 and 1,010 kc. With a gain of 10 at 1,000 kc, the amplifier output is 10 mv at this frequency. However, the tuned circuit in the amplifier provides only 71 per cent of maximum output at the off-resonance frequencies of 990 and 1,010 kc. As a result, the amount of amplifier gain is 7,

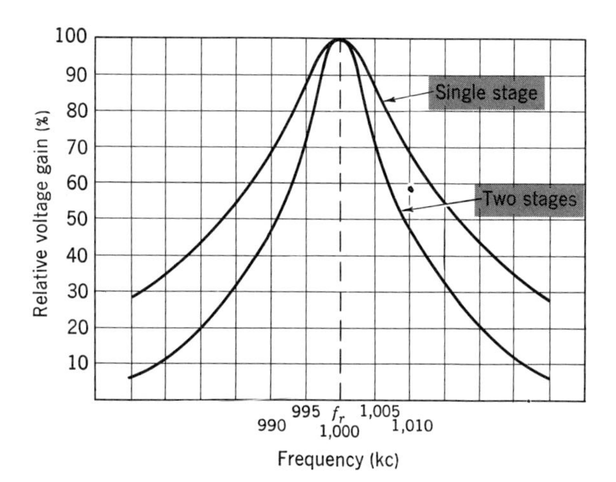


*Fig. 10·8 Effect of multiple-tuned stages on r-f selectivity.*

and the output 7 mv at these frequencies. When the output is coupled to the second stage, the signal frequencies are again amplified with the same relative response. The gain of the second stage is also 10 for the resonant frequency, producing 10 × 10, or 100 mv at 1,000 kc. At 990 and 1,010 kc, the output is 49 mv, however, since the original 1-mv signal is amplified twice, each time by a factor of only 7. As a result of the multiplication of the individual gain values, the combined response of multiple-tuned stages increases the over-all selectivity to provide sharper frequency response.

The number of multiple-tuned stages that can be used has practical limitations. One important restriction is the fact that the selectivity curve may become too sharp, with insufficient bandwidth for the sideband frequencies of the modulated carrier. Excessive sideband cutting would result, producing severe distortion of the signal. When more than one tuned stage is used, the combined effect of all the stages must provide an over-all response with the required bandwidth. Another trouble that may develop with multiple stages tuned to the same frequency is that excessive feedback can make the amplifier oscillate.

## *10 · 5 Regeneration in r-f amplifiers*

When several stages are tuned to the same frequency, some of the signal output from one stage may feed back to a previous stage and provide enough regeneration to make the amplifier operate as an oscillator. The sources of feedback in the amplifier are usually stray capacitive or magnetic coupling, the plate-to-grid capacitance of the tube, and coupling through some mutual impedance common to two or more amplifier stages, particularly the B supply. If the r-f stages oscillate, the receiver will either whistle or chirp. This audio output is caused by audio modulation of the r-f oscillations. In the detector output, the modulation provides audio output. Very often, the audio modulation results from intermittent oscillations, which go on and off at an audio rate. Because of the regeneration problems, it is generally not practical to use more than three or four cascaded amplifiers tuned to the same frequency.

Pentodes are generally used for the r-f amplifier stages because their very small plate-to-grid capacitance minimizes this as a troublesome source of feedback. The placement of connecting leads in r-f amplifiers is also very important. This is called *lead dress*. The input and output leads of one stage, or signal leads from two different stages, should be kept away from each other because capacitive or magnetic coupling between the leads can feed back enough signal to produce oscillations.

To minimize magnetic coupling between coils, the following precautions are helpful. Very often each coil is mounted in its own metal shield can. The shield should be a good electrical conductor, such as copper or aluminum, to provide low resistance for the induced eddy currents. This confines the magnetic field of the coil within the shield can. Coils that are physically small may omit the shield cans, since they can be positioned far apart compared with the coil size. It is helpful to mount unshielded coils at right angles to each other so that their magnetic fields are in different planes. Also, mounting one coil on top of the chassis and another below utilizes the metal chassis as a shield between the coils.

Mutual coupling through a common B supply is another source of regeneration. A decoupling network, such as $R_5C_5$ in Fig. 10·7, helps to isolate the stage from the mutual impedance of the B-supply output filter capacitor. Although $C_5$ is a large-capacitance electrolytic filter capacitor of 8 to 40 $\mu$f, its coiled construction provides enough inductance to develop appreciable r-f signal voltage. It may be necessary to shunt the large electrolytic with a small paper or mica capacitor of about 0.005 $\mu$f to act as an r-f bypass. In the screen-grid circuit, a voltage-dropping resistor and bypass capacitor combination, such as $R_4C_6$ or $R_2C_3$ in Fig. 10·7, also serves as a decoupling filter. If the screen bypass opens, regeneration can result from common coupling through the B supply.

Another source of mutual coupling between r-f stages may be the metal chassis itself, or a long chassis return lead that is common to several circuits. A wire only a few inches long has enough inductance to develop signal voltage at high radio frequencies. In order to minimize such mutual coupling, it is important to avoid connections that allow a common return path through the chassis for the input and output circuits of the amplifier. The return leads to chassis should be short. It may also be necessary to return capacitors in the plate and screen circuits directly to cathode. The higher the frequency, the more important these precautions become in eliminating regeneration in r-f amplifiers.

## *10·6 Detectors*

The function of the detector stage in a receiver is to recover the audio modulation of the modulated r-f signal by rectifying the modulated carrier wave and filtering out the r-f variations. The result is the desired audio signal. As illustrated in Fig. 10·9, for amplitude modulation the r-f input signal has amplitude variations with an envelope corresponding to the required audio voltage. In this form, however, the AM signal cannot provide audio signal, because the envelope is the same, top and bottom, for positive and negative polarities of the r-f carrier wave. With respect to the a-f variations, then, the average carrier voltage is zero. In order to obtain the audio envelope voltage, the r-f carrier must be rectified. Either polarity

*Fig. 10·9 Function of the detector.*

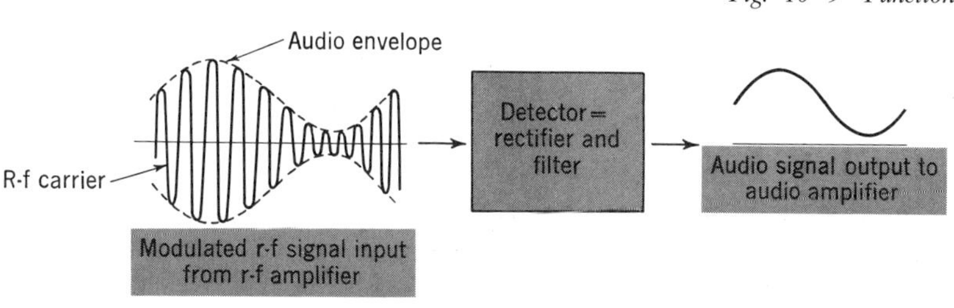

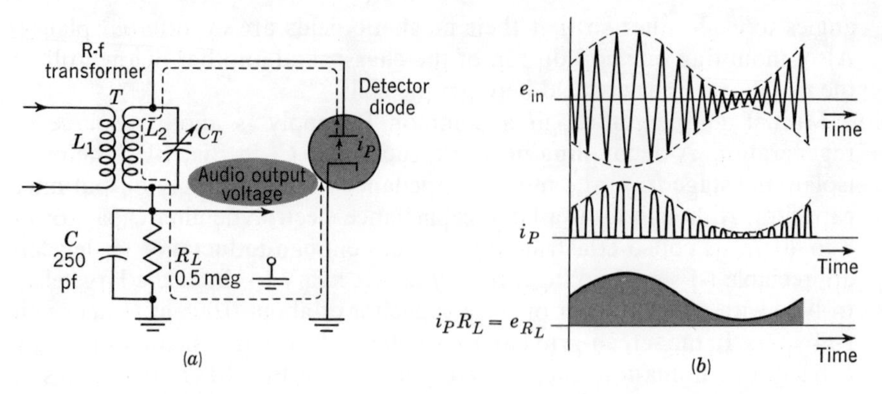


*Fig. 10·10 Diode detector. (a) Circuit diagram. (b) Current and voltage waveshapes.*

of the r-f signal can be rectified because the envelope is the same, top and bottom. After the r-f carrier has been rectified, either its peak or average value varies at the audio rate. Filtering out the r-f variations with a bypass capacitor causes the output of the detector to become just the desired audio signal without any r-f variations.

**Diode detector.** As shown in Fig. 10·10*a*, the modulated r-f signal supplies input voltage to the diode rectifier. When this a-c voltage makes the diode plate positive, plate current flows. Then $i_P$ can return through the d-c path of $L_2$ and $R_L$ back to the cathode. $R_L$ is the diode load resistor that provides rectified output voltage. In the waveforms in *b*, the top wave is the a-c input voltage, which is the modulated r-f carrier signal. The $i_P$ waveform below is the pulsating direct current consisting of half-cycles of the r-f signal, which results from the half-wave rectification. Note that $i_P$ flowing through $R_L$ produces the same waveform of pulsating d-c voltage. This is the rectified output voltage from the high side of $R_L$ to chassis ground.

As a pulsating d-c voltage, the output across $R_L$ varies at the audio rate. However, the r-f half-cycles are filtered out by the r-f bypass capacitor *C* so that no r-f signal is coupled to the audio amplifier. The result, then, is just an audio voltage across $R_L$ corresponding to the desired audio envelope of the modulated signal. The waveshape in *b* labeled $e_{R_L}$ is the $i_PR_L$ voltage drop across $R_L$. This audio output of the diode detector is coupled to the next stage through a coupling capacitor to block passage of the average d-c level of the detector output voltage; this could affect the bias on the first audio amplifier. Note that the r-f bypass *C* across $R_L$ also serves as an a-c return to cathode so that all the r-f signal voltage developed by the $L_2C_T$ tuned circuit is applied across the diode.

There is no B+ voltage for the diode detector. The r-f signal supplies the a-c input voltage that makes the diode conduct. The pulsating d-c output voltage across $R_L$ is the rectified a-c signal input. The detector then is just a rectifier for the a-c carrier wave. Suitable tube types are low-power

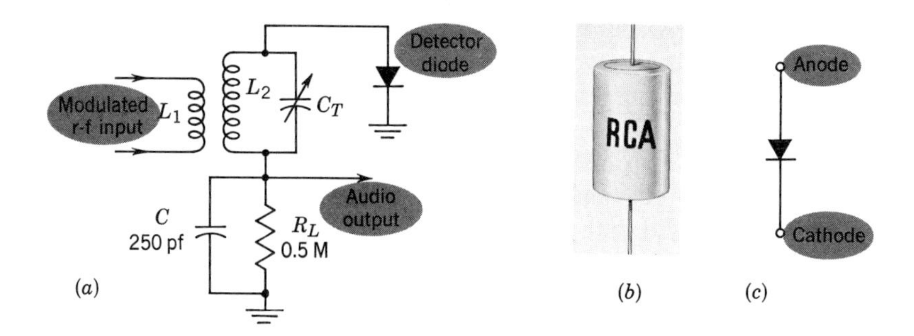


*Fig. 10·11 Crystal diode detector. (a) Circuit. (b) Typical diode. (c) Schematic symbol.*

diode rectifiers, such as the 6H6 and 6AL5 duodiodes. The 6SQ7, 12SQ7, 6AV6, and 12AV6 are duodiode triodes where the diodes are often used for the detector stage. Although duodiodes are common, the detector needs only one diode as a half-wave rectifier. Generally, either one diode plate is used or the two plates are connected together to operate as one diode.

Since germanium or silicon crystal diodes, such as the 1N34, are low-power rectifiers, these are often used in diode detector circuits. A typical circuit is shown in Fig. 10·11. The diode rectifier is the most common detector circuit for receivers because it has the least amount of distortion with an r-f signal level of 2 to 20 volts.

**Grid-leak-bias detector.** The type of detector shown in Fig. 10·12 uses the grid-cathode circuit of a triode or pentode as a diode rectifier for the r-f signal input, and the resultant audio voltage in the grid circuit is amplified in the plate circuit. Initially, without signal input, there is no bias. Then the positive half-cycles of r-f input voltage drive the grid positive to produce grid current. The result is grid bias across the $R_gC_g$ grid-leak combination. $C_g$ bypasses $R_g$ for radio frequencies so that the bias cannot vary at the r-f rate. However, the bias voltage can follow the a-f variations.

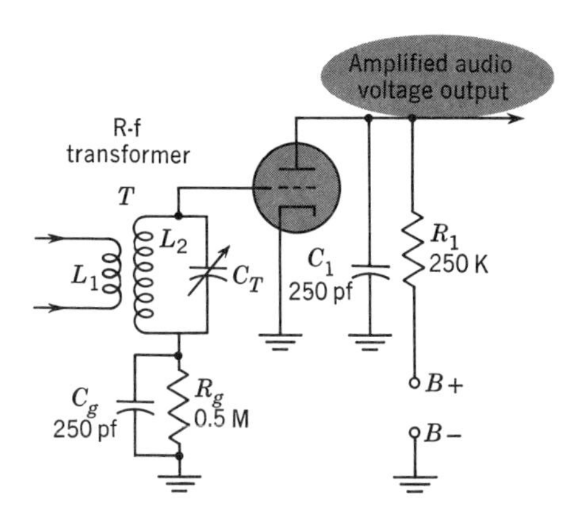


*Fig. 10·12 Grid-leak-bias detector circuit.*

As the r-f signal input varies in amplitude with the audio modulation, the amount of rectified output varies with the envelope. The rectified signal voltage across $R_gC_g$ then serves as a d-c bias voltage for the triode amplifier, where the amount of bias varies with the audio envelope of the modulated signal. More negative grid bias results in less plate current for the triode; less negative bias produces more plate current. With the current through $R_L$ varying at the audio rate, the output plate voltage is the amplified audio signal. $C_1$ is an r-f bypass to make certain that no r-f signal is coupled to the next audio amplifier.

Note that the grid circuit of this detector is the same as the diode detector in Fig. 10·11. Essentially, the grid-leak-bias detector consists of a diode detector in the grid circuit, d-c-coupled to the triode audio amplifier. Because of its amplification, the grid-leak-bias detector has the advantage of greater sensitivity for weak signals, but it has excessive distortion for strong signals of 2 to 20 volts.

**Regenerative detector.** As shown in Fig. 10·13, this circuit is similar to the grid-leak detector but with regenerative r-f feedback. $L_3$ in the plate circuit is inductively coupled to $L_2$ in the grid circuit, with the amount of feedback varied by $C_f$. The advantage is greater sensitivity. $L_c$ with $C_1$ and $C_2$ serve as an r-f filter for the detected audio output.

**Superregenerative detector.** This circuit is similar to Fig. 10·13 but has higher sensitivity, as the feedback is increased to the point of oscillations. However, the oscillations are blocked or *quenched* at a rate of 20 to 200 kc to allow detection of the desired audio signal. A separate quenching oscillator can be used or $R_gC_g$ in the detector circuit can be adjusted for the desired quenching rate. Normally, these regenerative detector circuits are used with only the simplest VHF receivers for weak signals.

**Plate detector.** The plate-detector circuit in Fig. 10·14 has a very high value of cathode resistance, so that the cathode bias is close to the grid-cutoff voltage of the triode or pentode amplifier. This bias results from the average value of plate current and is present with or without signal input. $C_k$ is large enough to bypass $R_k$ for audio frequencies, keeping the bias

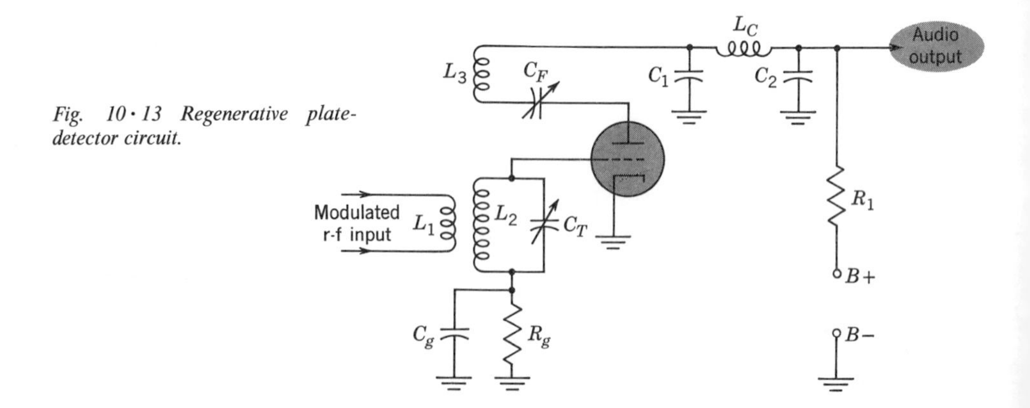


*Fig. 10·13 Regenerative plate-detector circuit.*

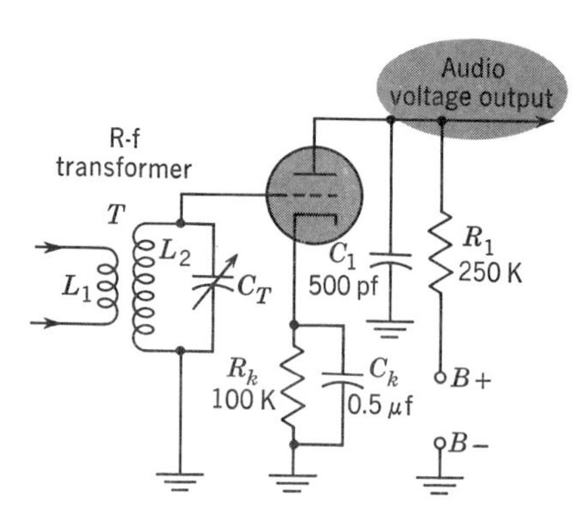


*Fig. 10·14 Plate-detector circuit.*

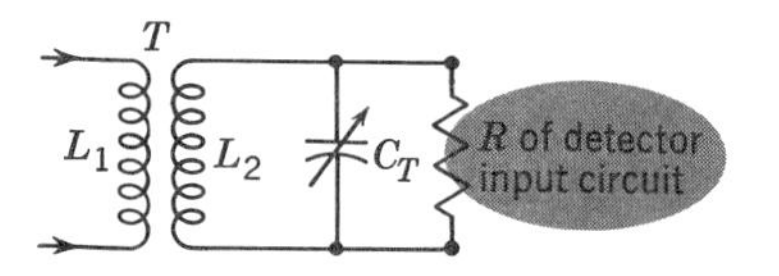


*Fig. 10·15 Detector as equivalent shunt R across tuned input circuit.*

constant. With the grid biased practically to cutoff, the amplifier operates essentially class B. The positive half-cycles of r-f input voltage produce rectified half waves of r-f plate current, while the negative half-cycles of the grid-input voltage are cut off. Fundamentally, the triode is serving as a half-wave rectifier for the modulated r-f signal input. When the r-f amplitude increases with the audio envelope, larger pulses of plate current are produced; lower amplitudes in the grid signal result in smaller plate-current pulses. Since the plate current through $R_1$ is varying at the audio rate, the result is audio-output voltage in the plate circuit. The plate-bypass capacitor $C_1$ in Fig. 10·14 bypasses the r-f variations.

The plate detector amplifies the signal, but it produces more distortion than a diode detector. The main feature of a plate detector is the fact that its input circuit has a very high resistance because there is no grid current. Sometimes this is called an *infinite-impedance* detector.

**Input resistance of detector.** When the rectified detector current results directly from rectifying the signal, the input circuit is equivalent to a low-resistance load on the r-f circuits supplying the input signal. For example, a 6AL5 diode conducting 100 μa with 10 volts of signal applied has an equivalent resistance of 100,000 ohms. This resistance of the detector is in parallel with the tuned r-f input circuit, as shown in Fig. 10·15. The lower the shunt resistance, the lower the $Q$ of the tuned circuit; higher resistance allows higher $Q$ with less detuning. The detector input circuit should draw minimum current, therefore, for maximum input resistance to prevent loading down the tuned input circuit. In the diode-detector circuit, the loading effect is minimized by using a high value for $R_L$ of 0.5 to 2 M.

## *10·7 Volume controls*

The amount of sound output from a loudspeaker depends on the power supplied by the audio amplifier. Varying the amount of audio signal therefore adjusts the volume. In a receiver there are several possible methods. Adjusting the gain of the r-f amplifier to vary the sensitivity changes the amount of detected audio signal. Or in the audio amplifier, the level of audio-signal voltage can be varied. Finally and most common is the

*Table 10 · 1 Comparison of detector circuits*

| *Type* | *Linearity* | *Sensitivity* | *Input volts* | *Input resistance* | *Remarks* |
|---|---|---|---|---|---|
| Diode | Good | Low | 3 or more | Low | No B+ voltage; most common detector |
| Grid-leak | Poor | High | Less than 1 | Low | Weak-signal detector |
| Plate | Poor | High | 1 to 3 | High | Infinite-impedance detector |

method of using a potentiometer for the diode-detector load resistance, so that the control can adjust the amount of detected signal coupled to the audio amplifier. This arrangement is shown in Fig. 10 · 16 with a typical volume control.

**Diode load as volume control.** In Fig. 10 · 16*b*, $R_L$ is a 0.5-M potentiometer. This is the fixed load resistance of the detector between lugs 1 and 3, regardless of the position of the variable arm. Therefore, the full detected audio-output voltage is developed from lug 3 to chassis ground. Assume 3 volts of audio signal. If the variable arm connected to lug 2 is set to the center of $R_L$, the voltage across lugs 2 and 1 will then be one-half the total voltage, resulting in 1.5 volts coupled to the audio amplifier between the variable arm and chassis ground. Or if the variable arm moves to lug 3, all the detector output of 3 volts is coupled to the audio amplifier for full volume. The control is wired so that rotating the shaft to the right increases the volume. Turning the control completely to the left moves the variable arm to lug 1 with zero resistance for the output circuit and no output voltage.

Such volume controls are available in sizes of 0.5 to 2 M, which are typical values for the diode load resistance. Very often the volume control has attached to it the power on/off switch operated by the same shaft. However, the control and the switch are in two separate circuits. Generally

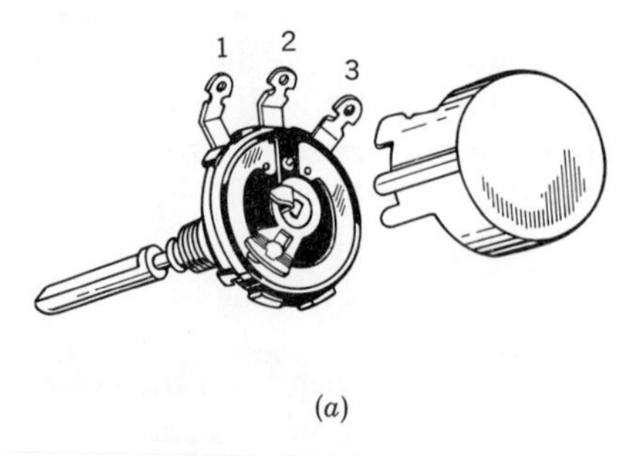


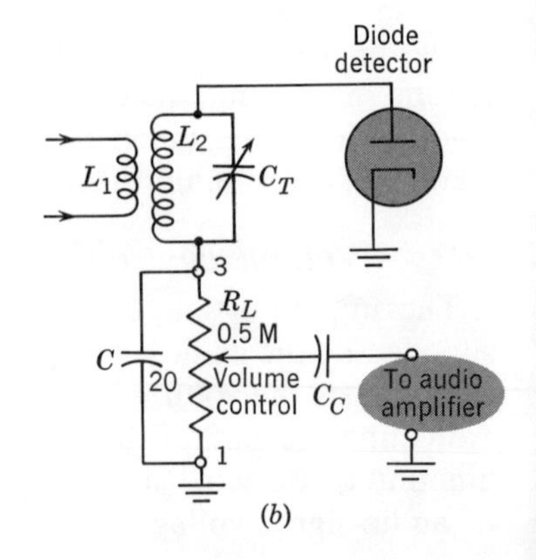


*Fig. 10 · 16 Diode-detector load as volume control. (a) Construction of potentiometer $R_L$. (b) Circuit.*

there is a push button or a rotary switch mounted on the back of the control. Pushing the shaft in or turning to the right operates the switch. Then the control is adjusted to vary the volume.

**Audio level control.** In Fig. 10·17 the 0.5-M potentiometer serves as the grid resistor for $V_2$, with adjustable output voltage to vary the volume. The idea is the same as the previous volume control. When the variable arm is at terminal 3 for maximum resistance, all the audio-signal voltage from $V_1$ is coupled to the grid of $V_2$ for maximum voltage. As the variable arm is moved toward terminal 1, the volume decreases. Note that the resistance in the grid circuit remains fixed at 0.5 M, however, allowing $V_1$ to produce its normal output voltage across $R_g$, regardless of the volume control setting. The idea here is the same as the volume control in Fig. 10·16*b*, with $R_g$ corresponding to $R_L$. However, varying the diode-detector load resistor is preferable for a receiver to prevent overloading the first audio amplifier on strong signals.

**Sensitivity control.** The volume control $R_k$ in Fig. 10·18 varies the r-f gain. Increasing the resistance of $R_k$ results in more cathode bias; more bias on the amplifier reduces the sensitivity and gain. This leads to less output of modulated r-f signal. If less signal is fed into the detector, less audio signal is developed. Decreasing $R_k$ to reduce the bias increases the r-f sensitivity, resulting in more audio and greater volume. The small additional cathode resistor $R_m$ provides a minimum bias of about 3 volts when $R_k$ is at its zero resistance setting. The cathode resistance is unbypassed here in order to allow degeneration. Then, adjusting $R_k$ varies both the bias and degeneration for better control of the gain.

Note that the resistance variation of $R_k$ is opposite from an audio control. The volume is increased by decreasing the resistance of $R_k$ for less bias and less degeneration. In addition to having much less resistance than an

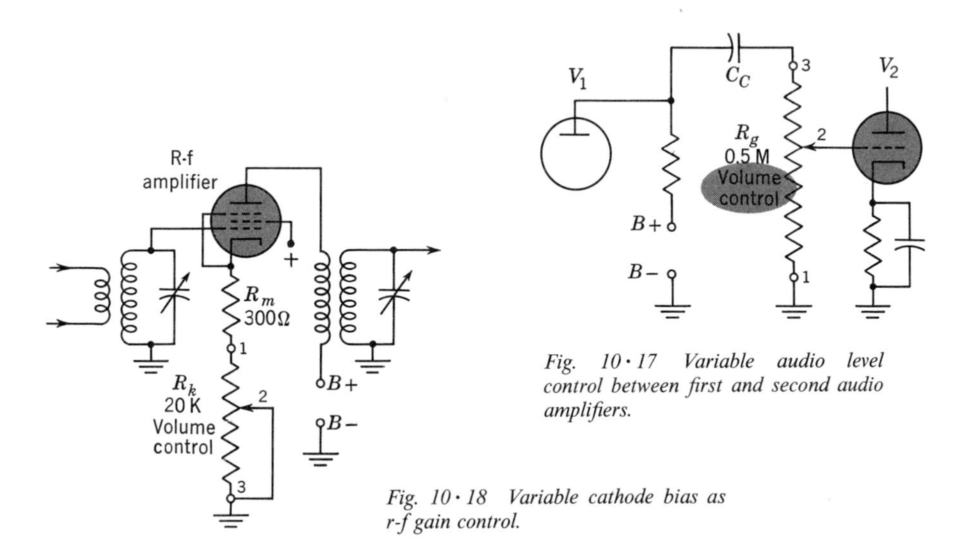


*Fig. 10·17 Variable audio level control between first and second audio amplifiers.*

*Fig. 10·18 Variable cathode bias as r-f gain control.*

audio control, therefore, the cathode-bias control must be wired so that rotation to the right for more volume decreases its resistance. $R_k$ is a 20,000-ohm potentiometer used as a rheostat by shorting the outside terminal 3 to the variable-arm terminal 2.

An r-f amplifier with variable bias should be a variable-$\mu$ tube, such as the 6BA6, which has a remote-cutoff grid voltage of $-20$ volts. The variable-$\mu$ and remote-cutoff characteristics enable the tube to operate at reduced gain as the negative bias is increased and without excessive amplitude distortion. Still, the sensitivity control has the disadvantage of overload distortion on strong signals unless the gain is set low, which is incorrect for weak signals. This method is used only when there is no automatic gain control (see Chap. 11) for the r-f signal. Otherwise, a potentiometer as the diode-detector load resistance is the best volume control.

**Audio-volume-control taper.** A potentiometer used for audio volume control has its resistance variation tapered to provide more uniform changes in volume. At low volume settings, the resistance should change by small

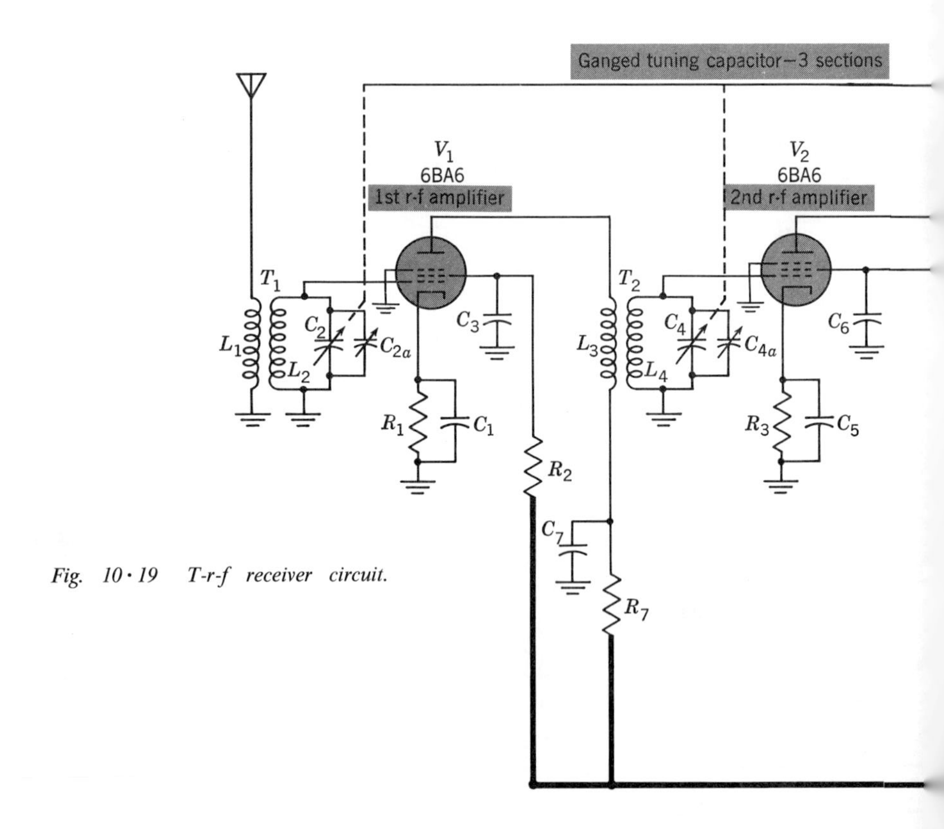


*Fig. 10·19 T-r-f receiver circuit.*

amounts because the ear is now more sensitive to an increase or decrease in sound level. When the volume is high, greater changes are needed for the same effect. Therefore, an audio volume control has a gradual change at the low-resistance end. When the control is turned up all the way to the right, the resistance changes are greater to provide a bigger change in audio level.

## *10·8 T-r-f receiver*

The result of combining a t-r-f amplifier and detector with an audio amplifier is a complete receiver, as shown in Fig. 10·19. It is called a t-r-f receiver because each r-f circuit is tuned to the radio signal frequency. The two-stage t-r-f amplifier with $V_1$ and $V_2$ is the same circuit as in Fig. 10·7. The amplified r-f output is coupled to the diode detector by the diode section of $V_3$. This is the detector circuit of Fig. 10·16*b*. The detected output is then amplified by the two-stage audio amplifier, which uses the triode section of $V_3$ with the audio power-output stage $V_4$ driving the

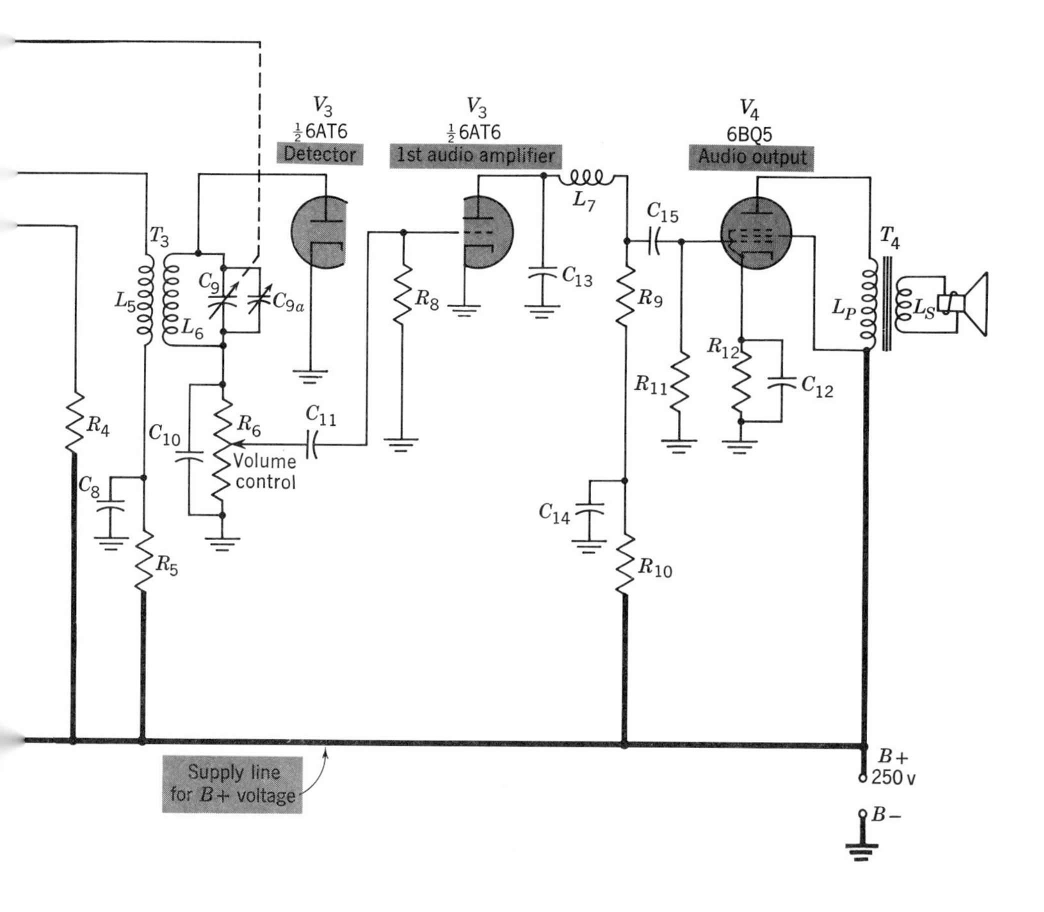

loudspeaker. Maximum undistorted audio power output from the 6BQ5 is 5.7 watts at full volume. The r-f and a-f amplifier stages all operate class A for minimum distortion.

Although not shown in the diagram, the power supply provides heater power for all the tubes and B+ voltage for the r-f and a-f amplifier stages. There are just two receiver controls. Tuning is varied by the three-section gang capacitor $C_2$, $C_4$, and $C_9$ for the r-f section. The volume control $R_6$ is a potentiometer serving as the diode-detector load resistance. The on/off switch for the power supply is mounted at the back of the volume control. Actually, most receivers use the superheterodyne circuit explained in the next chapter, but the simplified receiver here illustrates some important fundamentals for all types of receiver circuits.

**Signal path.** Starting at the antenna circuit, $T_1$ couples the received r-f signal to the first resonant circuit $L_2C_2$. Tuning this circuit selects the desired signal from the many carrier frequencies present in the antenna circuit. The selected signal is amplified and then coupled into the second r-f amplifier stage by $T_2$. Its resonant secondary circuit $L_4C_4$ is tuned to the same frequency as $L_2C_2$. Therefore the selected signal is amplified by $V_2$. With enough signal-output voltage from $V_2$, it is coupled to the detector stage by $T_3$. Note that the r-f input circuit to the detector is also tuned to the selected frequency; this makes three tuned circuits here.

There may be one to four r-f stages for the modulated carrier signal ahead of the detector, to provide the required selectivity and sensitivity. More tuned circuits provide better selectivity. More r-f gain allows greater sensitivity. This means a weaker r-f signal at the antenna can provide the voltage needed at the detector. If each of the two t-r-f stages shown here has a gain of 20, the over-all r-f gain is 400. Thus, a 10-mv antenna signal provides 4 volts for the detector.

The AM signal applied to the diode section of $V_3$ is rectified to produce the desired audio signal. In the output circuit, $R_6$ is the diode load resistor that provides the audio-output signal. $C_{10}$ is the r-f bypass capacitor. The audio coupling capacitor $C_{11}$ blocks the steady d-c component of the detector-output voltage and couples the a-c audio-signal variations to the grid circuit of $V_3$. The amount of audio signal coupled to $V_3$ depends on the setting of the volume control $R_6$. If 4 volts of r-f signal is fed to the detector, about 4 volts of a-c audio-output voltage appears across $R_6$. When the variable arm of the 500,000-ohm potentiometer is at 50,000 ohms for a low-volume setting, one-tenth the total voltage, or 0.4 volt, is coupled to the audio amplifier.

The triode section of $V_3$ is a typical $RC$ amplifier serving as the first audio stage. In the plate circuit, $L_7$ is an r-f choke and $C_{13}$ an r-f bypass to filter out any r-f signal from the detector. Although the reactance of $L_7$ is high for radio frequencies, it has little opposition for audio frequencies, allowing the amplified audio voltage to be developed across the plate-load resistor $R_9$. This output signal is coupled by $R_{11}C_{15}$ to the audio-output stage $V_4$. The primary of $T_4$ is the plate-load impedance of $V_4$. Plate-current variations in the primary induce a voltage in the secondary to provide the current needed in the voice coil of the loudspeaker.

**Alignment of t-r-f receiver.** In order to obtain maximum selectivity and sensitivity, the ganged r-f tuning circuits must be aligned. For a t-r-f receiver, the alignment procedure consists of adjusting the individual tuned circuits to make them all resonant to exactly the same frequency. To align the receiver circuit in Fig. 10·19, for instance, trimmer capacitors $C_{9a}$, $C_{4a}$, and $C_{2a}$ are adjusted, in that order, for maximum signal output. This is generally done for a frequency near the high end of the tuning range, usually 1,400 kc for the AM radio broadcast band. The tuning for other frequencies in the band can be equalized by bending the slotted rotor end plates on the ganged capacitor, aligning at approximately 1,100, 850, 700, 600, and 550 kc when there are five slotted sections. If there are no slotted sections to be adjusted, the stages can be aligned with the trimmers at 1,000 kc in the middle of the band. The alignment is always done by starting at the detector stage and working back to the antenna, so that a mistuned circuit will not cause a mistake in interpreting maximum output.

## *10·9 The B-supply line*

In the receiver circuit of Fig. 10·19, the plate and screen leads returning to B+ are marked in heavy lines to emphasize the importance of these B-supply connections. Although illustrated here for a t-r-f receiver for AM radio, the same principles apply in any radio receiver because all amplifier stages must have d-c operating voltages from the B supply. Physically, the B-supply line may consist of connections at several convenient tie points, all wired to the B+ output from the power supply.

The same idea applies to transistor amplifiers but with lower values of collector supply voltage. Also the required polarity of the supply voltage may be either positive or negative. Remember that the collector needs polarity opposite from its letter symbol for reverse bias voltage. This means that an N collector needs positive voltage while a P collector needs negative voltage. It should be noted, though, that the collector is often operated at ground potential, with inverted polarity of supply voltage applied to the base-emitter circuit.

Taking one stage at a time in Fig. 10·19, the plate circuit of the r-f amplifier $V_1$ has a d-c path to B+ through $L_3$ and $R_7$. The inductive reactance of $L_3$ provides the plate-load impedance needed for $V_1$ to amplify the r-f signal voltage. The decoupling resistor is not part of the a-c plate-load impedance, because it is bypassed for r-f signal variations by $C_7$. However, direct current must flow in the plate circuit or there will be no signal output, since it is the a-c component of the pulsating direct plate current that carries the program intelligence. The d-c path in the plate circuit includes the d-c resistance of $L_3$ and the decoupling resistance. If either one is open, the plate circuit is open. Then there is no plate voltage for $V_1$, the plate current is zero, and there is no output.

$R_2$ provides the d-c path for the screen-grid circuit of $V_1$. If $R_2$ opens, there is no screen-grid voltage and no screen current. Without screen-grid voltage in a pentode, there is no plate current, even when the plate has its supply voltage. The screen-bypass capacitor $C_3$ bypasses the r-f signal around $R_2$, but if $C_3$ becomes shorted, the d-c screen voltage is zero, and the amplifier cannot function. Also, if r-f bypass $C_7$ of the plate decoupling filter shorts, the plate voltage will be zero, and there will be no output.

Last but not least, the cathode must have a d-c path to chassis ground, which is B− here, in order to complete the circuit for the d-c operating voltages applied to plate and screen. If the cathode resistor $R_1$ opens, there can be no plate or screen current. Although the plate and screen voltages can be measured with respect to chassis ground, they must be returned to cathode for the tube to operate.

The second r-f amplifier stage $V_2$ is like $V_1$ and has the same B-supply requirements. The diode detector in $V_3$, however, does not need B+ for operation.

In the first audio amplifier, the triode section of $V_3$ has its plate returning to B+ through the r-f choke $L_7$, plate-load resistor $R_9$, and plate-decoupling resistor $R_{10}$. All three provide a continuous path for direct current. If any one opens, there will be no plate current and the amplifier cannot operate. However, only $R_9$ is the plate load that provides amplified audio-signal voltage, since $L_7$ has negligible reactance for audio frequencies and $R_{10}$ is bypassed by $C_{14}$. To complete the circuit for d-c plate-cathode voltage, the cathode is tied directly to chassis ground, which is B−.

The plate of the audio power-output tube connects to B+ through the primary of the output transformer $T_4$. The screen grid of $V_4$ is tied directly to B+ for maximum power output. In the plate circuit, the primary winding $L_p$ has little d-c resistance, but its reactance for audio frequencies provides the required plate-load impedance. If $L_p$ opens, there will be no plate current and no audio output. Finally, the cathode of $V_4$ returns to B− through the cathode-bias resistor $R_{12}$. Just as in $V_1$ and $V_2$, an open cathode return means zero plate and screen current, and the amplifier cannot function.

## SUMMARY

1. A t-r-f stage is an r-f amplifier with a resonant circuit that can be tuned to the desired r-f carrier frequency.
2. Sensitivity means high gain in the receiver so that it can provide useful audio output from a weak r-f signal at the antenna.
3. Selectivity is the ability of the receiver to tune to one r-f carrier frequency while rejecting other frequencies. The more tuned stages there are, the better the selectivity.
4. For the AM radio broadcasting band, the r-f circuits must tune through 535 to 1,605 kc. This is a 1:3 frequency range, approximately, requiring a 1:9 change in tuning capacitance. With a 250-$\mu$h inductance, the variable capacitance required is about 40 to 360 pf.
5. A ganged tuning capacitor has two or more sections, one for each tuned circuit of cascaded t-r-f stages.
6. Alignment means adjusting the trimmer capacitance in each tuned circuit of cascaded stages to resonate each at exactly the same frequency for maximum selectivity and sensitivity.

7. A detector rectifies the modulated r-f signal to recover the audio modulation. The three main types of detector circuits for AM radios are the diode detector, grid-leak detector, and plate detector. The diode detector is the most common.
8. A receiver's volume can be controlled by variable-cathode bias for the r-f amplifier, by adjusting r-f sensitivity, or most often by using a potentiometer as the diode-detector load resistor. The potentiometer then is an audio level control to adjust the amount of detector output coupled to the audio amplifier. This audio level control needs an audio taper for smooth control of volume.
9. In a t-r-f receiver each tuned stage is resonant at the r-f carrier frequency.
10. All the amplifier stages in a receiver connect to the B supply for plate and screen-grid d-c operating voltages. Cathodes return to B− (usually the chassis).

## SELF-EXAMINATION

Here's a chance to find out how well you have learned the material in this chapter. Work the exercises; then check your answers against the Key at the back of this book. These exercises are for your self-testing only.

1. Tuning in the desired station is the function of the (*a*) audio section; (*b*) detector; (*c*) r-f section; (*d*) power supply.
2. Sensitivity and selectivity result from (*a*) tuned amplifier stages; (*b*) high-gain audio amplifiers; (*c*) use of a diode detector; (*d*) use of a low B-supply voltage.
3. A 250-μh inductance resonating with 200-pf capacitance will provide maximum signal voltage at the carrier frequency of (*a*) 710 kc; (*b*) 1,000 kc; (*c*) 1,410 kc; (*d*) 1,605 kc.
4. In a t-r-f stage the plate-load impedance for amplified r-f signal is the (*a*) primary of the r-f interstage transformer; (*b*) plate decoupling resistance; (*c*) internal plate resistance of the pentode amplifier; (*d*) reactance of the bypass capacitor in the plate decoupling filter.
5. The diode detector is commonly used in receivers because it provides (*a*) maximum amplification with weak r-f signal input; (*b*) minimum distortion; (*c*) minimum fidelity; (*d*) a good load for the B supply.
6. In Fig. 10·19, with 40 ma through a 250-ohm $R_{12}$, the cathode bias on $V_4$ equals (*a*) 1 volt; (*b*) 3 volts; (*c*) 10 volts; (*d*) 30 volts.
7. When the receiver volume control is the diode-detector load resistor (*a*) maximum volume occurs for the lowest resistance to chassis ground; (*b*) maximum volume occurs for the highest resistance to chassis ground; (*c*) the load resistance for the diode varies with the volume setting; (*d*) the r-f bypass across the volume control must be varied.
8. In Fig. 10·19, if the primary winding $L_p$ in $T_4$ opens, the (*a*) screen-grid voltage of $V_4$ will decrease; (*b*) plate voltage of $V_4$ will be zero; (*c*) audio output will be distorted; (*d*) diode detector cannot operate.
9. In Fig. 10·19, typical values for $R_{12}$ and $C_{12}$ in the cathode circuit of the audio-output tube are (*a*) 250 ohms, 100 μf; (*b*) 1 M, 100 μf; (*c*) 25 ohms, 250 pf; (*d*) 10,000 ohms, 0.001 μf.
10. In the receiver circuit diagram in Fig. 10·19, which of the following statements is *not* correct? (*a*) $R_3$ is a cathode-bias resistor; (*b*) $C_8$ is a bypass capacitor; (*c*) The tuning capacitor has three sections; (*d*) $R_{10}$ is the plate load for the $V_3$ triode.

## QUESTIONS AND PROBLEMS

1. Draw the schematic diagram of a two-stage t-r-f amplifier with ganged tuning.
2. Give the function for each of five components in the diagram of Question 1.
3. Draw schematic diagrams showing a volume control varying (*a*) audio level, (*b*) r-f gain.
4. Draw the schematic diagram of a diode-detector circuit indicating where (*a*) r-f signal is applied, (*b*) audio signal is taken out, (*c*) a d-c voltmeter can measure d-c output.
5. Describe briefly how to align the t-r-f receiver in Fig. 10·19 at a frequency near the high end of the band.
6. Redraw the transistorized r-f amplifier circuit in Fig. 10·3, but with the collector circuit returned to chassis ground. Use a 12-volt supply connected to the base-emitter circuit, with the same amount of forward bias and reverse bias.

7. Explain briefly why a volume control with audio taper changes its resistance more gradually at the low-resistance end.
8. What would be the maximum-to-minimum capacitance ratio needed to tune through the 88- to 108-Mc range? (This is the commercial FM broadcast band.)
9. List three methods of reducing the possibility of regeneration in a receiver.
10. Refer to the receiver diagram in Fig. 10 · 19, and assume that the screen-bypass capacitor $C_6$ shorts in the $V_2$ stage. (*a*) How will this affect the r-f output of $V_2$? (*b*) How will this affect the audio output of the receiver? (*c*) How much current will flow through $R_4$ with a value of 10,000 ohms? (*d*) How much power will be dissipated in $R_4$? (*e*) Why will $R_4$ burn open?
11. Calculate the $C$ needed with 125-$\mu$h $L$ for resonance at 535 kc, and at 1,605 kc.
12. Calculate the $L$ needed with 25-pf $C$ for resonance at 98 Mc.

# Chapter 11 Superheterodyne receivers

Practically all modern receivers use the superheterodyne circuit because of its greater gain and selectivity than the t-r-f circuit. This applies to all receiver applications, including frequency modulation and television. In superheterodyne receivers, the desired station's r-f carrier wave is converted to a new frequency that is the same for all stations. This is the intermediate frequency of the receiver. In effect, the superheterodyne circuit shifts the r-f signal frequencies to make them all fit the frequency of the receiver i-f amplifier, instead of trying to make the receiver suit the different signal frequencies. The topics in this unit are as follows:



## *11·1 How the superheterodyne receiver operates*

The distinguishing feature of the superheterodyne circuit is changing the frequency of the incoming r-f signal to a lower intermediate frequency. This is done by mixing the incoming r-f signal with the output of an r-f oscillator in the receiver. The process of mixing or combining two waves

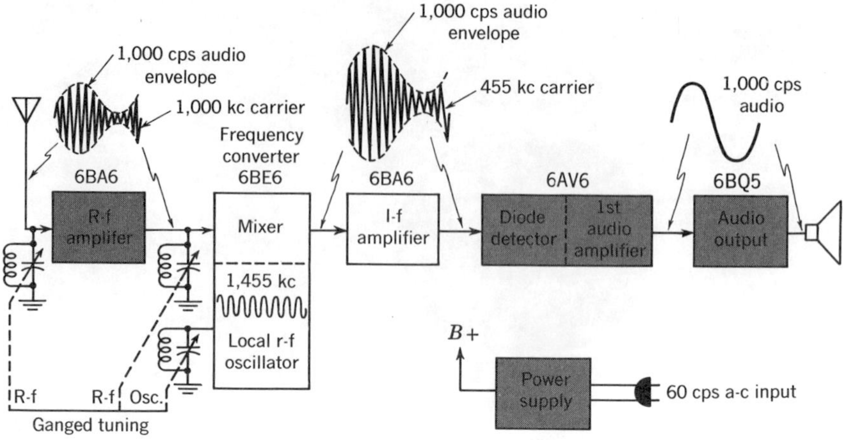

*Fig. 11·1 Block diagram of a superheterodyne receiver.*

of different frequencies to produce a new frequency is called *beating,* or *heterodyning.* Usually the heterodyning is done to produce a new frequency equal to the difference of the two input frequencies. In this case, the beat frequency is lower than the original frequencies. For example, when an r-f carrier wave at 1,000 kc is heterodyned with the oscillator output at 1,455 kc, the difference equals 455 kc. This is the intermediate frequency. The term *superheterodyne* applies here because the difference frequency is superaudible, or above the a-f range.

The block diagram in Fig. 11·1 illustrates the operation of a superheterodyne receiver. The r-f stage tunes to the desired carrier frequency, as in a t-r-f receiver. In fact, this buffer amplifier between the antenna and frequency converter is often called the *t-r-f stage.* In the example here, the r-f stage is tuned to 1,000 kc. This amplifier r-f signal is then coupled to the frequency-converter stage.

If the input signal is strong enough, the t-r-f stage can be omitted because it is not required for the heterodyning process. As a matter of fact, in many superheterodyne receivers the r-f signal from the antenna is coupled directly to the grid of the converter.

Also coupled into the converter is the output of the receiver local oscillator. The oscillator has a frequency that differs from the input signal by an amount equal to the intermediate frequency. In this example, the oscillator is operating at 1,455 kc, or 455 kc above the r-f signal frequency of 1,000 kc. Because of the heterodyning action in the frequency converter, its output is the intermediate difference frequency at 455 kc. The remaining resonant circuits in the receiver are tuned to the intermediate frequency, not the incoming-signal frequency. This is the i-f section of the receiver. There may be one, two, or three tuned i-f amplifiers. These i-f stages use tuned circuits, but they are not varied. Instead the local oscillator is varied as the receiver is tuned to produce the same intermediate frequency for *all* stations.

As shown by the waveforms in Fig. 11 · 1, the i-f signal is still a radio-frequency signal at 455 kc. No audio signal is produced until the i-f signal is rectified in the second detector stage. This is the same detector as in a t-r-f receiver, demodulating the carrier to produce the audio signal needed for the audio amplifier. It is called the second detector here because the frequency converter stage is considered to be the first detector in a superheterodyne receiver, since rectification is necessary for the heterodyning action that produces the intermediate frequency. The only effect of the frequency conversion in the first detector, however, is to shift the carrier frequency of the modulated r-f signal to a lower value equal to the intermediate frequency. The i-f output has the same modulation as the r-f input signal. After the modulation on the i-f signal is removed by the second detector, the resultant audio signal is amplified in the audio section of the receiver.

To analyze the superheterodyne operation further, suppose that the receiver is tuned to a station at 600 kc. Now the tuned circuits for the r-f amplifier and frequency converter (or mixer) are set for 600 kc to receive the desired station. More important, though, the tuned circuit of the local oscillator also varies since it is ganged with the r-f tuning circuits. Then the local oscillator frequency will be set to 1,055 kc, which is 455 kc above the r-f signal at 600 kc, so that the receiver can still produce the same intermediate frequency of 455 kc. The station at 600 kc will be received, therefore, because this frequency can beat with the oscillator at 1,055 kc to produce the difference frequency of 455 kc.

It is the local oscillator frequency that determines which station is tuned in. The r-f circuits tuned to the signal frequency amplify the r-f signal, but unless this frequency is converted to the intermediate frequency, it cannot produce any audio output, because most of the receiver's gain is in the i-f amplifier. Therefore only the signal frequencies that beat with the oscillator to produce the correct intermediate frequency receive enough gain to produce an output. Thus, the local oscillator actually tunes in the different stations for the receiver.

The receiver tuning for all stations in its r-f range is accomplished by making the local oscillator track 455 kc above the r-f signal circuits. See Fig. 11 · 2. For any r-f signal in the bottom row, the oscillator operates

*Fig. 11 · 2 Illustrating how the local oscillator frequencies track 455 kc above the r-f signal frequencies.*

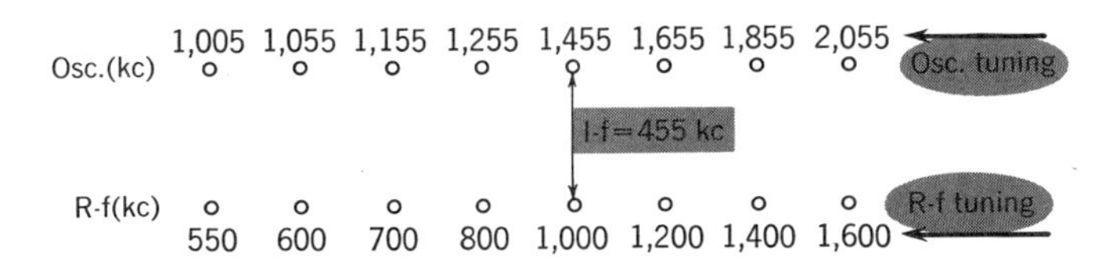

455 kc higher, with the values listed in the top row. There is a constant difference of 455 kc between the oscillator and the r-f signal circuits as both are tuned throughout their ranges.

It should be noted that the intermediate frequency is not always 455 kc. For AM radios, 455 kc, 456 kc, or 468 kc are common, with 262.5 kc used in auto radios. FM broadcast receivers and television receivers operating at higher frequencies use i-f values of 10.7 Mc and 41.25 Mc for the FM sound signal. However, the superheterodyne principle of converting all r-f signals to one intermediate frequency by tuning the local oscillator still applies. Furthermore, the local oscillator can beat below the frequencies of the r-f signal instead of above and still produce the desired difference frequency for the i-f signal.

## 11·2 Advantages of the superheterodyne receiver

The process of changing the signal frequencies to an i-f value adds to the complexity of the receiver, but the advantages gained by doing most of the r-f amplification with fixed-tuned circuits are very important. Compared with a t-r-f receiver, the superheterodyne provides (1) more gain, (2) better selectivity, (3) more uniform gain and selectivity for all stations over the tuning range of the receiver, and (4) better circuit stability with less tendency to produce oscillations as the receiver is tuned to different stations.

As a comparison, consider a t-r-f receiver with five tuned circuits, requiring a five-section tuning capacitor. This receiver would be expensive and bulky, requiring elaborate shielding to prevent oscillations because of feedback between the many r-f stages tuned to the same frequency. Furthermore, the receiver's gain and selectivity would vary through the tuning range as the $L/C$ ratio of the resonant circuits changes with each setting of the tuning capacitor. Double-tuned circuits would not generally be used because of their tendency to oscillate at some frequencies. All these problems are practically solved by using the superheterodyne circuit. In an i-f amplifier, both the primary and the secondary of the coupling transformer are tuned to the intermediate frequency, which provides a double-tuned circuit. There is little danger of oscillations as long as these circuits are not misadjusted. As a result, one i-f stage has four tuned circuits, counting the input and output coupling transformers. In addition, with high $Q$ for the fixed-tuned circuits, the i-f amplifier can have much more gain than a t-r-f stage.

The fact that the intermediate frequency is lower than any signal frequency automatically increases the receiver's selectivity. Suppose that the receiver is tuned to 1,000 kc and there is an interfering station at 1,050 kc. This 50-kc difference between the two frequencies is only 50/1,000, or 5 per cent, of the resonant frequency in a circuit tuned to 1,000 kc. At the intermediate frequency of 455 kc, however, the same 50-kc separation between the stations is 50/455, or approximately 10 per cent. With a greater percentage difference between the desired frequency at resonance and the interfering frequency off resonance, the relative response is reduced for the undesired frequency. As a result, the ability of the receiver to reject such interfering frequencies is improved, increasing the adjacent channel selectivity.

## *11 · 3 Heterodyning*

The idea of producing a new frequency by the process of heterodyning is similar to the more familiar example of beats commonly heard with musical sounds. When two tones of nearly the same frequency are produced at the same time, the ear can detect a regular rise and fall in the intensity of the resulting sound. This beat is produced as the two individual sound waves alternately reinforce and then cancel each other. With a small phase difference that is continuously changing between two waves of slightly different frequency, after a number of cycles one wave will be either completely out of phase or in phase with the other. When the two waves are in phase, they reinforce to produce maximum amplitude of the

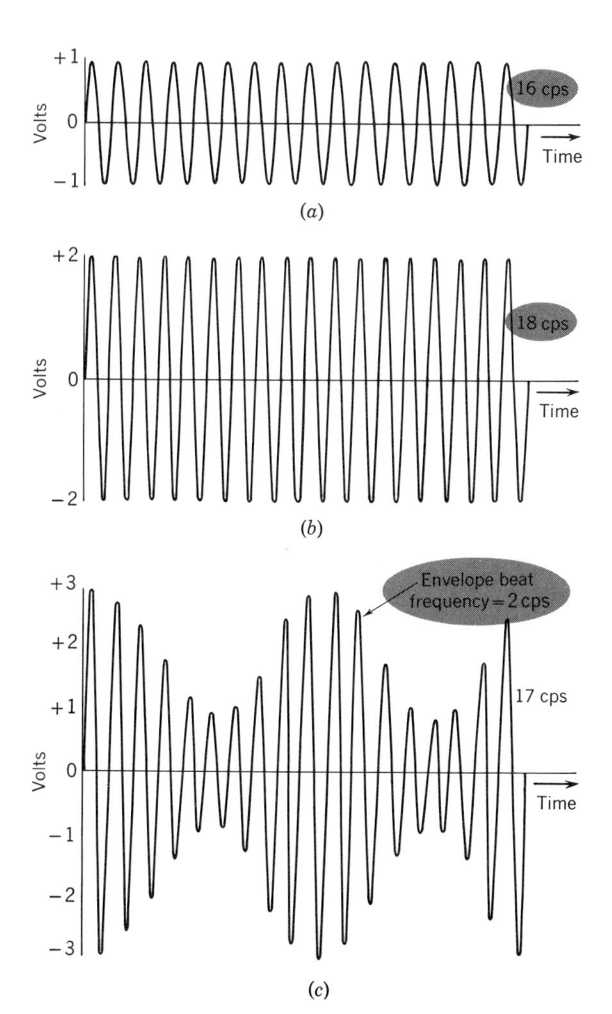


*Fig. 11 · 3 (a) Wave at 16 cps. (b) Wave at 18 cps. (c) The resultant beat, with envelope varying at the difference frequency of 2 cps.*

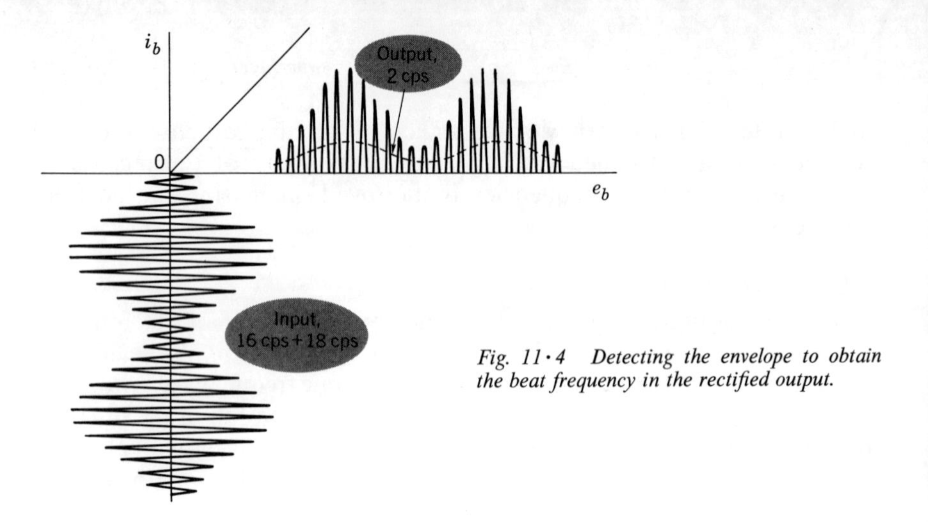


*Fig. 11·4 Detecting the envelope to obtain the beat frequency in the rectified output.*

resultant wave. Later, the two waves will be exactly out of phase and cancel each other. The rate at which the amplitude of the resultant wave rises and falls is called the *beat frequency*. It is exactly equal to the difference in frequency between the two original waves, since this determines how often they can reinforce and cancel each other. When the two frequencies are nearly equal, therefore, we can hear the pulsating intensity of the beat note in addition to the original tones.

**Producing the beat frequency.** The same beating action can be produced with electric waves by combining two voltages or currents in a circuit where the amplitude of the output depends upon the instantaneous values of the two input waves. An example of two waves with their resultant beat is shown in Fig. 11·3. Assume that wave *a* has a frequency of 16 cps, while wave *b* has a frequency of 18 cps. These two a-c voltages may be considered in series with each other in a circuit where the two waves can combine to produce the resultant wave *c*. When the voltages are in phase, as at the starting point at the left in the illustration, the voltages add to produce a larger resultant amplitude. A little later, the higher frequency wave will be one half-cycle ahead, and the two waves will cancel because they are then in opposite phase. As a result, the amplitude of the resultant wave in *c* varies at the difference frequency, which is 2 cps in this example. This is the beat frequency.

**Detecting the beat.** Since the voltage variations of the envelope at the beat rate are both positive and negative in the same amount at the same time, the average value of the resultant wave is zero. In order to produce the beat frequency, the resultant wave must be rectified. The reason is illustrated in Fig. 11·4, in terms of the operating characteristic of a diode detector. The resultant wave, equal to the sum of the 16-cps and 18-cps components, is applied to the diode as signal voltage, causing plate current to flow only for the positive half-cycles. As a result, the output consists of the rectified half-cycles only and is the current flowing in the load of the diode detector.

Notice that the average value of the rectified output, indicated by the dotted wave in the illustration, varies at the beat rate. The peak values of the output also vary at the difference frequency. This means that the output circuit can have current or voltage variations at the beat frequency. As a result, a new frequency equal to the beat or difference frequency is present in the rectified output, although it was just the envelope of the input wave. The undesired frequencies are filtered out, leaving just the beat-frequency signal as the output voltage.

The conclusion is therefore that in addition to the mixing of two frequencies, detection is necessary to produce the heat frequency. This is why the frequency converter stage in the superheterodyne receiver can be called the *first detector*. Broadcast receivers do not use a diode for the first detector, but triodes or pentodes can be operated as a plate detector.

Another application of the heterodyne principle is called *zero beat*. In this technique, two voltages of the same frequency are heterodyned to produce zero output as the difference when the frequencies are exactly the same and in opposite phase. The usefulness of the zero beat is that it indicates when a known frequency is the same as an unknown frequency. If the unknown frequency is close to the standard frequency, the beat note will be a low-pitched audio signal equal to the slight difference in frequencies. As the frequencies come closer together, the beat note frequency becomes lower and lower until there is a null with no output at zero beat when the frequencies are exactly equal.

**Producing the intermediate frequency.** The rectified current in the first detector of the superheterodyne receiver includes many frequencies. Those with the greatest amplitude are the signal and oscillator input frequencies, their difference frequency, and their sum frequency. In addition, there can be other frequencies that are combinations of the fundamentals and harmonics of the signal and oscillator frequencies. For a receiver with an intermediate frequency of 455 kc tuned to a station at 1,000 kc, the main frequencies produced in the first detector would be:

| | |
|---|---|
| Signal frequency | 1,000 kc |
| Oscillator frequency | 1,455 kc |
| Sum frequency | 2,455 kc |
| Difference frequency | 455 kc |

Of these, the difference, or beat, frequency of 455 kc is the desired one. The difference is generally used as the intermediate frequency because it is the lowest frequency. This allows more selectivity and gain with increased stability in the i-f section of the receiver. Obtaining 455 kc as the desired frequency is just a case of tuning the plate circuit of the converter stage to the intermediate frequency. Then there is practically no load impedance, and therefore, no gain for any frequencies except 455 kc.

## *11 · 4 Effect of heterodyning on the modulated signal*

Although the response of the i-f section of the receiver is made sharp to reject undesired frequencies, the bandwidth is great enough to accept the sideband frequencies associated with the modulated carrier. This bandwidth must be provided because the desired modulation is still present in the i-f signal. Note that the envelope of the i-f signal shown in Fig. 11·1 still has the same audio modulation as the r-f signal.

The reason the modulation of the r-f carrier wave passes through the converter to provide the same modulation at the intermediate frequency can be illustrated by some numerical values. Consider a modulated carrier of 1,000 kc in terms of its sideband frequencies, extending ±5 kc from 995 to 1,005 kc. When this signal is coupled into the converter, the local oscillator beats with all the sideband frequencies to produce a new difference frequency for each. For reference, these values are tabulated here:

| *R-f signal, kc* | *Local oscillator frequency, kc* | *Difference frequency, kc* |
|---|---|---|
| 1,005 | 1,455 | 450 |
| 1,000 | 1,455 | 455 |
| 995 | 1,455 | 460 |

Notice that the i-f signal still has sideband frequencies 5 kc below and 5 kc above the new i-f carrier at 455 kc. Although these values are for the extreme sideband frequencies farthest removed from the carrier, the same action occurs for all the sideband frequencies in the r-f signal. They are just converted to lower frequencies with the same separation from the i-f carrier as in the original r-f input signal. The effect of the frequency conversion with a modulated signal, therefore, is to shift the original carrier wave to a new frequency, but with the same modulation.

An interesting effect of the heterodyning is the inversion of the upper and lower sidebands. Referring to the tabulated values, you can see that the upper sideband frequency at 1,005 kc in the original r-f signal is converted to a frequency lower than the i-f carrier, while the lower sideband frequency of 995 kc in the r-f signal becomes the higher sideband frequency in the i-f signal. This inversion is only a result of operating the local oscillator above the incoming signal frequencies. As a result, the upper sideband frequencies of the r-f input signal are closer to the local oscillator frequency, producing a lower difference frequency. The modulation is not changed by the inversion, however, because both sidebands have the same information.

## *11·5 Frequency-converter circuits*

This section of the superheterodyne receiver is an arrangement where the plate current can vary at both the frequency of the r-f signal and that of the local oscillator. Then the resultant beat frequency is detected so that it can be selected by the converter plate circuit tuned to the intermediate frequency. Any one of several circuits can be used:

1. Separate tubes for the mixer and local oscillator. This arrangement is illustrated in Fig. 11·5. The oscillator is generally a triode. A conventional circuit is used, generally a Hartley, Colpitts or tuned-grid oscillator.

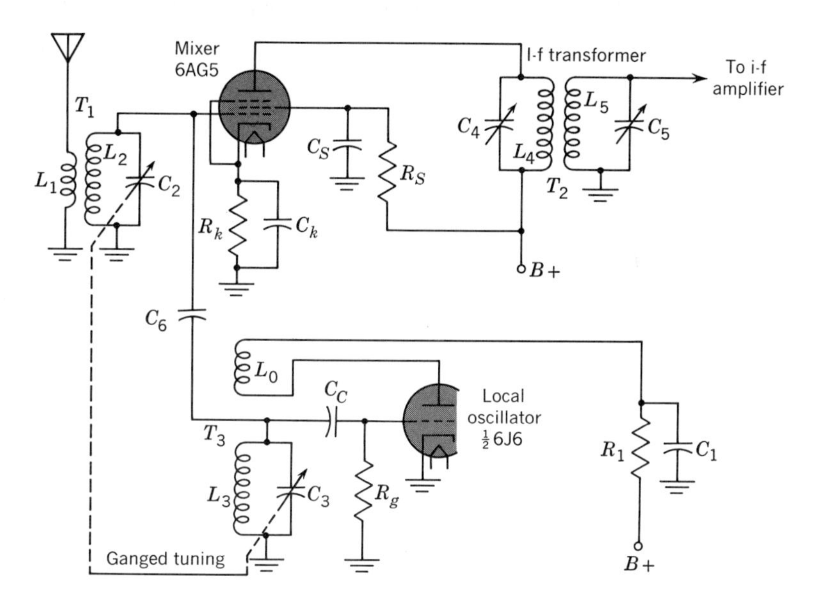


*Fig. 11·5 Frequency-converter circuit with pentode mixer and separate oscillator stage. Trimmer capacitors not shown.*

The mixer can be a pentode for more gain or a triode for less tube noise. A crystal diode mixer is generally used in the UHF range for minimum noise.

2. Mixer and local oscillator stages in one tube envelope, as illustrated by the triode-pentode converter in Fig. 11·6.
3. A pentagrid converter tube with five grids where the oscillator and mixer functions are combined in one stage, as illustrated by the 6BE6 converter in Fig. 11·7 and the circuit in Fig. 11·8. The pentagrid converter arrangement is used most often because it is compact and economical.
4. Autodyne circuit. One stage is used for both the oscillator and mixer functions (Fig. 11·9). The autodyne circuit is commonly used with a triode transistor for frequency conversion, in order to save a stage.

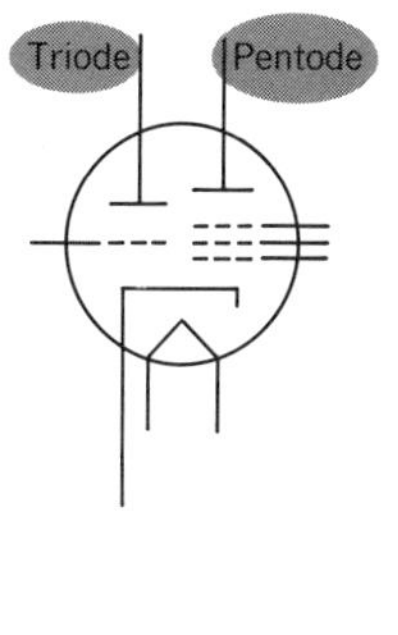


*Fig. 11·6 Triode-pentode frequency converter.*

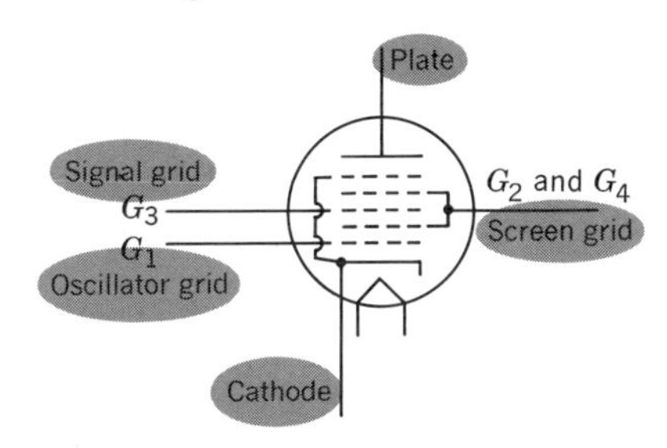


*Fig. 11·7 Pentagrid converter.*

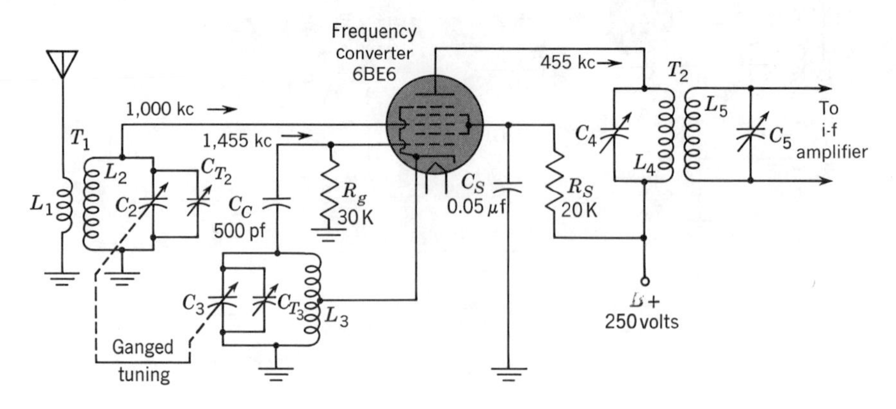


*Fig. 11·8 Pentagrid frequency-converter circuit for intermediate frequency of 455 kc.*

**Plate detector with separate local oscillator.** In Fig. 11·5, a triode serves as a tuned-grid plate-feedback oscillator. In the pentode mixer circuit, the input to the control grid is the desired r-f signal voltage across the tuned circuit $L_2C_2$. The oscillator voltage is capacitively coupled to the mixer grid by means of $C_6$. As a result, both input voltages vary the mixer plate current to produce the intermediate beat frequency. The mixer is biased close to cutoff so that the beat frequency can be obtained in the rectified output. Since the tuned-plate circuit $L_4C_4$ is resonant at the intermediate frequency, it selects the desired difference frequency by providing the greatest load impedance and maximum gain for the tube at this frequency. The i-f output is then coupled to the i-f amplifier by $T_2$.

The process of coupling the local oscillator output voltage into the mixer is called *oscillator injection*. In Fig. 11·5 the injection is accomplished by the loose capacitive coupling of $C_6$. However, the oscillator voltage can be injected by inductive coupling instead. Also, the oscillator injection voltage can be coupled into the mixer grid or cathode. In any case, there must be enough oscillator voltage injected to produce the heterodyning action. This is about 5 to 15 volts. Too much injection voltage is undesirable, however. The oscillator signal then may be radiated by the antenna to cause interference in nearby receivers. Also, close coupling may detune the oscillator and t-r-f circuits.

**Pentagrid converter circuit.** The five grids of the 6BE6 in Fig. 11·7 enable this tube to function as both oscillator and mixer. Grid 1 is called the *oscillator grid* because it is used as the control grid for a triode oscillator circuit. Grids 2 and 4 are connected internally to serve as the screen grid of the mixer. This dual grid also functions as the anode of the triode oscillator, operating with grid 1 and the cathode. Grid 3 between the two electrodes of the screen grid is called the *signal grid* because it is the control grid for the r-f input signal. Grid 5 is the suppressor grid connected internally to the cathode. The plate is for the pentode mixer function.

Note that there are two control grids to vary the plate current: the oscillator grid 1 and the signal grid 3.

A typical circuit using the pentagrid converter is shown in Fig. 11 · 8, for an AM radio receiver with an intermediate frequency of 455 kc. The Hartley oscillator uses just the single-tapped coil $L_3$. This inductance resonates with the oscillator tuning capacitor $C_3$ to provide the desired frequency for the oscillator output. Since the screen is bypassed by $C_S$ to chassis ground for radio frequencies, it is effectively connected to the grounded side of the oscillator inductance $L_3$. Therefore the screen grid serves as the oscillator anode. The opposite side of $L_3$ is coupled to the oscillator grid. $C_C$ is the oscillator grid-coupling capacitor, and $R_g$ is the grid resistor, providing grid-leak bias for the oscillator. Note that the cathode is connected to a tap on the oscillator coil. The oscillator anode current, returning from the screen grid to the cathode through the bottom section of $L_3$, induces the required feedback voltage across the grid-cathode section of the inductance. In this circuit, the control grid and cathode have oscillator signal voltage. With the oscillator anode at r-f ground potential, one side of the oscillator coil $L_3$ and the rotor of the tuning capacitor can be grounded. This eliminates the problem of hand capacitance in tuning.

The r-f tuned circuit $L_2C_2$ feeds the desired r-f signal voltage to grid 3 of the converter. Note that tuning capacitor $C_2$ in the signal circuits is ganged with the oscillator tuning capacitor $C_3$. Therefore the oscillator frequency will always differ from the signal frequency by the intermediate frequency as the receiver is tuned to different stations. As a result of the heterodyning action in the converter, the output in the tuned plate circuit $L_4C_4$ is the desired i-f signal at 455 kc. This is coupled to the i-f amplifier by the converter plate transformer $T_2$, which is tuned to the intermediate frequency.

*Fig. 11 · 9 Transistorized autodyne frequency converter circuit.* (From GE "Transistor Manual.")

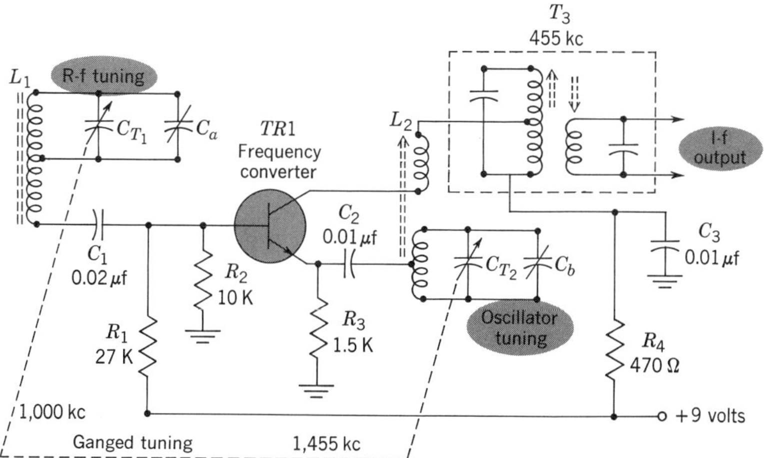

**Transistorized autodyne circuit.** In Fig. 11·9, TR1 is used for both oscillator and mixer, as an autodyne frequency converter. For the oscillator function, $L_2$ is the oscillator coil with feedback from emitter to collector. $R_3C_2$ provide emitter signal bias, similar to grid-leak bias in a tube. $L_1$ is the antenna coil for r-f input signal to the base. $T_3$ is the i-f transformer for the 455-kc output from the collector. Note that the frequencies are indicated for tuning in 1,000 kc. With this r-f input signal, the oscillator is at 1,455 kc to produce the i-f output at 455 kc. For any other r-f signal frequency to be tuned in, the oscillator will beat 455 kc above to produce the i-f output.

**Converter tube characteristics.** The characteristic of conversion transconductance determines the performance of a converter tube in the same way as the grid-plate transconductance characteristic for an ordinary amplifier tube. The conversion transconductance is defined as

$$g_c = \frac{\text{i-f current in converter output}}{\text{r-f signal-voltage input}} \tag{11·1}$$

For typical operating conditions the conversion transconductance of the 6BE6 tube is approximately 400 $\mu$mhos. In general, the conversion transconductance is about 30 per cent of the maximum transconductance of the tube acting as a straight amplifier without any frequency changing. The conversion conductance is less because the oscillator voltage swings the $g_m$ to very low values on its negative half-cycle in order to obtain rectification of the i-f beat.

Except for diode mixers, the converter always has gain at the intermediate frequency because the beat note is amplified in addition to the detection. The ratio of the i-f voltage output to the r-f signal voltage input is called the *conversion gain.* In a 6BE6 converter stage the gain may be about 50 for an intermediate frequency of 455 kc.

*Fig. 11·10 Two-gang capacitor with smaller section for oscillator tuning.*

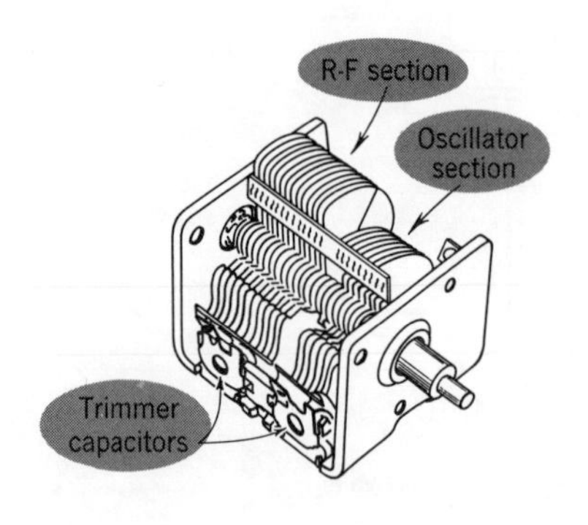


*Fig. 11·11 Mica compression-type oscillator padder capacitor.* (J. W. Miller Co.)

## 11 · 6 The local oscillator

For efficient heterodyning and minimum distortion of the r-f signal modulation, the oscillator must inject about 10 volts of c-w output into the mixer. Note that 10 volts rms equals 28 volts peak-to-peak. With typical converter tubes, nonlinear operation results with this wide swing. In addition, the oscillator injection voltage must be much greater than the r-f signal voltage because the resultant beat note has the modulation of the weaker signal. In normal operation, this requirement is automatically satisfied with 10 volts oscillator output and a typical r-f signal level in the millivolt range.

**Oscillator tracking.** The oscillator frequency must always differ from the signal frequency by an amount equal to the intermediate frequency. Maintaining this constant difference frequency means the oscillator tracks with the t-r-f circuits. If we assume that the oscillator operates above the r-f signal frequencies, the tracking is essentially a problem of making the oscillator tune through a higher frequency range. This is done by reducing the oscillator tuning capacitance and inductance. The oscillator capacitance can be decreased by using a tuning gang that has a smaller section, like the one in Fig. 11 · 10. A tuning capacitor having identical sections can be used, however, with a series capacitor in the oscillator circuit.

For series capacitance, an individual mica capacitor is used, such as the one in Fig. 11 · 11. This is called a *padder* because it adjusts the maximum capacitance in the circuit. Remember that the reciprocal formula applies to series capacitors. For the example shown in Fig. 11 · 12, if $C_1$ and $C_p$ are each 400 pf, their combined series capacitance equals 200 pf. The value of 400 pf for the oscillator-tuning capacitor $C_1$ is its maximum value in complete mesh at the low-frequency end of the tuning range. By adjustment of the capacitance of the padder $C_p$, the maximum capacitance is set for the lowest oscillator frequency, while the parallel-trimmer capacitance $C_T$ is used to adjust the minimum capacitance with the tuning capacitor out of mesh for the high end of the band.

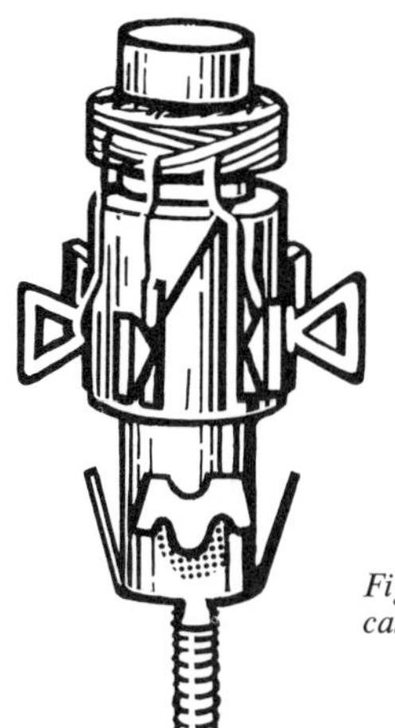

*Fig. 11 · 12 Trimmer and padder adjustments in oscillator tuning circuit.*

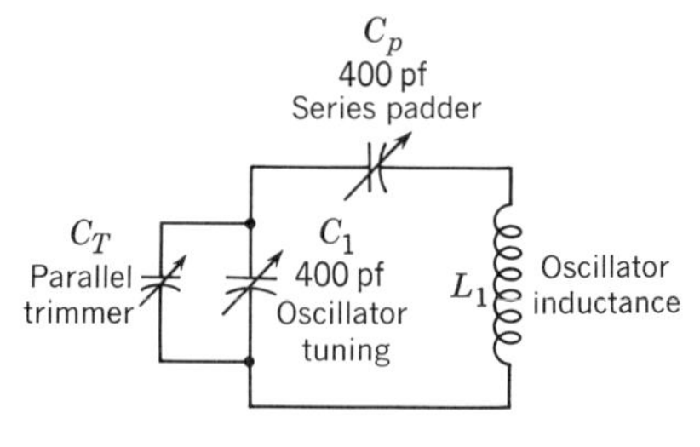


*Fig. 11 · 13 Oscillator coil for AM broadcast band.* (J. W. Miller Co.)

The oscillator inductance $L_1$ is also decreased to raise the oscillator frequency. For example, $L_1$ can be approximately 130 $\mu$h, while the inductance of a t-r-f coil is about 250 $\mu$h. When $L_1$ is fixed, the lowest resonant frequency is set by the padder capacitor. If the oscillator coil has an adjustable slug to vary its inductance, the padder capacitor is omitted. Figure 11 · 13 shows an adjustable oscillator coil for the AM broadcast band.

In a receiver for the AM broadcast band, the padder is adjusted for maximum receiver output with the tuning dial set to 600 kc. The trimmer capacitor is adjusted at 1,400 kc. The actual oscillator frequencies are above the frequency of the r-f signal by the amount of the intermediate frequency, but adjusting for maximum i-f output automatically provides the correct oscillator frequency.

With an oscillator-tuning capacitor smaller than the t-r-f section, there is no padder. Then the oscillator-tuned circuit is designed to track over the lower part of the band just for the specific intermediate frequency of the receiver. If necessary, the slotted end plates of the oscillator tuning capacitor can be bent to improve the tracking at the low-frequency end of the band. The shunt trimmer is adjusted as usual at 1,400 kc. If the oscillator inductance is variable, it can be adjusted for proper tracking at 600 kc on the dial. In multiband receivers, the oscillator coil for each range usually has its own trimmer and padder capacitors. Each of these is then switched across the tuning capacitors for the desired range.

The purpose of the oscillator-tracking adjustments is to align the oscillator with the r-f signal circuits for maximum sensitivity and to make the oscillator-tuning range fit the dial calibration of the receiver. As the tuning gang is varied, the oscillator will always bring in the station whose carrier frequency can beat with the oscillator to produce the intermediate frequency. Most of the receiver sensitivity is in the i-f amplifier. Therefore the oscillator frequency determines the station tuned in, even if the r-f signal circuits are resonant at a different frequency.

Correct oscillator tracking insures that the station pulled in by the local oscillator is the frequency to which the r-f circuits are tuned. Also, the oscillator tuning determines the dial settings at which different stations can be received. The oscillator trimmer is adjusted to make the receiver tune to the frequency of 1,400 kc at this point of the dial, while the padder adjustment is made at 600 kc. Assuming that the dial scale designed for the receiver is used, the frequencies between these two values will be received at their proper readings on the dial. If the oscillator does not track, the tuning range of the receiver will be compressed or extended, and it will not match the dial scale. Then all the stations are received too close together over part of the dial, or only a few stations can be received throughout the entire dial.

**The oscillator tuning range.** The local oscillator generally operates above the signal frequencies in order to reduce the tuning range of the oscillator, which simplifies the tracking problem. Just why the oscillator tuning range is smaller can be seen from some numerical values. Consider a receiver with an intermediate frequency of 455 kc. To cover the range of 535 to 1,605 kc, the receiver's local oscillator must tune through 990 to 2,060 kc when it operates higher than the r-f circuits. This is a tuning range of 2,055/1,005, or approximately 2 to 1, for the local oscillator. If the oscillator operates below the signal frequencies, it would tune through 80 to 1,150 kc, a range of approximately 15 to 1. Not only would the tracking problem be very difficult, but this tuning range could not be obtained with a single coil and capacitor combination.

**Oscillator bias.** Grid-leak bias is generally used for the local oscillator, as illustrated by $R_gC_C$ in Fig. 11·8. Since the a-c grid-feedback voltage develops the d-c bias, it provides a convenient way of checking the oscillator to see if it is operating. Just measure the bias at the oscillator grid with a d-c voltmeter. When the grid-leak bias is there, the oscillator must be operating. A typical bias value is approximately 10 volts, with the grid side negative. This should be measured with a high-resistance voltmeter, of at least 20,000-ohms-per-volt sensitivity. If the voltmeter lowers the oscillator grid resistance appreciably, the bias reading will be reduced, or the oscillator may not operate at all.

If there is no bias, the oscillator is not operating. Without the oscillator injection voltage for heterodyning, the r-f input cannot be converted to the i-f signal. As a result, there will be no audio output. Sometimes the oscillator may operate at the lower frequencies in the tuning range but fail at the high-frequency end of the band because of increased losses at the higher frequencies. Then the receiver will operate only for the low-frequency stations, with no output at the high end of the band.

## *11·7 Spurious responses*

The main disadvantage of the superheterodyne receiver is the fact that any voltage at the intermediate frequency coupled into the i-f amplifier will be amplified. This should be the desired r-f signal converted to an i-f signal, but there are other possibilities. Any false signal that does not have the correct radio frequency as listed on the dial is called a *spurious response*. The most important spurious frequencies are those close to the intermediate frequency of the receiver, harmonics of the intermediate frequencies, harmonics of the desired r-f signal, and image signals. An image frequency differs from the oscillator frequency by the intermediate frequency just as the desired signal does, but the image is above the oscillator instead of below. The effects of the spurious responses may be reception of two signals at the same time, whistles caused by an a-f beat note between signals, receiving one station at two points on the dial, or receiving r-f signals that are not on the dial.

**Image frequencies.** The most important of the spurious responses is the image-frequency signal. For example, assume that a receiver with an intermediate frequency of 455 kc is tuned to a station at 600 kc, which is the desired signal. The oscillator frequency is set at 1,055 kc, beating 455 kc above the signal to receive this station. At the same time, any undesired signal frequency at 1,510 kc that may be coupled into the converter can also beat with the oscillator frequency of 1,055 kc to produce the same difference frequency of 455 kc.

Assuming the oscillator operates above the signal frequencies, the image frequency of any desired station is equal to its r-f carrier frequency, plus twice the intermediate frequency. For instance, the image of a station at 600 kc in the broadcast band is 600 kc plus 910 kc, or 1,510 kc, with an intermediate frequency of 455 kc for the receiver. For 800 kc, the image frequency is 1,710 kc, which is outside the broadcast band.

**Undesired i-f signals.** Radio services at or near the intermediate frequency of the receiver produce r-f signals that may reach the grid of the converter when there is not enough r-f selectivity to reject this frequency. Interference of this type can usually be eliminated, however, by the use of a *wave trap.* The trap is tuned to reject the intermediate frequency, but it is connected into the r-f circuits before the converter. The desired r-f signals are not rejected, because they are changed to the intermediate frequency only in the output of the converter.

**I-f harmonics.** Whistles may be produced in a superheterodyne receiver when harmonics of the intermediate frequency produced in the output of the second detector are coupled back into the r-f section. This feedback can then beat with the incoming signal for some stations to produce an audible beat note. The remedy is to provide adequate filtering in the output of the second detector, arrange the wiring for minimum coupling between the second detector and the r-f section, or use a slightly different intermediate frequency for the receiver.

**R-f harmonics.** In some cases harmonics of the local oscillator can beat with r-f signals much higher than the normal frequency range of the receiver to produce a spurious response that is completely unrelated to the receiver's dial. As an example, suppose that the local oscillator is operating

*Fig. 11 · 14 I-f transformers. (a) Unshielded with mica trimmers. (b) I-f can with mica trimmers at top. (c) Permeability tuning at top and bottom of i-f can.* (J. W. Miller Co.)

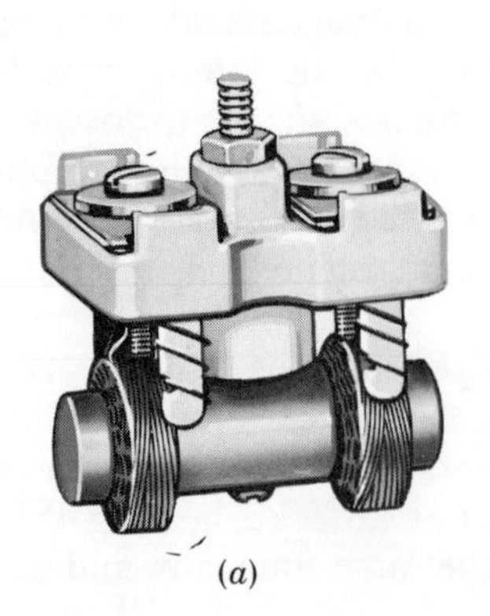

(*a*)

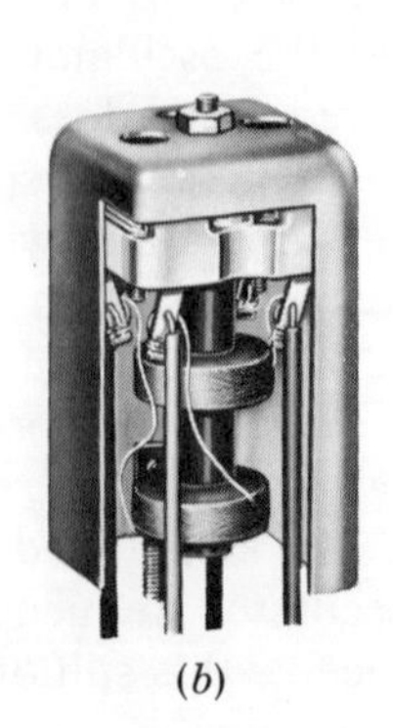

(*b*)

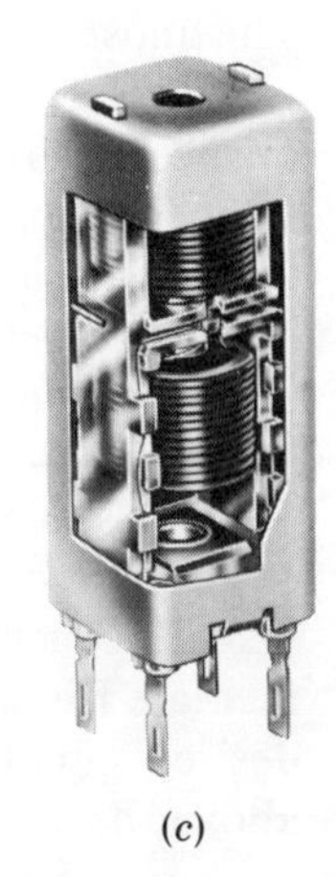

(*c*)

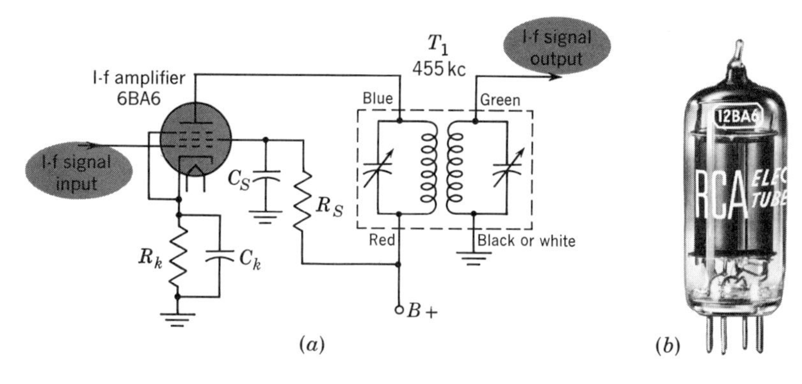


*Fig. 11·15 (a) Circuit of i-f amplifier stage. (b) Typical i-f amplifier tube.*

at 10 Mc to beat with a 9-Mc signal for an intermediate frequency of 1 Mc in a short-wave receiver. The sixth harmonic of the oscillator at 60 Mc can also beat with any spurious signal at 59 or 61 Mc to produce the same 1-Mc i-f output from the converter. Similarly, the fundamental frequency of the local oscillator can beat with harmonics of the r-f input signal, resulting in spurious responses below the tuning range of the receiver.

**R-f preselector.** The main factor in minimizing spurious responses in the superheterodyne receiver is the r-f gain ahead of the converter. For this reason, the t-r-f amplifier between the antenna and converter is often called a *preselector* stage. Its r-f gain and selectivity are the factors that allow only the desired signal to be converted to the intermediate frequency. Without a preselector stage, there is generally insufficient r-f selectivity in the antenna circuit to prevent spurious signals from reaching the converter-grid circuit. The preselector also has the advantage of improving the signal-to-noise ratio of the receiver.

## *11·8 The i-f amplifier section*

This section of the receiver includes one, two, or three stages tuned to the intermediate beat frequency produced in the output of the converter. Consider an AM radio receiver with an intermediate frequency of 455 kc. Each i-f stage is usually a transformer-coupled amplifier with both the primary and secondary tuned to this intermediate frequency. The i-f amplifier operates class A because it is amplifying a modulated signal. A double-tuned transformer can be used for the coupling circuit because the resonant frequency is the same for all stations. Several i-f transformers are shown in Fig. 11·14. Note the provisions for tuning the transformer to the intermediate frequency. Either mica trimmer capacitors are available across the primary and secondary, or permeability tuning is used. In aligning the receiver, the i-f transformers are adjusted for maximum output.

A conventional i-f amplifier circuit is shown in Fig. 11·15. The i-f

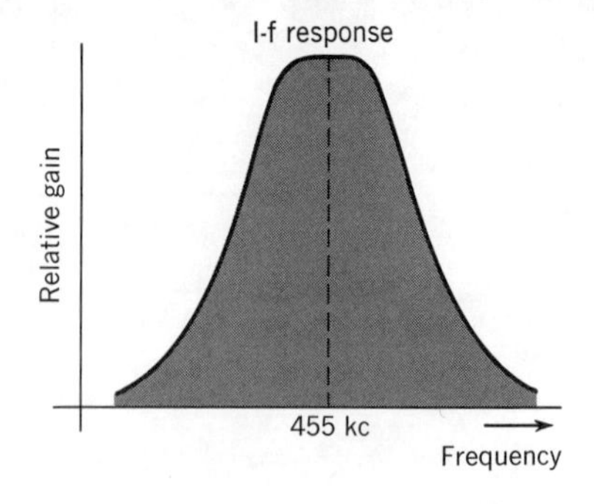


*Fig. 11·16 Response curve of i-f amplifier tuned to 455 kc.*

input signal to the grid is amplified by the pentode amplifier and then inductively coupled to the next stage by transformer $T_1$. Both the primary and the secondary of $T_1$ are tuned to the intermediate frequency. The required d-c screen voltage is supplied by the screen-dropping resistor $R_S$ and bypass capacitor $C_S$. The $R_kC_k$ combination in the cathode provides an operating bias of 3 volts. The gain of an i-f amplifier at 455 kc is 100 to 200. With approximately 40 mv of signal from the converter, a single i-f stage having a gain of 150 would provide 6 volts output, which is enough signal voltage to drive the second detector.

**The i-f response curve.** A typical i-f response curve is shown in Fig. 11·16. The important requirements are uniform gain over a band wide enough for the sideband frequencies of the modulated signal, and a sharp slope for the side responses off resonance. This band-pass characteristic is produced by double-tuned i-f circuits. With critical coupling in the transformer, wide band pass can be obtained. The total bandwidth may be 6 to 20 kc for AM radio receivers that have an intermediate frequency of 455 kc, or 200 kc for FM receivers with an intermediate frequency of 10.7 Mc. The slope of the side responses is sharp because two tuned circuits having a comparatively high $Q$ are used in one i-f coupling circuit.

The i-f response provides practically all the receiver selectivity, which is its ability to reject frequencies close to the desired r-f signal. With sharp skirts (that is, sides) on the i-f response curve, the only frequencies amplified are those that can beat with the local oscillator to produce the intermediate frequency of the receiver. There is very little gain for frequencies outside the i-f response.

**The i-f transformer.** The air-core transformers in Fig. 11·14*a* and *b* have the primary and secondary coils mounted on a cardboard dowel form, about ¾ in. from each other to provide the desired coupling. The metal shield can, usually aluminum, minimizes regeneration at the intermediate frequency. Two mica trimmer capacitors are mounted with the coils to tune the primary and secondary to the intermediate frequency. These can be adjusted with a screwdriver through the holes at the top of the can. The i-f transformer in Fig. 11·14*c* uses permeability tuning, utilizing adjustable slugs for tuning the primary and secondary inductances. Usually the top slug is the primary adjustment and the bottom one is the secondary.

For a 455-kc i-f transformer, the primary and secondary inductances are about 800 $\mu$h each, to tune with a capacitance of approximately 150 pf. The d-c resistance of either winding is generally about 15 ohms. The $Q$ of the primary and secondary is approximately 100. When the transformer has color-coded leads, they are arranged as shown in Fig. 11 · 15.

Two i-f transformers are normally employed for a single i-f amplifier stage, to tune both the input and output circuits. The i-f transformer receiving the signal from the converter is called either the *converter plate transformer* or the *input i-f transformer.* The unit coupling i-f signal into the second detector is the output i-f transformer. These differ slightly in their characteristics. The input transformer is designed to couple into the grid circuit of the first i-f amplifier, which is a very high resistance, compared with the load of the second detector on the output i-f transformer.

**I-f tubes.** The tubes used for the i-f amplifier are always r-f pentodes, like the 6SK7 or 6SH7 octal tubes, or miniature glass types, such as the 1U4, 6AG5, and 6BA6. Pentodes are used because the grid-plate capacitance is small. Also their high plate resistance is needed to avoid shunting the i-f tuned circuits with a comparatively low resistance that could reduce their $Q$.

**The intermediate frequency.** With the development of superheterodynes, the intermediate frequency used in AM radio receivers has steadily increased from early values of 50 to 100 kc up to the present popular intermediate frequency of 455 kc. The reasons for this change illustrate some important characteristics of the superheterodyne receiver. Actually, the receiver intermediate frequencies may be any frequency between the highest audio frequency and 535 kc in a broadcast-band receiver, which is the lowest signal frequency in the tuning range.

The advantages of using a lower intermediate frequency for the receiver are the increased selectivity, gain, and stability possible in the i-f amplifier. However, a low intermediate frequency has the serious disadvantage of increasing the number of image-frequency signals in the tuning range of the receiver. Also, the image and the desired signals are not widely separated in frequency, making it more difficult for the r-f circuits to reject the image frequency. A higher intermediate frequency, therefore, improves the image rejection of the receiver. Other advantages of having a higher value for the intermediate frequency are: (1) a reduction in the tuning range of the local oscillator when it operates above the signal frequencies, (2) less trouble with spurious responses caused by i-f harmonics, and (3) a reduction in the amount of oscillator radiation produced by oscillator signal coupled to the antenna through the r-f signal circuits. The oscillator radiation is reduced because of the greater separation in frequency between the signal and oscillator circuits. In any case, the exact intermediate frequency should not coincide with the frequency of any powerful radio service, nor with its harmonics.

Taking all these factors into account, the 455-kc frequency has become the most common intermediate frequency in broadcast-band receivers.

This is almost the highest possible intermediate frequency that can be used, providing all the advantages of a high intermediate frequency, especially with regard to image rejection. Sufficient gain can be obtained at this frequency without much difficulty. Most broadcast-band receivers use an intermediate frequency of 455 kc, 456 kc, 465 kc, or 468 kc. Automobile receivers generally use 262.5 kc, in order to obtain the increased gain and selectivity possible with a lower intermediate frequency. These i-f values apply whether tubes or transistors are used. A typical transistorized i-f amplifier circuit can be seen in the receiver schematic of Fig. 12 • 6 in the next chapter.

**I-f alignment.** The alignment of the i-f amplifiers is a process of tuning all the i-f transformers to the receiver's intermediate frequency. Starting with the secondary circuit coupled to the second detector, each i-f circuit is tuned, working back to the transformer primary in the converter plate circuit. After this is done, the oscillator and r-f stages are aligned. Misalignment of the i-f circuits can reduce the sensitivity or result in no reception at all. However, these effects are much more likely to be produced by other more common receiver troubles, because once the i-f circuits have been aligned they should seldom need readjustment. If the receiver howls or chirps and has poor sensitivity, though, the trouble may be the i-f alignment.

## *11 • 9 Automatic volume control (AVC)*

The AVC circuit in a receiver automatically controls the gain according to the strength of the carrier signal received. Less receiver gain is needed for a strong signal than for a weak signal. Therefore the AVC circuit reduces the gain by increasing the negative bias on the i-f and r-f amplifiers. All that is needed is a negative d-c bias voltage proportional to the signal strength. This is obtained by rectifying the signal itself. The advantage of automatic volume control is that it provides easy control of volume in the audio section without blasting on strong stations, while the AVC bias controls the gain in the i-f and r-f stages to prevent overload and minimize fading. A disadvantage is the fact that the AVC bias reduces the receiver gain. However, the superheterodyne circuit provides plenty of gain so that practically all receivers use automatic volume control.

A typical AVC circuit is shown in Fig. 11 • 17. The first requirement is producing the negative AVC bias voltage. This is done by the diode $V_3$, which rectifies the signal. $R_L$ is the diode load resistor in the plate circuit. The voltage across $R_L$ is a d-c voltage proportional to the signal strength. The polarity of this d-c voltage is negative because $R_L$ is in the plate return to chassis ground. Then the rectified signal voltage across $R_L$ must be filtered to eliminate r-f and a-f variations, since a steady bias is required. This is the function of the AVC filter $R_1C_1$. It has the relatively long time constant of 0.01 sec so that the d-c voltage across $C_1$ cannot vary at the audio rate. The voltage across $C_1$ is the filtered, negative d-c voltage that serves as AVC bias.

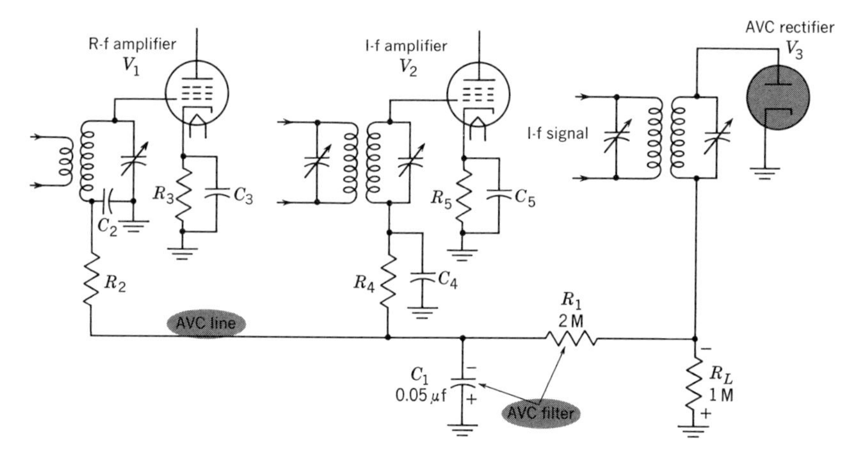


*Fig. 11 · 17 Simple AVC circuit.*

The i-f and r-f stages that need the AVC bias return to $C_1$ through the AVC line. Then each controlled stage has its bias controlled by the voltage across $C_1$. Since the AVC bias is negative, it is applied to the control grid. The method shown here is series feed, with each grid circuit returned to the AVC line instead of chassis ground. Then the AVC voltage determines the grid bias. The controlled stages also have minimum cathode bias.

To review the AVC action, suppose that a strong station is being received. Then the i-f signal input to $V_3$ may be enough to produce $-5$ volts AVC bias. This amount of bias reduces the i-f and r-f gain so that only 3 volts of audio signal is available from the second detector. When the receiver is tuned to another station that is weaker, less AVC bias will be produced and the audio output can still be 3 volts. The AVC circuit adjusts the gain, therefore, to provide about the same audio output from the detector for strong stations and weak stations.

The function of each part of the AVC circuit can be summarized as follows:

1. The AVC rectifier produces the negative d-c bias voltage. More signal produces more AVC bias; less signal results in less bias. It should be noted that the diode detector in a receiver can also function as the AVC rectifier. The output across the diode load resistor in the plate circuit is a negative d-c voltage which can be filtered to provide the AVC bias.
2. The AVC filter removes the a-c variations from the rectified signal voltage to provide a steady AVC bias. The filter time constant is generally 0.1 to 0.05 sec. A shorter time constant will not filter out the audio variations; a time constant that is too long will not allow the AVC bias to change fast enough when the receiver is tuned to different stations.
3. The AVC line is connected to the control grid circuit in each controlled stage, to provide the AVC voltage as grid bias. The amount of negative

bias voltage on the AVC line may be 1 to 10 volts, depending on the amount of signal.

4. Decoupling filters, such as $R_2C_2$ and $R_4C_4$ in Fig. 11·17, isolate each grid circuit from the common AVC line, minimizing feedback between stages.
5. The tube used in an r-f or i-f stage controlled by AVC bias should be a remote-cutoff (variable-$\mu$) type. A remote-cutoff grid characteristic minimizes amplitude distortion in the amplifier when its bias is varied over a wide range by the AVC voltage.

**Delayed automatic volume control.** The AVC circuit reduces the gain more for strong signals than for weak signals. However, there is still some bias and loss of gain for weak signals just when maximum gain is needed. For this reason, some receivers have provision for shorting out the AVC bias circuit when necessary to allow maximum receiver gain. Or, a delayed AVC circuit can be used. In this method, the AVC bias is not produced until there is sufficient signal voltage. Actually the delay is in amplitude, not in time, since the AVC rectifier is biased out of conduction for weak signals.

An example of a delayed AVC circuit is shown in Fig. 11·18. The i-f signal is coupled to AVC rectifier $D_1$ and to the second detector $D_2$ by coupling capacitor $C_2$. With delayed automatic volume control, there must be separate rectifiers. The AVC rectifier has a delay bias to prevent conduction on weak signals, but the detector must conduct for even the weakest signal. Furthermore, the delay bias is obtained from a battery or the power supply, not from the signal, because the delay must be available for weak signals. In this circuit, there is +3 volts on the cathode of the AVC rectifier. This means that the signal must drive the plate more than 3 volts positive in order for the tube to conduct. For weak signals, then, the AVC rectifier does not conduct, and there is no AVC output voltage. When the signal is strong enough to exceed 3 volts, the AVC rectifier conducts. Then it produces AVC voltage proportional to the signal strength, as in a simple AVC circuit. However, the AVC circuit is biased out of operation for weak signals; this permits maximum gain in the receiver.

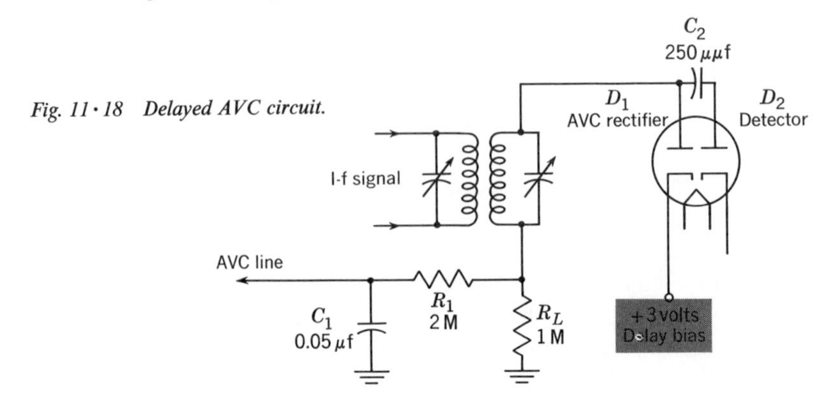


Fig. 11·18 Delayed AVC circuit.

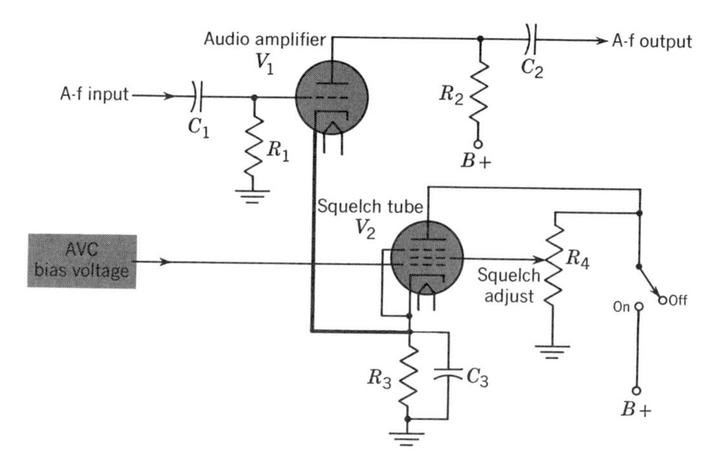


*Fig. 11·19 Squelch circuit for cutting off audio amplifier to eliminate noise between stations.*

**Amplified automatic volume control.** Some receivers have a stage to amplify the AVC bias voltage. In this way, a small change in signal level can provide a large change in AVC bias. This circuit usually has an AVC level adjustment. Set this control to eliminate overload distortion on the strongest signal.

**AVC voltage for transistors.** The AVC voltage is used as negative grid bias for tubes to reduce plate current in the r-f and i-f stages. Similarly, for transistors the corresponding idea is to use the AVC voltage to reduce the forward bias for less collector current. Note that negative AVC voltage opposes positive forward bias on a P base. Also, positive AVC voltage can be used on an N base. Either method for transistors is called *reverse AVC,* which reduces collector current with more signal strength, similar to the AVC bias for tubes.

## *11·10 Squelch circuits*

A receiver with automatic volume control operates at maximum gain between stations because then there is no signal input and no AVC bias. In a receiver with two or more i-f stages, the resultant amplification of the converter noise can produce a loud hissing sound between stations. When a signal is tuned in, the receiver quiets down as the AVC bias reduces the receiver gain. Still, for high-gain receivers it may be desirable to eliminate the noise between stations. This can be done by cutting off the audio amplifier when no carrier signal is being received. Such a circuit is called a *squelch, silencer,* or *quiet AVC* circuit. The AVC voltage indicates whether a signal is being received. Therefore the AVC bias can be used to operate a squelch or silencer circuit that cuts off the audio amplifier between stations but allows normal operation with a carrier signal.

One type of squelch circuit is illustrated in Fig. 11·19. Notice that the audio-amplifier cathode bias depends on the current in both $V_1$ and $V_2$, since both tubes have $R_3C_3$ in common. Furthermore, the current of the squelch tube is controlled by the AVC bias voltage directly coupled to its grid. Assume that a strong signal is being received, which produces enough negative AVC bias to cut off the squelch tube. Then the audio amplifier has its normal bias produced by plate current through $R_3$. Between stations, however, when the AVC bias is very low, the squelch tube conducts maximum current and the increased current raises the bias across $R_3C_3$ enough to cut off the audio amplifier. The tubes that are used allow the squelch tube to conduct with this bias, while the sharp-cutoff audio-voltage amplifier cannot conduct.

The sequence of operation is as follows:

1. When the squelch tube conducts, it cuts off the audio amplifier.
2. When the squelch tube is cut off, the audio amplifier conducts.

There are two provisions in the squelch circuit to make certain that weak stations can be received. The adjustment of screen-grid voltage determines how much AVC bias will cut off the squelch tube. Also, the squelch tube can be disabled completely by switching off its B+ voltage. To adjust the circuit, tune in the weakest usable signal with the squelch off; then turn it on and set the squelch adjustment at the point where this station can be heard.

## *11·11 Tuning indicators*

The AVC voltage indicates when a station is tuned in. Furthermore, this voltage increases to maximum as the receiver is tuned exactly to the signal frequency. Thus, the AVC bias voltage can be used as a tuning indicator. Three of the most common methods are:

1. A d-c voltmeter to read the negative AVC bias voltage. This voltage is highest for maximum carrier signal when the tuning is correct. A VTVM is preferable because of the high resistance in the AVC circuit. This

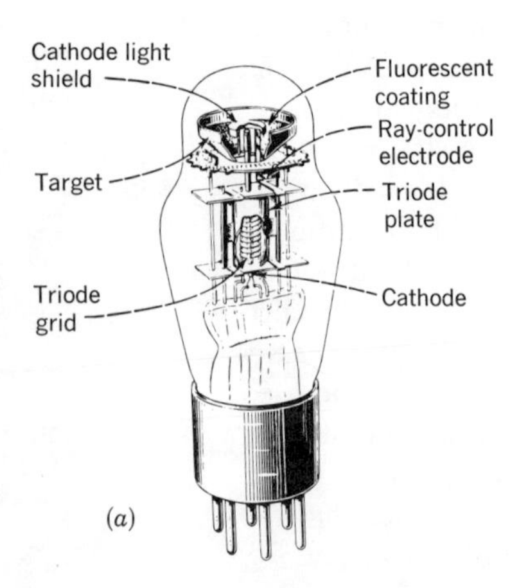


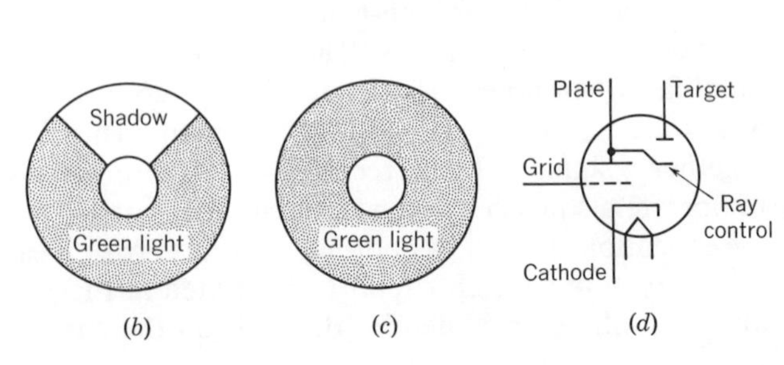


*Fig. 11·20 6E5 magic-eye tuning indicator tube. (a) Structure of tube. (b) Eye open with maximum shadow. (c) Eye closed. (d) Schematic symbol showing ray-control electrode tied internally to triode plate.* (RCA.)

method is convenient only when you are working on the chassis and the AVC bias is available for measurement.

2. A d-c milliammeter to read average plate current of the r-f and i-f amplifier stages controlled by AVC bias. This type of tuning meter dips to a minimum value for maximum signal. This surprising fact results because the meter reads d-c plate current in the class A amplifiers. The average plate current depends only on the bias, not the a-c signal. When there is more signal, the higher AVC voltage makes the bias more negative. This, in turn, reduces the amount of plate current.
3. An electron-ray indicator, or magic-eye, tube, as shown in Fig. 11·20. This method is very popular because the operator can watch the indicator mounted over the receiver dial. The magic-eye tube is controlled by the AVC bias voltage to make the shadow close the most for maximum signal. Actually, the amount of shadow indicates the relative value of the AVC bias voltage.

The construction of a magic-eye tube is shown in Fig. 11·20*a*. Its principal feature is a circular target at the top with a fluorescent coating that glows green when excited by electrons. The cathode of the tube supplies the electrons. The target is connected to B+, its positive potential accelerating electrons which make it fluoresce green. However, between the cathode and target at the top of the tube is a thin metal rod called the *ray-control electrode*. This electrode controls the number of electrons from the cathode which can strike the target directly behind it. The electron flow depends on the potential of the ray-control electrode. With the same voltage as the target, the electrode has little effect, and the entire target area glows. If the electrode is less positive than the target, electrons are prevented from striking the target. The dark area then results in a shadow just behind the rod.

A circuit for opening or closing the shadow of the eye of a 6E5 magic-eye tube is shown in Fig. 11·21. The target has the constant B+ voltage. The triode plate voltage, however, depends on the current through the plate-load resistor *R*. When the plate current is zero, the plate voltage equals the 250 volts of the B supply. Maximum plate current drops the plate voltage to 50 volts. Note that the potential of the ray-control electrode is identical with the triode plate voltage because they are connected internally.

The variation in shadow area depends on the fact that the target voltage

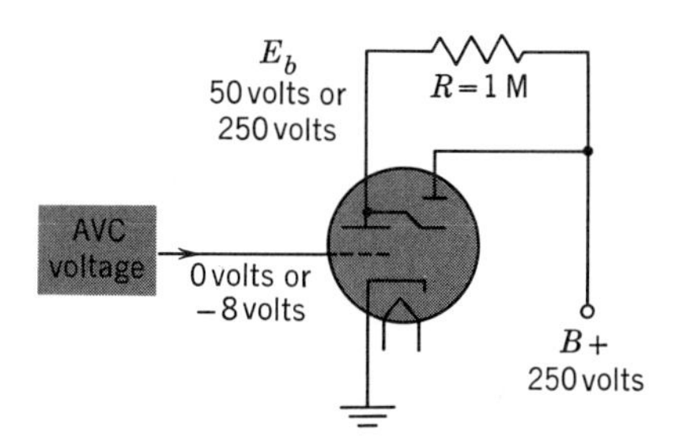


*Fig. 11·21 Circuit for 6E5 magic-eye tube.*

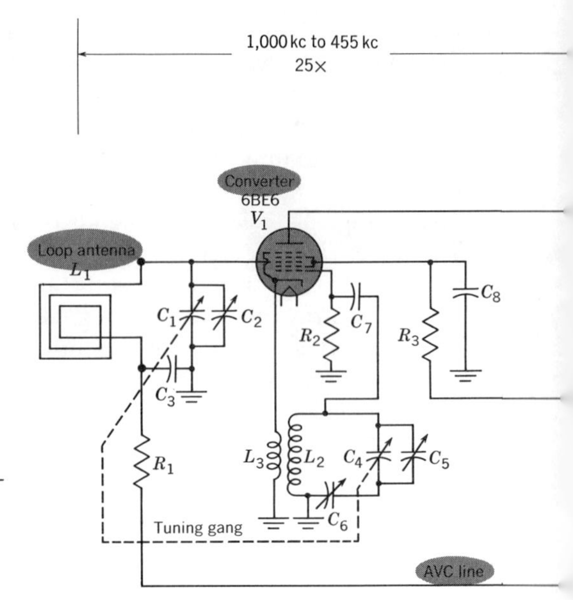


*Fig. 11·22 Schematic diagram of superheterodyne receiver. Power supply not shown.*

is constant at B+ but the potential of the ray-control electrode varies with the triode plate voltage. With 250 volts on the plate, the electrode and the target have the same potential and there is no shadow. When the ray-control electrode is at the 50-volt value, however, it is 200 volts less positive than the target. This causes the eye to open to its maximum shadow.

The amount of triode plate current is controlled by the AVC voltage at the grid. This is obtained directly from the AVC line. Zero AVC voltage at the grid allows maximum plate current in the triode, dropping the plate voltage to 50 volts. At the opposite extreme, an AVC voltage of −8 volts cuts off the triode plate current. Now the plate and the ray-control electrode have the full B+ voltage, and the eye closes. Since the AVC bias increases with more signal, the eye closes as a station is tuned in. This same principle of using the AVC bias to control fluorescence applies to all types of electron-ray tubes used to indicate tuning.

## 11·12 *Typical superheterodyne receiver*

Figure 11·22 illustrates how the r-f, i-f, and audio circuits are contained in a superheterodyne receiver. This circuit is for a compact table-model radio with four tubes, not including the power-supply rectifier. There is sufficient gain, however, because of the superheterodyne circuit. The numbers along the top of the diagram indicate how much gain each section provides. Notice the high gain of 100 in the single i-f stage. Gen-

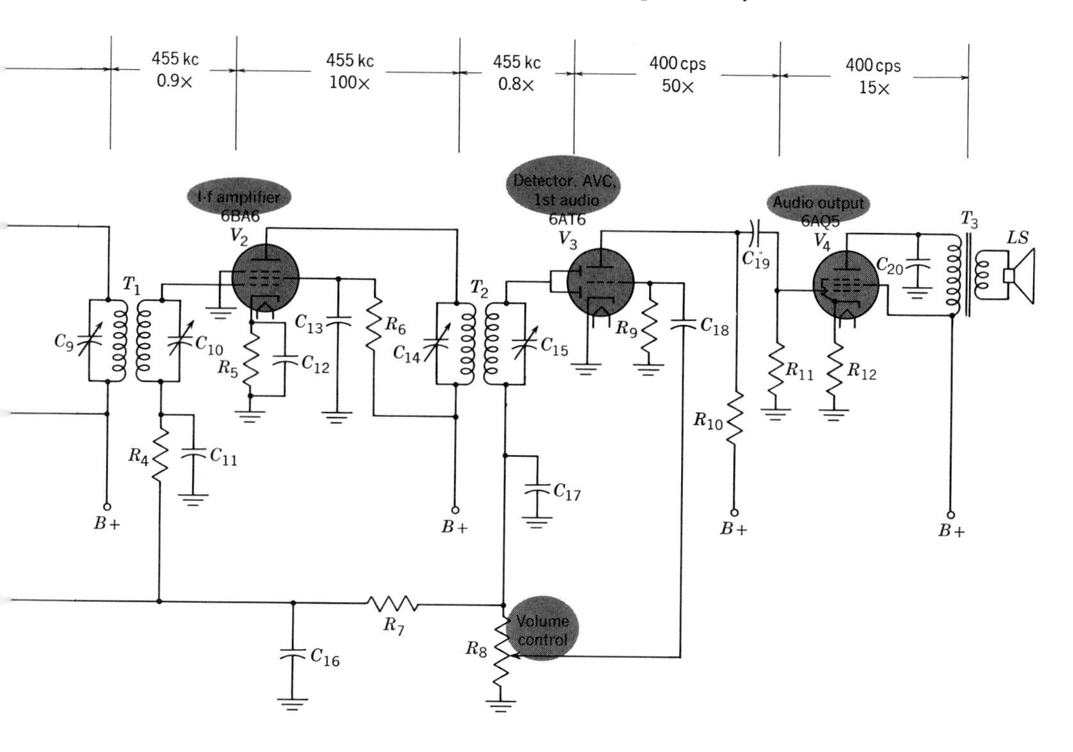


erally, table-model radios use either a loop antenna mounted in the cabinet or a compact ferrite loopstick at the top of the chassis. These antennas are shown in Fig. 11·23 and Fig. 12·3. Since these antennas are directional, moving the receiver can affect the amount of signal pickup.

The r-f signal picked up by the antenna is coupled to the signal grid of the converter. There is no t-r-f preselector stage. Also, the inductance of the loop antenna $L_1$ resonates with $C_1$ for r-f tuning, eliminating an r-f transformer. The low side of $L_1$ is bypassed to chassis ground by $C_3$. A direct connection cannot be used, because it would short the AVC bias voltage. However, the bypass joins $L_1$ and $C_1$ for r-f signals so that they can tune to the desired station. $C_2$ is the trimmer for the r-f tuning capacitor $C_1$.

The oscillator section of the 6BE6 converter stage consists of the cathode, grid 1, and the screen grid, which is bypassed to chassis ground by $C_8$. A tuned-grid oscillator is used with feedback obtained from the tickler coil $L_3$ in the cathode circuit. Grid-leak bias is produced by $C_7$ and $R_2$. The oscillator-tuned circuit includes $L_2$, the tuning capacitor $C_4$ with its shunt trimmer $C_5$, and series padder $C_6$. Tuning capacitors $C_4$ and $C_1$ are on the same shaft of a two-section gang, so that the oscillator can track 455 kc above the r-f signal. Because of the heterodyning action between the oscillator and r-f signal in the converter, the required i-f signal is produced in the converter-plate circuit tuned to 455 kc.

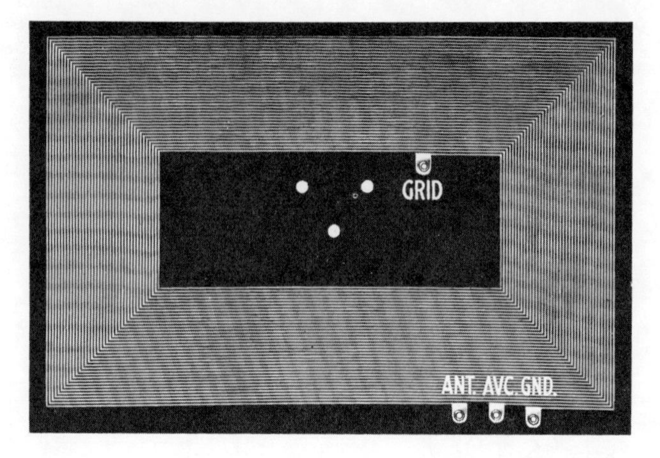


*Fig. 11 · 23 Typical loop antenna.*

$T_1$ is the converter-plate transformer, with primary and secondary tuned to the intermediate frequency. This is the input i-f transformer, coupling the 455-kc signal to the grid of the i-f amplifier. $C_9$ and $C_{10}$ are the trimmer capacitors to align the i-f transformer. The one i-f stage has enough gain to supply several volts of signal to the diode detector. The two diode plates in $V_3$ are tied together to form one diode that functions as detector and AVC rectifier. When a signal is coupled to the diode plates by the output i-f transformer $T_2$, rectified signal current flows to produce detected output voltage across $R_8$. The i-f bypass capacitor $C_{17}$ is actually in parallel with $R_8$, allowing only audio-voltage output. $R_8$ serves as the audio volume control. The desired amount of audio voltage is coupled by $C_{18}$ to the grid of the triode amplifier in $V_3$. This is the first audio amplifier. It drives the audio-output stage $V_4$, which is transformer-coupled by $T_3$ to the loudspeaker. In the plate circuit of $V_4$, $C_{20}$ is a bypass capacitor for radio frequencies and high audio frequencies.

In the detector circuit, besides serving as the volume control, $R_8$ is a diode load resistor that provides negative d-c voltage proportional to the signal strength. This rectified voltage is negative because $R_8$ is in the plate-to-ground circuit. Therefore the voltage across $R_8$ can be used for AVC bias. $R_7$ and $C_{16}$ form the AVC filter to remove i-f and a-f variations. The voltage across $C_{16}$ is the filtered AVC bias voltage for the AVC line. Although no audio voltage can be developed across $C_{16}$, it is isolated from $R_8$ by the high resistance of $R_7$, so that the volume control provides the desired output for the audio amplifier. In the AVC circuit, the r-f mixer section of the converter and the i-f amplifier are controlled by the AVC bias. $R_1C_3$ forms a decoupling filter for $V_1$, while $R_4C_{11}$ is the i-f decoupling filter, isolating each stage from the common AVC line.

In a receiver with automatic volume control, the manual volume control is in the audio circuits. This is set for the desired amount of audio signal while the AVC circuit automatically controls the gain of the r-f and i-f amplifiers to maintain the same volume for different stations.

**Push-button tuning.** Some radios have push buttons for automatically tuning to different stations. Either a mechanical or an electrical arrangement can be used. With mechanical push buttons, you can see the station indicator move across the dial. Each button just turns the shaft of the tuning gang to a preset stop. To set each button, the stop is loosened, the radio tuned manually to a station, and then the stop is tightened. Thereafter, pushing the button tunes in this station. With the electrical arrangement, the push button is a switch that disconnects the manual tuning circuit and substitutes a separate *LC* circuit for each station. Under the button there is usually a screw to adjust the oscillator. The frequency at which the oscillator is set determines which station is tuned in by the push button. Each button can be used to tune in any station within the range of its oscillator frequency adjustment.

**Alignment.** The alignment of a superheterodyne receiver consists of two parts: i-f alignment and r-f alignment. First the i-f transformers must be tuned to the intermediate frequency by adjusting the trimmer capacitors on the can or the slugs for coils that are permeability-tuned. The last i-f transformer in the input circuit of the detector is aligned first, working back to the converter-plate circuit. Correct alignment produces maximum output.

For the r-f alignment, the main requirement is setting the local oscillator frequency to tune in the station frequencies on the dial. The oscillator padder is adjusted to tune in 600 kc at the low end of the dial. Then the shunt trimmer is adjusted for 1,400 kc at the high end of the dial. If there is no series padder, the i-f alignment must be at the correct frequency for oscillator tracking. After the oscillator trimmer is set for 1,400 kc on the dial, the r-f trimmer is adjusted for maximum signal at the same frequency.

## SUMMARY

1. The superheterodyne circuit uses a local oscillator to beat with the r-f signal, converting the signal to the intermediate frequency. The local-oscillator frequency is varied to convert all stations to the same intermediate frequency in the receiver.
2. In the heterodyning action, two waves of slightly different frequencies are mixed in one circuit and rectified to produce the beat equal to the difference between the two input frequencies. This function is performed by the converter stage, which combines the mixer and local-oscillator functions. Common tube types are the 6BE6 and 6SA7 pentagrid converters.
3. It is the local oscillator that tunes the receiver to different stations, as the oscillator frequency tracks above the r-f signal by a constant amount equal to the intermediate frequency. Generally the oscillator beats above the r-f signal to reduce the tuning range required for the oscillator.
4. An i-f stage is a pentode, class A amplifier, generally transformer-coupled, with primary and secondary tuned to the intermediate frequency. Because of the double-tuned coupling and the i-f value lower than the frequency of the r-f signal, the i-f amplification provides most of the receiver sensitivity and selectivity. Higher i-f values improve the image rejec-

tion; lower frequencies allow more gain and selectivity. In radios for the 535-1,605-kc broadcast band, 455 kc is the most common i-f value.

5. The superheterodyne receiver has the disadvantage of allowing spurious responses, which are undesired frequencies that can beat with the oscillator to produce the correct intermediate frequency.
6. The AVC circuit automatically controls the gain of the r-f and i-f stages by providing bias proportional to the signal strength, providing relatively constant audio-signal output from the detector. The stronger the signal is, the more the negative AVC bias and the less gain in the receiver.
7. In delayed automatic volume control, the AVC rectifier is cut off by the delay bias for weak signals.
8. In quiet AVC, or squelch, circuits, the AVC bias operates a squelch tube that cuts off the audio amplifier to quiet the receiver between stations when no signal is being received.
9. The AVC bias voltage indicates when a station is tuned in. Maximum AVC voltage means maximum signal. A magic-eye tuning indicator is controlled by the AVC voltage to make the eye close as an indication of maximum AVC voltage.
10. In aligning a superheterodyne receiver (*a*) the i-f circuits are tuned for maximum i-f output, (*b*) the oscillator padder is adjusted to bring in 600 kc on the dial, (*c*) the trimmer is adjusted for 1,400 kc, and (*d*) the r-f trimmer is adjusted for maximum signal at 1,400 kc.

## SELF-EXAMINATION

Here's a chance to find out how well you have learned the material in this chapter. Work the exercises; then check your answers against the Key at the back of this book. These exercises are for your self-testing only.

1. In superheterodyne receivers, the (*a*) t-r-f preselector stage usually operates 455 kc above the r-f signal; (*b*) converter needs r-f signal input and oscillator signal; (*c*) power supply uses series heaters; (*d*) oscillator frequency is usually double the i-f value.
2. A converter for 455 kc has output signal at just this frequency because the (*a*) oscillator supplies 455-kc signal; (*b*) r-f signal is always a harmonic of 455 kc; (*c*) converter plate circuit is tuned to 455 kc; (*d*) converter grid circuit is tuned to 455 kc.
3. With a 455-kc intermediate frequency and the oscillator operating above the signal frequencies, the oscillator frequency to tune in 1,200 kc is (*a*) 455 kc; (*b*) 745 kc; (*c*) 1,200 kc; (*d*) 1,655 kc.
4. For the same receiver, the image of a station at 1,200 kc is (*a*) 910 kc; (*b*) 1,290 kc; (*c*) 2,110 kc; (*d*) 2,400 kc.
5. Which of the following i-f values will provide the best sensitivity and selectivity? (*a*) 262 kc; (*b*) 455 kc; (*c*) 465 kc; (*d*) 468 kc.
6. In an AVC circuit, (*a*) more signal produces less bias; (*b*) the highest bias is produced between stations; (*c*) the gain of the audio-output stage is controlled by the AVC bias; (*d*) maximum signal produces maximum AVC bias.
7. In a typical squelch circuit, the (*a*) squelch tube eliminates r-f interference when a strong signal is received; (*b*) squelch tube cuts off the audio amplifier between stations; (*c*) audio amplifier operates at maximum gain between stations; (*d*) AVC bias is practically zero when a strong signal is received.
8. If the local oscillator does not operate, the (*a*) sound will be distorted; (*b*) AVC voltage will be maximum; (*c*) receiver cannot tune in any stations; (*d*) audio amplifier is cut off.
9. To measure the grid-leak bias voltage on the local oscillator to see if it is operating, the best meter to use is (*a*) an r-f voltmeter; (*b*) a VTVM on the negative d-c volts position; (*c*) a 1,000-ohms-per-volt meter to measure positive voltage; (*d*) a VTVM on positive d-c volts position.
10. In the receiver diagram in Fig. 11 · 22, the AVC filter capacitor is (*a*) $C_{17}$ with a value of 250 pf; (*b*) $C_{20}$ with a value of 0.02 $\mu$f; (*c*) $C_{16}$ with a value of 0.05 $\mu$f; (*d*) $C_{19}$ with a value of 0.05 $\mu$f.

## QUESTIONS AND PROBLEMS

1. Give two advantages and one disadvantage of the superheterodyne receiver compared with a t-r-f receiver.
2. Give the function for each of the following sections in a superheterodyne receiver: r-f amplifier, mixer, local oscillator, i-f amplifier, second detector, and audio amplifier.
3. A receiver with an intermediate frequency of 262 kc is tuned to different stations at 660 kc, 880 kc, 1,010 kc, and 1,560 kc. For each of these, give the resonant frequencies for the (*a*) r-f signal circuits, (*b*) local oscillator, (*c*) i-f amplifier.
4. What is a typical value of grid-leak bias voltage for the local oscillator? How would you measure it?
5. How would you adjust the oscillator padder and trimmer capacitors?
6. Describe briefly two types of tuning indicators.
7. Give three requirements of an AVC circuit.
8. What is meant by delayed automatic volume control? By quiet automatic volume control?
9. Give the function and typical value for the following components in the receiver circuit in Fig. 11·22. $C_{20}$, $C_{19}$, $C_{17}$, $C_{16}$, $C_1$, $R_2$, $R_8$, $R_9$, $R_{12}$, $R_{10}$.
10. Referring to Fig. 11·9: (*a*) How much is the voltage across $R_2$ for the base? (*b*) Then how much is the emitter voltage $V_E$ for 0.3-volt forward bias? (*c*) With 8-ma $I_C$ how much is the voltage drop across $R_4$? (*d*) Then how much is the collector-emitter voltage $V_{CE}$?

# Chapter 12 Receiver circuits

Practically all modern receivers use the superheterodyne circuit. This unit shows how the superheterodyne principle is applied in many types of receivers, including transistor receivers, television receivers, and FM receivers. For FM signals, the additional requirements of a limiter in the i-f amplifier and discriminator for the FM detector are explained. Also, the sections of a television receiver are described, showing how they all work together to reproduce the picture. The topics are as follows:



## *12 · 1 A-c/d-c receivers*

Probably the most common type of radio receiver is the small a-c/d-c set (see Fig. 12 · 1). These receivers operate on either alternating or direct current. To achieve such operation, the receiver must operate without a power transformer, since direct current cannot pass through a transformer. Just what is done with the transformer gone will be seen as we analyze the operation of a typical a-c/d-c radio receiver.

**Frequency converter.** The schematic circuit of this receiver is shown in Fig. 12 · 2. The first stage is a frequency converter, using a 12BE6, which receives the incoming signal at grid No. 3. This signal is picked up initially by a loop antenna mounted at the rear of the chassis. It is also possible to use a ferrite-core antenna in place of the loop with equally good results. In such antennas, a number of turns of wire are wound around a long ferrite core (Fig. 12·3).

The oscillator signal required by the converter tube is developed by the circuit between the cathode and first grid. Energy is coupled from $L_1$ to the grid by means of a few turns of wire wound around $L_1$. These few turns are labeled $L_3$ in Fig. 12·2, and the voltage developed in them is the voltage applied to grid No. 1. By leaving one end of $L_3$ open, we obtain the same effect as a capacitance in the grid circuit and this enables us to develop d-c grid bias here. A bias of $-10$ volts is indicated, although somewhat lower values will also be found.

Before we leave the converter stage, it should be noted that AVC bias is applied to the signal grid (grid No. 3) via $R_2$ and the loop antenna. The purpose of this voltage is to regulate the gain of the stage so that a steady output is obtained for received signals of different strengths.

**The i-f stage.** The signal and oscillator voltages mix in the converter stage and produce an i-f signal of 455 kc. The signal is then applied to a 12BA6 i-f amplifier, where it is further strengthened and then passed on, by $T_2$, to a diode detector. The i-f tube is operated with very little grid bias, here $-0.9$ volt, and this is supplied by the AVC line. The only other d-c voltages brought to this stage are the screen-grid and plate voltages, the screen grid connecting directly to the B+ line while the plate receives its voltage through the primary winding of $T_2$.

Both i-f transformers $T_1$ and $T_2$ use small movable cores to tune each winding to resonance. When all the cores are properly adjusted, the 12BA6 receives maximum signal from the converter stage and transfers maximum signal to the following diode detector.

**Second detector.** A simplified circuit diagram of the detector is shown in Fig. 12·4. Current flows from the cathode to the plate in the tube, then through the transformer secondary winding to the volume control, through the resistance of this control to ground, and then back to the cathode of

*Fig. 12·1 Typical table model a-c/d-c receiver.* (Zenith Radio Corp.)

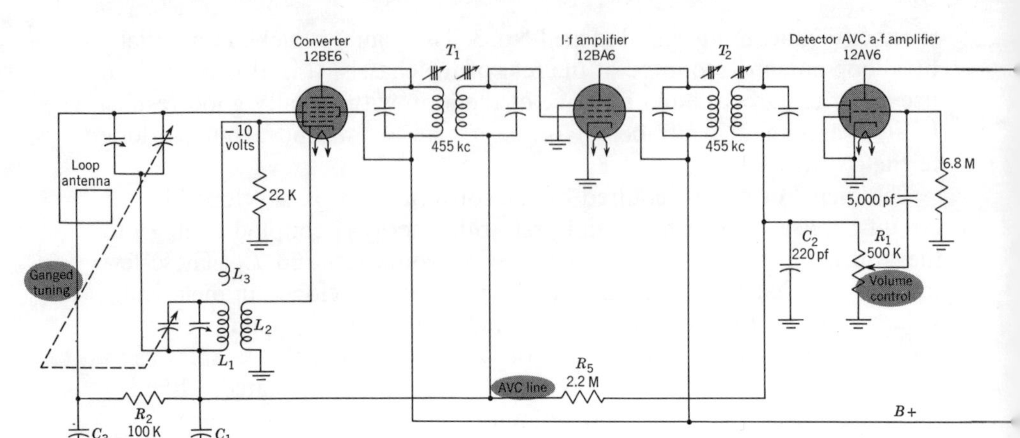


*Fig. 12 · 2 Schematic diagram of typical a-c/d-c receiver.*

the diode again. $C_2$, a 220-pf capacitor, bypasses the i-f carrier voltage away from $R_1$, preventing it from developing any voltage across the control. What does appear across $R_1$ is the audio signal and a d-c voltage proportional to the average level of the i-f carrier.

**Audio system.** The center arm of the volume control taps off as much audio voltage appearing across $R_1$ (Fig. 12 · 2) as required by the audio system and applies the voltage to the control grid of the 12AV6. Here, the signal is amplified, passed on to the audio-output tube, a 50C5, amplified again, and then sent to the loudspeaker through $T_3$. $R_3$ is the load resistor for the 12AV6, while $R_4$ is the input grid resistor for the 50C5. $C_4$ is a coupling capacitor, designed to transfer the signal from $R_3$ to $R_4$, at the same time preventing the B+ voltage present at the plate of the 12AV6 from reaching the control grid of the 50C5.

R-f bypass capacitors $C_5$ and $C_6$ serve to remove any i-f signal that may pass through the triode section and reach this point in the circuit. Finally, $C_7$ is designed to attenuate some of the higher audio-signal frequencies in order to give the sound produced by the loudspeaker a more mellow tone.

**The AVC network.** Since a d-c voltage is developed across $R_1$ which is proportional to the strength of the incoming signal, this voltage can be used to control automatically the gain of the frequency converter (12BE6) and the i-f amplifier (12BA6). $R_5$ and $C_1$ filter out the audio portion of the voltage obtained from $R_1$, permitting only the d-c voltage to be fed to the controlled tubes. $R_2$ and $C_3$ provide additional filtering for the AVC line.

**Power supply.** The power supply for this receiver is typical of power

*Fig. 12 · 3 Ferrite loopstick antenna.* (J. W. Miller Co.)

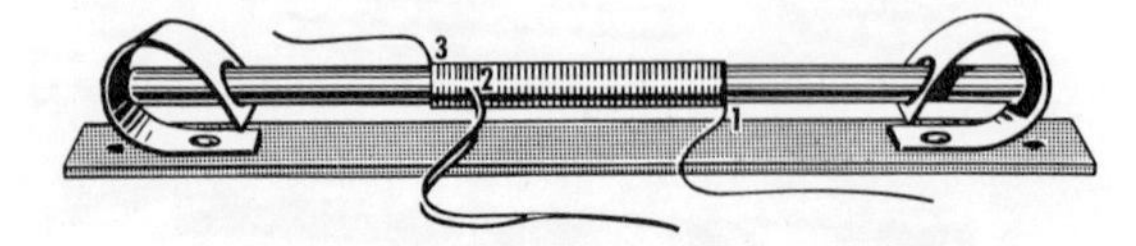

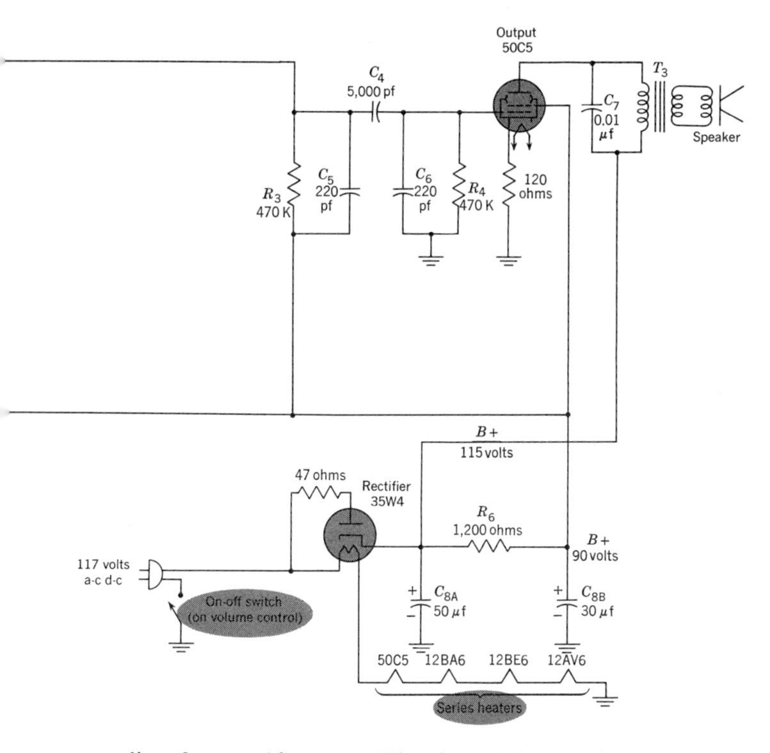


supplies for a-c/d-c sets. The 35W4 is a half-wave rectifier, permitting current to flow through the tube when the plate is positive with respect to the cathode. During the time the tube does conduct, electrons flow from the negative side of the power line to the ground conductor and from here to one plate of filter capacitor $C_{8a}$. The pile-up of negative charge on this plate forces electrons from the other plate of $C_{8a}$ to flow to the cathode of the 35W4, from here to the plate of the tube, and then back to the power line, completing the circuit. In this way, capacitor $C_{8a}$ becomes charged up.

During the half-cycles of the applied line voltage when the 35W4 does not conduct because the plate is negative with respect to the cathode, $C_{8a}$ tends to charge up $C_{8b}$. This is possible because $R_6$ and $C_{8b}$ are connected

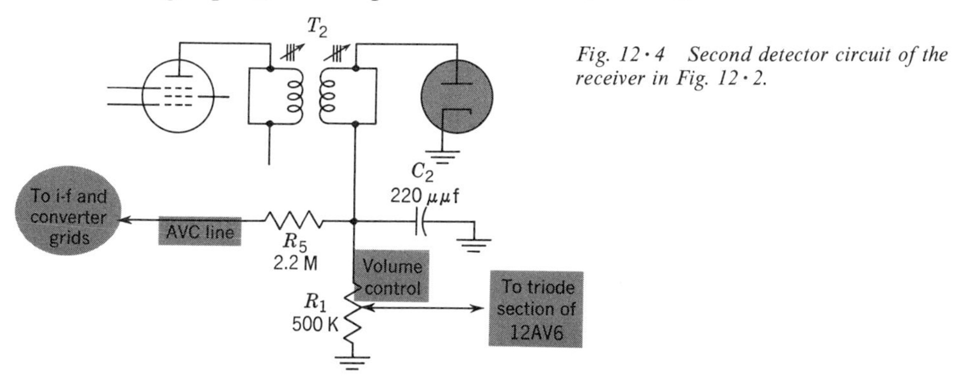


*Fig. 12·4 Second detector circuit of the receiver in Fig. 12·2.*

across $C_{8a}$. Since $C_{8a}$ has a charge, electrons will flow from $C_{8a}$ to $C_{8b}$, causing $C_{8b}$ eventually to become charged to a voltage close to that of $C_{8a}$. In the present circuit, Fig. 12·2, the voltage across $C_{8b}$ is 90 volts, while the voltage across $C_{8a}$ is about 115 volts.

Each half cycle, when the 35W4 conducts, it recharges $C_{8a}$ and through this $C_{8b}$, replacing the charge drawn off from these capacitors by the rest of the receiver circuit. In this way, the voltages across $C_{8a}$ and $C_{8b}$ remain constant. Incidentally, $C_{8a}$ and $C_{8b}$ are given similar capacitor numbers because they are housed in the same case.

**Series tube filaments.** In an ac/dc circuit, with no power transformer, all the filaments are connected in series. When their voltages add up to the line voltage, they are simply connected across the power line. Thus, in Fig. 12·2, we have a total of

35 for 35W4
50 for 50C5
12.6 for 12AV6
12.6 for 12BA6
12.6 for 12BE6
122.8 volts

This is close enough to the normal line voltage of 117 volts to permit satisfactory operation. (With a panel lamp, total voltage drop across the 35W4 heater is 32 volts.)

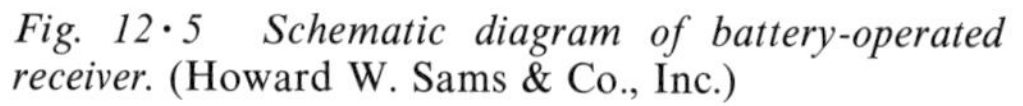

*Fig. 12·5 Schematic diagram of battery-operated receiver.* (Howard W. Sams & Co., Inc.)

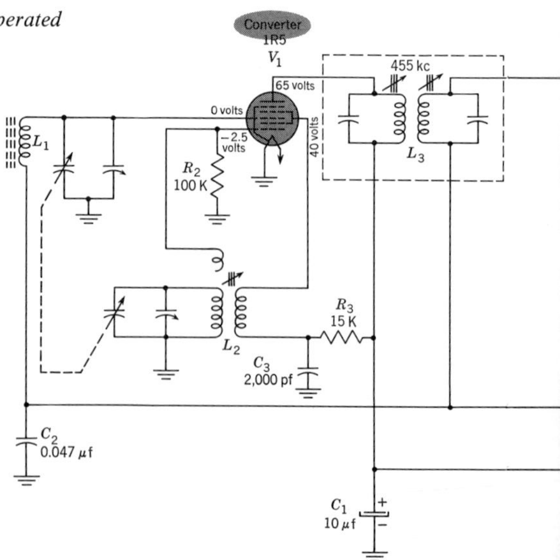

If, however, it happens that the total filament voltage required is less than 117 volts, then a voltage-dropping resistor is placed in series with the tube filaments. To illustrate how the resistance value would be determined, assume that all the filaments of a certain receiver require only 85 volts. The difference between 85 and 117 is 32 volts, and the series resistor is needed to absorb this difference. To compute the resistor value, we need to know how much current will be drawn by the tube filaments. From a tube manual, we determine that the tubes in this circuit require a current of 0.15 amp. Then, using Ohm's law,

$$E = IR$$

or

$$R = \frac{E}{I} = \frac{32}{0.15} = 213 \text{ ohms}$$

Actually, a 215-ohm resistor would be used here, rated at 10 watts.

In any series arrangement, each tube filament must operate with the same amount of current. Tubes are designed specifically with this requirement in mind so that simple series arrangements can be used. If one or more tubes require different filament currents, shunt resistors would have to be used for the excess current.

## 12·2 Portable tube receivers

The field of portable receivers is now completely dominated by transistors, and tube-operated sets are, for the most part, no longer being manufactured.

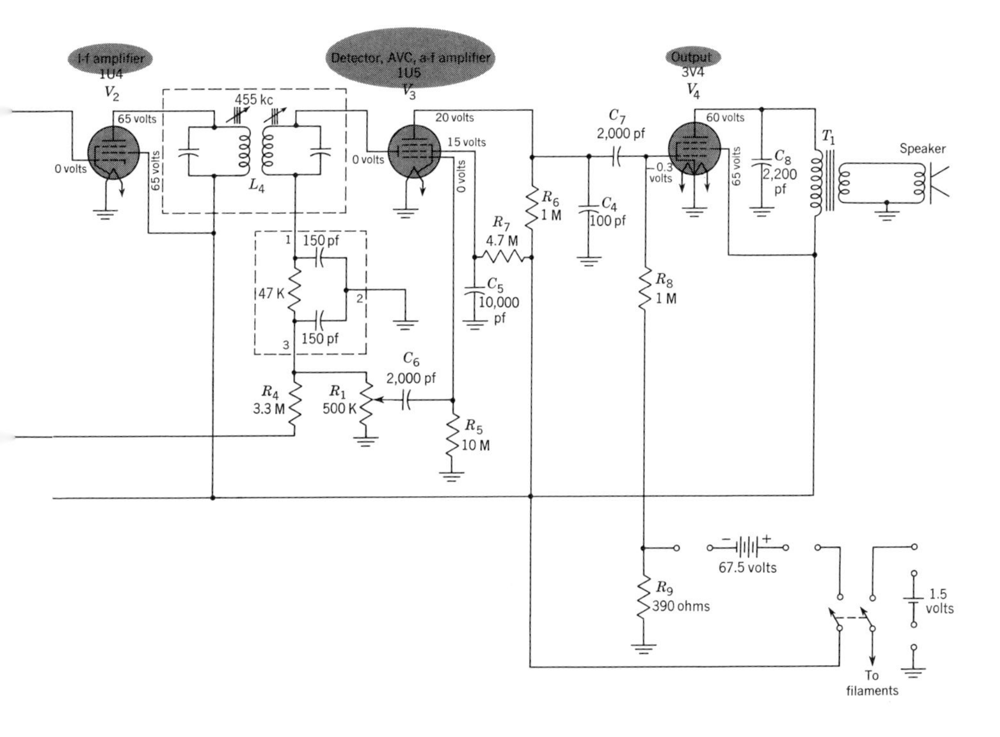

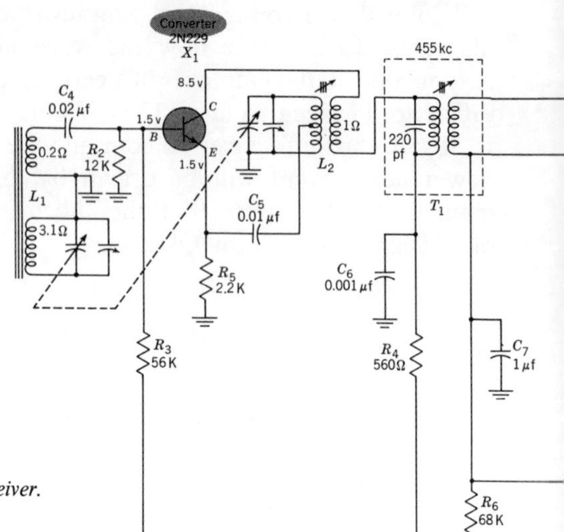


*Fig. 12·6 Schematic diagram of transistor receiver.* (Howard W. Sams & Co., Inc.)

However, since there are a substantial number of portable tube receivers still in use, let us briefly examine one such typical unit to see the arrangement of its circuits and the tubes which are used.

The schematic diagram of a completely battery-operated receiver is shown in Fig. 12·5. The sequence of stages is similar to Fig. 12·2 except for the tube types. These require very low filament voltages and only 50 ma of current. The tubes possess no cathodes because these would have to be heated by the filaments, necessitating higher filament voltages in order to bring the cathodes to a suitable operating temperature. Also, by eliminating cathodes, we shorten the time required by a tube to reach an operating condition. That is why battery receivers respond almost as soon as the power is turned on.

To power the circuit of Fig. 12·5, two batteries are used. One battery provides the voltages for the plates and screen grids of the several tubes. It is known as a *B battery,* with voltages ranging from 67½ to 90 volts.

The other battery, the *A battery,* has a terminal voltage of 1½ volts, and it powers the tube filaments. All the tubes shown require 1.4 volts, which is close enough to the 1½ volts of the battery to permit its use. This is true even of the 3V4, where the filament is constructed in two sections. Each section requires 1.4 volts, and they can be connected in series or parallel.

## *12·3 Portable transistor receivers*

The portability feature is further enhanced by the fact that only a small battery is required for power. Filament-heating batteries, which vacuum

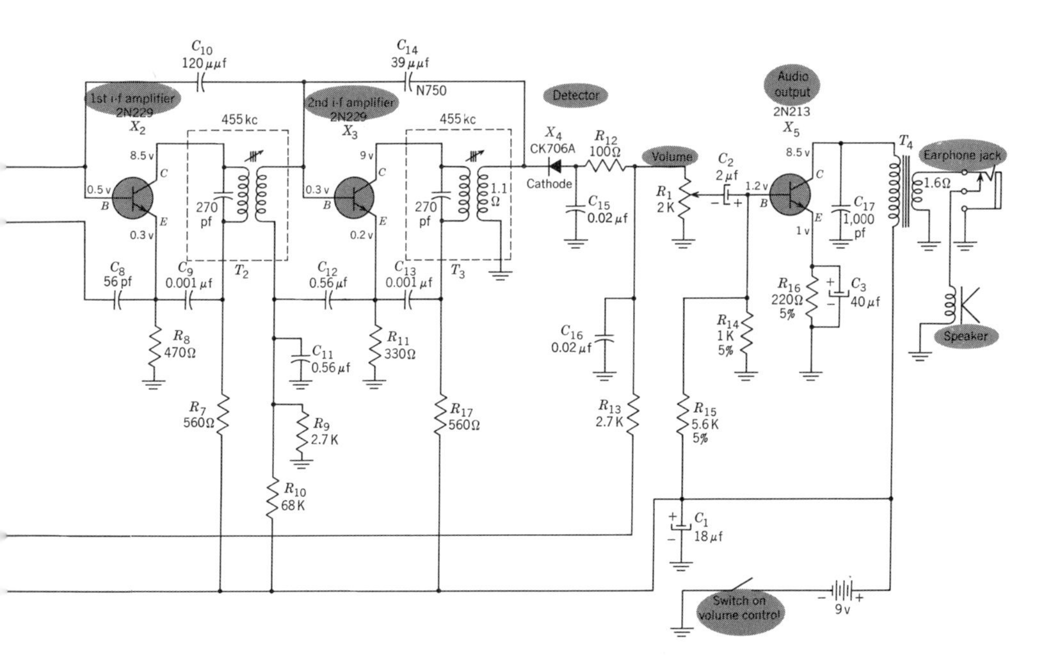


tubes require and which occupy a fair amount of space, are not needed with transistors.

The schematic diagram of a typical receiver is shown in Fig. 12·6. There are four transistors and five stages. The extra stage is the second detector, and its function is performed by a germanium diode, serving as a diode rectifier. The transistors are the NPN variety, and two types are used for the converter, i-f, and audio stages.

**Converter.** The first stage, containing transistor $X_1$, is essentially a self-oscillating converter. The input signal is picked up by a tuned ferrite-core coil which possesses a high $Q$. A low-impedance winding on the antenna coil couples the signal to the base of $X_1$.

Local oscillations are generated by a parallel-resonant circuit in the emitter circuit which is inductively coupled to a coil in the collector circuit. The low-impedance emitter is tapped down on the tuned circuit in order to provide the proper impedance match without lowering the $Q$ of the circuit.

The foregoing oscillator arrangement is a fairly common one. (Its equivalent vacuum-tube circuit is shown in Fig. 12·7.) With the incoming signal and the local oscillator voltage both being applied to the converter transistor, the appropriate i-f signal is formed and then fed to transformer $T_1$ and the i-f stages beyond.

A 2,200-ohm resistor is placed in the emitter circuit to provide d-c stabilization against temperature changes and variations among different replacement transistors. The positive voltage which the emitter current

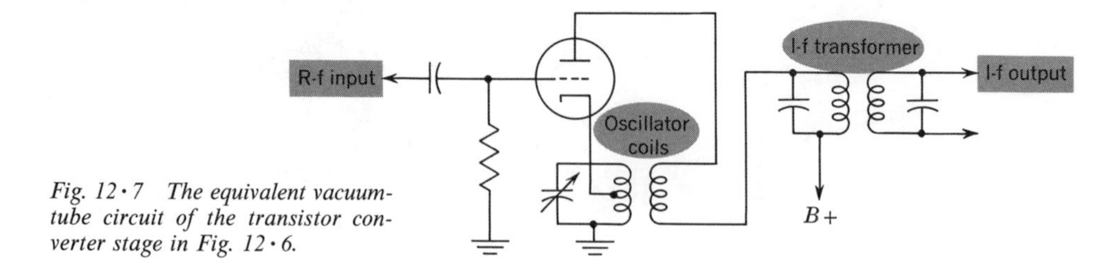


*Fig. 12·7 The equivalent vacuum-tube circuit of the transistor converter stage in Fig. 12·6.*

develops across $R_5$ is counterbalanced by a positive voltage fed to the base from the battery. The actual voltage difference between these two elements is on the order of 0.1 volt.

The proper biasing voltage for the collector of $X_1$ is obtained from a 560-ohm resistor which is tied to the 9-volt B line. A 0.001-$\mu$f bypass capacitor $C_6$ keeps the signal currents out of the d-c distribution system.

**I-f system.** There are two stages in the i-f system, and both are essentially similar. The primary of each i-f transformer is tuned with a fixed capacitor, while the secondary is untuned in order to match the higher collector impedance of the preceding stage to the low input impedance of the following stage. Peaking of each i-f coil is achieved by varying the position of an iron-core slug.

Each i-f stage is neutralized by feeding back a voltage from the base of the following stage to the base of the preceding stage. The feedback occurs through a small series capacitor. In one case this series capacitor is 120 pf, and in the second stage it is 39 pf. Whether an i-f stage will require neutralization depends upon the collector-to-base capacitance of the transistor being used. In special high-frequency transistors, this internal capacitance may be small enough so that the neutralization may not be needed, especially at the lower radio or intermediate frequencies. Where this capacitance is large enough to cause noticeable feedback, however, neutralization, as shown in Fig. 12·6, must be used. The intermediate frequency of the receiver is 455 kc.

Automatic volume control is applied to the first i-f stage only. A negative voltage is obtained from the second detector and applied to the base of $X_2$. Its purpose is to regulate the emitter and collector currents and, with this, the stage gain. When the incoming signal becomes stronger, the negative AVC voltage rises, reducing the collector current of $X_2$ and, with it, the gain. The opposite condition prevails when the signal level decreases. This method is quite effective and provides a wide range of control.

The base bias for the second i-f stage is obtained from $R_{10}$ and $R_9$. This bias voltage is heavily bypassed by $C_{11}$ and $C_{12}$.

Both emitters have d-c stabilizing resistors. If it were not for the presence of $C_7$, $C_8$, $C_{11}$, and $C_{12}$, signal degeneration would occur. As it is, only the direct portion of the current passes through $R_8$ and $R_{11}$. The 470-ohm resistor in the emitter leg of $X_2$ is a compromise between good

AVC action and the d-c stability of the amplifier. A high value of $R_8$ is desirable for stability purposes, but the degeneration that produces the stability would result in reduced gain-control action.

Each of the collectors of $X_2$ and $X_3$ receives its operating voltage through 560-ohm dropping resistors. $C_9$ and $C_{13}$, at the top end of the resistors, serve as decoupling and bypass capacitors.

**Second detector.** The second detector follows the second i-f stage, and its function is performed by a germanium diode. The load resistor for the detector is the volume control. Note the impedance of the control, 2,000 ohms; this low value is needed to match the input impedance of the audio-output stage $X_5$.

**Audio amplifier.** The final amplifier is operated with the emitter grounded through a 220-ohm resistor. Base bias is obtained from the voltage-divider network formed by $R_{14}$ and $R_{15}$. The output transformer matches the 10,000-ohm collector impedance of $X_5$ to the low voice-coil impedance of the miniature speaker. Diameter of the speaker is only 2¾ in. Provision also exists for a small earphone plug, which can be inserted into a small jack on the side of the receiver. When the earphone is in use, the speaker is disconnected.

The total power for the receiver is furnished by a hearing-aid-type 9-volt battery. Total current drain is on the order of 4 ma.

More powerful portable receivers are also available, receivers capable of utilizing exceedingly weak signals to produce a high-volume output. In such receivers there will be found an r-f amplifier stage, several audio-frequency amplifiers, and a push-pull output stage. If short-wave signals, as well as the standard broadcast band, are also to be received, then the converter is usually replaced by a separate r-f oscillator and a separate mixer. Such separation of functions provides more stable operation at frequencies above the standard broadcast band.

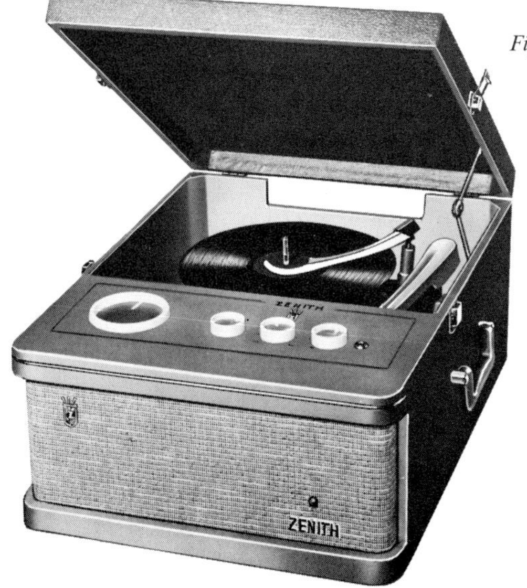


*Fig. 12·8 Radio phonograph.* (Zenith Radio Corp.)

## 12·4 Radio-phonograph combinations

A radio-phonograph combination (Fig. 12·8) brings together a radio receiver and a record changer in one cabinet. This enables the user to listen either to the radio or to his own records. For the latter use, the signal which is developed by the record is fed through the two audio amplifiers of the radio receiver and then made audible by the loudspeaker.

To see how this combination appears electrically, let us examine the circuit diagram of a radio-phonograph combination. With the switch in the radio position, as it is in Fig. 12·9, we have a conventional a-c/d-c receiver. (This switch is located in the center of the diagram.) When it is moved up, it is in the RADIO position; when moved down, it is in the PHONO position. Since this is the point where the circuit departs from a conventional radio receiver, let us examine it in greater detail.

The circuit directly concerned with the phonograph switch is shown in Fig. 12·10. This switch is located between the output load resistor $R_6$ of the diode second detector and the volume control $R_1$. The load resistor has developed across it the audio signal from the incoming broadcast, plus a d-c voltage proportional to the strength of the r-f carrier. This latter

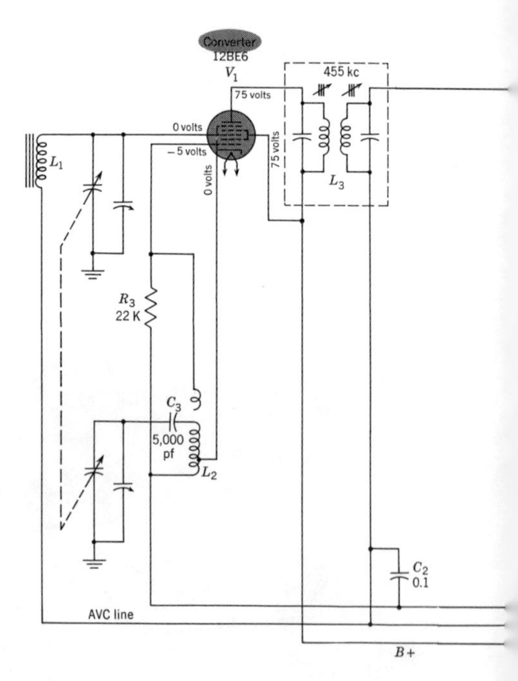


*Fig. 12·9 Schematic diagram of a typical radio phonograph.* (Howard W. Sams & Co., Inc.)

voltage, and its variations produced by changes in signal strength, is used by the AVC circuit ($R_5$ and $C_2$) to control the gain of the i-f amplifier and converter stages.

Let us consider the circuit when the RADIO-PHONO switch is in the RADIO position (Fig. 12 · 10). Tracing out the circuit, we see that the top or ungrounded end of $R_6$ connects to the volume control through the left side of the switch. The right side of the switch grounds the bottom of the oscillator coil $L_2$. This ground enables the oscillator section of the converter circuit to function, so that the proper oscillator frequencies will be developed and fed to the converter for mixing with the incoming signal to produce the intermediate frequencies. Capacitor $C_2$ (not shown in Fig. 12 · 10), is also grounded, and this ground enables the i-f signal to reach the control grid of the 12BA6 i-f amplifier.

Thus, with the switch in the RADIO position, the front-end stages are activated and the demodulated audio signal is transferred from the detector load resistor $R_6$ to the volume control $R_1$. Thereafter, the signal goes to the audio system, where it is made powerful enough to drive a loudspeaker.

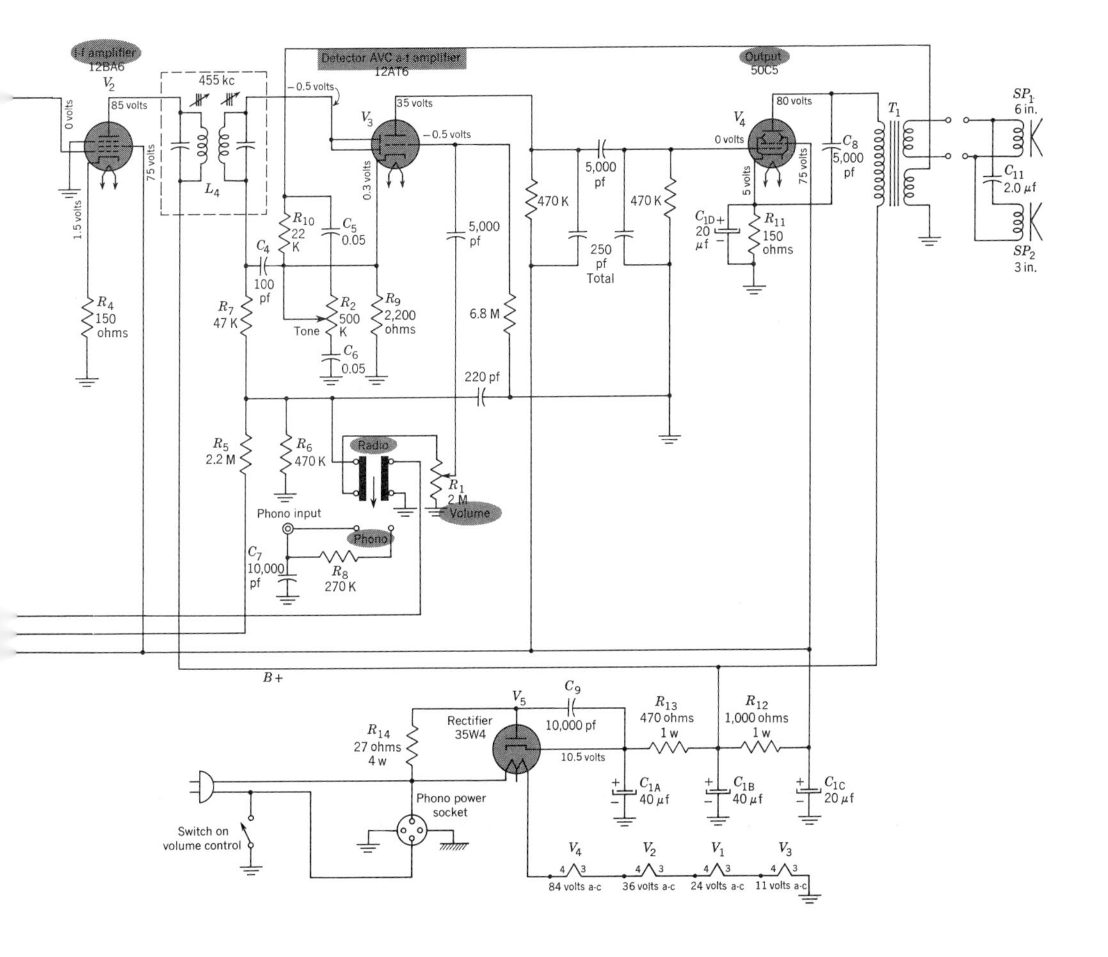

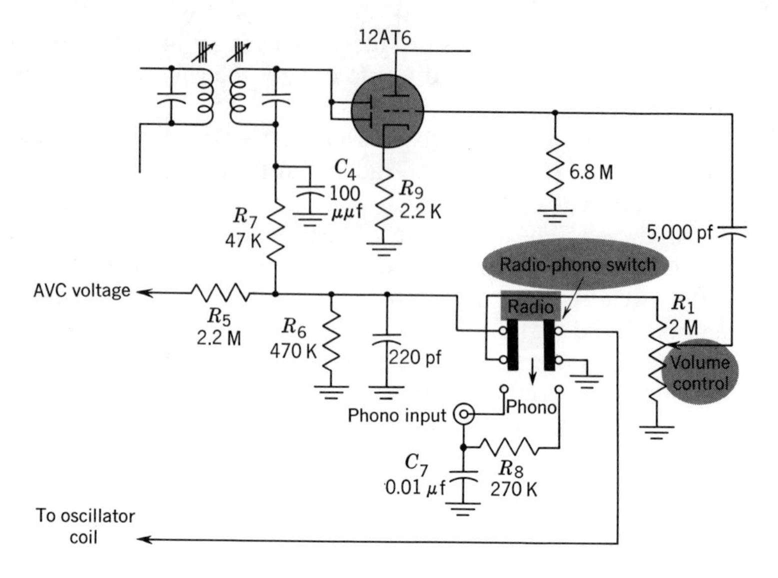


*Fig. 12·10 The* RADIO-PHONO *switch in the* RADIO *position. This is part of the full circuit shown in Fig. 12·9.*

In Fig. 12·9, there are two loudspeakers. One, labeled $SP_1$, is a 6-in. speaker to reproduce the lower frequencies. The other, $SP_2$, is a 3.5-in. speaker for the higher frequencies. To make certain that only the higher audio frequencies reach $SP_2$, a 2.0-μf capacitor, $C_{11}$, is placed in series with the $SP_2$ voice coil. This capacitor attenuates the low-frequency signals more than the high-frequency signals, in essence feeding more high-frequency voltages to $SP_2$. The other speaker, $SP_1$, receives all signals, but since it has a heavier cone, it cannot respond as well to the faster-changing higher audio frequencies and so does not reproduce them as well. Between both speakers, a fairly even response is maintained over a wide range of audio frequencies.

When the RADIO-PHONO switch is shifted to the PHONO position, the circuit appears as shown in Fig. 12·11. Now the path between $R_6$ and the volume control is open, and no broadcast signals can get through to the audio system and the loudspeaker. As further insurance along these lines, the ground connection to the oscillator coil $L_2$ is broken, disabling the oscillator. This effectively inactivates the front-end section of the receiver.

At the same time that the front-end stages of the receiver are being disabled, the input lead from the phonograph is brought into the circuit so that its signal reaches the volume control $R_1$ and from here to the rest of the audio system and the loudspeaker.

The audio signal of the phonograph is developed by a cartridge in the arm of the record changer as the needle rides in the grooves of the record. A crystal cartridge is employed in this system because it can develop enough voltage to drive the audio stages for full power output.

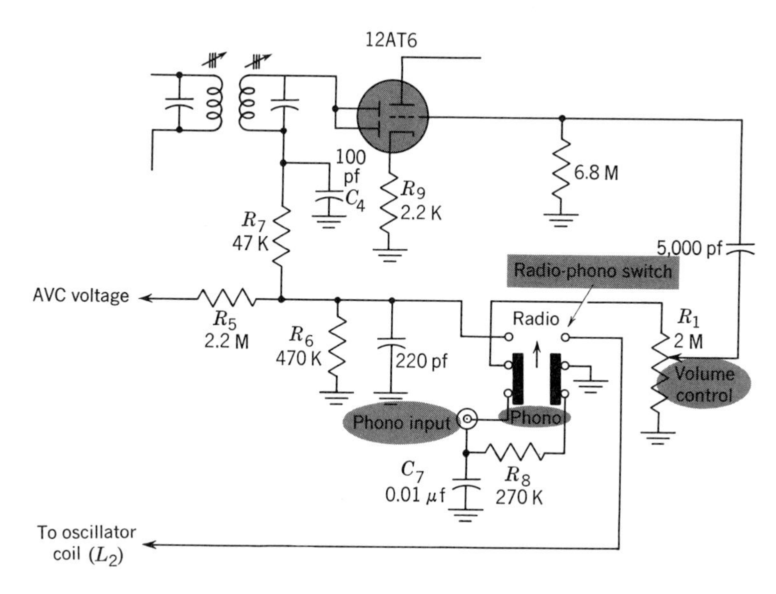


*Fig. 12·11 The* RADIO-PHONO *switch in the* PHONO *position. This is part of the full circuit shown in Fig. 12·9.*

A-c power to drive the motor of the record changer is obtained from the same line that brings power to the receiver power supply. A separate switch is provided so that the motor can be turned on when it is needed.

## *12·5 AM-FM receivers*

Another popular type of combination receiver is the AM-FM receiver (see Fig. 12·12). This set is designed to receive AM radio stations on the normal broadcast band, 535 to 1,605 kc, and FM signals on the 88- to 108-Mc band. In short, we have here two entirely different receivers which operate in different fashions on two entirely different signals. In the ensuing discussion, we will examine an AM-FM receiver to see what circuitry is required by each type of signal and how one over-all receiver design is achieved.

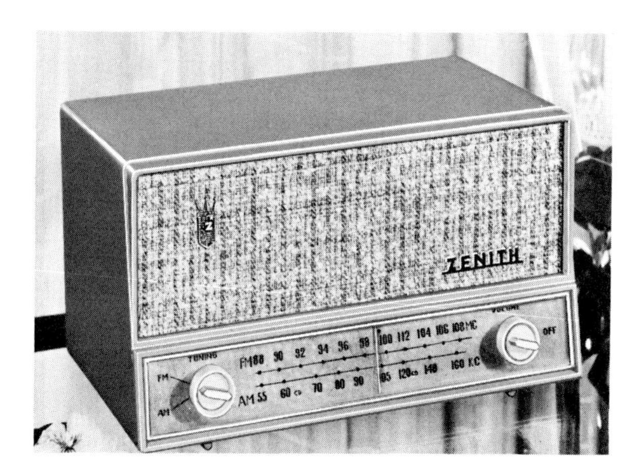


*Fig. 12·12 An AM-FM receiver. Note the two separate frequency scales.* (Zenith Radio Corp.)

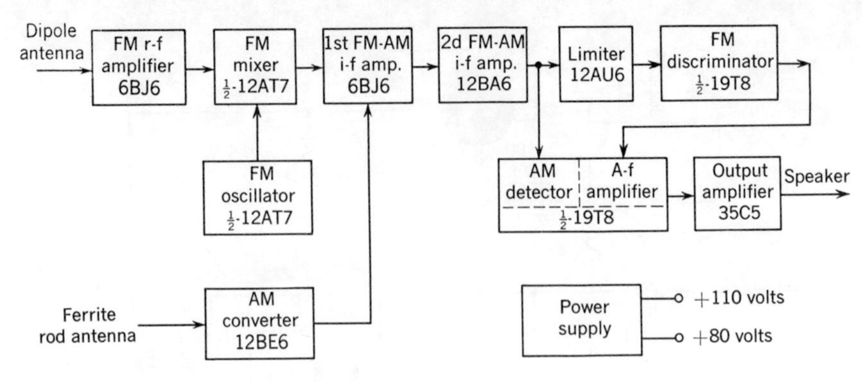


*Fig. 12 · 13 Block diagram of the AM-FM receiver circuit shown in Fig. 12 · 14.*

**Block diagram.** Before we consider the actual receiver circuit, let us examine a block diagram (Fig. 12 · 13). We see that the FM signal is received at an r-f amplifier, then transferred to a mixer stage. At the same time, a locally generated signal is developed in a separate oscillator and fed to the mixer where it combines with the incoming signal. The result of this mixing is a difference signal which represents the i-f voltage. The frequency of this i-f signal is generally 10.7 Mc for FM receivers. Thus far, the FM signal has been treated, in all respects, like an AM signal

*Fig. 12 · 14 Schematic diagram of AM-FM receiver. A full explanation of its operation is given in the text. Note that points 1, 2, 3, and 4 on facing pages are connected.*

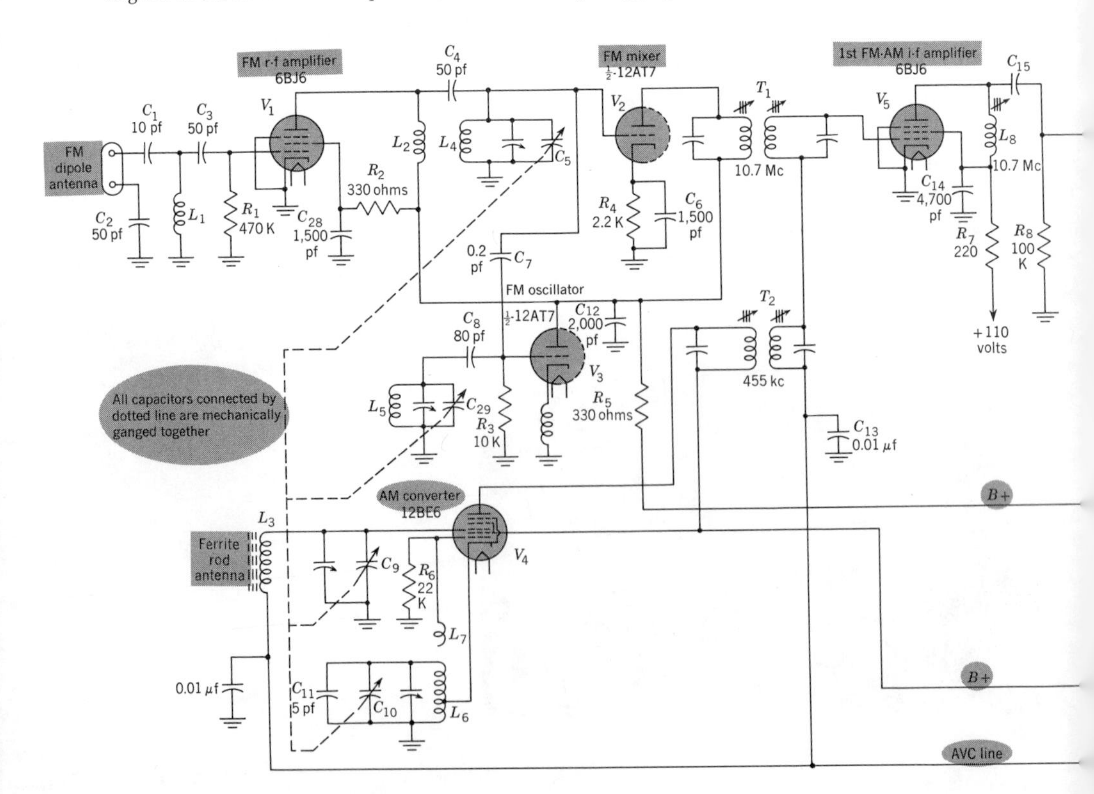

except that the operating radio frequency is much higher, being in the range from 88 to 108 Mc.

The FM i-f signal is now passed through two i-f amplifiers and then through a limiter, where any amplitude variations in the signal are removed. Next, the FM signal is applied to an FM detector, where it is converted back to audio. After this, it is fed to a conventional audio voltage amplifier, then to an audio power amplifier, and finally to the loudspeaker.

Now, consider the AM portion of this receiver. The incoming AM signal is picked up by a loop or ferrite-rod antenna and applied to a converter tube, where an intermediate frequency of 455 kc is produced. This signal is then fed to the same two i-f stages, where it receives additional amplification. In spite of the difference in frequencies between the AM and FM i-f signals, the same tubes serve each satisfactorily, thereby reducing the cost compared with separate i-f systems.

At the output of the second i-f amplifier, the AM signal is fed to a diode detector, where the audio is extracted from the signal. Then the audio signal is applied to the same audio voltage-and-power amplifiers as the FM signal and, from here, to the same loudspeaker.

Note, then, that the AM and FM signals use four stages in common, a feature which greatly aids in reducing the over-all cost of this combination. Furthermore, all stages use a common power supply and this, too, provides a saving.

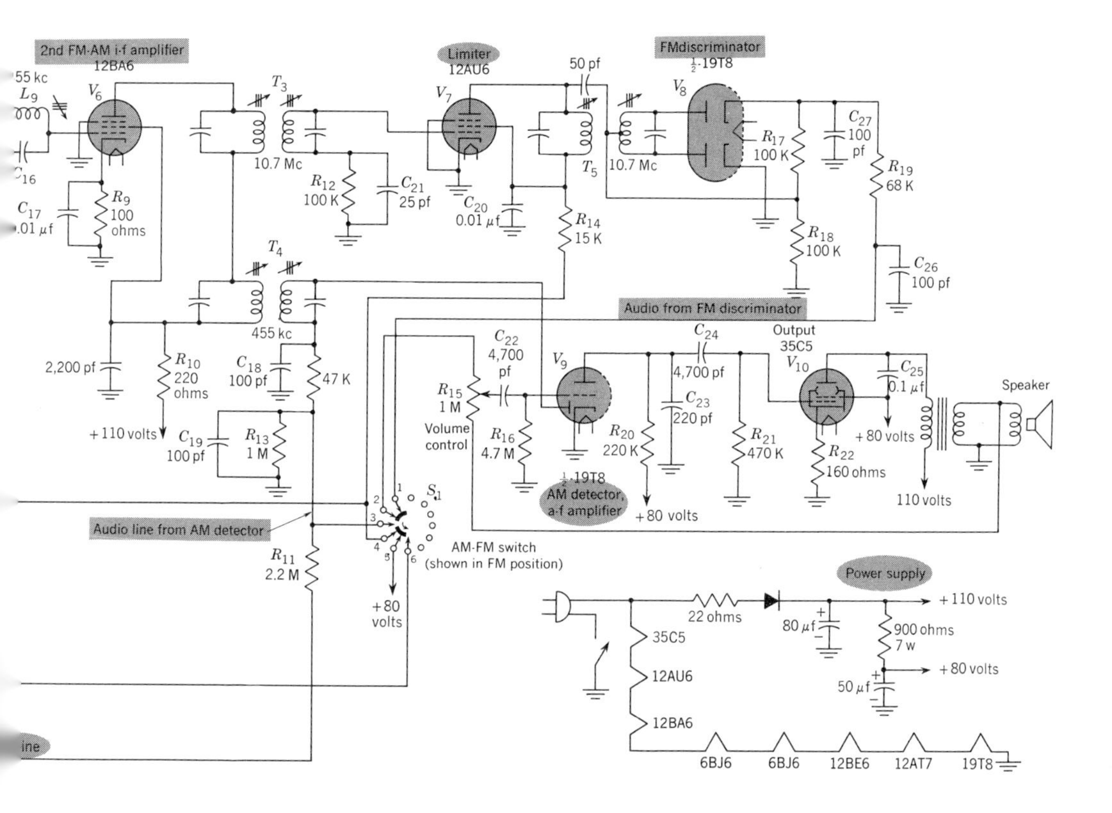

It should be understood that both systems do not operate at the same time. Only one functions at a time, so that no interference develops between them.

**AM section.** The schematic diagram for this AM-FM receiver is shown in Fig. 12 · 14. The various stages are labeled, and these should be tied in to the previous block diagram for clarification and to better follow the paths taken by either signal (AM of FM) through the system. As a start, let us consider the AM section first, since it is the more familiar of the two. The AM signal is picked up by a ferrite-rod antenna $L_3$ and applied to grid No. 3 of the 12BE6 converter. At the same time, the oscillator portion of this tube, formed by grid No. 1, the cathode, grid 2, and grid 4, provides an oscillator voltage which mixes the incoming signal to produce a 455-kc i-f signal. This voltage is fed to the primary of $T_2$ and then inductively coupled to the secondary.

The voltage which is developed across the secondary of $T_2$ is applied to the control grid of $V_5$. It does this through the secondary winding of the FM interstage transformer $T_1$, located just above $T_2$. This latter unit is tuned to 10.7 Mc, which means that it contains fewer windings than transformer $T_2$, owing to the fact that inductive reactance ($X_L$) is equal to $2\pi fL$, and, to maintain the same reactance as we increase the frequency, we must lower the coil inductance ($L$). Thus a coil which has a high inductive reactance at 10.7 Mc will possess very little reactance at 455 kc; consequently, the 455-kc signal on the secondary of $T_2$ passes through $T_1$ quite readily without any loss of voltage.

The first i-f stage $V_5$ amplifies the AM signal and passes it on, by means of a tuned circuit, to the control grid of the next i-f amplifier $V_6$. Between $V_5$ and $V_6$, there are two tuned circuits, one for 10.7 Mc and the other for 455 kc, and neither interferes with the other. At $V_6$, the AM signal is again amplified, and then it appears across the primary of $T_4$. Note that to reach $T_4$, the signal must pass through the primary of the 10.7-Mc transformer $T_3$, but for the same reasons stated for $T_1$, this is done quite readily.

From the primary of $T_4$, the AM signal is inductively coupled to the secondary and from here to the diode plate in the 19T8 tube. The AM signal is now detected, and the resultant audio voltage is transferred from the diode load resistor $R_{13}$ to the volume control through switch $S_1$.

Note that the 19T8 is a multipurpose tube. Two diode sections, each with its own cathode, are employed in the FM detector stage $V_8$. Another diode plate, using the same cathode as the triode $V_9$, serves as the AM detector. The remaining section, a triode, is the first audio amplifier. Thus, three separate functions are performed by a single tube.

The audio voltage obtained from the AM signal is fed from the volume control to the grid of the triode section of the 19T8. Here the signal is amplified, then transferred to the 35C5 output amplifier, and from this stage to the loudspeaker. Thus, the AM signal is treated in the same fashion in this receiver as it is in a purely AM set, except that it receives amplification from two i-f stages instead of the normal one. It is possible in some AM-FM combinations, however, to find that the AM signal is passed through only one i-f stage. Hence the fact that two i-f stages are employed in Fig. 12·19 should not be taken as a general rule for AM-FM combination receivers.

**FM section.** We are now ready to consider the FM section of this receiver. The FM carrier, with a frequency between 88 and 108 Mc, is generally received by a dipole antenna, either indoor or outdoor, and brought to the control grid of the r-f amplifier $V_1$. The input circuit is not tuned in this receiver, and the r-f amplifier amplifies all the FM signals that may be present on the antenna. Signal selection is actually done by $L_4$ and $C_5$ in the grid circuit of the following mixer $V_2$, as the signals pass from $V_1$ to the control grid of $V_2$.

Capacitors $C_5$ and $C_{29}$ are rotated in unison so that the frequency generated by the oscillator $V_3$ will possess the proper value to mix with the incoming signal and produce a difference frequency of 10.7 Mc. These two stages must keep in step with each other; otherwise, the correct intermediate frequency will not be obtained.

It will be noted that the r-f oscillator in the FM section employs a separate tube, whereas in the AM section, one tube functions as the mixer and the oscillator. The use of separate tubes for the mixer and oscillator (or completely separate sections in a multipurpose tube) is a common practice in high-frequency circuits. This is because it leads to greater oscillator stability, preventing the mixer from influencing either the frequency generated by the oscillator or the amplitude of this voltage. Whatever voltage the oscillator must transfer to the mixer (for the mixing operation), it transfers through a small capacitor, such as $C_7$ in Fig. 12·14. This capacitor value is only 0.2 pf, and a conventional capacitor may not even be used. Instead, a wire from the control grid of the oscillator may simply be twisted around a wire connecting to the control grid of the mixer. Since the two wires are both covered with insulation, no direct electrical contact is made but because the wires are twisted about each other, a capacitance is set up between them. It is this capacitance which is represented by $C_7$.

After the 10.7-Mc signal is developed in the mixer, it is fed to transformer $T_1$ and, from here, to the first i-f stage $V_5$. So far as the high-frequency FM signal is concerned, $T_2$ is not even present, because the high-frequency currents pass easily through the capacitor which shunts the secondary winding of $T_2$. Thus, as we saw before, $T_1$ does not hinder the 455-kc AM signal when the latter appears across $T_2$, and $T_2$ does not interfere with the 10.7-Mc signal when it is present across $T_1$. In this way, signals are able to use the same amplifier (although not at the same time) without mutual interference.

The FM 10.7-Mc signal is amplified by $V_5$ and is passed on to $V_6$, using $L_8$ as its resonant circuit, which tunes with the stray capacitance. $V_6$ amplifies the signal and transfers it to $V_7$, a limiter, through transformer $T_3$. The limiter stage is an i-f amplifier for the 10.7-Mc FM signal but has special operating characteristics.

**AM-FM changeover switch.** A special switch (located to the left of the volume control in the schematic diagram, Fig. 12·14) serves to switch the receiver between amplitude modulation and frequency modulation as desired by the listener. For example, the switch is presently shown in the FM position. For this condition, contact 1 is making connection with contact 2, and this brings the audio voltage output of the FM discriminator to the volume control. Contact 3 is open, but contact 4 is making connection with contact 5. This brings 80 volts, B+, to the FM r-f amplifier, the FM mixer, the FM oscillator, and the limiter. These are all stages solely concerned with the FM signal. Note that contact 6 is open, and since this contact, when it makes connection with contact 5, brings B+ to the 12BE6 converter, the latter stage is rendered inoperative. In this way, AM signals are prevented from entering the receiver.

For AM operation, the switch is turned in the direction of the small arrow, at the center of the switch. This arrow points in the counterclockwise direction. When the changeover is made, you must mentally move the internal switch contacts counterclockwise one position. (NOTE: Not 1, 2, 3, 4, 5, and 6, but the two small metal shaded stubs.) When this is done, contact 2 now connects to contact 3 instead of contact 1. (Contact 1 is open.) This brings the audio output of the AM detector instead of the output from the FM detector to the volume control. Furthermore, contact 5 now connects to contact 6 instead of 4, and hence B+ is fed to the AM converter. By the same token, it is removed from $V_1$, $V_{2a}$, $V_3$, and $V_7$, inactivating these stages which deal with FM signals only and preventing any FM signal from reaching the system.

In this way, either AM or FM signal operation is obtained.

## *12·6 Limiters*

In the limiter, the signal receives some amplification, but the principal function of a limiter is to clip off (or smooth out) any amplitude variations in the signal itself. Since the FM detector (which follows the limiter) is sensitive to amplitude variations in the FM signal, these amplitude changes will cause a voltage to appear at the output of the FM detector which is not part of the original audio signal transmitted by the broadcast station. Hence this additional voltage represents distortion and it must be minimized as far as possible.

The FM receiver illustrated in Fig. 12·14 uses a limiter stage because the FM detector is a discriminator circuit which responds to amplitude

*Fig. 12·15 A limiter stage removes the amplitude variations from the input signal.*

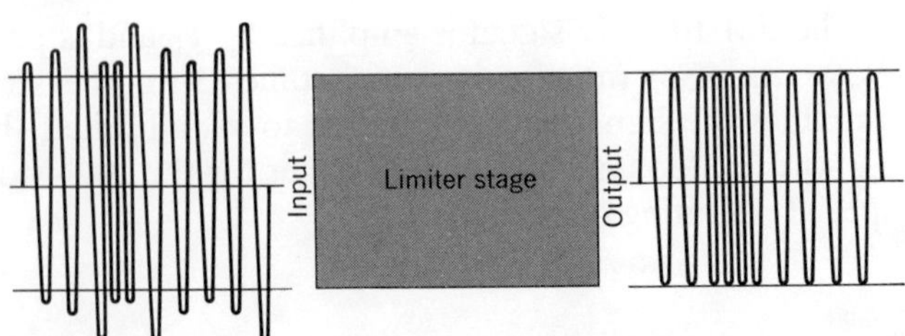

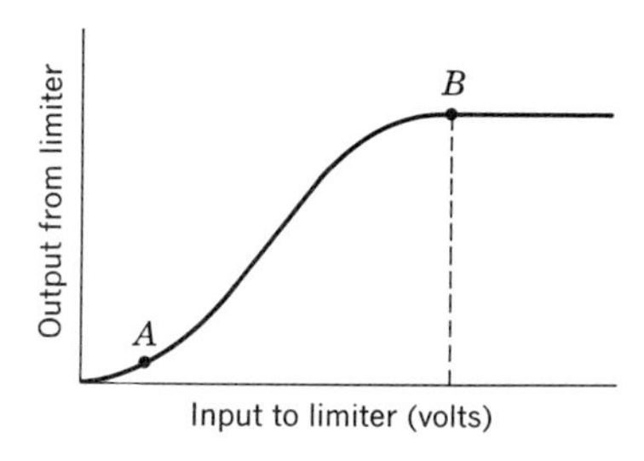

*Fig. 12 · 16 For proper limiting action, the signal reaching the limiter must be strong enough to operate the limiter beyond point B.*

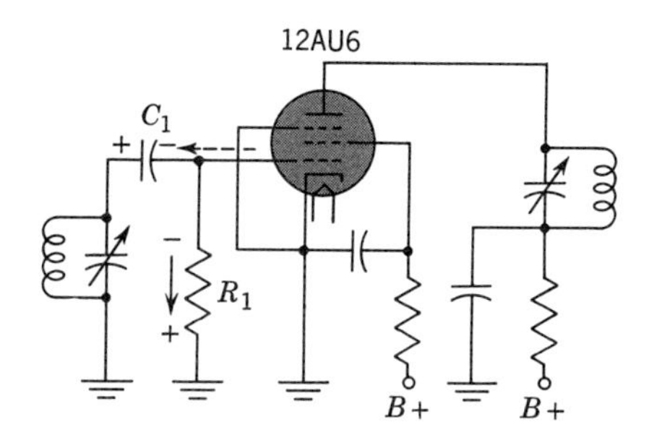

*Fig. 12 · 17 A limiter stage using grid-leak bias.*

variations in the FM signal. However, other types of FM detector circuits do not require a limiter stage.

**Limiter operation.** A limiter amplifier performs its function of removing amplitude modulation by providing a constant amplitude-output signal for a comparatively wide variation in input voltages. A simple illustration is shown in Fig. 12 · 15, where the forms of the input and output voltages of a limiter are indicated. A more exact representation of the ability of a limiter to remove amplitude modulation from a wave is given by the curve shown in Fig. 12 · 16. For all signals possessing more than a certain minimum input voltage at the antenna, the limiter produces a substantially constant output. In this region, starting at point *B* and extending to the right, the stage is purely a limiter in its action.

In the region from *A* to *B*, however, the stage is not driven into saturation, and different input voltages produce different output voltages. Any signal too feeble to drive the limiter beyond (or to the right of) point *B* will cause amplitude variations in the discriminator output and therefore distortion in the FM signal. For proper operation, it is necessary at all times that sufficient amplification be given an incoming signal in order that, when it reaches the limiter, it is strong enough to drive the tube beyond point *B*.

Limiting action in a tube is obtained by using a sharp-cutoff tube together with grid-leak bias and low plate and screen voltages. It is possible to obtain partial limiting using grid-leak bias or low electrode voltages separately, but best results are obtained by combining these two methods.

**Grid-leak-bias limiting.** The principle of operation of a grid-leak limiter is not difficult to understand. Initially, there is no bias, and the tube is operating with zero voltage on its control grid. Upon the application of a signal, the grid is driven positive and current flows in the grid circuit, charging up $C_1$ (Fig. 12 · 17). The capacitor continues to charge throughout the entire positive half of the incoming signal.

During the negative portion of the input cycle, the charge on $C_1$ begins to force its way through $R_1$, back to the other side of the capacitor. The capacitor must discharge because it represents a potential difference and, as long as a complete path is available, current will flow. This is funda-

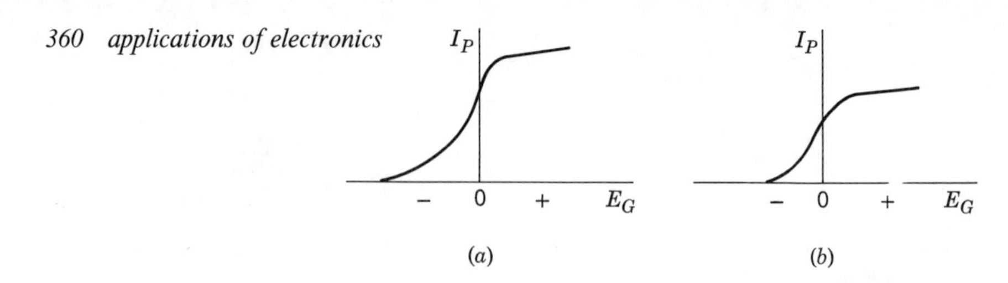


*Fig. 12·18 (a) Normal characteristic curve for a sharp-cutoff tube, such as a 12AU6. (b) Modified curve obtained with lowered plate and screen-grid voltages.*

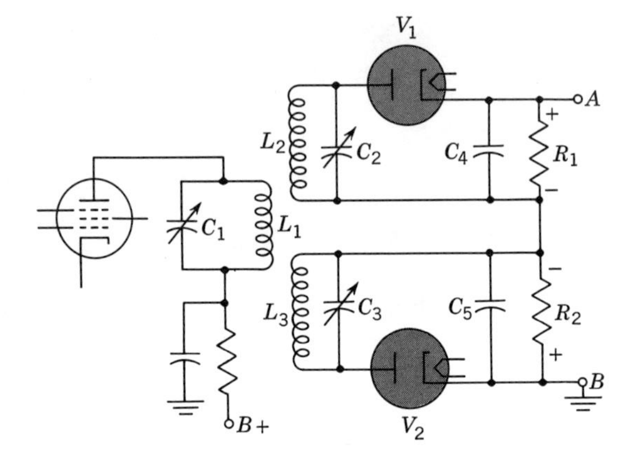


*Fig. 12·19 A triple-tuned FM discriminator. This circuit preceded the Foster-Seeley discriminator.*

mental to all electric circuits. The electrons passing through $R_1$ develop a potential difference that has the polarity indicated in Fig. 12·17. From the moment the input voltage departs from its positive values to the beginning of the next cycle, the discharge of $C_1$ continues.

At the start of the next cycle, there will be some charge remaining on $C_1$. Hence the grid does not draw current again until after the input voltage has risen to a positive value sufficient to overcome the residual negative voltage on $C_1$. When the input voltage becomes sufficiently positive, current flows, recharging $C_1$. This sequence will recur as long as the input voltage is active. Because of the fairly long time it takes $C_1$ to discharge through $R_1$, not all the charge on $C_1$ will disappear during any one cycle of input voltage. Hence each succeeding cycle of input voltage will add a little to what remains from the previous cycle. After a few cycles, a point of equilibrium is reached, and the voltage across $R_1$ remains constant. This establishes the operating bias, and the input voltage will fluctuate around this value.

Limiting is achieved by the way the grid-leak bias varies as the incoming signal varies. For example, when the signal strength increases, so does the value of grid bias developed. This, in turn, makes the grid of the tube more negative, lowering its gain. Consequently, the stronger effect of the signal is counteracted.

On the other hand, when a weaker than normal signal is received, less grid-leak bias is developed, and the negative grid voltage drops. This raises the gain of the tube and acts to produce a stronger output, again counteracting the signal decrease. In this way, various amplitude signals at the input tend to produce the same output signal.

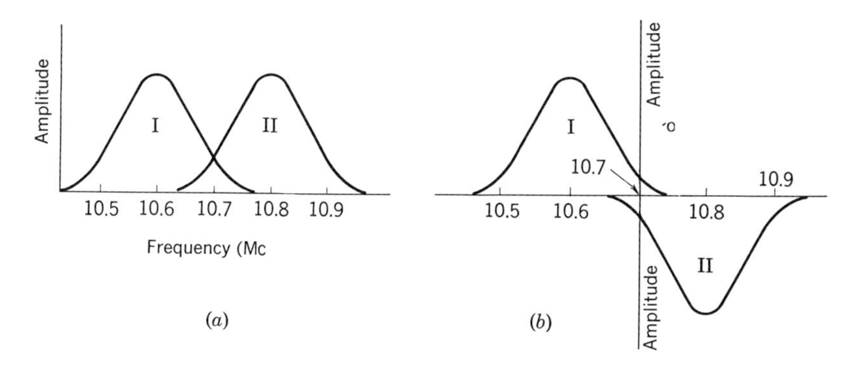


*Fig. 12·20 (a) The secondary response curves of discriminator. (b) The same curves rearranged according to the polarity of the voltages developed across their load resistors.*

**Limiting with low tube voltages.** The saturation or limiting effect of low plate and screen voltages is best seen by reference to the curves of Fig. 12·18. In Fig. 12·18*a*, we have the normal extent of the $E_G I_P$ curve of a sharp-cutoff pentode, say a 12AU6. A sharp-cutoff tube is necessary to remove fully the negative ends of the wave. It would be quite difficult to obtain sharp cutoff with a remote-cutoff tube, and amplitude variations would be present on the negative peaks of the wave.

When we lower the plate and screen voltages, the extent of the characteristic curve is diminished (Fig. 12·18*b*). It requires much less input signal now to drive the plate current into saturation. An FM receiver containing such a limiter would be capable of providing good limiter action with weaker signals than if the tube were operating with full potentials.

In the limiter in Fig. 12·15, we employ low plate and screen voltages plus grid-leak bias to obtain efficient limiting action.

## *12·7 Discriminators*

Beyond the limiter is the FM detector where the frequency variations in the received signal are converted to audio voltages. This stage is frequently called a *discriminator*, and its operation is best understood by first considering a discriminator that preceded the present circuit.

**The triple-tuned discriminator.** The triple-tuned discriminator was an early forerunner of the modern FM detector. It is seldom used today, but it is useful in explaining the operation of the more complex Foster-Seeley discriminator, shown in Fig. 12·14.

The triple-tuned FM discriminator is shown in Fig. 12·19. The $L_1C_1$ circuit of the stage preceding the detector is tuned to the carrier i-f value and inductively coupled to the two secondary circuits $L_2C_2$ and $L_3C_3$. $L_2C_2$ is tuned to a frequency lower than $L_1$ and $C_1$, and $L_3C_3$ to a higher frequency than $L_1C_1$. Each secondary circuit contains a diode rectifier, with a load resistor. The d-c voltages developed across these load resistors are of opposite polarity; consequently, the net voltage taken from the output terminals $AB$ will be the difference between the voltage across $R_1$ and the voltage across $R_2$.

Now, consider how the voltage across $R_1$ changes with the frequency of the applied voltage. If the applied frequency varies from a point below the resonant frequency of $L_2C_2$ to a point above it, the curve of the voltage across $R_1$ with respect to frequency appears as curve I of Fig. 12·20*a*. Maximum voltage is produced across $R_1$ at the resonant frequency of $L_2C_2$. The voltage falls off above and below this frequency.

The voltage across $R_2$ possesses a similar characteristic except that its peaking point is at another frequency (see curve II, Fig. 12·20*a*). The polarity of the d-c voltage across $R_1$ is opposite to that across $R_2$, so the curve of the voltage across $R_2$ with respect to frequency can be placed as curve II of Fig. 12·20*b*. The voltage at terminals $AB$ is equal to the voltage across $R_1$, plus the opposite-polarity voltage across $R_2$.

Combining the two individual graphs, we obtain the S-shaped curve of Fig. 12·21. It is this latter curve that can be called the over-all characteristic of the entire circuit, and it is the response curve of a discriminator.

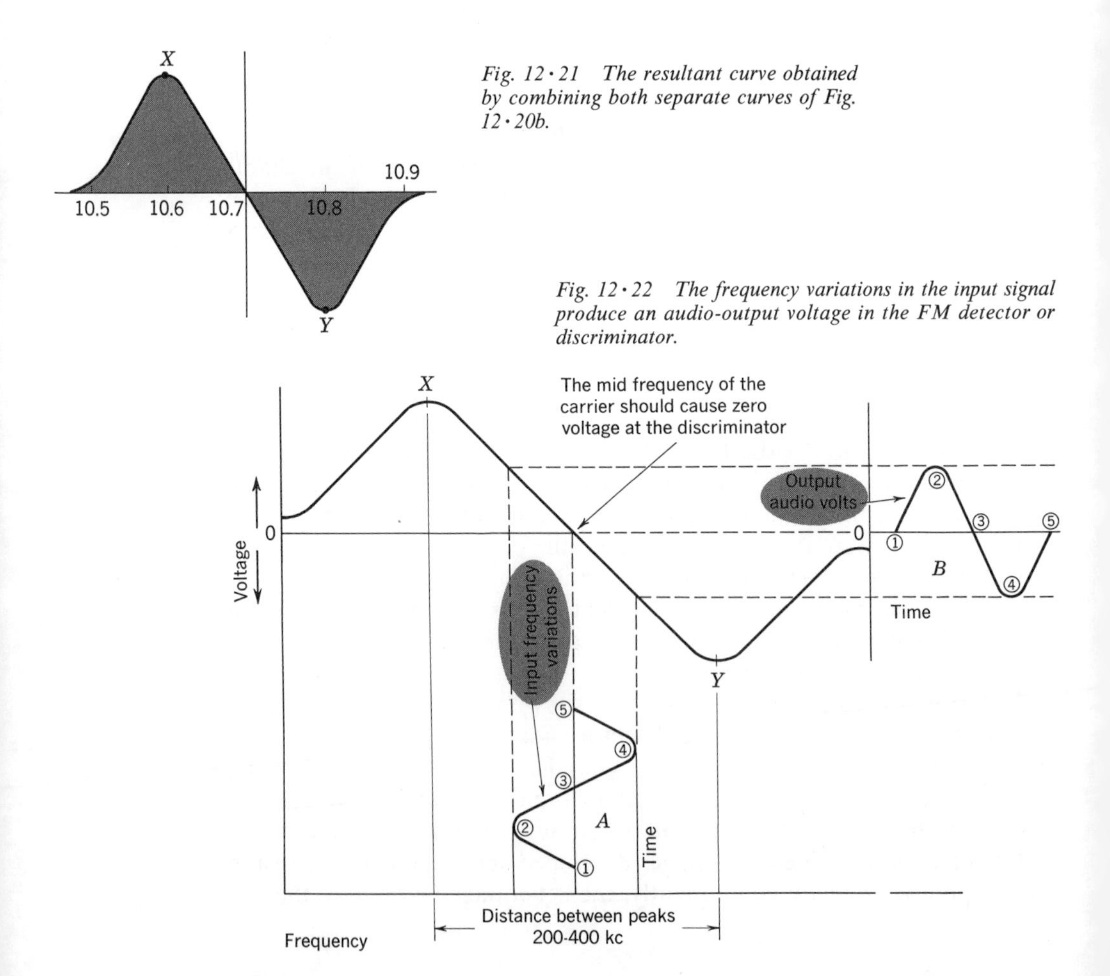


*Fig. 12·21 The resultant curve obtained by combining both separate curves of Fig. 12·20b.*

*Fig. 12·22 The frequency variations in the input signal produce an audio-output voltage in the FM detector or discriminator.*

It indicates how the voltage output changes as the frequency of the incoming signal changes. If an FM signal is applied to the discriminator, the output-voltage curve (*B* in Fig. 12·22) will have the same characteristic as the frequency-varying input signal (*A* in Fig. 12·22). Thus, as the frequency decreases from 1 to 2 in *A*, the voltage output of the discriminator increases from 1 to 2 in *B*. As the frequency increases from 2 to 4 in *A*, the output voltage decreases from 2 to 4 in *B*.

It might be well to repeat that, in an FM signal, the different positions of the signal frequency represent the audio modulation. Hence, if the frequency changes in a sine-wave manner, the output-audio voltage should likewise be a sine wave, as shown in Fig. 12·22.

The discriminator circuit is adjusted so that the mid-frequency of the FM signal will cause zero voltage at the discriminator output. Peaks *X* and *Y* are 200 to 400 kc apart, each equidistant from the mid-frequency of the FM carrier. Thus, if the mid-frequency of the carrier is at an i-f value of 10.7 Mc, one peak would be at 10.6 Mc and the other at 10.8 Mc, making the peaks 200 kc apart. In television receivers, the i-f value is 4.5 Mc. The peaks of the resonant circuits would then be spaced about this center frequency.

**The Foster-Seeley discriminator.** Now let us consider the discriminator which is employed in the receiver of Fig. 12·14. This discriminator, shown separately in Fig. 12·23, is known as a *Foster-Seeley discriminator.* It differs from the circuit of Fig. 12·19 because it has only one tuned circuit in the secondary of the coupling transformer instead of two. Furthermore, $L_1$ and $C_1$ are resonated to the same frequency as $L_2C_2$. At the resonant frequency (generally 10.7 Mc), voltages $E_2$ and $E_3$ are 90° out of phase with primary voltage $E_1$. Also, $E_2$ and $E_3$ are equal. This means that diodes *V8A* and *V8B* have equal voltages applied to them, producing equal voltages across $R_{17}$ and $R_{18}$. The net result of this action, at the frequency, is zero output voltage.

*Fig. 12·23 The FM discriminator employed in the circuit of Fig. 12·14. This is the Foster-Seeley or phase-shift discriminator circuit.*

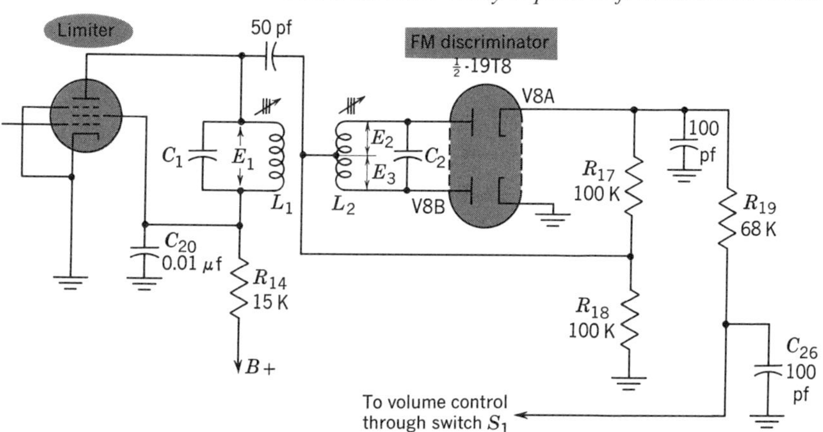

For frequencies above or below 10.7 Mc, the phase between $E_2$, $E_3$, and $E_1$ shifts from 90°, causing one diode to receive more voltage than the other. (For frequencies above 10.7 Mc, one diode, say $V_{8a}$, will receive more signal voltage. For frequencies below 10.7 Mc, the other diode will receive more signal voltage.) Under these conditions, the voltages across $R_{17}$ and $R_{18}$ will *not* be equal, producing resultant output voltages. This action is identical with that which occurred in the previous discriminator, and the net outcome is the same.

Once the audio signal is developed, it is transferred to the volume control, which is now set to receive this signal rather than the audio signal that it previously received from the AM system. Then the audio signal passes through the triode audio amplifier of $V_9$, the power-output amplifier $V_{10}$, and, finally, the loudspeaker (Fig. 12·14).

It is important to note that the discriminator is just one type of FM detector circuit. Furthermore, the discriminator is sensitive to amplitude variations. Other types of circuits are capable of providing AM rejection while detecting the FM signal. Two examples of FM detectors that do not need a limiter are the ratio-detector circuit, which uses a duodiode like the discriminator, and an FM detector circuit that uses a gated-beam tube, such as the 6BN6 or 6DT6.

## 12·8 FM stereo multiplexing

Within recent years, stereophonic records, tapes, and reproducing systems have become exceedingly popular. The added sound dimension greatly enhances the listener's enjoyment, giving him a sensation of immediacy and involvement that he could not achieve with monaural sound. The same desire for stereophonic reproduction has also extended to broadcast reception, and the FM broadcasting industry, in cooperation with the Federal Communications Commission, has developed a method by which such stereophonic reception can be achieved. This system is technically known as FM multiplex (or multiplexing) since it combines or meshes two sets of signals without detriment to either one. As we shall see presently, an FM multiplex signal will produce a conventionally monaural output from a monaural receiver and a stereo output from a stereo receiver. This compatibility is most important since the new signal must not disrupt the operation of the tens of thousands of FM receivers designed to operate only on monaural signals. This is the single signal we are so long familiar with.

In stereo reproduction, an illusion of three-dimensional sound is developed. The listener is provided with nearly the same aural effect in his living room that he would obtain were he sitting in the auditorium where the sound is being developed. In an auditorium, sound produced at the left-hand side of the stage reaches the listener from this section of the stage. By the same token, music formed by instruments on the right-hand side of the stage reaches the listener coming from this direction.

In monaural reproduction, all the sound produced, no matter where the

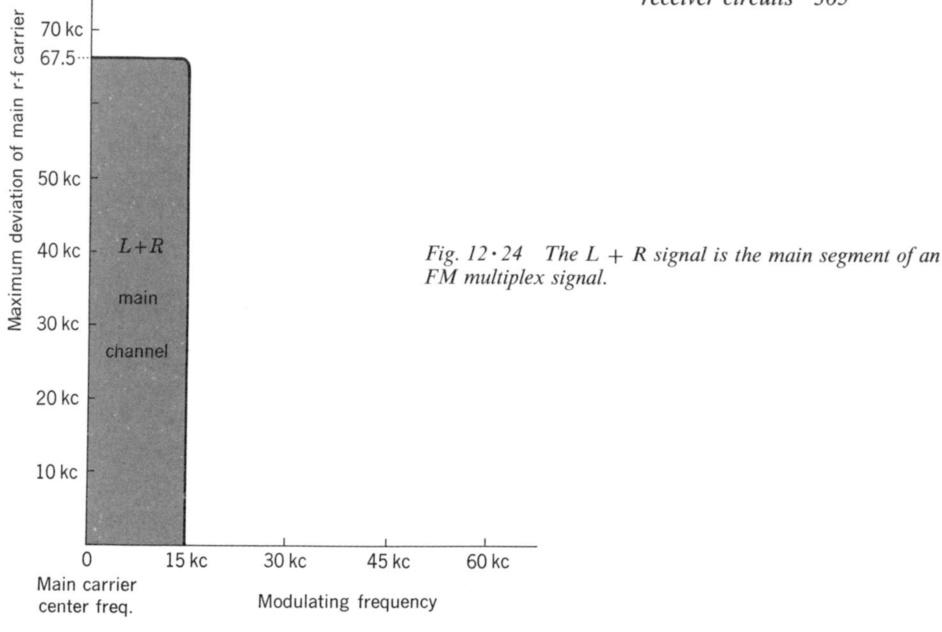


*Fig. 12·24 The L + R signal is the main segment of an FM multiplex signal.*

source may be located, is fed into one microphone and dealt with as a whole by the transmitter. At the receiver, only this sound is reproduced whether it comes from one speaker or many speakers. Under these conditions, it is not possible to develop the illusion of three-dimensional sound. In other words, it is impossible to re-create the same spatial effect at the receiving end that a listener would experience were he present where the sound originated. This is the limitation of monaural sound reproduction that stereo reproduction seeks to overcome. And stereo reproduction is made possible by use of FM multiplexing.

**FM multiplex signal.** Stereo reproduction utilizes two signals, nominally referred to as the left and right signals indicating the sources from which these audio signals were obtained. Separate microphones pick up each signal and separate audio systems amplify and otherwise process these signals until they are ready to modulate the r-f carrier for transmission through the atmospere to the receiver.

To appreciate how the left (L) and right (R) signals are combined in this final step, it is important to remember that this stereo signal must be usable by a conventional monaural receiver. Furthermore, the monaural set must have available to it *both* the left and right signals in combined fashion, since this is the way monaural signals are ordinarily received. That is, they obtain *all* the sound that is produced at the studio.

Thus, if we examine the first part of a stereo signal, we find that both the L and R signals are present; furthermore, they are present to 15,000 cycles. This is the full audio range.

Figure 12·24 shows this first segment of an FM multiplex signal. The combined L + R audio signals frequency-modulate the r-f carrier to produce the conventional FM output signal that the receiver picks up with its antenna. The only difference between what is done here and in regular monaural broadcasting lies in the fact that instead of modulating the r-f carrier ±75 kc, the modulation swing is now restricted to a maximum of ±67.5 kc, and possibly as little as ±60 kc. The reasons for these restrictions will be indicated presently.

So much, then, for the way in which the FM stereo signal maintains a normal relationship with monaural receivers. These sets will receive the L + R signal and produce the normal output.

For stereo reception, we now require some method of separating the L + R signals which are combined in this main channel. And the way this is done is both simple and ingenious. The two original L and R signals are, at the transmitter, also subtracted from each other (rather than added as above) to provide another signal, the L − R signal. This L − R signal, called the difference signal, then amplitude-modulates a 38-kc carrier. The resultant sidebands then extend from 38 kc minus 15 kc, or 23 kc, to 38 kc plus 15 kc, or 53 kc. This is the conventional action in amplitude modulation where the sidebands extend above and below the carrier by the value of the audio-modulating frequencies. The highest audio-modulating frequency is 15,000 cycles, and so the frequency spread noted above is developed.

There is this one difference, however. The modulation is achieved in such a way that the 38-kc carrier itself is suppressed. This is readily accomplished with a balanced modulator which produces the upper and lower sidebands but does not permit the carrier to reach the output. The

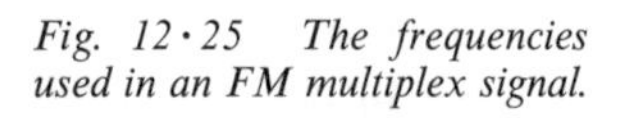

*Fig. 12·25 The frequencies used in an FM multiplex signal.*

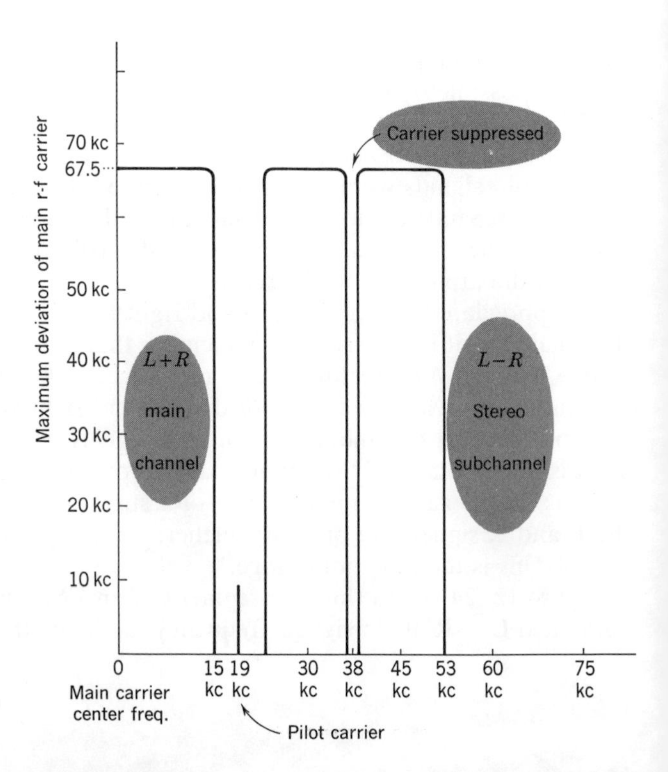

L − R carrier, known as the subcarrier, is suppressed to insure that there will be no stereo-subchannel output when a monaural signal is being broadcast. (The same condition is produced in a color television broadcast when only a black-and-white signal is transmitted.)

After the 38-kc sideband signals are formed, they and the main channel signal, plus a special 19-kc pilot signal (to be explained presently) frequency-modulate the r-f signal of the transmitter to provide the full stereo signal. See Fig. 12·25.

The 19-kc pilot carrier is required because the 38-kc carrier of the L − R signal is suppressed. In order to properly demodulate the L − R signal at the receiver, information concerning the suppressed 38-kc carrier is required. The 19-kc pilot carrier provides this information. At the receiver, the 19-kc pilot signal is received, then fed to a doubler where 38-kc is developed. This signal then goes to a detector where it combines with the subcarrier sidebands and the original L − R audio signals appear at the detector output.

At this point, the reader may wonder what a stereo receiver does with the L + R and L − R signals to obtain independent L and R signals for its stereo speakers. The action, mathematically, is quite simple and serves to illustrate the ingenuity of this approach.

If we add the (L + R) and the (L − R) signals, in an adder stage, we have

$$\begin{aligned}(L + R) + (L - R) &= L + R + L - R \\ &= 2L\end{aligned}$$

In other words, adding (L + R) to (L − R) gives us the left signal alone.

By the same token, subtracting (L + R) from (L − R) gives

$$\begin{aligned}(L + R) - (L - R) &= L + R - L + R \\ &= 2R\end{aligned}$$

Thus, only the right signal is developed by this subtraction. Now the left signal can be fed to its speaker and the right signal to its speaker and stereo reproduction can be achieved.

There is still another signal that can be sent with the stereo signal, although it really has no relationship to stereo. Background music, or SCA[1] transmission, has been in operation for several years and is provided by a number of FM stations. This SCA signal is placed at 67.5 kc where it is well out of the way of the two parts of the stereo signal. The 67.5-kc SCA carrier is modulated ±7 or 8 kc about the 67.5-kc point. When this SCA service is present, it is combined with the two parts of the stereo signal to modulate the r-f carrier of the station. The latter will, of course, fall somewhere in the FM broadcast range, 88 to 108 Mc.

[1] SCA is an abbreviation for Subsidiary Communication Allocation.

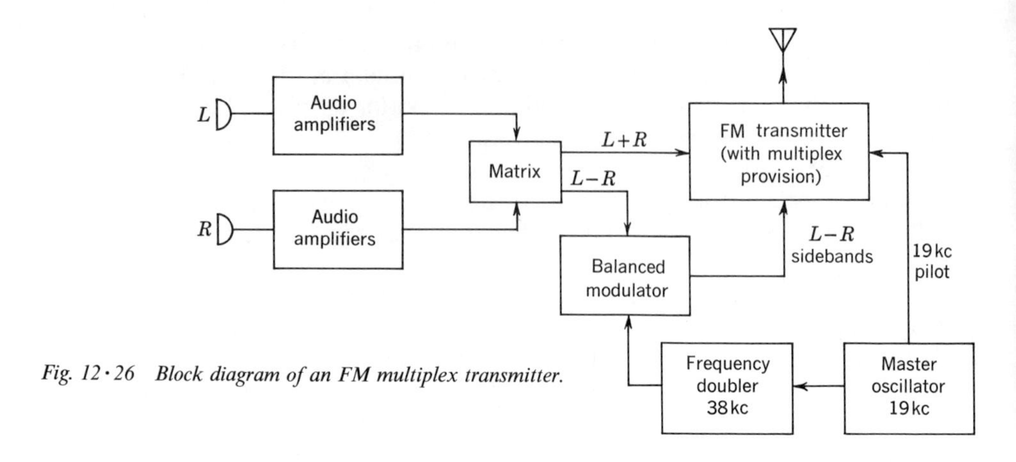


*Fig. 12·26 Block diagram of an FM multiplex transmitter.*

Before we examine briefly the arrangement of the stages at the transmitter and receiver, let us see what happens at a monaural receiver when a stereo signal is received. The main channel (L + R) component is handled in the normal manner to produce a conventional output. The (L − R) subcarrier, at 38 kc, does not produce any output because all frequencies above 15 kc are suppressed at the output of the FM detector. The same is true of the 19-kc pilot carrier, as well as any SCA signal that might be present. Furthermore, since 19-kc signals and all those above it are inaudible, nothing would be heard by the listener even if these should reach the speaker.

**FM multiplex transmitter.** A simplified block diagram of an FM multiplex transmitter is shown in Fig. 12·26. Separate microphones and audio amplifiers pick up and amplify the left and right signals. These are then fed to a mixer (or matrix) where separate (L + R) and (L − R) signals are developed. The (L + R) signal remains as it is whereas the (L − R) signal is shunted to a balanced modulator where it amplitude-modulates the 38-kc subcarrier. In the modulator, the 38-kc subcarrier is suppressed, but the (L − R) sidebands are developed and transferred to the FM transmitter. In the transmitter, the (L + R) signal, the (L − R) sideband signal, and the 19-kc pilot signal frequency-modulate the station carrier. This FM signal is now broadcast.

Note that a 19-kc master oscillator provides the 19-kc pilot carrier as well as serving as the driving source for the 38-kc frequency doubler. It is from this latter stage that the balanced modulator receives its 38-kc carrier. By this arrangement, the 19-kc pilot signal can properly serve the stereo receiver to recreate the original (L − R) audio signal.

**FM multiplex receiver.** The FM multiplex receiver is best considered in two sections. The first section, shown in block-diagram form in Fig. 12·27*a*, contains the r-f tuner, the i-f section, the i-f limiter, and the FM detector. The line-up of stages is representative of every FM receiver, whether it is

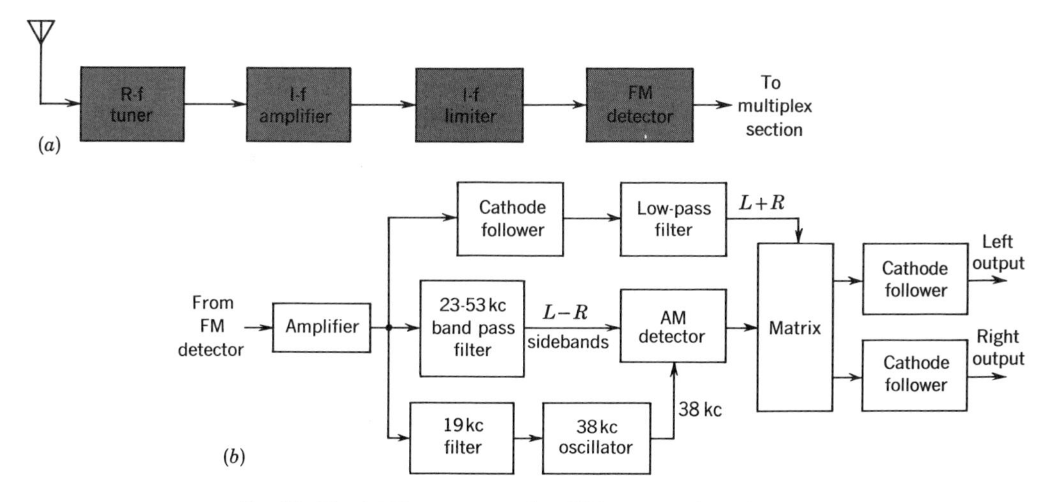


*Fig. 12·27 (a) Tuner section of an FM receiver, for either monaural or stereo. (b) Multiplex section of an FM stereo receiver.*

being employed to receive monaural or stereo signals. It is only beyond the FM detector that a significant difference occurs and this can be readily seen from Fig. 12·27*b*. Let us go through this section of the receiver, stage by stage.

In Fig. 12·27*b*, the full output signal from the FM detector is applied to an amplifier where the signal is further strengthened. At the output of the amplifier, the signal is applied to three points: a cathode follower, a 23- to 53-kc bandpass filter, and a 19-kc filter.

The cathode follower, at the top of Fig. 12·27*b*, is followed by a low-pass filter that permits signals with frequencies up to 15 kc to pass, but which sharply attenuates all others. This low-pass filter thus permits only the (L + R) segment of the demodulated signal to pass. This signal is fed to the matrix or mixer stage.

The second output of the input amplifier reaches a band-pass filter that permits only the (L − R) sidebands to pass through, but which attenuates all other segments of the demodulated signal. Finally, the third amplifier output feeds a 19-kc filter where the 19-kc pilot carrier is retrieved. This carrier then synchronizes a 38-kc oscillator to insure that the 38-kc carrier required to demodulate the (L − R) sidebands possesses the proper phase.

The generated 38-kc signal and the (L − R) sidebands are combined in the AM detector. The output of the detector is then the (L − R) audio signal, as well as a −(L − R) signal. The two signals are then appropriately combined with the (L + R) signal in the matrix network according to the fashion previously indicated to provide a true L output and a true R output. These two signals then pass through separate cathode followers and perhaps even amplifiers (not shown) and are applied to separate speakers. In this manner, the stereo effect is produced.

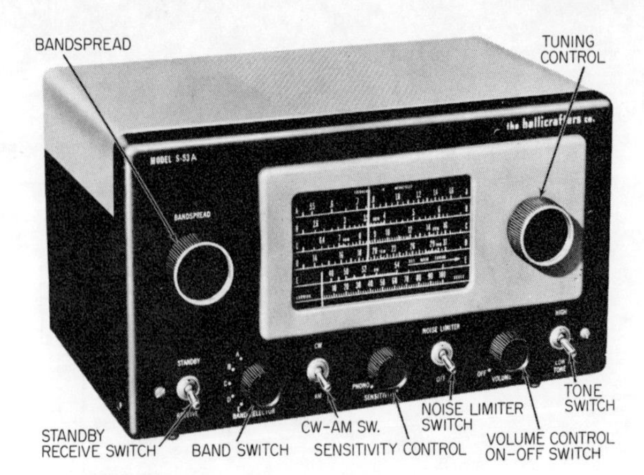


*Fig. 12·28 A popular communications receiver.* (Hallicrafters.)

## *12·9 Communications receivers*

Another type of receiver that is used extensively is the communications receiver. It is almost always a superheterodyne receiver which can receive signals over a wide range of frequencies extending well beyond the radio broadcast band. Thus, a typical communications receiver, such as the unit shown in Fig. 12·28, can operate from 540 kc (which is the start of the radio broadcast band) to 54.5 Mc. In that range will be found a wide variety of broadcasting services, the more important of which are: short-wave stations, both commercial and governmental, amateur stations, radio-navigational signals for aircraft and ships, and frequencies for mobile equipment and for industrial electronic equipment, such as diathermy machines. This particular set skips some frequencies in this range, as we shall see, but this practice varies with the receiver. Some sets cover a range completely, while others select only those frequencies which are most needed.

A communications receiver, if it employs a superheterodyne circuit, possesses the same sequence of stages as a superheterodyne radio receiver, modified to some extent by the nature of the signals to be received. A block diagram of the receiver shown in Fig. 12·28 is given in Fig. 12·29. The input signal is applied to the mixer stage; a separate oscillator also feeds its signal to the mixer and, as a result of the mixing of these two signals, an i-f voltage is developed. In this set, the i-f value is 455 kc, although this

*Fig. 12·29 Block diagram of the communications receiver shown in Fig. 12·28.*

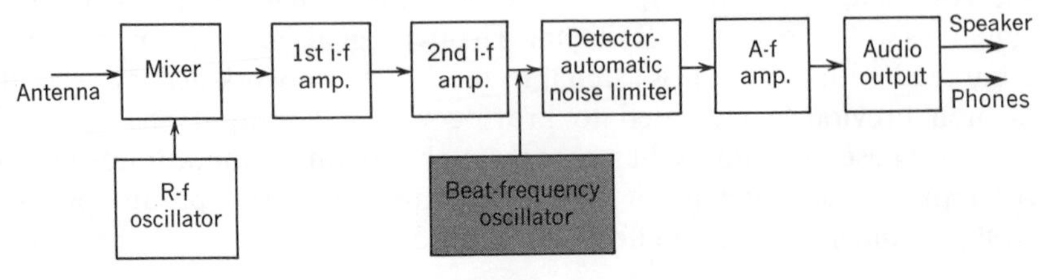

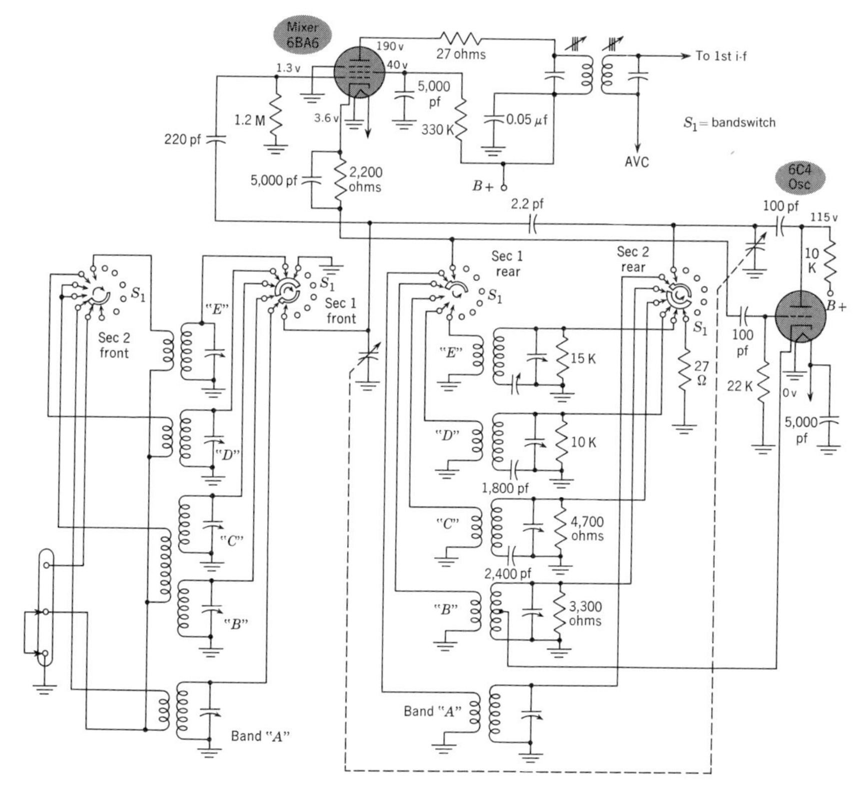


*Fig. 12·30 The bandswitching arrangement used in the circuit of Fig. 12·28. The coils for each of the five bands are indicated.*

figure may vary with different receivers. Thereafter, the signal is amplified by two i-f stages, detected by a diode detector, amplified by two audio stages, and then made available to a loudspeaker or a pair of headphones.

**Bandswitching.** Communications receivers are designed to operate over a wide range of frequencies, so wide, in fact, that one tuning circuit would not be able to cover all the frequencies involved. To solve this problem, a number of resonant circuits are used, each tuned to a different frequency, and when a desired range of frequencies is to be received, the corresponding tuning circuit is switched into the circuit.

For example, the receiver of Fig. 12·28 covers the range from 540 kc to 54.4 Mc in five bands as follows:

| | | | | |
|---|---|---|---|---|
| Band A | 540 | to | 1,630 | kc |
| Band B | 2.5 | to | 6.3 | Mc |
| Band C | 6.3 | to | 16 | Mc |
| Band D | 14 | to | 31 | Mc |
| Band E | 48 | to | 54.5 | Mc |

Note that the coverage is not continuous, with frequency skips between Band A and Band B and between Band D and Band E. This receiver

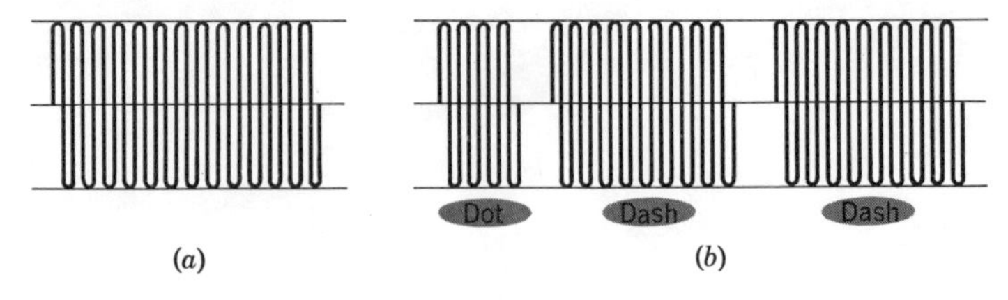


*Fig. 12·31 (a) A c-w signal. (b) A c-w signal broken up into dots and dashes.*

is designed for amateur and short-wave work, and the frequency ranges not covered do not carry such signals. More extensive coverage can be obtained with more expensive receivers, but the nature of the circuitry remains the same.

A special rotary switch, labeled BAND SWITCH on the receiver front panel, enables the set user to place the receiver in the desired band. The dial, too, has separate markings for each band so that the user can identify the frequency (or wavelength) of a received signal. Internally, the various tuning circuits for each band appear as shown in Fig. 12·30. Note that all front-end tuning circuits must be changed in unison; in the present diagram, this concerns the input tuning circuit to the mixer and the tuning circuits in the oscillator stage. If an r-f amplifier is used, its tuning circuits will have to be changed as well.

**C-w operation.** One of the signals a communications receiver is called on to receive is a c-w signal. The letters c-w stand for continuous wave, which is an r-f carrier without any modulation (see Fig. 12·31*a*). Now, when code is to be sent, this r-f continuous carrier is broken up into a series of short and long pulses (Fig. 12·31*b*) called dots and dashes. In the receiver, this r-f signal is picked up, converted in the mixer to the intermediate frequency, passed through the i-f system, and then fed to the detector. However, because the signal has no amplitude or frequency modulation, no audio signal will appear at the detector output.

In order to make dots and dashes audible, a special oscillator, called a beat-frequency oscillator (BFO), is added to the circuit at the detector stage or just prior to it. This oscillator is tuned to a frequency that is 1 kc less than the intermediate frequency. When the dots and dashes arrive, they mix in the detector with the signal from the beat-frequency oscillator. The result is a 1-kc audio note whenever a dot or a dash arrives, but no output when no c-w signal is being received.

Since the beat-frequency oscillator would produce a disturbing 1-kc note for voice or music broadcasts, a special switch on the front panel enables the set user to turn the oscillator on or off, as desired. The switch, in Fig. 12·28, is labeled CW-AM.

**Standby operation.** Another feature of most communications receivers is the STANDBY-RECEIVE switch. With this switch in the STANDBY position, all the B+ voltage is removed from the tubes and only the filament power is left on. This keeps the tubes in readiness for instant signal reception simply by flipping the switch to the RECEIVE position. Persons using communications receivers like to have their sets available for immediate operation, yet for the long periods when the set is not in use, removal of the B+ voltage lowers the power requirement of the circuit.

**Bandspreading.** A BANDSPREAD control (Fig. 12·28) enables the set listener to separate signals that are so close in frequency that it would be quite difficult to tune them in separately with the normal tuning control. In some sets, such as this one, the BANDSPREAD control is mechanically geared to the regular tuning control. However, the gearing between the two controls is such that it takes a number of turns of the BANDSPREAD control to rotate the normal tuning control once. In this way, very fine adjustment of the tuning circuits can be made easily.

Another method of spreading stations apart is by electrical means. One of the simplest ways of doing this is to place a small variable capacitor in parallel with the regular tuning capacitor. Then, when a point in the band is reached where several stations are crowded together, rotation of the small capacitor (or capacitors, if there are several) will shift the frequency of the tuning circuits by small amounts, enabling the listener to tune in each station separately.

**Sensitivity control.** Most communications receivers contain sensitivity controls which vary the gain of the r-f or i-f stages to suit the level of the incoming signal. For example, when very strong signals are being received, the sensitivity of the receiver is cut down; when the signals are weak, the sensitivity is raised. This special control is added because most communications receivers are very sensitive and the normal AVC system is not

*Fig. 12·32 The automatic noise-limiter circuit employed in the communications receiver of Fig. 12·28. The limiter circuit uses half of the 6AL5; the AM detector uses the other half.*

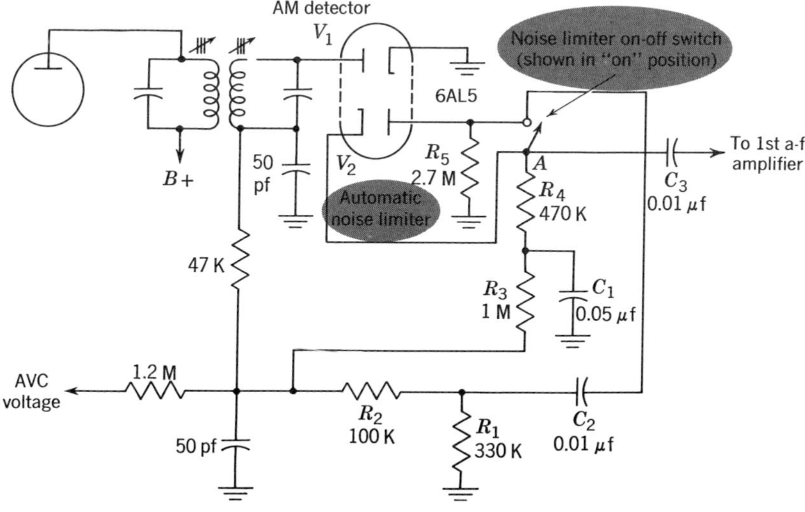

capable of effectively dealing with the wide range of signal levels that the set is called on to receive. By adding the auxiliary sensitivity control, greater flexibility is achieved.

Note that automatic volume control is still present, and once the sensitivity of the set has been established (by the sensitivity control), the AVC circuit acts to keep the signal level constant. Some sets have a switch to short out the automatic volume control for maximum sensitivity when necessary.

**Noise limiters.** One final circuit of importance commonly found in communications receivers is the noise-limiter circuit. Its purpose is to reduce the interference caused by sharp noise pulses due perhaps to ignition systems, electric storms, or motors of various types. It does this by limiting the amplitude of such pulses, causing the receiver to become inoperative (that is, silencing it) for the second or so that the pulse is present. This temporary cutout is usually better than permitting the noise to pass through undiminished, producing a loud, sharp noise in the speaker or the headphones.

A typical noise limiter, such as that employed in the receiver of Fig. 12·28, is shown in Fig. 12·32. One half of the double diode (such as a 6AL5) is employed as the second detector for the receiver, developing its output audio signal across $R_1$. The noise limiter uses the second diode of the tube and functions as follows. The incoming carrier produces a negative d-c voltage across $R_1$ and $R_2$. This voltage is filtered by $R_3$, $R_4$, and $C_1$ and applied to the cathode of the noise limiter $V_2$. The plate of $V_2$ is

*Fig. 12·33 The block diagram of a modern television receiver.*

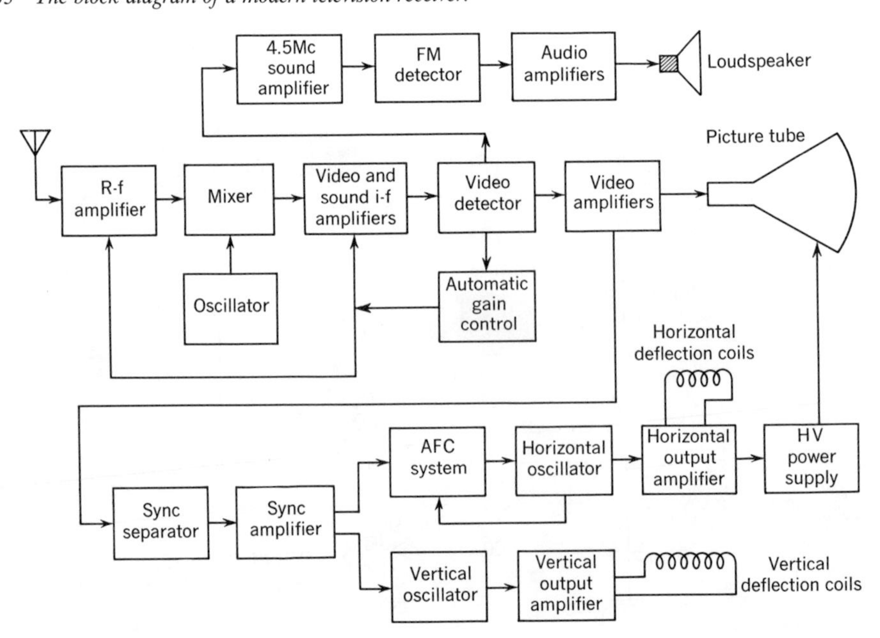

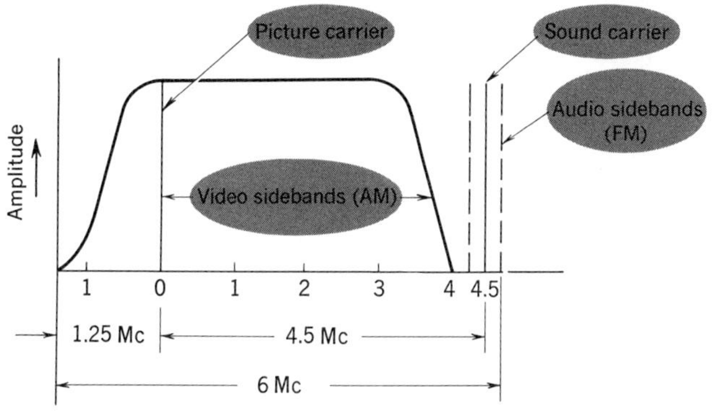


*Fig. 12·34 A television channel (video and sound) occupies a bandwidth of 6 Mc.*

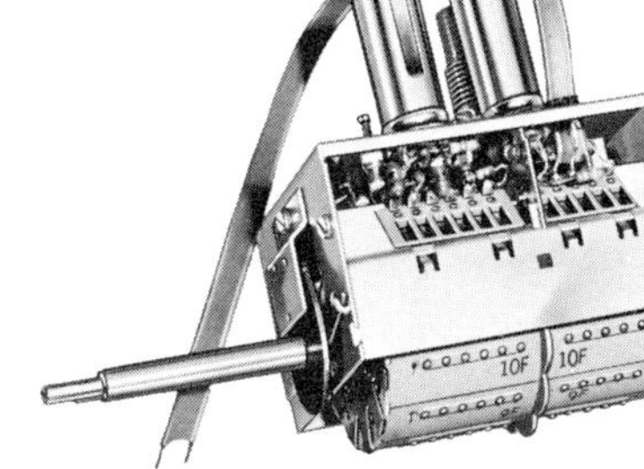

*Fig. 12·35 A typical turret-type television tuner.* (Kollsman Instrument Corp.)

essentially at ground d-c potential because it connects to ground through $R_5$. Since the cathode is negative with respect to the plate, the tube conducts. Hence an audio signal which is present across $R_1$ is fed through $C_2$ to the diode, passing through this tube and appearing at point $A$. From here, it is led by $C_3$ to the volume control and the audio system.

Now, suppose a strong noise pulse arrives. The negative d-c voltage across $R_2$ and $R_1$ rises sharply. Because of the large values of $R_3$ and $C_1$, however, this sharp rise is prevented from reaching the cathode of $V_2$, but the negative pulse is brought to the plate of $V_2$ by capacitor $C_2$, driving the diode quickly to cutoff. The tube remains in this condition as long as the pulse keeps the plate negative with respect to its cathode, and during this time no signal passes through $V_1$ to the audio amplifier. Only when the pulse ends does $V_2$ resume normal operation.

Some noise signal does get through, but the greatest portion of it is kept from reaching the loudspeaker.

A switch is provided, so that the noise-limiter circuit can be rendered inoperative, if desired. The OFF position may be necessary to prevent limiting modulation peaks on the desired signal.

## *12·10 Television receivers*

A television receiver is, in a sense, an expanded radio receiver in that it must not only do what a radio receiver does, that is, receive the sound portion of the broadcast, but it must also develop a picture of the scene which is being transmitted at the same time. Obviously, this added function will require more stages, so that a television receiver can never be as simple as a radio receiver. The block diagram of a typical television set is shown in Fig. 12·33.

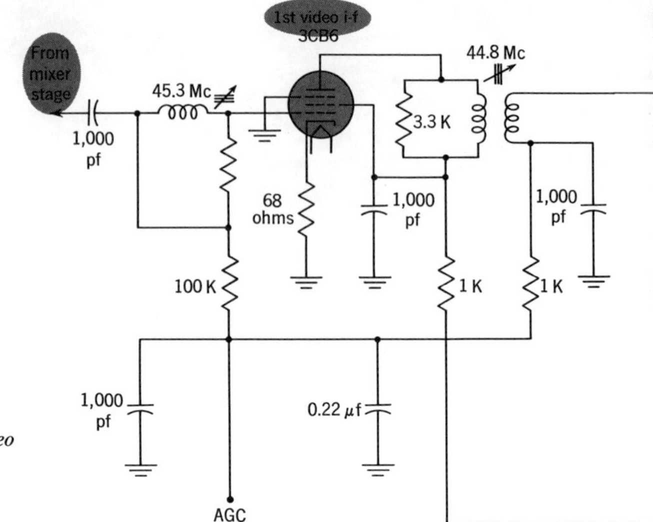


*Fig. 12·36 A typical three-stage video i-f system.*

**Front-end stages.** The first stage in the television receiver is the r-f amplifier. Such a stage is almost always present because it not only makes the set more sensitive so that weaker signals can be picked up, but it also improves receiver selectivity because the r-f circuits are always tuned. This selectivity reduces the amount of interference that might otherwise pass through the receiver and show up on the television screen. The eye is a critical judge of what it sees, much more so than the ear of what it hears, and every effort must be made to keep everything but the proper picture off the screen. A well-designed r-f amplifier will help do this, besides strengthening any weak television signals that may arrive. Without the assistance of the amplification by the r-f amplifier, the incoming signal would often not possess enough strength to overcome the electrical noise which is developed in the antenna and the tubes and other components of the receiver circuitry. As a result, both signal and noise would be amplified together in succeeding stages of the receiver, producing a picture full of noise spots called *snow.* If, however, the arriving signal is given a boost by the r-f amplifier, it can be made powerful enough to override this circuit noise and develop a clear, sharp picture on the cathode-ray-tube screen.

Television signals are very broad, occupying a channel 6 Mc wide (see Fig. 12·34). To tune in this wide channel properly requires a special tuning mechanism. Several types have been developed, but the most widely used tuner has the turret-type construction shown in Fig. 12·35. The tuner has 12 positions, one each for the 12 VHF television channels (54 to 88 Mc and 174 to 216 Mc). In addition, there are 70 more channels in the UHF region, from 470 to 890 Mc. If these are to be received also, then an auxiliary tuning assembly is attached to the turret tuner.

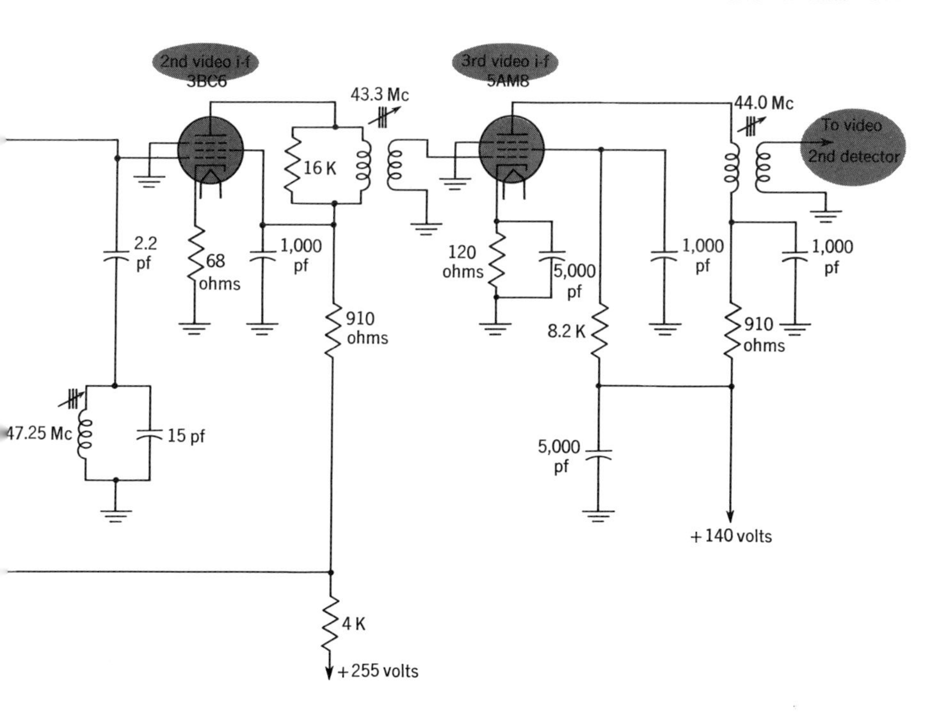


All tuners contain the circuits for the r-f amplifier, the oscillator, and the mixer stage. The incoming signal which the r-f amplifier receives is passed on to the mixer, where it is combined with a signal from the r-f oscillator. This mixing produces an i-f signal which represents the difference between the incoming radio signal frequency and the oscillator signal frequency. For example, if channel 3 is being received, then the frequencies in the signal will extend from 60 to 66 Mc. For this channel, the r-f oscillator signal will be about 107 Mc, so that their difference frequency will fall between 41 and 47 Mc. More specifically, the sound carrier will be found at 41.25 Mc, and the video carrier at 45.75 Mc. Most of the frequency space between these two values will be occupied by the video signal and its sidebands.

**I-f system.** The signal now enters the video i-f channel, where three or four amplifiers are generally found (see Fig. 12·36). Each stage is tuned, generally, to a different frequency. However, the net effect of this staggering of the resonant frequencies is to produce a uniform band pass about 3.5 to 4.0 Mc wide. This is wide enough for the video signal and the sound carrier, while at the same time it attenuates any signals which lie outside the band.

It is customary to control the gain of the first stage or two of the video i-f system by means of automatic gain control. The AGC system provides a small negative signal which varies with the strength of the incoming signal. That is, if the signal gets stronger, the negative AGC voltage increases; if the signal gets weaker, the AGC voltage becomes less negative.

When this voltage is applied to the control grid of a video i-f amplifier tube, it tends to even out the strength of the amplified i-f signal for the detector.

Beyond the video i-f system is the video detector, where the video and sound i-f carriers are removed and the modulating signals themselves remain. In the video signal, the frequency range extends from 0 to 4.0 Mc. In the sound signal, the situation is slightly different in that a signal is developed which has a central frequency of 4.5 Mc, while the audio intelligence is contained in sidebands on either side of the 4.5 Mc. Furthermore, the sound signal is frequency modulated, while the video signal is amplitude modulated. Note that 4.5 Mc is the difference between the picture and sound carrier frequencies. This method of detecting the sound as a 4.5 Mc FM signal is called an *intercarrier sound* circuit.

**The sound system.** Tracing the path of the sound signal in Fig. 12·33, we find that it leaves the video signal generally at the video detector and moves on to the sound system. Here, the first stage is a 4.5-Mc amplifier, where the signal is further strengthened. Then the signal is passed on to an FM detector, where the sound intelligence of the broadcast is removed. Thereafter the sound signal is amplified first by an audio voltage amplifier and then by a power amplifier, after which it is applied to the loudspeaker and audible sound is produced. This latter portion of the receiver is identical to the audio-amplifier stages of a radio receiver.

**Video stages.** With the sound signal followed to its destination, let us return to the video detector and trace the path of the video signal in Fig. 12·33. This signal, which contains all the picture information, is applied first to a video amplifier (one or two stages) that is so designed that it will pass a band of frequencies from near 0 to 3.5 or 4.0 Mc. This is necessary because the video frequencies extend over this wide a band. Then the signal goes to the picture tube, where it varies the strength of the electron beam striking the fluorescent screen at the end of the tube. The video signal voltage is coupled to the control-grid–cathode circuit of the picture tube. If a bright patch of light is desired, the video signal permits maximum beam current to strike the screen. If a weaker or dimmer spot is wanted, the video signal reduces the number of beam electrons striking the screen and less light is obtained.

The intensity of the picture-tube beam varies very rapidly as the beam travels back and forth and up and down across the screen of the picture tube, and thus a picture is produced. Beam movement is very rapid; actually 30 full pictures are reproduced every second. Because the rate is so fast, the individual pictures are never seen. Rather, the eye tends to combine them all, and the viewer sees motion.

**Beam scanning in picture tube.** It was stated above that the beam in the picture tube moves back and forth across the screen, and at the same time travels from top to bottom, producing 30 full images or pictures a second. In order to produce this combined motion, two separate magnetic fields are set up in the picture tube. One field causes the electron beam to travel across the screen horizontally, while the other field causes it to move vertically. In somewhat greater detail, here is how the scanning process occurs.

At the start, let us assume that the beam is striking the screen at the upper left-hand corner, point *A* in Fig. 12 · 37. Then, under the influence of one magnetic field, the beam travels toward the right side. While this is happening, however, the other magnetic field is also slowly forcing the beam downward. The result of these combined forces is to move the beam to the right and slightly downward until it reaches point *B*. At this moment, a blanking pulse arrives in the video signal and drives the picture tube to cutoff. During the short interval that the beam is cut off, a synchronizing pulse in the video signal causes the beam to move rapidly to point *C*. When this has been accomplished, the blanking voltage decreases to zero (that is, it ends), and the electron beam again starts moving to the right.

It will be noted from Fig. 12 · 37 that, in retracing from right to left, the beam returned to position *C*, thereby skipping a line. This is done purposely and is followed throughout the scanning. The beam scans every odd line in the image, returns to the top, and scans every even line; then the process starts all over again, scanning odd lines, then even lines, then odd lines, and so on, for as long as the receiver is in operation.

With this interlaced system of scanning, it is possible to achieve a flickerless image at the receiver, without using a very high scanning rate. Experience has shown that if all the odd lines in an image are scanned in 1/60 of a second and all the even lines in 1/60 of a second, the reproduced image in the receiver does not flicker, but changes smoothly from scene to scene. All the odd and even lines combine to form a complete image. Therefore, one full image is received every 1/30 of a second.

In each picture there is a total of 525 lines. Each field, then, contains half

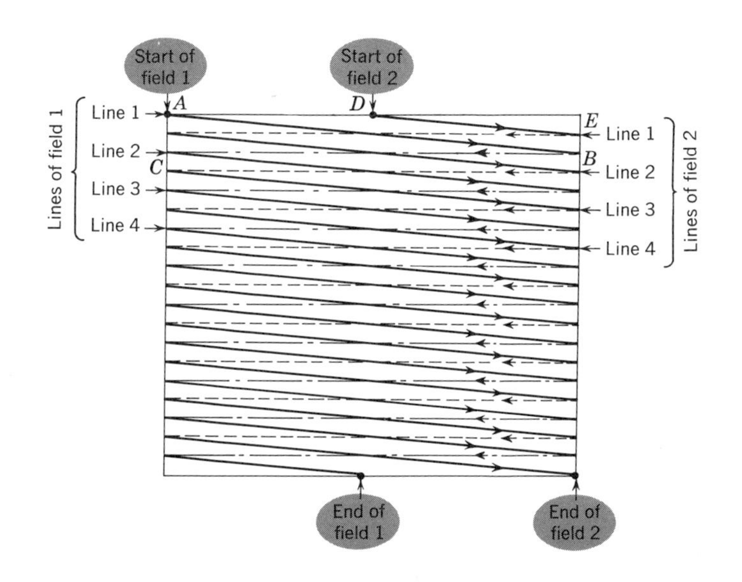


*Fig. 12 · 37 The motion of the electron beam over a picture screen in tracing out the image.*

this amount, or 262½. Further, since there are 30 complete pictures a second, there are 30 × 525, or 15,750 total lines a second, and this is the *horizontal scanning rate.* The *vertical scanning rate,* or the number of fields (odd plus even), is 60 per second.

Now, the next question that might be asked is, "How does the electron beam in the picture tube know when it has reached the end of a line or the bottom of the picture?" The answer lies in the synchronizing pulses which are sent along with the video signal. Several lines of a video signal, after detection by the video detector, are shown in Fig. 12·38. One line lasts for 1/15,750 of a second, and during this time the video information for that one line must be sent, plus a sharp pulse, called a *synchronizing pulse,* at the end of the line.

It can be seen, from Fig. 12·38, that the horizontal sync pulse is on a different level from any of the video-signal variations. After the video detector, some of the video signal is fed to a sync separator stage where, as its name suggests, the sync pulses are separated from the rest of the signal. All these pulses then feed to a horizontal sweep oscillator through an automatic frequency-control circuit.

**Horizontal sweep system.** Here is how this particular combination operates. The horizontal oscillator has a natural frequency of about 15,750 cps. However, this oscillator and its frequency must be securely locked in step with the horizontal sync pulses arriving with the incoming signal from the broadcast station. It is the function of the sync pulses to achieve this lock-in. One of the ways is by triggering or pulsing the oscillator every time pulses arrive, that is, every 1/15,750 of a second. This is the direct approach, but it possesses one great disadvantage. Any noise or interference pulses that are received with the signal will also reach the oscillator and cause it to trigger off beat. This will disrupt the operation of the beam and act to jumble up the picture, producing the visual effect shown in Fig. 12·39.

A more effective approach is by means of automatic frequency control. The AFC circuit consists of a phase detector and a long-time-constant filter. The incoming sync pulses feed into the phase detector, where the timing is compared with the wave present at the output of the horizontal sweep oscillator. The purpose of the long-time-constant filter between the phase detector and the horizontal oscillator is to remove the effects of any

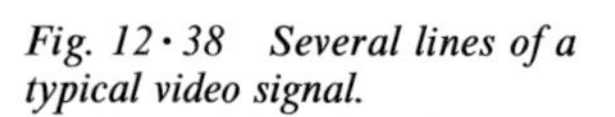

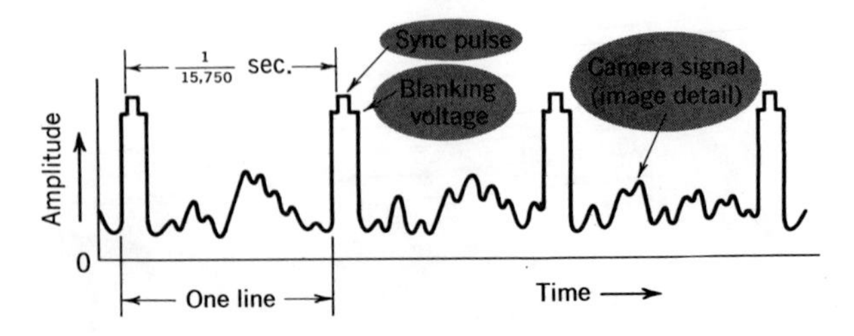


*Fig. 12·38 Several lines of a typical video signal.*

*Fig. 12·39 A picture out of horizontal synchronization.*

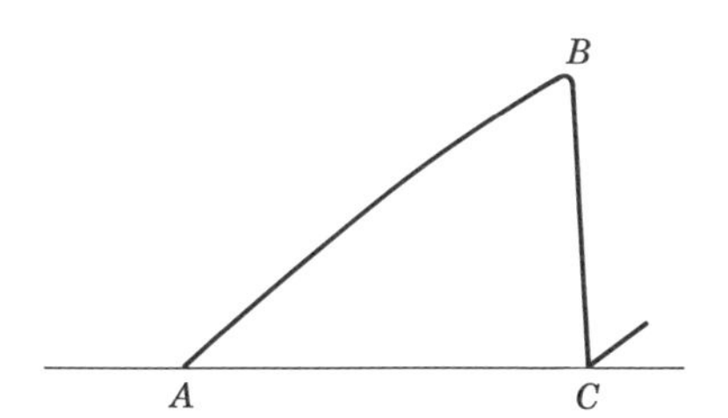


*Fig. 12·40 The deflection wave developed by the horizontal oscillator.*

noise or other interference pulses that may arrive with the signal. By this method, we can stabilize the oscillator and prevent it from falling out of sync and causing the visual effect shown in Fig. 12·39.

The wave at the output of the horizontal oscillator possesses the form shown in Fig. 12·40. From *A* to *B*, it rises at a steady rate, and during this period the electron beam is moving evenly from the left-hand to the right-hand side of the screen. At point *B*, the voltage drops quite rapidly, causing the electron beam to return rapidly to the left-hand side of the screen, in position for the next line. Then the sequence starts all over again.

The wave of Fig. 12·40 is amplified by the horizontal output amplifier and then transferred via a transformer to the horizontal windings of a special yoke which mounts over the neck of the picture tube (see Fig. 12·41). The magnetic field of the windings in the yoke extend through the neck of the tube, and, as the electron beam passes through the field (on its way to the screen), it is moved back and forth in the manner already described.

An interesting feature of this horizontal system is the additional function served by the transformer (mentioned above) located between the output stage and the deflection yoke. Every time the sawtooth wave of Fig. 12·40 drops down sharply from *B* to *C*, it causes a large voltage to develop in the transformer windings. This effect is known as an *inductive kick* and occurs whenever the current flowing through any inductance changes rapidly. The large voltage that develops is rectified by a special diode and is used to charge a capacitor to a value of 10,000 to 16,000 volts. This voltage is then applied to the picture tube and serves to bring the electron beam all the way down the neck of the tube and to the fluorescent screen. Therefore, horizontal scanning is necessary to produce the high voltage for the picture tube.

**Vertical sweep system.** So far we have shown how the beam motion back and forth across the screen is produced and controlled. Still neces-

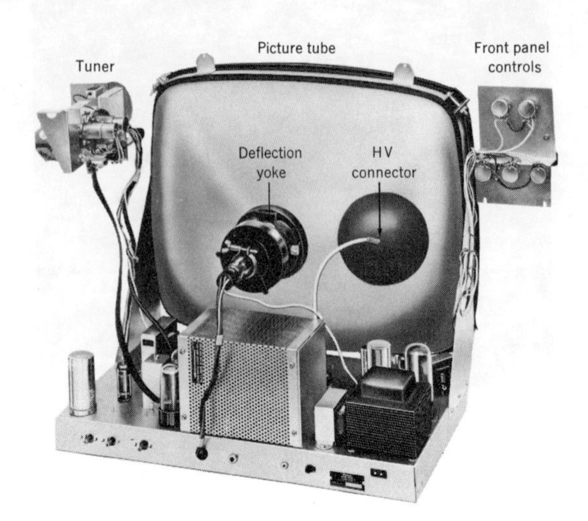


*Fig. 12·41 The position of the deflection yoke on the neck of the picture tube. The yoke serves to swing the electron beam back and forth and up and down over the face of the screen.*

sary is some force to move the beam from the top of the screen to the bottom at an even and steady rate, and then to return the beam quickly to the top of the screen for the next field. This action is provided by a separate vertical system controlled by vertical sync pulses, which are also contained in the incoming video signal.

We have previously seen that a horizontal pulse occurs at the end of each line. For control of the vertical motion of the beam, a special vertical pulse is inserted at the end of every field, that is, after every 262½ lines. The form of this vertical pulse, actually a collection of pulses, is shown in Fig. 12·42. The vertical pulse is broken up into the form shown so that the horizontal system may be kept in synchronism while the vertical is active.

The same sync separator (Fig. 12·33) that strips the horizontal sync pulses from the video signal also clips off the vertical pulses whenever they appear. These vertical pulses are then amplified and fed to the vertical oscillator where they control the frequency of this stage. Note that the vertical oscillator is not preceded by an AFC system because the vertical system has a much lower frequency and is not affected by most noise, which is high frequency.

The vertical oscillator, kept on frequency by the vertical sync pulses, produces an output sawtooth wave of the form shown in Fig. 12·40. This voltage is amplified by an output amplifier and then transferred to the vertical windings of the deflection yoke. The magnetic field set up by these windings moves the electron beam in the picture tube slowly downward and then quickly upward sixty times a second. In this way, beam motion is controlled both vertically and horizontally to scan out an image on the screen. The action occurs very quickly and yet very accurately; otherwise, the high-quality television pictures we now obtain would not be possible.

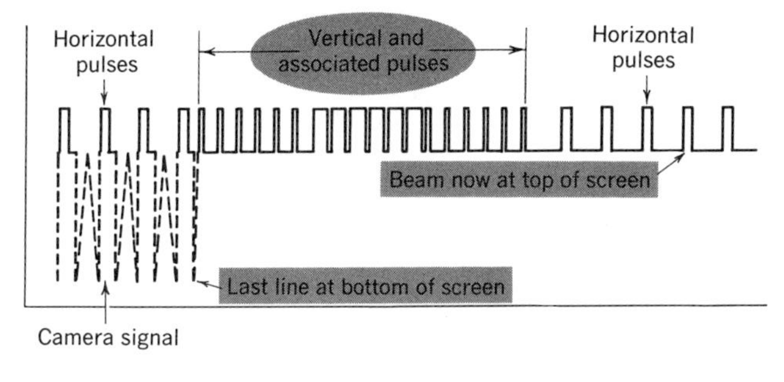


*Fig. 12·42 The form of the vertical synchronizing pulses.*

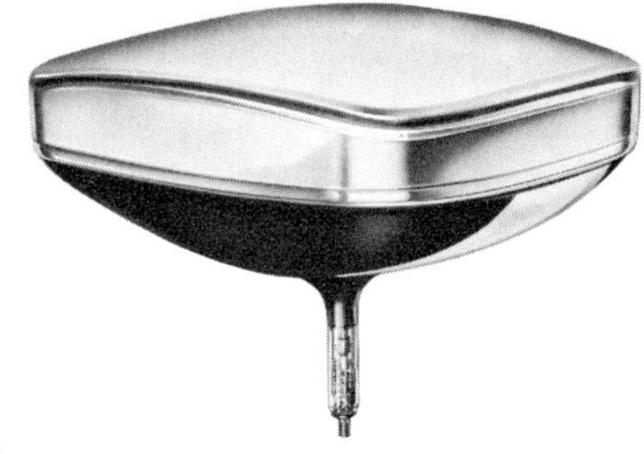

*Fig. 12·43 A television picture tube.* (RCA.)

**Television picture tubes.** A picture tube has a rectangular screen set in a glass body, which gradually narrows down to a neck about 2 in. in diameter (see Fig. 12·43). At the far end of this neck is a base with pins to which a socket can be firmly attached. Through this socket, the filament voltage, the video signal, and some of the lower B+ driving voltages are brought to the electron gun in the neck. The beam is formed by this gun and then projected down the neck to the fluorescent screen (see Fig. 12·44). A very high accelerating voltage, between 10,000 and 16,000 volts, is connected to an inner conductive coating of the tube by a metal button recessed in the side of the tube bulb. The inner coating extends from a point near the screen to the electron gun. As a result, the electric field set up by the high voltage applied to the coating impels the beam toward the screen. Whenever the beam strikes the screen, a dot of light appears; as the beam moves across the screen, the succession of light dots forms a visible line. If the video signal now varies the number of electrons in the beam, the dots of light will vary similarly in intensity, and the appropriate variations over the screen form an image.

Mounted along the neck of the tube are several devices designed to help produce the desired image (see Fig. 12·44). Closest to the screen end of the tube is the deflection yoke to deflect the beam horizontally and vertically. Next is a centering device to place the image squarely on the screen and prevent it from moving off to one side. Sometimes a focus magnet will follow the centering device, although in the newer tubes, focusing is achieved by placing the proper voltage on a special grid in the electron gun. Finally, there may be an ion-trap magnet. Its purpose is to prevent any gas molecules that may be present from striking the screen. Gas ions will be found even though the tube has been evacuated. It is impossible to attain a perfect vacuum, and the few gas ions that may

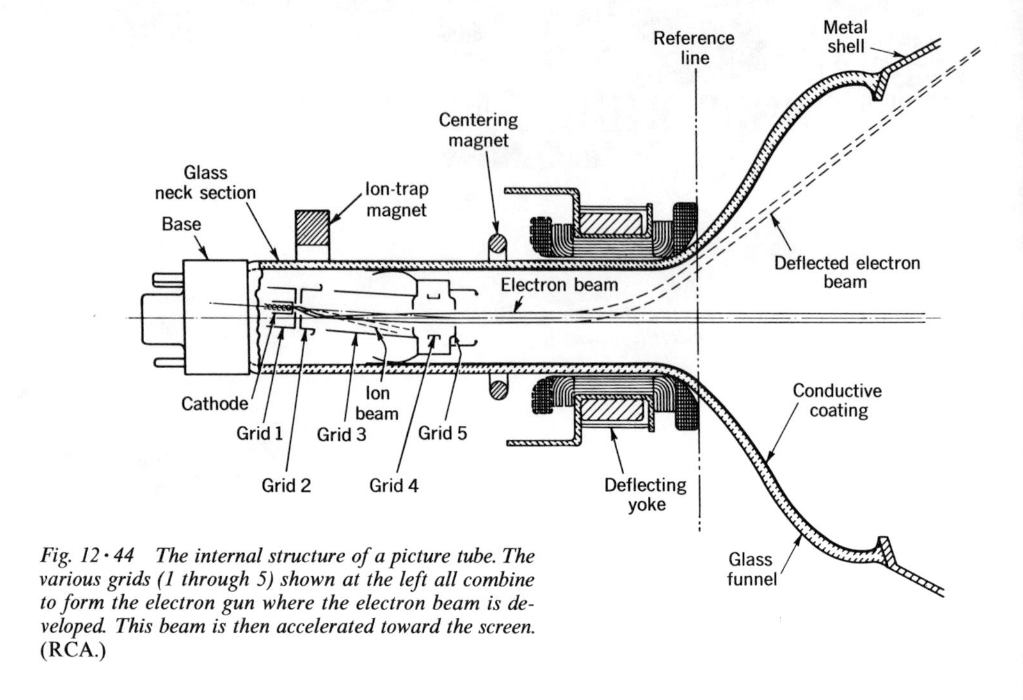


*Fig. 12·44 The internal structure of a picture tube. The various grids (1 through 5) shown at the left all combine to form the electron gun where the electron beam is developed. This beam is then accelerated toward the screen.* (RCA.)

remain can damage the screen if they are drawn to it at high velocity. The ion-trap magnet is set for maximum brightness on the screen.

In the newer tubes, ion traps are unnecessary because the inner side of the fluorescent screen is coated with a microscopically thin layer of aluminum. This stops the relatively large ions, but it permits the smaller electrons to pass through. This type of picture tube is referred to commercially as having an *aluminized* screen.

**Television-receiver controls.** Television receivers possess more operating controls on the front panel than radio receivers. These are as follows:

*Contrast,* or *picture,* regulates the contrast between the dark and light portions of the picture. It is similar to the volume control on a sound receiver. The setting depends upon the amount of light in the room and the personal preference of the viewer. This control varies the amount of video signal for the picture tube.

*Background,* or *brightness,* controls the brightness of the scanning beam. Actually, within the circuit, it fixes the d-c operating bias for the cathode-ray tube. If it is turned up too high, the beam saturates easily, and the image appears bright, thin, and watery. If it is turned down too low, the general appearance of the image is dark. The normal position of the control is such that, with no incoming signal, the scanning roster is just visible.

*Hold* (one for the horizontal and one for the vertical sweep system) varies the free-running frequency of the vertical and horizontal sweep oscillators. It is adjusted until the picture is locked in with the sync pulses of the incoming signal.

*Fine tuning* varies, within narrow limits, the frequency of the front-end r-f oscillator. This is necessary to counteract slight drifting of the oscillator frequency during operation.

*Station-selector* switch controls the channel to which the set is tuned. This controls the r-f circuits, including the local oscillator.

In addition to these, there are the usual sound controls (volume and tone), plus the ever-present on/off switch. If the receiver is capable of receiving other standard broadcast bands, then a further addition, in the form of a tuning dial, will be found. This represents the maximum number of controls found on any commercial television receiver. On some of the sets, even some controls listed previously are found on the rear panel. However, the usual controls in the back of the chassis are:

*Linearity.* The beam should travel across the screen at an even rate. If it does not, the picture will tend to crowd together. To correct this condition, two controls are available. These are the vertical and horizontal linearity controls. If the picture is crowded on the left or right side, the horizontal linearity control is adjusted. If the picture is crowded together at the top or bottom, the vertical linearity control is adjusted.

*Height* determines the height of the picture and also affects the vertical linearity.

*Width* determines how wide the picture is. Both height and width controls are properly adjusted when the picture fits its allotted frame.

The foregoing represent the basic controls. Special designs may produce additional controls, and when these are encountered, their purpose will generally be indicated by their names.

## SUMMARY

1. A-c/d-c receivers employ a transformerless power supply. The filaments are series-connected across the power line.
2. Portable receivers may rely solely on battery power, or they may operate either from batteries or the a-c or d-c power line.
3. The ultimate in portability is achieved with transistor receivers. Here no filament voltage at all is required, and only a small battery powers the transistors themselves. The small size of transistors permits extensive miniaturization of the various circuit components. The circuit arrangement using transistors is similar to the stage sequence in vacuum-tube receivers. Because transistors at the present time are capable of less gain, however, more transistors may be employed.
4. A radio-phonograph combination brings together a radio receiver and a record changer in one cabinet. The receiver has all the stages normally used for this purpose; the phonograph feeds its signal into the audio system.
5. AM-FM receivers are another popular combination, with one chassis housing both types of receivers. One audio system serves both receivers; furthermore, many such sets employ the same i-f tubes, although with different resonant circuits because of the wide difference in intermediate frequencies.
6. A limiter stage removes amplitude variations that may be present in the FM signal.
7. The purpose of the FM discriminator is to convert the frequency variations in the FM signal into corresponding audio voltages.
8. With FM multiplex operation, stereo transmission and reception is made possible. The main carrier of the FM multiplex signal combines the left (L) and right (R) audio signals. The 38-kc subcarrier sidebands contains the (L − R) component. In addition, a 19-kc pilot carrier is also transmitted.

9. A communications receiver operates over a wide range of frequencies and deals with a variety of broadcasting services. Because of the wide frequency coverage, band-switching is employed. To receive c-w code signals, a beat-frequency oscillator (BFO) is included.
10. An automatic noise limiter is a circuit which reduces the interference caused by strong noise pulses.
11. Television receivers are more complex than radio receivers because they must develop a picture of the scene being transmitted as well as reproduce the sound of the program.
12. The r-f section of a television receiver requires a special tuning mechanism to receive the 6-Mc-wide sound and video signal. The most common tuners are the turret-type unit and the rotary-switch type for the 12 VHF channels. If the UHF band is to be received also, an auxiliary tuning assembly is required. Then the fine tuning control for VHF becomes the main tuning for UHF.
13. Sound and video signals remain together until the video second detector. Then the sound signal, which at this point is an FM signal with a center frequency of 4.5 Mc, is fed to a separate sound system, where it is converted to its corresponding audio voltages and then applied to a loudspeaker.
14. The video signal, after the second detector, is amplified by one or more video amplifiers, then sent to a picture tube. Here it varies the beam intensity to reproduce the picture on the screen.
15. To move the electron beam in the picture tube vertically and horizontally, special deflection voltages are produced by the vertical and horizontal sweep systems.
16. Vertical and horizontal synchronizing pulses are sent along with the video signal so that the beam's up-and-down and back-and-forth motion is geared to the way the picture is developed originally at the transmitter.

## SELF-EXAMINATION

Here's a chance to find out how well you have learned the material in this chapter. Work the exercises; then check your answers against the Key at the back of this book. These exercises are for your self-testing only.

1. The most common sequence of stages in an a-c/d-c receiver is (*a*) mixer, oscillator, i-f amplifier, detector, audio amplifier, and output amplifier; (*b*) converter, two i-f amplifiers, detector, audio amplifier, and output amplifier; (*c*) converter, one i-f amplifier, detector, audio amplifier, and output amplifier; (*d*) r-f amplifier, converter, two i-f amplifiers, detector, and three audio amplifiers.
2. Transistor receivers require less battery power than vacuum-tube receivers because (*a*) transistors are larger in size than tubes; (*b*) transistors have no filaments to be heated; (*c*) fewer transistors than tubes are used; (*d*) transistor filaments are connected in parallel, while tube filaments are placed in series.
3. The phonograph of a radio-phonograph combination requires only which sections of the receiver circuit? (*a*) Audio stages; (*b*) detector and audio stages; (*c*) i-f and audio stages; (*d*) audio-output stage.
4. FM receivers utilize separate r-f oscillators and mixer stages because (*a*) of the nature of FM signals; (*b*) FM signals operate at high frequencies, so that a separate mixer and oscillator provide greater stability; (*c*) above a certain frequency, it is not possible to use the same tube to mix signals as well as generate the necessary oscillator voltage; (*d*) a single converter would be more expensive.
5. The same i-f section can be employed for FM and AM signals because (*a*) the signals use different elements in the tubes; (*b*) the same tuned circuits take care of both signals; (*c*) FM signals require very little amplification; (*d*) AM signals can easily pass through FM tuning circuits, and FM signals can easily pass around AM tuning coils.
6. In FM multiplex reception, a monaural receiver uses only (*a*) the 19-kc pilot signal; (*b*) the (L − R) sidebands; (*c*) the (L + R) carrier signal; (*d*) all three of the above components.
7. Bandswitching is required in communications receivers because (*a*) of the wide range of

audio frequencies received; (*b*) of the wide range of radio frequencies covered; (*c*) signals are to be received; (*d*) stations are frequently situated very close together.

8. A noise limiter in a communications receiver (*a*) kills the noise as it enters the set; (*b*) produces a voltage which is out of phase with the noise voltage, thereby canceling out the latter; (*c*) reduces the aural effect of noise pulses, without completely removing them from the circuit; (*d*) becomes active only when FM signals are received.
9. In a television receiver, the vertical and horizontal sweep frequencies are, respectively, (*a*) 30 and 15,750 cycles; (*b*) 60 and 15,750 cycles; (*c*) 60 and 525 cycles; (*d*) 30 and 525 cycles.
10. The video and sound signals received by a television receiver possess which forms of modulation? (*a*) AM for sound and FM for video; (*b*) FM for sound and FM for video; (*c*) AM for sound and AM for video; (*d*) FM for sound and AM for video.

## QUESTIONS AND PROBLEMS

1. Why are the tube filaments in an a-c/d-c receiver connected in series? What must be added when the total required filament voltage is less than 117 volts?
2. How do the filaments receive their voltage in a three-way receiver? Illustrate your answer for battery, a-c, and d-c operation.
3. Explain how AVC is achieved in a transistor receiver.
4. What happens in a radio-phonograph combination when the system is switched from radio operation to phonograph operation? Use Fig. 12·9 as your example.
5. How does the limiter stage in an AM-FM receiver operate? Use the limiter in Fig. 12·4 as your illustration.
6. Explain four differences between the AM and the FM sections of an AM-FM receiver.
7. Indicate the purpose of each of the front-panel controls used in the communications receiver of Fig. 12·28.
8. Which stages in a television receiver are not found in either AM or FM radios? Describe briefly the purpose of these stages.
9. Describe the manner in which an electron beam scans the face of a picture-tube screen. Illustrate your answer.
10. Name five controls found on the front of a television receiver. Explain what they do.

# Review of chapters 10 to 12

## SUMMARY

1. A t-r-f receiver consists of one or more t-r-f stages followed by a detector and audio amplifier. Each t-r-f stage is resonant at the r-f carrier frequency.
2. Sensitivity in a receiver means high gain to produce audio output from the detector with weak antenna signal.
3. Selectivity of a receiver is the ability to tune in just the desired carrier frequency. The more tuned circuits in the receiver, the better is its selectivity.
4. A ganged tuning capacitor has two or more sections, one for each tuned circuit in cascaded stages.
5. Alignment means adjusting each tuned circuit of cascaded stages to the correct resonant frequency.
6. A detector rectifies modulated r-f signal to recover the audio modulation. Common types are plate detector, grid-leak detector, and diode detector, the last being the most common.
7. Using a potentiometer of 0.5 to 2 M for the diode-detector load resistance is the most common method of volume control.
8. The superheterodyne receiver uses a local oscillator to beat with the r-f signal, converting the station's carrier frequency to the intermediate frequency of the receiver. The local oscillator frequency is varied when you tune the receiver to convert all stations to the same intermediate frequency.
9. Pentagrid converter tubes, such as 6BE6 and 6SA7, are commonly used for the frequency conversion.
10. A t-r-f stage may be used as a buffer ahead of the converter for better signal-to-noise ratio, less oscillator radiation from the antenna, and better r-f selectivity to reduce spurious responses.
11. The image frequency is equal to the r-f signal frequency, plus twice the intermediate frequency.
12. An i-f stage is a pentode, class A amplifier, generally with double-tuned transformer coupling. The stage is fixed-tuned at the intermediate frequency. Most of the receiver's selectivity is provided by the i-f section. Typical i-f values are 455 kc, 456 kc, and 468 kc in AM radios, or 10.7 Mc in FM receivers for the 88- to 108-Mc band.
13. The AVC circuit automatically controls the i-f and r-f gain according to signal strength. Stronger signal produces more negative AVC bias and less receiver gain.
14. In delayed automatic volume control, a delay bias cuts off the AVC rectifier for weak signals.
15. In quiet AVC, squelch, or silencer circuits, the AVC bias operates a squelch tube that cuts off the audio amplifier to quiet the receiver between stations when no signal is being received.
16. Maximum AVC bias indicates maximum signal for exact tuning. A magic-eye tuning indicator is controlled by AVC voltage to close the eye.
17. To align a superheterodyne receiver, adjust: (*a*) i-f circuits for maximum output, (*b*) oscillator padder at 600 kc, (*c*) oscillator trimmer at 1,400 kc, (*d*) r-f trimmer at same setting of 1,400 kc.
18. All the amplifier stages in the receiver are connected to the B supply for positive plate and screen voltages. Cathodes return to B−.
19. A-c/d-c receivers use a transformerless power supply, with series heaters.

20. Portable receivers are battery-operated; three-way receivers can also operate from the power line. Transistors are used for miniature portable receivers.
21. In radio-phonograph operation, the receiver audio section is used to amplify output of the phonograph pickup.
22. In AM-FM receivers, separate r-f and frequency-converter stages are needed. Generally the same i-f stages are used with individual tuned circuits for 455 kc and 10.7 Mc. The FM detector is generally a ratio detector or discriminator, using two diodes in a balanced detector circuit. These detector circuits recover the audio modulation from the frequency variations in the FM signal. The audio section of the receiver amplifies the detector output for either AM or FM reception.
23. A communications receiver has high sensitivity for weak signals, with multiple bands for wide frequency coverage. To receive c-w code signals, the receiver often includes a BFO. A noise-limiter circuit may be included to reduce interference from strong noise pulses.
24. A television receiver includes an AM section for the AM picture signal, FM section for the FM sound signal, with deflection and synchronizing circuits for scanning the screen of the picture tube.
25. The r-f section of a television receiver uses a separate r-f tuner, with provision for 6-Mc bandwidth on each channel to include the r-f picture-carrier signal and r-f sound-carrier signal.
26. Since the picture- and sound-carrier frequencies are always separated by 4.5 Mc, intercarrier sound receivers use this 4.5-Mc beat for automatically tuning in the sound signal when the picture is tuned in.
27. The detected AM picture-carrier signal provides video signal, corresponding to the desired picture information. After being amplified in the video amplifier, the video signal is coupled to the picture-tube grid-cathode circuit.
28. Deflection of the electron beam in the picture tube is produced by vertical and horizontal deflection circuits in the receiver. The vertical deflection circuits produce 60-cycle sawtooth output for vertical scanning; the horizontal deflection circuits produce 15,750-cycle sawtooth output for horizontal scanning. The timing of these deflection circuits in the receiver is controlled by synchronizing pulses in the video signal from the station.

## REFERENCES (*Additional references at back of book.*)

### *Books*


Ghirardi, A. A., and J. R. Johnson, *Radio and Television Receiver Circuitry and Operation,* Rinehart & Company, Inc.

Grob, B., *Basic Television,* 3d ed., McGraw-Hill Book Company.

Henney, K., and G. A. Richardson, *Principles of Radio,* 6th ed., John Wiley & Sons, Inc.

Hickey, H. V., and W. M. Villines, *Elements of Electronics,* 2d ed., McGraw-Hill Book Company.

Kiver, M. S., *TV Simplified,* D. Van Nostrand Company, Inc.

———, *FM Simplified,* D. Van Nostrand Company, Inc.

Levy, A., and Frankel, M., *Television Servicing,* McGraw-Hill Book Company.

Sheingold, A., *Fundamentals of Radio Communication,* 3d ed., D. Van Nostrand Company, Inc.

Slurzberg, M., and W. Osterheld, *Essentials of Radio-Electronics,* 2d ed., McGraw-Hill Book Company.

Van Valkenburgh, Nooger, and Neville, Inc.: *Basic Electronics,* John F. Rider, Publisher, Inc.

Watson, H. M., H. E. Welch, and G. S. Eby: *Understanding Radio,* 2d ed., McGraw-Hill Book Company

Zbar, P. B., and P. W. Orne, *Advanced Servicing Techniques,* John F. Rider, Publisher, Inc.

## REVIEW SELF-EXAMINATION

Here's another chance to check your progress. Work the exercises just as you did those at the end of each chapter and check your answers.

Answer true or false

1. The preselector stage ahead of the converter is an example of a t-r-f stage.
2. The i-f amplifier operates class C for maximum gain.
3. Critical coupling in the i-f transformer will provide maximum bandwidth without double peaks.
4. I-f stages often have AVC bias.
5. With an intermediate frequency of 455 kc and the local oscillator at 1,855 kc, the receiver will tune in a station at 1,400 kc.
6. For Question 5, the image frequency is above the standard broadcast band.
7. With an intermediate frequency of 10.7 Mc, to tune in an FM radio station at 100 Mc, the local oscillator can be at either 110.7 Mc or 89.3 Mc.
8. About 3 volts input signal is a typical value for a diode detector.
9. Selectivity measures how much gain the receiver has.
10. A receiver will produce 1-volt output from the detector with 50-$\mu$v input signal at the antenna. This measurement indicates receiver sensitivity.
11. A diode detector usually has B+ voltage to make the rectifier conduct.
12. Rectified output voltage from a detector indicates i-f signal input.
13. In a two-gang tuning capacitor with one smaller section, this section is for oscillator tuning.
14. A typical volume control serving as detector load resistance is 10,000 ohms, for a vacuum-tube diode.
15. With stronger signal, the AVC bias becomes more negative.
16. If you align the i-f stages for maximum i-f output, the output of the detector will increase.
17. A squelch circuit silences converter noise between stations by cutting off the audio section.
18. About 1-volt output from a phonograph pickup can drive the audio section in a typical receiver.
19. If one heater is open in a series string, the full-line voltage will be across the open heater.
20. If the primary of the audio-output transformer is open, the plate voltage on this stage will be zero.
21. If the AVC bias is shorted to chassis, the receiver gain will be maximum.
22. In the 6BE6 pentagrid converter the first control grid next to cathode is the oscillator grid.
23. The Hartley circuit cannot be used for the local oscillator in a receiver.
24. An *LC* circuit resonant at 10.7 Mc has maximum impedance at 455 kc.
25. The capacitance of 50 pf has 1-M reactance at 10.7 Mc.
26. Transistor receivers need a voltage-doubler power supply.
27. Miniature portable transistor receivers often use 6 to 8 transistors for enough gain.
28. The Foster-Seeley discriminator is balanced at center frequency.
29. If a radio-phonograph receiver operates normally on phonograph position but not on radio, the trouble must be in the audio-output stage.
30. The ratio detector and gated-beam detector are two examples of FM detector circuits.
31. The horizontal scanning and synchronizing frequency in television receivers is 15,750 cps.
32. The vertical scanning and synchronizing frequency is 60 cps.
33. The video signal in television receivers provides the desired picture information.
34. In television broadcasting the separation between r-f picture and sound carriers is 6 Mc.
35. Intercarrier sound receivers have a 4.5-Mc-sound i-f section.
36. The FM sound signal in television broadcasting has 25-kc maximum deviation.
37. The local oscillator in the r-f tuner of a television receiver tunes in the desired channel.
38. The BFO in a communications receiver is used for c-w radiotelegraph emission.
39. In an i-f amplifier with AVC bias, plate current decreases with more signal.
40. A magic-eye tube for tuning indicates AVC bias voltage.

# Chapter 13 Test instruments

Test instruments represent the most useful tool a technician can have in his work with electronic circuits. When a piece of equipment breaks down, test instruments will reveal not only which section of the equipment is at fault, but actually the component within that section that is causing the trouble. The same instruments will also help the technician select a suitable substitute when an exact replacement part cannot be obtained. New equipment frequently requires a number of preliminary adjustments before proper operation is achieved, and here, again, test instruments play a vital role. In fact, there is very little maintenance, design, or service work that can be performed on electronic equipment without suitable test instruments. In view of their importance, this chapter will examine those instruments which the technician will use most extensively. The topics are as follows:



## *13 · 1 Volt-ohm-milliammeters*

A volt-ohm-milliammeter, or VOM for short, is an instrument that can measure voltage, resistance, and current. In the latter category, it is customary to find that the current measurements extend into the ampere range also. All quantities are d-c, with the exception of the volt range where a-c volts can be measured as well.

Externally, a typical VOM appears as shown in Fig. 13 · 1. Internally, the instrument consists essentially of separate voltage, resistance, and current-measuring circuits, with only the meter movement in common. A selector switch may be employed to set up the desired circuit for a certain measurement, or the switch may be dispensed with, and a large number of pin jacks made available on the front panel. The instrument user will then insert the test leads in the desired jacks and make the measurement. The VOM shown in Fig. 13 · 1 employs the first approach, thereby reducing the number of needed pin jacks to just a few. This approach is more convenient for the user, although it tends to increase the price of the instrument somewhat.

**Ohmmeter section.** The ohmmeter section of a typical VOM is shown in Fig. 13 · 2. The illustrations have been set up to show the actual circuit when the selector switch is in the $R \times 1$, $R \times 100$, and $R \times 10{,}000$ positions. If we examine the first of these circuits, Fig. 13 · 2*a*, we see that when the positive and negative terminals are shorted together, the full 1.5 volts of the internal battery appear across the 11.5-ohm resistor. This voltage drop, in turn, causes a current to flow through the meter movement and the three resistors in series (1,138 ohms, 21,850 ohms, and the 10,000-ohm zero-adjust potentiometer). The ZERO ADJUST control is rotated until the

*Fig. 13 · 1 Typical volt-ohm-milliammeter.* (Simpson Electric Co.)

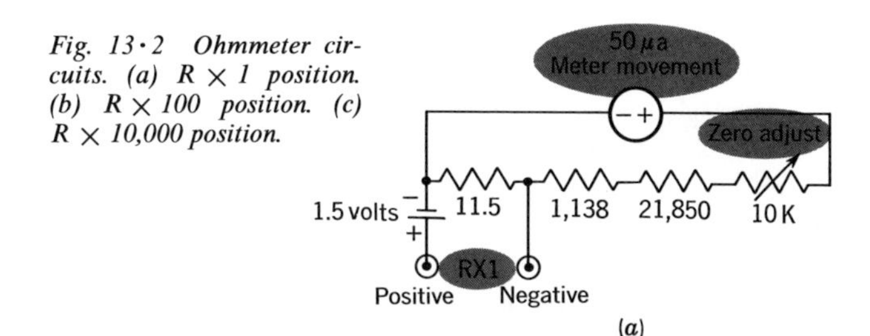


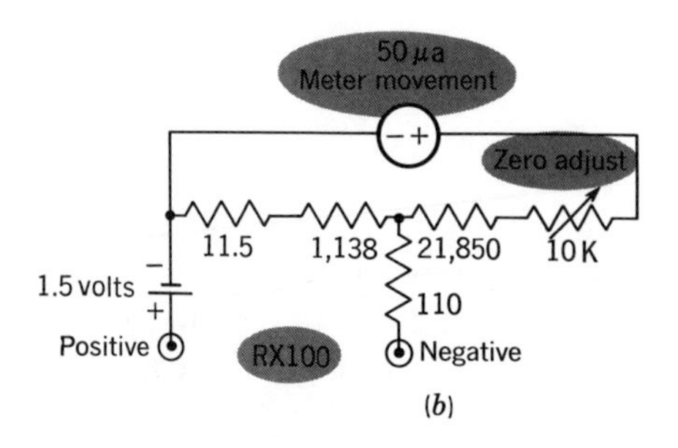


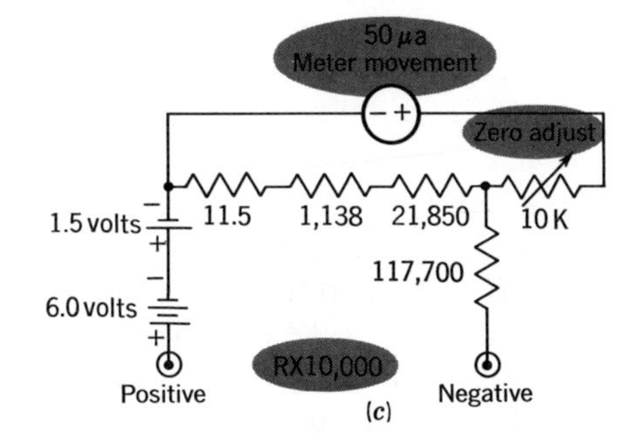


*Fig. 13 · 2 Ohmmeter circuits. (a) R × 1 position. (b) R × 100 position. (c) R × 10,000 position.*

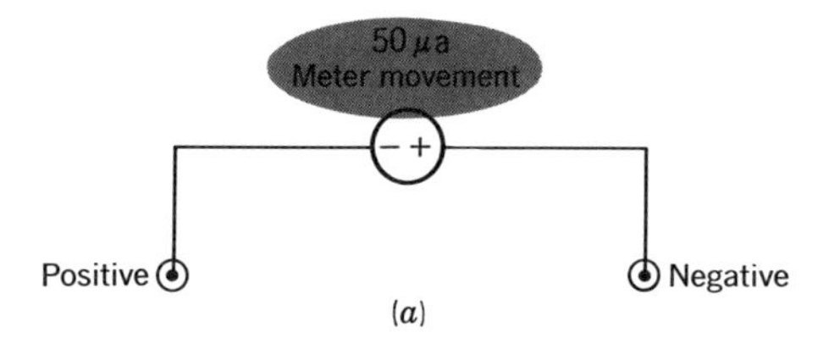


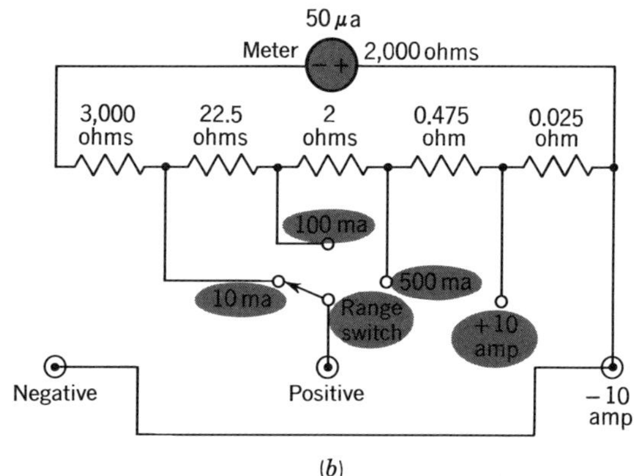


*Fig. 13·3 Current ranges of VOM. Meter movement itself has resistance of 2,000 ohms. (a) 50-μa range. (b) Milliampere and ampere ranges.*

meter needle is over the zero-ohms marker on the far right-hand side of the scale.

If we insert an 11.5-ohm resistor externally between the positive and negative terminals, we will reduce the voltage drop across the internal 11.5-ohm resistor by half, causing the meter needle to drop to mid-scale. At this point the value of 11.5 (actually 12) ohms should be indicated on the scale. By increasing the value of any external resistance between the positive and negative terminals, less and less of the 1.5 volts will appear across the internal 11.5-ohm resistor, and the meter needle will deflect less and less to the right.

Comparison of Fig. 13·2*a* with Fig. 13·2*b* shows that two additional resistors are connected across the input for the $R \times 100$ position. The 110-ohm resistor is in series with the negative terminal, while the 1,138-ohm resistor provides more internal voltage drop to extend the range of the ohmmeter.

For the $R \times 10{,}000$ position (Fig. 13·2*c*), two new items are brought into the circuit. These are a 117,700-ohm resistor and a 6.0-volt battery. More voltage is needed to produce the same meter current with higher values of resistance.

**D-c milliammeter and ammeter circuit.** The current-measuring section of the instrument is shown in Fig. 13·3. The smallest current range is 50 μa. When the instrument is set to this range, all the current flows through the meter. There are no shunts. Then 50 μa produces full-scale deflection because this is a 50-μa movement. Less current will produce proportionately less needle deflection along the scale.

The total current applied to this meter movement must never exceed 50 μa. This means that as the current coming into the system rises, more and more of it must be diverted through the shunt path. If you examine Fig. 13·3*b*, you will note that with each higher current position, the resistance of the shunt path decreases. These ranges are for direct current. Conventional multimeters do not have provisions for reading alternating current, although they do measure a-c voltage.

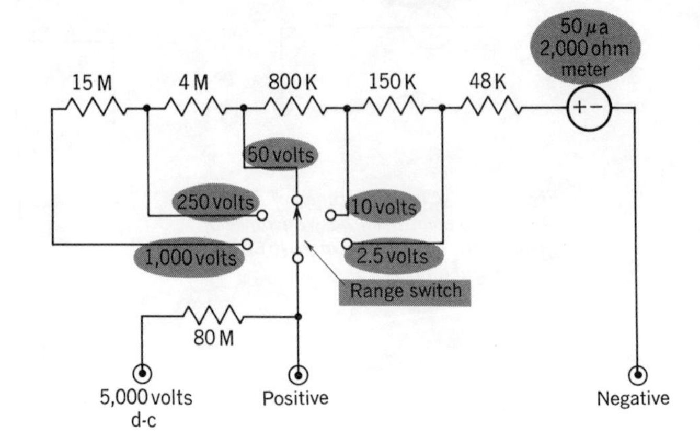


*Fig. 13 · 4 D-c voltmeter section of VOM.*

**D-c voltmeter circuit.** A simplified diagram of the d-c voltmeter portion of a typical VOM is shown in Fig. 13 · 4. This consists basically of the meter movement and a number of series resistors, called *multipliers.* Their purpose is to limit the maximum current in the circuit to 50 μa or less for any applied voltage within a specified range. Thus, when the selector switch is in the 50-volt position, there are three resistors in the circuit. These are 800,000, 150,000, and 48,000 ohms. In addition, the meter movement itself possesses an internal resistance of 2,000 ohms; thus the total series resistance in the circuit is 1,000,000 ohms. A current of 50 μa will be produced by application of 50 volts to such a circuit. Therefore, full-scale deflection corresponds to 50 volts in this range.

The same reasoning holds true for all the other d-c voltage ranges. Each multiplier limits the full-scale current to 50 μa for its range.

**A-c voltmeter circuit.** The a-c voltmeter section of the meter is shown in Fig. 13 · 5. With the exception of the rectifiers, the arrangement is basically the same as the d-c voltmeter circuit. The purpose of the rectifiers is to provide the meter movement with direct current, this being the only type of current to which it will respond. Only one half-cycle of current passes through the meter, and this occurs when the terminal marked POS is positive with respect to the negative, or common, terminal. Current then flows through the meter in the direction indicated by the solid arrows. During the reverse half-cycle, the current path is around the meter, as revealed by the dotted arrows. Shunt resistor $R_2$ and series resistor $R_1$ are precision-wound and calibrated for the rectifier with which they are used.

**Operating controls.** The front panel controls of most VOMs are relatively few in number and simple to manipulate. Two leads are provided for insertion in appropriate pin jacks. For example, if d-c voltages are to be measured, one lead (generally red) is plugged into the pin jack marked +. The other lead goes to the pin jack marked −. Then the selector switch is turned to the desired voltage level. For the VOM in Fig. 13 · 1, the small knob at the far left-hand side is turned to −DC or +DC, depending on the

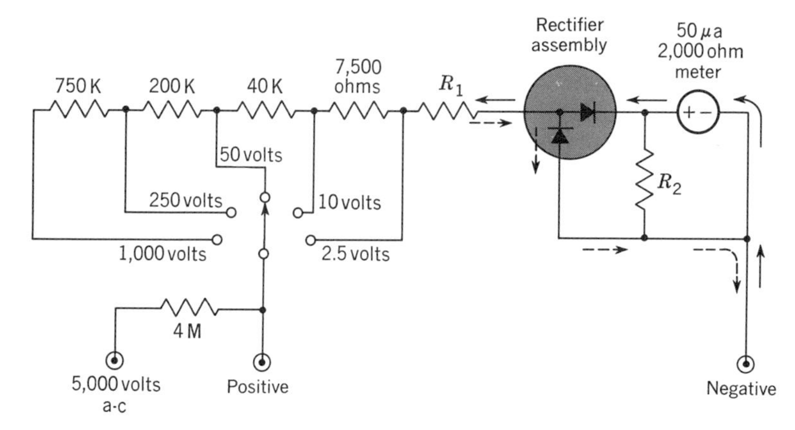


*Fig. 13·5 A-c voltmeter circuit of VOM. Polarity of input terminals does not matter for a-c measurements.*

polarity of the voltage to be measured. Thus, if the lead connected to the common terminal connects to the negative side of a voltage and the lead in the POS terminal goes to the positive side of a voltage, then the small knob should be set to the +DC position. If the same leads connect to opposite polarity voltages, however, the meter needle will attempt to travel to the left, off the scale. To have the needle read up-scale, in the proper manner, one of two actions can be taken. Either the small knob can be turned to the −DC position, or the lead positions can be reversed.

When setting the selector control, make certain that the scale chosen will accommodate the level of the voltage to be checked. *If the approximate value of this voltage is not known, start with the highest voltage range and work down.*

Another precaution to observe is to read the proper scale when a reading is to be made. Note that both a-c and d-c voltages use the same 0–10, 0–50, and 0–250 scales. For readings in the 0–2.5-volt range, the 0–250-volt markings are used, and each value is divided by 100 to bring it to the proper level. For the 0–1,000-volt range, each reading on the 0–10 scale is multiplied by 100. Finally, for the 0–5,000-volt range, each reading on the 0–50 scale is multiplied by 100.

For a-c voltage measurements, the same two front-panel pin jacks are employed and the same voltage ranges. For the VOM in Fig. 13·1, the small knob at the left-hand side of the panel is turned now to the a-c position. On other multimeters the range switch may have separate positions for a-c voltage measurements.

Note, though, that when d-c voltages between 1,000 and 5,000 volts are to be measured, the pin jack for the positive lead is the one marked "D.C. 5000 V." For a-c voltages in this range, the same lead is inserted in the pin jack marked "A.C. 5000 V." This is just to the left of the 5,000-

volt d-c pin jack. For this measurement, the large central selector switch is set to the position labeled "1000 V." (NOTE: Just beneath this number, "5000 V." is indicated in small letters.)

The negative lead (for direct current) remains in the same common pin jack. For a-c voltage measurements, where polarity is not important, the other lead also remains here.

The − and + pin jacks are employed again for resistance measurements although the polarity markings have no significance now. The large selector switch is rotated to the proper mark ($R \times 1$, $R \times 100$, or $R \times 10{,}000$), and the resistance scale readings are multiplied by the factor indicated.

It is important, however, before any resistance measurements are made, that the ends of the two test leads be shorted together and the ZERO OHMS knob at the right adjusted until the meter needle is directly over the zero level. When this has been done, the leads are separated and resistance measurements begun. The ohmmeter zero adjustment must be made for each range.

Direct-current measurements up to 500 ma are made by connecting the leads to the same two pin jacks and the selector switch set to the proper current range (0–1 ma, 0–10 ma, 0–100 ma, or 0–500 ma). For the 10-amp range, the selector switch is rotated to the 10-ma position. (NOTE: This is also marked as 10 amp.) However, one test lead is inserted in the pin jack marked +10A, while the other lead goes to the pin jack marked −10A. This is done to prevent heavy current from passing through the low-current circuits, such as the selector switch.

The foregoing discussion outlines the procedure for setting the front-panel controls to ohms, volts, milliamperes, and amperes. Before the leads actually make contact with the circuit or component to be checked, always check carefully the setting of the center selector switch and the small left-

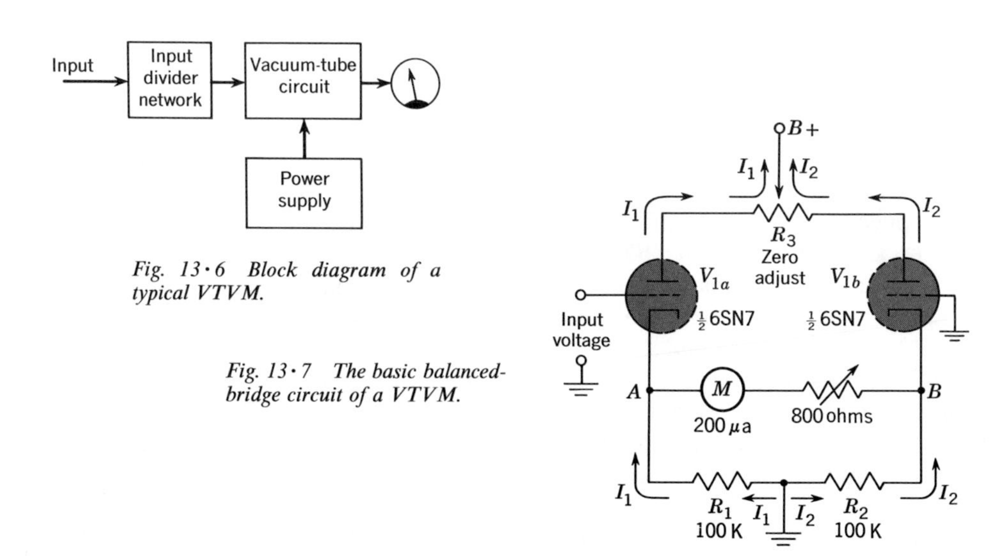


*Fig. 13·6 Block diagram of a typical VTVM.*

*Fig. 13·7 The basic balanced-bridge circuit of a VTVM.*

hand control. This is particularly important when voltages are to be measured. For resistance checks, the ZERO OHMS knob should be remembered.

## 13 · 2 Vacuum-tube voltmeters

The VTVM is an instrument which performs essentially the same measurements as a VOM. The VTVM, however, because of its design, can cover a wider range of voltages (a-c and d-c) and resistances, and, further, it does this with less loading effect on the circuit under test. To achieve this greater flexibility, the VTVM requires a source of power, either alternating current from the power line or direct current from a battery.

Basically, the VTVM consists of an input-divider network, a vacuum-tube circuit, which is generally a balanced-bridge network, and a power supply (see Fig. 13 · 6). In the discussion that follows, the role that these circuits perform in voltage and resistance measurements will be examined.

**Balanced-bridge circuit.** The circuit which is most widely employed in the VTVM is the balanced bridge shown in Fig. 13 · 7. The current through each tube flows in the path indicated by the various arrows. For $V_{1a}$, current $I_1$ flows from the plate through part of $R_3$ to $B+$ and from ground through $R_1$ back to the cathode of the tube. This current, in flowing through $R_1$, develops a voltage drop which places point $A$ at some positive value above ground.

If the grid of $V_{1a}$ is grounded (to place it at the same potential as the grid of $V_{1b}$), we might expect the potentials at points $A$ and $B$ to be equal, and no current would flow through meter $M$. If the currents in both paths are not identical, some difference in voltage will exist between points $A$ and $B$. In this case, current will flow through meter $M$, and its needle will deflect. To zero the meter and thus bring about a balance between both branches of the circuit, variable resistor $R_3$ is provided. Through its adjustment, the currents through $V_{1a}$ and $V_{1b}$ can be varied until points $A$ and $B$ possess identical positive potentials. $R_3$ is the knob on the front panel of the VTVM and is labeled ZERO ADJ.

To employ this circuit for the measurement of voltages, a voltage is applied between the grid of $V_{1a}$ and ground. If this voltage is positive, the current through $V_{1a}$ will increase, and point $A$ will become more positive than it was. Point $B$, on the other hand, will remain unchanged since the grid of $V_{1b}$ is grounded.

With a very definite difference of potential existing now between points $A$ and $B$, current will flow through meter $M$ from $B$ to $A$. Just how much current will flow will depend upon the value of voltage applied to the grid of $V_{1a}$; consequently, the meter dial can be calibrated directly in volts.

When a negative voltage is applied to the grid of $V_{1a}$, the current through this tube decreases, causing point $A$ to become less positive. Current will flow through meter $M$ from point $A$ to point $B$. This will force the meter pointer to move from zero toward the left. Since the zero position on most instruments is already as far to the left as the pointer normally goes,

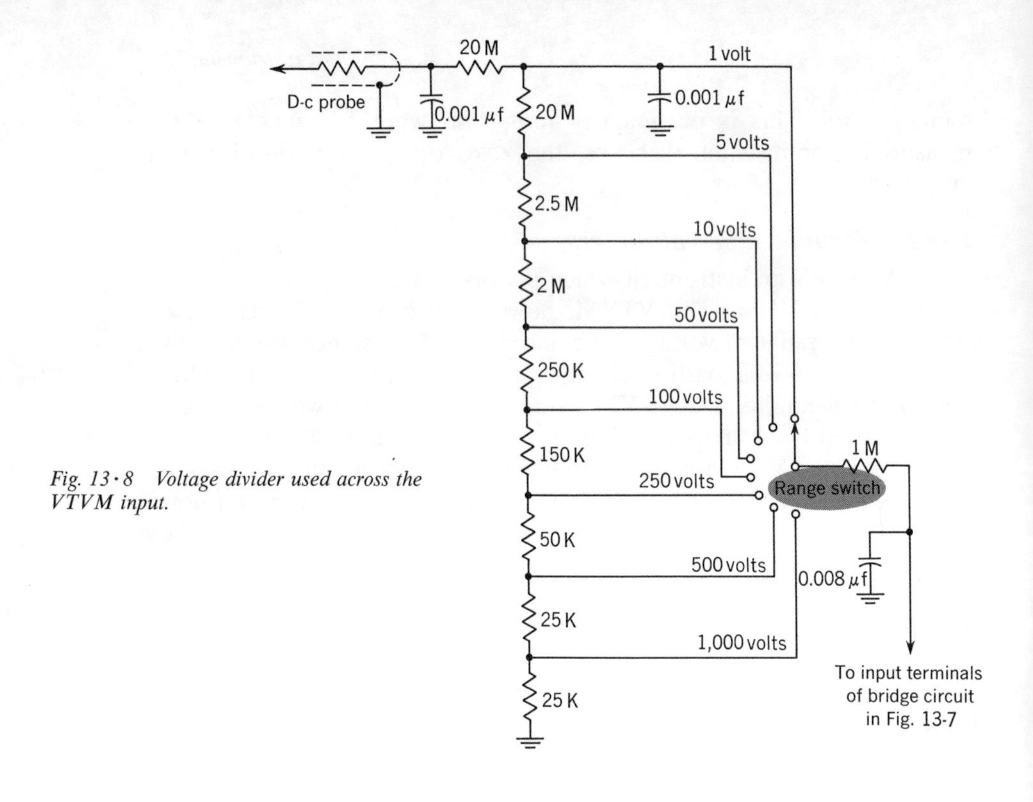


*Fig. 13·8 Voltage divider used across the VTVM input.*

applying a negative voltage to the VTVM will drive the pointer off scale. To overcome this limitation, we may either reverse the test leads or incorporate a switch which will accomplish the same thing by reversing the meter connections. This switch is known by a variety of names, but the most widely used is +DC and −DC.

In order to permit the VTVM to measure a variety or range of voltages, a voltage-divider circuit is placed across the input to the meter, as shown in Fig. 13·8. The total value of the resistances in this string is 50 M; and for the voltage ranges shown (1 volt to 1,000 volts). Therefore 50 M is the input impedance of the VTVM. With an input impedance this high, it can readily be appreciated why the VTVM scarcely disturbs the circuit into which it is connected to measure voltages. This is one of the major advantages of the VTVM.

**A-c voltage measurement.** The measurement of a-c voltages with a VTVM is based upon: (1) the rectification of the a-c voltage by a diode (sometimes by a copper-oxide rectifier), (2) the subsequent application of this voltage to the grid of the input triode of the bridge, and (3) the functioning of the bridge circuit. In the a-c positions, the meter will indicate the rms value of the voltage.

The rectifier diode may be contained in a special probe, or it may be situated within the instrument case and a conventional test prod or probe may be used for the a-c measurements. In some meters, there is a separate plug-in jack to which a-c voltages are applied and a separate plug-in jack to which d-c voltages are applied. In other models, both voltages are brought in through the same terminal. Note, however, that in all vacuum-tube voltmeters the range switch has a position for alternating current and a separate position for direct current.

The measurement of a-c voltages follows exactly the same procedure as that for d-c voltages. The only precautions to observe are that the proper probe is being used and that the selector switch has been shifted from d-c volts to a-c volts.

**Resistance measurement.** The ohmmeter section of a VTVM is shown in Fig. 13·9. A small battery (such as the Mallory RMBZ4 1.34-volt unit) is used to supply the potential. This potential, when applied to the grid of one triode section of the 6SN7 tube, is sufficient to cause full-scale deflection of the meter. A variable control, marked OHMS ADJ on the front panel, permits the operator to position the meter needle accurately so that it will stop directly over the final right-hand marking of the OHMS scale. This is done with no resistor connected between the OHMS and COMMON terminals of the meter and with the leads from these terminals not touching. The needle position at the other end of the scale should also be checked by the procedure previously outlined (with the meter leads shorted together). If necessary, the ZERO ADJ knob can be used to bring the needle directly over the zero line. These adjustments should hold for all ranges.

When the resistance under test is connected between the COMMON and OHMS test leads, a voltage-divider circuit is produced. It consists of the 1.34-volt battery in series with one of the standard resistors $R_1$ to $R_7$ and

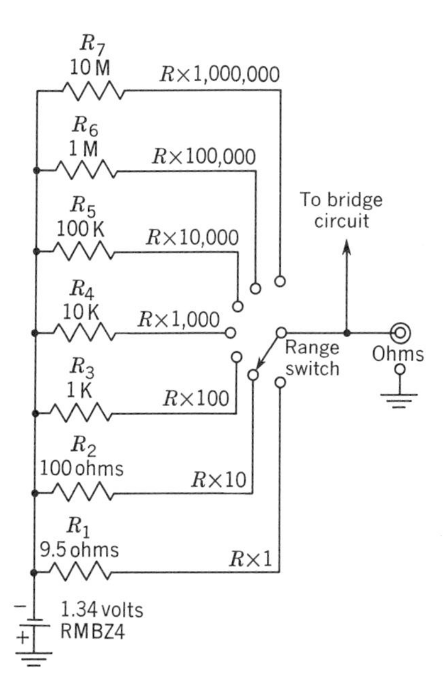


*Fig. 13·9 Ohmmeter section of a VTVM.*

the resistor under test. The voltage across the unknown resistor is proportional to its resistance. This voltage is applied to the grid of one section of the bridge circuit, and a meter deflection proportional to the unknown resistance is produced.

Modern vacuum-tube voltmeters are capable of measuring resistances up to 1,000 M (some units even go beyond this). This is 1 billion ohms and is more than sufficient for any normal service work.

When the ohmmeter is not in use, the meter needle remains at the extreme right side of the scale at the INF mark. This is opposite to its resting position for the volts scales.

**D-c section.** Most VTVMs do not provide for measuring direct current, but if this facility is present, the circuit employed is identical to the current-measurement circuit of a VOM (see Fig. 13 · 3). The vacuum-tube circuits do not enter into this measurement, so that there is no need to plug the instrument into the power line. Merely set the function selector switch to the proper current position, and connect the meter leads into the circuit where the current is to be measured.

**Operating controls.** While there are a number of vacuum-tube voltmeters commercially available, an examination of the operating controls of the unit shown in Fig. 13 · 10 will demonstrate what to expect from most units.

The VTVM in Fig. 13 · 10 has on its front panel two adjustment knobs (ZERO ADJ and OHMS ADJ), two switch controls, and three plug-in terminals. For d-c voltage measurements (both low and high voltages), a cable is connected to the DC VOLTS terminal. Another lead would be plugged into

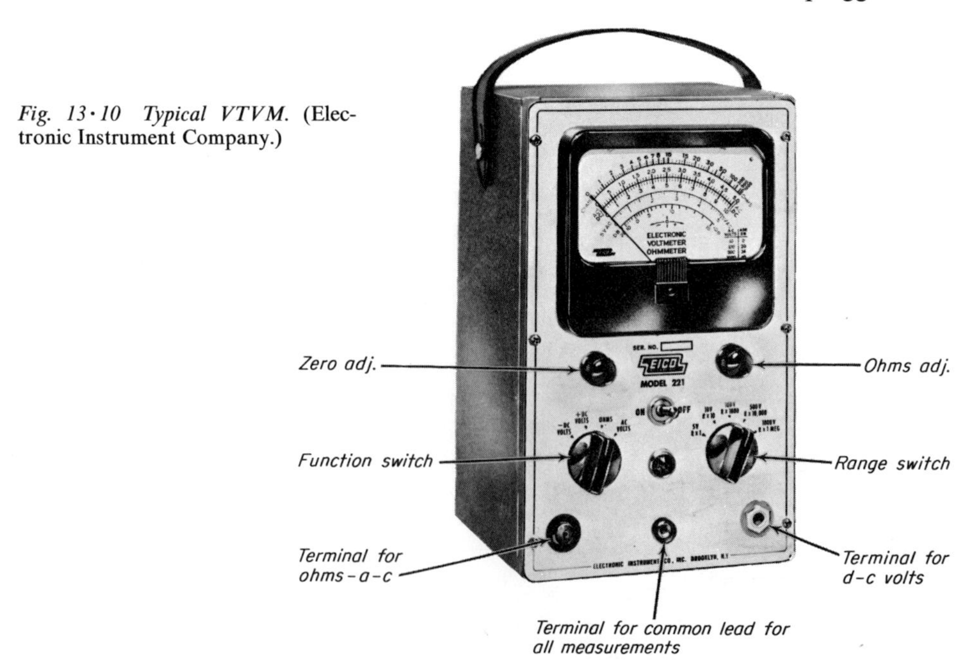


*Fig. 13 · 10 Typical VTVM.* (Electronic Instrument Company.)

the COMMON terminal, thus providing two prods or probes for voltage measurement. The same COMMON terminal is employed when a-c volts and ohms are to be measured. In these instances, however, the second lead is plugged into the terminal marked OHMS-AC.

Besides using the proper set of terminals for any given measurement, it is also necessary to move the FUNCTION and RANGE switches to the proper positions. The FUNCTION switch chooses the electrical quantity to be measured. At −DC VOLTS, the meter will record the values of any negative d-c volts applied to the d-c volts probe. At +DC VOLTS, a positive voltage should be applied to the d-c probe. Always connect the common lead to the low side of the voltage to be measured (usually chassis ground) and use the switch for reversing the polarity. The two remaining positions of the FUNCTION switch are for ohms and a-c volts.

The right-hand switch is the RANGE switch. Its position determines *how much* of the quantity to be measured can be safely handled by the meter. For example, when this switch is set to 250 volts, no more than 250 volts (a-c or d-c) can be safely measured by the instrument. Any higher voltage would drive the needle off scale and might possibly damage the instrument.

Before any work is done with this VTVM, the FUNCTION and RANGE switches should be properly set. This is an important rule to observe.

The ZERO ADJ control enables the operator to balance the meter's internal circuit and to compensate for any changes that may have occurred to upset this balance. After the meter has been placed in operation by turning on the power for a sufficient warmup period and the RANGE switch rotated to the desired voltage measuring position, the ZERO ADJ knob should be rotated to the right or left until the needle pointer is directly over the zero indicator on the meter scale. It is good practice to repeat this procedure whenever the setting of the FUNCTION switch is changed.

On a-c and d-c volts scales, only the ZERO ADJ knob need be checked. For the measurement of resistance, however, on the ohms scale, the OHMS ADJ is also employed.

## *13 · 3 The oscilloscope*

The cathode-ray oscilloscope is one of the most useful and versatile of test instruments. It is essentially a device for displaying graphs of rapidly changing voltage and current, but it is also capable of giving information concerning frequency values, phase differences, and voltage amplitude. The oscilloscope is used to trace test signals through radio receivers and audio amplifiers, to measure percentages of modulation in transmitters, and to localize the sources of distortion and hum voltages in communication equipment. It is used to measure peak a-f and r-f voltages, to measure audio amplifier gain and fidelity, to make over-all frequency response curves, and to study dynamic tube characteristic curves. These are a very few of its many applications.

The principal components of a basic oscilloscope (Fig. 13 · 11) include a cathode-ray tube, a sweep (sawtooth) oscillator, deflection amplifiers (hori-

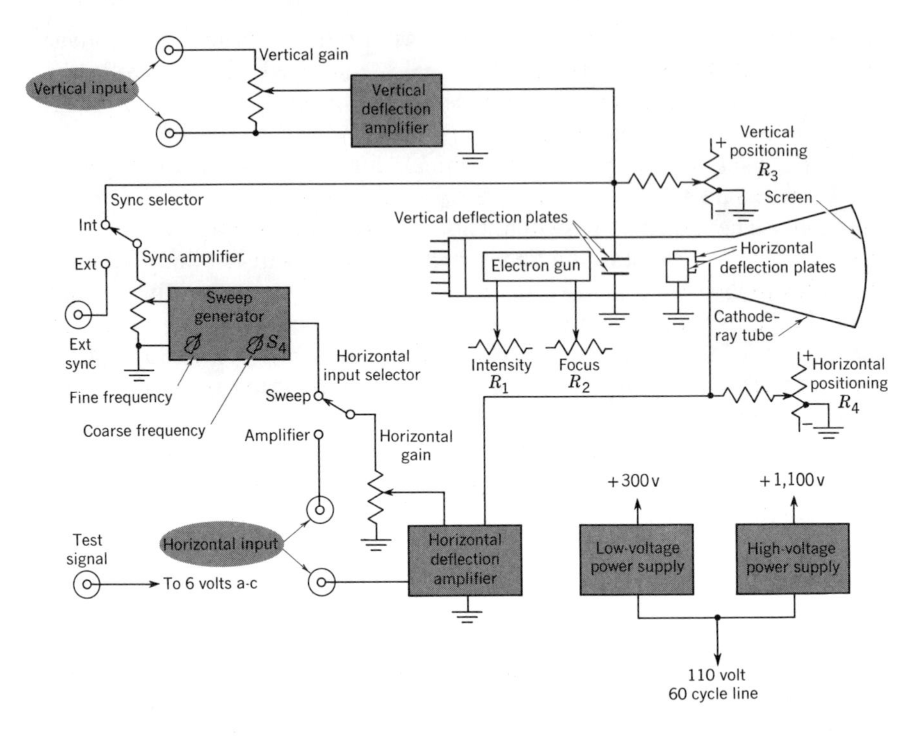


*Fig. 13 · 11 Block diagram of a cathode-ray oscilloscope.*

zontal and vertical), and suitable controls, switches, and input terminals for the proper operation of the unit.

**Cathode-ray tube.** The heart of an oscilloscope is the cathode-ray tube (CRT). This is a special type of electron tube (see Fig. 13 · 12) in which electrons emitted by a heated cathode are focused and accelerated to form a narrow beam having high velocity. The beam is then controlled in direction and allowed to strike a fluorescent screen, whereupon light is emitted at the point of impact and produces a visual indication of the beam position.

The electronic process of forming, focusing, accelerating, controlling, and deflecting the electron beam is accomplished by the following principal elements of the CRT: the electron gun consisting of a heated cathode, a grid, a focusing anode, and an accelerating anode; a deflection system, for controlling the direction of the beam emanating from the electron gun; a fluorescent screen, for visually indicating the movement imparted to the electron beam; and an evacuated glass bulb, which contains all the above elements of the CRT. Partially covering the inside of the glass bulb is an Aquadag (graphite) coating which provides a return path for electrons and at the same time serves to shield the electron beam electrostatically.

**The electron gun.** The electron gun, shown in block form in Fig. 13 · 11 (expanded in Fig. 13 · 12), provides a concentrated beam of high-velocity electrons. The cathode is an oxide-coated metal cylinder which, when properly heated, emits electrons. These electrons are attracted toward the accelerating and focusing anodes by the high positive potential of the anodes in relation to the cathode. In order to reach the anodes, however, the electrons are forced to pass through a control grid (a cylindrical piece of metal, closed at one end except for a tiny circular opening), which concentrates the electrons and starts the formation of a beam.

Electrons leaving the grid aperture are strongly attracted by the positive charge on the focusing (anode No. 1) and accelerating (anode No. 2) anodes. These electrodes are cylindrical in shape and have small openings to permit the beam to pass through. Between these two anodes, an electrostatic field exists. This field serves as an electron lens which focuses the electrons in somewhat the same manner as an optical lens focuses a beam of light. The electron lens differs from the optical lens, however, in that its focal length can be changed by simply changing the ratio of potentials between the first and second anodes. This ratio is changed by varying the potential on anode No. 1 by means of the FOCUS CONTROL, $R_2$ (Fig. 13 · 11), which is a potentiometer located on the front panel of the oscilloscope. The potential on the second, or accelerating, anode remains constant. The intensity of the beam (number of electrons constituting the beam) is varied by potentiometer $R_1$ (INTENSITY CONTROL), which changes the grid potential with respect to the cathode, thus permitting more or fewer electrons to flow.

**Electrostatic deflection system.** After the emitted electrons have been accelerated and focused to form a high-velocity beam, the electrons continue their travel toward the viewing screen until they strike the screen, which causes the screen to fluoresce, or give off light, within the region bombarded, forming a spot of light at or near the center of the tube. Other areas of the tube screen may be similarly activated by deflecting the beam from its center path. The beam may be deflected by either electrostatic or EM means. The electrostatic system is the predominant method for oscilloscopes.

Electrostatic beam deflection is accomplished through the use of two

*Fig. 13 · 12 Internal construction of a CRT.*

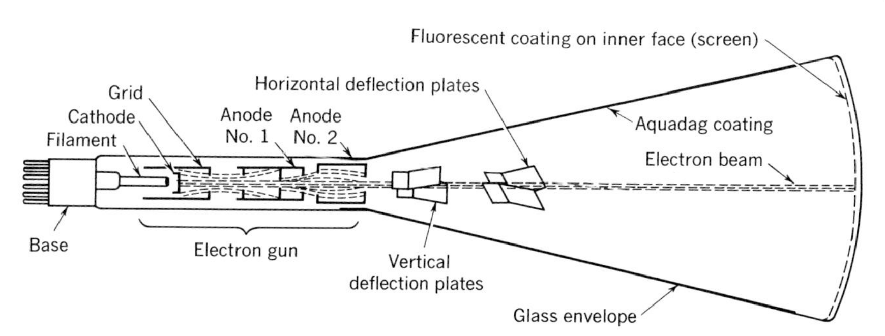

pairs of parallel plates that straddle the path of the beam. The second pair is perpendicular to the first; thus the electrons must pass between each set of deflection plates (see Fig. 13 · 12). If no electric field exists between the plates of either pair, the beam will follow its normal straight-line path, and the resulting spot will be at or near the center of the screen. A voltage potential applied to one set of plates will cause the beam to bend toward the plate that has the positive potential and away from the plate that has the negative potential. Deflection of the beam occurs virtually instantaneously, since it possesses an infinitesimal mass, and the bending is in direct relationship to the amount of voltage applied to the plates. The second pair of plates influences the beam in the same manner, except that the bending occurs in a plane perpendicular to the first. A voltage that keeps varying with time, when applied to either set of plates, will move the spot back and forth across the screen in a straight line. The movement of the spot across the screen will appear as a solid line when its motion is fast enough to exceed the persistence of human vision or the persistence of the phosphor material forming the screen. The required frequency is about 16 cps or more.

**Resultant motion of electron beam.** In nearly all applications of the oscilloscope, voltages are applied independently and simultaneously to each set of deflection plates. Thus, the electron beam is continually acted upon by two forces at right angles to each other. Figure 13 · 13 illustrates what happens under these conditions. When the sliders of potentiometers $R_3$ and $R_4$ are at ground potential, all four deflection plates are at ground potential and the spot appears at $o$ on the screen. If the slider on $R_4$ is moved toward the negative voltage, horizontal-deflection plate $HP_2$ becomes negative with respect to $HP_1$, and the electron beam is repelled by the negative voltage on $HP_2$. Under the action of this repulsion, the spot moves to point $x$ on the screen. If the slider on $R_3$ is moved toward the positive voltage, the vertical-deflection plate $VP_1$ becomes positive with respect to $VP_2$, causing the beam to be attracted from point $x$ to point $m$. Now, if the above adjustments of $R_3$ and $R_4$ are made simultaneously and at the same rate, the spot will move to point $m$ along the

*Fig. 13 · 13 The resultant motion produced by independent deflection voltages applied at the same time to the vertical and horizontal deflection plates.*

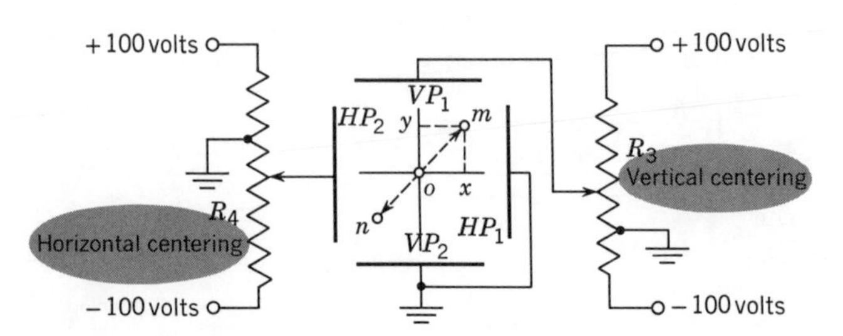

line *om*. If the same control setup is turned in the other direction, the spot moves along the new resultant *on*. Thus, when two voltages are applied simultaneously, one to each pair of deflection plates, the position of the spot at any instant is proportional to the resultant of the forces exerted upon the beam at that instant.

**Positioning (centering) controls.** Since structural imperfections in the manufacture of CRTs may cause the beam to strike at some point other than the center of the screen when no signals are applied, it is necessary to provide some means of positioning the beam. This is usually done by applying small d-c potentials to the deflection plates by means of potentiometers, such as $R_3$ and $R_4$ in Fig. 13 · 13. Centering, or positioning, controls are also useful whenever the enlargement of a waveform to permit examination of minute characteristics is so great that the portion of interest moves off the CRT screen. In such cases, the centering controls may be used to change the position of the waveform so that the desired portion is visible.

**Fluorescent screen.** In order to convert the energy of the electron beam into visible light, the screen is coated with a phosphor chemical which, when bombarded by electrons, has the property of emitting light. This property is known as *fluorescence*. The intensity of the spot on the screen depends upon the speed of the electrons in the beam, and the number of electrons that strike the screen at a given point per unit of time. In practical cases, the intensity is controlled by varying the number of electrons that are allowed to reach the screen.

All fluorescent materials have some afterglow, which varies with the screen material and with the amount of energy expended to cause the emission of light. The length of time required for the light output to diminish by a given amount after excitation has ceased is defined as the *persistence* of the screen coating. The general classification of screen materials is in terms of long, medium, or short persistence.

Various phosphors are used in oscilloscope work, each for specific applications. White and blue-white phosphors of short and very short persistence are used where photographic records are taken of screen patterns. For general service work where visual observation is most important, a green phosphor having medium persistence is used.

**Aquadag coating.** As previously described, the fluorescent screen of a CRT is bombarded by a beam of electrons. If these electrons were allowed to accumulate upon the screen, the screen would soon acquire a negative charge that would effectively repel and disperse the electron beam, thus blocking the tube in its primary function. This does not occur, however, because the beam, upon striking the screen, dislodges electrons from its surface (a process known as *secondary emission*). These dislodged electrons must be returned to the power supply. When the number of secondary electrons conducted away from the screen equals the number of electrons that return to, or are delivered to, the screen, there is no accumulation of charge. Present-day CRTs have a coating of graphite painted upon the

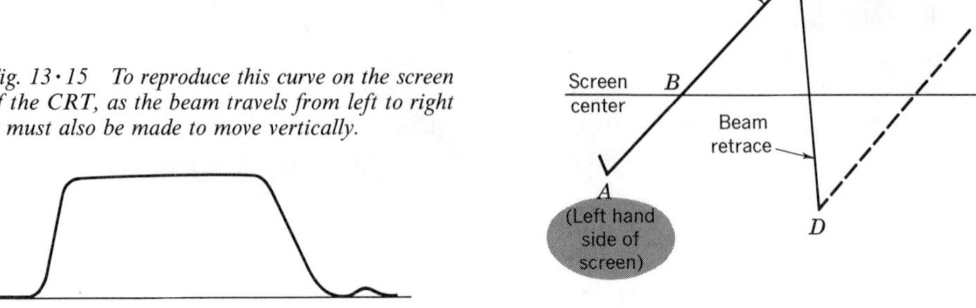


*Fig. 13·14 Sawtooth voltage wave for horizontal deflection.*

*Fig. 13·15 To reproduce this curve on the screen of the CRT, as the beam travels from left to right it must also be made to move vertically.*

inner glass surfaces but not connected with the screen. The functions of this coating are to (1) collect and return secondary electrons to the power supply, (2) serve as an electrostatic shield against external electric fields, and (3) in some tube types, act in addition as an accelerating anode.

**Power supplies.** If the oscilloscope is to function as a complete unit, both high-voltage and low-voltage power supplies are necessary. The output voltage of the high-voltage power supply is usually over 1,000 volts d-c, depending upon the size of the CRT, and the output voltage of the low-voltage power supply is usually 300 volts d-c. The high-voltage power supply is necessary to operate the CRT, and the low-voltage power supply provides the necessary voltages for the associated circuits.

**Vertical and horizontal amplifiers.** The deflection plates of a CRT require voltages on the order of several hundred volts for full-scale deflection. Therefore it is necessary to utilize amplifiers between the input terminals and the deflection plates so that test signals of low amplitudes may be effectively presented. When an amplifier is used between the signal source and the deflection plates, the signal is faithfully reproduced only if the limitations of the amplifier are not exceeded. These limitations include frequency response (in the amplifier and also in the input attenuator circuit), phase distortion, and the maximum allowable input voltages (both d-c and peak a-c). The frequency response and phase distortion are determined by the grid and plate circuits of the amplifiers. The maximum input voltage is limited by the input coupling capacitors and by the dynamic range of the amplifier.

The test signal whose waveform is to be viewed is applied to the vertical deflection system. For most oscilloscope applications the voltage applied to the horizontal deflection plates provides for the horizontal movement of the spot at a uniform rate with respect to time. The most common type of voltage which is applied to the horizontal deflection plates is the sawtooth wave shown in Fig. 13·14. From $A$ to $C$ the voltage rises steadily and linearly, moving the beam across the face of the oscilloscope screen at an even rate. At point $C$, it drops sharply, returning to the same level as point $A$. This drop causes the electron beam to retrace rapidly.

Half of the applied sawtooth wave is negative (points *A* to *B*). While this portion of the wave is active, the beam is at some point to the left of center. At *A* the beam is farthest to the left, but as the sawtooth voltage gradually rises, the beam is drawn in toward the center, reaching this point when the voltage reaches point *B*. As the voltage continues to rise, the forward motion of the beam brings it to the far right-hand section of the screen when the sawtooth voltage reaches point *C*. From point *C* to point *D*, the sawtooth voltage drops sharply, causing the electron beam to retrace quickly back to the left-hand side of the screen again.

We see from this sequence that the application of a sawtooth voltage to the horizontal deflection plates moves the beam first one way across the screen, and then the other way. If this back-and-forth motion is repeated often enough per second, the traces blend into each other, producing a steady horizontal line (also known as a *base,* or *axis*) of uniform intensity.

To reproduce a certain waveform, say, the curve shown in Fig. 13·15, then, as the beam travels on its way from left to right, we also want it to move vertically (or up and down). This can be accomplished by applying the wave to be reproduced to the vertical deflection plates. When the voltage at the vertical plates increases, the beam moves up; when it decreases, the beam moves down. In this way, the beam moves up and down as it travels across the face of the CRT, and the waveshape of any voltage fed to the vertical deflection plates is traced out.

**Oscilloscope operating controls.** The oscilloscope shown in Fig. 13·16 contains what might be termed a typical number of controls. The FOCUS control is used to adjust the sharpness of the trace or point of light on the screen. The INTENSITY control enables the operator to adjust the brilliance of the spot or trace. To move the beam vertically or horizontally, VERT

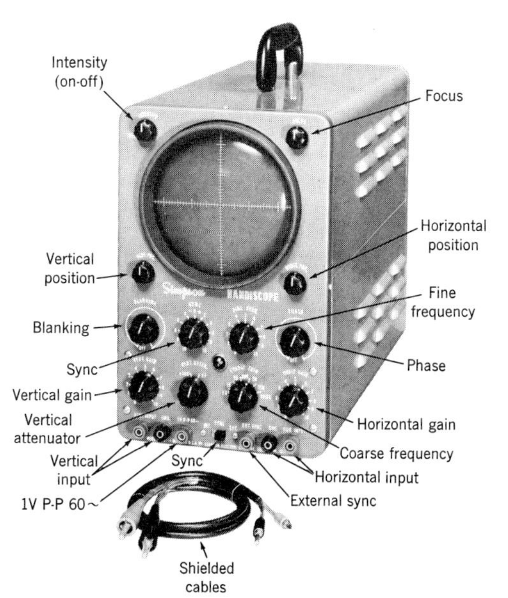


*Fig. 13·16 Operating controls on a typical oscilloscope.* (Simpson Electric Company.)

POSITIONING and HORIZ POSITIONING controls are provided. With the aid of these two controls, the pattern can be positioned anywhere on the face of the tube.

The three controls in the center of the panel, SYNC, FINE FREQ, and COARSE FREQ, are all concerned with the sawtooth generator contained in this instrument. There are other names by which these controls are known. Thus, the SYNC control is also labeled SYNC LOCK and LOCKING control. Substitute names for FINE FREQ include RANGE FREQ, FREQ VERNIER, and VERNIER. For COARSE FREQ there is SWEEP FREQ, SWEEP RANGE, and STEPS CONTROL.

The simplest of these three controls, and actually the one to be set first, is the COARSE FREQ control. This is a six-position switch (here) with each position identified by a different number. The sequence of numbers is: 15, 90, 500, 2,000, 15,000, and 100,000. With the switch in the first position, rotation of the FINE FREQ control will cause the horizontal sawtooth frequency to vary from 15 to about 90 cycles. When the FINE FREQ control is in its extreme counterclockwise position, the generated sawtooth frequency will be lowest; when it is turned completely clockwise, the sawtooth frequency will be at the highest point of that range.

Turning the COARSE FREQ control to the second position, marked 90, will cause the generated sawtooth frequency to rise. By rotating the FINE FREQ control, the sawtooth frequency can be varied from a point below 90 cycles to a value near 500 cycles. To insure that a continuous range of frequencies will be attained from the low end to the high end, sufficient overlapping occurs between successive positions of the coarse-frequency control. The high end of this range is somewhat above 100,000 cycles, and the low end somewhat below 15 cycles. This means that any wave with a frequency above 100,000 cycles will develop more than one cycle on the oscilloscope screen; any wave having a frequency less than 15 cycles will develop *less* than one cycle during one forward trace of the electron beam.

To use the COARSE and FINE FREQ controls to observe one or more cycles of any wave applied to the vertical input terminal (and ground, of course), set the COARSE FREQ switch to the range within which the signal frequency falls. Then rotate the FINE FREQ control until one cycle (or two) appears on the oscilloscope screen.

If you do not happen to know the approximate frequency of the applied signal, it takes only a minute to try each of the five range positions of the COARSE FREQ switch until you find the best range to use.

In order to work with any pattern obtained on the screen, the pattern should be held stationary. With the FINE FREQ control, it is possible, with some patience, to adjust the frequency of the sawtooth generator until it exactly equals (or is an exact fraction of) the frequency of the applied vertical signal. But, unless this control is constantly adjusted, the frequency of the sawtooth generator will change (even if only a few cycles) and the pattern will drift.

To keep the trace or pattern steady without frequent recourse to the

FINE FREQ control, a portion of the incoming signal is fed to the sawtooth generator and serves as a synchronizing pulse to lock the generator in step with the vertical-input frequency. The SYNC control enables the operator to vary the amount of synchronizing pulse or signal fed to the sweep oscillator. The optimum position for this control is at that point where the smallest amount of sync signal causes the pattern to become stationary. Thus, you start with the SYNC control at zero, slowly turn it to the right (clockwise) until the pattern locks in. It is important before using the SYNC control to adjust the FINE FREQ until the pattern is close to being stationary.

If it is desired to feed an external signal to the horizontal system, in place of the normal sawtooth wave, then the single-pole double-throw toggle switch located below the COARSE FREQ knob is used. When this switch (labeled SYNC) is flipped to the left, the internally generated sawtooth wave is applied to the horizontal deflection amplifiers. When it is snapped to the right, marked EXT, the sweep voltage must come from an external source, usually a generator. In sweep alignments, this external voltage is obtained from the sweep generator and is generally a 60-cycle voltage. However, it could be any type of voltage. This is applied to the horizontal input terminals.

Two gain controls are available, one each for regulating the gain of the two amplifier systems (that is, vertical and horizontal). For the vertical system, there is the VERT GAIN control at the left-hand side. This control adjusts the amplitude of the signal wave fed into the first vertical amplifier stage and hence controls the *height* of the pattern on the viewing screen. For the horizontal system, there is the HOR GAIN potentiometer at the right. This control adjusts the input to the horizontal amplifier to produce the desired pattern *width* on the screen. This control is effective whenever any voltages, external or internal, are applied to the horizontal amplifier.

Just beneath the VERT GAIN control is the VERT INPUT terminal with a corresponding GND terminal beside it. A similar set of terminals for horizontal input signals is located beneath the HOR GAIN control. These terminals would be used only when the HOR INPUT switch was in the EXT position.

A SYNC toggle switch is located to the right of the vertical input terminal. This switch has two positions: INT (internal) and EXT (external). When the switch is in the INT position, a portion of the signal being fed in at the vertical input terminal is fed to the sweep oscillator through the SYNC AMP control. If the frequency of the applied signal is near the frequency of the sweep, or is some multiple of it, the pattern can be made stationary on the screen of the CRT. This process is called *syncing* in the pattern and is achieved by the method described above when we were discussing the correct method of setting the COARSE and FINE FREQ controls. Note how we adjusted these two controls and then slowly advanced the SYNC control until the pattern became stationary. The SYNC AMP control determines how much of the synchronizing voltage is fed to the sweep oscillator to lock it in.

If we wish to obtain a synchronizing signal from some external source, we shift the SYNC toggle switch to the EXT position and then feed this external sync voltage in at the binding post located just to the right of the switch. This is labeled EXT SYNC. Note that the sawtooth sweep voltage is still active in the oscilloscope, and that the setting of the SYNC control still determines how much of this external sync voltage reaches the sawtooth oscillator.

All the sync signals are effective only when the internally generated sawtooth wave is functioning. They are not effective when an external sweeping voltage is being employed.

To the right of the VERT GAIN control will be found a control marked VERT ATTEN, standing for vertical attenuation. This switch has two positions, one marked ×1 and the other marked ×100. It is the purpose of this network to cut down strong signals which may be applied to the vertical input terminals of the oscilloscope, in order to prevent overloading the input vertical amplifier. When the switch is set at the ×100 position, any large input vertical voltages are reduced to one-hundredth of their value. If a small input voltage is applied to the vertical input terminals, the vertical attenuation switch should be set to the ×1 position. This applies the full signal to the vertical input amplifier. In other oscilloscopes, this divider or attenuator may have additional ratios. Common ratio values are 1, 10, 100, and 1,000. In the 1 position, the full value of any applied signal is fed to the vertical system. In the 10 position, only a tenth of the input voltage reaches the first stage, etc. In this way, we prevent the vertical deflection system from being overloaded. The overload distortion usually produces clipping and linking of the input waveform.

A similar attenuation network may be installed in the horizontal system for the same purpose. Sometimes in place of "attenuator" these switches are labeled "sensitivity controls;" the function, however, remains the same.

Just above the vertical gain control is the knob labeled BLANKING. When the oscilloscope is operated with an internal sweep, this control is adjusted until only the forward trace of the pattern appears on the oscilloscope tube face; the return trace will not be seen. When operating with an external horizontal input signal, this control is placed in the off position so that the entire input waveform may be viewed. Note then that this control is designed to remove any visible portions of the retrace that may occur ordinarily and which would be seen on the screen in the absence of this particular facility.

At the right-hand side of the oscilloscope front panel there is a knob labeled PHASE. This control adjusts the phase relationships between an internal sine-wave sweep at the line frequency and the signal which is providing vertical deflection. This particular combination is generally used for FM and television visual alignment response curves. To operate the internal sweep oscillator, the PHASE control is placed in the off position.

Finally, there is a knob at the bottom of the oscilloscope control labeled 1V P-P 60 CYCLES. From this knob it is possible to obtain 1 volt of the

60-cycle line voltage for use at either the vertical or the horizontal input terminals. When applied to the vertical input terminals, it will permit the measurement of the peak-to-peak value of any waveform appearing on the screen by comparison with the 1-volt reference.

**Oscilloscope probes.** Oscilloscopes generally use three separate probes for different applications. One is a direct probe, which is just a length of shielded cable. The shielding is important to prevent stray pickup of interfering signals. Stray pickup of hum can cause the pattern on the screen to bend; stray r-f pickup may result in hash that makes the trace look too thick. The low-capacitance, or isolation, probe is for use in circuits where the capacitance of the direct lead can cause problems. This probe has series decoupling resistors to isolate the circuit being tested from the input capacitance of the oscilloscope and its connecting lead. In r-f circuits, the low-capacitance probe should be used to prevent detuning. Also, this probe should be used when checking nonsinusoidal wave-shapes, since excess capacitance can distort the waveform observed. The decoupling resistor in the low-capacitance probe reduces the amount of signal input for the oscilloscope, usually to one-tenth the measured voltage. The third probe is a detector probe, primarily for sweep alignment of individual tuned stages without using the receiver detector.

## *13·4 Signal generators*

A signal generator is an instrument which generates an a-c signal suitable for test purposes. Actually, it is a miniature radio transmitter and can be made to generate signals for any desired frequency.

These generated signals can be modulated or unmodulated and are used for:

1. *Receiver alignment.* Adjusting the i-f and r-f tuned circuits to their correct frequencies.
2. *Receiver performance testing.* Checking receiver sensitivity and a-f response, selectivity, and signal-to-noise ratio.
3. *Receiver servicing.* Trouble shooting a defective receiver.

**Types.** When signal generators are classified according to frequency, they can be either a-f or r-f generators. A-f generators usually are called *audio oscillators.* They are capable of producing signals in the audio range from 20 to 20,000 cps. R-f generators generate signals over any specified range of frequencies above 20,000 cps, but no single generator will cover all the r-f ranges used in radio and radar. Different r-f generators are available, each covering a specified frequency range. Many of these also have an audio output available. The audio output can be variable over part or all of the audio range, or can be a fixed frequency, usually 400 cps.

The classification of r-f and a-f signal generators can also be subdivided according to their signal output. Audio oscillators can have either a sine-wave or a square-wave output. R-f signal generators can provide either c-w, r-f, or amplitude-modulated signals, frequency-modulated signals, or a pulse-modulated output. In AM generators, an audio signal is mixed with the radio frequency and the r-f amplitude varies with the amplitude of the audio signal. In FM generators, the radio frequency is varied in frequency at a rate determined by the amplitude of the audio signal. In pulse modulation, the r-f signal is provided in the form of pulses.

In the following discussion, we shall investigate the operation of audio, AM, and FM signal generators. We shall not examine pulse-modulated generators, because they are encountered less frequently.

## *13 · 5 Audio generators*

At the low frequencies encountered in a-f generators, few of the basic oscillator circuits are practical, principally because of the large size and expense of the inductive and capacitive components necessary for the tuned circuits. Therefore, special types of audio-oscillator stages are used to provide signals of the required amplitude and frequency. These circuits generally utilize Wien-bridge oscillators and beat-frequency oscillators.

**Wien-bridge oscillator.** An oscillator in which a frequency-selective bridge circuit is used as the *RC* feedback network is called a *Wien-bridge* oscillator. One widely used circuit for this type of oscillator is shown in Fig. 13 · 17*a*, where the feedback circuit is drawn to show that the phase-

*Fig. 13 · 17 Wien-bridge oscillator. (a) Actual circuit. (b) Circuit redrawn to illustrate its operation.*

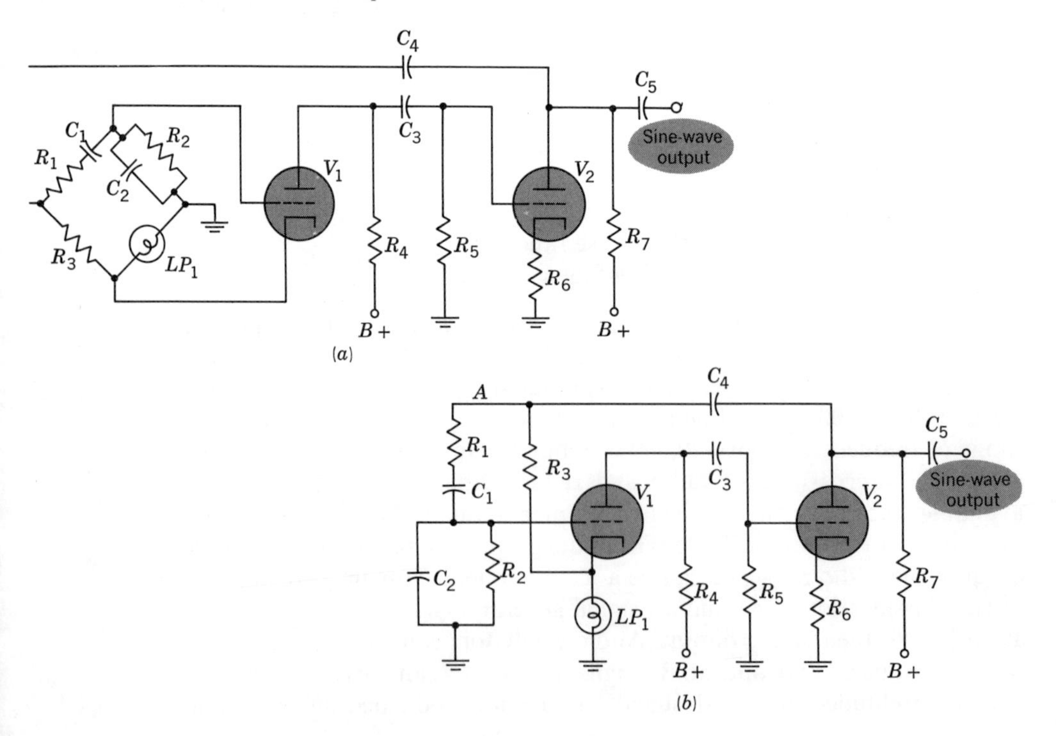

shifting element of the circuit is a frequency-selective bridge. It is simpler, however, to use the circuit as shown in Fig. 13·17*b* for purposes of discussion, since the feedback paths are revealed more clearly.

Tube $V_1$ is the oscillator. Tube $V_2$ acts as an amplifier and inverter. Thus, even without the bridge circuit, this system oscillates, since any signal that appears at the grid of $V_1$ is amplified and inverted by both $V_1$ and $V_2$. The voltage feedback to the grid of $V_1$ must reinforce the initial signal, which causes oscillations to be set up and maintained. However, the system amplifies voltages over a very wide range of frequencies. Voltages of any frequency or of any combination of frequencies can cause oscillation. The bridge circuit is used, then, to eliminate feedback voltages of all frequencies except the single frequency desired in the output.

The bridge allows a voltage of only one frequency to be effective in the circuit because of the degeneration and phase shift provided by this circuit. Oscillation can take place only at the frequency $f_0$ which permits the voltage across $R_2$, the input signal to $V_1$, to be in phase with the output voltage of $V_2$, and for which the positive feedback voltage exceeds the negative feedback voltage. Voltages of any other frequency cause a phase shift between the output of $V_2$ and the input of $V_1$ and are attenuated by the high degeneration of the circuit. Then the feedback voltage is not adequate to maintain oscillations at a frequency other than $f_0$.

A degenerative feedback voltage is provided by the voltage divider consisting of $R_3$ and lamp $LP_1$. Since there is no phase shift across this voltage divider and since the resistances are practically constant for all frequencies, the amplitude of the negative feedback voltage is constant for all the frequencies that may be present in the output of $V_2$.

The positive feedback voltage is provided by the voltage divider consisting of $R_1$, $C_1$, $R_2$, and $C_2$. If the frequency is very high, the reactance of the capacitors is almost zero. In this case, resistor $R_2$ is shunted by a very low reactance, making the voltage between the grid of $V_1$ and ground almost zero. On the other hand, if the frequency is reduced toward zero, the current that can flow through either $C_2$ or $R_2$ is reduced to almost zero by the very high reactance of $C_1$. Therefore the voltage between the grid and $V_1$ and ground falls almost to zero. At some intermediate frequency the positive feedback voltage is a maximum. It is at this frequency that the oscillator oscillates.

The voltage across $R_2$ is in phase with the output voltage of $V_2$ if $R_1C_1 = R_2C_2$. If the frequency of the output of $V_2$ increases, the voltage across $R_2$ tends to lag the voltage at the plate of $V_2$. If the frequency decreases, the voltage across $R_2$ leads the output voltage of $V_2$.

The frequency at which the circuit oscillates is $f_0 = 1/(2\pi R_1C_1)$. This assumes that $R_1 = R_2$ and $C_1 = C_2$, with $f_0$ in cycles, $R_1$ in ohms, and $C_2$ in farads. At this frequency, the positive feedback voltage on the grid of $V_1$ just equals or barely exceeds the negative feedback voltage on the cathode, and the positive feedback voltage is of the proper phase to sustain oscillation. At any other frequency, the negative feedback voltage is

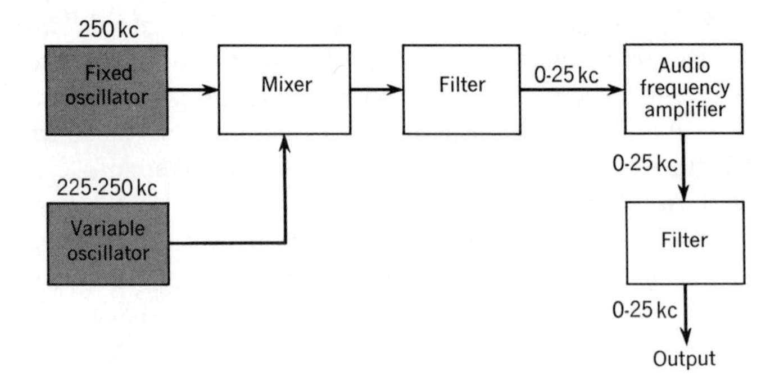


*Fig. 13 · 18 Block diagram of a beat-frequency audio signal generator.*

larger than the positive, so that the resultant degeneration of the amplifier suppresses these frequencies.

Frequency variation is achieved most frequently by varying the two capacitors, $C_1$ and $C_2$, in unison.

Lamp $LP_1$ is used as the cathode resistors of $V_1$ in order to stabilize the amplitude of oscillation. If, for some reason, the amplitude of oscillation tends to increase, the current through the lamp tends to increase. When the current increases, the filament of the lamp becomes hotter, making its resistance greater. A greater negative feedback voltage is developed across the increased resistance of the hotter lamp filament. Thus, more degeneration is provided, which reduces the gain of $V_1$ and thereby holds the output voltage at a nearly constant amplitude. Since the waveform is sinusoidal only at a small amplitude of output from $V_1$, the lamp serves also to prevent distortion of the sinusoidal waveform of the output.

In commercial a-f generators using the Wien-bridge oscillator, the output of the oscillator is amplified by several stages and then made available at an output terminal.

**Beat-frequency a-f generator.** The second most widely employed type of a-f generator develops its output signal by beating two r-f signals together and taking their difference for the desired a-f signal.

To illustrate the approach employed in this method, consider the typical block diagram shown in Fig. 13 · 18. A fixed oscillator operating at 250 kc and a variable oscillator whose frequency can be varied from 225 to 250 kc feed their signals into a mixer stage. Here the two signals beat against each other, producing a difference frequency. When the variable oscillator is set to 250 kc, the difference is zero. This difference increases until it reaches a value of 25 kc when the variable oscillator is operating at 225 kc.

The audio signals produced by the mixer are passed through a low-pass filter that prevents any of the high-frequency mixing signals from reaching the audio amplifier (1 or more stages). Here the desired audio signals are strengthened and then passed through another low-pass filter to the output terminal of the instrument.

The frequency of the variable oscillator is changed by a variable capacitor which is operated from a dial on the front panel of the instrument. This dial is not marked in the actual frequencies generated by the variable oscillator but in the difference frequencies which each position of the dial produces. The beat-frequency audio oscillator must be calibrated after sufficient warm-up time (about 20 minutes). Usually the calibration is done at 60 cps.

## *13·6 AM signal generators*

The AM generator, whatever frequency range it covers, consists basically of an r-f oscillator whose output frequency can be varied over a certain range. This signal is available as is (that is, unmodulated), or it can be combined with a low-frequency audio signal (that is, modulated). To achieve this amplitude modulation, the generator also includes an audio oscillator, operating at a frequency of about 400 cycles. Provision is always made to bring the modulation in, when desired, or to cut it off when only the r-f signal (or carrier) is wanted. For direct testing of the receiver audio system, the 400-cycle signal voltage is also made available at a front-panel jack.

The front panel of a typical AM generator is shown in Fig. 13·19. The dial face is seen to consist of a number of scales (here, seven), each scale covering a different set of frequencies. The lowest frequency scale is the outermost scale (labeled *A*). The frequencies across this scale range from 75 kc at the low end to 220 kc at the high end.

The second, or B, scale covers the frequency range from 200 to 600 kc. A comparison of this range with that of scale A reveals that the low end of scale B overlaps the high end of scale A. In other words, a portion of

*Fig. 13·19 Front-panel controls of an AM r-f signal generator.* (Electronic Instrument Company.)

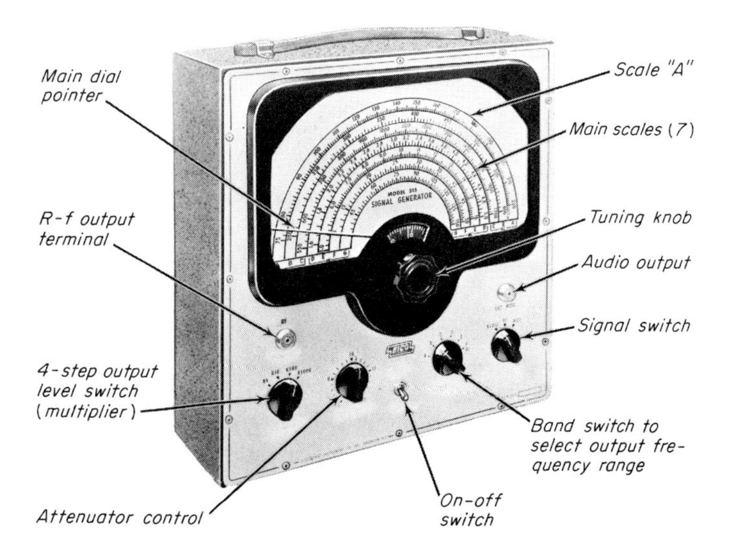

the highest frequencies of scale A reappear at the low end of scale B. This is another common practice and insures that the rise in frequencies is continuous. If the beginning of scale B started just at the end of scale A, it might happen under some circumstances that the frequencies generated by the instrument for scale A did not quite extend to the highest frequencies indicated on that scale. For example, change in value of a circuit capacitance, resistance, or inductance, a variation in the operating voltages, or aging of the oscillator tube itself might readily alter the frequencies generated by an oscillator. Especially critical in this respect are the end frequencies of any resonant circuit. This situation would cause a gap to appear between A and B and impair, to some extent, the usefulness of this generator. It is to prevent this from happening that the frequency ranges of the various scales overlap.

The range of the third, or C, scale of the generator extends from 550 to 1,700 kc. Writing 1,700 kc in megacycles produces a result of 1.7 Mc. Scale D starts below this, at 1.6 Mc, and rises to 5.0 Mc. Scale E ranges from 5.0 to 16 Mc. Next follows scale F with 10 to 50 Mc, and finally scale G with 45 as the first number and rising to 150 Mc. While 45 is the first number on scale G, the dial actually contains markings below this point. However, the section below 45 (on scale G) need not be used, since it repeats a portion of scale F.

The choice of a particular range is governed entirely by the switch marked BAND, located just below and slightly to the right of the dial knob. There is a separate position for bands A, B, C, D, and E. Note, however, that bands F and G both occupy the same selector-switch position. Also, if you look closely at scales F and G, you will discover that every number on scale G is exactly three times the number appearing directly above it on scale F. This means that one tuned circuit is being used to generate all frequencies from 10 Mc on up. The basic or fundamental frequencies being developed by the generator (with the BAND switch in the F, G position) are those indicated on scale F, namely, 10 to 50 Mc. As is true of nearly all oscillators, however, harmonics are generated together with the fundamentals. In this instrument we are interested in the third-order harmonics. Note, however, that there are also present the second harmonics and possibly some fourth, fifth, and even higher harmonics, although the harmonic amplitude falls off quite sharply as the harmonic order rises.

The range, then, of the generator shown in Fig. 13·19 extends from 75 kc to 50 Mc on fundamentals and from 50 to 150 Mc on harmonics.

**Other generator controls.** In addition to the ON-OFF switch and the BAND switch, there are three other controls and two signal outlets on the generator of Fig. 13·19. One control, the MULTIPLIER, is a four-way selector switch serving to regulate, in four steps, the intensity of amplitude of the signal delivered by the generator. In the $\times 1{,}000$ position, the maximum amount of signal is available at the RF OUTPUT terminal located just above this control. In the $\times 100$ position, approximately $1/10$ as much signal is permitted to reach the output terminal, the remainder of the signal being dissipated in resistive attenuation pads. In the ×10 position, the signal level is still further decreased, and in the ×1 position it is at its lowest point.

In each of the four positions, the level of the signal may be continuously varied by means of the ATTENUATOR control. This arrangement provides full control of the signal from maximum to minimum.

On the other side of the panel there is a SIGNAL control possessing three positions: AUDIO, wherein the audio signal (400 cycles) is made available at the AUDIO OUTPUT terminal; RF, when only an unmodulated r-f signal appears at the RF OUTPUT terminal; and MOD, when the r-f signal at the RF OUTPUT terminal is modulated by the audio note.

Control of the amplitude of the audio signal at the AUDIO OUTPUT terminal is also achieved by rotating the ATTENUATOR control. This knob attaches to a dual potentiometer, one section of which serves for the r-f signal and one for the audio signal.

To obtain the signal voltages present at either the RF OUTPUT terminal or the AUDIO OUTPUT terminal, a coaxial cable is provided with the instrument. One end of the cable screws onto the output terminals, while the other end contains two alligator clips. One clip attaches to the outer conductor of the cable and represents the ground-output connection of the generator. The other clip connects to the inner conductor of the cable and is the one usually referred to as the *hot* lead, since it carries the signal.

While most generators—in common with this one—contain an internal audio oscillator whose output can be used to amplitude-modulate the r-f signal, modulation by an external signal can also be achieved, if this is desired. In the instrument shown in Fig. 13·19, the external modulating voltage is fed into the instrument at the AUDIO OUTPUT—EXT MOD terminal. The SIGNAL control is placed in the RF position to disable the internal audio oscillator and prevent it from feeding its signal to the buffer tube at the same time.

## *13·7 FM Signal generators*

FM generators provide an r-f signal that varies in frequency. When an audio signal is mixed with a radio frequency whose amplitude is constant, the radio frequency changes at a rate determined by the instantaneous amplitude of the audio signal. On one alternation of the audio signal, the radio frequency is higher than the resting, or original, frequency; on the other alternation of the audio signal, the radio frequency is lower than the resting frequency. The radio frequency is no longer steady, but is rather a shifting one, and the amount of frequency shift from the center frequency is called the *deviation*. The amount of deviation depends on the amplitude of the audio signal, and the number of times per second it takes place depends on the frequency of the audio signal. The amount of deviation may vary from a few kilocycles to 10 Mc, or more.

The basic unit of the FM generator is an oscillator that generates continuous radio frequency at a given frequency. To obtain an FM signal,

*Fig. 13·20 An FM sweep generator.* (Allied Radio Corporation.)

there must be an audio signal that will cause the frequency of the oscillator to vary. One method of doing this is to take a 60-cps audio signal from the filament winding of a power transformer and feed it through a potentiometer to the voice coil of a small permanent-magnet loudspeaker. By varying the resistance of the potentiometer which is in series with the voice coil, the amount of vibration of the speaker cone can be varied. Now if a copper disk is attached securely to the speaker cone and mounted close to the coil of the r-f oscillator circuit, when the disk comes closer to the r-f coil, eddy currents are induced in it. The lines of force from the disk oppose the lines of force from the r-f coil. Therefore the effective inductance of the coil is reduced, and the frequency of the tuned circuit becomes higher. When the disk moves away from the r-f coil on the other alternation of the audio cycle, the frequency change is in the opposite direction. The amount of frequency change can be regulated by adjusting the amount of vibration of the speaker cone. This is done by controlling the amplitude of the 60-cps input to the voice coil.

Sweep signals are employed to reveal the response of a circuit or system to a band of frequencies. The sweep signal is fed into the equipment at the input to the section being tested, and the output is viewed on the screen of an oscilloscope. From the resulting pattern, we can tell at a glance the condition of the circuit. By combining an auxiliary marker signal with the sweep signal and then varying the frequency of the marker signal, we can determine where the response falls off and where it rises.

To sum up, a 60-cps audio signal causes the FM signal to vary on either side of the resting frequency 60 times a second. Also, the higher the amplitude of the audio signal, the greater the frequency deviation of the FM signal.

**FM generator controls.** An example of a low-cost sweep generator is shown in Fig. 13·20. (To bring out the various control markings better, the front panel is reproduced separately in Fig. 13·21.) It is designed to cover a range of frequencies from 300 kc (0.3 Mc) to 250 Mc. On the main tuning dial, there are four scales: 0 to 50 Mc, 40 to 130 Mc, 120 to 170 Mc, and 160 to 245 Mc. The markings are center sweep frequencies about which the frequency sweeping takes place. If the dial is set for 100 Mc, for example, and the SWEEP WIDTH control is set to maximum clockwise position, the output signal will sweep back and forth from 94 to 106 Mc. The approximate center frequencies of the vhf television channels are indicated on the scale in smaller numbers.

The various controls found on this instrument are as follows:

*Crystal.* The switch will select one of the two crystals plugged into the crystal socket and electrically connect it to the crystal oscillator circuit. Marker or identifying pips will be generated at the crystal frequency as well as harmonics, such as 2nd, 3rd, 4th, and so on. For example, a 4.5-Mc crystal will produce harmonics at 9.0 Mc, 13.5 Mc, 17.0 Mc, 22.5 Mc, and so on.

*Marker amplitude.* Controls the strength of the marker signal. This control should be advanced only as far as required to produce a visible pip on the sweep trace. Too much marker amplitude will distort the curve. This control will adjust the strength of the crystal marker or the marker from an external generator.

*Phase.* This control varies the phase of the modulating voltage applied to the oscillator of the sweep generator. The phase control and main tuning should be adjusted together so that a single symmetrical trace appears on the oscilloscope screen.

The sweep signal, after it has passed through a system, is applied to an

*Fig. 13·21 Front-panel operating controls of FM sweep generator in Fig. 13·20.*

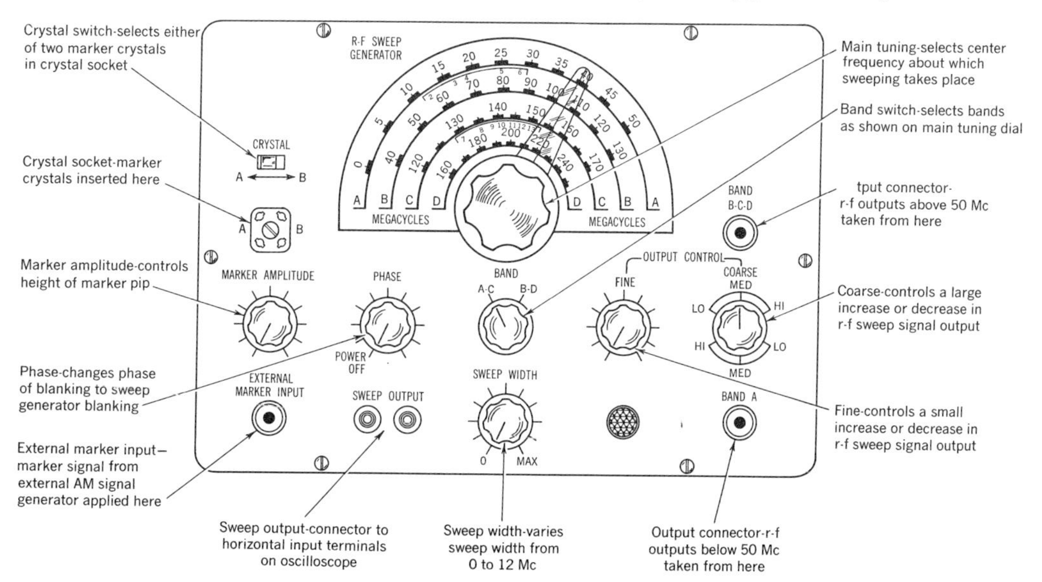

oscilloscope where we can observe what effect the system had on the sweep signal. To develop a single pattern on the oscilloscope screen, the motion of the tracing electron beam in the oscilloscope must be synchronized to the 60-cycle sweeping voltage in the signal generator. It is the function of the PHASE control to achieve this.

When this control is turned to the extreme counterclockwise position, the sweep generator is turned off.

*Band.* This switch has two positions, A–C and B–D, which correspond to the A, B, C, and D scales on the main tuning dial.

*Output control.* (1) *Fine:* Controls small increases or decreases of the output signal. Clockwise rotation of this control increases the output or amplitude of the sweep signal. (2) *Coarse:* This is a step-type output control and produces large increases or decreases of the output signal. This control is marked LO–MED–HI corresponding to low, medium, and high output. The top set of markings is for the output connector of bands B, C, and D. The lower set of markings is for the band A output connector. Set this switch according to the output connector being used.

*Band* B–C–D. Sweep output from bands B, C, and D is taken from this connector.

*Band* A. Sweep output from band A is taken from this connector. This connector is used when sweep frequencies between 0 and 50 Mc are required. The built-in filter eliminates undesired frequencies above 50 Mc.

*External marker input.* An external generator can be connected to this connector to provide the necessary marker pips on the sweep waveform. This eliminates a separate connection of the marker generator to the unit under alignment.

*Sweep output.* Horizontal deflection voltage for the oscilloscope is available from this pair of terminals.

*Sweep width.* This control varies the amount of sweep width about the center frequency indicated by the main tuning-dial pointer. Maximum sweep width is approximately 12 Mc. The sweep-width control is not calibrated. The markers will enable you to determine the bandwidth of the circuit being aligned.

## 13·8 Wavemeters

Wavemeters offer the simplest means of measuring the frequency of an r-f wave. These devices indicate the fundamental or harmonic frequencies of oscillators or harmonic generators on a calibrated dial. There are two basic wavemeters, both of which absorb part of the output power of the device whose frequency is to be measured. One is the reaction wavemeter, the other is the absorption wavemeter. The reaction wavemeter absorbs very little power. A current meter, located in the circuit of the device whose frequency is to be measured, usually serves as an indicator. Since the power absorbed is not sufficient to load the equipment being measured to any great extent, this wavemeter can be used to measure the frequency of low-power equipment.

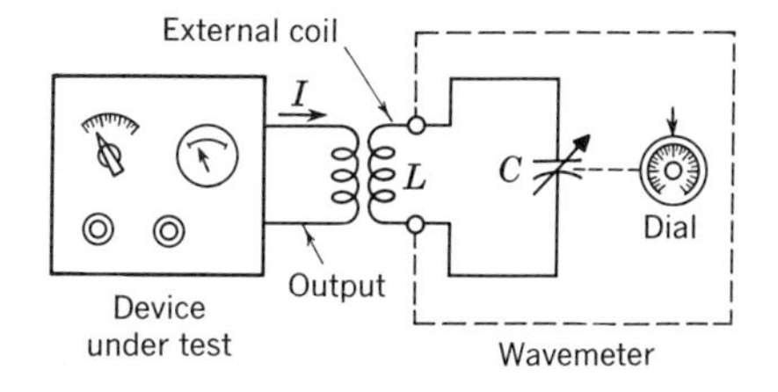


*Fig. 13·22 Basic circuit of a reaction wavemeter.*

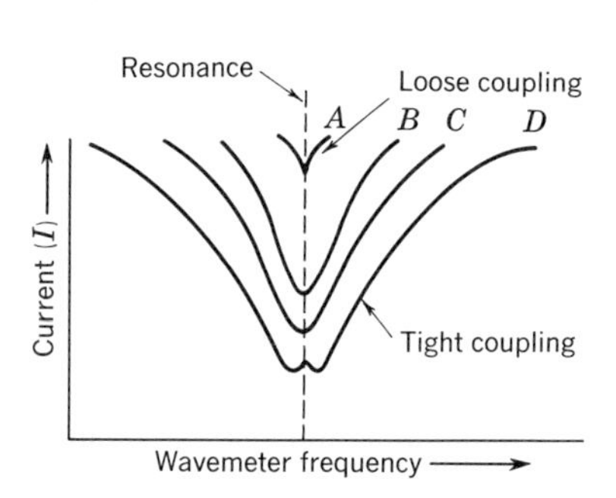


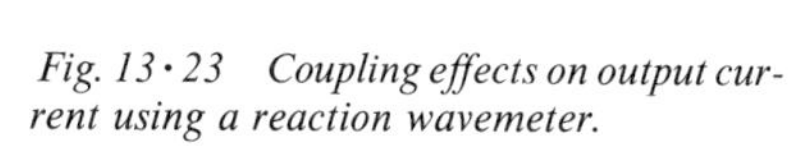

*Fig. 13·23 Coupling effects on output current using a reaction wavemeter.*

The absorption wavemeter is more accurate than the reaction wavemeter but absorbs slightly more power from the equipment whose frequency is being measured. It generally is used on high-power equipment only, since it tends to load the equipment. An ammeter, a lamp, or an earphone is used to indicate the unknown frequency.

**Reaction wavemeter.** The basic circuit of a reaction wavemeter, including a coil $L$ and a variable capacitor $C$, is shown in Fig. 13·22. The external coil $L$ is loosely coupled to the output coil of the device whose frequency is to be measured. The capacitor $C$ is then tuned until the resonant frequency of the wavemeter is equal to the frequency of the device under test. At this point, the current meter indicates resonance which can be either maximum or minimum, depending on where the ammeter is located in the device under test. The capacitor is operated by an accurately calibrated vernier dial, with the graduations in terms of some arbitrary unit. The frequency or wavelength is found by means of a calibration curve or chart, which relates the dial setting to either frequency or wavelength.

When the wavemeter is moved into the r-f field of the device under test, the coupling produces a change in the current and increases the load on the device. When the wavemeter is tuned to resonance, this load becomes maximum, and the indicating ammeter reads a minimum or maximum, depending on its locations. To prevent loading of the equipment being measured, the wavemeter is moved away until the least possible variation of the tuning capacitor causes a maximum deflection on the meter.

A plot of the output current vs. wavemeter frequency for various degrees of coupling is shown in Fig. 13·23. When the wavemeter is resonant, the load on the device is a maximum, and the current is a minimum. The dips in the curve indicate a dip in the reading of the meter. Curve $A$ indicates loose coupling (the wavemeter has been moved away from the equipment being measured). As the coupling is increased, the dip in primary current becomes greater. Curve $B$ represents tight coupling, curve $C$ critical coupling, and curve $D$ overcoupling. For accurate frequency measurements, the coupling is decreased until there is a very small dip in the current meter. Loose coupling is ideal for accurate frequency measurements, since its dip at resonance is very sharp.

**Absorption wavemeter.** Although the absorption wavemeter is similar to the reaction wavemeter, it contains in addition a resonant-frequency indicating device. Figure 13 · 24 shows an absorption wavemeter circuit using a lamp to indicate resonance. The functions of tuning capacitor $C_1$ and external coil $L$ are identical with those explained in the discussion of the reaction wavemeter. The lamp is at maximum brilliance when the wavemeter is at the resonant frequency of the device under test. The amount of brilliance that results depends on the voltage appearing across fixed capacitor $C_2$. The capacitive value of $C_2$ is much larger than that of $C_1$, and, since its reactance is negligible at the resonant frequency, $C_1$ and $L$ determine the resonant frequency of the wavemeter.

When the wavemeter circuit is tuned to the same frequency as the unknown frequency, maximum current flows in the wavemeter circuit. Since this current occurs at resonance and current is maximum at resonance, the lamp glows brightly. As $C_1$ is tuned to either side of resonance, the circulatory current becomes less, and the lamp grows dimmer.

In Fig. 13 · 25, the circulatory current is plotted against the wavemeter frequency for various degrees of coupling. For accurate frequency measurements, the external coil of the wavemeter is loosely coupled to the device under test. This accuracy is indicated by the sharpness of curve $C$ at resonance. Although overcoupling, as shown in curve $A$, produces a greater circulatory current and lamp brilliance, it results in inaccurate frequency measurements. Overcoupling results in a double-humped curve where a maximum circulatory current is obtained on either side of resonance.

In Fig. 13 · 26*a*, the headphones of the absorption wavemeter are coupled to the wavemeter circuit by means of small pickup coils and are used as the indicating device. Resonance is obtained when a *click* is heard in the headphones as $C_1$ is tuned slightly above resonance. The crystal rectifier serves to provide the direct current necessary to operate the headphones.

*Fig. 13 · 24 Basic circuit of an absorption wavemeter.*

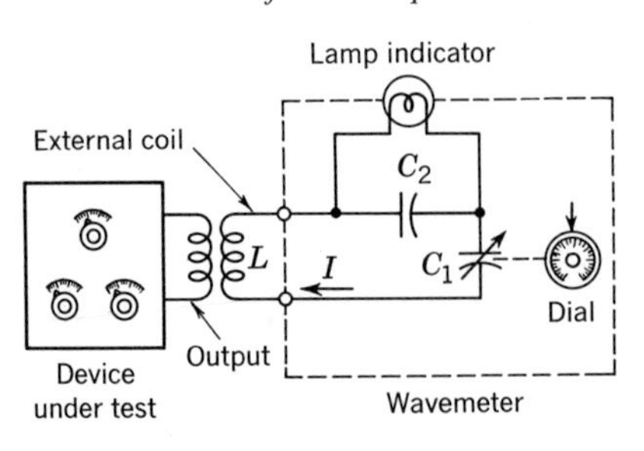


*Fig. 13 · 25 Effect of coupling on current in absorption wavemeter.*

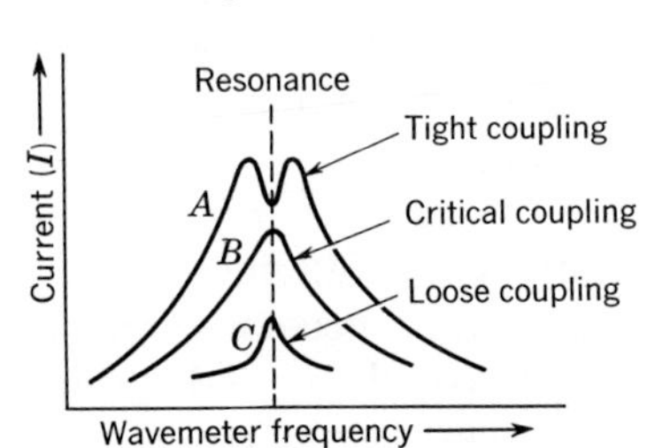

In Fig. 13·26*b*, a d-c meter movement replaces the headphones, a diode rectifier replaces the crystal, and the circuit operates as a simple vacuum-tube voltmeter. The diode is a filament-type tube operated by means of a small self-contained battery. Resistor $R_1$ is used to adjust the filament voltage to its proper value. Capacitors $C_2$ and $C_3$ bypass the r-f circulatory current around the ammeter and battery. The ammeter reads in direct proportion to the potential difference existing across tuning capacitor $C_1$. This potential difference results from the circulatory current in the wavemeter circuit and, at resonance, the ammeter reads maximum.

**Determining unknown frequency.** When using the reaction wavemeter, the external coil of the wavemeter is moved into the r-f field of the device whose frequency is to be measured. The dial knob on the wavemeter is slowly rotated through its frequency range until some reaction is noted on the indicating meter in the device under test. The wavemeter is moved slowly away from the oscillator until the deflection on the indicating meter is barely perceptible. The dial-setting knob is now adjusted for a maximum deflection on the meter. This point on the dial indicates the frequency of the circuit being tested.

In the absorption wavemeter, the frequency to be measured is determined in a similar manner. A lamp is used as an indicator, and the wavemeter is brought near the device under test. The dial-setting knob is turned slowly through its range until the indicator lamp just begins to glow. The knob is not turned for maximum lamp brilliance, which might cause it to burn out. The wavemeter is moved slowly away from the oscillator until the lamp glows more dimly or goes out. It is then tuned for maximum lamp brilliance. For accuracy, the wavemeter should be taken as far from the device as possible, where maximum brilliance is a faint glow.

Wavemeters usually contain several external coils of the plug-in variety. Each coil represents a specific frequency range, and if the approximate frequency to be measured is known, the selection of the proper coil is simplified. Where the approximate frequency is unknown, each coil must

*Fig. 13·26 Two circuit variations of the absorption wavemeter.*

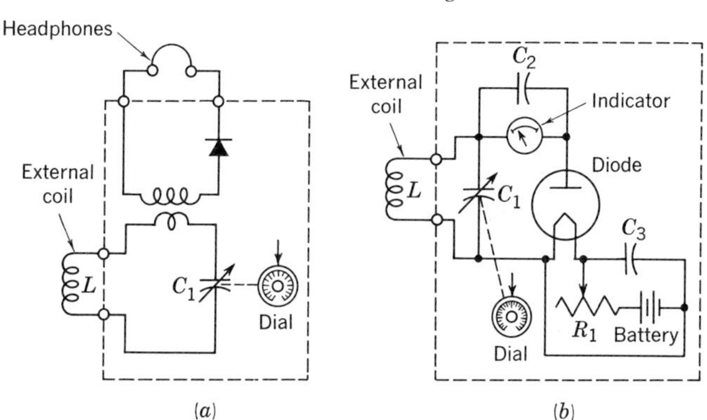

be tried separately to obtain resonant indications. The tuning capacitor is the air type, and the frequency range it covers determines the number of plug-in coils that are needed. A frequency standard, from which a fixed known frequency signal can be obtained, is used to calibrate the wavemeter.

The foregoing discussion has covered the circuitry and operation of the instruments which will enable the technician to perform over 90 per cent of the measurements he ordinarily encounters. For the remaining 10 per cent, he will need additional instruments, such as capacitor meters, *Q* meters, signal tracers, impedance bridges, and frequency meters. In all cases, the user is urged to read the instruction manual carefully before using the instrument.

## SUMMARY

1. A volt-ohm-milliammeter (VOM) measures a-c and d-c voltages and direct current without the need for external source of power. For resistance measurements, a small battery is enclosed within the instrument case. Each section functions independently, with only the meter movement serving all sections.
2. The vacuum-tube voltmeter (VTVM) performs essentially the same measurements as a VOM. However, the VTVM can cover a wider range of voltages and resistances. The heart of the VTVM is a balanced bridge which operates on the unbalance that occurs when voltages are applied to the control grid of one tube in the bridge.
3. Before any resistance measurements are made on any ohms range of the VOM, it is important that the leads be shorted together and the needle position adjusted until it is directly over the zero mark. This is the zero adjustment. For a VTVM, with the test leads separated, the OHMS ADJ control is rotated until the needle is set directly over the scale position which indicates an open circuit. This is usually designated as infinite ohms.
4. For all measurements, the selector switches must be carefully set. This includes not only the function, such as a-c volts, d-c volts, and ohms, but the range as well. If any doubt exists as to the size of a certain voltage to be measured, always start with the highest range, and carefully work down until the most suitable range is found.
5. The cathode-ray oscilloscope provides a visual picture of the voltage in a circuit. The oscilloscope can be considered as a high-impedance voltmeter.
6. The principal components of a basic oscilloscope include a CRT, a sweep (sawtooth) oscillator, and vertical and horizontal deflection amplifiers.
7. Sawtooth waves are used most frequently to move the electron beam in an oscilloscope back and forth across the screen. The traveling time of the beam from left to right (trace time) is usually much slower than it is from right to left. The latter time is known as the retrace time.
8. A signal generator is an instrument which generates an a-c signal suitable for test purposes. Actually, it is a miniature radio transmitter and can be made to develop signals for any desired frequency.
9. Audio, or a-f, generators produce signals in the audio range from 20 to 20,000 cps. The two main types use either the Wien-bridge oscillator or r-f oscillators to produce audio beat frequencies.
10. R-f generators usually provide c-w output or m-c-w, modulated at 400 cps. The audio output is generally available separately for testing audio circuits. When a wide range of radio frequencies is to be generated by a single generator, band switching is employed. Each band possesses its own tuning circuit.
11. FM generators provide an r-f signal that varies in frequency, generally at a 60-cycle audio rate. Thus, the radio frequency is no longer a steady frequency, but is rather a shifting one, and the amount of frequency shift from the center frequency is called the deviation.
12. Wavemeters are used to measure the frequency of an r-f wave.

## SELF-EXAMINATION

Here's a chance to find out how well you have learned the material in this chapter. Work the exercises; then check your answers against the Key at the back of this book. These exercises are for your self-testing only.

1. The purpose of the OHMS ADJ control in a VTVM is to make certain that (*a*) the meter reads correctly at both ends of the ohms scale; (*b*) when the two leads are shorted together, the meter needle is exactly over the zero mark; (*c*) when the two leads are apart, the meter needle is directly over the final right-hand marking of the ohms scale; (*d*) the meter reads correctly when in the exact center of the scale.
2. One advantage which a VTVM has over a VOM is (*a*) greater portability; (*b*) low power consumption; (*c*) a lower input impedance; (*d*) the ability to measure wider ranges of voltage and resistance.
3. In an oscilloscope, the voltage to be viewed is applied to the (*a*) vertical input terminals; (*b*) horizontal input terminals; (*c*) SYNC AMP terminal; (*d*) vertical or horizontal input terminals, depending on its frequency.
4. A pattern on the scope screen can be made to stand still by carefully advancing the (*a*) focus control; (*b*) sync-amplitude control; (*c*) coarse-frequency control; (*d*) horizontal-positioning control.
5. In the beat-frequency method of developing an audio signal, (*a*) two low-frequency signals are added together; (*b*) a high- and a low-frequency signal are mixed; (*c*) the difference of two r-f signals is taken; (*d*) the sum of two r-f signals is taken.
6. The purpose of the lamp in the cathode circuit of $V_1$ in Fig. 13·17 is to (*a*) stabilize the amplitude of the generated signal; (*b*) indicate the circuit is operating; (*c*) act as a fuse; (*d*) regulate the frequency of oscillation.
7. Which of the following controls is not found on the front panel of an AM generator? (*a*) Attenuator control; (*b*) sweep-width control; (*c*) multiplier switch; (*d*) r-f output.
8. In using reaction or absorption-type wavemeters, it is best to (*a*) keep the coupling between the wavemeter and the equipment being measured as loose as possible; (*b*) keep this coupling as tight as possible; (*c*) vary the distance between wavemeter and the unit being checked until a maximum indication is obtained; (*d*) position the meter for critical coupling.
9. To align an i-f amplifier at 455 kc for maximum audio output across the loudspeaker, the signal generator must supply (*a*) FM sweep at the rate of 60 cps; (*b*) audio output at 400 cps; (*c*) r-f output modulated with 400 cps; (*d*) FM output with a deviation of 455 kc.
10. Which of the following is the best meter to use for measuring the AVC bias voltage in a receiver? (*a*) VOM, with a sensitivity of 1,000 ohms per volt; (*b*) VTVM; (*c*) reaction wavemeter; (*d*) high-impedance a-c voltmeter.

## QUESTIONS AND PROBLEMS

1. Explain how to set the controls on the VOM shown in Fig. 13·1 to measure: −25 volts d-c, 25 volts a-c, 1,500 ohms.
2. What control balances the bridge circuit of Fig. 13·7? How does it become unbalanced in the normal course of making a d-c voltage measurement?
3. Why can a VTVM, using only a 1.34-volt battery, measure higher resistances than a VOM?
4. Draw the block diagram of an oscilloscope, and explain briefly the purpose of each section.
5. List the electrodes in a CRT, and give the function of each.
6. Explain the purpose of the following oscilloscope controls: INTENSITY; SYNC AMP; VERT GAIN; INT-EXT switch.
7. Explain briefly how an FM generator operates.
8. How is an audio signal developed by the beat-frequency method?
9. What is the purpose of each of the controls on the AM generator shown in Fig. 13·19?
10. How does an absorption wavemeter operate? How would it be used?

# Chapter 14 Pulse circuits

A pulse is an abrupt surge of voltage or current of short duration, as shown in Fig. 14·1*a*. The pulses may occur at irregular intervals or have a definite frequency called the *pulse repetition rate.* Also, the pulse amplitude may be positive or negative (Fig. 14·1*b*). The pulse duration $t_d$ is the length of time the amplitude remains at its maximum value. This is a short interval, usually in microseconds. There are many ways of obtaining the pulses. For example, one class of waveshaping circuits changes sinusoidal input waveforms to nonsinusoidal output pulses. In this chapter, we shall discuss the special group of circuits which generate output pulses directly, without a sinusoidal input. Such pulse circuits are used in industrial control systems, radar, television, computers, and other applications where precise timing is important. These circuits are described in the following sections:



## 14·1 Pulse generators

A circuit that delivers nonsinusoidal pulses at its output is called a *pulse generator*. Among the many types of pulse generators is a group of circuits called *relaxation oscillators.* These circuits are capable of generating a continuous output of pulses at a specific pulse repetition rate (*prr*) without any external signal. In relaxation oscillators, the output frequency or *prr* is determined by a capacitor-resistor or inductor-resistor combination rather than by a conventional inductor-capacitor tuned circuit. Energy builds up in the capacitor until the capacitor is charged to a certain voltage level. The circuit then "relaxes" and the capacitor discharges; hence the term relaxation oscillator.

In general, pulse generators can be considered in two classes. The first type is a *free-running* or self-oscillating circuit, which is basically a relaxation oscillator. No external triggering signal is needed. This group includes plate-coupled and cathode-coupled multivibrators and blocking oscillators. The second type consists of *trigger* circuits which produce output pulses only when external triggering signals are applied. Typical trigger circuits are the Eccles-Jordan circuit (or *flip-flop,* as it is commonly called) and the Schmitt trigger circuit.

The operation of these free-running oscillators and trigger circuits is described in Secs. 14·2 to 14·6, but first we can consider some definitions related to stability. Fundamentally, the method of providing for discharge in a relaxation oscillator is to use a tube or transistor as a conducting path for the capacitor. Conduction is the on condition; cutoff is the off condition. The sharp changes between cutoff and conduction produce the pulses in the output circuit. When the tube or transistor can remain in one state, either on or off, without external signal, this condition is a *stable state.* However, in many cases the circuit can change from on to off, or vice versa, without any external signal. This condition is called *quasi-stable* because it can change between the on and off states by itself.

## *14·2 Multivibrator circuits*

A multivibrator (MV) has two stages, with the output of one driving the input of the other. Because multivibrators are so commonly used in pulse circuits, the different types are often classified as follows. In terms of the circuit arrangement, the two most common types are the plate-coupled MV in Fig. 14·2 and the cathode-coupled MV in Fig. 14·4. Both these circuits

*Fig. 14·1 (a) Oscilloscope photograph of typical pulses used in electronic circuits. (b) The amplitude E and duration $t_d$ of pulses.*

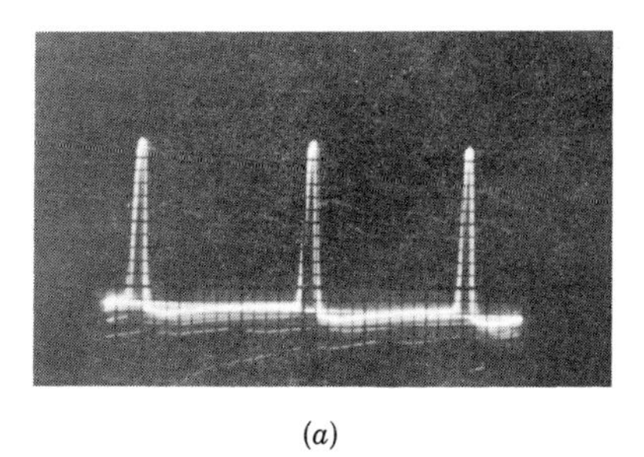

(*a*)

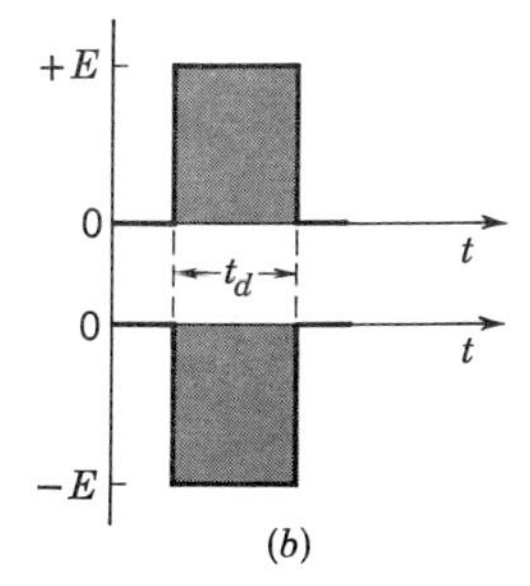


(*b*)

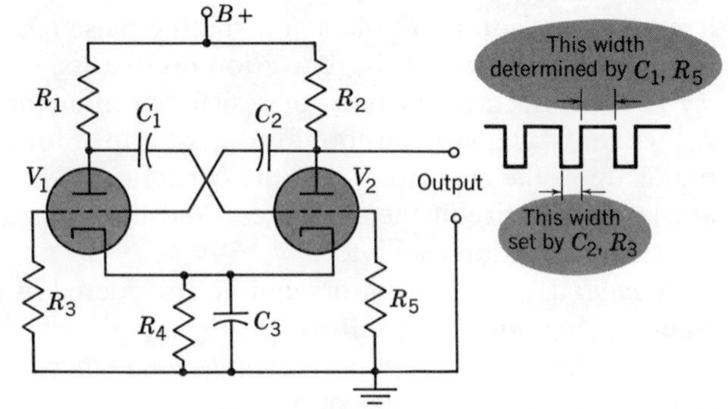

*Fig. 14·2 A plate-coupled multivibrator.*

are free-running relaxation oscillators. The Eccles-Jordan circuit in Fig. 14·6 and Schmitt trigger in Fig. 14·10 are actually multivibrator circuits, but they require external triggering signals to produce output pulses. All these pulse circuits are shown using tubes and then the corresponding transistor circuits are illustrated.

In terms of stability, multivibrators are considered *monostable, bistable,* or *astable.* Remember that a multivibrator has two stages. When both are stable, with one off and the other on, the circuit is bistable. With both stages quasi-stable, the result is an astable multivibrator, meaning it does not have any stable state. Finally, a monostable multivibrator has one stable state and one quasi-stable state. In their applications, the astable multivibrator is simply a free-running relaxation oscillator. The bistable and monostable multivibrators are trigger circuits, as they need input signal to upset the stable state.

**Plate-coupled vacuum-tube multivibrators.** As shown in Fig. 14·2, a plate-coupled multivibrator is simply a two-stage $RC$-coupled amplifier. It is called a plate-coupled circuit because the output at the plate of $V_2$ is coupled back through $C_2$ to the grid of $V_1$. The plate-coupled multivibrator is an astable circuit having two quasi-stable states; hence it has a continuous pulse output.

In the circuit of Fig. 14·2, the combination of $R_4$ and $C_3$ serves only as a bias source. The cutoff time of $V_2$ is determined by $C_1$ and $R_5$, and the cutoff time of $V_1$ is determined by $C_2$ and $R_3$. Thus, the output frequency of this circuit is controlled by $RC$ circuits as mentioned earlier. The square-wave pulse output waveform is also shown in the figure.

To follow the action, let us start with $V_1$ cut off and $V_2$ conducting. At this time, $C_2$ is discharging, with its left-hand plate losing electrons. The discharge current for $C_2$ flows through $R_3$. As the discharge of $C_2$ nears completion, the negative voltage across $R_3$ decreases, and $V_1$ starts to conduct. The current of $V_1$, flowing through $R_1$, causes the B+ at the plate of $V_1$ to drop. The decrease is transmitted to the grid of $V_2$ through capacitor $C_1$, appearing across $R_5$ as a negative voltage. This voltage

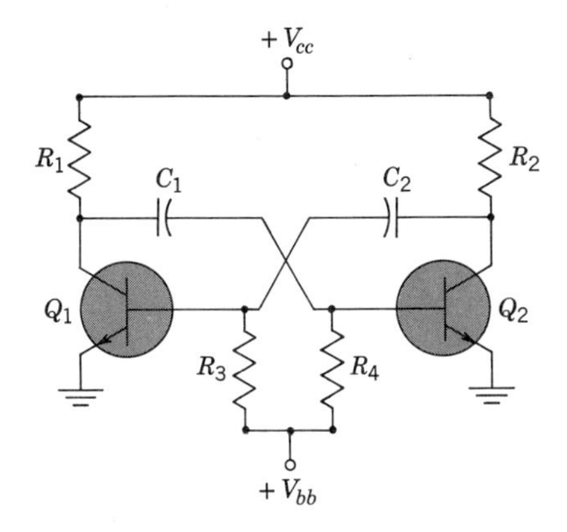


*Fig. 14·3 A collector-coupled multivibrator, similar to plate-coupled circuit in Fig. 14·2.*

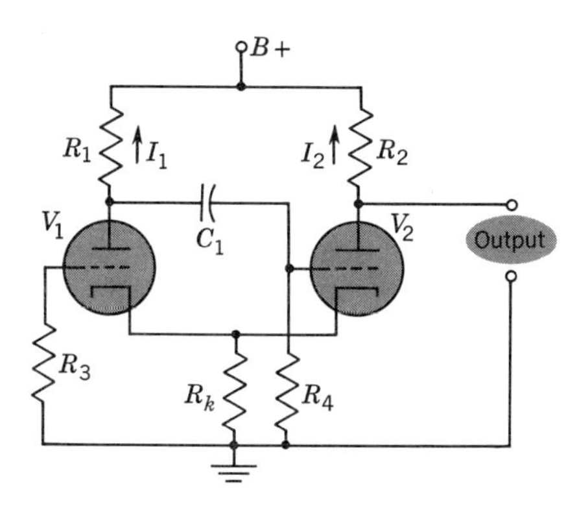


*Fig. 14·4 A cathode-coupled multivibrator. Frequency depends on $R_4C_1$ time constant.*

causes the plate current of $V_2$ to drop and the plate voltage on this tube to rise. This rise is transmitted to the grid of $V_1$ by capacitor $C_2$, further increasing the current flowing through this tube. The action continues, with the current flowing through $V_1$ increasing and the current through $V_2$ decreasing until $V_1$ is conducting strongly and $V_2$ is cut off. The system will remain in this condition while $C_1$ is discharging. When the discharge of $C_1$ nears completion, the negative voltage across $R_5$ decreases and permits $V_2$ to conduct again. When this happens, the entire sequence repeats itself.

**Collector-coupled transistor multivibrators.** The transistor counterpart of the vacuum-tube plate-coupled multivibrator is the collector-coupled multivibrator, shown in Fig. 14·3. This is essentially a two-stage, *RC*-coupled, common-emitter amplifier. It uses two NPN transistors, and the output at the collector of $Q_2$ is coupled back through $C_2$ to the base of $Q_1$. The base forward bias $V_{bb}$ is applied to $Q_1$ and $Q_2$ through resistors $R_3$ and $R_4$, and the collector reverse bias $V_{cc}$ is applied through $R_1$ and $R_2$. The output pulse frequency of this multivibrator is controlled by the *RC* circuits $R_3C_2$ and $R_4C_1$.

**Cathode-coupled vacuum-tube multivibrators.** The circuit of a cathode-coupled multivibrator is shown in Fig. 14·4. Before the power is applied to the circuit, capacitor $C_1$ is not charged, and the grids of both $V_1$ and $V_2$ are at ground potential. When the B+ voltage is applied, both tubes start to conduct. Since the plate current of both tubes flows through $R_k$, the cathode potential of both $V_1$ and $V_2$ rises above ground, producing a bias voltage which tends to limit the magnitude of the current flowing in the tubes. The flow of $I_1$, the plate current of $V_1$, through $R_1$ reduces the voltage at the plate of this tube. The plate voltage, it can be seen, is applied to the series combination of $C_1$ and $R_4$. Since the voltage across $C_1$ cannot change instantaneously, the full drop that takes place at the plate of $V_1$

appears across $R_4$ as a negative voltage. This further reduces the current through $V_2$. This reduction of current decreases the voltage developed across $R_k$, so that $I_1$ increases. This increased current causes the plate voltage of $V_1$ to drop further, and the conduction of $V_2$ is decreased even more. The action keeps building up, finally culminating with the current in $V_2$ reduced to zero and the current in $V_1$ at maximum.

The foregoing sequence occurs very rapidly, and $V_2$ is driven beyond cutoff almost instantaneously.

$V_2$ will remain cut off only as long as the negative voltage remains across $R_4$; and this, in turn, is determined by the $R_4C_1$ time constant. When the voltage across $R_4$ falls low enough, $V_2$ starts to conduct. Its current, flowing through $R_k$, increases the positive voltage here, which is equivalent to a greater negative bias voltage for $V_1$. This causes the current through $V_1$ to decrease, producing a rise in voltage at its plate. This positive increase is coupled to the grid of $V_2$ by $C_1$ and $R_4$, causing $V_2$ to conduct even more. This again raises the voltage across $R_k$, increasing the negative bias on $V_1$ and further lowering the current through this tube. (It also raises the bias of $V_2$, but this tube is receiving positive grid drive from the increasing plate voltage of $V_1$.) The foregoing action continues until $V_1$ is driven to cut off by its negative grid drive. However, $V_2$ is conducting maximum plate current.

At this time, capacitor $C_1$ charges through the flow of grid current in $V_2$ until the grid current falls to zero. This decreases the plate current flowing through $V_2$ and, consequently, the negative bias on $V_1$. Then the plate current of $V_2$ decreases and the grid voltage on $V_1$ increases until $V_1$ begins to conduct again, and the cycle repeats itself.

The on-off reversal (switching) of $V_1$ and $V_2$ can be summarized as follows:

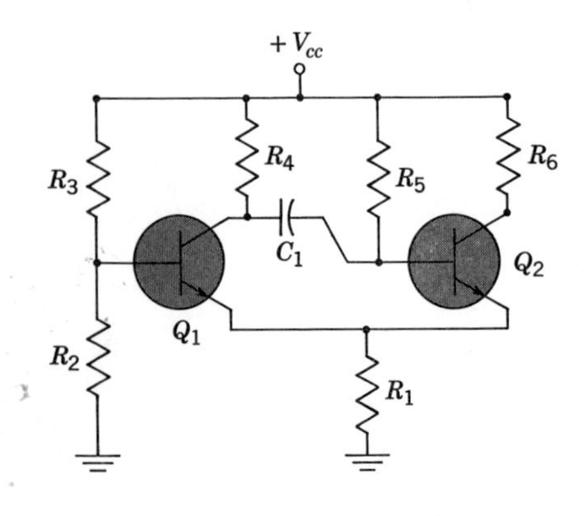


*Fig. 14·5 An emitter-coupled multivibrator, similar to the cathode-coupled multivibrator in Fig. 14·4.*

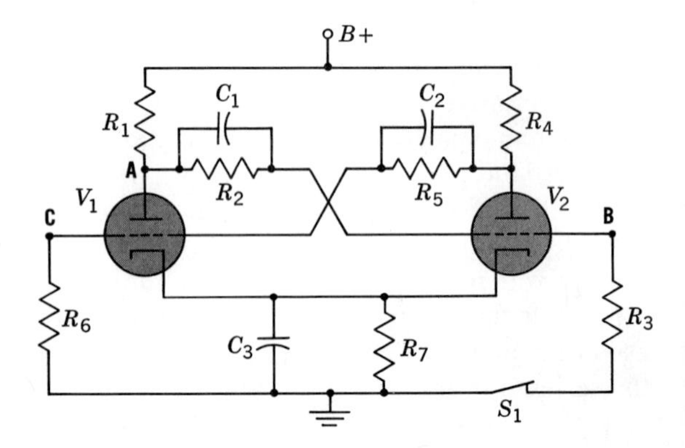


*Fig. 14·6 The Eccles-Jordan circuit is a bistable trigger circuit which can be triggered by several methods.*

1. $V_1$ conducting (on) cuts off $V_2$ (off). The time $V_2$ is off depends on how long it takes $C_1$ to discharge through $R_4$ when the $V_1$ plate voltage drops with conduction.
2. $V_2$ conducting cuts off $V_1$. The plate current of $V_2$ must now make the voltage across $R_k$ high enough to cut off $V_1$. This depends on how long it takes for $C_1$ to charge through the grid-cathode circuit of $V_2$.

The output pulses from either plate provide an unsymmetrical square wave, as the plate voltage drops quickly when the tube conducts and then rises sharply to the B+ value when the tube is cut off.

**Emitter-coupled transistor multivibrators.** The emitter-coupled multivibrator, shown in Fig. 14·5, is the transistor counterpart of the cathode-coupled multivibrator of Fig. 14·4. In this emitter-coupled circuit, two NPN transistors are used. The collector of $Q_1$ is coupled to the base of $Q_2$ through capacitor $C_1$. There is a common-emitter resistor, $R_1$, which provides bias for the two transistors.

The operation of this circuit is similar to that of the vacuum-tube circuit. When $Q_2$ conducts ($Q_1$ not conducting), $C_1$ charges through the low-resistance base-to-emitter path of $Q_2$ and causes a decrease in the forward bias on $Q_2$. As the forward bias on $Q_2$ decreases, the collector current from $Q_2$, flowing through $R_1$, decreases. Hence, the reverse bias due to the current through $R_1$ also decreases. Because of the voltage-divider action of $R_2$ and $R_3$, the positive voltage at the base of $Q_1$ overcomes the negative bias due to $R_1$, and $Q_1$ becomes forward-biased. When $Q_1$ conducts, its collector voltage drops, and this negative-going voltage is coupled to the base of $Q_2$ through $C_1$. $Q_2$ is then cut off; but it remains cut off only as long as it takes $C_1$ to discharge. As the discharge of $C_1$ nears completion, $Q_2$ becomes forward-biased again and begins to conduct. Its collector current again flows through $R_1$, causing a voltage drop here that is sufficient to cut off $Q_1$. The entire sequence then repeats itself.

## *14·3 Eccles-Jordan circuit*

The Eccles-Jordan circuit, shown in Fig. 14·6, is a trigger circuit that has two stable states (bistable multivibrator). One stable state exists when $V_1$ conducts, and the other when $V_2$ conducts. In this circuit, $R_7$ and $C_3$ provide cathode bias for both tubes. The plate of each tube is d-c coupled to the grid of the other by resistors $R_2$ and $R_5$, respectively. Capacitors $C_1$ and $C_2$ across $R_2$ and $R_5$, respectively, are *speed-up* or *commutating* capacitors. They improve the speed with which the circuit changes state by offering a low impedance to high frequencies.

Recall that a stable trigger circuit cannot change state by itself; it will change state only when triggered by an external influence. For example, if we momentarily short-circuit $R_6$, tube $V_2$ will conduct; tube $V_1$ will now be cut off. By the same token, if we short-circuit $R_3$, tube $V_1$ will conduct and $V_2$ will be forced into cutoff. This is one way to trigger the circuit so that it changes state. Another way is to insert a switch in the grid-to-

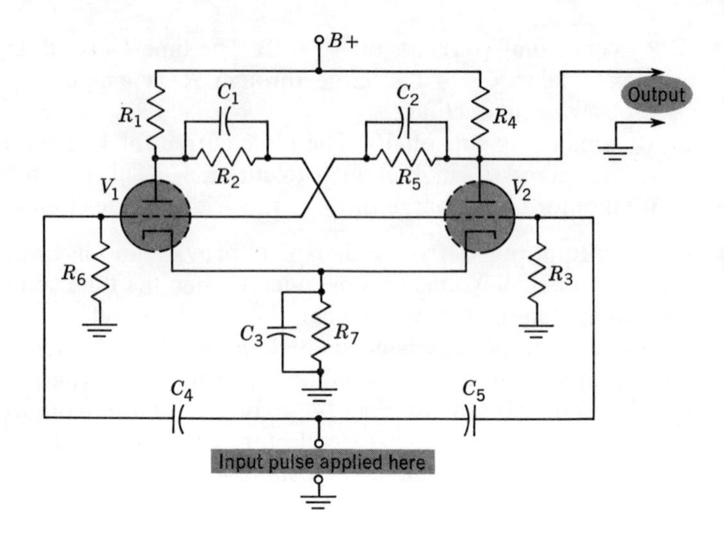


*Fig. 14·7 Input negative pulses are applied to both tubes in this Eccles-Jordan circuit. However, only the conducting tube is affected.*

ground circuit. For example, if $S_1$ in Fig. 14·6 is opened, point $B$ will become positive as the $V_2$ grid is d-c coupled to the $V_1$ plate. Then $V_2$ will begin to conduct and $V_1$ is cut off. To reverse this situation, we need a similar switch in the grid-to-ground circuit of $V_1$.

The foregoing methods of triggering a bistable circuit are very slow and inefficient. However, the use of triggering pulses from an external source provides a very practical method for triggering the Eccles-Jordan circuit. Let us now see how this can be done.

**Grid coupling.** Two means of coupling pulses into an Eccles-Jordan are shown in Figs. 14·7 and 14·8. In the first diagram, Fig. 14·7, capacitors $C_4$ and $C_5$ couple negative pulses to the grids of $V_1$ and $V_2$. A negative pulse to a tube in cutoff will have no effect, but a strong negative pulse to a tube in conduction will drive that tube into cutoff.

Assume $V_2$ is conducting. A negative pulse is applied to the input terminals and coupled through $C_4$ and $C_5$ to the grid of each tube. The pulse fed to $V_1$ has no effect, since the tube is already cut off. However, $V_2$ is driven into cutoff by the negative pulse it receives from $C_5$. With no tube current, the voltage drop across $R_4$ drops to zero, causing the plate voltage of $V_2$ to rise. This increase is coupled through $R_5$ and $R_6$ to the grid of $V_1$, driving it more positive. $V_1$ conducts, and a voltage is developed across $R_1$. This action lowers the plate voltage of $V_1$ and the grid voltage of $V_2$, through the coupling resistors $R_2$ and $R_3$. $V_2$ remains cut off until another input pulse is applied. A stable-state transition has occurred from $V_2$ conducting to $V_1$ conducting. Since output is taken from $V_2$ plate, a positive pulse will appear in the output at this time.

Let us now apply a second negative input pulse and see what happens.

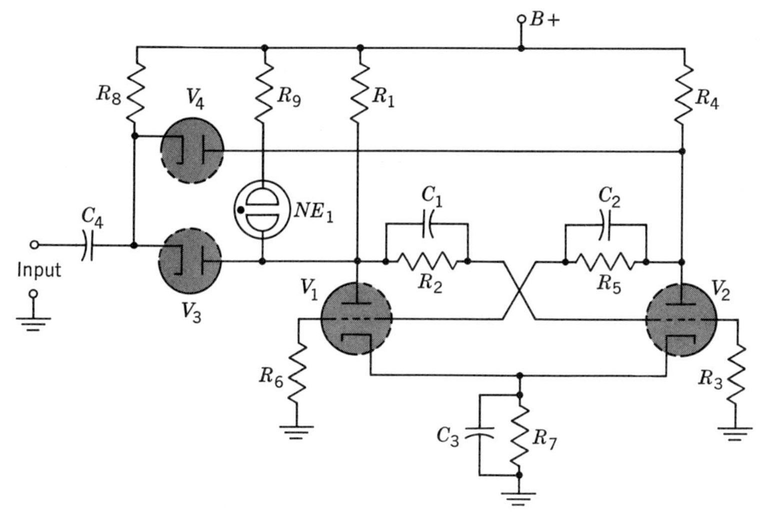


*Fig. 14·8 Faster response to input pulses can be achieved by using diodes $V_3$ and $V_4$ to couple input pulses to $V_1$ and $V_2$.*

When the second pulse is applied, only $V_1$ is conducting. Therefore it will be driven into cutoff, producing a positive signal for the grid of $V_2$ and causing $V_2$ to conduct. The conduction of $V_2$ will produce a negative output pulse at its plate. Thus two negative input pulses are needed to obtain one negative output pulse. This is a reduction of 2 to 1, or a division by two. Hence, the Eccles-Jordan circuit may be used as a frequency divider.

**Diode coupling.** Capacitive coupling for the input pulses works successfully when response speed is not too great. When high-speed pulse operation is desired, however, diode input coupling is employed. A pulse-input circuit which uses diodes is shown in Fig. 14·8. Diode coupling improves response time over the simple capacitor coupling method because it does not add any capacitance at the grid. With diode coupling, a negative pulse applied to the diode cathodes will cause one to conduct. Since the plate of diode $V_3$ is directly connected to the plate of $V_1$ and the plate of diode $V_4$ is attached to the plate of $V_2$, the diode plate connected to the nonconducting triode will have a higher positive voltage. If $V_1$ is conducting, then the plate voltage of $V_2$ is almost at the B+ value, while $V_1$ is about 150 volts below B+. Under these circumstances, only the plate of $V_4$ is sufficiently positive to permit this tube to conduct. A negative input pulse causes $V_4$ to conduct through $R_4$, producing a negative pulse across $R_4$. This pulse is coupled through $R_5$ and $C_2$ to the grid of $V_1$, driving this tube into cutoff. The next input pulse will pass through $V_3$, since the plate of $V_1$ is more positive, and through the coupling circuit, causing $V_2$ to cut off. In this way, the circuit flips back and forth and, because of this, is frequently called a *flip-flop* circuit. Diode coupling offers the advantages of polarity selection, grid isolation, and fast response.

An indicating device is often necessary to reveal which stable condition exists. One of the simplest indicators is a neon lamp and a series current-limiting resistor. The neon lamp is shown as $NE_1$ in Fig. 14·8, and $R_9$ is the current-limiting resistor. The neon lamp is useful because of its conduction characteristics. It requires 65 volts to light, but after the lamp ionizes, the voltage across it settles down to 50 volts. A reduction of the applied voltage to less than 50 volts will cause the lamp to deionize. When $V_1$ is cut off, the voltage drop across $R_1$ is about 47 volts because of the current passing through the coupling resistor $R_2$ and the grid-to-ground resistor $R_3$. When $V_1$ is conducting, the tube current through $R_1$ results in 150 volts appearing across $R_1$. As a result, the neon lamp across $R_1$ will not light when $V_1$ is cut off, but it will light when $V_1$ conducts.

**Transistorized Eccles-Jordan circuit.** The circuit in Fig. 14·9 is a transistorized Eccles-Jordan circuit in which the triggering pulses are applied to the base of each transistor. This type of coupling corresponds to grid coupling in the vacuum-tube circuit. The operation of the transistor circuit is essentially the same as that of the tube circuit.

Assume that $Q_2$ is conducting and that a negative input pulse is applied to the base of each transistor through $C_3$ and $C_4$. Since $Q_1$ is cut off, the negative pulse has no effect on this transistor. But $Q_2$ is driven into cutoff by the negative triggering pulse. As a result, the collector voltage of $Q_2$ rises and this positive voltage is coupled to the base of $Q_1$ through $R_2$ and $R_3$. Therefore, $Q_1$ becomes forward-biased and it conducts. $Q_2$ remains cut off and $Q_1$ keeps conducting until the next triggering pulse is applied. The next triggering pulse causes $Q_2$ to conduct and $Q_1$ to be cut off; the action is the same as just described for the first triggering pulse. The output is taken from the collector of either transistor, and each transistor generates an output pulse for every other input triggering pulse.

*Fig. 14·9 A transistorized Eccles-Jordan circuit. The triggering pulses ($e_{in}$) are applied to each transistor base.*

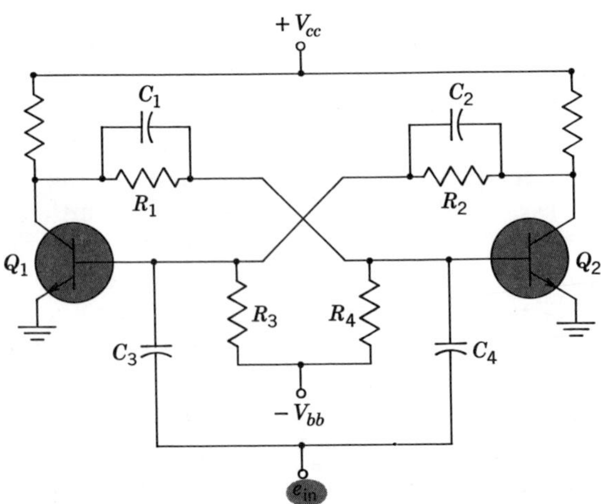

## 14·4 Schmitt trigger

The Schmitt trigger, shown in Fig. 14·10*a*, is a cathode-coupled bistable trigger circuit. This circuit is similar to the Eccles-Jordan circuit except that the coupling network from the plate of $V_2$ to the grid of $V_1$ has been eliminated, and a common-cathode resistor $R_k$ has been added. Because it is bistable, the Schmitt trigger has two stable states, in either of which it will remain indefinitely unless triggering is applied. One stable state exists when $V_1$ conducts and $V_2$ is cut off; the other exists when $V_1$ is cut off and $V_2$ conducts.

The switching from one stable state to the other can be accomplished by two methods: by applying a triggering pulse to the grid of $V_1$, and by raising or lowering the grid voltage of $V_1$. In the latter method, a sinusoidal input signal can be used to raise and lower the grid voltage.

Suppose we use an arbitrary input signal such as shown in Fig. 14·10*b*. Let us start with $V_2$ conducting and $V_1$ cut off by the bias developed across $R_k$ due to the plate current of $V_2$. The circuit remains in this condition until $e_{in}$ becomes sufficiently positive to drive $V_1$ out of cutoff. This happens at $e_{in} = e_1$ in the figure. When $V_1$ begins to conduct, its plate voltage drops and a negative voltage is coupled to the grid of $V_2$ by the voltage-divider network consisting of $R_2$ and $R_4$. Then $V_2$ is driven into cutoff, and the voltage output taken from the plate of $V_2$ increases. After the circuit has been triggered, any further increase of the input voltage does not affect the output voltage; hence the output stays at a constant level $E$.

$V_1$ continues to conduct until $e_{in}$ decreases. But this time, triggering occurs at a lower voltage level $e_2$, as shown in Fig. 14·10*b*. This discrepancy in the triggering voltage level occurs as a result of a form of *hysteresis* (or *backlash*) that exists in the circuit when the loop gain is greater than

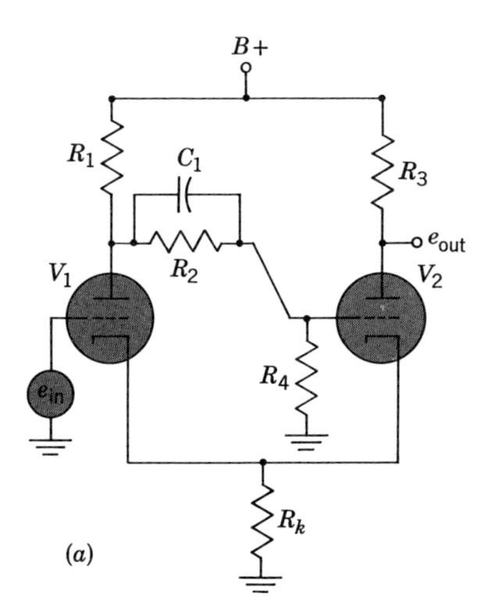


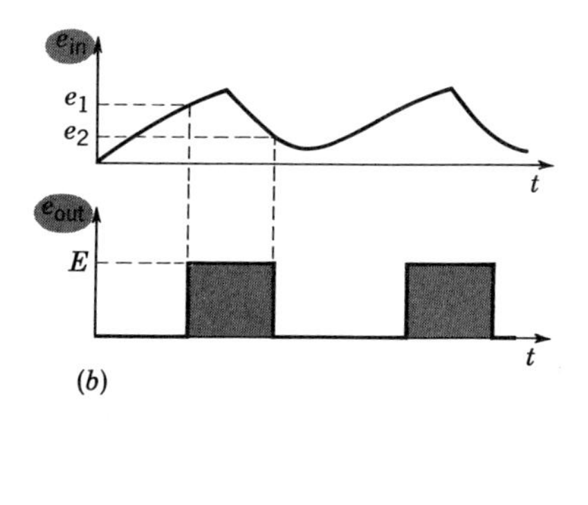


*Fig. 14·10 (a) A Schmitt trigger circuit. (b) The input triggering waveform and output pulses. There are two different triggering voltage levels in this circuit.*

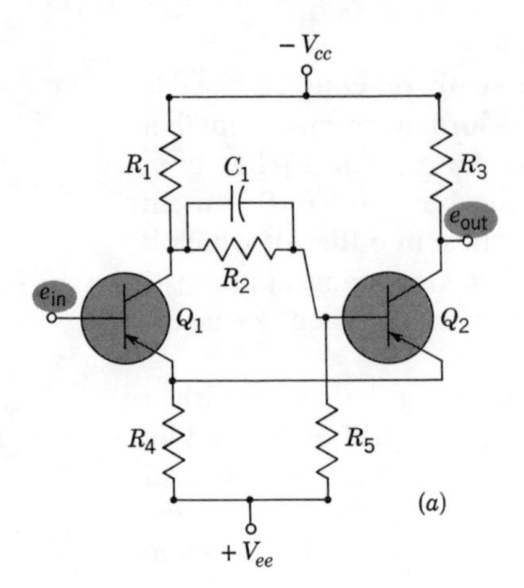


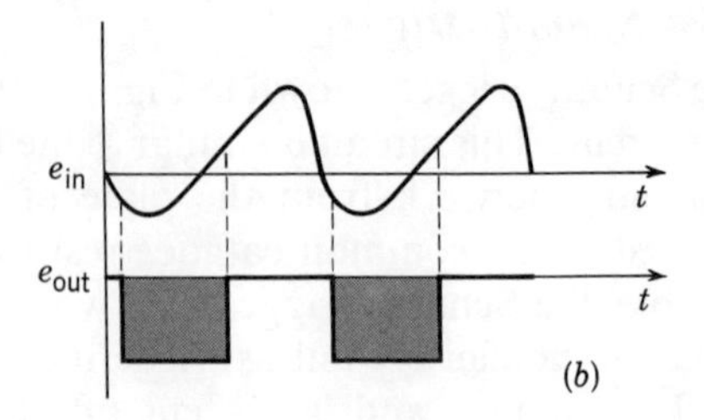


*Fig. 14·11 (a) A transistorized Schmitt trigger circuit. (b) Its input and output waveforms. This is an emitter-coupled bistable circuit, which produces negative pulses at the collector of $Q_2$.*

unity. The hysteresis effect can be minimized if a resistor, connected in series with the cathode of $V_2$, is used to limit the loop gain to unity.

When the input voltage decreases, the plate current from $V_1$ decreases, and the grid voltage of $V_2$ becomes less negative because of the decrease in current flowing through $R_k$. Then, at $e_{in} = e_2$, $V_2$ begins to conduct again and $V_1$ is cut off. Further triggering by $e_{in}$ causes the cycle to repeat again.

Regardless of the shape of the input voltage, the output of $V_2$ is always a square wave. Because of this characteristic, the Schmitt trigger is often used as a squaring circuit.

The transistorized Schmitt trigger shown in Fig. 14·11*a* uses two PNP transistors. Notice the change in the polarity of the bias voltages. $R_4$ is the common-emitter resistor, and the collector of $Q_1$ is coupled to the base of $Q_2$ through the $R_2C_1$ network. In the stable state when $Q_1$ is cut off ($Q_2$ conducts), its collector voltage is equal to $-V_{cc}$. This negative voltage is applied to the base of $Q_2$ through the coupling network, and it forward-biases $Q_2$. The collector current flows through $R_4$ and causes $Q_1$ to be reverse-biased.

The circuit remains in this stable state until a negative-going input signal of sufficient amplitude is applied to the base of $Q_1$. See Fig. 14·11*b*. This signal triggers the circuit by forward-biasing $Q_1$ into conduction. When $Q_1$ conducts, its collector voltage drops (becomes less negative). This voltage change, which is coupled to the base of $Q_2$, reverse-biases $Q_2$ and drives it into cutoff. Now, $Q_1$ is conducting and $Q_2$ is cut off; this is the other stable state. The circuit is triggered again when the amplitude of $e_{in}$ increases (becomes less negative) and $Q_1$ becomes reverse-biased. When this happens, $Q_1$ is cut off, its collector voltage rises (becomes more negative), and $Q_2$ is forward-biased and conducts again. The cycle repeats itself as further triggering is applied. The output of the transistorized Schmitt trigger is taken from the collector of $Q_2$; the output waveform is shown in Fig. 14·11*b*.

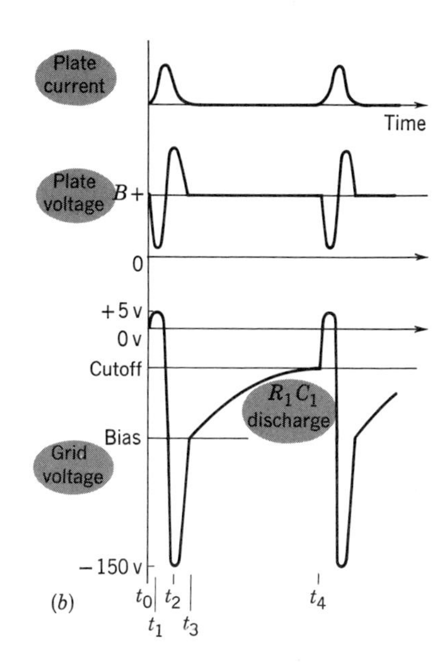


*Fig. 14 · 12 A blocking oscillator and its voltage and current waveforms.*

## 14 · 5 Blocking oscillators

The free-running blocking oscillator, shown in Fig. 14 · 12*a*, is a relaxation oscillator. One difference between this circuit and the relaxation oscillators discussed earlier is that, here, one of the vacuum-tube stages is replaced by a transformer. Blocking oscillators are essentially transformer-coupled oscillators with grid-leak bias; they are called "blocking" oscillators because of the way the grid-leak bias *blocks* the oscillatory action of the circuit.

The transformer provides grid feedback voltage with the polarity required to reinforce the grid signal and start the oscillations. When the oscillator feedback drives the grid positive, grid current flows to develop grid-leak bias. This regenerative circuit could oscillate with a continuous sine-wave output at the natural resonant frequency of the transformer, depending on its inductance and stray capacitance. However, several factors enable the oscillator to cut itself off with high negative grid-leak bias. A large amount of feedback is used. Also, the $R_1C_1$ time constant is made long enough to allow the grid-leak bias to keep the tube cut off for a relatively long time. Finally, the transformer has a high internal resistance (or low $Q$) so that after the first cycle, the sine-wave oscillations do not have enough amplitude to overcome the negative bias.

The tube remains cut off until $C_1$ can discharge through $R_1$ to the point where the grid-leak bias voltage is less than cutoff. Then the plate current can flow again to provide feedback signal for the grid, and the cycle repeats itself at the blocking rate. Therefore, the circuit operates as an intermittent or blocking oscillator. The tube conducts a large pulse of plate current for a short time and is cut off for a long time between pulses. Refer to the waveforms in Fig. 14 · 12*b*.

**Triggered blocking oscillators.** The triggered or monostable blocking oscillator is shown in Fig. 14·13. This circuit is very similar to that of the free-running blocking oscillator, except for the large negative voltage $-E_{cc}$ which is applied to the grid of the tube. This voltage keeps the tube normally cut off (the stable state). Since the circuit is monostable, it has only one stable state. The quasi-stable state occurs when the tube conducts. This state exists for a short length of time, after which the circuit automatically reverts back to its stable state without any external triggering.

When a positive triggering pulse $e_{\text{in}}$ of sufficient amplitude to overcome the negative bias is applied, the tube is driven into conduction. The increasing plate current flows through the primary winding of the transformer, which is connected so that it will provide regenerative feedback. Therefore, the grid is made increasingly positive, and plate-current saturation is quickly reached. As the plate current increases toward saturation and the grid is driven positive, grid current flows and capacitor $C_1$ charges.

As saturation is approached, the plate current increases more slowly, and the amount of positive feedback decreases. When the plate current starts to decrease, the polarity of the feedback voltage changes and decreases the plate current still further. The grid voltage reaches the cutoff level and continues to swing negative to some maximum value. When the cutoff level is reached, $C_1$ begins to discharge. Finally, the grid voltage rises to the fixed bias level and the plate voltage returns to B+. The circuit is again in its stable state where it will remain until the next triggering pulse is applied. As you can see, the operation of this circuit is the same as that of the free-running circuit except for the biasing arrangement and necessary triggering.

**Transistorized blocking oscillators.** Transistorized blocking oscillators operate in essentially the same manner as do their vacuum-tube counterparts. The transistorized circuits may be either free-running or triggered (monostable), depending on the biasing arrangement. In the free-running blocking oscillator, forward bias is applied to the base-to-emitter junction

*Fig. 14·13 A monostable blocking oscillator will produce an output only when triggered by a positive pulse, $e_{in}$.*

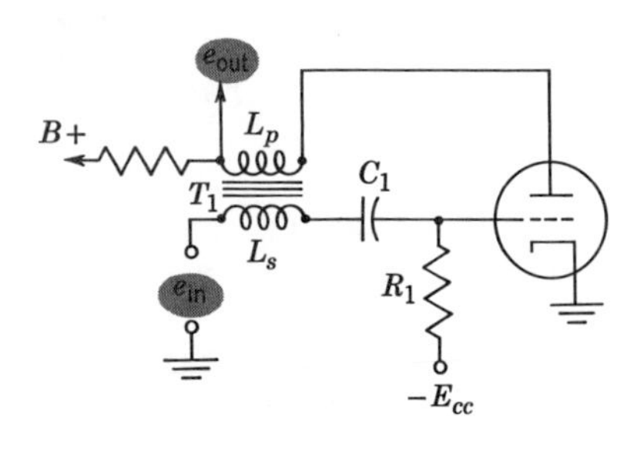


*Fig. 14·14 A transistorized monostable blocking oscillator. The diode $D_1$ suppresses the large negative voltage swing and prevents damage to the transistor.*

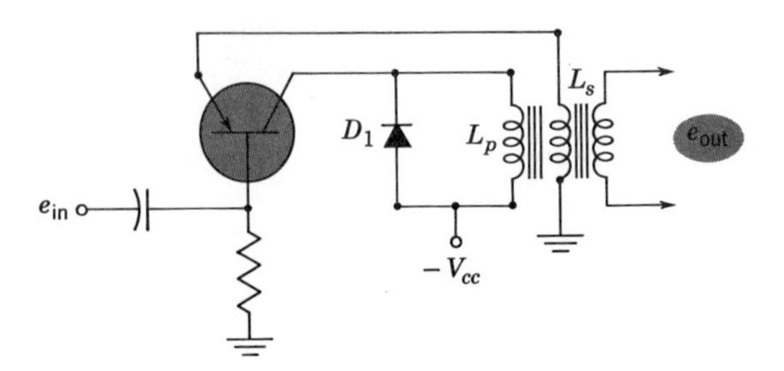

and the circuit operates as soon as power is applied to the circuit. In the monostable blocking oscillator, reverse bias is applied to the base-to-emitter junction and the transistor is held at cutoff until a triggering pulse is applied. Except for the difference in biasing, the actual operation of the two transistorized blocking oscillators is the same; hence only the monostable circuit will be described.

The circuit of the transistorized monostable blocking oscillator is shown in Fig. 14·14. Before triggering is applied to the base of the transistor, the circuit is in its stable state (cut off), and the emitter-base bias is zero. When a negative triggering pulse $e_{in}$ is applied, the transistor is forward-biased and begins to conduct. The increasing collector current flows through the primary winding of the transformer, causing a voltage to be induced in the secondary. The polarity of this induced voltage is such that it increases the forward bias on the emitter. As this action continues, the transistor approaches saturation, and the rate of change of collector current decreases.

When the collector current becomes constant at saturation, the induced voltage in the transformer secondary drops to zero. The transistor becomes reverse-biased and stops conducting. Ordinarily, the voltage in the secondary would continue to decrease below zero to some peak negative value, and then rise again to zero. However, this effect is not desirable in the transistor circuit because the negative swing can exceed the collector breakdown voltage. The diode $D_1$, connected across the transformer primary, is used to suppress the negative voltage swing and, hence, prevents damage to the transistor. The circuit is again in its stable state, in which it will remain until the next triggering pulse is applied. The output $e_{out}$ then is a single cycle, repeated for every input pulse.

## *14·6 Gas-tube relaxation oscillators*

**Gas-diode oscillators.** A cold-cathode gas tube has a definite ionization voltage which starts tube conduction. After a gas tube fires, the voltage across it drops to a sustaining value and remains there until it is interrupted or reduced. In addition, a minimum current of 5 ma is required to maintain ionization of the gas in most tubes. The resistor in series with a gas diode controls the behavior of the tube. When its value is such that the sustaining voltage is reached with 5 ma or more, the tube will remain fired. If the sustaining voltage is reached and the current is less than 5 ma, the tube will extinguish itself.

In many gas tubes, the difference between ionization and sustaining voltages is about 15 volts. This voltage difference can be used to develop pulses with a minimum of circuitry. One tube used for pulse forming in this manner is the OA3. Its ionizing voltage is 90 volts, the sustaining voltage is 75 volts, and the minimum ionizing current is 5 ma. In Fig. 14·15, the value of $R_1$ is made large enough to cause the plate voltage to drop below the sustaining voltage with a minimum of current. Conduction will cease when these conditions are met.

The action is made positive by the selection of a current well below the

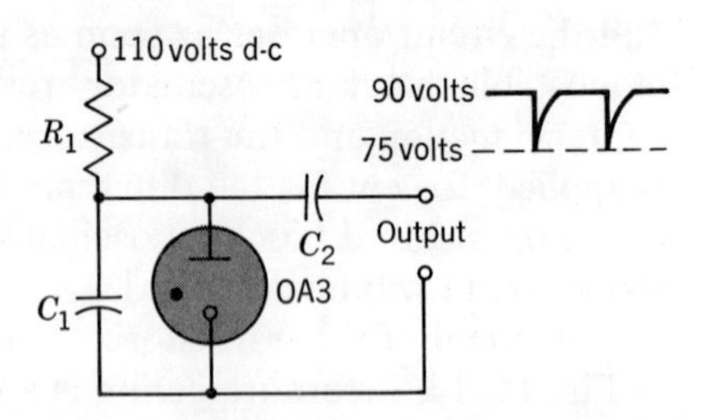


*Fig. 14·15 A simple pulse generator which relies on the difference in ionizing and sustaining voltages of a gas tube.*

sustaining value. For instance, with 1 ma of current and $R_1$ equal to 35,000 ohms, the voltage across $R_1$ will be 35 volts. The OA3 cannot maintain conduction with only 1 ma; thus the tube must stop conducting after a momentary conduction. This terminates the leading edge of the output pulse.

Here is how the circuit of Fig. 14·15 functions. With 110 volts applied, current flows through $R_1$ to charge $C_1$. When the voltage across $C_1$ reaches 90 volts, the OA3 fires, discharging the capacitor to 75 volts. However, the value of $R_1$ is made large enough so that minimum current (5 ma) for sustaining the tube does not develop. Hence the OA3 tube extinguishes itself. This leaves the capacitor, which now has a potential of 75 volts, to charge again to 90 volts. When this point is reached, the OA3 fires again, discharges $C_1$ to 75 volts, and extinguishes itself. The sequence thus repeats itself over and over.

The number of pulses formed in Fig. 14·15 depends on the time required to charge $C_1$ from 75 to 90 volts through $R_1$. The B+ voltage will also affect the time necessary to accomplish this charge. If the voltage is made higher, the capacitor charges more quickly to the firing potential; this means an increase in frequency. Conversely, when the voltage is lower, say 95 volts, it takes $C_1$ longer to charge to 90 volts (from 75 volts), and consequently the pulse repetition rate is lower.

Frequency control can be readily achieved if we change the value of the charging capacitor $C_1$ with a rotary selector switch. Another simple form of control is to insert a variable resistor $R_1$. Large values of resistance lower the repetition rate, whereas small values raise it. With this latter form of control, it is important that $R_1$ always remain large enough to reduce the current below the sustaining value so that the tube may be extinguished.

**Gas-triode oscillators.** A thyratron is a hot-cathode gas triode whose firing potential depends on the grid bias voltage. However, once this gas tube has fired, the grid has absolutely no control over the tube current, which is controlled, for the most part, by the circuit external to the tube. When the tube has fired, the voltage across it drops to a sustaining value and remains essentially constant. The tube can be extinguished only if we reduce the tube current below the minimum value required to maintain ionization. When the gas tube stops conducting the grid regains control, and the grid bias determines the amount of plate voltage needed to fire the tube again.

The basic circuit of a thyratron relaxation oscillator used as a sawtooth sweep generator is shown in Fig. 14 · 16. When switch $S_1$ is closed, power is applied to the circuit and $C_1$, which is initially uncharged, begins to charge through $R_3$. The voltage developed across $C_1$ is positive with respect to ground and it is applied to the plate of the thyratron. Before $C_1$ can charge to the value $E_{bb}$, it reaches the firing potential of the tube which is determined by the grid bias. The tube then starts to conduct and $C_1$ discharges very quickly through the tube and the series resistor $R_2$. When the voltage across $C_1$ falls below the sustaining voltage, the tube stops conducting. $C_1$ then begins to charge again, and the cycle repeats itself.

The output waveform appears across $C_1$, and as shown in Fig. 14 · 16, it is a sawtooth voltage. The frequency of this output is dependent on the plate voltage, the grid bias, $R_3$, and $C_1$. Usually, though, $E_{bb}$ and $E_{cc}$ are kept constant, and the frequency is varied by changes in the values of $R_3$ and $C_1$. Thus, the amplitude and linearity of the voltage waveform can be kept constant. The value of $R_3$ must always be large enough to prevent the plate supply voltage from providing the tube with enough current to maintain ionization; if this should happen, the circuit would stop oscillating.

## *14 · 7 Synchronization*

It is often desirable to control oscillators within very narrow frequency limits, since variations in tube characteristics and circuit components may shift the frequency. In many cases, the pulses must be generated at exactly the same instant as another generator thousands of miles away in order to convey information. Television pictures are possible only because the transmitter and receiver use synchronization circuits that maintain simultaneous operation. A ship-to-shore message can be coded on the ship, and only a station which can synchronize with the ship's output pulses will be able to lock in and decode the message.

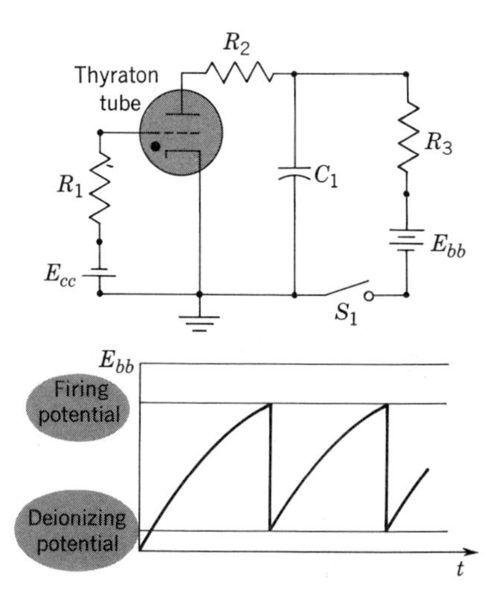


*Fig. 14 · 16 The basic circuit of a gas-triode oscillator. The waveform across $C_1$ is a sawtooth.*

Triggering and synchronization are very similar. However, triggering is used to drive a circuit into conduction so that an output voltage is produced; whereas synchronization is used to make the already existing output frequency of a relaxation circuit conform to the reference synchronizing frequency. (If a continuous synchronizing signal is applied to a bistable or monostable circuit, the circuit simply acts as if it is being continuously triggered. Astable circuits generate an output with or without a synchronizing signal.)

Methods of synchronizing two separate oscillators rely on the slow approach toward capacitor discharge which is made by a timing circuit. The curve in Fig. 14 · 17 follows the voltage discharge curve seen across a grid resistor when a timing capacitor discharges through the resistor. Since all the timing circuits described previously rely on a capacitor discharge, the basic synchronizing action is the same for all. Assume the natural discharge of a capacitor extends for nine divisions as shown in Fig. 14 · 17. This discharge time can be shortened, however, by the application of a pulse at any time after division No. 7, as shown. What the applied pulse does is bring the tube out of cutoff earlier than the normal pulse-spacing time. This is called *triggered synchronization.* The division line at which synchronization takes place depends on the amplitude of the trigger (synchronizing) pulse.

Figure 14 · 18 shows a cathode-coupled multivibrator which accepts positive synchronizing pulses. The pulses are applied to the $V_2$ grid. When

*Fig. 14 · 17 Triggered sync. The pulse must arrive just before conduction time, as shown.*

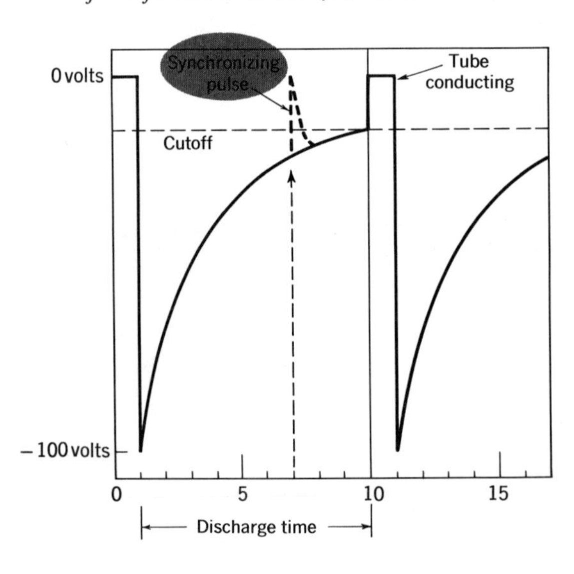


*Fig. 14 · 18 A cathode-coupled multivibrator responds to a positive pulse applied to the grid of $V_2$. The pulse will bring $V_2$ into conduction before $C_1$ has discharged completely.*

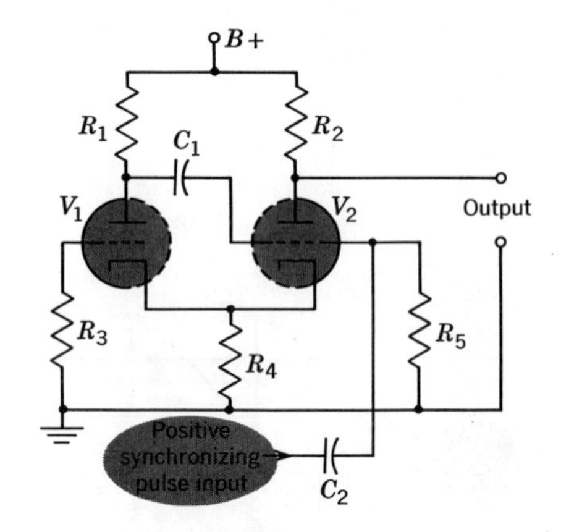

a pulse arrives, the tube is driven out of cutoff and $V_1$ is forced into cutoff —*before* this action would normally occur in the circuit. In Fig. 14·19, however, a negative sync pulse is applied to the grid of the opposite tube $V_1$. This pulse will initiate a circuit change as readily as positive pulses in Fig. 14·18. When a negative pulse is applied to $V_1$, its current flow is sharply reduced. This causes less voltage drop across $R_4$ and permits $V_2$ to come out of cutoff.

Figure 14·20 illustrates methods of applying synchronizing pulses to a blocking oscillator. At *A*, when a positive sync pulse is applied to the grid, it brings the tube out of cutoff early, and the cycle begins.

Sync pulses can also be brought into the circuit through an extra winding on the transformer. The pulse sends a current through this extra winding, developing a positive voltage in the grid winding, which brings the tube out of cutoff.

Synchronizing principles are used in a great many electronic fields, such as telemetering, computers, radio communications, and television. In television, two synchronizing signals are required to completely lock in the receiver with the transmitter. One pulse at 60 cycles controls the vertical sweep, and one at 15,750 cycles controls the horizontal sweep.

## *14·8 Gating*

Gating is the technique of controlling tube conduction by the application of pulses to bring a tube out of cutoff. When several signals are applied

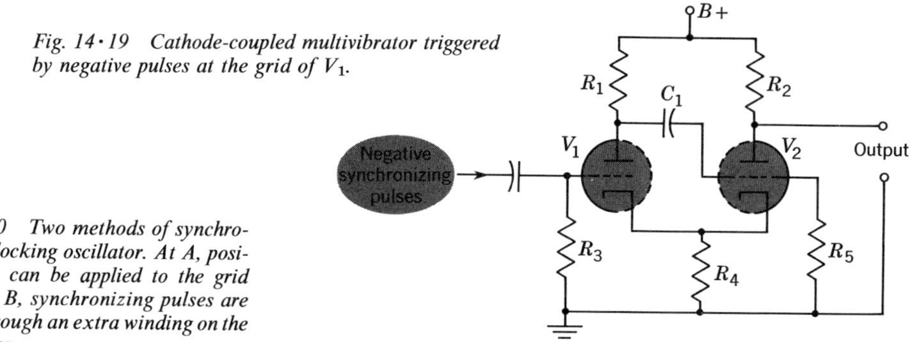


*Fig. 14·19 Cathode-coupled multivibrator triggered by negative pulses at the grid of $V_1$.*

*Fig. 14·20 Two methods of synchronizing a blocking oscillator. At A, positive pulses can be applied to the grid circuit; at B, synchronizing pulses are applied through an extra winding on the transformer.*

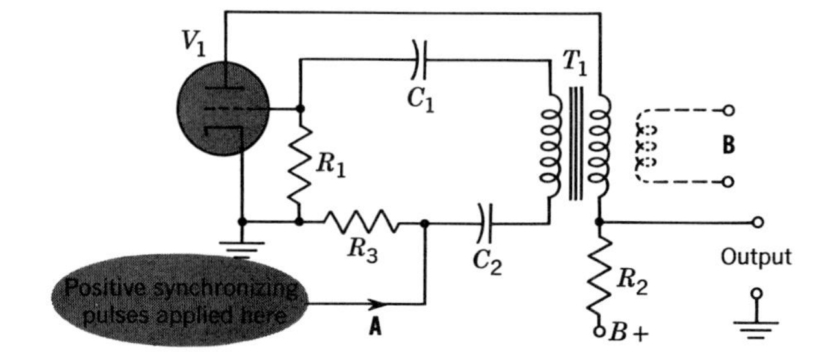

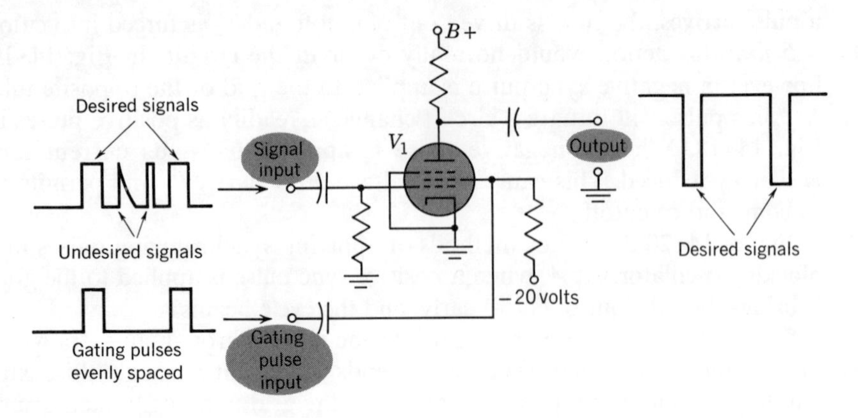


*Fig. 14·21 A gated amplifier. The screen voltage holds the tube cut off until a positive gating pulse is applied to the screen grid. Then the desired pulses in the signal at the control grid can be amplified.*

to a tube (or transistor), and only certain ones are to be amplified, selection is possible because the tube is permitted to conduct only when the desired signal appears at the input. Tube (or transistor) conduction is prevented at all other times.

Figure 14·21 illustrates a gated amplifier circuit. The input waveform contains the desired signal and some unwanted signals. A pentode screen grid does have control over tube conduction, and −20 volts bias on this grid will hold the tube in cutoff. No signal can pass through a tube in cutoff; therefore, any signals applied to the control grid during cutoff will not appear at the plate output. When the gating pulses appear, they momentarily add positive voltage to the screen-grid bias and permit the tube to conduct. The control-grid input signals that appear during this brief moment are amplified. Comparison of the input signals with the gating pulses reveals that the desired signals and gating pulses occur at the same time. Therefore, the desired signal is amplified while the remainder of the input signal is suppressed.

**Telemetry uses.** Telemetering is a method of transmitting information concerning a commodity or process over a distance. Information is obtained from a metering unit in the form of modulated pulses whose shape, duration, amplitude, or time of occurrence bears a relationship to the quantity to be measured. Generally, more than one signal from one variable is transmitted on an r-f carrier. All the different information pulses are combined into a series of pulses which will modulate the carrier.

Power companies measure voltage and current at a distant substation and transmit the information concerning these variables back to a central control station. The signals obtained from the substation are compared with desired values. A difference between the value indicated by the signal received and the desired value requires corrective action. Central control will transmit a correction signal to the substation to change the value of supplied current or voltage. Subsequently, supplied power is altered to compensate for the original variation.

The heart of this system is a decoding device which separates pulses transmitted on a common r-f carrier. A pulse train containing amplitude-modulated signals from two different sources is shown in Fig. 14 · 22. A measured voltage at the substation is used to develop an amplitude-modulated pulse signal, shown in the top line of Fig. 14 · 22. Current demand is measured and converted to an amplitude-modulated signal, as shown in the center row. By careful synchronizing, these two signals are transmitted sequentially, that is, first a pulse of the top line, then a pulse of the second line, etc. The composite pulse train, including narrow sync pulses, is shown at the bottom. To separate these signals at the receiver, a circuit is required which passes one pulse through one amplifier and the next pulse through another amplifier. A synchronized multivibrator at the receiver supplies amplifier gating pulses to achieve this.

Telemetering has many other applications, including rockets and missiles, where it is used to record the various conditions (such as temperature, pressure, vibration, and speed) existing in these vehicles as they speed through space. This information, obtained electronically from a variety of different measuring devices, is combined in a previously established sequence and radioed back to the testing base. Here, the many bits of information are separated from each other and individually analyzed.

*Fig. 14 · 22 A series of pulses from two different sources A and B are intermixed to produce the output waveform shown in (c).*

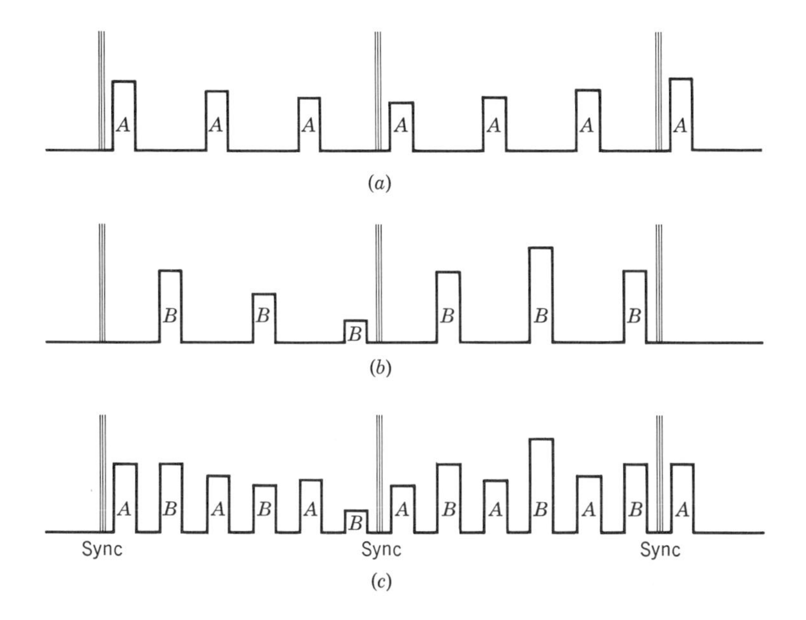

*Fig. 14·23 A mechanical register capable of counting pulses up to 1,000 per second.* (Production Instruments Co.)

## *14·9 Pulse counters*

**Electromechanical counters.** Counting pulses at a slow rate is performed by an electric counter which has wheels similar to the automobile speedometer. Each pulse develops a magnetic field around a coil with a movable center iron core. This core then responds to the magnetic field by moving toward the center of the coil. The core is physically linked to the units wheel by a mechanical ratchet and pawl, so that each core movement advances the units wheel one number. With every 10 pulses, the units wheel makes one complete revolution and a small extension protruding from the units wheel engages the tens wheel. In this manner, the tens wheel is caused to advance one number for every complete revolution of the units wheel. Similarly, every wheel advances one number after the next lower-order digit wheel has made a complete turn.

Figure 14·23 identifies the units, tens, hundreds, and thousands of an electric counter. This unit is designed for wall or external panel mounting, but a variety of other shapes and mounting arrangements are also available. Most counters respond to a maximum of 20 to 50 counts per minute; however, the one shown in Fig. 14·23 is designed to operate at 1,000 counts per minute, or 1,600 counts per minute with an electronic driver. Driving a counter may be as simple as closing a switch to complete the circuit to a power source or as involved as using several tubes to control the counter action. Counter solenoids (coil and core) are built to suit a wide variety of requirements in voltage and current. Common voltage values are 120 volts a-c, 24 volts d-c, and 6 volts d-c, at currents ranging from ½ amp down to several microamperes. When the repetition rate of input pulses exceeds the response of a mechanical counter, several electronic units will extend the range of the basic counter.

**Electronic counters.** The need for counting pulses at faster and faster rates has led to the development of electronic circuits which respond to high-speed input pulse repetition rates and provide a lower-speed pulse repetition-rate output. To achieve pulse speed reduction, an electronic unit must respond to each input pulse but produce only one output pulse after a specific number of input pulses. The relation between input pulse rates and output pulse rates is called the *scale factor*. For example, a circuit that can respond to 10 input pulses and produce one output pulse has a scale

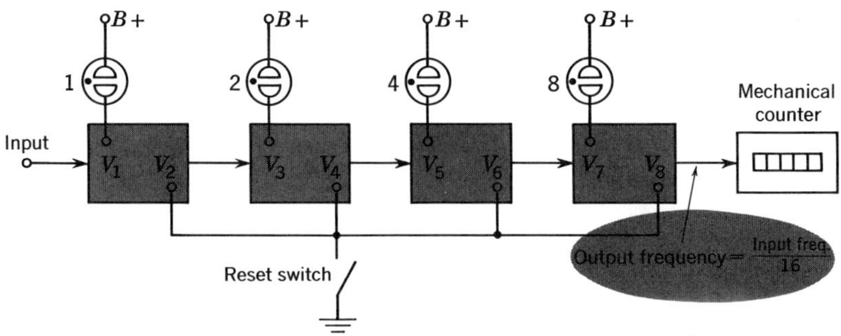


*Fig. 14·24 Four binary stages, connected together, provide a count division of 16.*

factor of 10. Ten is the most desirable scale factor, but very few electronic units are able to provide this scale factor directly.

Electric-counter speeds could be double if a simple scale-factor-of-2 circuit is placed before the counters. This scale-factor-of-2 circuit must produce one output pulse for every two input pulses. A circuit that meets these requirements is the Eccles-Jordan circuit discussed earlier in this chapter.

Scale factors greater than 2 can be obtained if we connect several Eccles-Jordan stages in series. The output pulse of one stage becomes the input pulse for the next stage. In this manner, the first stage, with a scale factor of 2, divides the input pulses by 2 and produces half as many output pulses. The second stage also has a scale factor of 2, and the output pulses from the first stage are divided by 2. The total scale factor for the two stages in series is 4. A third stage following this provides an over-all scale factor of 8. Most plug-in units in computers are made up of four stages to provide a total scale factor of 16.

Figure 14·24 shows a block diagram of four cascaded Eccles-Jordan stages with indicating lamps. The lamps are numbered according to the number of input pulses required to light them. The first lamp, No. 1, will come on when the first input pulse causes $V_2$ to cut off and $V_1$ to conduct. The second lamp is No. 2, since the first stage produces one output pulse after receiving two input pulses. The third lamp is No. 4 to correspond to the output pulse produced by the second stage after four first-stage input pulses.

Table 14·1 shows the conducting tubes after each number of input pulses. Note that for the fourth input pulse, only lamp No. 4 will light. This is because $V_5$ is the only conducting tube which has a light across its plate-load resistor. Pulse 7 will be indicated by all three lamps, 1, 2, and 4. The lamps are cumulative; that is, all lighted lamps must be added together to find the actual number of input pulses applied. When all the lamps are lit, the total count of input pulses is 15, since 1, 2, 4, and 8 add up to 15. One pulse applied after No. 15 will turn all the lamps off. This is the same condition as the zero count.

*Table 14·1 Conduction sequence of each tube, for scale of 8 in Fig. 14·24*

| Input pulses | Conducting tubes | | | | Input pulses | Conducting tubes | | | |
|---|---|---|---|---|---|---|---|---|---|
| 0 | $V_2$ | $V_4$ | $V_6$ | $V_8$ | 8 | $V_2$ | $V_4$ | $V_6$ | $V_7$ |
| 1 | $V_1$ | $V_4$ | $V_6$ | $V_8$ | 9 | $V_1$ | $V_4$ | $V_6$ | $V_7$ |
| 2 | $V_2$ | $V_3$ | $V_6$ | $V_8$ | 10 | $V_2$ | $V_3$ | $V_6$ | $V_7$ |
| 3 | $V_1$ | $V_3$ | $V_6$ | $V_8$ | 11 | $V_1$ | $V_3$ | $V_6$ | $V_7$ |
| 4 | $V_2$ | $V_4$ | $V_5$ | $V_8$ | 12 | $V_2$ | $V_4$ | $V_5$ | $V_7$ |
| 5 | $V_1$ | $V_4$ | $V_5$ | $V_8$ | 13 | $V_1$ | $V_4$ | $V_5$ | $V_7$ |
| 6 | $V_2$ | $V_3$ | $V_5$ | $V_8$ | 14 | $V_2$ | $V_3$ | $V_5$ | $V_7$ |
| 7 | $V_1$ | $V_3$ | $V_5$ | $V_8$ | 15 | $V_1$ | $V_3$ | $V_5$ | $V_7$ |
| | | | | | 16 | $V_2$ | $V_4$ | $V_6$ | $V_8$ |

## *14·10 Decade counters*

Mechanical counters would be easy to read if the electronic scale factor were 10. Decade electronic counters can be formed by reducing the scale of 16 (four-stage Eccles-Jordan) by 6 counts. The reduction method used is pulse feedback. That is, a pulse is obtained from one stage and routed back to a preceding stage to make this prior stage believe it has received still another pulse from the input. So far as this prior stage is concerned, each feedback pulse looks the same to it as any incoming pulse. If we now apply 10 pulses at the input to the system and feed back 6 pulses from one stage to a prior one within the system, then 16 pulses in all will have passed through the system and reached the counter. For these 16 pulses, the counter will record a value of 1, but this 1 stands for 10 rather than 16.

Feedback employed as shown in Fig. 14·25 results in a reduction from 16 to 10. One feedback carries a positive pulse from the third stage ($V_6$ plate) back to the second stage ($V_3$ grid). This pulse causes the conduction of $V_3$, after the fourth applied pulse, without applying another input pulse, thus reducing the required number of pulses by two. Since it normally takes two input pulses to change stable states in the second stage, the first feedback loop eliminates two input pulses. A second feedback is shown from stage 4 ($V_8$ plate) to stage 3 ($V_5$ grid), causing $V_5$ conduction after six input pulses without additional input pulses. Four pulses are required at the input to cause a transition of stage 3; therefore, the second feedback loop reduces the number of required input pulses by four. When both loops are considered, the scale factor is shown to be 16 minus 2 (first feedback loop) minus 4 (second feedback loop), or 10.

**Gas-glow tubes.** A counter tube using a gas discharge arc responds to input pulses by transferring a discharge arc from one set of electrodes to another. Figure 14·26 shows a four-decade gas-glow tube counter capable of recording 9,999 pulses. A comparable unit using feedback Eccles-Jordan counters would be many times larger. Thus glow-transfer counters save space. Although the response speed of glow tubes ranges from 1 kc to 100 kc, which is less than Eccles-Jordan counters, they are gaining recognition for easy reading, conservative space requirements, and servicing simplicity.

A circuit for two decade counters appears in Fig. 14·27. Pulses are amplified and shaped before being applied to PULSE INPUT, where $V_{1a}$ amplifies them. Opening reset switch $S_1$ forces the arc in the counter tube to the zero (0) cathode by removing the ground connection from all other cathodes. Once the tube is ionized, its plate voltage drops to a sustaining value which is below the ionizing voltage. As a result, no other cathode can form an arc when the reset switch is closed. Input pulses are coupled from $V_{1a}$ through capacitors $C_1$ and $C_2$ to transfer grids $G_1$ and $G_2$. The input pulses to $V_{1a}$ are positive, so that when they appear in the plate circuit of the tube, they have been reversed to a negative form.

The negative pulse from $V_{1a}$ will reach $G_1$ first (through capacitor $C_2$), and since it makes $G_1$ quite negative, an arc will form between the anode and $G_1$. A short time later, the same negative pulse will reach transfer grid $G_2$, via $C_1$, $R_3$, $R_4$, and $C_3$. The latter network is purposely inserted in the circuit to delay the passage of the negative pulse to $G_2$. The arc now skips to $G_2$, then to the next clockwise cathode, which in this case would be

*Fig. 14·25 Block diagram with two feedback circuits to reduce the scale factor by 6. The first feedback loop reduces the scale factor by 2, while the second loop reduces it by 4.*

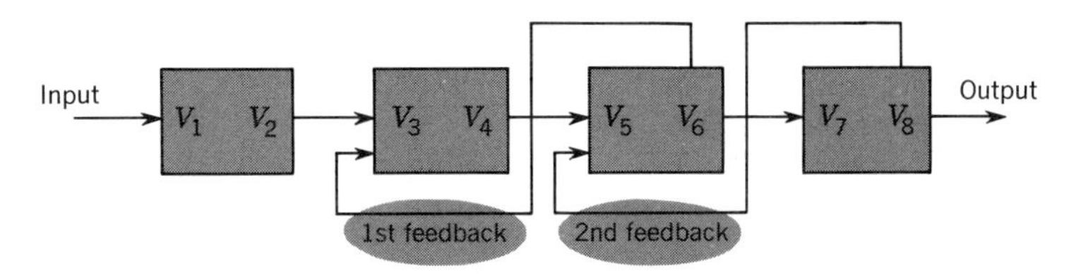


*Fig. 14·26 Gas-glow discharge tubes provide compact scales of 10. Four tubes provide a scale factor of 10,000. This unit is used in process control and radioactive-particle counts.* (Electro Pulse, Inc.)

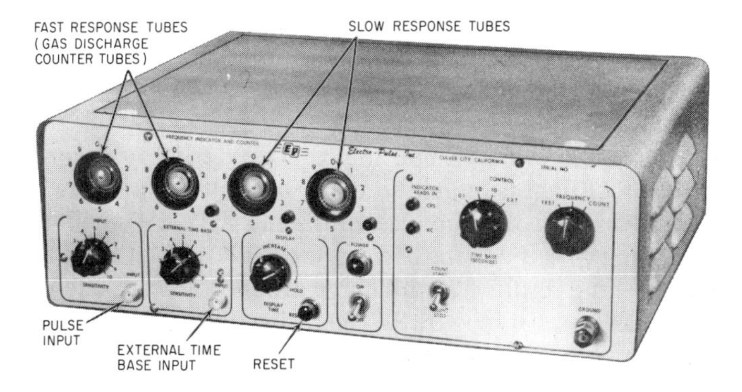

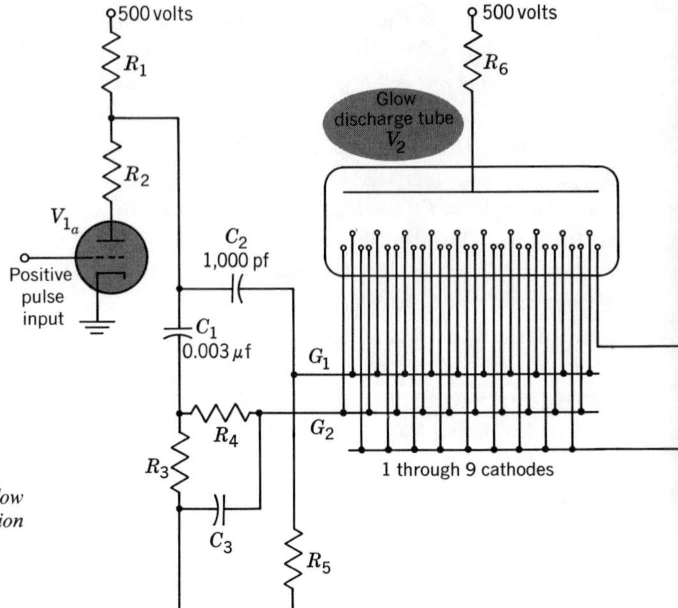


*Fig. 14·27 Decade counter using glow discharge tube. See text for description of circuit.*

cathode No. 1. Thus each input pulse advances the discharge arc around the tube. When 10 input pulses have been applied, the zero cathode will pick up the arc and generate a positive output pulse across $R_7$. When pulse 100 is applied, a positive pulse will appear at the second tube output across $R_{14}$.

The glow discharge tubes are all slow-speed counters when compared with counter circuits capable of a response up to 4 Mc. However, when considering glow tubes only, it is permissible to refer to low-frequency response tubes (up to 5 kc) and high-frequency response tubes (over 20 kc). The reason for the difference in response speeds lies in the glow-tube design. High-response glow tubes contain three or four transfer grids and must have low interelectrode capacitance despite additional tube elements. These requirements result in a much more expensive tube than the simpler design requirements of the glow tubes in the low-response area.

**Nixie.** A single-envelope indicating device has been developed which presents the numbers 0 through 9 by means of a glow discharge. When a tube is filled with gas, a glow will surround the cathode during tube conduction. This production of light is limited to the immediate cathode area. Thus an irregular cathode shape will produce an irregular glow. The Nixie tube has 10 cathodes, each with the shape of a single digit. Completing only one cathode circuit at a time results in a discharge around that cathode in the shape of a particular number. Figure 14·28 shows the multicathode Nixie and an actual number size for each tube.

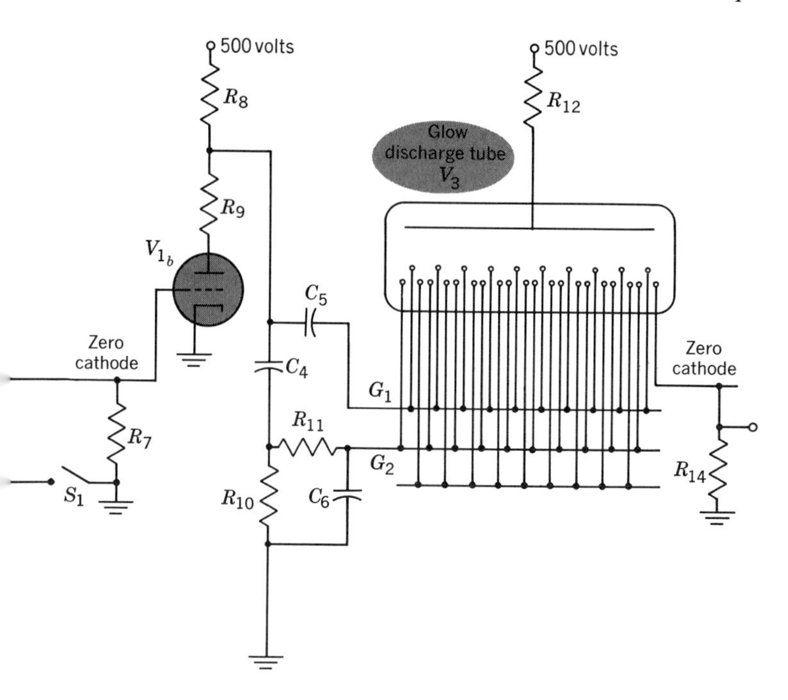


## 14·11 Computers

The two classes of computers are digital and analog. In general, those devices which add, multiply, subtract, and divide by pulse techniques are called *digital* computers.

The second type, the *analog* computer, relates parts of a mathematical equation to a physical or electrical quantity. For example, an automobile speedometer is actually an analog computer. Speed is measured in miles per hour; hence, the type of information required is distance and time.

*Fig. 14·28 The Nixie tube provides the numbers from 0 to 9 by means of a gas arc surrounding a cathode that is shaped like the number.* (Burroughs Corp.)

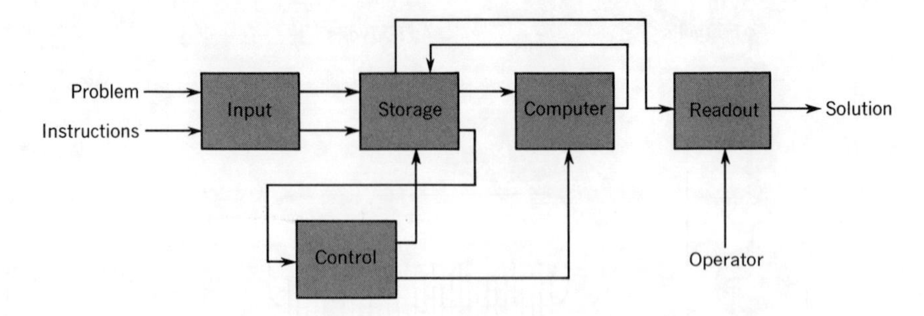


*Fig. 14·29 The functional sections of a computer are input, control, storage, computer, and readout. Information is applied to the input and stored. The control unit routes the problem through the computer and back to the storage unit. The readout unit samples the store on demand from the operator.*

Distance traveled is determined at the drive shaft, but time is not directly available. Time is introduced through an analogous magnetic-field equation. Each revolution per second of the speedometer cable causes a small magnetic deflection of the needle. The needle will be deflected in proportion to the number of revolutions occurring in 1 sec.

Since digital computers use the pulse circuitry described in this chapter and analog computers generally do not, the following description is restricted to an analysis of digital machines.

A computer is composed of several functional sections, as outlined in the block diagram of Fig. 14·29. The input unit receives the instructions and parts of a problem from punched cards, tapes, or switch settings. The control unit responds to the instructions from the input unit and selects those mathematical functions required to solve the problem. The sequence of operation is determined by the order in which the instructions are placed in the input. The storage unit provides a memory for the instructions, initial problem, intermediate solutions, and final answer. The computer unit contains the mathematical circuits required to solve a problem. Signals from the control unit select the parts of the computer required and their sequence of operation. A readout unit receives the solution from the storage unit and puts it on tape, prints it, or displays it visually.

Where computers can perform mathematical operations in millionths of a second, human mathematicians require hours to translate a problem into machine language. Each part of a problem must be reduced to a single operation. It is not enough to tell the machine to add two numbers. Every detail must be supplied as an instruction. Thus, to add two numbers, the machine must be told to store the first number in a particular part of the storage unit (let us call this $A$) and to store the second number in another part of the storage unit (let us call this $B$). The machine instructions are usually placed in another section of the storage unit ($C$) reserved especially for them. The computer circuits are then selected by the control unit and

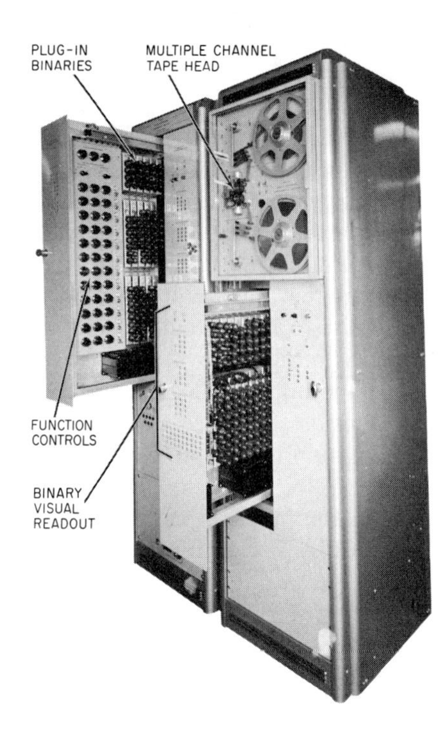


*Fig. 14·30 A complete data-processing computer requires a great many scale-of-16 and scale-of-10 plug-in units. Data storage is supplied by magnetic tape with several channels available for information.* (Potter Instruments Co.)

told to add the number in store *A* to the number in store *B*. The computer unit begins the addition function, while the control unit samples store *C* for the next instruction. This instruction is to place the sum from the computer unit in store *D*. The computer unit complies with the instruction and is ready for another problem. Readout occurs when the operator instructs the machine to present the information contained in store *D*.

The computer unit is the heart of the machine. Here are found all the mathematical and logic circuits. The circuits for addition of numbers are formed of rows of two-stable-state multivibrators, which receive pulses from a controlled oscillator. Subtraction is accomplished by adding the complement of the subtrahend to the minuend. Thus subtracting 13 (subtrahend) from 90 (minuend), when the maximum number which can be indicated by the computer is 99, is actually performed by adding 87 (the complement of 13) to 90, and the result is 177. However, since the machine can indicate only two numbers, the result shown is 77, which is the same result as subtracting 13 from 90. Discussion of the number of digits which can be indicated by the machine determines what the complement is. If the machine in the example above could indicate three digits, the complement of 13 would be 987 and the answer would appear as 077.

Multiplication by circuitry alone requires a great many components; therefore most present-day computers have multiplication tables built into them. The tables are not self-sufficient but do reduce the circuit complexity. Each number in the multiplier is handled separately with special circuits for carry and shift. The partial products are added, and each subsequent partial is shifted to the left. The total product is the sum of each of the partials and is stored until readout is desired.

Division is the most complex of all mathematical operations and is usually performed in some devious manner. The circuitry is large and performance is slow if division is handled in a straightforward manner. A possible first-digit quotient is tested by multiplying the quotient by the divisor and comparing the product with the dividend. When the maximum quotient is found, the product of the quotient and the divisor is subtracted from the dividend, and the next digit in the dividend is brought down to the right of the difference. The whole process is repeated until the degree of exactness desired is reached. Other methods of division require less circuitry or perform the operation faster, but their description is beyond the scope of this book.

Figure 14·30 shows the external view of a computer. Input is supplied by switch settings, and computer and control by Eccles-Jordan circuits. The memory or storage unit uses magnetic tape, and readout may be obtained on adding tape or by electric typewriter.

## SUMMARY

1. Electronic circuits which generate pulses are usually relaxation oscillators. Common circuits are the multivibrator, either plate-coupled or cathode-coupled, and the blocking oscillator. Pulses from such sources are often used for synchronization, gating, and testing pulse circuits.
2. Monostable and bistable multivibrators are usually trigger circuits; they produce an output only when triggered by an external source. Astable multivibrators have two quasi-stable states and they are self-oscillating.
3. A circuit that uses the relaxation principle has an output frequency that depends on an *RC* network. The relaxation principle involves the charging of a capacitor to a certain value, and the discharge of the capacitor through a resistor when the circuit "relaxes."
4. The Eccles-Jordan trigger circuit is one form of bistable multivibrator, plate-coupled.
5. The Schmitt trigger circuit is a bistable form of cathode-coupled multivibrator.
6. The blocking oscillator uses a transformer for feedback but the oscillations are blocked at a rate determined by the grid *RC* time constant. This circuit can operate either free-running or triggered, depending on the bias.
7. Multivibrators and blocking oscillators are often synchronized by input pulses to make the oscillator run at the frequency of the triggering pulses or an exact submultiple for frequency division.
8. Storage and pulse-counting circuits are the heart of computers and industrial process controls. An Eccles-Jordan counting circuit which has two stable states provides a division by two.
9. Refinement of a series of Eccles-Jordan stages is accomplished by the use of feedback loops to reduce the scale factor. Since our numbering system is based on 10 digits, feedback is used to provide decade scaling units from a scale of 16.
10. Pulse techniques are used in many electronic fields; some of them are nuclear, computer, industrial controls, and radar.

## SELF-EXAMINATION

1. A pulse occurs (*a*) in sine-wave oscillators only; (*b*) only in Eccles-Jordan circuits; (*c*) whenever there is an abrupt change in voltage or current; (*d*) when a tube is cut off, never when it starts to conduct.

2. A relaxation oscillator (*a*) cannot produce a series of pulses; (*b*) has some tube cutoff during a portion of each cycle; (*c*) does not require synchronization; (*d*) produces output pulses at the cathode only.
3. Three stages of Eccles-Jordan circuits provide a scale factor of (*a*) 3; (*b*) 6; (*c*) 8; (*d*) 16.
4. The tube current in a thyratron sweep generator may be decreased by (*a*) a decrease in grid voltage; (*b*) a decrease in the firing potential; (*c*) a decrease in the voltage across the tube; (*d*) an increase in the grid bias voltage.
5. The blocking oscillator is unique because it (*a*) uses degenerative feedback; (*b*) obtains feedback from a triggering signal; (*c*) obtains feedback through a coupling transformer; (*d*) is a monostable circuit.
6. Gating is a pulse technique which (*a*) allows a tube to conduct for an interval determined by a pulse; (*b*) advances the number stored in a counter circuit; (*c*) changes the shape of a pulse; (*d*) amplifies the counted pulse.
7. A two-tube bistable circuit (*a*) operates like a pulse amplifier; (*b*) provides two output pulses for every input pulse; (*c*) can produce one negative output pulse after three negative input pulses; (*d*) Divides the input pulse repetition rate by two.
8. Telemetering is a method of (*a*) transmitting pictures from one place to another; (*b*) counting pulses sent over long distances; (*c*) transmitting information concerning a commodity or a process over a distance; (*d*) synchronizing gating pulses.
9. A feedback loop in scaler circuits (*a*) improves high-frequency response; (*b*) reduces distortion; (*c*) reduces the output repetition rate in relation to the input repetition rate; (d) reduces the number of input pulses required to produce one output pulse.
10. The lights in a four-stage Eccles-Jordan counter represent the numbers (*a*) 0,1,2,4; (*b*) 1,2,3,4; (*c*) 1,2,4,8; (*d*) 2,4,6,8.

## QUESTIONS AND PROBLEMS

1. Draw the circuit of an astable (free-running) plate-coupled multivibrator.
2. Draw the circuit of a Schmitt trigger circuit using two NPN transistors.
3. Draw the circuit of a free-running blocking oscillator with a triode tube.
4. Draw an Eccles-Jordan trigger circuit using tubes.
5. Explain why capacitors $C_1$ and $C_2$ are used in the circuit of Fig. 14·6.
6. Connect feedback in a block diagram for three stages of Eccles-Jordan counters to provide a scale of six.
7. Describe two methods of applying input pulses to an Eccles-Jordan circuit.
8. What two classifications of computers are there? Describe one difference between them.
9. How do gas-glow tubes advance from one pulse to the next?
10. What would happen in the circuit of Fig. 14·14 if $D_1$ were removed?

# Chapter 15 Industrial electronics

Industrial electronics is one of the largest branches of electronics. This field includes industrial applications of amplifiers, oscillators, and rectifier circuits. In industry, electronic systems are used in a great variety of applications such as heating, measuring, and controlling. Some components that are widely used in these systems are gas tubes, silicon controlled rectifiers, zener diodes, photoelectric tubes, thermistors, and many others. These and other important topics are presented in the following sections:



## *15·1 Electronic heating*

Industrial electronic heating, known as *high-frequency* heating, is a process by which high-frequency alternating current develops heat within an object. This is in contrast to older methods in which heat, supplied by an external source, penetrates to the interior. This newer process has the advantage of providing heat which not only can be controlled accurately but also can be localized to desired areas of the object.

Two general types of heating methods have been developed for industrial use. One, known as *induction heating* and operating on the general plan of Fig. 15·1, employs a magnetic field, produced by an alternating current, for heating electrically conductive materials. The second, known as *dielectric heating* and operating on the general plan of Fig. 15·2, employs an electric field, developed between two plates by an a-c voltage, for heating nonconductive materials.

The heating effects of magnetic and electric fields have been known for years as hysteresis, eddy current, and dielectric losses, which impaired the operating efficiency of various devices, but their deliberate application to various industrial processing methods is of comparatively recent origin. In many operations where heat is required, production time has been reduced and the product improved in quality and uniformity through the use of electronic heating.

Early applications of dielectric heating were in drying or curing tobacco leaves, killing insects in grains or cereals, and producing artificial fever in various parts of the human body to help cure infections. Induction heating is used primarily in heat treating, melting, and other commercial processes associated with metal fabrication.

The eddy currents which develop the heat in metal are confined mainly to the surface of the work, a condition which becomes more pronounced as the frequency is increased. Therefore the required operating frequency depends upon the nature of the material to be heated and the kind of work that is to be done.

As a general rule, frequency varies inversely with the thickness of the metal to be heated. For example, a large volume of steel alloy may be melted efficiently with a current frequency of 1,000 cycles, whereas a 0.010-in.-thick strip of the same material requires a frequency of 100,000 cycles for proper heat treatment.

Recently, the tendency has been toward the use of higher frequencies, especially where shallow heating effects are desired, and the band from 1.5 Mc to 50 Mc has become important.

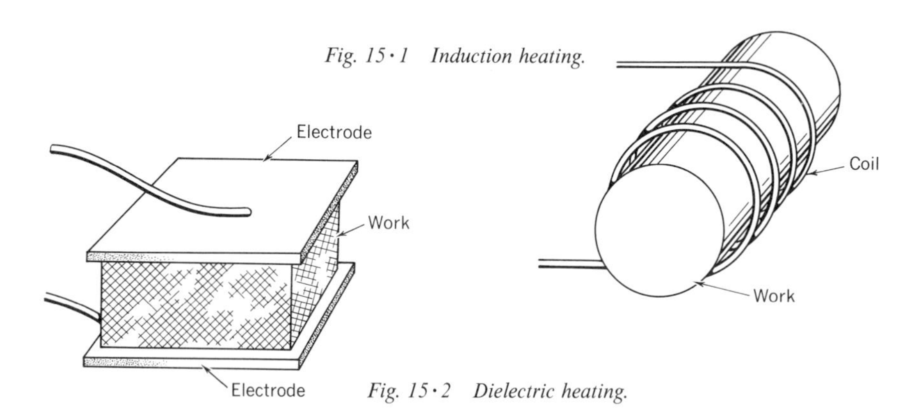


*Fig. 15·1 Induction heating.*

*Fig. 15·2 Dielectric heating.*

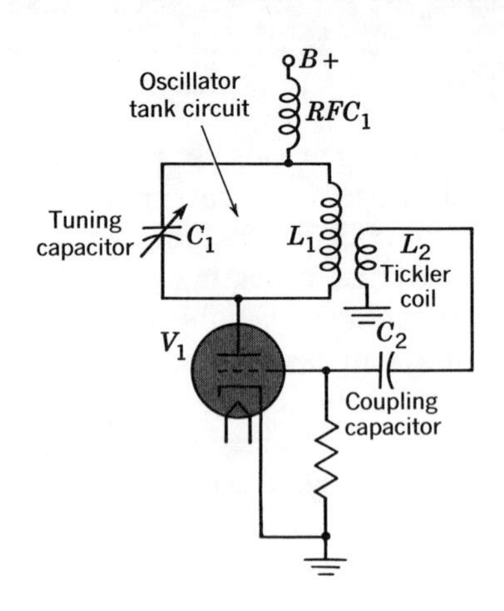


*Fig. 15·3 A typical high-frequency generator found in industrial equipment. This circuit is a tickler-coil oscillator.*

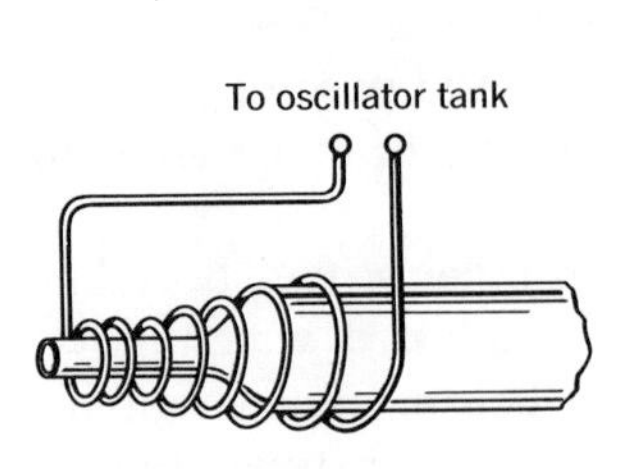


*Fig. 15·4 Power is coupled from the oscillator circuit to the work through a coupling coil.*

The frequencies from 15,000 cycles to 50 Mc are supplied by electronic generators, or oscillators, similar to those employed for high-power high-frequency radio transmission. Since this type of equipment is already developed to a high state of dependability, it needs only to be adapted to heating purposes. These high-frequency electronic generators are designed to operate at one or more fixed frequencies depending upon the class of service for which they are to be used.

A typical oscillator used to develop r-f power in industrial electronic equipment is shown in Fig. 15·3. This is a tickler-coil oscillator, plate-tuned with grid feedback. Frequency of oscillation is governed by the values of $L_1$ and $C_1$.

The basic difference between oscillators employed in communications receivers and those found in industrial equipment is the power developed by these circuits. In communications receivers, the oscillators develop very small amounts of output power, frequently less than 1 watt. In industrial equipment, many hundreds, even thousands of watts of power are often required. Consequently the components of such power oscillators are quite large and well insulated to withstand the high voltages involved.

## *15·2 Induction heating*

As previously mentioned, induction heating is employed for electrically conductive materials and uses the magnetic field set up by an alternating current. The material to be heated, known as the *work,* is placed within or adjacent to a coil which carries current. This is the *work coil,* and the varying magnetic field which it sets up induces a voltage in the work piece. This voltage, in turn, causes a current to flow in the surface of the work.

The action is essentially the same as that of a transformer, and the work coil can be considered as the primary, while the work itself serves as a single-turn secondary. Because of resistance losses ($I^2R$), the resulting current in the work develops the desired heat, which is confined mainly to the surface layer.

The amount of heat developed depends on the strength of the work current, which is controlled by the strength of the magnetic field. This depends on the value of current in the work coil and the closeness of coupling between the work and the coil. The heat developed by the resistance of the work is known as an *eddy-current* loss. In magnetic materials, the hysteresis loss will generate some heat, but it is generally small in value compared with the eddy-current loss and, except at the lower temperatures, can be neglected.

While the eddy currents appear mainly on the surface of the work, the depth to which they will penetrate and develop heat depends, among other factors, upon the frequency of the current in the work coil. At low frequencies, the eddy currents penetrate to greater depths. This action makes induction heating particularly adaptable for metal surface treatments, such as case hardening, soldering, and brazing.

**Case hardening.** Case hardening is a special form of treatment in which a piece of relatively soft but tough steel is rendered extremely hard on its outer surface. The operation consists of covering the surface of the steel with some powdered carbon and applying heat at the proper temperature for a certain period of time. Figure 15·4 shows the end of a shaft being treated. After the surface has been heated, the work is rapidly cooled to produce the case hardening.

**Soldering and brazing.** The localized heating zone of induction heating makes it ideally suited for the many industrial and manufacturing operations which require the soldering or brazing of separate parts.

For soldering, the heat can be generated in the metal itself, along the seams or other areas to be joined, so that the solder will flow freely but only where needed. Because of the high rate of energy transfer, as well as the concentration of the heat, excessive temperatures do not develop in other areas of the work, and there is little or no discoloration of the surfaces. This action is so pronounced that it is seldom necessary to immerse the work in water or wrap it in wet cloths to keep it cool.

For brazing, which requires the heating of dissimilar metals, the magnetic flux of the work coil must be concentrated on the slower heating metal, so that the joint will be at the correct temperature before the brazing alloy melts. Under these conditions, the brazing alloy will be drawn properly into the joint.

**Other applications.** An excellent illustration of the application of the surface action of induction heating to modern processing occurs in the manufacture of tin plate. Known best as the material of which the common "tin" can is made, tin plate consists of a sheet of steel covered with a thin coat of tin. During its manufacture, long sheet-steel strips, about 42 in. wide, are drawn through electrolytic plating tanks at speeds from 500 to 1,000 ft per min.

While in the tanks, the steel receives a thin coating of tin, but, because of the minute irregularities of the deposit, the final product has a dull matte finish and may be somewhat porous. Since the coating is designed to provide protection for the steel, no porosity can be tolerated, and the dull finish is not acceptable for many products.

Both of these objections are overcome with induction heating by passing the plated steel sheets through suitable work coils. The developed heat raises the temperature of the tin to a value slightly above its melting point, so that it flows freely and closes any open pores. Also, as there is no mechanical contact between the work coils and the sheet, the melted tin solidifies to form the smooth, shiny surface typical of modern tin cans.

## 15·3 Dielectric heating

We have just seen that induction heating depends upon the resistance the work piece offers to the electrical currents induced in it. Therefore this method is applicable only to the heating of materials which are good conductors.

In contrast, dielectric heating operates on the principle that when a substance is placed in an electric field, its molecules are subjected to a stress and are disturbed. By the application of an alternating electric field of high frequency, the molecules are made to vibrate, creating heat. This action can be considered dielectric hysteresis loss. Within limits, the higher the frequency of the electric field, the greater is the heat.

Thus, dielectric heating is applicable to materials which are electrical nonconductors or which present an extremely high resistance to electrical current. A high-frequency alternating voltage is impressed across a pair of plates, or electrodes, and the work material, placed between them, is subjected to the influence of the electric field. The arrangement can be considered as a conventional capacitor with the work material serving as the dielectric.

The dielectric may carry small conducting currents which produce $I^2R$ losses, but most of the heat is due to dielectric hysteresis. This corresponds to magnetic hysteresis in iron.

The power losses occur in the entire dielectric material, from one electrode to the other. Therefore the heating is distributed uniformly throughout the entire mass. However, because of surface heat losses, by radiation and conduction, the interior areas actually become somewhat warmer than the outer layers.

This ability to develop heat instantly and uniformly throughout a material is an important advantage of dielectric heating and makes it adaptable to the manufacture of all types of nonconducting materials that are formed or processed with heat.

## 15·4 Gas tubes and zener diodes

In industrial applications, two general types of tubes are used. One is the high-vacuum tube, which most readers are familiar with. The other is a gas-filled tube, which has many operational features not found in vacuum tubes.

In all tubes, the electrons leave the cathode and progress toward the plate, gaining momentum along the way. This fact is responsible for the success of the gas tube in high-current requirements. The electron builds up momentum after leaving the cathode just as it does in the vacuum tube. When gas atoms are present, however, these electrons collide with the atoms. Then the cathode electrons can dislodge electrons from the gas atoms.

This action of removing electrons from gas atoms is *ionization*. The released electron is a negative charge, while the atom becomes a positive ion. The plate current in a gas tube depends not only on cathode electrons but on ionization charges as well. This has the effect of increasing the current for any given plate voltage.

When voltage is applied to a gas tube, ionization does not occur until the cathode electrons have attained a definite energy. Ionization will occur at some plate voltage, known as the *ionization voltage* of the tube. After the gas in a tube has been ionized, the voltage across the tube drops to a value just sufficient to maintain ionization. This voltage is referred to as the *sustaining voltage*, or *arc drop*. Ionizing and sustaining voltages are controlled at the time of manufacture by the type and amount of gas used. These voltages are so stable that some gas tubes with common arc-drop values of 150 and 75 volts are used as voltage regulators. A variety of gases are used, but all are inert; that is, they will not combine chemically with other elements. Neon, argon, and mercury vapor are common examples.

**Gas diodes.** Mercury vapor is used generally in these power rectifiers because it is ionized easily. The tube can then conduct a large current with only 15 volts sustaining voltage. A warmup time of 20 sec or more is needed so that gas pressure in the tube rises to its normal value. Otherwise, positive ions can be attracted to the filament cathode and disintegrate the cathode. The droplets of mercury seen on the glass envelope and bottom of a cold tube must be heated and evaporated before the tube is used.

**Gas triodes.** The presence of gas in a triode changes its characteristics. A negative grid will hold the tube in cutoff when plate voltage is applied. But the grid will have no control over tube conduction once the gas has been ionized. The reason for the loss of grid control is this: The positive ions require an electron in order to return to their stable gas atom state. Positive atoms obtain the necessary electrons from the cathode unless there is a more negative tube element. By concentrating more electrons on the grid in an effort to cut the tube off, the grid is made more negative than the cathode. The positive ions are attracted to the negative grid, forming a positive shield that neutralizes the normally negative field of the grid.

**Thyratrons.** A thyratron is a gas-filled triode. When it is conducting, a thyratron can be brought to cutoff only by interrupting the cathode or anode path. Without conduction to maintain ionization, the gas ions quickly recombine, and the grid is able to regain control of the tube. This lack of grid control is very useful, where conduction is desired long after the initiating signal has passed. When this tube is used as a rectifier, we can control the output power by allowing conduction for only a fraction of a complete sine wave.

A common practice which allows grid control is the application of alternating current to the plate. When the applied voltage goes through its negative half-cycle, the gas deionizes. When the plate becomes sufficiently positive, during the next half-cycle, the tube fires again. In this way, the tube is turned on and off once each cycle.

Schematically, gas-filled tubes are identified by a black dot placed inside the tube circle. The tube elements, however, are presented in the same fashion as vacuum tubes.

**Thyratron timers.** Since thyratron conducting characteristics are sharply defined, thyratrons make excellent tubes for timing devices.

Figure 15·5 illustrates a timing circuit used in many commercial gas-tube timers. Alternating current can be applied directly, without using a transformer, if desired. Both halves of the applied alternating current perform useful functions.

At the bottom of the diagram there are two straight lines labeled USER SWITCH (do not confuse this with a capacitor which is represented by one straight line and one curved line). This switch is normally open (N.O.), but it may be closed if the user so desires. Note that a switch may also be normally closed (N.C.). To open these contacts, a specific action is required, either physically opening the contacts or actuating a relay which, in turn, accomplishes the same action. One such normally closed set of contacts is used elsewhere in this circuit. A closed set of contacts is represented by two straight parallel lines, with a diagonal line running through them.

Now let us consider the operation of this circuit when the user switch is open. If we assume that the negative half-cycle of the 110-volt line is active, then point *A* will be more negative than point *B*. The applied volt-

*Fig. 15·5 Timing circuit using thyratron for welding control.*

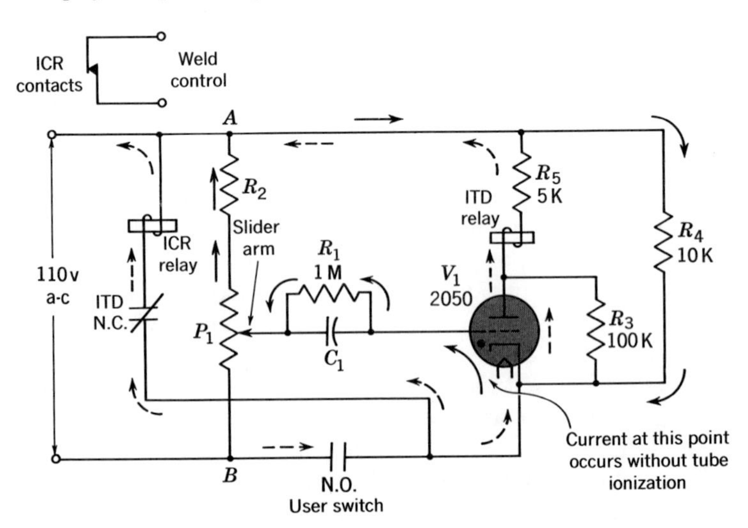

age will distribute itself between $A$ and $B$ in proportion to the resistance values of $R_2$ and $P_1$. Furthermore, the slider arm of $P_1$ will be more positive than point $A$ and the $V_1$ cathode, which is connected through $R_4$.

Initially, when the power is first applied, capacitor $C_1$ is not charged. Hence the full voltage between the slider arm of $P_1$ and point $A$ will be applied between the grid and the cathode of $V_1$. Since the grid is positive with respect to the cathode, a small current passes from cathode to grid. (The full path followed by this current is indicated by the solid arrows in Fig. 15 · 5.) This current develops a voltage across $R_1$, and $C_1$ becomes charged to this voltage.

Although grid current flows, $V_1$ does not ionize because its plate voltage is zero. Notice that both the plate and cathode return to $A$ when the user switch is open.

During the positive half-cycle of the a-c line voltage, the slider arm of $P_1$ will be negative with respect to point $A$. This will apply a negative voltage between grid and cathode of $V_1$, and no current will flow through the tube. During this half-cycle, therefore, the excess of electrons at the left-hand plate of $C_1$ will flow through $R_1$ to the right-hand plate. The value of $R_1$ is high, however, and the capacitor will not be able to discharge completely before the next half-cycle appears.

During its positive intervals, the grid will again be driven positive, current will again flow from cathode to grid, and the voltage across $C_1$ will build up. After the circuit has been in operation over a number of cycles, an average voltage will develop across $C_1$. This voltage will make the grid negative with respect to the cathode.

Now let us consider the operation of this circuit when the user-switch contacts are closed. When the switch is closed, the cathode is connected to point $B$ through the switch. Now when $A$ is negative and $B$ is positive, the cathode is more positive than the grid. The grid is connected, through $R_1$ and $C_1$, to the slider of $P_1$, which is negative with respect to $B$. In addition, the voltage drop across $C_1$ makes the grid negative with respect to the cathode. Thus the grid is negative not only because of the slider of $P_1$ but also because of $C_1$. As a consequence, no current will flow through the tube.

During the next half-cycle, point $A$ goes positive with respect to $B$, with the thyratron plate also becoming positive. With the user switch closed, the full line voltage appears between cathode and plate. The tube will conduct provided the grid is more positive than cutoff. The slider of $P_1$ is positive with respect to $B$; however, negative voltage across $C_1$ is still present. The thyratron will fire as soon as the negative voltage across $C_1$ equals or becomes less than the positive voltage from the slider to point $B$.

Notice the relay coil labeled ICR and the relay coil in the plate circuit of the thyratron labeled ITD. These coils are labeled according to their function in the circuit. Thus ICR is a contact relay whose only purpose is to control an external circuit, while ITD is a time-delay relay energized only after the thyratron has been fired.

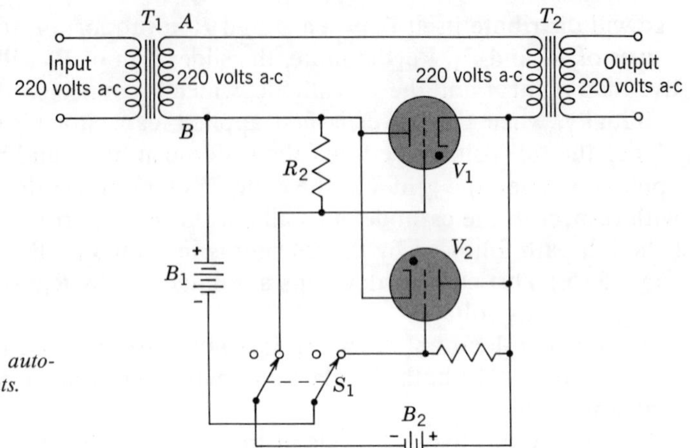


*Fig. 15·6 Thyratrons as automatic switch for heavy currents.*

The complete timer action starts when alternating current is applied. After a few cycles, $C_1$ develops an average voltage and maintains it until the user switch is closed. The timed cycle begins as soon as the user switch is closed. ICR energizes immediately and causes the contacts labeled WELD CONTROL to close, thereby initiating whatever sequence the device is controlling. A short time later, the grid voltage on $V_1$ becomes positive enough to fire the tube. This ends the controlled time delay. When the thyratron fires, relay ITD is energized, opening its normally closed contacts in the ICR current path. ICR no longer has a complete path from $A$ to $B$ and deenergizes. The controlled circuit thus is interrupted, ending the timed sequence. The user switch must be opened before another timing event can take place. Timing starts when the user closes the switch and ends when the tube fires.

Chemical or physical processes are often timed and controlled by one timer. A tank could be filled with liquid and required to remain full for a specified time. A simple timer and control circuit could open the input valve, fill the tank, time the interval, and empty the tank. Every industry has need for dependable timers, some with 10 to 15 sequences.

**Thyratrons in power control.** Thyratrons are built to carry currents up to 50 amp for control and power rectification. A mechanical switch arcs each time it is opened or closed; and metal contactors would soon wear out under this punishment. A diagram of an all-electronic switch is shown in Fig. 15·6. $S_1$ controls the bias on a pair of *inverse-parallel* thyratrons. The inverse-parallel arrangement means that the plate of one tube connects to the cathode of the other, but the two control grids are operated in parallel.

Points $A$ and $B$ in Fig. 15·6 will reverse polarity every half-cycle of the a-c input voltage. In order to see what happens in the circuit, assume $A$ is positive but neither thyratron has fired. The plate of $V_2$ is connected to point $A$ through $T_2$, and the cathode is connected to point $B$. This means that $V_2$ can conduct if its grid will permit. To trace the grid circuit, start at the $V_2$ grid, assuming $S_1$ to be closed, as shown. The circuit extends from the grid, through $S_1$, through battery $B_1$ to the cathode of $V_2$. The bias battery makes the grid negative and prevents the conduction of $V_2$. When $S_1$ is moved to the open position, the grid of $V_2$ connects only to the plate. It now has positive bias, and the tube conducts.

Over the next half-cycle, point $A$ is negative with respect to $B$. The cathode of $V_1$ connects to $A$ through $T_2$, while the plate connects to $B$. $V_1$ will conduct provided the grid is not biased to cutoff. With $S_1$ closed, a bias battery ($B_2$) is connected between the grid and cathode which holds the tube in cutoff. When $S_1$ is open, however, the battery is removed, and now the tube will conduct.

$S_1$ controls the 220-volt line current through $T_2$, either preventing or permitting the conduction of $V_1$ and $V_2$. When $V_1$ and $V_2$ are permitted to conduct, power appears at the output of $T_2$. When $V_1$ and $V_2$ are kept cut off, no power appears at the output of $T_2$.

The combination of the two thyratrons actually functions as a switch. $S_1$ does not carry a large current, and therefore presents no special problem. Power loss in the thyratrons is small because of their low internal voltage drop. A 10-amp current through the thyratrons will result in a power loss of 150 watts, while the load receives 2,050 watts.

**Ignitrons.** A big brother to the thyratron, the ignitron has the same characteristics but its construction is different. Instead of a solid, heated, cathode, the ignitron uses a pool of mercury. As shown in Fig. 15·7, an igniter needle is used as a starting electrode. This silicon carbide ignitor penetrates just far enough into the surface of the cathode to form the small gap necessary for forming an arc so that the mercury can become ionized. Then current can flow from cathode to anode when anode voltage is applied. The anode-cathode voltage drop is about 8 volts.

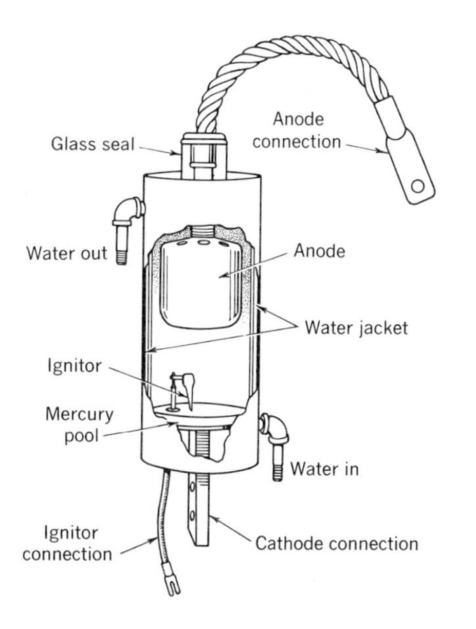


*Fig. 15·7 The complete ignitron contains an arc-forming rod (igniter tip), a large plate (anode), and a water-cooled jacket which is at cathode potential.*

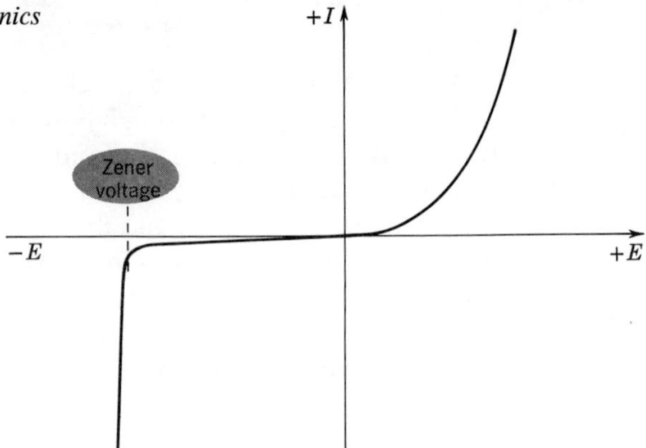


*Fig. 15 • 8 When the reverse voltage applied to a zener diode reaches the zener voltage level, a high reverse current flows through the diode.*

These powerful tubes carry hundreds of amperes for heavy-industry control. Figure 15 • 7 shows the internal construction in a cutaway view. Water is used to cool the ignitron, and the water jacket, which is at cathode potential, extends completely around the tube. Electrically, the ignitron operates in the same fashion as a thyratron.

**Zener diodes.** A typical volt-ampere curve for a semiconductor diode is shown in Fig. 15 • 8. When a diode is reverse-biased ($-E$ is increased), the flow of current through the diode is very small. But when a certain value of reverse bias is reached, a high reverse current begins to flow. This critical voltage is called the "breakdown" or *zener* voltage; and silicon diodes specifically designed to operate in this region of the characteristic curve are called *zener diodes.*

Ordinarily, the diode that has the least amount of reverse current is considered the best diode. However, for applications such as voltage regulation, the high reverse current that flows at the zener voltage level is very desirable. As shown in the curve of Fig. 15 • 8, for a large variation in reverse current, the voltage drop across the zener diode remains almost constant. Because of this characteristic, the zener diode can be used in place of the glow tube in voltage-regulator circuits, as shown in Fig. 15 • 9. The diode is *reverse-biased,* and the input voltage must be slightly higher than the zener voltage rating of the diode for proper circuit operation. The operation of this circuit is basically the same as that of the glow-tube regulator.

In general, zener diodes are more versatile than their gas-tube counterparts. They are manufactured with voltage ratings that range from 1 volt to several hundred volts and have power ratings up to 100 watts. Other common applications of zener diodes are in waveshaping circuits, voltage reference sources, and switching circuits.

## *15 • 5 Silicon controlled rectifiers*

The silicon controlled rectifier (SCR) is a semiconductor device that has become increasingly important in industrial electronics. In fact, in many

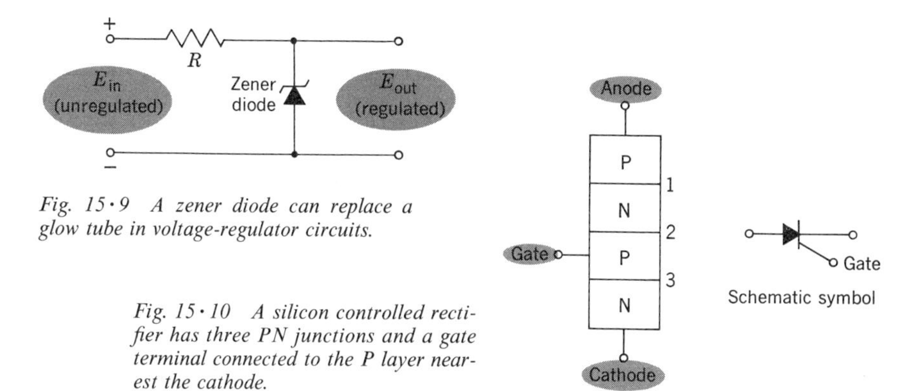


*Fig. 15·9 A zener diode can replace a glow tube in voltage-regulator circuits.*

*Fig. 15·10 A silicon controlled rectifier has three PN junctions and a gate terminal connected to the P layer nearest the cathode.*

applications, it has successfully replaced power transistors, thyratrons, vacuum tubes, fuses, relays, and many other devices. Basically, the SCR consists of three PN junctions arranged like the PNPN sandwich shown in Fig. 15·10. The upper P layer is the anode, and the lower N layer is the cathode. The third terminal, connected to the lower P layer, is the *gate* terminal. In many ways, thc SCR is similar in operation to a thyratron, and its gate element serves essentially the same purpose as the thyratron grid. The schematic symbol for the SCR is also shown in Fig. 15·10.

Let us assume that the gate terminal is open-circuited and that a negative voltage is applied to the anode. Under these conditions, junction 1 is reverse-biased, junction 2 is forward-biased, and junction 3 is reverse-biased. Hence, current flow is blocked by the two reverse-biased junctions and only a small leakage current exists. This leakage current for a negative anode voltage is shown in the volt-ampere curve of Fig. 15·11.

Now assume that a positive voltage is applied to the anode. This time

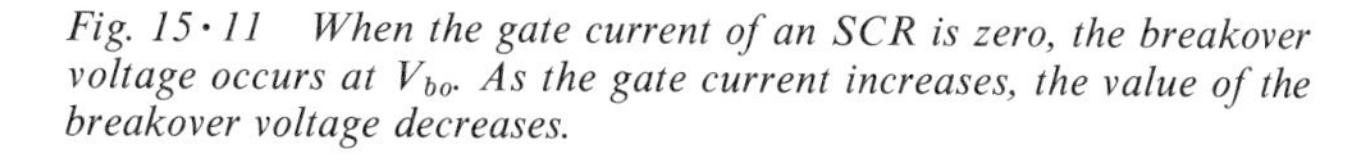

*Fig. 15·11 When the gate current of an SCR is zero, the breakover voltage occurs at $V_{bo}$. As the gate current increases, the value of the breakover voltage decreases.*

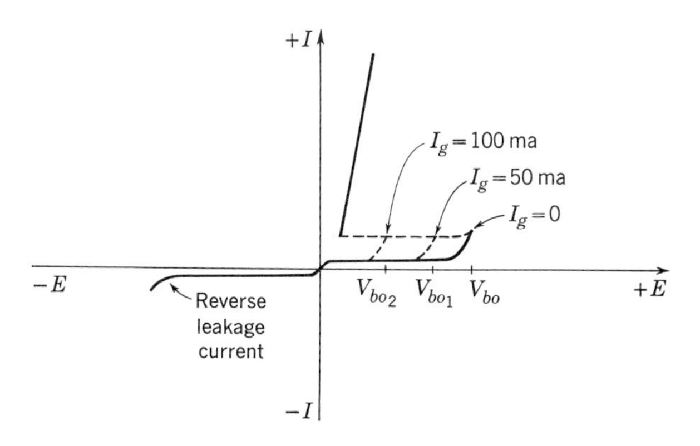

junction 1 is forward-biased, junction 2 is reverse-biased, and junction 3 is forward-biased. Again, current flow is limited to the leakage current because of the reverse-biased junction 2. However, if the positive anode voltage is increased, a "breakdown" suddenly occurs; the voltage across the SCR drops to a low value, and the current through the SCR rises very rapidly. The voltage at which this phenomenon occurs is called the *breakover* voltage, and it is indicated as $V_{bo}$ in Fig. 15 · 11. After $V_{bo}$ has been reached and conduction occurs, the internal resistance of the SCR becomes very small. Therefore, very small changes in voltage result in large changes in current. To return the SCR to nonconduction, it is necessary to remove the applied voltage (or reverse it) so that the amount of current flow is too small to sustain the "breakdown" effect that occurred at $V_{bo}$. In this respect, the operation of the SCR is similar to that of the thyratron.

In the foregoing discussion, we considered the action of an SCR with no gate voltage applied. If we apply a small positive voltage to the gate terminal, we increase the forward bias on junction 3, and gate current flows. This current causes a greater accumulation of electrons in the lower P layer, which makes it possible for the SCR to conduct at a *lower* breakover voltage. If we further increase the positive gate voltage, the gate current increases, and the SCR will begin to conduct at a still lower breakover voltage. Therefore, we see that the gate element can control the amount of anode voltage needed to make the SCR conduct. The effect of the gate current on the value of $V_{bo}$ is illustrated in Fig. 15 · 11. When $I_g$ is 50 ma, the breakover voltage is $V_{bo1}$; when $I_g$ is 100 ma, the breakover voltage decreases to $V_{bo2}$.

The sole function of the gate element is to control the value of the SCR breakover voltage. Once the SCR has begun to conduct, the gate loses all control. Again, the similarity between the SCR and the thyratron is apparent. As mentioned earlier, the SCR will cease conduction only when the anode voltage is removed or reversed.

The amount of gate current needed to bring the breakover voltage point from its value at zero gate current to almost zero volts is on the order of milliamperes. By contrast, the anode current of a practical SCR can be 20 amp (rms) or higher. Hence, it is evident that the current gain of an SCR is fairly high. Typical gate currents used to fire the SCR range from a few milliamperes to about 2 amp; and peak inverse voltages range from about 50 to 500 volts.

**SCRs in power control.** One of the principal uses of silicon controlled rectifiers in industrial equipment is for power control. Since it is a rectifier, an SCR in an a-c circuit can conduct only during one half of a cycle. Furthermore, an SCR can be made to conduct anywhere from a fraction of the half-cycle to the entire half-cycle by the proper application of gate current. Thus an SCR can control the power in a circuit by controlling the average value of the load current.

The circuit in Fig. 15 · 12*a* is a basic half-wave rectifier employing an SCR. $E$ is the a-c input voltage and $E_g$ is a pulse generator that sup-

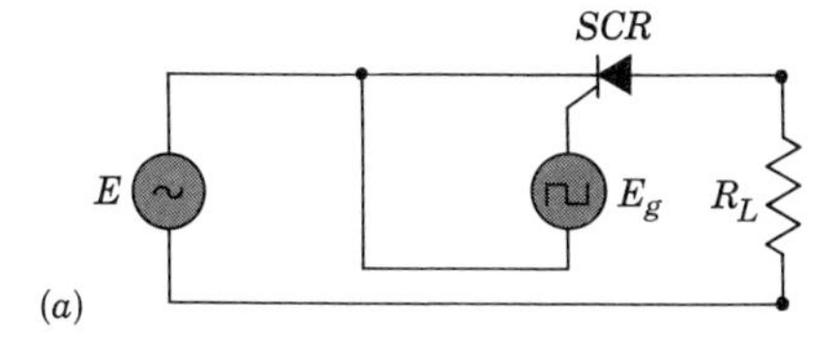


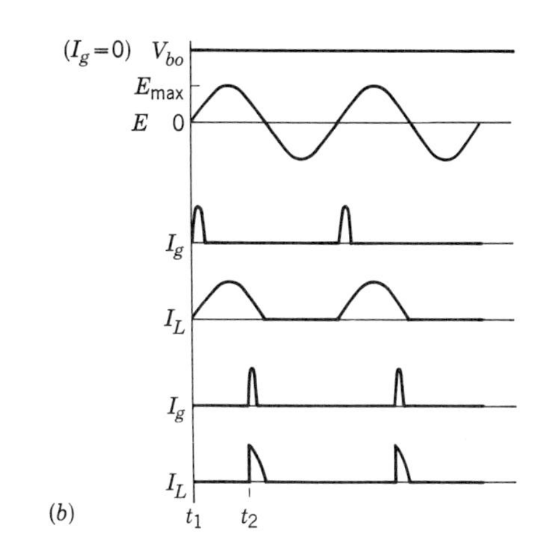


*Fig. 15·12 (a) A silicon controlled rectifier used in a half-wave circuit can control the power delivered to the load. (b) The waveforms of current when $I_g$ occurs at different times in the cycle.*

plies the gate voltage. In this circuit, the maximum value of $E$ is less than the breakover voltage of the SCR so that conduction can occur only when a gate signal is applied. Although it is possible to control the load current through $R_L$ by varying the amplitude of the gate current, it is usually more practical to use constant-amplitude gate pulses and vary the phase difference between the input signal and the gate signal, as is the case in this circuit. The effect of varying the phase shift between the two signals is illustrated in Fig. 15·12*b*. Current waveforms are shown for the case when $I_g$ occurs at the very beginning of the positive half-cycle ($t_1$) and for the case when $I_g$ occurs near the end of the positive half-cycle ($t_2$). In both cases, the amplitude of $I_g$ reduces the value of $V_{bo}$ enough to permit the SCR to conduct almost as soon as the input voltage goes positive. As shown in the figure, when $I_g$ occurs at $t_1$, $I_L$ flows for essentially the entire half-cycle; when $I_g$ occurs at $t_2$, $I_L$ flows for only a very small part of the cycle. From this discussion, we can see that the average load current is decreased when the phase shift is increased and, conversely, it is increased when the phase shift is decreased.

Because only one-half of the a-c cycle is rectified in this half-wave circuit, maximum power control is not obtained. However, when SCRs are used in full-wave or bridge circuits, maximum power control is obtained because both halves of the a-c cycle are used. One of the greatest advantages in using silicon controlled rectifiers in power control is that they provide excellent control while causing very little power loss. They also offer better reliability and are more rugged than the electronic and electromechanical devices which they can replace.

## *15·6 Photosensitive transducers*

A photosensitive transducer is an electronic device that responds to light by producing electrical impulses or by changing its resistance. Such transducers are used in industrial electronic systems to detect, measure, and control light.

**Photoemission.** When light strikes certain metal compounds they release electrons from the surface. Cesium oxide is an example. The emitted electrons are *photoelectrons,* which can be collected by a positive anode. Many phototubes operate on this principle of photoemission. Sometimes they are called "electric eyes."

**Photoconduction.** This effect causes a variation of resistance in a photosensitive material exposed to light. As examples, selenium and lead sulfide have high resistance in the dark but their resistance decreases with light. This type of photoelectric cell needs an external voltage source to produce current, which is controlled by the variable resistance of the photosensitive material.

**Photovoltaic effect.** Some combinations of semiconductor material deposited on a metal plate can generate voltage at the junction when exposed to light. The exposure meter for photography is a common example of such a photoelectric cell.

The construction of a voltaic photocell is shown in Fig. 15 · 13. A very thin layer of gold admits light to the light-sensitive material bonded to a steel plate. The plate is one terminal for voltage output, while the opposite terminal is a contacting ring around the gold window.

The active element exposed to light is often selenium or copper oxide. Selenium is used in the photovoltaic cell for the light meter of Fig. 15 · 14. The voltage-producing part of the light meter is the same as in Fig. 15 · 13. Since the current meter completes the path for electron flow between the two voltage-output terminals, the meter current indicates light intensity. More light on the selenium generates more output voltage to increase the current through the meter.

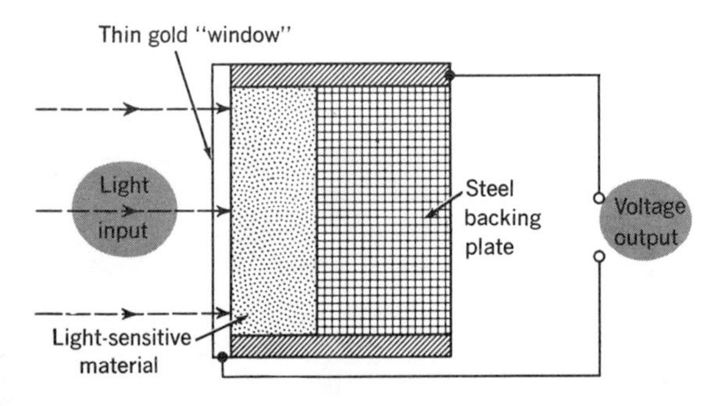


*Fig. 15 · 13 Basic construction of a voltaic photocell.*

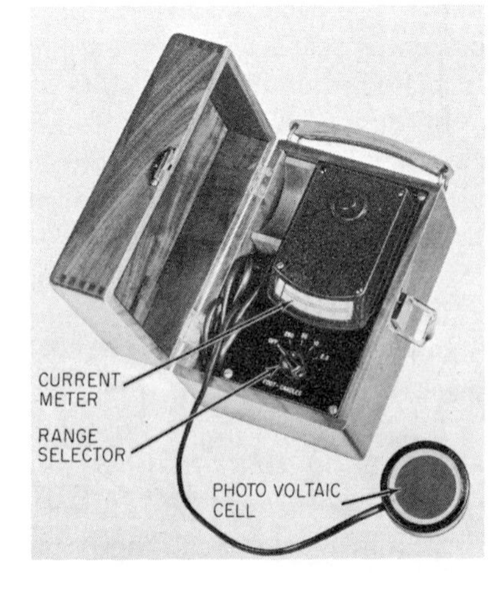


*Fig. 15 · 14 A photographic light meter.* (Photovolt Corp.)

**Photoconductive cells.** The resistance of some materials will change when they are illuminated by light. The additional energy imparted by the light weakens the internal bond between electron and nucleus in the atom and increases electron flow through the material. Since the photoconductive cell does not produce its own power, as the voltaic cell does, it requires an external power source and a current meter or amplifier. A typical construction is shown in Fig. 15 · 15.

An application of a lead-sulfide photoconductive cell is shown in Fig. 15 · 16. A bridge arrangement of three resistors and the cell form a sensitive circuit to convert resistance changes to voltage changes suitable for amplification. Light from a small lamp is directed onto the photoconductive cell after passage through the sound track of a motion-picture film. The sound appears on the film as light and dark variations along a narrow strip parallel to one edge. As the amount of light reaching the cell is varied by the passage of the film, changes in cell resistance occur. This acts to unbalance the bridge circuit, which, in turn, varies the voltage fed to the control grid of the amplifier tube $V_1$.

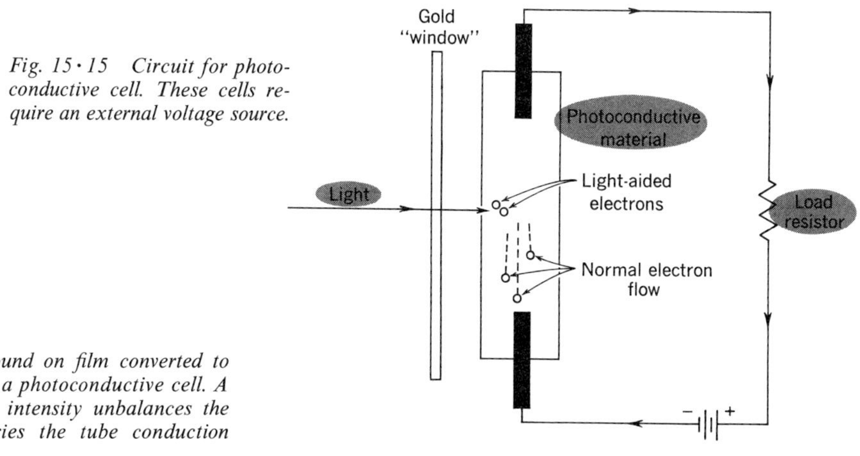


*Fig. 15 · 15 Circuit for photoconductive cell. These cells require an external voltage source.*

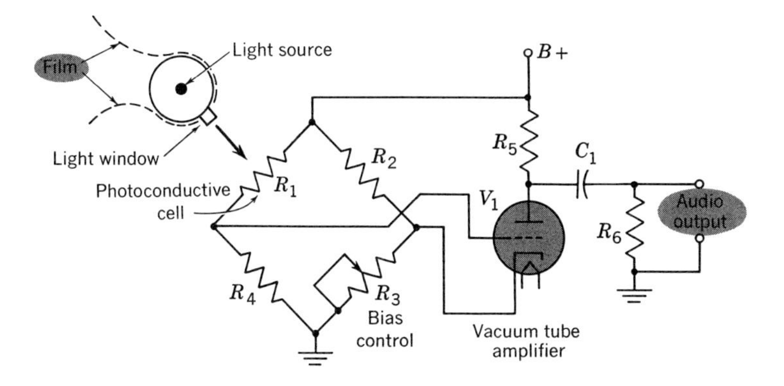


*Fig. 15 · 16 Sound on film converted to audio signal by a photoconductive cell. A change in light intensity unbalances the bridge and varies the tube conduction proportionately.*

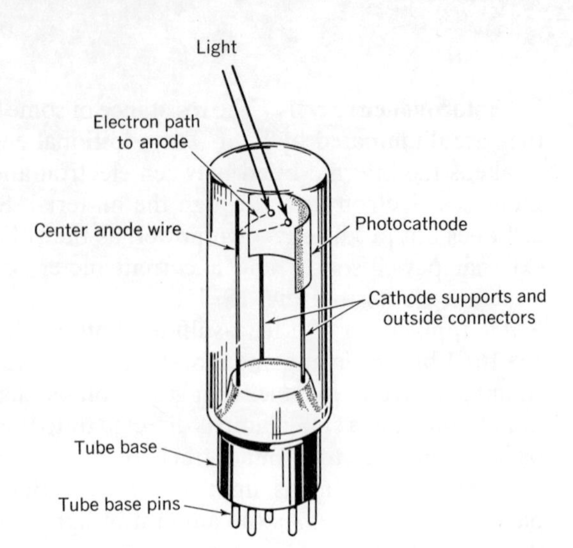


*Fig. 15 · 17 A photoelectric tube or PE cell. Light on the photocathode causes electron emission. These electrons are then attracted to a positive center wire, which is the plate or anode.*

The response of lead sulfide to light variations is fast enough to make this arrangement practical for the audio range of frequencies.

**Photoelectric tubes.** The emission of electrons from a light-sensitive material is the most widely used of the photoelectric effects. The impinging light rays impart enough energy to the electrons in the sensitized material to enable them to escape from the surface. These are then collected by a positive electrode. The released electrons are called *photoelectrons,* and their source is a *photocathode.*

A photoelectric tube contains an electron-emitting element called the *cathode* and a positive center wire called the *anode.* In this respect, it is similar to the more familiar vacuum tube; however, the cathode is not heated, since the tube depends entirely on light energy for electron emission. The number of electrons emitted by the cathode with a given light intensity is limited by the area of the cathode. In all photoemissive tubes, the cathode is made quite large; in addition, it is curved toward the center wire (Fig. 15 · 17). Electrons leave the surface at right angles; therefore every electron is aimed right at the plate wire by the cathode curvature.

Light causes cathode emission of photoelectrons from the photocathode, but the plate must be positive to attract the electrons after they become free of the cathode. Since there is an electron displacement in the tube, an external circuit must complete the path from the plate, through a load resistor, to a power supply, and back to the cathode. When the electron path is complete and voltage is applied between cathode and plate of this phototube, the tube is ready to conduct.

Figure 15 · 18 shows the circuit of a control unit which provides controlled voltage whenever there is an absence of light at $V_2$. Transformer taps are provided on $T_1$ for a 115-volt or 230-volt source. This is often necessary in industrial units, since many manufacturing plants are wired for 230 volts. The secondary voltage is applied to a divider consisting of a fixed resistor

$R_1$ and a variable potentiometer $P_1$. The cathode of $V_1$ is connected to the slider of $P_1$ to obtain control bias.

The grid of $V_1$ connects to the lower end of $R_2$ at the junction of $R_2$ and $V_2$. When light is applied to the cathode of $V_2$, it causes electron emission. The flow of electrons through $R_2$ develops a voltage which makes the grid of $V_1$ negative with respect to the plate of $V_1$. The current through $V_2$ increases as light intensity increases. Thus sufficient current can flow through $R_2$ to drive the grid negative with respect to its cathode. The exact light intensity required to do this can be controlled by adjustment of $P_1$, which varies the voltage point of $V_2$ cathode along the voltage divider.

The relay is deenergized when the light reaching $V_2$ becomes intense enough to produce a grid-to-cathode voltage difference greater than cutoff. Conversely, the relay is energized during the time that the incident light intensity on $V_2$ is low enough to permit $V_1$ to conduct.

When tube current flows, it travels through the relay, causing the relay contacts to close. This completes the circuit between the a-c line and the controlled unit (see $CR_1$ contacts at top of Fig. 15 · 18). Capacitor $C_1$ across the control-relay coil provides filtering action to prevent relay chatter. Thus $V_1$ will deenergize the relay when light of modcrate or strong intensity reaches $V_2$. Thc relay will be energized when little or no light reaches $V_2$. What this means to the controlled circuit depends on the type of contacts which the relay possesses. If these are normally closed, then with no light on $V_2$, they will open and remove the a-c power from the controlled equipment, whatever it may be. When there is light, power will be reapplied.

Conversely, if the relay contacts are normally open (with the relay not energized), the opposite action will occur.

The use of an a-c supply voltage for $V_1$ and $V_2$ does not present a

*Fig. 15 · 18 A photoelectric-tube control circuit.*

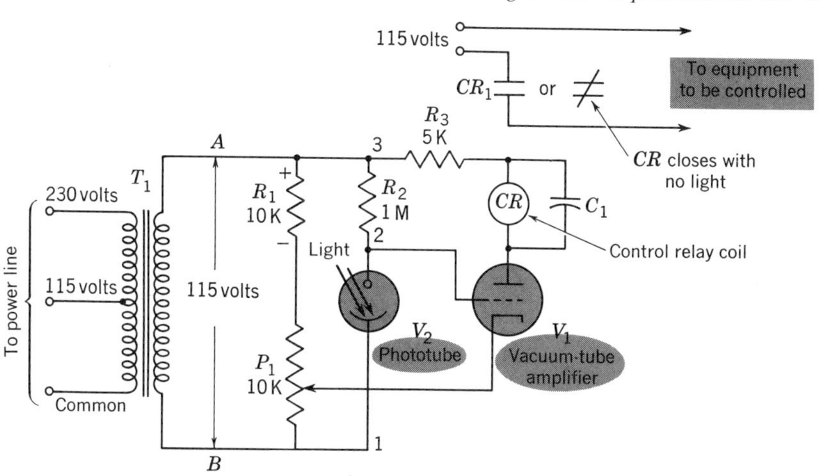

problem, since they will conduct only on the half-cycle which makes $A$ positive with respect to $B$.

**Photomultiplier tubes.** The very small current available from the standard phototube is a limitation in many circuits. One method of increasing the electron yield for a given light intensity is based on an action called *secondary emission.* In Fig. 15 · 19, electrons from the photocathode travel toward a metal plate labeled *first dynode.* The electrons increase in momentum in moving toward the dynode.

When the photoelectrons arrive at the dynode, they strike this metal plate with sufficient force to knock off other electrons. Most of these displaced electrons leave the surface of the metal and move toward a more positive second dynode. This action, of one electron causing the emission of several other electrons, is called secondary emission. After leaving the first dynode, the electron stream contains many more electrons than were emitted by the photocathode originally. The secondary emission amplification continues through the eight remaining dynodes, producing a final-plate electron flow that is many times larger than the original photocathode emission.

The photocathode, final plate, and every dynode are connected to separate base pins. Starting at the cathode, every tube element must have a more positive voltage than the preceding element. Thus, dynode 1 is more positive than the cathode, dynode 2 is more positive than dynode 1, and so on. Finally, the plate must be more positive than dynode 9.

## 15 · 7 Thermoelectric transducers

A thermoelectric transducer is a thermosensitive device that responds electrically to heat. When such devices are exposed to heat, they react by

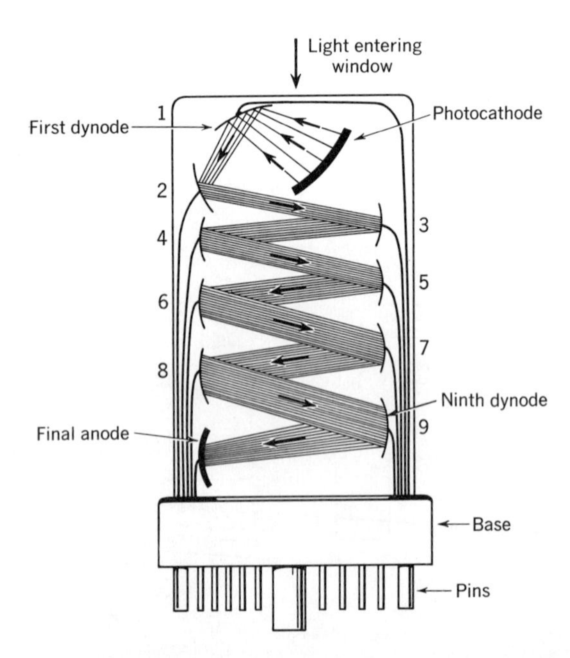


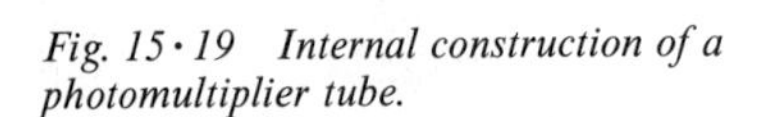
*Fig. 15 · 19 Internal construction of a photomultiplier tube.*

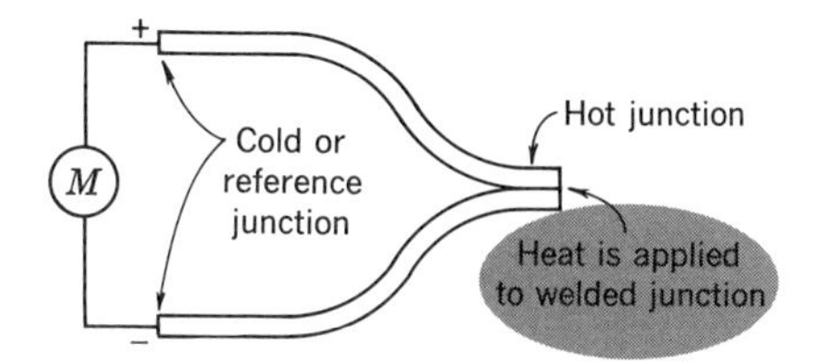


*Fig. 15·20 A basic thermocouple is made of two dissimilar metals welded together at one end. When heat is applied to the welded junction, a potential difference occurs at the reference junction.*

producing a voltage or current, or by changing in resistance value. In industrial systems, it is often necessary to detect, measure, and control temperatures accurately; and there are many thermoelectric devices which perform these functions over a large range of temperatures. For example, a thermoelectric transducer used in an industrial oven may measure temperatures from 660 to 2600°F. And a transducer in an industrial cooling system may measure temperatures of −200°F and lower. Examples of thermoelectric transducers are thermocouples, thermistors, and PN junctions.

**Thermocouples.** A thermocouple is made of two different metals welded together at one end (or sometimes both ends). When heat is applied to the welded junction, a potential difference appears between the other ends of the metals. If a meter or other device is connected between the two free ends, thereby closing the circuit, a current proportionate to the amount of heat at the junction will flow. The welded end of the thermocouple is called the hot or high-temperature junction, and the opposite end is called the cold or reference junction. These are shown in Fig. 15·20.

When a metal is heated at one end, electrons leave that end and concentrate in the cooler end, causing the metal to become polarized. The amount of polarization differs from metal to metal. Therefore, if heat is applied to a welded junction of two dissimilar metals, their opposite ends are polarized to a different degree, and a potential difference exists. The amount of voltage across the reference junction, or the current through it, can be calibrated in degrees.

The metals used in a thermocouple usually depend on where it will be used. Thermocouples made of copper and constantan have a useful temperature range from about −400 to 1100°F and will not rust or corrode easily. However, chromel-alumel thermocouples which are used for temperatures between −300 and 2400°F will become contaminated if used in hydrogen at high temperatures. As a final example, platinum and platinum-rhodium thermocouples may be used for temperatures from 32 to about 3000°F; these become contaminated easily in atmospheres containing metallic vapors, carbon, and hydrogen. This thermocouple also has a low output.

A thermocouple, in conjunction with a meter, measures temperature directly. In another application, a thermocouple connected to a recording device comprises a unit that can trace on graph paper a curve of the temperature variation over a period of time. A frequent application of thermocouples in industrial equipment is in automatic temperature control. Figure 15 · 21 shows a simplified temperature-control circuit using a thermocouple to sense heat variations in a system. The purpose of the circuit is to turn the power supply to the heat generator on and off, thus controlling the temperature of the system. When the temperature begins to rise, the thermocouple current increases. This current passes through the moving coil which is similar to those used in d'Arsonval meter movements. Attached to this coil is a pointer which travels across a temperature-indicating scale. When the thermocouple current is zero, the pointer points to zero, and when the current increases the pointer moves upscale. Should the system temperature exceed the desired value, the pointer moves past the desired control point, and a flag attached to the pointer interrupts the beam of light shining on the photocell. This photocell controls the relay. When the beam of light is interrupted, the resistance of the photocell increases, causing a decrease in the current flowing through the relay coil. This causes the relay load contacts to open and turns off the heat supply.

When the system temperature decreases sufficiently, the thermocouple detects the decrease, and less current flows through the moving coil. Consequently, the pointer moves downscale, and the flag moves out of the light path. When light once again falls on the photocell, the resistance of the cell decreases, a large current flows through the relay coil, and the relay contacts close. Thus, the heat generator is turned on again.

*Fig. 15 · 21 A simplified temperature-control circuit that uses a thermocouple to sense heat variation in a system.*

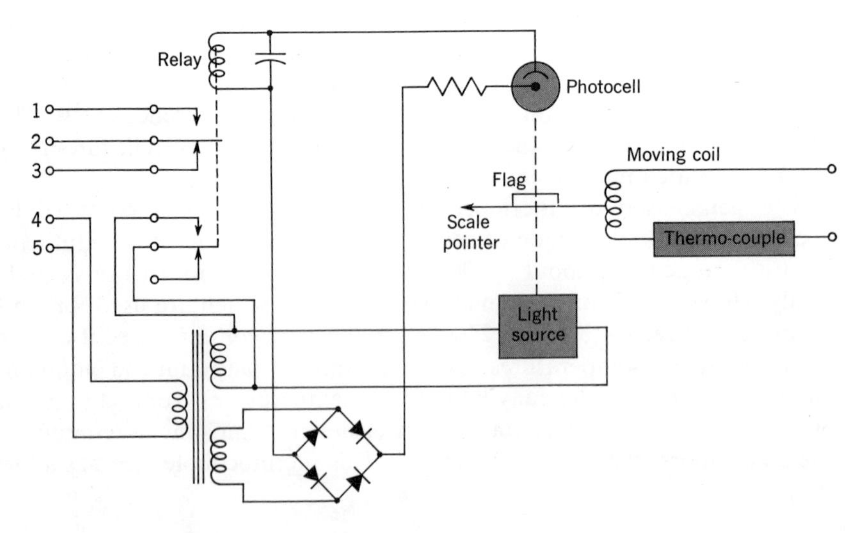

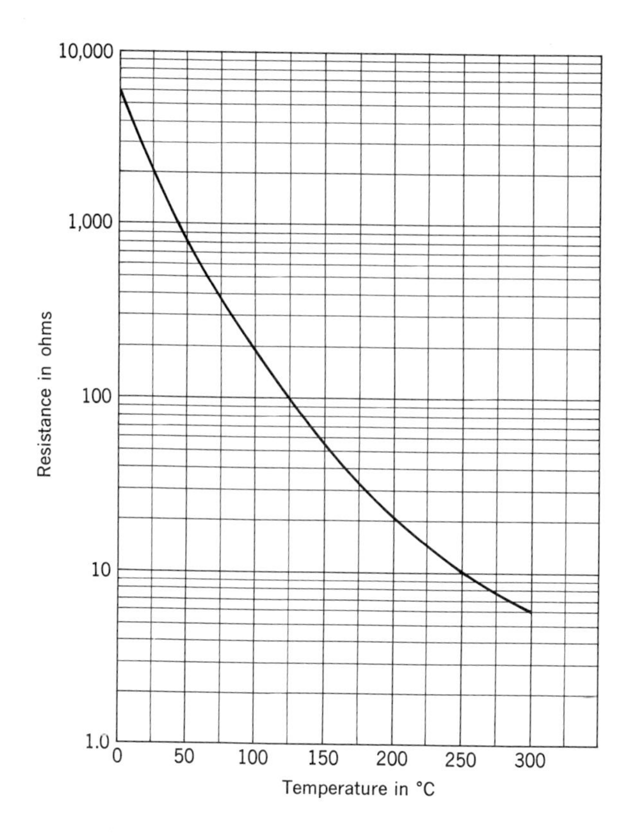


*Fig. 15 · 22 A curve showing how the resistance changes with temperature in a typical thermistor.*

**Thermistors.** A thermistor is a thermoelectric transducer whose resistance changes very rapidly with small changes in temperature. Actually, a thermistor is a specially constructed resistor made of metallic oxides such as manganese, cobalt, nickel, and others. These oxides are processed at very high temperatures and are pressed into shapes such as beads, tubes, disks, and washers. Most thermistors have leads for electrical connections, but some like the washer type are mounted on a bolt.

A curve showing how the resistance of a typical thermistor changes with temperature is given in Fig. 15 · 22. The resistance of this thermistor decreases as the temperature increases, which means that the thermistor has a negative temperature coefficient. Thermistors are also made with positive temperature coefficients.

One of the most common applications of thermistors is in transistor circuits where they are used to compensate for loss in gain caused by an increase in the temperature of a transistor. Thermistors are also often used as electronic thermometers for accurate temperature measurement. The simple thermistor bridge circuit of Fig. 15 · 23 is an example of a temperature-measuring circuit. Such circuits are extremely sensitive and can measure temperature differences of fractions of a degree. The circuit is balanced at a known normal temperature, and no current flows through the meter. When the temperature changes, even slightly, the resistance of

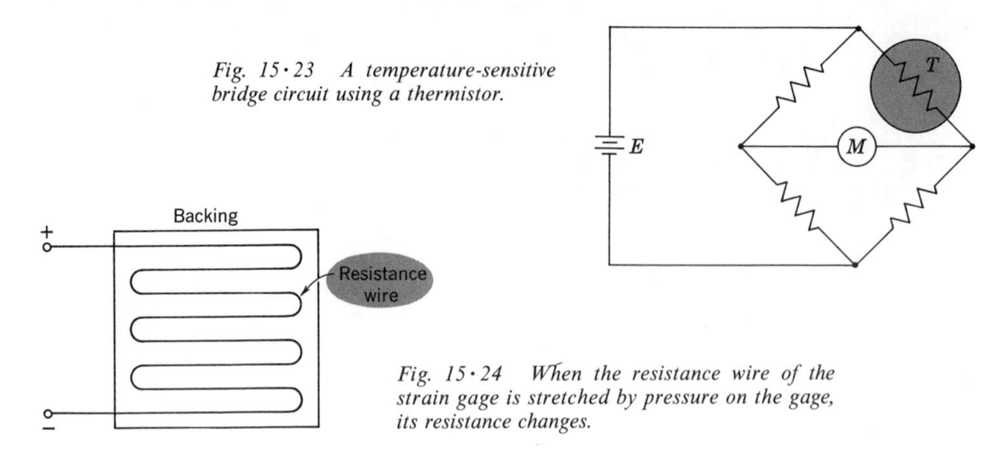


*Fig. 15·23 A temperature-sensitive bridge circuit using a thermistor.*

*Fig. 15·24 When the resistance wire of the strain gage is stretched by pressure on the gage, its resistance changes.*

the thermistor changes and causes the bridge to become unbalanced. When this happens, current flows through the meter which is calibrated in degrees, and the change in temperature from the normal temperature is measured. Thermistor bridge circuits can be used for sensing and measuring temperature changes in automatic temperature-control systems.

**PN junctions.** A PN junction is another device that responds electrically to heat. In thermoelectric applications, the P-type and N-type materials are heavily doped with impurities to increase their efficiencies. Because the metals used in bimetallic thermocouples are good conductors of heat, the heat applied to the hot junction is rather quickly conducted to the reference junction and, therefore, the efficiency of thermocouples is low. Semiconductors, however, are not good conductors of heat and, consequently, provide much higher efficiencies.

Although PN junctions may be used to change heat into electrical energy, they are more commonly used to change electrical energy into heat or for cooling. When a direct current is passed through a PN junction, electrons are forced across the junction. When they move across the junction into a higher level of energy, they absorb heat energy at the junction; when they move into a lower level of energy, they release heat energy at the junction. The absorption of energy causes cooling at the junction and the release of energy causes heating there. Furthermore, the same junction is cooled when the current flows in one direction and heated when the current flow is reversed.

The heating and cooling effects in a PN junction are being used in many industrial applications. For example, they are used in refrigeration and air-conditioning systems, and for controlling the temperature of electronic components (e.g., transistors) to increase their life and sensitivity. PN junctions used as thermoelectric devices are often more advantageous than devices they can replace because they offer quiet operation, light weight, and easy reversibility from cooling to heating; also, there is no chance of mechanical failure because they have no moving parts.

## 15 · 8 *Mechanical transducers*

**Strain gage.** A strain gage is a mechanical transducer whose resistance changes as a result of applied pressure. This pressure may be in the form of bending, twisting, tension, etc. Strain gages are made of thin resistance wire looped back and forth as shown in Fig. 15 · 24. The wire is usually bonded to a backing material such as paper or cloth, or it may be bonded to the object undergoing the strain or pressure. When pressure is applied, the gage wire stretches, becoming longer and thinner. This stretching produces an increase in the wire resistance. The strain gage is usually connected as one of the branches of a bridge circuit, and when its resistance changes, the bridge becomes unbalanced. As a result, current flows through the meter in the circuit, and the amount of pressure can be measured very accurately on the properly calibrated meter.

**Bourdon gage.** The Bourdon gage is another mechanical transducer that responds to pressure. As shown in Fig. 15 · 25, this device may assume different shapes. One end of the gage is sealed and connected to a dial pointer, or to a variable resistance, capacitance, or other electronic component. The other end is fixed and not sealed. When the gage is under pressure internally, it tries to straighten out or untwist. The amount of movement at the sealed end of the gage depends on the amount of internal pressure; and this movement is coupled to the dial pointer or to the electronic component to which the Bourdon gage is connected. Bourdon tubes are thin-walled tubes usually made of brass. They are used extensively in industrial systems to measure and control the pressure of liquids and gases.

## 15 · 9 *Basic control systems*

Perhaps the most important single factor in the rise of automation was the development of many sophisticated electronic devices and systems, some of which have been described in this chapter. These devices make possible the fast, accurate, completely automatic control systems that are now used in every branch of industry. Industrial processing requires many different control operations. These can, however, be classified in two general categories, open-loop and closed-loop systems. In the first group are switches, valves, and other simple actions which require an operator to make changes in the control setting. The higher class of control is the

*Fig. 15 · 25 Bourdon gages try to straighten or untwist when internal pressure is applied.*

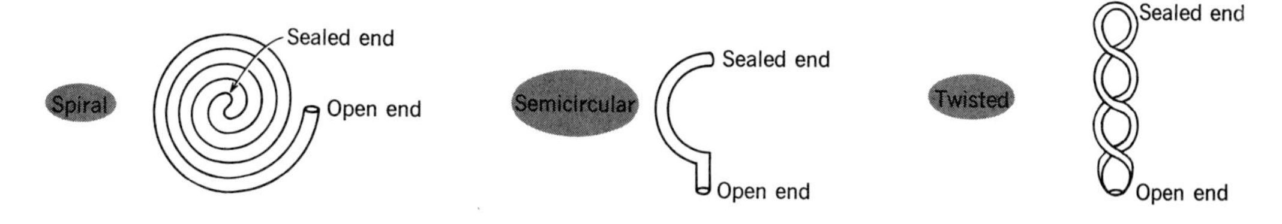

closed-loop system. It includes almost all automatic, self-correcting systems from home heating controls to large automatic production machines.

**Open-loop systems.** In the open-loop system, an action is initiated, but there is no way for the system to determine whether the action really took place or not. An example is the common light switch in home lighting. The control circuit is an open loop, since the switch can be turned to the ON position, but the observer must determine whether the light actually lit or not. The system itself cannot do this.

Figure 15·26*a* shows a functional block diagram of an open-loop system which applies to a great many circuits. The initiating action starts the control device and also identifies the action which is to be performed. In the previous example of a light switch, the switch starts the action required to turn the light on and also identifies what is to be accomplished. The second block represents the unit which actually performs the operation required to satisfy the command given by the initiating function. The last block is the end result: the completion of the initiating-function command. In the case of the light switch example, the end result is light.

**Closed-loop control systems.** The simple control circuit can be altered to provide closed-loop control. For this, an additional sensing device is required which would turn on a second light in the event of a failure in the first light circuit (for instance, if the first light burned out).

A block diagram of a closed-loop system is shown in Fig. 15·26*b*. Notice that two blocks have been added in a feedback loop: the sensing device and a comparer. The sensing device measures some characteristic of the end result and produces a signal representing that measurement. A comparer receives two signals, one fed back from the sensing device and another from the initiating action. If the sensing device signal and the command signal do not agree, the comparer generates a correction signal which changes the setting of the control device.

As an example of closed-loop control, consider the home heating system. The wall thermostat contains the initiating action, reference signal, sensing device, and comparer. The initiating device is the temperature control, which is set at the temperature desired in the home. Moving this dial or control also provides the reference signal for the comparer. The sensing

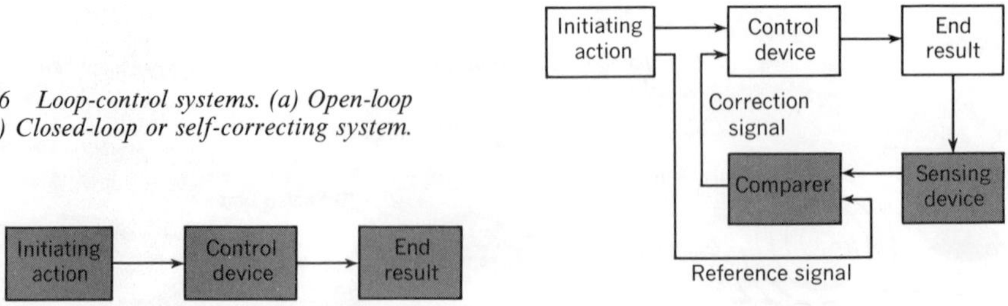

*Fig. 15·26 Loop-control systems. (a) Open-loop circuit. (b) Closed-loop or self-correcting system.*

Initiating action
Control device
End result

(*a*)

Initiating action
Control device
End result
Correction signal
Comparer
Sensing device
Reference signal

(*b*)

device is a thermometer which provides a signal proportional to the actual room temperature. If there is a difference between the desired heat (thermostat setting) and the actual measured temperature, the comparer will start the heating plant and stop it only after the desired and actual temperatures are the same. This is a closed-loop control, since no operator is needed to maintain a constant end product.

## *15 · 10 D-c motor operation*

An electric motor converts electrical energy to rotational energy by using the properties of magnetic forces. Because of this importance of magnetic forces in motor operation, let us briefly review some basic facts concerning magnetic fields.

Current in a wire produces a magnetic field proportional to the magnitude of the current. Thus large currents produce large magnetic fields, and small currents produce small magnetic fields. The strength of the magnetic field for a given current increases as the square of the turns when the wire is wound in the shape of a coil. If one coil is formed with two turns and another coil is formed with four turns, the turns ratio between the two coils is 2. However, the magnetic field when both coils carry the same current is four times greater in the larger coil.

Recall that magnetic poles act just like electric charges. That is, like charges (or poles) repel, while unlike charges (or poles) attract. A coil of wire with current flowing through it produces a magnetic field which has both a north and a south pole. When two magnets are placed close together, the poles tend to line up, with a north pole touching a south pole.

Now, all that is needed to make a d-c motor is a U-shaped soft-iron core and a straight piece of soft iron. Iron carries the magnetic lines of flux just as copper carries electron flow. The magnetic force is directed across the open end of the U-shaped core, producing magnetic flux that extends from one open end across to the other. Soft iron does not remain magnetized; therefore, a means of providing a magnetic force is required. A coil of wire with a constant electron flow will provide such a constant magnetic force. As shown at the top of Fig. 15 · 27, the wire is coiled around the closed end of the iron core. This combination, coil and iron core, is referred to as the motor *stator,* since it does not move.

Located between the open ends of the stator core is the second magnet required for two opposing or attracting magnetic fields. The second magnet is composed of soft iron also. Hence provision must be made to develop a magnetic field. This is accomplished by placing a coil around the iron piece. This assembly is supported at each end of a nonmagnetic shaft which runs at right angles to the soft-iron piece. The coil, soft-iron piece, and nonmagnetic supporting shaft are called, collectively, the motor *armature.* The armature is so mounted that it is free to rotate within the open end of the stator. This presents a problem.

How can direct current be supplied to the armature coil when the armature is free to rotate? The wire ends of the coil are placed parallel to the

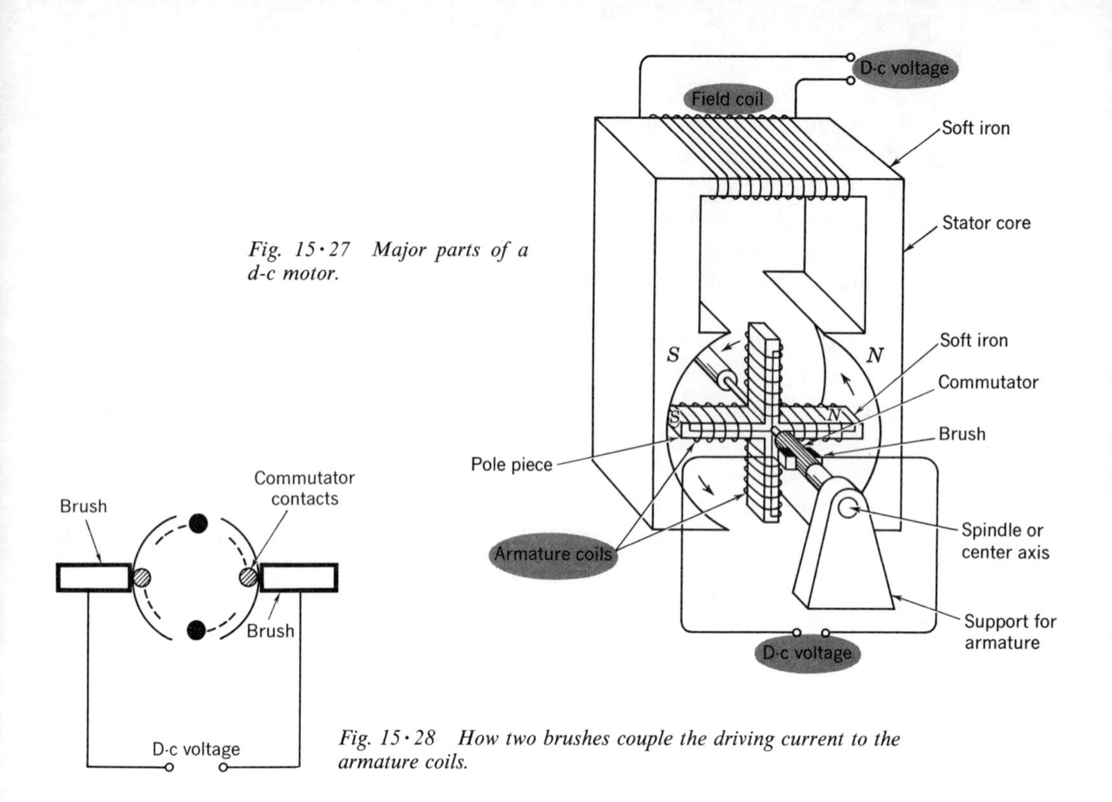


*Fig. 15·27 Major parts of a d-c motor.*

*Fig. 15·28 How two brushes couple the driving current to the armature coils.*

shaft and anchored to the shaft. Two curved *brushes,* made of copper, graphite, or other conductive material, are placed on the rotating shaft. When the coil wires touch the brushes, current flows through the brushes to the armature coil. This current then produces the magnetic field.

Applying the proper polarity of direct current at the brushes results in a magnetic field which places the armature north pole near the stator north pole. This causes the two poles to repel each other and the armature rotates. As the armature rotates, the wires reverse under the brushes, and the end of the armature coil now approaching the stator north pole changes its polarization from south to north.

Similarly, the pole that was previously north and that is now approaching the south pole of the stator is also changed to south. In Fig. 15·28, the armature contains two soft-iron pieces with two separate coils. Each coil is connected to the d-c source through the brushes. The brushes contact only two wires at a time. This action provides direct current only to the coil nearest the ends of the stator. The armature receives maximum repelling force when two like poles are closest to each other. When the armature poles are equidistant between the stator poles, the force is greatly reduced; therefore current is applied only when the maximum push is obtainable.

The brushes and coil-wire ends which are brought out from the coil form a unit called the motor *commutator.* The minimum requirements for a d-c motor are a stator, an armature, and a commutator. A commercial motor

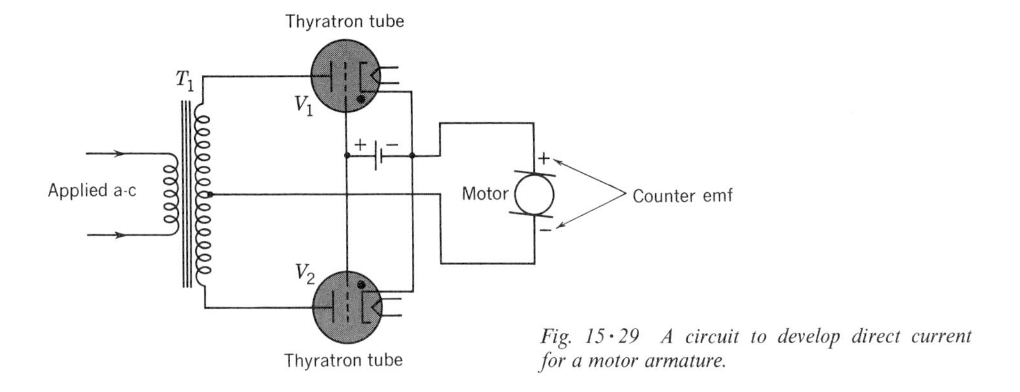


*Fig. 15·29 A circuit to develop direct current for a motor armature.*

has a stator (just under the motor housing), an armature with a great many separate coils, and a commutator having two brushes.

**Counter electromotive force.** While the direct current flowing in the armature creates a magnetic force which helps to drive that armature, the armature coils are moving through the stator magnetic field. Wires cutting through a magnetic field have a voltage induced in them. This induced voltage is a counter electromotive force. The induced voltage polarity is such that it opposes the armature current. The induced voltage is positive at the positive brush and negative at the negative brush. The counter emf reduces the effective voltage applied to the armature, causing a smaller current in the armature.

**Speed-control circuit.** To develop the direct current required by the armature, it is not unusual to use thyratron tubes to rectify the a-c line voltage to direct current. Such a circuit, in its simplest form, is shown in Fig. 15·29. What we have here is a full-wave rectifier in which each thyratron will conduct over the portion of the cycle when its plate is positive with respect to its cathode. The conduction does not last for a full half-cycle per tube, as the thyratron receiving the positive plate voltage does not fire until the plate is positive by an amount equal to the ionizing voltage of the tube, plus the counter emf developed by the motor armature.

A small battery is connected between the control grid and cathode of each thyratron in Fig. 15·31 to enable each tube to fire when its plate goes positive with respect to its cathode. When automatic motor control is employed, this grid-to-cathode voltage is part of the control system.

When there is little or no load on the motor, armature speed is high, and so is the counter emf. A high counter emf reduces the conduction time of the thyratrons, which is to be expected, since very little power is required by the armature.

Increasing the load on the motor slows down the armature rotation. Since the armature wires cut the stator magnetic field at a slower rate, the counter emf decreases. Thus, reducing the motor speed reduces the counter

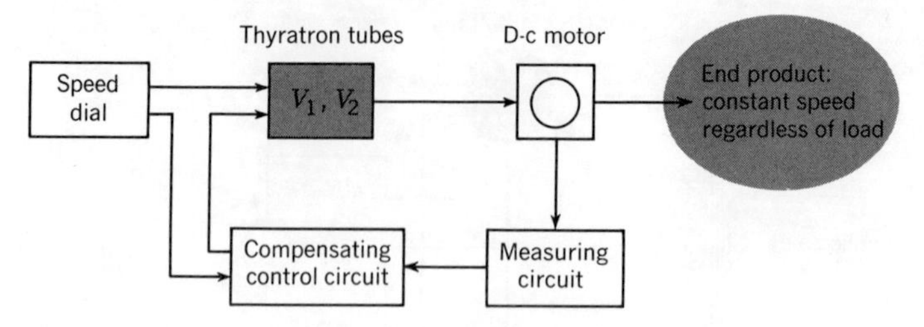


*Fig. 15·30 A block diagram of the motor-control circuit. Tubes $V_1$ and $V_2$ are the armature thyratrons shown in Fig. 15·29.*

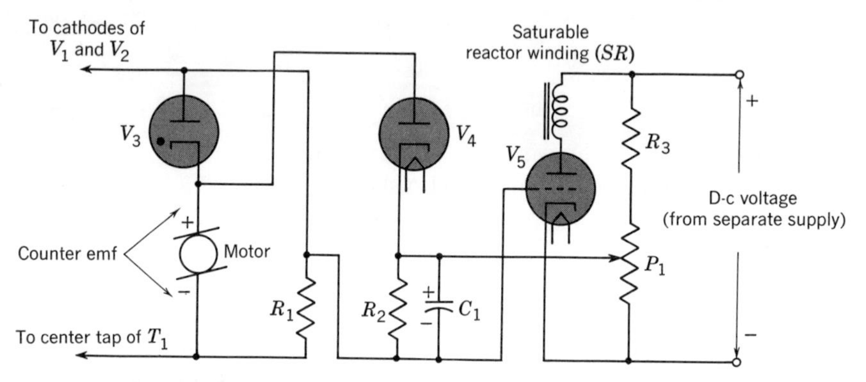


*Fig. 15·31 The circuit which provides the power required by the motor armature as the load varies.*

emf. The speed control can increase the power supplied to the armature to increase the armature rotation and, in this manner, restore the counter emf. In doing this, the speed control must allow the thyratrons to conduct for a longer time.

Thus, motor speed can be determined by the counter emf, and the load can be determined by the conduction time of the thyratrons. These two characteristics provide the variables needed for automatic motor control. Speed is held constant by measuring counter emf and developing a control signal from it. Load changes can be compensated by responding to the conduction time and developing a control signal from this variation in time.

## 15·11 D-c motor control

A d-c motor control is a closed-loop control. The operator sets the speed dial. Then the measuring and compensating circuits act to maintain the desired speed. A block diagram of the motor control is shown in Fig. 15·30. The speed dial sets the initial conditions, the measuring device senses load and speed changes, and the compensating circuits develop correction signals to maintain the desired end product.

**Sensing circuit.** A typical measuring circuit is shown in Fig. 15·31. Gas diode $V_3$ is added to the thyratron conduction path to prevent the voltage spikes produced by thyratron conduction from appearing in the sensing voltage. In this manner, only counter emf will be measured.

It is important to note that when the armature thyratrons $V_1$ and $V_2$ shown in Fig. 15·30 conduct, the plate of $V_3$ in Fig. 15·31 is positive with respect to its cathode, and it conducts. Also, when $V_1$ and $V_2$ are cut off, the plate of $V_3$ is negative with respect to its cathode because of the counter emf.

The counter emf causes the plate of $V_4$ to go positive when $V_3$ is not conducting. $V_4$ conducts through $R_2$, $C_1$ and $R_1$. During this conduction, a charge is built up across $C_1$ with the cathode side positive and the $R_1R_2$ junction negative. This charge on $C_1$ will provide a current through $R_2$ after $V_4$ conduction stops. The time required for $C_1$ to discharge completely depends on its capacitance and the resistance of $R_2$. Larger values of either component will prolong the time required for $C_1$ to lose its charge.

A reduction in counter emf will reduce the voltage across $R_2$. An increase in conduction time of $V_1$ and $V_2$ will allow the voltage across $R_2$ to decrease. The capacitor begins to discharge the moment $V_4$ ceases conduction. $V_4$ conducts only when $V_3$ and the armature thyratrons do not conduct because when $V_3$ conducts, it appears as essentially a short circuit across $V_4$ and keeps $V_4$ from conducting. Therefore, a longer thyratron conduction means a shorter $V_4$ conduction. And, a shorter $V_4$ conduction means less voltage across capacitor $C_1$. This, in turn, produces less voltage across $R_2$ since $R_2$ and $C_1$ are in parallel. Thus, the voltage across $R_2$ is related to the speed of the motor. This is important because the voltage from $R_2$ will regulate the current flow through the control tube $V_5$. In essence, $R_2$ is "telling" $V_5$ the condition of the motor speed. Now let us see how $V_5$ uses this information to bring the motor speed back to a pre-set value when this speed changes value.

**Control amplifier.** Control tube $V_5$ has a separate power supply and uses the voltage across $R_2$ as a bias only. The positive side of $R_2$ is connected to the cathode of $V_5$ through the slider of $P_1$. The grid of $V_5$ connects directly to the negative side of $R_2$. Thus the voltage developed across $R_2$ makes the grid $V_5$ more negative with respect to the slider of $P_1$. As the counter emf increases, the conduction of $V_5$ decreases. As the motor load increases, the conduction of $V_5$ increases.

$R_3$ and $P_1$ serve to provide an additional positive bias voltage for the cathode of $V_5$. This is in addition to the positive voltage developed across $R_2$, which is also applied to $V_5$.

**Compensation signal.** Figure 15·32 shows the complete armature control circuit. The saturable-core transformer has two other windings in addition to *SR*. One supplies alternating current to the transformer, while the other winding supplies a low a-c voltage to a rectifier whose output controls the grid bias of the armature thyratrons.

The speed and power requirements of the motor are measured by $V_4$, $R_2$, and $C_1$ as described. Current through $R_2$ is proportional to the counter emf and armature-thyratron conduction time. $V_5$ conduction is controlled by the voltage developed across $R_2$. In Fig. 15·32, the $SR$ winding is shown with the rest of the transformer. Current through $V_5$ (controlled by the voltage across $R_2$) flows through $SR$. Thus changes due to speed or load cause similar changes in the current through $SR$. The saturable-core transformer is a variable-ratio transformer. The step-down voltage varies from about 15 volts to a minimum of 3 volts. With no current in $SR$, the transformer provides maximum voltage to winding $S_1$. For any current through $SR$, the voltage induced in $S_1$ is reduced proportionately. Effectively, the turns ratio is varied by the current through $SR$. When a smaller voltage is induced in $S_1$, a smaller voltage is developed across $R_4$.

**Thyratron control.** The bias for thyratrons $V_1$ and $V_2$ is the voltage developed across $R_4$. A large current through $SR$ will result in a small voltage across $S_1$ and longer conduction of $V_1$ and $V_2$. A small $SR$ current allows a large voltage across $S_1$, which drives the grids of thyratrons $V_1$ and $V_2$ negative. Conduction times of $V_1$ and $V_2$ now decrease.

The complete sensing and control loop begins at the motor. Thus, as the speed decreases, the counter emf decreases. The voltage across $R_2$ decreases, and tube $V_5$ conducts more. A larger current through $SR$ causes less coupling between the primary and $S_1$ of the power transformer. The voltage applied to $V_6$ is smaller; therefore $V_1$ and $V_2$ bias is smaller, and the tubes conduct longer. The increased conduction speeds up the motor and restores the counter emf. Conversely, a motor-speed increase produces the opposite sequence, with less conduction through $V_1$ and $V_2$. This will send less current through the motor armature, and the motor speed will be reduced.

*Fig. 15·32 The complete circuit diagram of the motor-speed-control circuit represented by the block diagram in Fig. 15·26.*

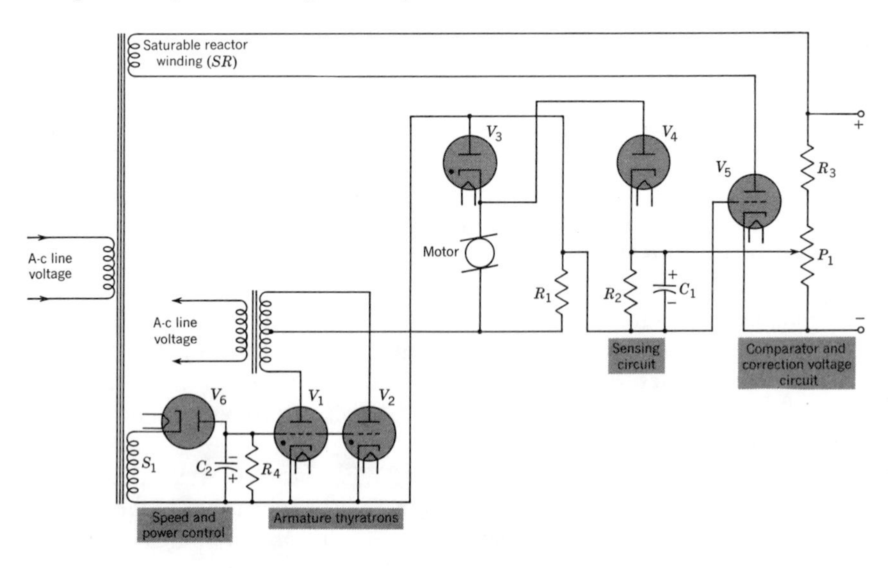

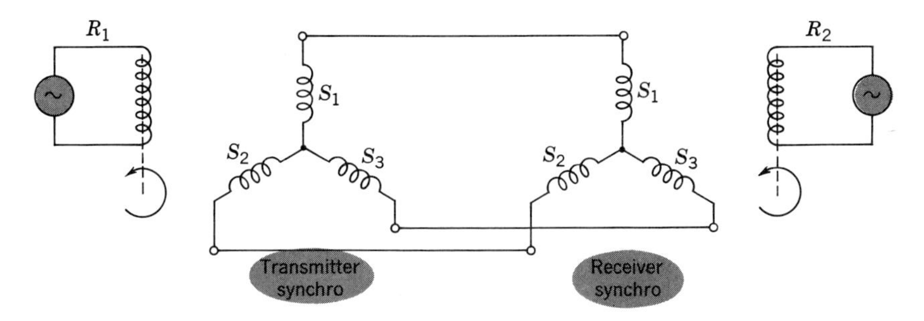

*Fig. 15·33 A generator synchro and a receiver synchro make up a basic synchro unit that can be used as a positioning control.*

The circuits just described are simplified versions of those found in the industrial field. In addition to armature control, there is motor field control, acceleration control, dynamic braking, reversing control, maximum armature-current limiting control, and other refinements available in commercial units. Each of these contributes to the close-loop system, but none is vital to the motor operation.

## *15·12 Synchros*

A basic synchro unit, shown in Fig. 15·33, consists of two motorlike devices, one of which is used as the transmitter, and the other as the receiver. The synchros are electrically connected to each other; and when the shaft of the transmitter synchro is rotated to a new position, the receiver shaft rotates to the same relative position. There is actually a master-slave relationship between the two synchros, and they are effectively synchronized even though they are not coupled mechanically.

The windings labeled $S_1$, $S_2$, and $S_3$ are the stator windings of each synchro, and they are Y-connected and placed 120° apart. The rotor windings are labeled $R_1$ and $R_2$, and each is connected to a power supply. A-c and d-c synchros are used, and the operating principles are the same for both types. In both synchros, when the rotor winding is in alignment with $S_1$, the shaft is in the zero-degree position, and maximum voltage is induced across $S_1$. The voltages in $S_2$ and $S_3$ are at some lower value because these windings are at an angle to the rotor windings. When the rotor winding is at right angles to $S_1$, no voltage is induced in $S_1$, but voltage is still induced in $S_2$ and $S_3$. The same discussion applies to $S_2$ and $S_3$.

Suppose the two synchro shafts are initially in the zero-degree position, and that the rotors are aligned with $S_1$. The same voltages appear across the corresponding windings in each of the synchros. Now, if the shaft of the transmitter synchro is rotated 60° in a counterclockwise direction, the transmitter rotor becomes aligned with $S_3$ and maximum voltage is induced in $S_3$. The changes in voltage that appear across each pair of terminals in the transmitter synchro are coupled to the corresponding terminals in the receiver synchro, causing current to flow in the receiver windings. As a result of this current, a magnetic field is produced in the receiver synchro stator windings and it interacts with the magnetic field of the receiver rotor. This causes the rotor to turn until it is in the same relative position as the transmitter rotor. When the receiver rotor has turned 60° counterclockwise, the voltages across each pair of its terminals are the same as the voltage across the transmitter terminals. Hence, there is no longer any voltage difference, and the current in the receiver stator windings is zero. As a result, the stator magnetic field is reduced to zero, the rotor stops turning, and the receiver shaft is in the same position as the transmitter shaft.

Because they are not mechanically connected, synchro units are commonly used to control position from a distance. For example, an operator can turn a knob on a transmitter synchro in one part of a building, and the shaft of a receiver synchro in another part of the building will rotate to the same relative position. In another application, a synchro receiver pointer knob on an indicating panel could continually display the position of a remotely located transmitter shaft. Though synchro units find many applications in industrial systems, they can only be used for very light loads.

**Servomotors.** In a closed-loop control system, the end result is frequently the output of an electric motor, a synchro, or some similar device, which is being controlled by the preceding stages. In some applications where the load is light, devices such as synchro units can be used satisfactorily. But for heavier loads, more powerful devices are needed. Such devices are generally called servomotors.

Typical servomotors include d-c shunt motors, d-c series motors, and a-c induction motors. The choice of a motor for a particular control system depends on such factors as allowable error, available power source, the amount of output power and torque needed, and environmental conditions such as temperature, humidity, and vibration.

In general, d-c motors are advantageous because they offer less weight for the same amount of output power, and they have much higher starting torques than do a-c motors. A-c motors, however, offer greater reliability because they are free from commutation problems such as noise and wear.

## SUMMARY

1. An induction heater creates a high-frequency radio wave that is coupled through mutual inductance to a metal where heat is generated by induced currents.
2. Dielectric heating is a means of generating heat in an insulating material. High-frequency waves produce heat because of dielectric losses in the insulator.
3. Gas is added to a tube at the time of manufacture to increase maximum current ratings and provide a smaller voltage drop across the tube. The gas changes the grid-control characteristics from continuous current control to an on-off function. A thyratron is a gas tube with a control grid.
4. Ignitrons are cold-cathode tubes which have great current-carrying capacity. The cathode is a pool of mercury, which provides gas molecules when an arc is formed with an electrode positioned very near the mercury.

5. The zener voltage level is the value of reverse bias for which a diode will conduct a high reverse current. Diodes designed to operate especially around the zener voltage level are called zener diodes.
6. A silicon controlled rectifier contains three PN junctions and has a gate element that controls the value of the breakover voltage. The breakover voltage of an SCR is similar to the ionization potential of a gas tube, and the gate terminal is similar in operation to the grid of a thyratron.
7. Photovoltaic cells generate a voltage when light strikes the sensitive element. Photoconductive cells change their resistance when activated by light. Photoelectric tubes are the most widely used light-sensitive devices. Light striking a photocathode causes emission of photoelectrons that are attracted to the positive anode. The resulting current develops a control voltage across a resistance placed in series with the tube. The current may be increased by dynode secondary emission in a photoelectric tube called a photomultiplier tube.
8. When heat is applied to the hot junction of a thermocouple, a voltage appears at its reference end. Thermistors respond to temperature variations by changing resistance value. PN junctions act as cooling devices when current is made to flow through the junction; changing the direction of the current through the junction causes the PN junction to act as a heating device.
9. Process-control circuits are grouped in two classifications: open-loop systems and closed-loop systems. The open-loop system is started by an initiating action, such as a switch. Any action is performed by a control circuit, which should produce the desired end result. However, the circuit includes no method of determining the effect of the control system on the end result.
10. The closed-loop system responds to an initiating signal, then proceeds through a control circuit to a process function. The end product is monitored, and a signal representing some aspect of the end product is sent back to a comparator circuit. The end-product signal and a reference signal obtained from the initiating switch are compared. When a difference exists, a signal is sent to the control circuit to alter the end product. In this manner, the end-product signal is changed to coincide with the reference signal. A d-c motor-control circuit is an important example of a closed-loop system.
11. Two motorlike synchros which are electrically coupled to each other are used as positioning controls over reasonable distances. When the shaft of the transmitter or master synchro is rotated, voltage changes are coupled to the receiver or slave synchro. These voltage changes cause the shaft of the receiver synchro to rotate until it is in the same relative position as the transmitter shaft. Synchro units can only be used for very light loads. For heavier loads, servomotors are used.

## SELF-EXAMINATION

Here's a chance to find out how well you have learned the material in this chapter. Work the exercises; then check your answers against the Key at the back of this book. These exercises are for your self-testing only.

1. A metal bar may be heated electronically by (*a*) emission heating; (*b*) dielectric heating; (*c*) induction heating; (*d*) convective heating.
2. An electronic heater uses (*a*) a detector; (*b*) an oscillator; (*c*) an amplifier; (*d*) a pulse-counter circuit.
3. Adding gas to a tube to form a thyratron causes (*a*) a decrease in current capacity; (*b*) greater voltage drop across the tube; (*c*) the loss of grid control above cutoff; (*d*) the tube to be useless in control circuits.
4. The arc drop of a conducting thyratron is about (*a*) 8 volts; (*b*) 15 volts; (*c*) 24 volts; (*d*) 150 volts.
5. An SCR is conducting and its anode current $I$ is 20 amp. If $I_g$ is reduced by one-half, (*a*) $I = 10$ amp; (*b*) $I = 40$ amp; (*c*) $I = 20$ amp; (*d*) $I = 0$ amp.
6. A photocell is most sensitive to (*a*) AM radio waves; (*b*) sound waves; (*c*) FM radio waves; (*d*) light waves.

7. A thermistor reacts to changes in temperature by (*a*) generating a voltage; (*b*) stretching or twisting; (*c*) generating a current; (*d*) changing its resistance.
8. A control that corrects itself for any error in the end product is (*a*) an open-loop system; (*b*) a closed-loop system; (*c*) an amplifier; (*d*) an initiating control.
9. The d-c motor control senses (*a*) armature current; (*b*) counter emf; (*c*) field voltage; (*d*) the voltage across a resistor in series with the armature current.
10. A disadvantage in using d-c servomotors is that (*a*) they can only be used for very light loads; (*b*) they have low starting torques; (*c*) they are very heavy; (*d*) they may develop commutation problems.

## QUESTIONS AND PROBLEMS

1. Describe the heat-producing action caused in a metal when it is placed in a magnetic field.
2. Draw the circuit of a tickler-coil oscillator.
3. Compare a thyratron and a vacuum-tube triode with respect to: (*a*) amount of plate current; (*b*) control of plate current by the grid.
4. Give two similarities between an SCR and a thyratron.
5. Describe briefly the operation of a photomultiplier tube.
6. Explain why an external power supply is not needed for a photovoltaic cell circuit, but is needed for a photoelectric tube.
7. If the resistance of a thermistor is 100 ohms at 125°C, what is the *change* in thermistor resistance when the temperature reaches 250°C? Refer to Fig. 15·22.
8. Explain how the heating and cooling effect at a PN junction takes place.
9. List the basic parts of a closed-loop-control block diagram.
10. Describe the conduction time and current path of $V_4$ in Fig. 15·31.

# *Review of chapters*  *to* 

## SUMMARY

1. A volt-ohm-milliammeter tester (VOM) measures a-c or d-c voltages and direct current without using any external power. For resistance measurements, a small battery is enclosed within the instrument case.
2. The vacuum-tube voltmeter (VTVM) performs essentially the same measurements as a VOM. However, the VTVM can cover a wider range of voltages and resistances.
3. For either instrument, the selector switches must be carefully set. This includes not only the function, but the range too.
4. The cathode-ray oscilloscope provides an image of the waveshape of current or voltage in a circuit. The principal components of a basic oscilloscope include a cathode-ray tube, a sweep (sawtooth) oscillator, and vertical and horizontal deflection amplifiers.
5. A signal generator is an instrument which generates an a-c signal suitable for test purposes. A-f (audio) generators produce signals from 20 to 20,000 cps. R-f generators produce c-w or AM signals above 20,000 cycles. The modulating signal is generally a single audio note, either 400 or 1,000 cycles.
6. FM generators provide an r-f signal that varies in frequency, generally at a 60-cycle rate. Thus, the radio frequency is no longer a steady frequency, but rather one that shifts back and forth about a center frequency.
7. In all signal generators, when a wide range of frequencies is to be covered, bandswitching is employed.
8. Wavemeters offer the simplest means of measuring the frequency of an r-f wave. Two general types are available: reaction wavemeters and absorption wavemeters.
9. A pulse is a fast variation in voltage or current of short duration. Pulses are nonsinusoidal in shape, but they are repeated at a regular rate and do have a definite frequency.
10. Electronic circuits which generate pulses are usually relaxation oscillators, such as a multivibrator or blocking oscillator. The pulses are often used for synchronization, gating, instrument calibration, and testing of pulse-actuated circuits.
11. A monostable multivibrator has one stable and one quasi-stable state; a bistable multivibrator has two stable states; and an astable multivibrator has two quasi-stable states.
12. Multivibrators are usually divided into two categories: free-running of self-oscillatory oscillators which are relaxation oscillators, and trigger circuits which produce output pulses only when external triggering signals are applied.
13. A widely employed method of counting pulses combines a mechanical counter with an electronic scaling circuit. The circuit most often used for this purpose is an Eccles-Jordan counter.
14. Pulses are coupled into an Eccles-Jordan circuit by grid capacitors or diodes. The latter method is capable of a faster response time.
15. Scale factors greater than 2 can be obtained by connecting several Eccles-Jordan stages in series.
16. Ten-stable-state devices which rely on gas discharge are being used increasingly in computer design.
17. Two classes of computers are digital and analog. In general, those devices which add, multiply, subtract, and divide by pulse techniques are called digital computers. Analog computers relate parts of a mathematical equation to a physical or electrical quantity.
18. In induction heating, a high-frequency wave is coupled by mutual inductance to a metal where heat is generated by large circulating induced currents.

19. Dielectric heating is a means of generating heat in a material which is a poor conductor of electricity.
20. Gas added to a tube tends to alter its operating characteristics appreciably. The gas changes the grid control from a continuous function to one which is either on or off.
21. Photo-operated devices alter their resistance, voltage, or current when exposed to a beam of light.
22. Thermoelectric transducer such as thermocouples and thermistors are useful for measuring, sensing, and controlling heat because they respond electrically to variations in temperature.
23. Process-control circuits can be classified into open-loop or closed-loop systems. The open-loop system is simpler, but it is not self-correcting. The closed-loop system will correct itself if it is not operating properly.

## REFERENCES FOR SUPPLEMENTARY MATERIAL

### *Books*

Annett, F. A., *Practical Industrial Electronics,* McGraw-Hill Book Company.
Blitzer, R., *Basic Pulse Circuits,* McGraw-Hill Book Company.
Chute, George M., *Electronics in Industry,* 3d ed., McGraw-Hill Book Company.
Henney, Keith, and James D. Fahnestock, *Electron Tubes in Industry,* 3d ed., McGraw-Hill Book Company.
Johnson, J. Richard, *How to Use Signal and Sweep Generators,* John F. Rider, Publisher, Inc.
Kloeffler, R. G., *Industrial Electronics and Control,* John Wiley & Sons, Inc.
Miller, Robert E., *Maintenance Manual of Electronic Control,* McGraw-Hill Book Company.
Philco Technological Center, *Servomechanism Fundamentals and Experiments; Electronic Precision Measurement Techniques and Experiments,* Prentice-Hall, Inc.
Platt, Sidney, *Industrial Control Circuits,* John F. Rider, Publisher, Inc.
Ruiter, Jacob H., Jr., *Modern Oscilloscopes and Their Uses,* Rinehart & Company, Inc.
Swallow, T. P., and W. T. Price, *Elements of Computer Programming,* Holt, Rinehart and Winston, Inc.
Turner, Rufus P., *Basic Electronic Test Instruments,* Rinehart & Company, Inc.
Zbar, P. B., *Electronic Instruments and Measurements,* McGraw-Hill Book Company.

## REVIEW SELF-EXAMINATION

Here's another chance to check your progress. Work the exercises just as you did those at the end of each chapter and check your answers.

1. The section of a radio receiver which does not require alignment is the (*a*) t-r-f stage; (*b*) i-f amplifier; (*c*) antenna input circuit; (*d*) audio amplifier.
2. An isolation transformer is required for (*a*) FM receivers only; (*b*) a-c/d-c receivers; (*c*) all AM sets; (*d*) sets operating on alternating current only.
3. When voltage measurements are made in the AVC circuit (*a*) only a 1,000-ohms-per-volt VOM should be used; (*b*) all i-f tubes should be removed; (*c*) no signal should be tuned in; (*d*) only a VTVM or high-input-resistance VOM should be used.
4. The FUNCTION control on a VTVM (*a*) establishes the operating range of the instrument; (*b*) prevents the meter movement from being damaged by excessive voltage; (*c*) sets up the instrument to measure a certain electrical quantity; (*d*) is nothing more than an ON-OFF switch.
5. In receiver alignment, the oscilloscope is used in conjunction with (*a*) an FM signal generator; (*b*) a VOM; (*c*) a VTVM; (*d*) an audio signal generator.
6. A 1,000-cps wave would be produced by (*a*) an r-f signal generator; (*b*) an FM signal generator; (*c*) a Wien-bridge oscillator; (*d*) a wavemeter.
7. A marker signal is used to (*a*) identify a certain circuit; (*b*) identify a range of voltages; (*c*) identify the frequency of a specific point on a response curve; (*d*) peak a double-tuned circuit.

8. To determine the current in a circuit without an ammeter, we could (*a*) measure the voltage drop across a tube; (*b*) measure the voltage drop across a known resistor; (*c*) add all the resistance values in the circuit and divide by the number of resistors; (*d*) divide the total circuit resistance by the total circuit voltage.
9. To measure screen-grid voltage on an i-f amplifier, the meter should be on (*a*) ohms; (*b*) a-c volts; (*c*) + d-c volts; (*d*) − d-c volts.
10. To measure continuity of an i-f coil, the meter should be on (*a*) highest ohms range; (*b*) lowest ohms range; (*c*) highest d-c volts range; (*d*) lowest a-c volts range.
11. Induction heating depends on (*a*) dielectric hysteresis; (*b*) eddy-current losses; (*c*) $I^2R$ losses; (*d*) capacitive reactance.
12. A cold-cathode gas tube is capable of developing pulses because (*a*) the sustaining voltage is higher than the ionizing voltage; (*b*) the sustaining and ionizing voltages are equal; (*c*) the ionizing voltage is higher than the sustaining voltage; (*d*) its minimum current is zero.
13. Pulses are *not* generated in which one of the following circuits? (*a*) Computers; (*b*) telemetering equipment; (*c*) television receivers; (*d*) intercom systems.
14. The primary purpose for using additional electronic stages before a counter is to (*a*) increase counting speed; (*b*) reduce counting speed; (*c*) raise counter accuracy; (*d*) prevent motorboating.
15. A counter tube using a gas discharge arc possesses the advantage of (*a*) greater speed than an Eccles-Jordan counter; (*b*) greater compactness; (*c*) greater reliability; (*d*) very high division factors.
16. To achieve good dielectric heating action, we require (*a*) a good conductor for the dielectric; (*b*) a poor conductor for the dielectric; (*c*) ionized gas; (*d*) an inert gas.
17. A substance whose resistance changes when illuminated by light is known as (*a*) photovoltaic; (*b*) a photomultiplier; (*c*) photoconductive; (*d*) photoelectric.
18. A device whose resistance changes when it is subjected to pressure in some form is (*a*) a Bourdon gage; (*b*) a thermistor; (*c*) a strain gage; (*d*) a zener diode.
19. A device used in control systems for controlling the position of small loads only is (*a*) an a-c induction motor; (*b*) a d'Arsonval meter movement; (*c*) a synchro unit; (*d*) a d-c shunt motor.
20. Automatically operated systems possess (*a*) open-loop circuits exclusively; (*b*) no feedback paths; (*c*) ignitron switches; (*d*) closed-loop circuits.

# Chapter 16 *Military electronics*

One of the largest users of electronic equipment is the United States Armed Forces. Communications, radar, fire-control, sonar, guidance, and infrared systems are used to exchange messages and orders, detect aircraft, control aiming and firing of guns, detect presence and location of ships and submarines, direct missile flights, and locate areas of infrared radiation, such as factories, planes, and ships. The communications systems used by the military services are similar to those operating about us every day. Telephone, radio, telegraph, and television are a few of the most common ones. Of greater interest here is the use of electronics in systems built primarily to perform military functions. The topics are as follows:

## *16 · 1 Radar*

The word *radar* is derived from the phrase "radio detection and ranging." In the radar system, high-power r-f energy, usually hundreds or thousands of megacycles, is transmitted from a directional antenna in a narrow beam. The r-f wave strikes objects in the beam and is reflected back to the antenna in somewhat the same way that an echo returns to the source of the sound. A receiver detects the reflected wave and compares it with the original transmission to determine certain facts about the target.

Radar waves, like all radio waves, travel through air at the speed of light, approximately 186,000 statute, or land, miles a second. This corresponds to 162,000 nautical, or sea, miles per second. It takes about 12.4 $\mu$sec for a radar wave to travel 1 nautical mile to a target, be reflected, and return to the source. This 2-mile path is called a *nautical radar mile.* Three different systems have been used in radar: frequency modulation, frequency shift (or continuous wave), and pulse radar.

**FM radar.** Figure 16 · 1 is a plot showing frequency vs. time in an FM radar system. At time $t_0$, the frequency is 1,000 Mc. The frequency is increased linearly and in 12.4 $\mu$sec ($t_2$) becomes 1,100 Mc. At this time, it is abruptly changed back to 1,000 Mc, where it once again starts to increase to 1,100 Mc in 12.4 $\mu$sec. Suppose the signal leaves the transmitting antenna at $t_0$ and travels to a target where it is reflected back to the source. It arrives at the source at time $t_1$ (3.1 $\mu$sec). The frequency being transmitted at this time is 1,025 Mc. Since the radar wave travels 1 radar nautical mile in 12.4 $\mu$sec, the distance can be computed as 3.1/12.4, or ¼, radar nautical mile.

If a receiver located near the transmitter receives a signal from the transmitter by a direct path at the same time that it receives the reflected signal, a 25-Mc beat note will be produced at the receiver. If the total time taken for one complete frequency sweep and the number of megacycles of deviation are known, the 25-Mc beat note can be used to determine the distance to the target. The frequency of the beat increases as the target moves away from the antenna and decreases as the target approaches the antenna.

Figure 16 · 2 shows a block diagram of an FM radar system. A trigger

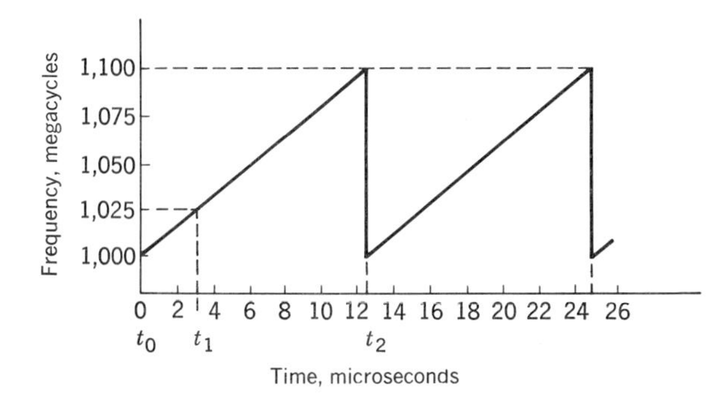


*Fig. 16 · 1 Frequency vs. time in an FM radar system.*

pulse starts the frequency sweep of the FM transmitter. This transmitter feeds a signal to the receiver and the antenna. The signal leaves the antenna, is reflected by the target, and returns to the receiver. At the time the reflected wave reaches the receiver, a signal of different frequency arrives at the receiver directly from the transmitter. The beat produced by these two signals is fed to the calibration unit, where it is deciphered and converted into miles.

**Frequency-shift radar.** Frequency-shift, or continuous-wave (c-w), radar operates on the Doppler effect created as a signal leaves a moving object. This phenomenon is the same as that produced by the horn of an automobile approaching a pedestrian. As heard by the pedestrian, the note of the horn changes frequency as the car approaches and passes. In frequency-shift radar, a continuous single frequency leaves the antenna and strikes the object or target. The signal reflects from the target, and the echo returns to the source where it is picked up by a receiver. If the target is leaving or approaching the antenna, the frequency of the returned signal varies in proportion to the target speed. The receiver compares the fixed c-w frequency generated by the transmitter with the Doppler echo to determine target speed.

There is one disadvantage to this system. An object moving crossways to the antenna produces no Doppler signal, and consequently it remains undetected. Similarly, a stationary object cannot be detected. There are times, however, when this system can be used to good advantage. In mountainous areas, other types of radar are virtually useless. They present the fixed land masses in such large proportion to moving planes that these craft are obscured. Under such conditions, Doppler radar can be used to locate the moving object without detecting the mountains.

**Pulse radar.** The most common system of radar uses pulsed r-f energy. The transmitter is turned on for short periods to produce pulses or bursts of radio frequency. It is then turned off for longer periods of time. During the time the transmitter is off, the r-f pulses leave the antenna and strike the object, and the echo returns to the receiver. The distance to the target in nautical miles can be determined by dividing the echoing time of the pulse by 12.4 $\mu$sec.

A cathode-ray tube (CRT) is used to display visually the original pulse and its echo. Figure 16·3 shows these two signals. The initial pulse that

*Fig. 16·2 Basic components of an FM radar system.*

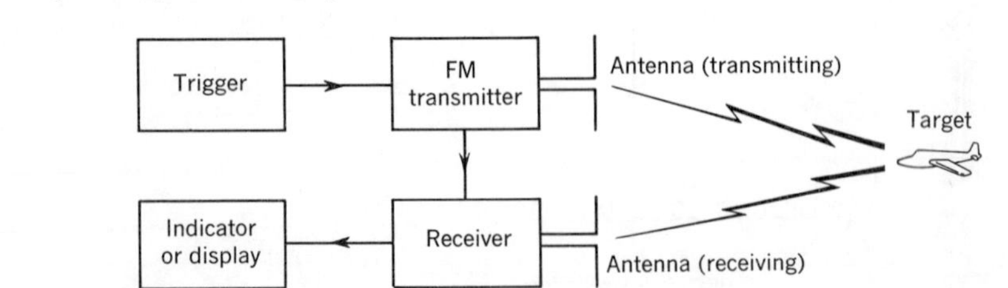

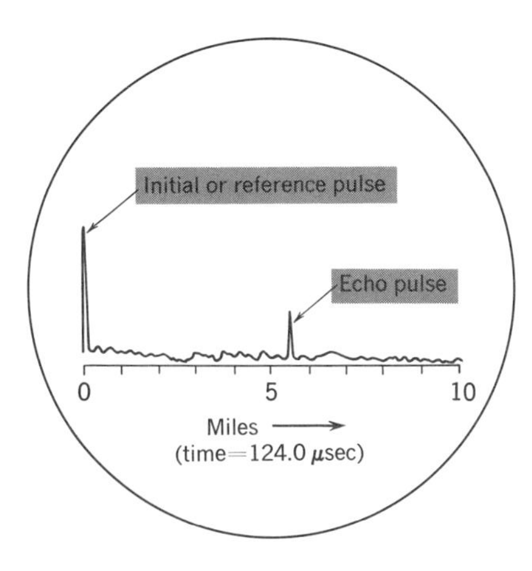


*Fig. 16·3 Cathode-ray-tube display of a radar signal and echo. The distance between the initial or reference pulse and the echo pulse indicates the distance from the transmitting antenna to the detected object.*

starts the system into one cycle of operation is shown at the left of the CRT. This pulse also starts the sweep circuit that provides the base line on the screen. By proper selection of the components in the sweep circuit, the exact time of the base-line sweep can be set. In this case a 124.0 μsec period is required for the sweeping dot to produce the complete base line. Thus the base line may be divided by 10 equally spaced vertical lines to represent a total distance of 10 nautical miles. The echo pulse shown in the middle of the screen falls at the 5½-mile marker or at the 68.2-μsec position ($5\frac{1}{2} \times 12.4 = 68.2$). In addition to the antenna, a pulse-radar system consists of: a trigger, a modulator, a transmitter, a receiver, and an indicator. See the block diagram in Fig. 16·4.

Operation of this equipment is as follows:

1. Trigger starts system by sending pulse to indicator and at the same time turning on modulator for a short period.
2. The modulator applies voltage to transmitter for the same period as the trigger pulse.
3. The transmitter supplies radio frequency to the antenna, and the signal travels in a narrow beam to the object.
4. An echo from the object returns to the antenna and goes to the receiver.
5. The receiver sends this echo pulse to an indicator, where it is compared in time with the original pulse.
6. The space between the initial pulse and the echo pulse is used to compute the distance to the object.

*Fig. 16·4 A pulse radar system.*

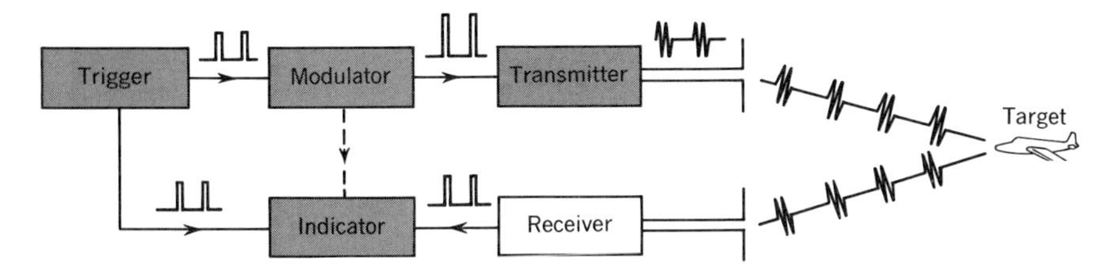

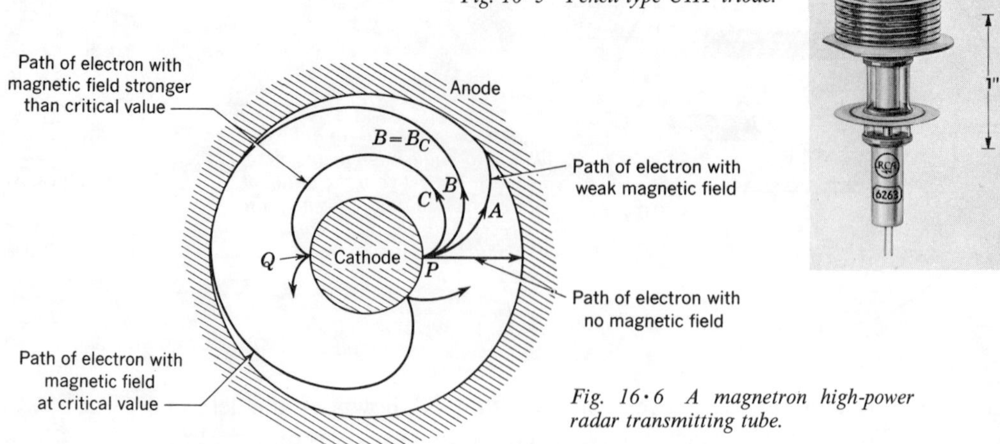


*Fig. 16·5 Pencil-type UHF triode.*

*Fig. 16·6 A magnetron high-power radar transmitting tube.*

## *16·2 Radar transmitting tubes*

The basic r-f generator in a radar transmitter usually consists of an oscillator stage with no amplification. In some of the lower-frequency radar systems (in the range of hundreds of megacycles), triode oscillators are used. At higher frequencies (1,000 Mc and up), lighthouse tubes, klystrons, and magnetrons perform this function. Another type, shown in Fig. 16·5, uses a coaxial pencil-type structure.

One of the simpler radar oscillator tubes, the lighthouse, is nothing more than a ruggedly constructed triode. It is specially built, however, so that the plate, grid, and cathode elements are positioned very close to each other. Their connections are brought out through sturdy metal rings instead of wires or socket pins. This type of coaxial construction eliminates frequency instability when the tube is used in a mechanically rigid oscillator circuit. One disadvantage of the tube is its low power-handling capacity. It does provide good voltage gain at radar frequencies, and it can be employed for receiver r-f amplifiers as well as local oscillators in superheterodynes.

Klystrons (see Sec. 5·9), or velocity-modulated tubes, contain their own frequency-determining or tuned circuit. This circuit is in the form of a resonant cavity, which replaces the tuning capacitor and coil normally associated with oscillators. The tube is similar in some ways to a multigrid vacuum tube, with the exception of the voltage applied to the plate. The klystron's plate voltage is minus with respect to the other elements. The reaction between this negative plate and the positive grids sustains oscillations in the tube.

The magnetron is the most common high-power radar r-f oscillator. As discussed in Sec. 5·10, the tube consists of a filament and a circular anode block, physically mounted between the poles of a powerful permanent magnet (Fig. 16·6). Groups of small cylindrical resonant cavities

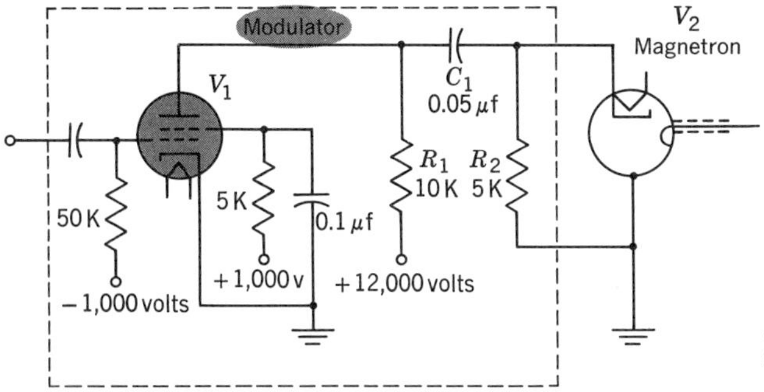

*Fig. 16·7 Schematic diagram of a radar pulse modulator.*

within the tube determine its operating frequency. The magnetic field between the cathode and plate of a magnetron is perpendicular to the electric field. These fields sustain oscillation in the magnetron through their effect on electrons traveling from cathode to anode. Magnetrons are capable of delivering very large quantities of r-f power for short periods when operated under pulsed conditions.

## *16·3 Radar modulators*

The radar modulator applies operating power to the oscillator in the transmitter in short bursts or pulses. This pulse of power turns the transmitter on for a few microseconds and initiates the r-f signal that is radiated from the antenna. The on time for a transmitter is usually about 2 or 3 μsec, after which the oscillator is cut off for about 1,250 μsec. Thus, the frequency of modulation is about 800 cps. As the oscillator tube in the transmitter is functioning only about 1/500 of the time, it may be operated at extremely high peak power. The average power under these conditions is 1/500 of the peak power. Magnetrons may be pulsed to produce hundreds of kilowatts (or even megawatts) of peak power. Since the tube is off far longer than it is on, heating in the tube is low enough so that many hours of operation can be realized.

Vacuum tubes or spark gaps can be used as the basic part of the modulator. Either device acts as a switch that discharges a large capacitor across the magnetron. Since the transmitter is to be on for a few microseconds only, this discharge must occur rapidly. Also, the pulse must be synchronized with the signal that is sent to the indicator, since it is the reference signal for measuring echo time or distance to the target. The modulator tube, in most cases, is a high-power transmitting-type tetrode.

A typical modulator circuit is shown in Fig. 16·7. In the normal or off state, $V_1$ is cut off by the 1,000-volt negative bias on the grid. Capacitor $C_1$ is charged by the 12,000-volt source through $R_1$ and $R_2$. In this condition, there is no voltage across the magnetron, and it will not oscillate. To initiate the transmission, a high-voltage positive pulse, 2 μsec long, is applied to the input grid of $V_1$. This turns $V_1$ on, and the tube conducts heavily, becoming a virtual short circuit. This conduction, in turn, grounds the left terminal of $C_1$. The plate of the magnetron is already grounded, and its cathode is connected to the other terminal of $C_1$. $C_1$ therefore discharges through the magnetron to produce an r-f pulse that lasts as long as $V_1$ is conducting (2 $\mu$sec).

It is possible to use a rotating spark gap in place of $V_1$ if this device is synchronized with the indicator.

## *16 • 4 Radar transmission lines*

The transmission line that couples the r-f output of the radar transmitter to the antenna differs from the transmission line used for receivers. Some radar systems operating at frequencies up to 5,000 Mc use coaxial cable for the transmission line. This cable consists of a heavy center conductor located within a tubular shield or braid. Insulation in the form of solid plastic, glass or plastic beads, or air is used, and the center conductor is held so that it does not touch the outer shield. This insulation must be able to withstand the very-high-voltage pulses produced by the radar transmitter, without breaking down or arcing over. Coaxial line, or *co-ax,* is available in either rigid or flexible types. A disadvantage of co-ax is its relatively high attenuation, or loss per foot of length. This loss increases as the frequency is raised and becomes so great above 10,000 Mc that the line is rendered useless.

In the range of 3,000 Mc and up, the most efficient transmission line is a waveguide. A waveguide, sometimes called *microwave plumbing,* consists of a hollow tubular or rectangular pipe. This transmission line has several advantages over co-ax at higher frequencies. It is more durable physically, it has less loss per foot of length, more power can be handled, and it is simpler to construct. Figure 16 • 8 shows two sections of waveguide. The flanges serve to connect various-shaped pieces together.

There are some disadvantages to a waveguide that make it useless at lower frequencies. The guide must have certain minimum dimensions that depend on its operating frequency. That is, the lower the frequency, the greater must be the physical dimensions of the waveguide. To be used at 200 Mc, a waveguide would have to be 4 ft wide and at least 4 ft high.

*Fig. 16 • 8 Waveguides for use as transmission line at radar frequencies. These are very efficient and introduce exceedingly low losses.* (Andrew Corp.)

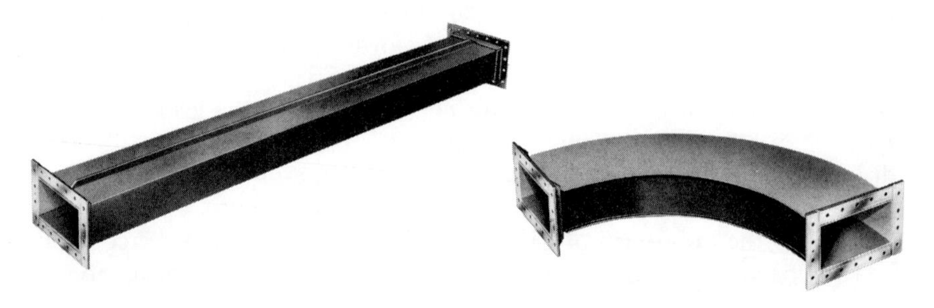

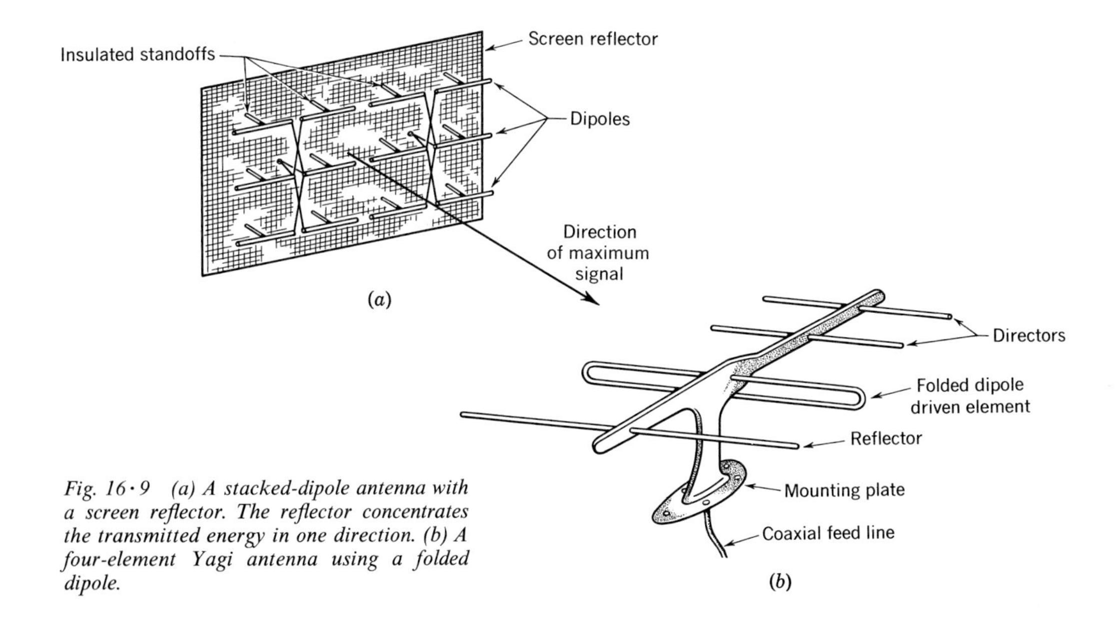


*Fig. 16·9 (a) A stacked-dipole antenna with a screen reflector. The reflector concentrates the transmitted energy in one direction. (b) A four-element Yagi antenna using a folded dipole.*

At a frequency of 1 Mc (the center of the broadcast band), it would have to be 400 ft wide.

Power is coupled to or from a guide by inserting a small loop or rod in the structure. "Plumbing" may be constructed in a variety of shapes. Rotary joints that connect to the antenna, flexible sections, U shapes, angles, and off-sets are some common ones (see Fig. 16·8).

## *16·5 Radar antennas*

A number of antennas are used in radar systems, but the most common are stacked dipoles with reflectors, Yagi arrays, and parabolas. Figure 16·9*a* illustrates a typical stacked-dipole array. The screen reflector serves two purposes. It reflects the received or transmitted wave much as a light reflector concentrates or focuses the light rays. This produces a narrow beam for transmission or reception and prevents side signals from affecting the radar display. It also prevents signals arriving from the rear of the antenna (possibly from other radar sets) from interfering with the system.

A Yagi antenna is shown in Fig. 16·9*b*. The signal in this antenna is fed to or taken from the driven element. The directors are shorter than the driven element, and the reflector is longer. This same type of antenna is used for television receivers, where a high-gain directional array is necessary.

While the two foregoing antennas are usually found at lower radar frequencies, the parabola is the most common antenna at higher frequencies. A typical parabola antenna is shown in Fig. 16·10. This name is taken from the fact that the large saucer shown here has the curvature of a parabola.

In some radar systems, separate antennas are used for transmission and reception of the signal. In other systems, a common antenna performs

Fig. 16 · 10 *A typical parabola antenna for radar use.* (D.S. Kennedy and Company.)

*Fig. 16 · 11 A T-R switch connected to furnish maximum signal to the antenna during transmission and to protect the receiver from large amounts of energy.*

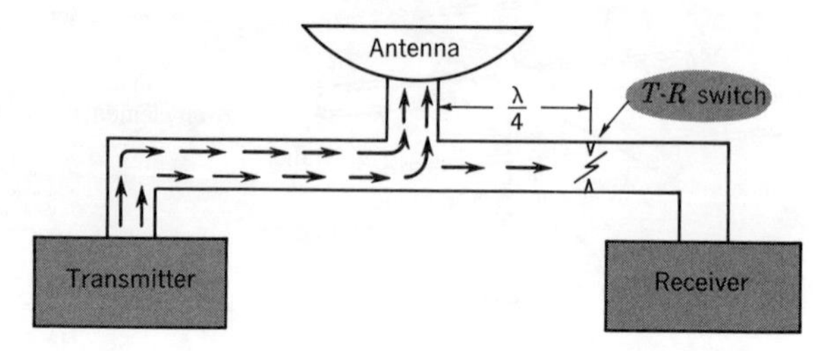


both functions. When the same antenna is used for transmitting and receiving, a changeover system is inserted to connect the antenna to the correct piece of equipment at the proper time.

**T-R (transmit-receive) switches.** As the time required for switching the antenna from the receiver to the transmitter and back to the receiver in a radar system must be very short, mechanical relays are useless and an electronic T-R switch is necessary. The T-R switch consists of two electrodes that form a gap very similar to a spark gap (see Fig. 16 · 11). This switch is mounted in the waveguide or line going to the receiver. Its position is usually about one-fourth of a wavelength from the T joint where the line from the antenna is split to feed the transmitter and receiver. When the large pulse generated by the transmitter starts toward the antenna, it reaches the T and can go either to the antenna or the receiver. Some of the energy goes down the line to the receiver but encounters the T-R switch first. The switch, being a gap, breaks down or arcs over in the presence of such great energy. It then becomes effectively a short circuit as the resistance across the electrodes drops to a low value. The transmitted energy then goes to the antenna, since it presents a better load. (Its impedance is more nearly that of the transmitter than is the shorted T-R switch, and consequently it receives the most power.) When the transmitter is turned off, the arc across the T-R switch extinguishes, and the weak echo signal can travel down the line to the receiver as the signal has insufficient energy to arc the switch.

## *16 · 6 Radar receivers*

Figure 16 · 12 shows the block diagram of a superheterodyne radar receiver. This receiver is basically very similar to a broadcast receiver with some exceptions. First, the receiver operates at extremely high frequencies. It must be more accurately designed than a broadcast receiver. Unlike a

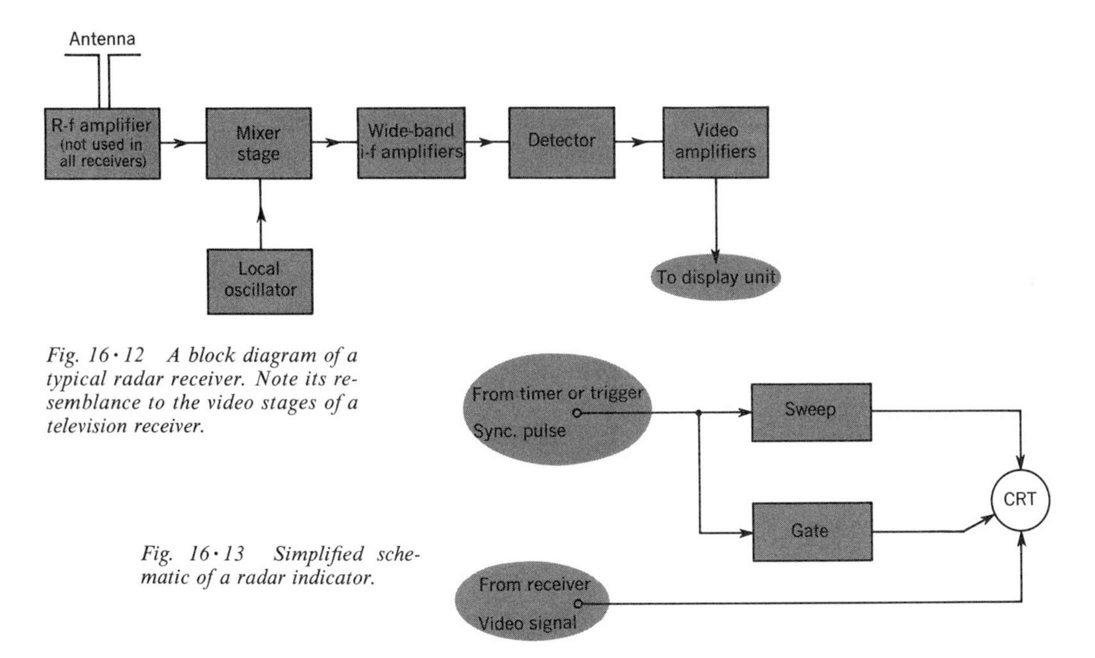


*Fig. 16·12 A block diagram of a typical radar receiver. Note its resemblance to the video stages of a television receiver.*

*Fig. 16·13 Simplified schematic of a radar indicator.*

broadcast receiver that passes only a few kilocycles to accommodate voice transmission, the radar receiver must have a bandwidth of many kilocycles or even megacycles. The information received in radar is so complex that the bandwidths of the tuned circuits are wide to pass the signal and its side bands. Where the broadcast receiver has an audio-amplifier stage that has a response up to about 10 or 15 kc, the radar receiver has a video amplifier. This is similar to the amplifier in a television set that feeds the picture tube. The output of the video amplifier is fed to the indicator or visual display device. Not all radar receivers have an r-f amplifier, since it is difficult to design a practical amplifier in the range of thousands of megacycles.

**Indicators.** Figure 16·13 is an expanded block diagram of a radar indicator. A sync pulse from the timer starts the sweep section of the indicator. The pulse may cause a dot to move across the screen, forming a single line, or a number of lines to form across the tube as in a television receiver, or a circle to develop on the tube face. The gate is used to supply a reference line or pulse on the display tube. A gate pulse is about 1 or 2 μsec wide and is generated in a blocking oscillator circuit. These circuits are capable of generating short time bursts with a fast rise-and-fall time. This gives a point of beginning or bearing to the information shown.

Figure 16·14 is the simplest type of presentation and has been previously discussed. The antenna of the set is stationary, and objects directly in front of it are shown as small pulses on the base line. The distance between the transmitted or reference pulse and the echo pulse is proportional to the distance to the object. If the speed at which the dot traverses

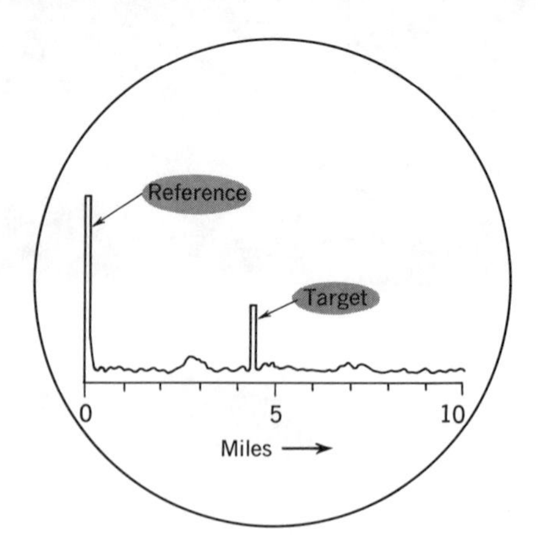


*Fig. 16 · 14 The A-scan type of presentation.*

the screen is known, the distance to the object can be computed. This distance is usually shown on a graph that covers the face of the CRT. Unless the exact position of the antenna is known and it is adjusted for maximum amplitude of the echo pulse, this system tells only the distance to the target, and very little about its altitude, direction, and speed can be determined. This type of display is called an *A-scan* presentation.

Figure 16 · 15 shows range-and-azimuth radar, or *B-scan* presentation. The range is the distance; azimuth is the direction in terms of angles. In this system, the antenna is moved back and forth through a certain angle of deflection, and the area of interest in a flat plane is shown. (In Fig. 16 · 15, the antenna moves through an angle of 75° on either side of center or dead ahead.) As the antenna swings through this angle, the beam in the CRT sweeps vertically with each pulse (see Fig. 16 · 15*a*). Hence, the position of the image measured vertically represents the range. Whether the target is left or right of the radar heading is also shown. A definite advantage of this system is that the direction in which the target is moving can be ascertained from the display.

The total image obtained, including targets, is shown in Fig. 16 · 15*b*.

Figure 16 · 16 is a *C-scan* presentation, or elevation-and-azimuth display. It is possible to tell if the target is directly ahead, high or low, left or right. This system is used primarily in aircraft to show locations of targets that are being approached. The disadvantage in this system is that it does not show the distance to the target.

A plan-position-indicator (PPI) scan is shown in Fig. 16 · 17. This system displays a map of the surrounding vicinity with the radar station located in the center. The antenna of a PPI system is rotated continuously through 360°. Zero degrees indicates the area dead ahead, 90° the area to the right, and so on. The sweep line in this display resembles a spoke in a wheel. It is placed between the center and the outer rim of the CRT and rotates about the tube, one end always remaining in the center, which is equivalent to the antenna position. This sweeping line follows the antenna heading and is in sync with it. The PPI scan is used in equipment designed for search, harbor control, convoy keeping, ground-controlled

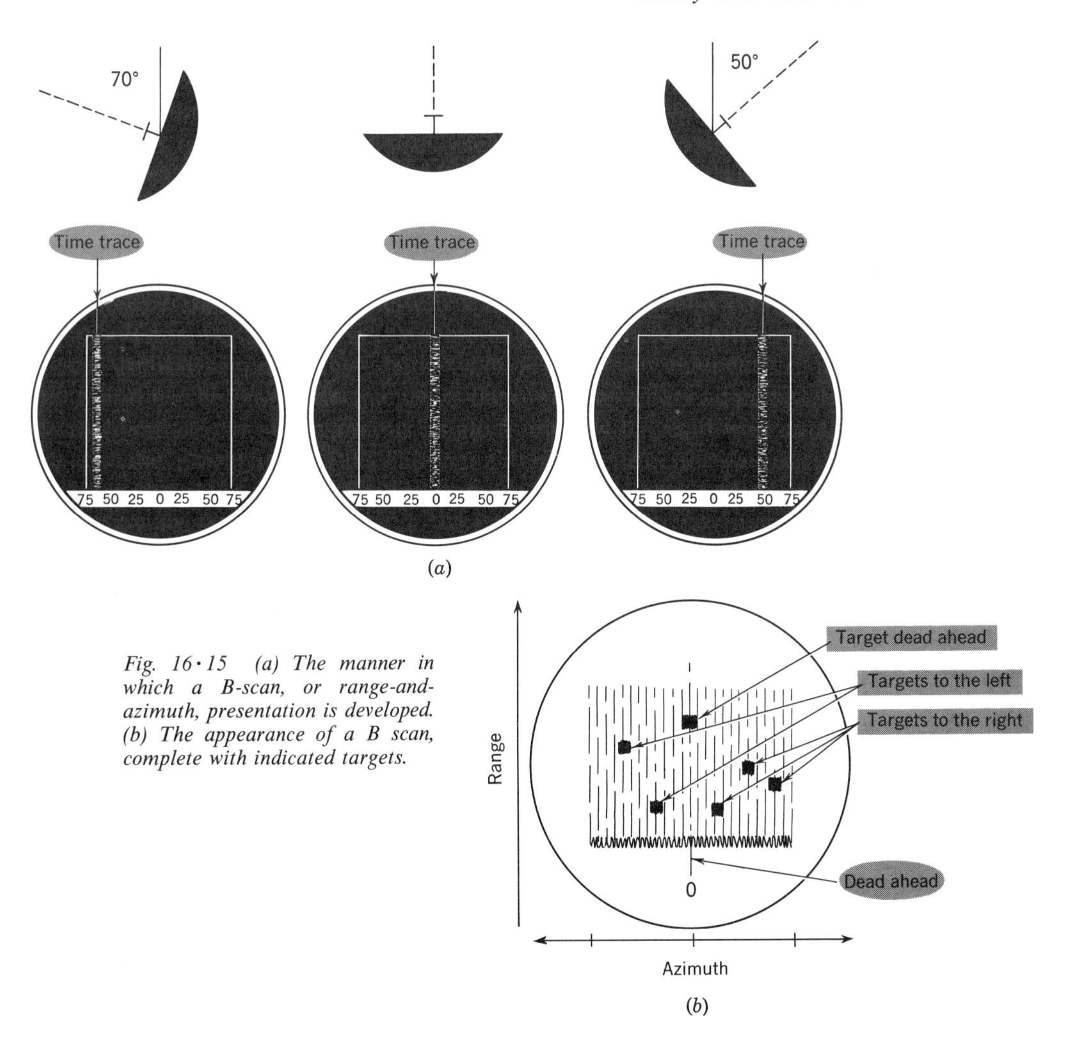


*Fig. 16·15 (a) The manner in which a B-scan, or range-and-azimuth, presentation is developed. (b) The appearance of a B scan, complete with indicated targets.*

interception (GCI), and navigation. Range and direction may be determined with this system.

**Timers, or synchronizers.** The timer, or synchronizer, provides the base or starting pulse in a radar system. This is the pulse that turns on the transmitter, references the indicator, initiates the sweep on the display tube, and in some cases controls antenna rotation. A block diagram of a timer is shown in Fig 16 · 18. In this diagram, a master oscillator, operating at 800 cps, generates a sine wave. A limiter circuit clips the peaks of the sine wave, and an overdriven amplifier changes its shape to a square wave. The square wave is differentiated, and negative-going pulses, 2 $\mu$sec wide, are formed. A differentiating circuit has a short *RC* time constant, to produce voltage peaks in the output across *R*. The pulses are fed to a cathode follower whose output feeds the 800-cps sync pulses to the transmitter and indicator.

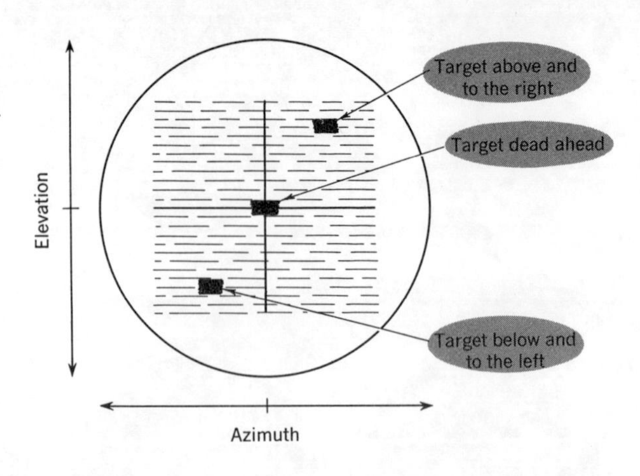


*Fig. 16·16 C-scan, or azimuth-and-elevation, presentation.*

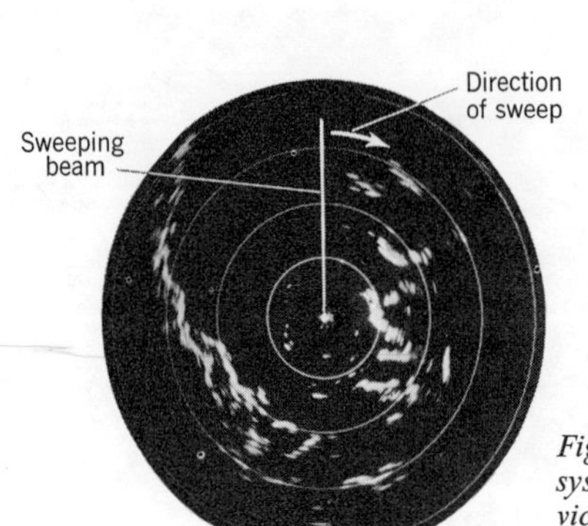


*Fig. 16·17 A PPI type of presentation. The system displays a map of the surrounding vicinity with the radar station located in the center.*

## *16·7 Radar altimeters*

An important item of electronic equipment on airplanes is the electronic, or radar, altimeter. This unit is a simplified radar system that measures the time required to bounce a signal from the ground back to the plane. The time is then used to give an indication of real altitude. Unlike the barometric-pressure altimeters, the radar type is not dependent on weather conditions and general changes in atmospheric pressure. Operating frequencies of altimeters are usually in the range of hundreds of megacycles. Antennas in use on such equipment are simple, and many times dipoles are used. Small standard vacuum tubes work efficiently at these frequencies, and equipment is usually compact and straightforward.

## *16·8 IFF equipment*

To afford protection to friendly aircraft during wartime, IFF (Identification, Friend or Foe) equipment is used in conjunction with radar and radar gun-aiming systems. Through the use of IFF equipment, craft can be identified as friendly without resorting to radio communications or visible insignia. In the IFF base or ground station, a transmitter emits a signal on a particular frequency that is received by a small set on the plane being questioned. The receiver on the plane in turn starts an accompanying transmitter that sends a coded signal back to the base station. If the base-station receiver is set to the particular coded pulse chain that is returned, it automatically signals gun operators that the plane is friendly. If no signal is returned or the code is incorrect, the plane is assumed to be unfriendly. The code can be changed periodically in the event the equipment falls into enemy hands. While this identification system could fail, a positive identification would almost surely indicate the plane was not one of the enemy's.

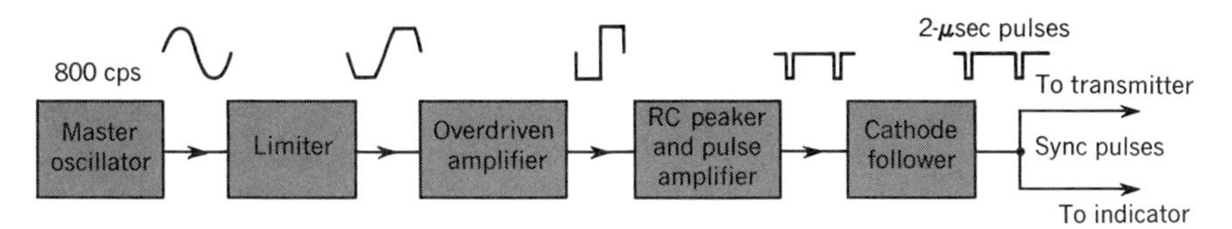


*Fig. 16·18 Block diagram of a radar timer.*

## 16·9 Navigational radar

All ships of appreciable size are now equipped with navigational radar. While this equipment is not always necessary in open sea, it is almost required when entering harbors, traveling in convoys, among floating ice blocks, and in darkness or fog. Aircraft are also beginning to use navigational or flight radar to prevent midair collisions. Ship navigational radar usually has only horizontal sweep and B scan. However, PPI presentations can be used to give all necessary information. C scan is generally used in aircraft.

## 16·10 Sonar

The word *sonar* is a contraction of the phrase "sound navigation and ranging." Sonar might well be called "underwater radar," since it is used by an ocean-going vessel to detect the presence of another craft either below or above the surface. Where the radar system uses r-f waves, the sonar system uses sound waves. Sound waves travel easily through water, while r-f waves are almost completely absorbed and rendered useless. Sound waves travel at a rate of 5,000 ft per sec in water compared with a speed of 186,000 miles per sec for radio waves in air. Two basic sonar systems are employed: the passive system uses only a receiver; the active system uses a transmitter and receiver. The type of craft and ultimate use determine if an active or passive system is to be used.

The passive sonar system is most frequently found on submarines. In this system, a sensitive microphone or group of microphones is placed in a chamber on the outside of the submarine. These mikes pick up the noise created by propellers and water turbulence of other craft. The noises are amplified and visually displayed on a CRT or are aurally reproduced by earphones. The microphone assembly is usually rotatable so that a bearing may be taken on the other craft by observing the direction of maximum signal.

As surface ships create a high volume of noise themselves, their use of a passive sonar system is limited. Submarines operate more quietly and therefore are adaptable to passive sonar. This system is advantageous where it is desirable to avoid detection by other craft. Active sonar generates a signal that may be picked up by enemy ships. It is therefore possible to use the sonar wave being emitted as a homing signal for missiles, shells, or torpedoes. By use of this signal to direct fire, deadly accuracy can be obtained.

The active sonar system is commonly found on surface ships where noise conditions make a passive system impractical. Active sonar consists of a timer, modulator, transmitter, transmitting and receiving transducer (audio equivalent of an antenna), receiver, and an indicator or display unit. These parts are connected to form a system, as shown in Fig. 16·19. As in radar, the timer produces the reference pulse and triggers the transmitter through the modulator. The transmitter then emits short bursts of energy in the range of 15 to 35 kc. This, of course, is above the limit of audibility of human ears and is frequently referred to as the *ultrasonic* range.

The transducer changes the electrical waves into sound waves. These waves leave the transducer in a narrow beam and travel through water at 5,000 ft per sec. If they strike an object in the water, some sound is reflected back to the transducer. This unit converts the audio waves back to electrical currents. A very-low-frequency receiver tunes, amplifies, and detects these signals, and feeds them to an indicator. The time for the complete trip is measured by comparing the original pulse with the echo, and distance is computed from this information.

**Sonar transmitters.** The simplest sonar transmitter is an audio oscillator capable of operating in the range of 15 to 35 kc. Most transmitters are slightly more extensive, but they are still much simpler than those used in television and broadcast stations.

A popular frequency for sonar operations is 24 kc. A transmitter for generating this frequency is shown in Fig. 16·20. This transmitter is somewhat unique in that the variable-frequency oscillator that determines the output frequency of this unit is also used as the local oscillator of the superheterodyne receiver in this system. The system operates as follows. The oscillator stage $V_1$ produces a stable 150-kc frequency which is amplified by $V_2$. $V_3$ is a tunable oscillator stage covering the range of 160 to 180 kc (at present set at 174 kc), while $V_4$ amplifies the output of $V_3$. $V_5$ is a mixer stage that combines the 174-kc signal from $V_4$ and the 150-kc frequency signal from $V_2$ to produce a difference frequency of 24 kc. Tubes $V_6$, $V_7$, $V_8$, and $V_9$ are push-pull power amplifiers that increase the 24-kc signal power to 150 watts, or more. Tuned transformer $T_1$ matches the amplifier impedance to the transducer (similar to the output transformer in a broadcast receiver). Relay $RL_1$ is the modulator that places the transmitter in the ON condition for the necessary millisecond periods.

*Fig. 16·19 A block diagram of an active sonar system.*

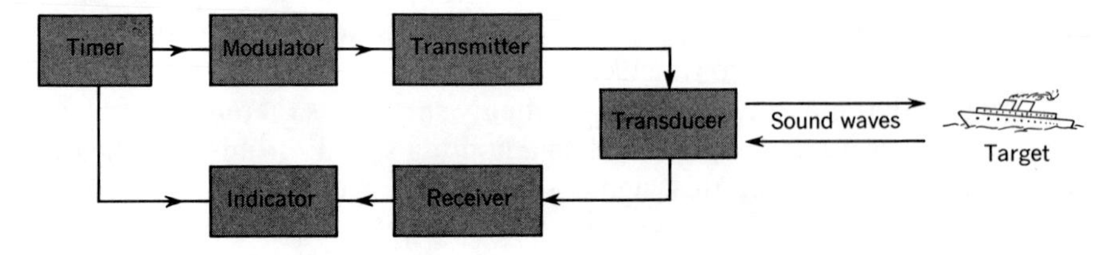

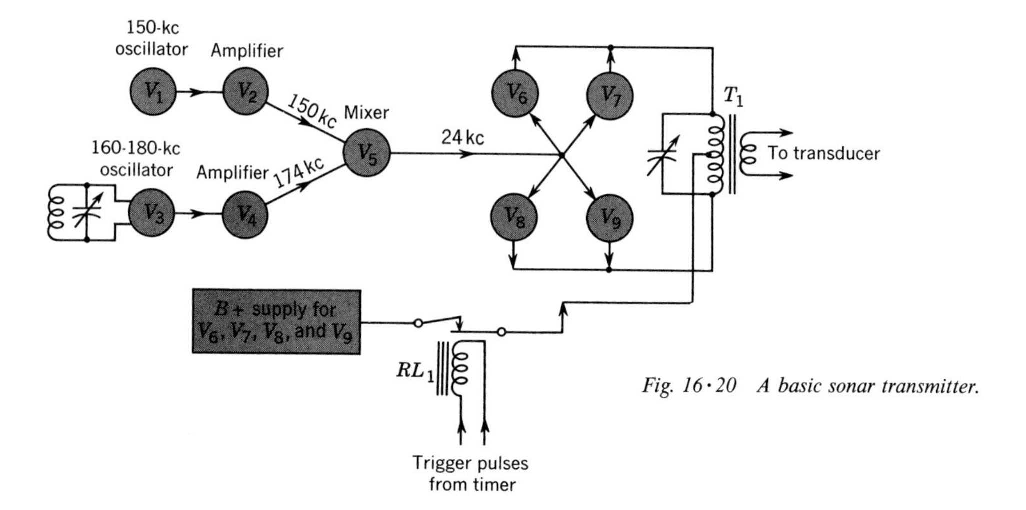


Fig. 16·20 A basic sonar transmitter.

**Sonar receivers.** Either t-r-f or superheterodyne receivers may be used in sonar. At these frequencies, a t-r-f receiver can be constructed with selectivity as good as a superheterodyne. A basic block diagram of the companion receiver to the transmitter of Fig. 16·20 is shown in Fig. 16·21. The echo signal received from the transducer is amplified in the r-f stage $V_{10}$ and sent to the mixer tube $V_{11}$. Here the 24-kc signal is heat against the 174-kc signal produced by oscillator $V_3$ (the same tunable oscillator used in the transmitter). The two signals produce a 150-kc intermediate frequency that is amplified by $V_{12}$ and $V_{13}$. $V_{14}$ is another beat-frequency oscillator (BFO) that is adjusted to be a few kilocycles higher or lower than 150 kc. This BFO produces an audible beat note, which is within the range of human hearing. $V_{15}$ is a detector, and $V_{16}$ and $V_{17}$ are audio amplifiers. These amplifiers feed speakers or phones and the indicator unit.

An interesting part of this receiver is the variable-oscillator circuit, common to both the transmitter and receiver. Should it be desirable to raise the operating frequency of the system to 26 kc (possibly because of interference from another sonar in the vicinity), the oscillator is returned to 176 kc. This produces a 26-kc beat note in the transmitter. The received echo now has a frequency of 26 kc also. The local oscillator in the receiver,

Fig. 16·21 A superheterodyne sonar receiver.

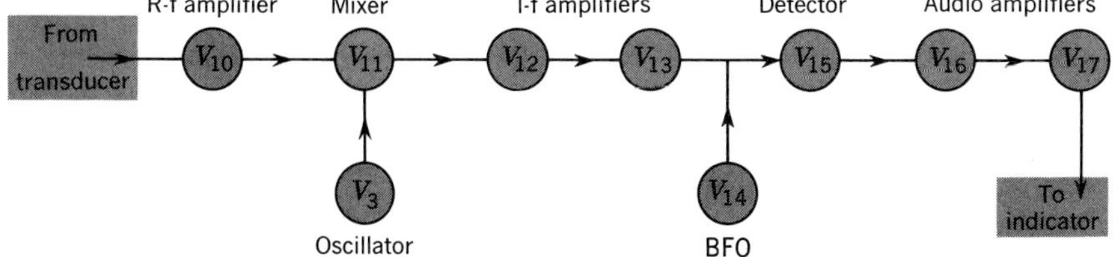

however, has been changed to 176 kc, and this still produces the 150-kc intermediate frequency (176 − 26 = 150). Therefore the transmitter and receiver track automatically.

**Sonar transducers.** The transducer in a sonar system is analogous to the antenna in a radar system. More simply, the transducer acts as a loudspeaker during transmission and a microphone during reception. Two basic types of transducers, magnetostrictive and piezoelectric, are used in sonar gear.

The piezoelectric effect is commonly used with crystal phonograph pickups, microphones, and earphones. Rochelle salts are common piezoelectric substances that generate small electric voltages when sound waves strike them. Conversely, when small electric voltages are applied to these crystals, they vibrate mechanically in accordance with the frequency and magnitude of the voltage. In a sonar transducer, several hundred of these crystal blocks are connected in a series-parallel arrangement to form the complete transducer.

Barium-titanate transducers are frequently used in sonar systems, especially the fathometer or depth-measuring types. These are essentially the same as ceramic phonograph cartridges. When barium-titanate crystals are deformed by the pressure set up in water by sound waves, they produce small voltages proportional to the pressure and consequently the magnitude of the sound. The crystals can be sealed in a pickup head to protect them from water and other corrosive chemicals. If greater output is desired, the crystals can be stacked in parallel.

Magnetostrictive transducers operate on the principle of shrinkage of certain metals when they are placed in a magnetic field. If nickel is placed in a strong magnetic field, the small atomic particles in the metal align themselves in an orderly manner and the material shrinks. This may be likened to a box of dominoes. If the small wood blocks are haphazardly tossed into a box made for them, they usually do not fit the allotted space, but, when they are neatly stacked in the box, end to end, they will fit. Likewise, if the magnetic field around the nickel is removed, the atoms will return to their haphazard relationship and take up more space.

If nickel is formed into small tubes and connected by a diaphragm, changing magnetic fields produced in the nickel by current fluctuations in a coil wrapped around it will change the size of the nickel. The nickel will in turn change the position of the diaphragm, and it will vibrate in accordance with the current.

It is interesting to note that the frequency produced by a magnetostrictive transducer is twice the driving-current frequency. The magnetic field that produces vibration is present whether the current is positive or negative. It is not present when the current is zero. Suppose, therefore, that a sine wave is placed across the transducer coils. At first it is zero, and no field is present. The metal is expanded, and the diaphragm is pushed away from the nickel rod. As the current increases, the field increases, and the metal shrinks. The diaphragm is now pulled toward the rod. When the current reaches zero again, the field collapses, the metal expands, and the diaphragm is again forced away. Now the current begins its negative half-cycle and the process repeats. For every cycle of supplied power, two contractions and expansions occur, and the frequency is doubled.

One disadvantage of this type of transducer is its narrow resonant frequency that limits its operating range. In a passive system, magnetostrictive transducers can pick up only a very narrow range of the sound and are useless. In active systems, if the transducer is resonant to the system, it works correctly.

**Sonar indicators.** Various indicators are used in sonar to reveal the presence of the enemy, the direction they are traveling, and their range or distance. For presence, a loudspeaker or earphones are used to produce the audible beep tone of the echo. Rotating transducers or a pair of transducers can be used to tell the direction the object is traveling, or a CRT can display the sonar signal to indicate presence, direction, and distance.

A novel distance indicator is the motor-driven type using a neon bulb (Fig. 16·22). In this indicator, a small synchronous motor drives a disk with a neon bulb attached to its rim. As the motor is synchronous, the exact time taken to rotate the disk once (directly or through a gear reduction) is known. Suppose that in Fig. 16·22 the time for the disk to make one revolution is 4 sec. Every time the neon light passes zero, contacts are closed by the disk, thus initiating a sonar pulse. The pulse is transmitted, the echo returns and is received, and the pulse is applied through slip rings to the neon light. When this pulse arrives at the light, it flashes. If the target is 5,000 ft away, the pulse will take 2 sec to go and return. In this time, the disk has moved one-half revolution. The neon light therefore flashes when it is at the top of the scale. This point could be calibrated "5,000 ft." A quarter of a revolution would be 2,500 ft, three-quarters would be 7,500 ft, etc. As the disk is rotating and the light is always flashing at the same spot, it appears to stand still, indicating the target distance.

## 16·11 The fathometer

As radar is used in aircraft altimeters for navigational purposes to indicate altitude, sonar is used in a fathometer to indicate the depth or

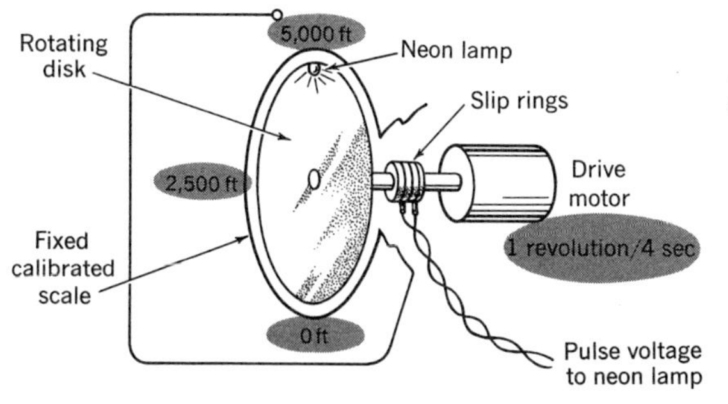


*Fig. 16·22 Sonar indicator using a synchronous motor drive.*

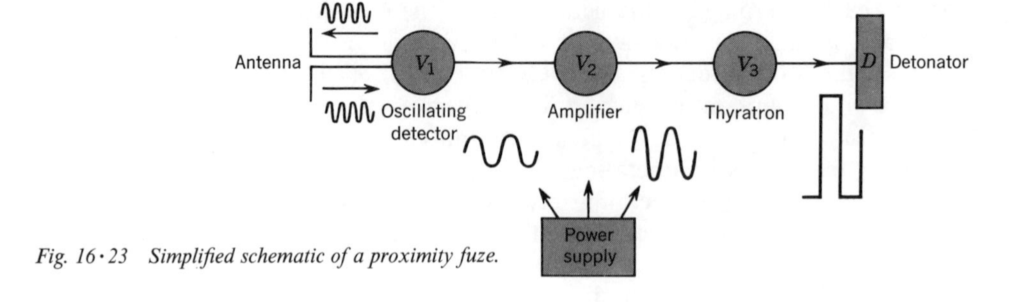


*Fig. 16·23 Simplified schematic of a proximity fuze.*

distance to the bottom of the ocean. That is, the sonar pulses are transmitted down. The system is relatively simple and usually synchronously driven neon lamps are used. Since good accuracy can be obtained, navigation and mapping can be carried out with such a device. Fathometers are produced for military and commercial ships.

## *16·12 The radio proximity fuze*

The proximity fuze is a simple miniaturized radar system that may be mounted on the nose of a rocket, projectile, or bomb. This unit detonates the charge when a target is approached. The fuze normally uses three vacuum tubes to perform all functions, and the transmitter and receiver have one stage in common. An oscillating detector in the receiver serves as the transmitter for the system. This stage generates a signal that is sent to the antenna, radiated to the target, reflected back to the antenna, and detected in this stage. As the strength of the echo increases, the output of the oscillating detector increases. An amplifier stage boosts this output, and it is then used to trigger or turn on a thyratron.

When the thyratron conducts, indicating proximity to a target or reflecting object, its plate current increases. This plate current is applied to a current-actuated detonator that triggers the explosive charge.

The basic block diagram of a proximity fuze is shown in Fig. 16·23. $V_1$ is the oscillating detector stage that generates the original signal and detects the echo. The waveforms are shown on the schematic. As the echo increases in amplitude, it adds to the signal already present in the detector. When this signal reaches a sufficient amount, it increases the output of the amplifier $V_2$ to the point where it turns on thyratron $V_3$. The plate current of $V_3$, flowing through the detonator $D$, heats the triggering element and fires the explosive material. This explosion in turn sets off the main charge.

The power supply is the source of filament and plate power for the system. Proximity-fuze power supplies are somewhat unique in use and operation. The fuze must be constructed so that it will work reliably after long months of storage. Also, for safety purposes, the fuze should not operate until after it is some distance from the gun, aircraft, or mortar. As batteries tend to deteriorate after extended periods of storage, they would be impractical in such a device. The solution to these problems is a

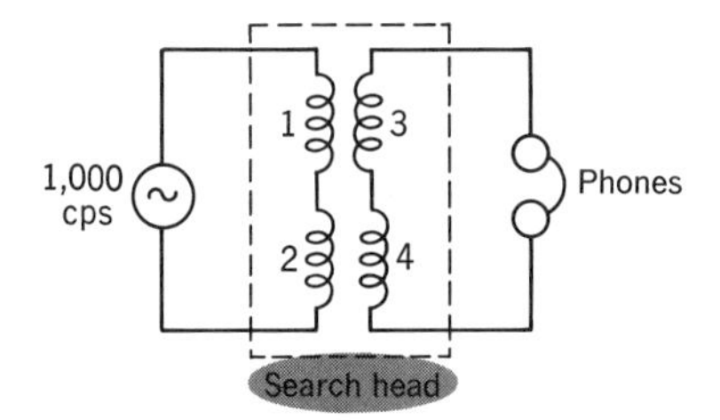


*Fig. 16·24 A Hughes balance circuit.*

propeller-powered a-c generator mounted in the fuze capsule. The blades rotate when the projectile moves through the air, and as they turn, they drive a shaft that works through a gear train to move an interrupter plate between the detonator and the charge. In the SAFE position, this plate prevents the charge from being detonated. The projectile, therefore, must move a certain distance through the air, and the blades must make a certain number of revolutions, before the charge can be detonated.

A small a-c generator is also driven by the propeller. This generator has two windings, one producing filament voltage, the other producing alternating current that is rectified for the B+ or plate voltage.

As the proximity fuze receives a large initial shock when the shell is fired from a gun, the components and assembly must be extremely rugged. This was one of the first uses of printed circuits and miniature packaged systems. The greatest deterrent to building such a device was the difficulty of combining small size with ruggedness and reliability. This difficulty was overcome by using miniature hearing-aid tubes and plotting the entire assembly.

## *16·13 Mine detectors*

Another electronic device used by the armed services is the mine detector, or buried-metal indicator. This is very similar to "buried-treasure" or ore detectors used by treasure hunters and prospectors. A popular model used by the army operates on the principle of a Hughes balance circuit, as shown in Fig. 16·24. In this circuit, the audio generator produces a 1,000-cycle signal that is impressed on coils 1 and 2. Coils 3 and 4 are constructed so that 3 is in phase with 1 and 4 is out of phase with 2. They are positioned so that the signal produced in coil 4 by coil 2 cancels the signal produced in coil 3 by coil 1.

The four coils form the search head, and this head is mounted on the end of a long handle and passed back and forth over the surface of the ground where the mine is believed to be. Should the head pass over buried metal, the coupling between the coils will change and the canceling signals will no longer cancel. When this occurs, the previously silent phones will emit a 1,000-cycle note. The larger the area of the buried metal, or the closer the head is to the metal, the louder the signal in the phones will be. An a-c voltmeter can be used in conjunction with the phones or to replace them if a visual indication is desired.

To increase the sensitivity of the mine detector, amplifiers are used between the pickup coils and the earphones. A three-tube model is shown in block-diagram form in Fig. 16·25. The system is self-contained with the exception of the pickup head. Batteries are used to supply the operating power. $V_1$ is a 1A6-GT battery-operated twin triode connected in a push-pull audio-oscillator circuit. It furnishes the 1,000-cycle signal to the search head coils. Two 1N5-GT tubes are used in the tuned audio amplifier. The pickup coil in the search head supplies the signal to $V_2$, where it is amplified and fed to $V_3$. A resonant circuit is used in the interstage coupling to increase the gain at 1,000 cycles. The output of $V_3$ feeds an indicating meter and earphones. All circuitry used in this system is very similar to that found in commercial radio receivers and test equipment.

## 16·14 Fire-control equipment

A fire-control system accurately aims and fires guns by electronic and mechanical units that are not subject to human error. Information about target speed, distance, heading, and direction are fed to a computer that predicts where the target and shell should meet. The information leaving the computer is then fed to gun-aiming motors that align the barrel so that the shells being fired will intercept the target at the predicted spot. When the gun is mounted in or on a moving vehicle (ship or aircraft), information about the direction of movement, speed, pitch, or roll of the vehicle is also fed to the computer to correct the aiming.

Some fire-control systems are so accurate that gun-operating personnel are no longer needed. This is especially true in some jet bombers. The copilot can operate the electronic system of the fire control and can shoot at craft attacking the rear of the plane that he cannot see.

Figure 16·26 shows the basic units of a fire-control system in use aboard a ship. Four different basic units comprise this system. The director locates the distance to the target. Target speed, direction, and heading are determined through an optical or radar system in the director. This information is fed to the computer where the proper aiming of the gun is determined through a mechanical or electronic system, or a combination of both. The stable element detects pitch or roll of the ship and sends com-

*Fig. 16·25 Block diagram of an army mine detector.*

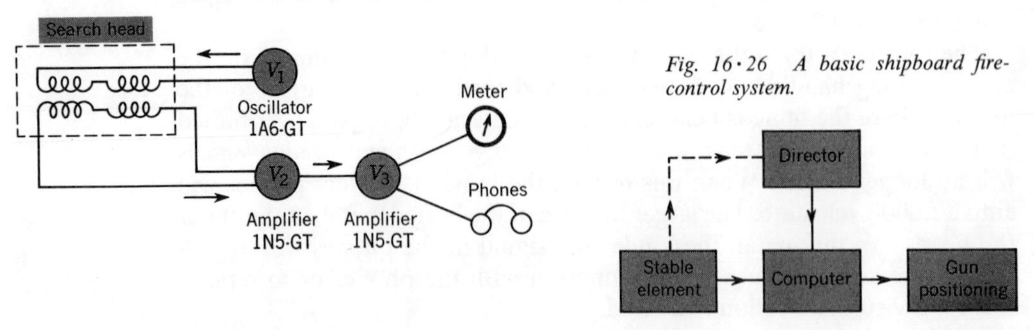


*Fig. 16·26 A basic shipboard fire-control system.*

pensating signals to either the director or computer. These signals are eventually used to keep the gun constantly pointed toward the same place regardless of rolling or other ship movement.

The electric signals from the computer are then fed to the aiming motors, and the gun is moved horizontally and vertically to the correct position. The electronic and electric parts of such a system are quite complex. Mechanical arrangements are used in each of these four units to perform many of the operations. The radar unit has been discussed previously. Mechanical computers are beyond the scope of this text as are the other mechanical sections of this system, but a number of servo systems that are primarily electronic are used in fire control, and these are of interest.

## *16·15 Servo systems*

While servomechanisms find a number of uses in fire-control systems, they are also useful for turning antennas, positioning a ship's rudder, and setting the flight of aircraft. A servo allows a small amount of torque or pressure, such as the turning of a knob, to control a considerable amount of energy at a remote or near place. The turning of a large gun on shipboard, for example, would be quite difficult if it had to be done manually. It would be even more difficult to turn the gun from a position many feet away. A motor, of course, can be used to turn the heavy gun through a gear reduction. One disadvantage of using only a motor to turn such a weight is that it cannot always be stopped at the exact desired place. It will almost invariably go past the spot where it should have stopped because of the momentum gained in turning. It is even possible that the gun will turn from the desired spot after it is reached because of pitching of the ship in a heavy sea. Figure 16·27 shows a servo system that will solve these problems.

In the schematic, two potentiometers of equal resistance have their fixed terminals connected in parallel. One of these, $P_1$, is the input-control potentiometer. The other, $P_2$, is the output monitoring or error-indicating potentiometer. The shaft of $P_2$ is connected directly to the shaft that rotates the gun. A high-power vacuum-tube amplifier is used to furnish operating

*Fig. 16·27 A gun-control servo system.*

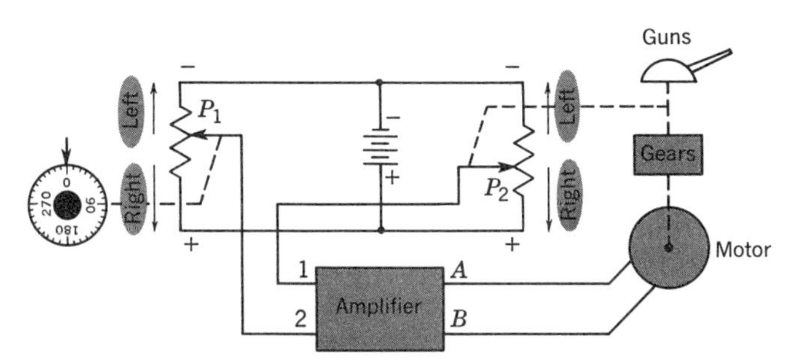

voltage to the motor $M$. The motor is connected to a gearbox that turns the gun.

In practice, two systems would be used to control the azimuth and elevation of the gun, but for the sake of simplicity only one is shown here. The input-control potentiometer $P_1$ may be turned by hand or by the fire-control computer. A scale on this potentiometer is calibrated in degrees. The battery supplies a reference voltage for the system and need not be an accurate voltage source.

If the wipers on both potentiometers are set at exactly the same degree of rotation, no voltage will be present across terminals 1 and 2 of the amplifier (the input terminals). If the gun was set to aim dead ahead, both potentiometers would be centered and the scale on $P_1$ would read zero degrees. Suppose that the computer took all available information about a certain target and found that to score a hit the gun should be pointed 10° to the right. If the computer was directly connected to the servo, it would turn $P_1$ to the right (down) to the scale marking 10°. If the computer was not mechanically linked to the servo, the gun operator would turn the dial right to 10°.

Since $P_1$ has been moved so that a more positive voltage appears at its rotor, while $P_2$ has remained stationary, the input voltage to the amplifier becomes more positive at terminal 2 than at terminal 1. If a d-c amplifier is used, it will increase the magnitude of this voltage and produce a potential across terminals $A$ and $B$ powerful enough to drive the motor. $B$ will then be positive with respect to $A$. The motor will now rotate, turning the gearbox, which will turn the gun in the required direction. As the gun rotates, however, $P_2$ will also turn in the same direction that $P_1$ was initially rotated. When the gun reaches the desired position, $P_2$ will have reached the same position as $P_1$, and the voltage developed by these potentiometers will be zero. Now, there is no longer a signal across terminals 1 and 2 of the amplifier. There is also no output voltage at terminals $A$ and $B$ of the amplifier because there is no input voltage. With no drive voltage, the motor stops, and the gun comes to rest at the desired spot.

Even with a servo system, the problem of "overshoot" (turning past the desired stopping point) exists. The gun is a heavy object that tends to keep moving once it is started. It is almost certain that after the motor is turned off, the gun will move a fraction of a degree past the desired stop point. This is no problem, however, since the servo recorrects this error. Should the gun go a few tenths of a degree past the 10° desired deflection, $P_2$ will also go past this point. This means that the rotor of $P_2$ is now at a position slightly more positive than the rotor of $P_1$. Therefore a positive voltage exists at terminal 1 of the amplifier (with respect to terminal 2). The amplifier increases this small positive signal, and output terminal $A$ becomes highly positive with respect to $B$. The polarity across the motor is reversed from what it was previously, and the motor reverses direction. It moves the gun and $P_2$ the few tenths of a degree to the left as required. When proper position is reached, the rotors on both potentiometers balance the voltage, and the system stops as before.

*Fig. 16·28 An infrared photograph of an airplane. Taken in mid-afternoon of a bright, hot day, the sun-heated ground is the hottest part of the picture. Under side of plane is hot because of reflection from ground. Upper surface reflects cold sky. The plane landed only a few minutes before, but turboprop-engine nacelles are cool.*

Depending on the constants of the system, this correcting may take place a number of times until the gun is set accurately. Each correction will be slightly smaller than before. This correcting process is called *hunting*. The voltage produced by $P_2$ when it is not in the same position as $P_1$ is called the *error signal.* Sometimes a friction brake is connected on the gun rotating shaft to make it stop soon after the motor stops. This brake is called a *damper,* and the action of braking is called *damping.*

One interesting thing about this type of control is that the more degrees of rotation required, the greater will be the potentiometer displacement, and the larger the voltage fed to the amplifier's input. If the input is larger, the output signal is larger, and the motor moves more rapidly to turn the gun. This is desirable in a gun-aiming system. It is possible to use a-c motors and transformer-coupled amplifier stages in a servomechanism. In an a-c system, some type of phase detection or other means to reverse motor direction is required. The servo amplifier uses high-power tubes to feed the motor and must have a wattage output equal to or larger than the required motor power.

## *16·16 Infrared systems*

All objects whose temperatures are higher than absolute zero (−273°C) radiate infrared waves. The lowest frequency in the infrared spectrum is about one million megacycles and the highest is in the vicinity of a few hundred million megacycles or nearly visible light. As the temperature of the object increases, the amount of infrared radiation increases. When these invisible waves strike such materials as lead sulfide or lead selenide, the material produces small voltages. These voltages are proportional to heat and can be amplified in low-frequency amplifiers to indicate the temperature of the object emanating the radiation. By means of a scanning disk to break the radiation into lines or by an image-converter tube, a thermal photograph of the object can be obtained. Figure 16·28 is such a photograph, taken on a hot day. The sun-heated ground is the hottest part of

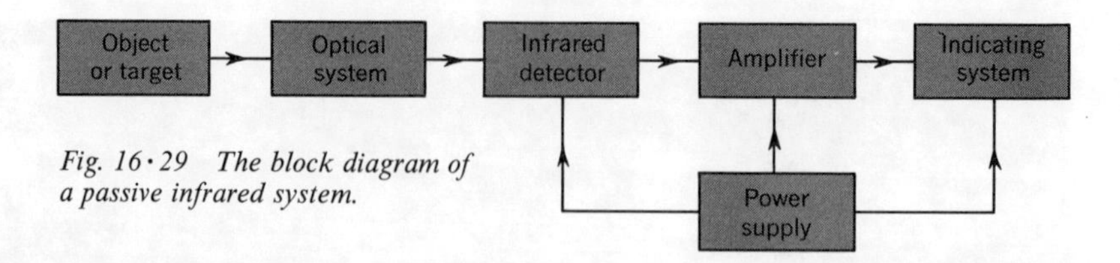


*Fig. 16·29 The block diagram of a passive infrared system.*

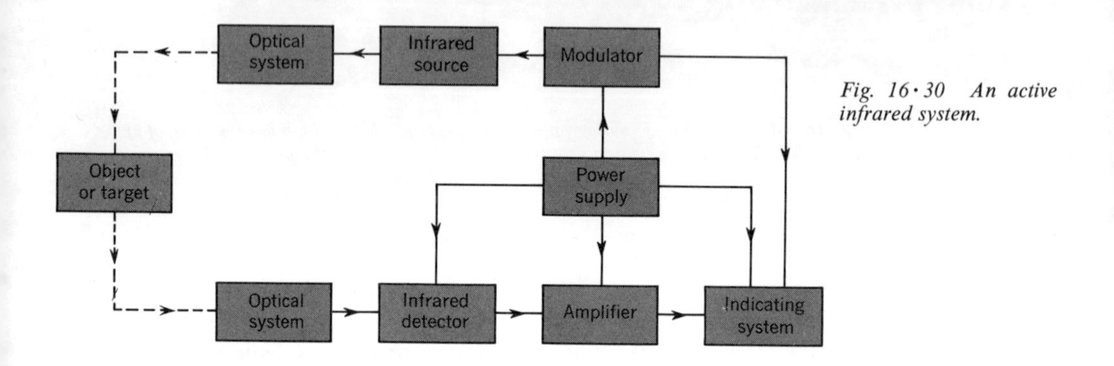


*Fig. 16·30 An active infrared system.*

the picture; the under side of the craft is hot from the reflection on the ground; while the upper surfaces reflect the cool sky.

Two types of infrared systems are in use. The most common of these, the passive system, utilizes only the radiation emanating from the object to supply the information necessary for a thermal picture. The more complicated one, the active system, irradiates the object of interest with an auxiliary source of infrared. One disadvantage of the active system is that it can betray the observer's location to the enemy. Active systems are, however, quite effective for point-to-point communications in total darkness. A radiator is modulated with speech or other information, and a detector at some other point picks up the signal, detects it, and restores the information to its original intelligence. This sort of communication is, of course, limited to line-of-sight.

Figure 16·29 shows the block diagram of a passive infrared system. The object or target radiates infrared waves through the atmosphere; these are picked up by the optical system. The optical system consists of a focusing mirror and special lens that concentrate the signal. A scanning device, usually a disk similar to those used in early television flying-spot scanners, breaks the radiation from the object into lines. The signal is now similar to a video signal used in television. This signal is amplified and then presented on a display or indicating system (possibly a CRT).

Passive infrared systems can be made extremely sensitive. By means of such a system to guide a missile, perfect homing on hot objects, such as aircraft, steel mills, industrial areas, or vehicles, can be obtained. The *sidewinder missile* uses infrared detection and is so accurate it can knock a flare from the wing of a drone plane. Infrared shows great possibilities as a replacement for radar. It is more difficult to jam and in some cases has greater definition.

Figure 16·30 is the block diagram of an active infrared system. An infrared source, optical system, and switching circuit have been added to a passive infrared circuit to convert it into an active device. Some active systems use optics common to both transmitting and receiving to simplify the equipment. The switching circuit turns the source on for short periods and then switches on the detector while the source is turned off. It can also be used to code the source where the system is being used for communications.

## *16·17 Guided missiles*

Guidance systems used in missiles encompass almost every phase of electronics. Servos, radar, infrared, and telemetering are a few of the important parts of the electronics used in a guided missile. The complexity of these circuits is in many cases the cause for failure of these guided space vehicles. Initial shock of firing is very great and casualties suffered by the components in the circuitry can affect the whole guidance system. To test and fire a missile properly, ground observers must have more than radar and visual sightings to inform them of in-flight progress.

A reliable telemetering system is almost always a necessity. A telemetering system is an answering device that supplies information to the engineers on the ground when it receives questions from them. The more complex systems consist of two transmitters and two receivers, with coding and decoding units.

The basic principle of a missile telemetering system is indicated in Fig. 16·31. A ground-based transmitter is controlled by observers to query the

*Fig. 16·31 A basic missile telemetering system.*

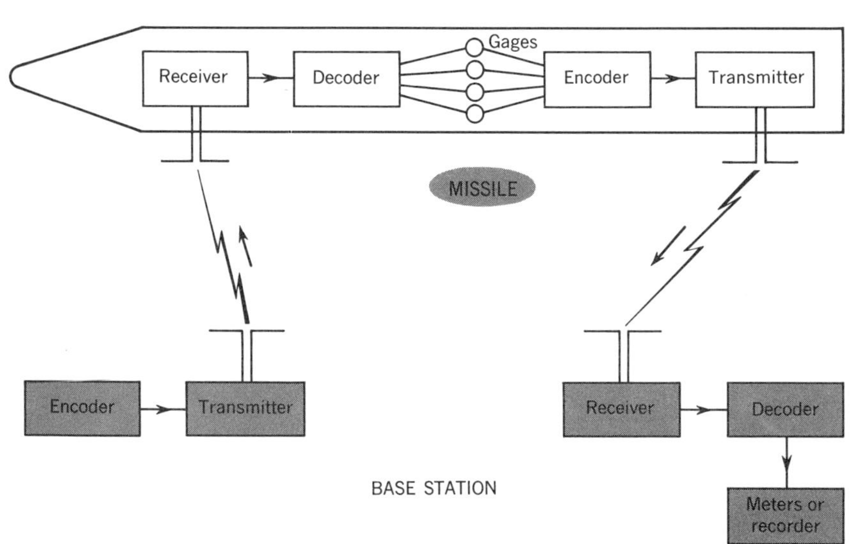

missile about certain operating conditions. These may include residual fuel, speed, direction, and air resistance. This transmitter sends out certain numbers of pulses or tones for certain pieces of desired information. A simple example would be air resistance.

Suppose a 400-cycle tone is used as the special signal for information on this quantity. The basic telemetering r-f signal will be modulated with this tone. A receiver on the missile will pick up the basic telemetering signal, which is possibly at 225 Mc. It will detect the signal, and the 400 cycles will be obtained. A special circuit on the missile that responds only to 400 cycles will then trip a relay that connects the gages measuring air resistance to the modulator circuit. This modulator circuit changes the values of air resistance into coded pulses, and the pulses modulate the transmitter in the missile. The transmitter radiates a coded r-f signal that is received at the observation station. The receiver at this point picks up the signal, then detects it, and feeds the pulses into a decoder. The decoder deciphers the pulsed information, and it is then fed into a printer or magnetic tape for a permanent record.

Timing pulses are impressed on all signals present, so that, at some later time, the engineers will know exactly when this measurement was taken. These timing signals are spaced over certain periods and do not interfere with or obscure the information being obtained. This system of telemetering can handle many channels of information, with different tones being employed to find out different operating conditions.

A simpler, but not quite so versatile, system of telemetering continuously transmits various pieces of information back to the observing station. From the time the missile is fired until it destroys itself, conditions on flight and internal functioning are transmitted in coded form to the ground receiver. This information is recorded and studied to determine why the missile flew correctly or failed.

If telemetered information indicates failure of one of the missile systems, a third portion of the telemetering system is sometimes used. Should it appear that the flight course has altered and a threat to nearby installations is encountered, a coded signal is sent to the missile that destroys it before damage is done.

By using the principles of radio control, the observers of a missile can determine through telemetering what measures should be taken to correct the flight, if necessary, and transmit information back to the vehicle to make these corrections. Radio control has been in use for a number of years and was once considered more of a novel than a useful device. In a radio-control system, certain pulses or tones trigger relays that are then used to actuate valves, motors, or other mechanical contrivances that change fuel injection, vane angles, or other flight-determining variables.

## SUMMARY

1. Radar determines the distance to a target, aircraft, or ship by bouncing a high-frequency radio signal off the object and measuring the echo time electronically.
2. Radar can also be used to determine direction of travel, speed, bearing, size, quantity, or other information about the target.

3. A complete radar system consists of a pulser or transmitter modulator, a transmitter, an antenna, a receiver, a display unit, and a synchronizer.
4. The types of radar display described here include A scan for distance or range, B scan for range and azimuth, C scan for azimuth and elevation, and the plan-position-indicator, or PPI, scan.
5. Sonar, operating on similar principles to radar, uses sound waves instead of radio frequencies and operates in water.
6. Passive sonar picks up sound generated by a ship or submarine to determine direction, size, speed, and other information. It has an advantage in that it cannot be detected by other sonar receivers.
7. Active sonar generates a sound wave that is bounced off the other ship or submarine, reflected back to a receiver, and then analyzed for information. This system has a disadvantage in that other sonar receivers can detect the presence of the transmitter.
8. The active sonar system consists of a synchronizer, modulator, sound generator or transmitter, transducer, receiver, and display unit.
9. The radio proximity fuze is a small radar system contained in the nose of a shell or projectile. It is capable of determining nearness of the target and can detonate the explosive charge.
10. Mine detectors usually work at audio frequencies and consist of a search head comprised of a large transformer with two or more windings, a sensitive receiving amplifier, and an audio generator. A disturbance in the vicinity of the search head caused by underground metal objects changes the coupling in the search-head transformer and consequently the tone in the receiver.
11. Fire-control systems take data obtained from radar, visual sighting, and other sources, automatically correlate this data in a computer, and then feed electric signals to gun-aiming motors to fire accurately on a moving target.
12. Infrared systems, operating like radar, determine the positions of objects by detecting the amount of heat leaving the object or the radiation in the infrared region. These systems may be either passive or active.
13. The active infrared system radiates the target or object under surveillance with these invisible waves and thus is not entirely dependent on other heat. Infrared waves are invisible, but their presence may be detected by certain crystals.
14. A missile-guidance system is able to control flight by computing target and flight information and feeding it to the missile-controlling members to steer them toward the target.
15. Radio control is the sending of pulses of information over r-f waves to control or operate a device at some distant point.

## SELF-EXAMINATION

Here's a chance to find out how well you have learned the material in this chapter. Work the exercises; then check your answers against the Key at the back of this book. These exercises are for your self-testing only.

1. The time a radar wave takes to travel one radar nautical mile is approximately (*a*) 12.4 μsec; (*b*) 25 μsec; (*c*) 32.2 μsec; (*d*) 1,100 μsec.
2. In FM radar, the variable that increases as distance to target increases is the (*a*) sweep rate; (*b*) basic frequency being transmitted; (*c*) beat produced by the echo and base frequency; (*d*) pulse repetition or triggering frequency.
3. In pulse radar, an echo time of 6.2 μsec fixes the target at (*a*) a distance of two radar nautical miles; (*b*) a distance of one-half of a radar nautical mile; (*c*) a position directly in front of the antenna; (*d*) an altitude of 5,280 feet.
4. A C-scan presentation system is useful, since it tells (*a*) distance and direction to target; (*b*) elevation and distance to target; (*c*) target speed and size; (*d*) target height and azimuth.
5. Sound, or sonar, waves travel in water at a speed of (*a*) 3,141 ft per sec; (*b*) 5,000 ft per sec; (*c*) 186,000 ft per sec; (*d*) 12.4 miles per sec.
6. Sonar-frequency operating range is usually between (*a*) 50 and 15,000 cps; (*b*) 535 and 1,605 cps; (*c*) 15,000 and 35,000 cps; (*d*) 454.5 and 455.5 kc.

7. A circuit in sonar systems that is sometimes common to both transmitter and receiver is the (*a*) audio-output stage; (*b*) oscillator; (*c*) power supply; (*d*) antenna.
8. Proximity fuzes are of interest, since they were one of the first users of (*a*) transistors; (*b*) printed and packaged circuits; (*c*) FM telemetering; (*d*) mercury batteries.
9. If an infrared source is viewed in total darkness (*a*) total blindness will result; (*b*) a violet cast is plainly visible; (*c*) nothing is visible; (*d*) the mixture of all parts of the spectrum produces white.
10. A servo system that is constantly overshooting or hunting probably requires a (*a*) potentiometer; (*b*) higher gain amplifier; (*c*) damper; (*d*) 400-cycle power source.

## QUESTIONS AND PROBLEMS

1. Draw the block diagram of a pulse-type radar system.
2. Give the function of each block in the diagram of Question 1.
3. Draw the image, as seen on the CRT, of three different types of representation in radar.
4. In a pulse radar system, the target is 4 miles away and 45° to the right. Show the target on B scan. Also compute the echo time in microseconds.
5. Draw the block diagram of a sonar transmitter, labeling all parts and showing the unit that may also be common to the sonar receiver.
6. The echoing time of a sonar signal is 3.8 sec. What is the target distance in feet? In miles?
7. The transmitted frequency in a sonar system is 20 kc. The receiver uses the same variable oscillator as the transmitter. The variable oscillator frequency is 190 kc. What is the probable intermediate frequency of the receiver?
8. Draw the block diagram of a proximity fuze, label all blocks, and explain its operation.
9. Explain hunting in a servo system.
10. Explain the operation of a telemetering system that could be used to indicate light and heat in an earth satellite.

# Chapter 17 Electronic navigational aids

Electronic controls and navigational aids are essential components of our military, commercial, business, and private aircraft. The trend is toward even wider use of electronic equipment to provide means for safely handling the rapidly increasing numbers of commercial jet aircraft and to establish methods for adequate air-traffic control. Such electronic navigational aids are described in the following topics:

- 17·1 Loop Antennas
- 17·2 Radio Direction Finders
- 17·3 Radio Ranges and Beams
- 17·4 VHF Omnidirectional Range
- 17·5 Distance-measuring Equipment
- 17·6 Tacan System
- 17·7 Radio Marker Beacons
- 17·8 Instrument Landing Systems
- 17·9 Loran

## *17·1 Loop antennas*

Most direction finders and radio compasses employed on aircraft use the loop antenna because it can be made quite small and compact. The direction finder locates the actual position of a ship or aircraft and is therefore of great importance in navigation. Before taking up the theory and operation of the direction finder, it will be useful to review the loop antenna and its characteristics.

The variation and intensity of the electromagnetic field in space, as it expands outward with the speed of light, can be represented by a sine curve, as in Fig. 17·1. The intensity of the field at any distance $X$ from the origin, or starting point 0, is represented by the length of a vertical line from the

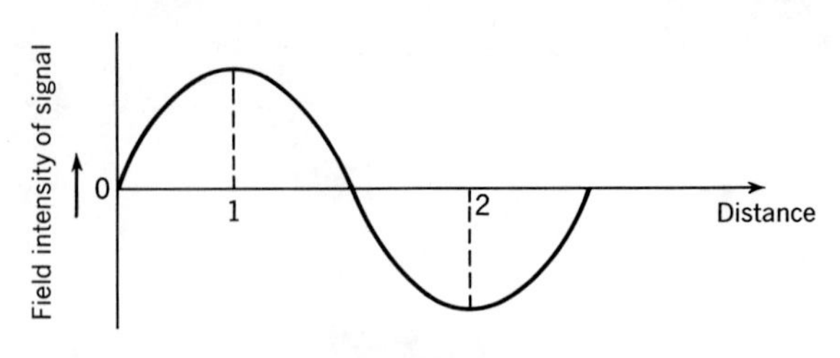


*Fig. 17·1 Sine-curve representation of an electromagnetic field in space.*

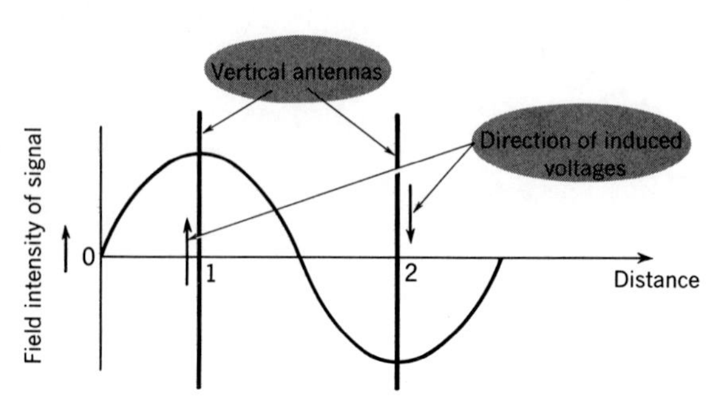


*Fig. 17·2 Antennas placed in an electromagnetic field to obtain maximum induced voltage.*

sine curve to the base line. Thus, at points 1 and 2, we have maximum fields, although their respective directions are opposite. If a vertical antenna is located at point 1 and another is located at point 2, both antennas will intercept and be cut by the maximum number of lines of force. We will therefore obtain the maximum induced voltage. Since the field polarities are exactly opposite, the induced signals will likewise be opposite, as shown in Fig. 17·2. Points 1 and 2 are one-half wavelength apart. As a general rule, any two points along a sine curve one-half wavelength apart will have equal and opposite voltage values. Furthermore, although points 1 and 2 are maximum points, this represents a condition existing only at one instant of time. At other instants, there will be other values at points 1 and 2 as the wave travels outward. But whatever these values may be, they will always be equal and opposite to each other.

Now connect the tops and bottoms of the vertical antennas, as indicated in Fig. 17·3. This forms a loop antenna. If, say, the bottom horizontal line is broken and the ends connected to a receiver input, a rather large signal can be introduced into the receiver. Since the loop and receiver input circuit constitute a closed circuit, a current will flow in the loop. The two vertical antennas are in series and aid each other in the circuit. This produces a large current flow in the loop. The field is parallel to the top and bottom horizontal loop elements; hence they will not be cut by the lines of force. Only the vertical sections are cut by the field, and only the vertical members provide an induced voltage. Note that the distance between points 1 and 2 is exactly one-half wavelength. Thus, for maximum signal, the width of a loop antenna should be one-half wavelength. At low frequencies, where the wavelength is extremely large, it is impractical to use this size loop. A practical loop is a small fraction of one-half wavelength.

Figure 17·4 shows the conditions for a loop whose dimensions are small compared with one-half wavelength. Fields of the same polarity cut both

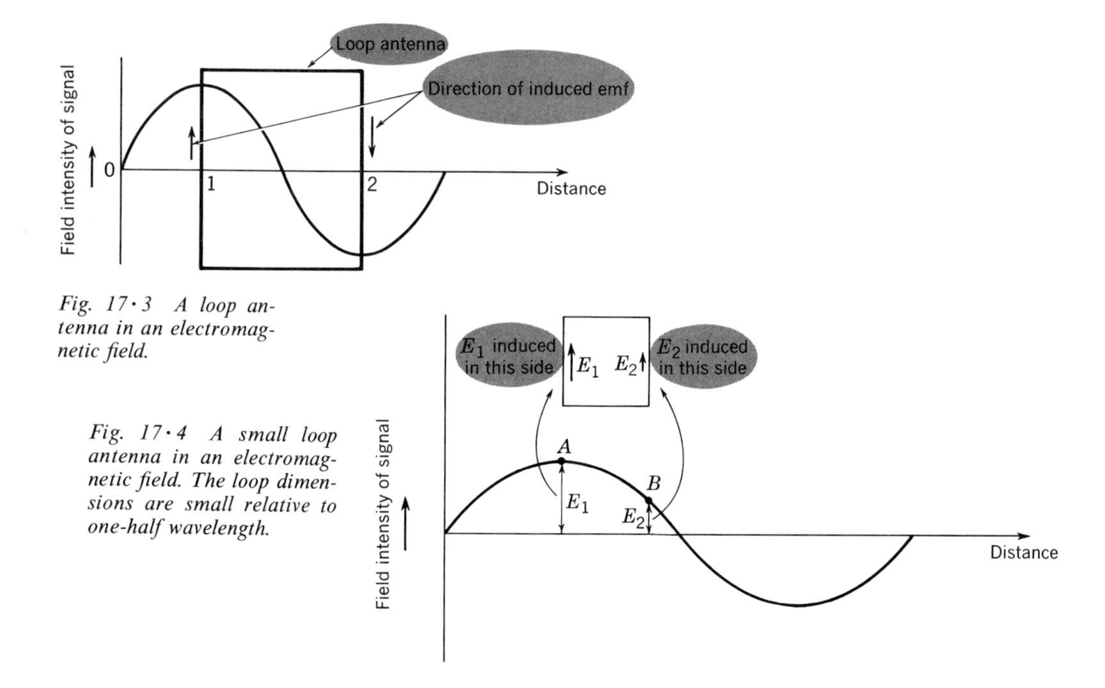


*Fig. 17·3 A loop antenna in an electromagnetic field.*

*Fig. 17·4 A small loop antenna in an electromagnetic field. The loop dimensions are small relative to one-half wavelength.*

sides of the loop. The voltages induced in the vertical sections of the loop are therefore of the same polarity, as indicated by the arrows of Fig. 17·4. The two voltages are not equal to each other, however, and when they subtract from each other, a very small net voltage remains. In Fig. 17·4, the voltage induced in the left-hand vertical side is $E_1$, and this is the voltage represented by point $A$ on the sine curve. The voltage induced in the right-hand vertical side is $E_2$, and this is the voltage represented by point $B$ on the sine curve. The difference between $E_1$ and $E_2$ is the *net* loop voltage.

The loop antenna is, in general, an inefficient device. Early loops were, of necessity, constructed of rather large dimensions in order to develop sufficient voltage to drive the low-sensitivity receivers then in use. If the loop is rotated from the maximum signal position, it is apparent that the

*Fig. 17·5 The net voltage in the loop antenna will vary with loop orientation. Maximum net voltage is achieved in (a), minimum net voltage in (c).*

Oncoming wave | Rotated 30° | Oncoming wave | Rotated 90° | Oncoming wave

(*a*) (*b*) (*c*)

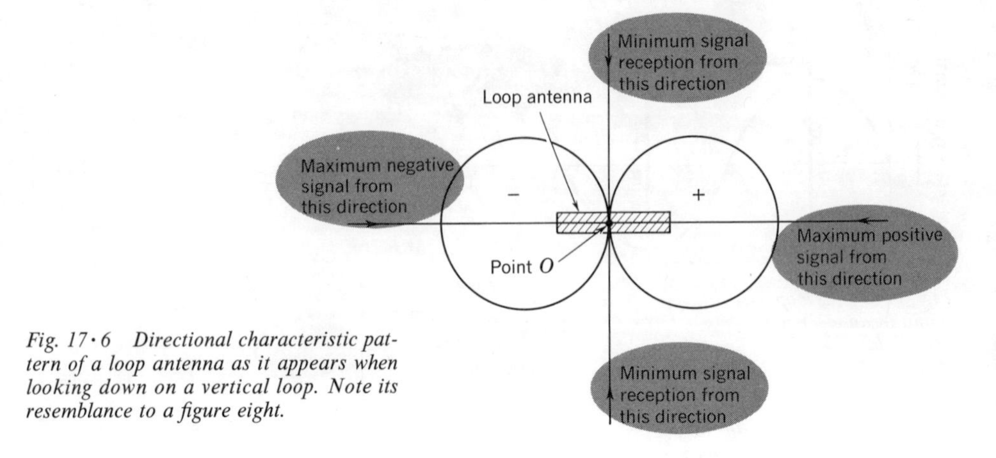


*Fig. 17·6 Directional characteristic pattern of a loop antenna as it appears when looking down on a vertical loop. Note its resemblance to a figure eight.*

distance the wave is required to travel between the two vertical sections is reduced (see Fig. 17·5). The net loop voltage is then decreased. If the loop is positioned at right angles to the direction of wave propagation, the same field cuts each of the vertical legs of the loop. This produces equal induced voltages, and the net voltage is zero. Thus the signal voltage varies from a maximum to a minimum, as the plane of the loop is changed from a position parallel to the wave motion to a position at right angles to it.

Figure 17·6 shows how the voltage induced in the loop varies as the direction from which the signal arrives is changed (assuming the loop's position remains fixed as shown). The length of a line drawn from point 0 to any point on the figure-eight curve represents the relative amount of signal voltage developed in the loop for various angles of signal arrival. The figure-eight directional pattern has two sections or lobes—one positive and one negative. Note that maximum induced voltage is obtained when the signal approaches either end, while minimum voltage is induced when the signal arrives broadside to the loop.

Instead of keeping the loop fixed in position and having the signal direction vary, it is more usual to rotate the loop, with the signal arriving from a fixed direction. Under these conditions, as the loop (and its directional pattern) rotate 360°, the induced loop voltage will experience two maxima and two nulls.

Figure 17·6 also indicates an important loop principle. When the loop is oriented for maximum signal, a relatively large angular rotation is required before an appreciable change in differential voltage is noted. With the loop in the minimum signal position, however, a small rotation will produce a large change in the loop-signal amplitude. This principle is of great importance in direction finders. Since the rate of change of the difference voltage is much greater near the minimum signal position, the loop is rotated to determine this minimum signal position, or *null.* In this minimum voltage position, the loop is at right angles to the direction of wave propagation (the broadside position).

Since the ear cannot accurately determine the minimum or null points, it is general practice to employ visual rather than aural means of detection. In this way, a loop antenna can be used to determine the direction of a transmitter signal.

As explained previously, the net loop voltage is very small. By using a number of turns of the conductor in the loop, the total voltage can be increased because the turns are connected in series. The loop voltage, and hence the sensitivity of the system, can be further increased by tuning the loop to resonate at the signal frequency.

The height of the vertical sections of the loop will also determine the differential voltage, since longer sides will be cut by more lines of force. Also, the wider the loop, the greater will be the signal voltage. Since the differential voltage depends upon the height and the width of the loop, it is a function of the loop area (height times width).

The width of the loop, if measured as a fraction of a wavelength, will vary for different frequencies. As frequency is increased, the wavelength decreases. The shorter the wavelength, the closer the loop will approach the ideal half-wavelength spacing between vertical sections.

The differential loop signal voltage therefore is a function of the field strength of the incoming wave, the total number of turns or conductors, the loop area, the wavelength, and the angle of loop orientation. Figure 17 · 7 illustrates five different types of loops used in radio direction finding.

Up to now we have considered only the ideal loop. Departures from these conditions occur in practice and give rise to an error in measuring bearing or location. This error deviation is of little consequence if it is constant and can be determined. Then suitable correction or calibration can be made to account for it. If the deviation changes, however, it becomes more difficult to obtain accurate bearings.

Deviation in direction-finding loops is the result of a number of causes. There is an important difference between the voltage induced in a loop antenna and that induced in an ordinary open antenna. The voltage at the loop terminals represents the difference between two induced voltages, while the open-antenna voltage is only the induced voltage. The maximum loop terminal voltage is obtained when the difference between the voltages induced in the two sides is the greatest. If a single-turn loop is cut at the center of the turn, as in Fig. 17 · 8*a*, and each end of the loop is connected to ground, currents in both sides will flow to ground, since this arrangement now constitutes two antennas. Rotating these two antennas

*Fig. 17 · 7 Various loop shapes employed in radio direction finders.*

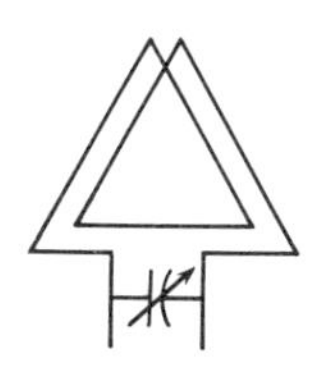
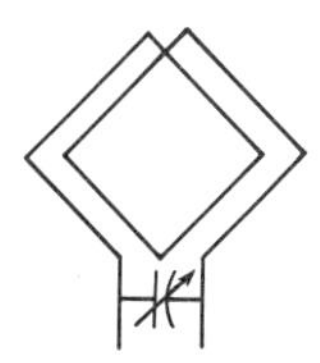
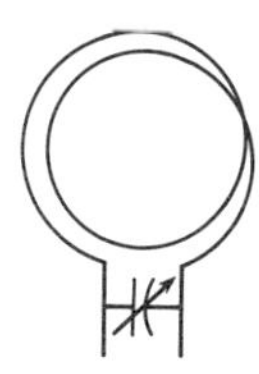
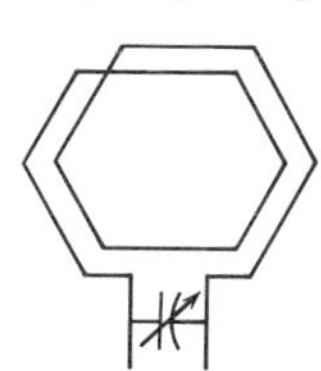
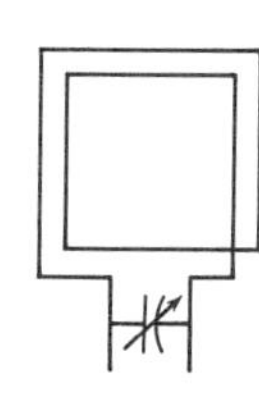

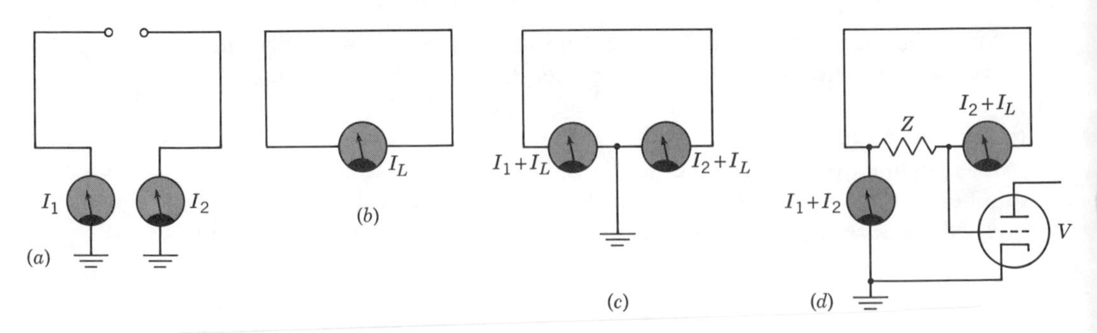


*Fig. 17·8 (a) Broken loop with grounded sides. $I_1$ and $I_2$ are currents from loop sides to ground. (b) Ungrounded closed loop has current $I_L$. (c) Grounded closed loop with both loop and ground currents. (d) Loop connected to input amplifier stage of receiver. Input voltage is developed across impedance Z by currents $I_2$ and $I_L$.*

will have no effect on the value of the currents that would flow. If the loop circuit is re-formed and its free ends are connected, as in Fig. 17·8*b*, a current flows in the loop.

If, however, the loop is closed and, in addition, the center of the lower horizontal section is grounded, the total loop current may be considered to be the result of two separate currents. This is illustrated in Fig. 17·8*c*. This type of loop connection is similar to that obtained when the loop antenna is attached to a radio receiver. One of the two current components is the loop current which flows completely around the loop and which is due to the difference voltage developed by the loop, as illustrated in Fig. 17·8*b*. The loop current varies with the position of the plane of the loop, since this determines the difference voltage. The second current is due to the fact that the right and left halves of the loop each have a different voltage with respect to ground, thus causing currents to flow from the top horizontal loop member down through both vertical members to ground. This current is not a function of the loop orientation or position. This second current component will flow even when the loop component is zero, thereby eliminating the null in the loop connection. It is known as *unbalanced current effect* and may result in an elliptical pattern instead of the ideal figure eight.

Figure 17·8*d* illustrates how this unbalanced current effect develops a voltage input for a receiver. A vacuum-tube circuit connected across the inserted loop load impedance $Z$ will have a grid-input voltage that is a function of $I_2$ and $I_L$. A true null cannot, therefore, be achieved.

A system for eliminating the unbalanced current effect is shown in Fig. 17·9. Two tubes ($V_1$ and $V_2$) are employed in a balanced-input circuit. The ground currents ($I_1$ and $I_2$) produce equal but opposite voltages across their respective load impedances. The grid-to-grid voltage due to these ground currents, therefore, is zero. Only the loop current $I_L$ provides an input voltage to the two tubes. This will produce well-defined null and maximum points as the loop is rotated.

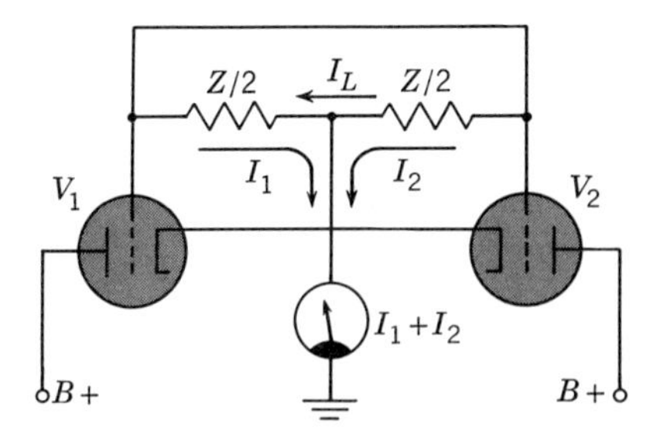


*Fig. 17·9 Balanced-input system cancels the effect of ground currents and utilizes only the loop current to provide receiver input signal.*

Another method of eliminating the effect of the unbalanced current in the loop antenna is achieved by shielding the loop. The shielded loop is extensively used in modern aircraft, since its mechanical construction meets the needs of aeronautical service. Figure 17·10 shows shielded loops. The electrical diagram of this type of loop is illustrated in Fig. 17·11. The shield is a metallic cover, having one break at the top. Because the shield is not a closed turn, no current will circulate around the shield. However, the arriving signal will induce voltages in the vertical sections of the shield. These voltages will cause ground currents $I_1$ and $I_2$ to flow. These currents will induce opposite currents ($I'_1$ and $I'_2$) in the loop. Since these induced loop currents are equal in magnitude but opposite in direction, they cancel and thus do not develop a net voltage across the load impedance $Z$. The normal loop difference voltage will produce a circulating loop current and its associated voltage output across $Z$.

The loop circuitry, as well as the direction-finding equipment, must be

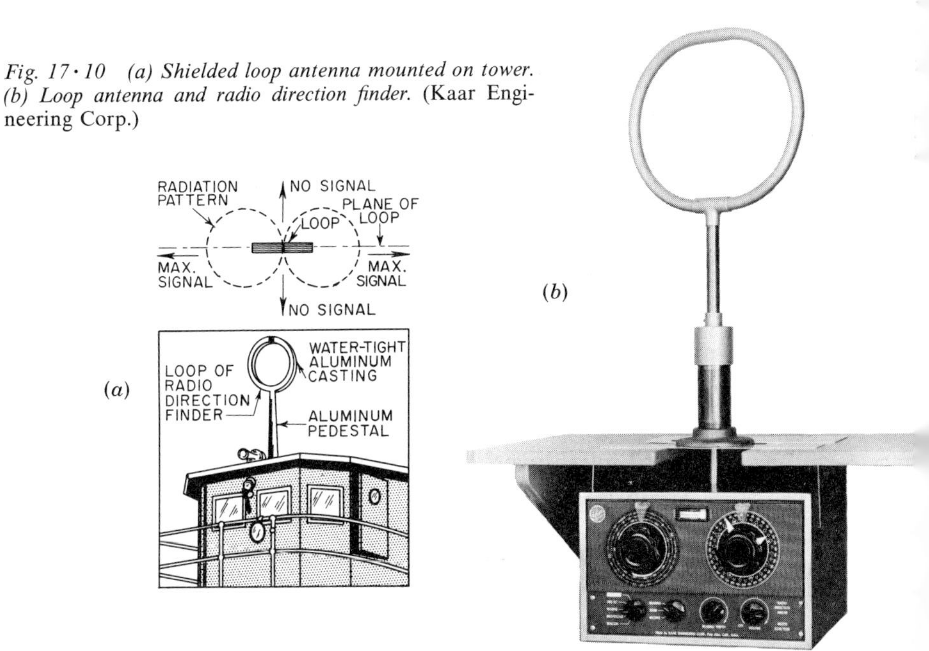


*Fig. 17·10 (a) Shielded loop antenna mounted on tower. (b) Loop antenna and radio direction finder.* (Kaar Engineering Corp.)

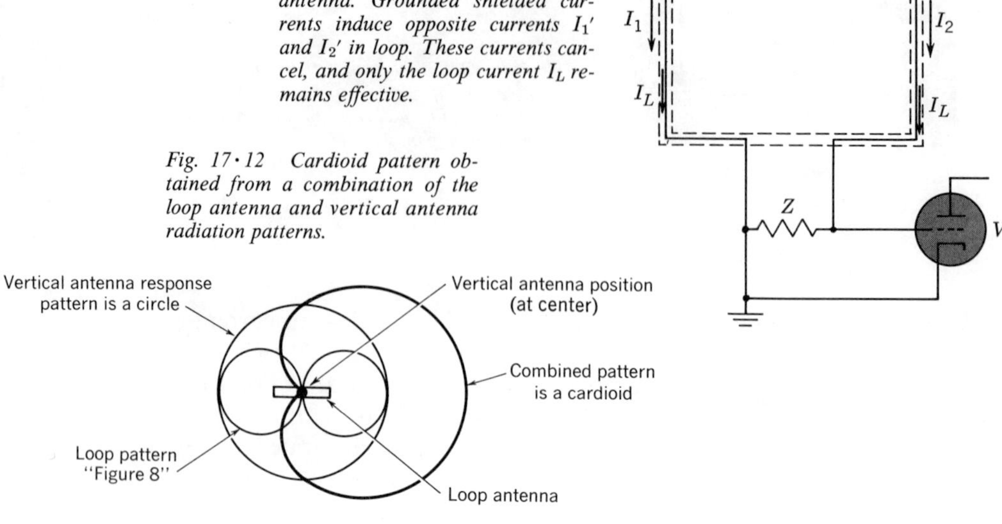


*Fig. 17·11 Typical shielded loop antenna. Grounded shielded currents induce opposite currents $I_1'$ and $I_2'$ in loop. These currents cancel, and only the loop current $I_L$ remains effective.*

*Fig. 17·12 Cardioid pattern obtained from a combination of the loop antenna and vertical antenna radiation patterns.*

extremely well shielded to eliminate any stray induced voltage that is not picked up by the loop itself. This stray signal will cause bearing error, since it obscures the null point. All power leads and interconnecting cables must also be well shielded to protect against these spurious voltages.

The distributed capacitance of the loop turns also tends to prevent a perfect null. *Night effect,* caused by reflected signals, usually from the ionosphere, also reduces the null sharpness because of the change of polarization of the waves during propagation. Refraction of electromagnetic waves will also cause polarization errors.

Because the basic loop antenna has two maximum and minimum positions, it is sometimes unsatisfactory as a direction-finding device. A *cardioid* (heart-shaped) field pattern, Fig. 17·12, is more desirable, since it yields only one maximum and one minimum point. A vertical antenna, used in conjunction with the loop, will provide such a cardioid response. With this pattern, one angle of rotation provides a maximum output, and one angle produces the null signal. The vertical antenna is known as the *sense* antenna. In a unidirectional bearing location system, the sense antenna may be cotinuously employed or may be switched into the antenna circuit by the operator.

## *17·2 Radio direction finders*

The directivity characteristics of loop antennas and vertical antennas are used in radio direction finders to determine the direction of a distant transmitter. This is done by detecting the direction of arrival of the oncoming r-f wave, using the directional antenna and a receiver.

**The aural-null direction finder.** The simplest form of aircraft direction finder consists of a radio receiver and its associated loop antenna. The loop is usually shielded and is used in conjunction with a loop azimuth indicator. The loop is rotated by the operator through a gearbox to achieve a very slow rotational speed. A flexible torsion shaft permits control of the rotation from the cockpit. The indicator is geared also to the torsion shaft, so that it is kept in step with the loop rotation and can be calibrated directly in degrees of rotational angle.

The direction finder provides a bearing on a fixed transmitting station. Two of these bearings, each using a different station, when plotted on a map, will intersect and determine the location of the aircraft. This procedure is called *taking a fix*. The null indication is used because it provides the greatest sensitivity and accuracy. It is possible to make the null of a loop extremely sharp, less than 1°.

**The right-left radio compass.** Figure 17 · 13 shows the indicator unit of a right-left radio compass. The indicator points to the center mark when the plane points to the radio station. It points to the right or to the left if the plane is off the course of the radio station being received.

For the antenna system a loop is mounted with its plane at right angles to the nose-tail axis of the aircraft; a vertical sensing antenna is also employed. A switch $S_1$, as shown in Fig. 17 · 14, can be used to reverse the antenna field pattern. The sense antenna and loop combination normally have a cardioid characteristic. The antenna voltages are fed to a special receiver which produces a rectified output signal. A meter $M$, with a pair of opposing windings, is connected to the receiver output through a switch $S_1$, ganged with $S$. As the loop reversing switch $S$ is changed from one position to another, the connections to $M$ are changed. The meter will give a right indication for one connection and a left indication for the other.

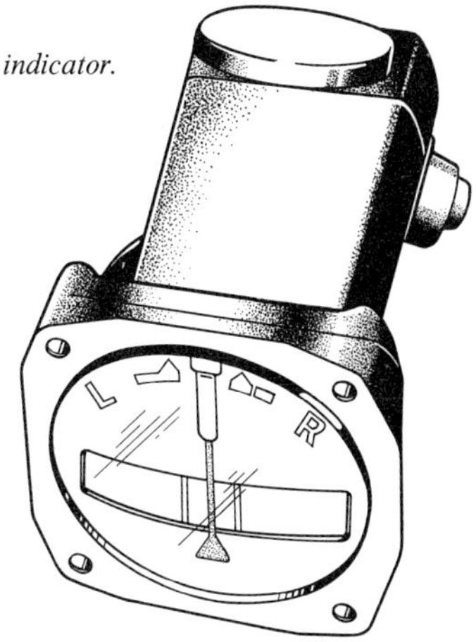

*Fig. 17 · 13 Right-left radio compass indicator.*

*Fig. 17 · 14 Block diagram of the right-left radio compass.*

Loop antenna
Vertical sense antenna
Mechanical coupling
$S$
$S_1$
$M$
Receiver

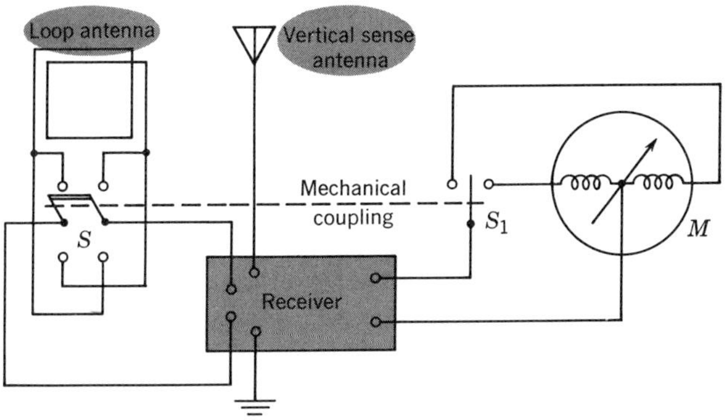

If switching is rapid enough, $M$ will deflect an amount which depends upon the differential output for the two loop senses.

Instead of using mechanical switches, the practical R-L compass employs an electronic switch. This type of direction finder is known as a *visual-null* device. It is a bidirectional system and therefore provides bearings when approaching or leaving the radio station. This is, at times, a distinct disadvantage.

**The automatic direction finder.** The manually operated direction finder, or radio compass, led the way to the development of completely automatic direction-finding equipment. The automatic direction finder (ADF) provides a continuous bearing indication. This is presented on a full 360° indicator and gives the pilot immediate bearing information, while freeing him to attend to other matters in the cockpit. In addition to the cockpit bearing indicator, ADF equipment employs a compass indicator on the output of the receiver, as shown in Fig. 17 · 15.

Electric contacts are positioned on either side of the indicator needle. When the aircraft is off course (for the station tuned in), the needle deflects and touches a contact which, in turn, operates a motor. The motor will rotate the loop until the loop plane is at right angles to the direction of the arriving signal. The rotation is always such as to reestablish a null. A flexible shaft connects the loop antenna and the bearing indicator in the cockpit to provide a continuous monitoring of bearing between the airplane's heading and the radio station. A combination loop-and-sense antenna is employed.

Electronic circuitry can be used instead of the mechanical contact mechanism. The ADF receiver has its controls on a panel in the cockpit. In some equipment, these controls are mounted directly on the bearing indicator unit. The receiver is tuned by a handle control, as shown in Fig. 17 · 16. Other receiver controls are also available on the indicator panel.

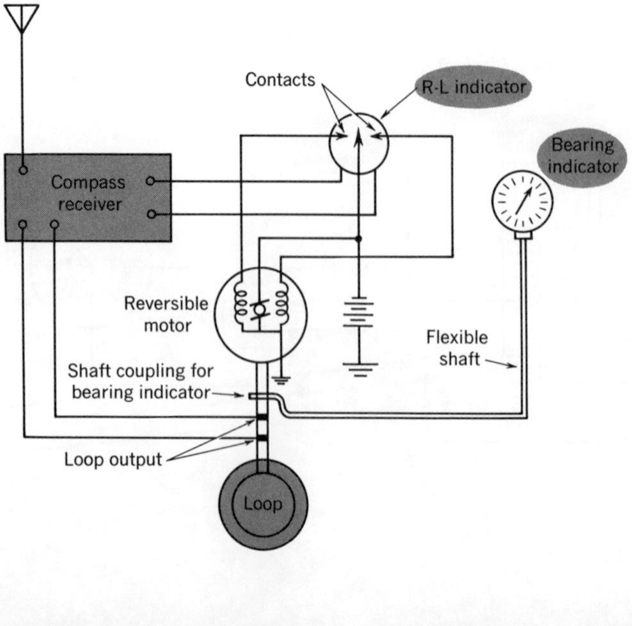

*Fig. 17 · 15 A simple automatic direction finder.*

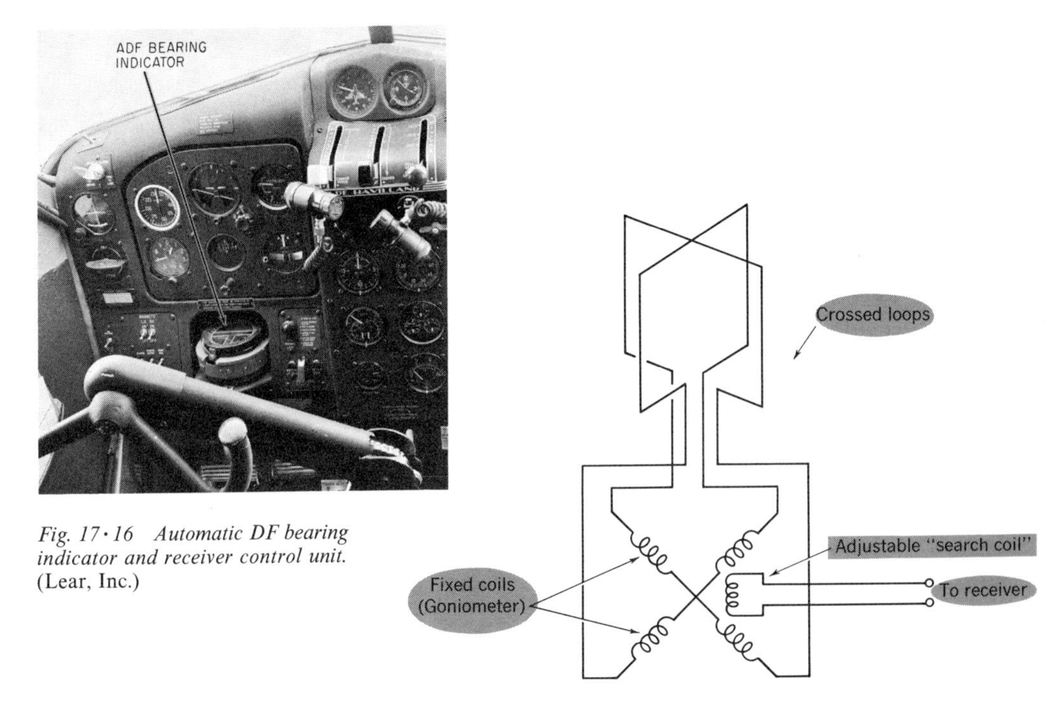


*Fig. 17·16 Automatic DF bearing indicator and receiver control unit.* (Lear, Inc.)

*Fig. 17·17 Basic circuit of the crossed-loop, or Bellini-Tosi, DF system.*

The simplicity and ease of operation of ADF equipment have been its greatest advantage.

**The dual automatic direction finder.** The basic ADF will provide the operator with an indication of a radio transmitter, but does not, in one step, obtain a fix on the location of the station. To obtain a fix, the operator must take a bearing on two transmitters. A dual ADF can take the two required bearings simultaneously. The dual unit consists of two separate ADF's. Each unit is tuned to a different transmitter and provides its respective bearing indication. Both bearings are displayed on the same indicator in the cockpit. This gives the pilot a visual presentation of the plane's direction relative to the two broadcasting stations.

A cathode-ray indicator has been devised for the dual direction-finder (DF) equipment. The outputs of the two receivers control two separate electron beams in a specially designed tube. The beams produce bearing lines on a fluorescent screen. The pilot can see his location immediately, since a map of the area also appears on the tube screen. The location is indicated by the intersection of the bearing lines.

**The Bellini-Tosi direction finder.** A common direction-finding system employed on surface vessels or ground DF stations is the Bellini-Tosi, or *crossed-loop,* system. This equipment uses two loop antennas mounted at right angles to each other. A simple crossed-loop unit is illustrated in Fig. 17·17. The two loops are fixed-mounted and receive signals which depend upon the angle at which the electromagnetic wave cuts the turns

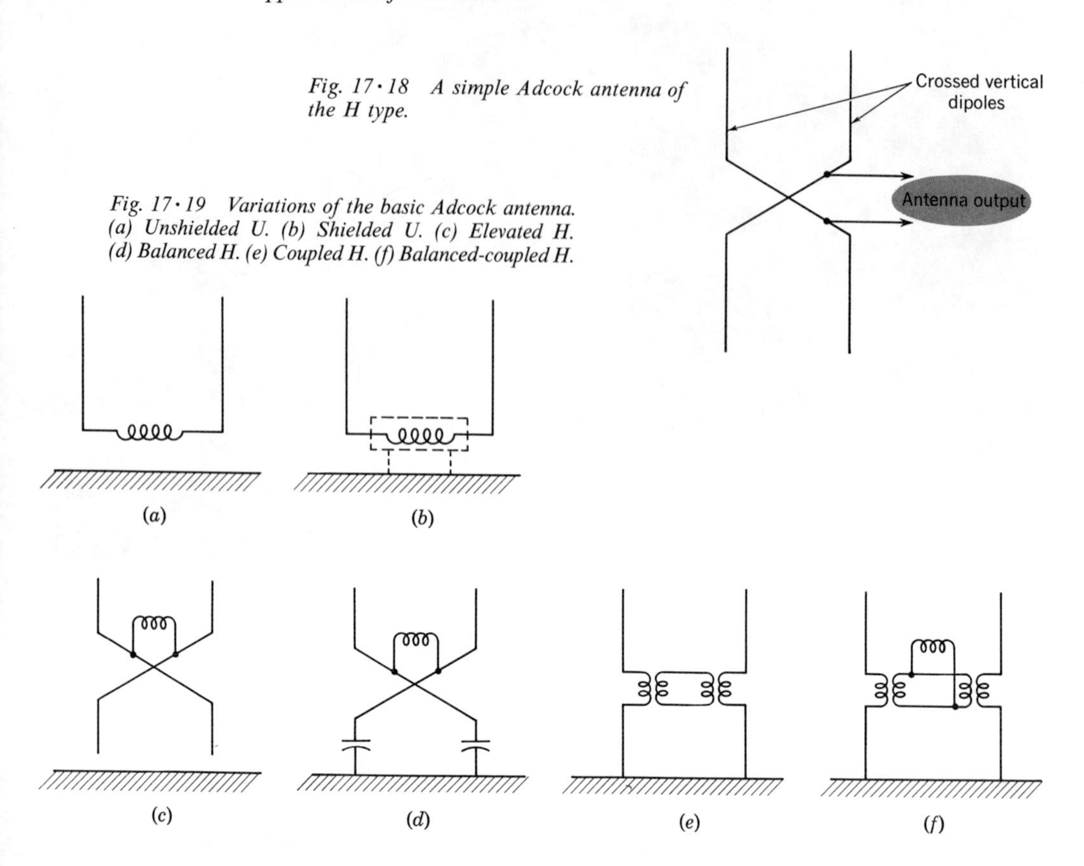


*Fig. 17·18 A simple Adcock antenna of the H type.*

*Fig. 17·19 Variations of the basic Adcock antenna. (a) Unshielded U. (b) Shielded U. (c) Elevated H. (d) Balanced H. (e) Coupled H. (f) Balanced-coupled H.*

of the loop. The loops are connected to a *goniometer,* as shown in Fig. 17·17. The goniometer is a coupling device having two fixed coils mounted at right angles to each other. An adjustable coil, placed inside the two coils, is free to rotate with respect to these coils. The voltage induced in the adjustable coil will vary with its angular position. Since the currents flowing in the fixed goniometer coils depend on the voltage induced in the loops, magnetic fields are set up around the fixed coils which have the same directional relationship as the received signal in the two loops.

The movable coil may then be rotated to find the position at which it is normal (at right angles) to the established magnetic field. When this condition is attained, the "search," or movable, coil will have zero induced voltage. Since the balance condition occurs (in a unidirectional system) at only one coil position, the goniometer can be calibrated to read directly in degrees of angular displacement or bearing. Certain errors, such as those imposed by the limitations of the goniometer circuitry, reduce the over-all accuracy of the system. Changes in the polarization of the propagated wave, due to the action of the ionosphere, can also cause errors in observation.

Because of the polarization, or night-effect, error, the Bellini-Tosi DF system is usually employed for distances up to approximately 500 miles.

The loop antennas are generally untuned, but the adjustable coil is tuned. A vertical antenna is used in conjunction with the crossed loops to provide unidirection indications.

**The Adcock direction finder.** Many electromagnetic waves are reflected from the ionosphere in such a way that their original polarization is changed. Direction-finding equipment operates almost exclusively on vertically polarized waves. When these vertical waves emerge from the ionosphere, they possess a considerable horizontal field component. This horizontal component will induce voltages in the horizontal arms of a loop antenna and prevent a sharp null indication. This results in errors and provides incorrect bearings. The effect is most prevalent at night, when the ionosphere is not under the influence of the sun, and for this reason is known as *night error*. It can be eliminated by using a different type of antenna, such as the Adcock. An Adcock antenna is simply a pair of vertical dipoles, placed close together and crossed over at their centers (see Fig. 17 · 18). The simple Adcock illustrated in the figure is in the form of the letter H.

The operating principle of this antenna is similar to that of the common loop for vertically polarized waves. Arriving waves induce voltages in the vertical dipole elements as in a loop. Horizontally polarized waves induce voltages in the horizontal members of the Adcock, but these voltages cancel out because of the crossed configuration of the antenna. The horizontal voltages are equal in magnitude, but opposite in phase, with respect to the output. The vertical members are not cut by the horizontally polarized waves. The resultant output voltage therefore is the difference between the voltages induced in the two vertical elements. The net voltage output is dependent, as in the loop, on the direction or orientation of the vertical dipoles with respect to the oncoming wave.

Since the dipoles are the equivalent of only a single-turn loop, the output voltage of the Adcock antenna is extremely small. This is a major disadvantage of the Adcock and is a limitation on its practical application. However, this antenna can provide accurate bearings when it is impossible to employ the loop. The Adcock will give very accurate bearings over distances as great as 1,000 to 6,000 miles.

If the dipole spacing is small compared with the wavelength of the received signal, the field pattern becomes the familiar figure eight. If the spacing is greater than one-half wavelength, the field characteristic departs appreciably from the figure eight and may be a four-lobed pattern. Variations of the H-type Adcock are commonly employed in direction finders. Figure 17 · 19 illustrates a few of the many possible types. In the coupled and the balanced-coupled types, the error due to night effect is almost nonexistent. Modern Adcock direction-finding equipment is employed in transoceanic navigation.

## *17 · 3 Radio ranges and beams*

Travel of any kind utilizes some well-defined trails, paths, roads, or courses. To proceed from one point to another, we depend upon such devices for direction and guidance. One of the great problems posed by air travel is a result of the lack of natural directivity devices or symbols during flight. Artificial highways through the sky must, therefore, be set up along all the routes used by commercial aircraft. By making use of the directional properties of antennas, radio roadways for aircraft have been developed and are now used extensively for marking the courses of flight. These radio trails employ a system of electronic equipment, known as *radio-range beacons,* or *radio beacons.*

**Radio-range fundamentals.** We have learned about the directional characteristics of the loop antenna and its figure-eight pattern. If two loop antennas are placed over the same center, but positioned at right angles to each other, the pattern of Fig. 17·20 is obtained. Note that each loop produces its own figure-eight pattern. The two patterns combine to produce four overlapping areas. Within these four intersection areas, a receiver tuned to the proper frequency would pick up signals from both loops. If the power input and frequency of the two loops are equal, they will generate equal-amplitude signals. In practice, the power output from a common transmitter is connected first to one loop, then to the other, alternating between them. A field-strength meter, or a receiver, placed on a line making an angle of 45° with the plane of either loop, will indicate no difference in output as the common transmitter is switched from one loop to the other. The radio beam employs this principle.

The crossed loops of the system produce four courses, including two reciprocal courses. The reciprocal courses are merely the negative of a course; that is, they point in the opposite directions. Along the four 45° lines, the courses are of equal intensity. An indicating device informs the airplane pilot whether the two loop signals picked up by a receiver are equal in magnitude, or if one is greater than the other. The pilot knows that, as long as the two signals are equally strong, he is on the proper course. He may determine a left or right deviation from the correct course by knowing which loop provides the greatest signal for the particular position of the plane.

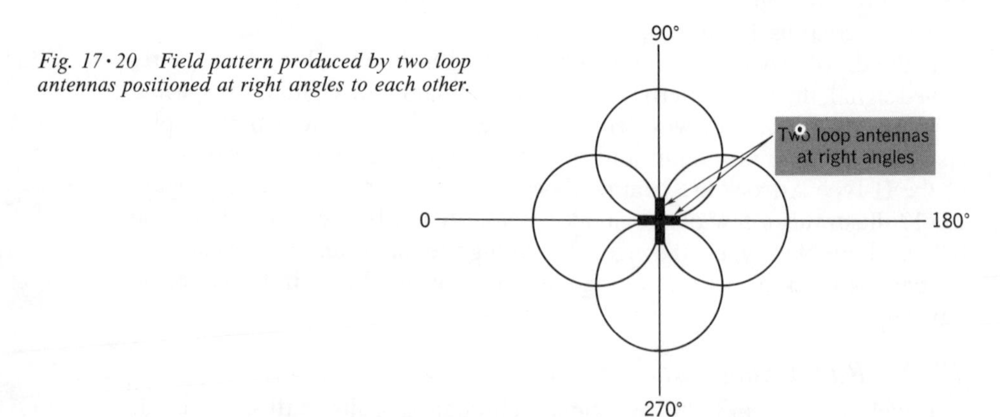


*Fig. 17·20 Field pattern produced by two loop antennas positioned at right angles to each other.*

In place of two loop antennas to provide the intersection figure-eight patterns, we may employ vertical dipoles. The directional pattern of two vertical dipoles can be made into a figure eight. Therefore, by employing four vertical antennas, with r-f energy being fed to them in pairs, we obtain the same resultant field pattern as we do with two crossed loops. The four dipoles are located on the corners of a square, with diagonally opposite antennas forming pairs. Energy is fed to each set of diagonally opposite towers 180° out of phase but with equal amplitude. This will produce the figure-eight pattern.

Four-course ranges permit only courses that are at right angles to each other and do not readily permit the navigation of other desired courses. By feeding different amounts of energy to the two loops, however, or the two pairs of dipoles, the orientation of the courses could be varied. Also, it was found that the addition of reflector antennas will alter the figure-eight pattern and enable the courses to be shifted. The basic principle of the radio beacon thus permits many variations. Right and left indicators have been employed to give a visual rather than an aural determination of the heading of the plane.

**The A-N aural range.** The accuracy of course indication is found to improve materially with the inclusion of interlocking code letters. In this system, the pilot, flying along a radio-range course, hears a continuous aural tone or note. This tone is produced by two interlocking signals and appears to be a continuous note. If the pilot goes slightly off his required range course, however, he hears one of the two interlocking tones with greater amplitude than the other. As the plane drifts more off course, this one tone will become increasingly dominant until the pilot hears only the single signal. If the two interlocking signals are different code groups or code letters, an off-course heading will bring one letter to the forefront, thus identifying which side of the required course the plane is on.

The International Morse Code letters A (dot-dash) and N (dash-dot) have been selected as standard for range-beacon systems. Figure 17 · 21

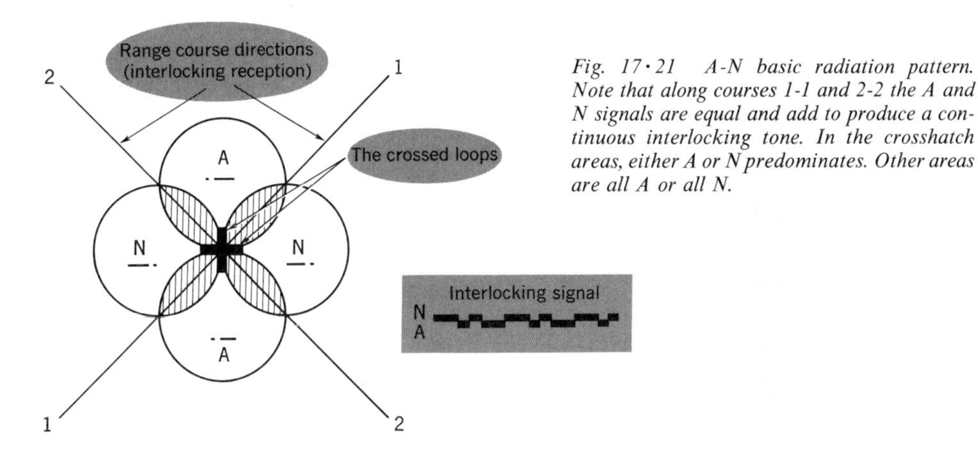


*Fig. 17 · 21 A-N basic radiation pattern. Note that along courses 1-1 and 2-2 the A and N signals are equal and add to produce a continuous interlocking tone. In the crosshatch areas, either A or N predominates. Other areas are all A or all N.*

illustrates the radiation pattern of a range station. In the areas marked A or N, the respective letter is received with greatest amplitude. Along the course lines, however, they interlock to provide a continuous tone. In the transmitters, the code letters are synchronized so that the dot of one letter fits into the space between the dot and dash of the other letter. The dash of each A occurs during the blank space present between the end of each N and the beginning of the next. The dash of each N appears during a similar period in the A transmission. Furthermore, the intensity of the predominating aural signal is a rough indication of the plane's distance from the range course.

On-course signals are actually heard by the pilot or navigator for a short distance on either side of the range-beacon course line. This is because the ear is unable to recognize very small differences in loudness. This zone of on-course signal is generally about a degree or two wide, becoming narrower as the plane approaches the beacon location. Directly above the beacon antennas, however, there is very little reception of the range signals. This dead area is known as a *cone of silence* and tells the pilot he is directly over the range station. The aural signal thus will rapidly increase in volume as the plane approaches the bacon and then disappear when the plane flies directly over the range antennas.

Since air routes are not necessarily 90° apart, some method is needed to obtain the required practical directional antenna patterns. Vertical antennas are used for this purpose. When fed in pairs and located at the corners of a square, so that diagonal antennas form pairs, four vertical radiators will produce the same beacon characteristic as the two crossed loops. However, the two pairs of dipoles can be arranged on corners of rectangles instead of squares, they can be fed different amounts of power, and their input phase can be varied. Additional dipoles can also be added to the basic array. These methods may all be employed to shift the directions of the range-course lines in accordance with navigational requirements. Because of the great flexibility, vertical antennas or towers are most commonly used in range systems.

The radio beacon system in the United States operates on frequencies from 200 to 400 kc. Range stations cover the entire United States, generally spaced at distances from 125 to 200 miles. A and N signals are transmitted for approximately 30 sec, after which a code signal is sent first from the N and then from the A antennas. This code signal identifies the particular beacon and permits the pilot to refer to a map and obtain the plane's position. It also alerts him, as the beacon signal gets weaker, to tune the range receiver to the next beacon on his course.

Since radio-range beacons operate on rather low frequency, sky waves are produced. These are extremely troublesome at night, because reflections from the ionosphere can cause an apparent change in the directional pattern of the beacon, thereby altering the on-course direction. This night effect is the same as that discussed for direction finders. Loop antennas, since they have horizontal members, are particularly subject to this effect.

Vertical dipoles can be operated so that the night effect is negligible. The towers are fed through tuned coaxial cables buried in the ground. This prevents vertical radiation and reduces the amount of sky wave. This range system is known as the *TL radio range,* TL being an abbreviation for tuned line. This type of range is sometimes called the *Adcock range.*

**The simultaneous radio range.** The original radio-range sytems installed in the United States served a dual purpose. In addition to providing the A and N range-course signals, the beacon transmitters would interrupt these interlocking code-letter signals to broadcast local weather information. As air traffic and air speed increased, the loss of the range signals during these short intervals became extremely critical. The problem was solved by simultaneously transmitting weather information and range signals. The receiver is designed to permit selection of either signal, as desired. To accomplish this in the TL range, a fifth vertical tower is added to the antenna system. It is placed at the center of the square, or rectangle, formed by the beacon antennas. Since this additional tower is nondirectional, its field radiation pattern is a circle.

In this system, called the *simultaneous radio range,* the range transmitter operates on a frequency 1,020 cycles higher than the weather-tower carrier. The range transmitters transmit a mechanically keyed pure sine-wave signal. In the aircraft range receiver, the radio frequency from the range transmitter beats with the radio frequency from the center tower, to produce the difference frequency, a 1,020-cycle note. The weather transmitter must transmit continuously; otherwise, the 1,020-cycle note is not available in the receiver. A wave filter is placed in series with the microphone used for weather broadcasting to eliminate any 1,020-cycle component from the speech signal (Fig. 17 · 22). In the receiver, two filters are employed. One filter passes only 1,020 cycles and is used for the beat frequency developed in the detector. This is the range signal. The second filter passes all audio frequencies except 1,020 cycles. This is the weather signal. By means of a switch, the pilot can connect his headphones to either filter, selecting either the weather or range direction information. The simultaneous range operates within the same band of low frequencies as the other TL ranges.

**The VHF radio range.** In order to overcome some of the problems of the low-frequency range, extensive investigation was made of the VHF characteristics as applied to radio-range systems. It was known that these relatively high frequencies would penetrate rather than be reflected by the ionosphere. This completely eliminated the night effect and also the necessity of employing only vertical antennas. In fact, it was found that, in the VHF spectrum, horizontal antennas are desirable. Since these high frequencies are easily reflected and refracted from certain obstructions, they are subject to the *mountain effect.* This phenomenom frequently occurs over mountainous terrain and is basically a distortion of the range-signal wave front by obstructions, such as hills and mountains. It is well to remember that VHF waves do not propagate beyond the line of sight,

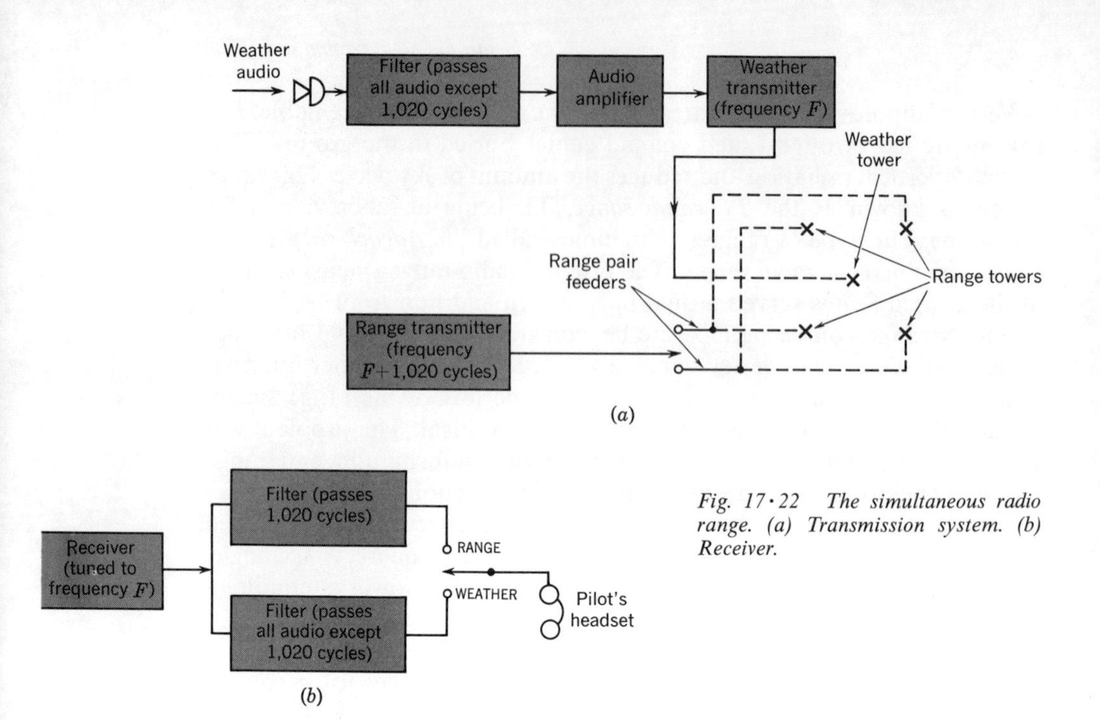


*Fig. 17·22 The simultaneous radio range. (a) Transmission system. (b) Receiver.*

or the horizon. The use of horizontal antennas to produce horizontally polarized waves substantially reduces the mountain effect. If low power output from the transmitter is employed, little signal will be received beyond the horizon, or line-of-sight distance.

While many types of VHF antennas are useful for range application, a special form of loop has been very successful for VHF ranges. Most horizontal antennas have a small amount of pickup of vertically polarized radiation and transmit some vertical component when radiating. An antenna free from these defects is the *Alford loop.* Two general forms are shown in Fig. 17·23. The length of the sides $L$ is made one-eighth wavelength or less. The folded ends are adjusted for maximum current at the center of each of the four sides. The field pattern of both types is identical to that of a simple vertical dipole, but horizontally polarized. The field pattern in the vertical plane is a figure eight, and the horizontal characteristic is a circle.

Reliable two-course VHF ranges are used today in air navigation. Remember that VHF antennas are much smaller physically than corresponding arrays for low frequency. Most range antennas are mounted on tall towers to achieve the proper height and to increase the horizon distance. These VHF ranges employ four or five Alford loops. The fifth loop is mainly for weather information.

## *17·4 VHF omnidirectional range*

*Omnirange,* short for omnidirectional range, provides extremely versatile facilities for radio-range air navigation. Several different systems, or variations of the basic type, are available.

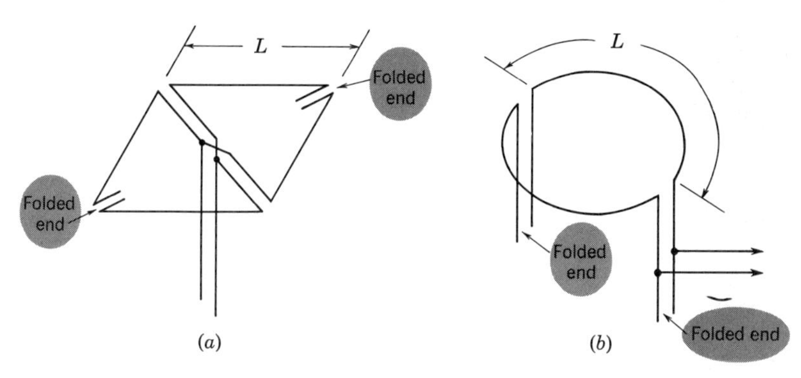


*Fig. 17·23 The Alford loop. (a) Transmitting antenna. (b) Receiving antenna.*

A definite advantage of the omnidirectional range (ODR) is its use of very high frequencies which are almost completely free of atmospheric and precipitation static. This is in contrast to the low-frequency facilities which are always cluttered with static when a pilot needs them most, say, during a severe thunderstorm.

Another important advantage is the greater number of available channels in the UHF band, compared with the low-frequency bands. There is also much less interference between channels, since transmission usually does not extend beyond the horizon at these high frequencies. Probably the most important advantage is that the ODR is not confined to just two or four courses. It will provide accurate navigational information on any course to stations which the pilot may select. This characteristic is the origin of the name (omnidirectional radio range). It is commonly called *VOR*, an abbreviation obtained from combining VHF and ODR.

The VOR furnishes definite track guidance between the chosen range station and any point up to about 50 miles (the horizon distance), when flying at minimum instrument altitude. The actual line-of-sight path depends, of course, on the terrain en route. At higher altitudes, this distance increases materially. At 14,000 ft, the horizon distance is approximately 150 miles.

Three basic instruments are used by the pilot when flying with the VOR. A cross-pointer instrument (Fig. 17·24*a*) indicates the plane's position relative to the desired course.

*Fig. 17·24 Dial faces of the three basic omnidirectional radio range instruments. (a) Cross-pointer course indicator. (b) Azimuth selector. (c) Sense indicator.*

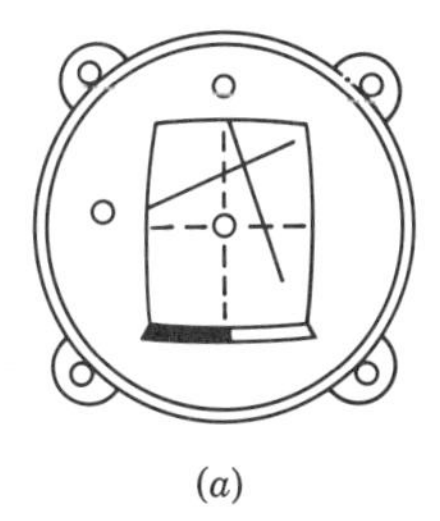


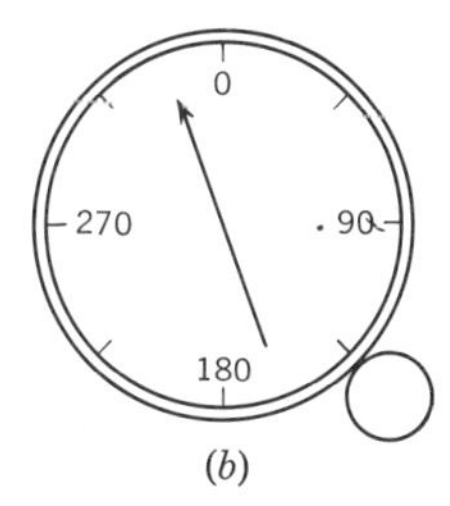


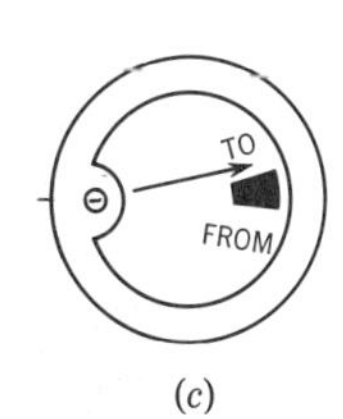

An azimuth selector (Fig. 17·24*b*) has a dial face that is similar in appearance to that of the ADF. The pointer of the azimuth selector can be rotated by means of an adjoining knob. This knob is employed as a manual course-selector control. After the cross-pointer needle has been centered, the position of the azimuth-selector pointer indicates the magnetic bearing of the plane, either to or from the station.

The third instrument is a sense indicator (Fig. 17·24*c*). It is similar in appearance to an ordinary panel meter, but is mounted so that the needle pivot is at the left side of the case, instead of at the bottom. The instrument is mounted with the word TO at the top of the meter scale and the word FROM at the bottom. A red sector between the TO and FROM markings indicates loss of signal or unsatisfactory receiver or transmitter operation.

In the operation of VOR navigation, the pilot selects the desired course by setting the pointer on the azimuth selector to the required magnetic bearing and then flies that course by reference to the cross-pointer instrument. The VOR navigation has a reciprocal course also. Since there is a reciprocal course, the sense indicator is included to indicate whether the aircraft is flying the desired or reciprocal course.

Once a course has been chosen and the sensing determined, the plane is always flown in the same direction as the deflection of the vertical cross-pointer indicator. When the aircraft is on the selected course, the vertical cross-pointer indicator will be directly vertical and over the row of dots on the cross-pointer dial face. If the sensing indicator reads TO, the plane is flying toward the station. If it reads FROM, the flight is away from the station.

To determine his position during flight, the pilot tunes his range receiver to one station in his general area and rotates the azimuth selector until the vertical needle of the cross-point indicator is centered. The azimuth bearing is then determined. Then, repeating this procedure for a second station, the pilot will have his bearings from two stations and can readily find his position on the map. The pilot does not have to determine his sensing in order to get a fix by the above method.

Each VHF ODR station is identified by a three-letter code signal, which is transmitted on the omnirange frequency.

**Theory of operation.** The directional information of the ODR station is radiated by means of two radio signals which add to form a single signal in the receiving antenna. One of these signals is nondirectional. The other is a rotating field obtained by means of a goniometer. The radiating system consists of five antennas, one at each of the four corners of a square and one in the center. The center antenna provides the nondirectional signal, while the other four are connected in pairs and are so fed that they form two figure-eight patterns at right angles to each other. In each pair, the r-f voltage to one antenna is 180° out of phase with that fed to the other antenna. This pattern is similar to that shown in Fig. 17·20.

The two directional antenna pairs are fed by means of a capacitive goniometer. Each antenna of a pair is connected to one side of the split-

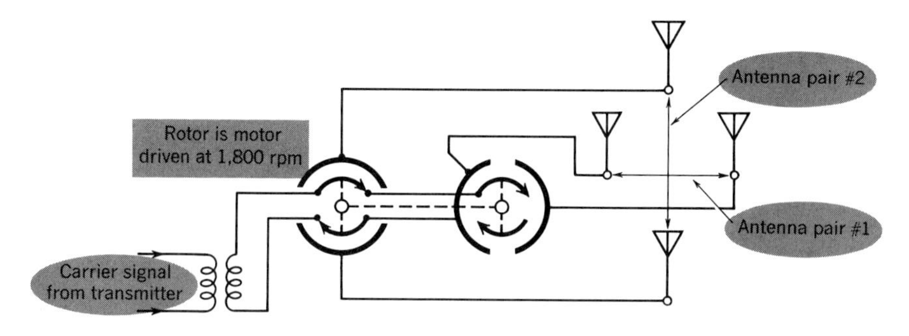


*Fig. 17·25 ODR capacitor goniometer operation. The two antennas of a pair are fed 180° out of phase.*

stator plates of a capacitor, as shown in Fig. 17·25. The split rotor of the capacitor is motor-driven at 1,800 rpm. The split rotor is connected, through a matching transformer, to the radio transmitter of the ODR station. As the rotor revolves, the voltage fed to the pair of antennas varies sinusoidally at the rate of 30 cps (1,800 rpm ÷ 60 = 30 cps). The other pair of directional antennas is fed by a similar capacitor, on the same shaft and in the same manner as the first, but with the split-station shifted by 90° with respect to the rotor. As the rotor revolves, the voltage fed to the second pair of antennas varies in the same manner as that fed to the first pair but leading it 90°. When the field due to pair No. 1 is maximum, the field set up by pair No. 2 is zero. When the field due to pair No. 2 is maximum, the field due to pair No. 1 is zero.

The two figure-eight field patterns add together at each instant to form a new over-all single figure-eight pattern whose position changes from instant to instant. This resultant field rotates through a full 360°. It rotates uniformly at the 1,800-rpm motor speed. This action is quite similar to that obtained in a two-phase motor.

The nondirectional circular field pattern of the fifth antenna is added to the rotating figure-eight field. Its r-f field is in phase with the field of one lobe of the rotating field pattern and out of phase with the other lobe. Thus, one lobe is strengthened while the other is partially canceled by the nondirectional field. The result is the familiar cardioid pattern.

The rotating field sweeps around all points of the compass. An aircraft receiver at any point picks up this signal. The signal varies in amplitude 30 times a second, or 1,800 times a minute. Also, the phase of this 30-cps modulation is different for each different direction or point of the compass. Thus, a plane receiving this signal from one direction will find that its phase is one value, while, if it picks up the signal from another direction, the phase will have some other value. This 30-cps modulation on the rotating r-f signal is called the *variable phase voltage.*

The nondirectional antenna transmits a signal containing a fixed 30-cps modulation called the *reference phase voltage.* In the receiver, the variable-voltage signal is compared with the reference voltage. Measurement of the phase angle between these two voltages is an indication of the bearing of the receiver from the ground station with respect to magnetic north or zero bearing.

In order to separate these two 30-cps signals for comparison in the receiver, a 10-kc FM subcarrier is used to carry the nondirectional reference-voltage signal. (Actually, the frequency is 9,960 cycles, or 9.96 kc.) In addition, the nondirectional r-f carrier contains voice modulation, so that weather and other information will be available with each range signal.

The operation of the ODR is therefore based on a comparison of the phase angle between two 30-cps audio signals. The difference in phase between the two received signals indicates the position of the receiver in the azimuth, thereby revealing the position of the aircraft relative to the ground station.

Figure 17·26 is a block diagram of a typical VOR transmitter station. The station equipment consists essentially of a VHF carrier transmitter and an auxiliary rack, containing a sideband generator in the form of a rotating goniometer, a reference generator, a carrier modulator, a modulator driver, an identification keying device, and a modulation eliminator. A remote-control rack contains a speech amplifier, a line equalizer, a voice-operated relay, a remote-control unit, and a rectifier. An antenna changeover and

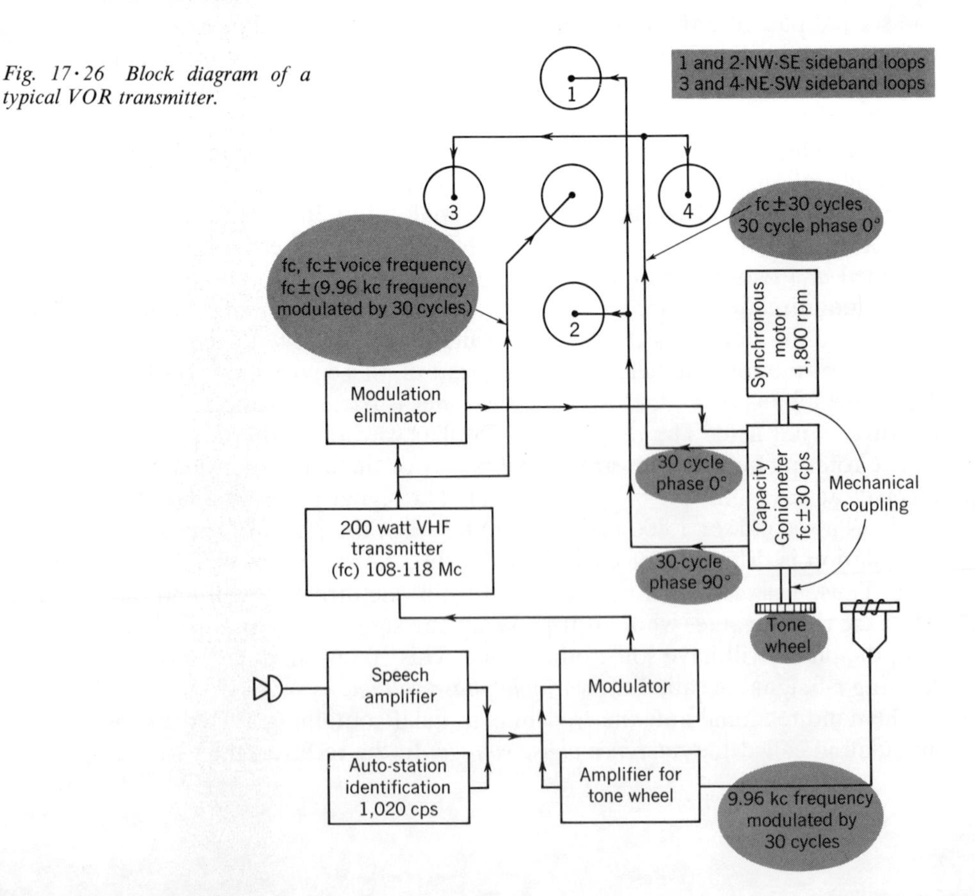


*Fig. 17·26 Block diagram of a typical VOR transmitter.*

phasing unit, a standby power plant, a voltage regulator, and a monitoring system complete the equipment list.

The crystal-controlled transmitter used in the VOR has an output power of 200 watts, and operates in the 108- to 118-Mc frequency range. The output is capable of 100 per cent modulation, but the transmitter is normally modulated 30 per cent by the 10-kc subcarrier and 10 per cent by a 1,020-cps identification signal. Voice modulation of about 30 per cent may be used simultaneously with the 10-kc modulation. The 10-kc subcarrier itself is frequency-modulated at 30 cps. The output from the final stage is fed to the center antenna of a five-loop array, and also to the input of the modulation eliminator.

The goniometer functions as a mechanical sideband generator. Goniometer output No. 1 delivers 30-cps sideband energy at 0° phase to a pair of antennas. Output No. 2 of the goniometer feeds a similar 30-cps sideband energy of 90° phase to a second pair of antennas.

On the same shaft with the goniometer is the reference-signal generator. This generator consists of a coil of wire wound over a permanent magnet, whose lines of flux pass through a toothed gear, called the tone wheel. As the teeth of the gear pass the pole piece, a varying magnetic field is produced around the magnet, inducing a voltage in the coil. The basic frequency of this voltage is dependent upon the number of teeth and the speed of rotation and is designed to be 9,960 cps. In addition, the spacing between the gear teeth is varied around the circumference of the wheel, so that the frequency varies ±480 cps from 9,960 cps as the wheel revolves. A tone wheel, having 332 teeth and a rotating speed of 1,800 rpm, produces an output of 9,960-cps frequency modulated at a 30-cps rate. This signal is referred to as the 10-kc frequency-modulated subcarrier.

Since the goniometer, which produces the variable-phase signal, is mounted on the same shaft with the tone wheel, it is necessary to have a means of properly phasing the signals they generate. This phase adjustment is accomplished by mounting the permanent magnet and coil assembly on a radial arm to amplitude-modulate the carrier. The modulation eliminator removes the amplitude modulation from that portion of the transmitter output which is fed to the goniometer. It is essential that the r-f input to the goniometer be practically free from modulation to prevent cross-modulation between the carrier and sideband signals.

The modulator provides facilities for combining three low-level audio-input signals and amplifying the resultant composite signal to a power level sufficient to modulate the VHF transmitter. Channels are provided for voice, the 10-kc subcarrier, and a 1,020-cps identification signal. The 1,020-cps signal unit consists of an oscillator-keyer combination. The oscillator operates at 1,020 cps, and a motor-driven keying unit interrupts this tone in any desired repetition sequence of dots and dashes. A relay in the output circuit is energized by a d-c voltage supplied through an externally connected control line and permits the output to be interrupted from a remote point, when it is desired to substitute voice modulation for the keyed 1,020-cps tone.

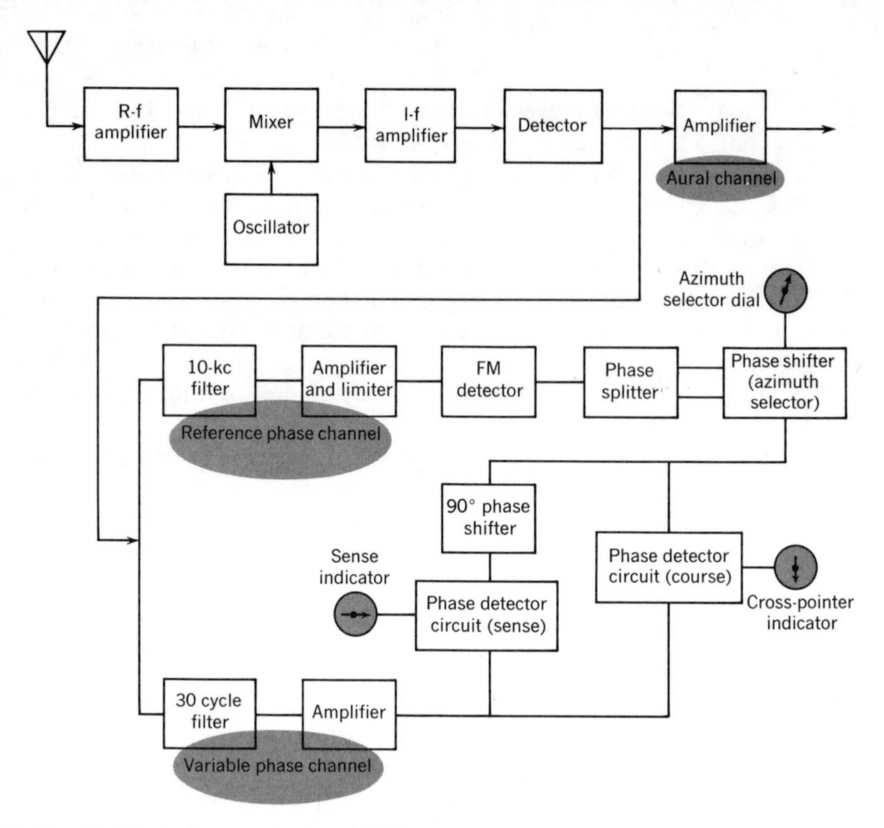


*Fig. 17·27 Simplified block diagram of a typical VOR receiver.*

**The omnirange receiver.** A simplified block diagram of a typical VOR receiver is shown in Fig. 17·27. The output of the detector contains a combination of the 10-kc subcarrier, 30-cps variable-phase signal, and either the 1,020-cps identification tone or voice frequencies. The voice frequencies and the 1,020-cps identification tone are passed through an audio amplifier and then to headphones. Actually, when the transmitter is being voice-modulated, the 1,020-cps identification is interrupted. The 10-kc frequency-modulated subcarrier portion of the detector output is fed to an amplifier, a limiter, and an FM discriminator to obtain a 30-cps reference signal, which is independent of the receiver bearing with respect to the station. The reference signal is fed through a phase-splitting stage to provide two equal voltages which have a phase difference of 90°. These two voltages are then passed through a manual phase-shifting control, which constitutes the azimuth selector in the aircraft. The resultant reference signal at the output of the azimuth selector is fed to a phase-detector circuit, where it is mixed and compared with a filtered and amplified 30-cps variable-phase signal to obtain plus or minus d-c indications. The polarity and amplitude of the direct current depend upon the phase relationship of the reference and variable signals.

The output of the phase-detector circuit is fed to a cross-pointer instrument. When the cross-pointer indicator is centered, the phase difference between the reference and variable signals is indicated by the azimuth selector. The phase-detector circuit is designed so that the cross-pointer instrument will indicate ON COURSE when the variable and reference signals are 90° out of phase upon arrival at the phase detector. The amount of phase shift required in the reference signal to result in the 90° phase angle is provided by the azimuth selector or phase shifter.

Since, for any given bearing from the station, there can be two reciprocal positions of the azimuth selector for which ON COURSE will be indicated, a sense indicator is included to identify the desired course. This sense indicator is connected to the output of a seperate phase-detector circuit, which operates at a phase relationship of 90° from the cross-pointer phase-detector circuit. The sense circuit merely determines whether the original reference signal, before phase-splitting, is in phase, or 180° out of phase with the variable voltage when the cross-pointer instrument is centered. The red sector of the sense indicator indicates loss of either or both the reference and variable signals.

## *17·5 Distance-measuring equipment*

A relatively new radio aid to navigation is called *distance-measuring equipment* (DME). It enables the pilot to obtain a fix by using only one VHF range station. This equipment sends out a signal from the aircraft, which is received by a DME receiver at the VOR station. It is then retransmitted to the aircraft. The total time in microseconds required for the signal to travel from the aircraft to the range station and back to the aircraft is measured. Since the speed at which the radio signal travels is accurately known, its travel time is a direct measure of the distance. The distance to the range station is automatically given by the DME. The radial to the range station may also be determined. The location of the aircraft can therefore be determined continuously when one VOR station is tuned in (Fig. 17·28).

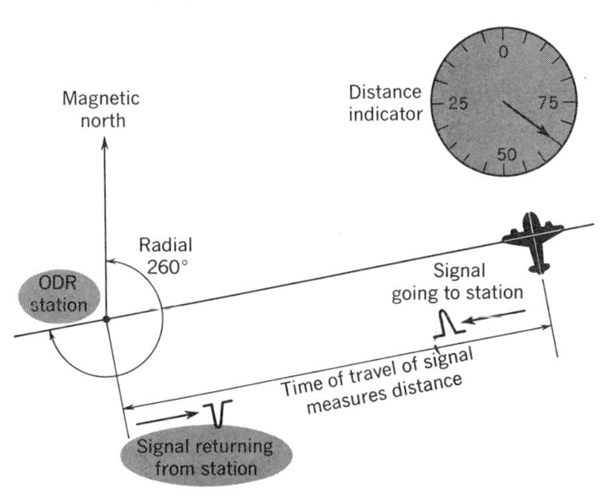


*Fig. 17·28 The method of measuring distance with DME. Note only one VHF ODR station is needed.*

**R-θ computer.** The fact that the distance and radial angle to the range station may be continuously available in the aircraft presents some very interesting possibilities. The location of a point may be determined if its distance from a fixed point is known and the magnetic bearing to the fixed point is known. The distance is called $R$, since it would be the radius of a circle drawn at that distance around the fixed point. The magnetic bearing is an angle which is often designated by the Greek letter $\theta$ (theta), as in Fig. 17·29.

Suppose that the pilot of an aircraft wishes to fly from point $A$ to point $B$, along the line $AB$ (Fig. 17·30). If he knows the distance from the ODR station that the plane should be at each instant, he can tell where he is in relation to his selected course. Of course, the pilot can determine the distance corresponding to several ODR radials along the line. His course would then be a zigzag one at best, as he made corrections. At modern aircraft speeds, the process would be rather unsatisfactory. A computer is designed which will make that calculation for him instantaneously and present the answer in usable form. The computer takes the radial indication from the navigation-system equipment and calculates instantly and continuously the distance from the ODR station. This places the aircraft on the straight line from $A$ to $B$. At the same instant, it takes the *measured* distance from the DME and compares it with the *calculated* distance. This gives an indication of positional deviation from the line $AB$ on the course-deviation indicator. The pilot can then fly the course from $A$ to $B$ by keeping a course-deviation indicator pointer centered.

The computer also continuously calculates the distance from the aircraft to point $B$ and indicates this "distance to go." When the distance meter reads zero and the course-deviation indicator is centered, the pilot knows he is over point $B$.

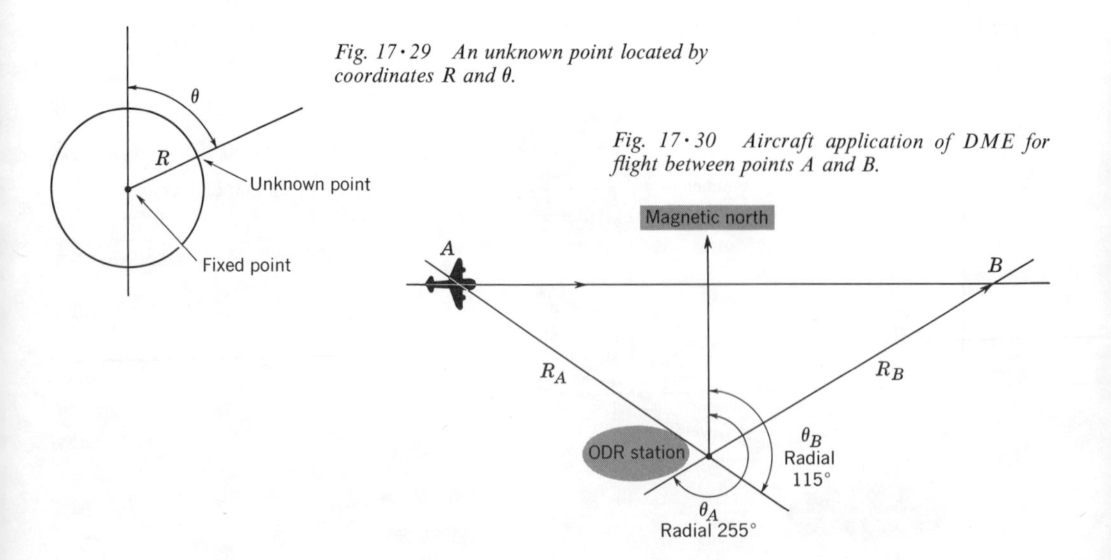


*Fig. 17·29 An unknown point located by coordinates R and θ.*

*Fig. 17·30 Aircraft application of DME for flight between points A and B.*

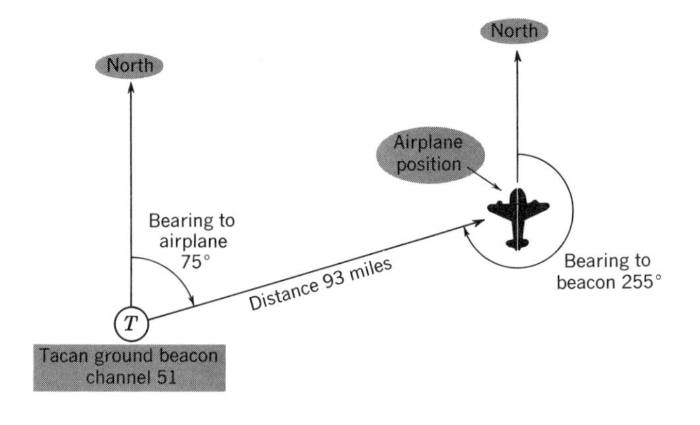


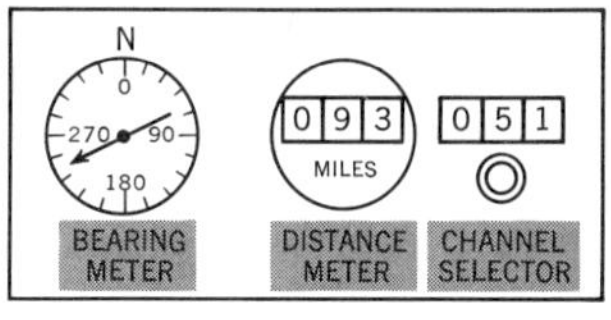

Cockpit display

*Fig. 17·31 Tacan navigation instrument displays.*

## *17·6 Tacan system*

Tacan (tactical air-navigation system) is one of the latest electronic aids to all-weather navigation. Developed for the U.S. Navy and Air Force, it is now being used for commercial and private aircraft. Tacan provides a bearing, or ODR, indication in degrees and a distance, or DME, indication in miles. These indications are obtained relative to the geographical location of a reference ground-beacon station. Knowing the bearing and distance from a specific geographic point, the pilot can fix his position on a chart. Operating in the 1,000-Mc band, this navigational system employs a single multichannel airborne receiver-transmitter. Pulse signals provide both the bearing and distance information. There are 126 clear-frequency two-way channels available for this purpose. These channels are spaced at 1-Mc intervals. In the 1,025- to 1,150-Mc band, 126 frequencies are available for air-to-ground transmission. Ground-to-air transmissions are handled on 63 frequencies in the 962- to 1,024-Mc band and 63 frequencies in the 1,151- to 1,212-Mc band.

Pulse coding, that is, the transmission of characteristic pulse groups, with some prearranged spacing between component pulses of the group, is used in the tacan system to increase the average radiated power and signal-to-noise ratio. In addition, the pulse coding serves the purpose of transmitting the bearing, or ODR, information on the same channel as the DME signal. Figure 17·31 illustrates the principles of the tacan navigational-information display.

Radar principles are employed in the DME portion of the system. The time of round-trip travel of pulses between two points is measured. The airborne transmitter sends out narrow, widely spaced "interrogation" pulses, which are picked up by the ground-beacon receiver. The receiver output triggers the associated transmitter, which sends out "reply" pulses on a different channel. The replies are picked up by the airborne receiver.

Automatic-timing circuits measure the round-trip travel time and convert this information into electric signals to operate the distance meter.

A single ground beacon may be interrogated simultaneously by a number of aircraft which are in the vicinity and which have tuned to the beacon's channel. The ground beacon will reply to all interrogations. Each aircraft will receive all replies on its channel.

To prevent interference, each aircraft's interrogation pulses occur at a rate that, within limits, is randomly wobbled. A search circuit in the receiver examines all reply pulses received by the aircraft, and within a short time finds the one series of reply pulses which has the identical repetition rate as the aircraft's own initiated interrogation pulses. This search process occurs only when a ground beacon is first tuned in. After this initial period, the circuit locks in to the proper series of reply pulses automatically and tracks them as their time delay slowly changes because of the aircraft's changing distance.

Pulse repetition rates are of the order of 30 per second (temporarily higher during the search process). The pulses are narrow, in the order of a few microseconds. The pulse signals are twin pulses with a prearranged spacing from 10 to 80 $\mu$sec. The receivers, ground and airborne, are followed by discriminators or twin-pulse decoders, which are set to pass only pulse pairs of the prescribed spacing. Isolated single pulses, or groups with some other spacing, will not pass through the decoder.

For bearing operation, the ground antenna produces 3,000 pulses per second. Some pulses are distance-reply pulses; others are random-filler pulses. Strength of the radiated signals from the antenna depends on the antenna direction pattern which, for tacan, is a cardioid.

By rotating this cardioid pattern, the signal received along any given direction from the beacon goes through corresponding variations in strength as a function of time. Since the cardioid is a single-lobed pattern, it takes one full turn of the antenna pattern in 1/15 sec for the received signal to go through one complete cycle of variation—say, maximum to minimum, back to maximum again. The airplane receives a 15-cps modulation, from which an audio sine-wave signal may be extracted by radio-detection methods.

The phase of this audio signal is important in determining the bearing. For example, consider the instant during which the maximum lobe points due north. At this moment, let us say that the 15-cps modulation wave is passing through zero; 1/60 sec later, the maximum lobe of the pattern is pointed due east and the phase of the 15-cps modulation is now passing through its 90° point. Then, 1/60 sec later, the maximum lobe points due south, and the 15-cps signal modulating the carrier has shifted another 90° in phase. For other directions around the beacon, the electrical phase of the 15-cps modulation is proportionately different. For the cardioid configuration, each degree of geographical bearing change corresponds to one degree change in electrical phase of the 15-cps signal. Ground station beacons are identified by keyed Morse characters, transmitted at regular intervals.

## 17·7 Radio marker beacons

As we have learned previously, a plane, when flying directly over the beacon transmitter, passes through a cone of silence. This zone of no-signal is an advantage, since it informs the pilot that he is passing over the range station. However, it is not an error-proof indication. The silence may be due to some other cause, such as an inoperative receiver or transmitter.

When the radio-range system was first devised, it was intended for the guidance of aircraft only during weather classified as "good for flying." It was presumed that the weather was favorable enough so that after the plane had employed the range to reach its destination, the pilot could recognize the airport and make a landing. The cone of silence served as a terminal marker for pilots for many years. The "approach procedure," or maneuver used by a pilot to guide him through an overcast and lead him to the airport, utilized these no-signal indications.

**The Z marker.** The cone, however, is affected by the presence of nearby objects and terrain. To correct for these difficulties, an entirely different type of terminal indication was developed. The new marker, known as the *Z,* or zone, *marker,* consists of an arrow cone-shaped radiation field, transmitted vertically at 75 Mc (4 meters). The very high frequency employed provides virtually static-free operation. A dual transmitter with a 5-watt output is used in order to insure continued operation should one transmitter unit become inoperative. The basic frequency of these transmitters is crystal-controlled at the fundamental frequency, 6.25 Mc. Both aural and visual indication are provided for in the Z-marker system. When the aircraft is flying over the narrow vertical field of the marker, a 3,000-cps tone is obtained in the receiver output, and a flashing white light on the instrument panel serves to inform the pilot. Z-marker fields are normally about 1 mile wide at an altitude of 1 mile.

**The fan marker.** Fan markers also operate on a frequency of 75 Mc. They radiate a fan-shaped field, which, at its highest point, is approximately 15 miles wide and 5 miles deep. Fan markers are located on each leg of a radio range and mark the point at which a new range station should be tuned in. The same aural indication (3,000 cps) is utilized. The cockpit light is also actuated. Keying by a series of dashes is used to identify the particular range leg.

Fan markers have been useful to indicate "hold" points, over which planes wait for permission to make a landing from the air-traffic control tower. The fan field extends to much higher altitudes than do Z-marker radiations. In order to accomplish this, fan transmitters have a 100-watt rating. The antenna used consists of four half-wave dipoles in the same plane. That is, each dipole is fed in phase with the dipole next to it. The entire array is situated one-quarter wavelength above a galvanized iron-screen reflector surface. This screen provides unidirectional (vertical) radiation.

## 17·8 Instrument landing systems

Navigational aids have been studied in the previous sections which permit pilots to fly planes where they wish to go. Upon arrival at their destination, it becomes necessary to land the aircraft under all weather conditions and under every possible degree of visibility. About a hundred instrument landing systems (ILS) have been proposed over a period of years to accomplish a truly "blind" landing.

Threefold guidance is necessary to accomplish instrument landing. Directional guidance to determine the runway direction may be obtained from a four- or two-course radio range, or localizer. Horizontal guidance provides an indication of the distance to the airfield. This is supplied by one or two markers placed along the approach course of the aircraft. The third, or vertical, guidance gives an indication of the aircraft's height and provides a path down to the runway. This radio trail is called a *glide path.*

**The diamond-dumore ILS.** This system, devised in 1928, employs a runway localizer, longitudinal guidance, and a glide path. The localizer operated on 278 kc, and the two markers used for longitudinal indication were on a frequency of 3,105 kc. The inclined glide path was obtained by using a special horizontally polarized antenna array on 90.8 Mc.

The signals transmitted by the localizer are displayed visually by a vertical pointer on a cross-pointer instrument. The glide path is indicated by the horizontal pointer. The marker indication is both aural and visual. In some early systems, passage over a marker was indicated aurally by its characteristic tone value and visually by the flickering of a white light on the instrument panel. Modern indicators use a purple light and a low tone for the outer marker and an amber light and high tone for the inner marker. The white light is now used for *Z* and fan markers.

**The FAA system.** Previous systems of instrument landing had difficulty in obtaining a suitable glide-path pattern. Curvature of the glide path is, of course, desirable. The FAA (Federal Aviation Agency) system provides an essentially straight-line glide path (see Fig. 17 · 32). Ordinary glide-path antenna arrays radiate a teardrop-shaped pattern. In the FAA system, the normal teardrop-shaped pattern is distorted so that it produces, all along the radio "corridor," an almost straight radio "floor." The array is located off the runway, to one side, and transmits its signal at an angle to the flight path of the aircraft. This kind of glide path is generally good enough for any pilot to make an instrument landing in complete safety.

ILS installations consist of a localizer unit just beyond the airport, a localizer monitor and antenna, the glide-path transmitter and its array, and the inner-marker equipment and antenna. An airport traffic-control tower contains a special instrument landing board on which colored lights monitor the landing-system operation. Meters on the control board give the operator visual indication of the localizer, the glide path, and the two marker transmitters. An automatic recorder provides a permanent record of the various units.

On the aircraft, a horizontally polarized receiving loop is mounted on top of the fuselage. The receiving equipment includes a localizer receiver, glide-path receiver, and marker receiver, as well as the cross-pointer indicator. Localizers operate on either 109.5, 109.9, or 110.3 Mc. For the glide

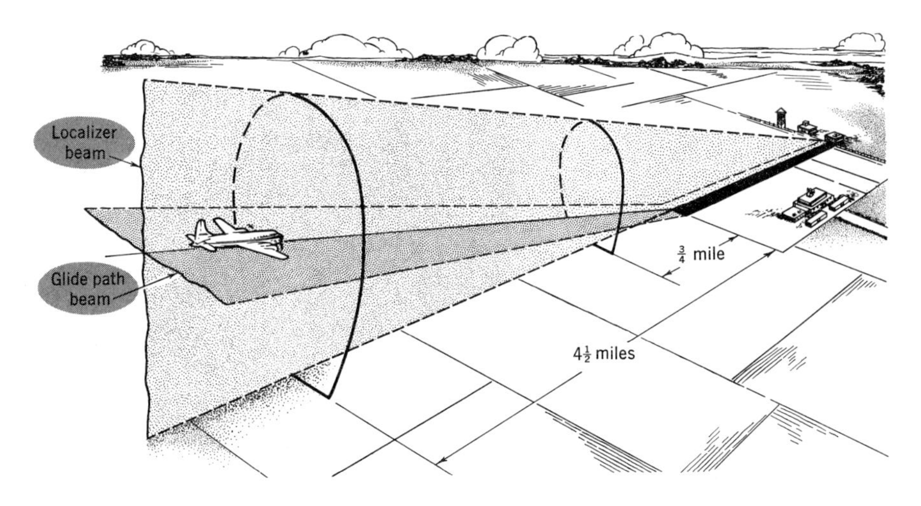


*Fig. 17·32 The guiding beams that are employed in an ILS system.*

path, 93.5, 93.9, and 94.3 Mc are used. The marker beacons operate on 75 Mc. They transmit fan-shaped patterns across the approach course. The outer marker is modulated with a 400-cycle note and flashes a purple light. The inner marker is 1,300-cycle-modulated and flashes the amber light.

**Ground-controlled approach.** A contribution to instrument landing made during World War II employs a ground-based scanning radar for obtaining accurate relative position of a landing plane with respect to a safe path of descent to a runway. A "talk-down" technique for giving information and instructions for landing to the pilot by radio is also used in this system, known as ground-controlled approach (GCA).

A self-contained mobile unit consists of two radars: a 10-cm search radar and a 3-cm sector-scanning system that covers a limited volume of space surrounding the final descent path. A single trailer houses the radars and operating positions for a six-man crew: four scope operators, a controller, and a supervisor. Also included is radio equipment for communicating with both the aircraft and the control towers.

The search system detects and identifies individual airplanes coming into the landing zone. Instructions to the pilot by radio help him guide the plane through an approach pattern into a rough alignment with the final descent path, some 4 to 10 miles from the field. The search set thus feeds the airplane into the region where the 3-cm scanning beams are operating and where the pilot receives more precise positional information.

Search data are presented on two identical 7-in. CRTs, used as position indicators and equipped with range scales of 7.5, 15, and 30 miles and with compass and heading lines. The two position-indicator operators, the controller, and the aircraft selector, although supplied with identical information, perform different functions. The controller observes the traffic entering the area and identifies those airplanes calling for instructions.

The controller normally guides the aircraft through a definite traffic pattern and then turns it over to the aircraft selector. The selector guides the planes into the narrow sector covered by the scanning system until the aircraft is being tracked by the precision operators, and the final approach controller takes over for final landing instructions.

## 17·9 Loran

Another important electronic aid to navigation is a system of long-range navigation known as *loran.* Developed during World War II by the Allies, it was put to immediate military use in guiding ships and aircraft in bombing raids and in general navigation to all parts of the globe. Shipboard and aerial navigators today use the loran system constantly, checking its accuracy occasionally with time-tested methods of sun and star sightings. Loran provides a reliable fix in just a few minutes, no matter how overcast the sky, how fogbound the surface, or how rough the sea.

The word *loran* is derived from the words "long-range navigation." It is a navigational system by which a pilot or navigator can determine his location or follow a plotted course by radio navigation. It is especially useful for long overwater flights or journeys.

**Operation.** Let us consider a radio receiver capable of measuring the very small time difference between the arrival of two signals at the receiving location. If the receiver is placed between two transmitters which are keyed at the same instant, as in Fig. 17·33*a*, it can measure the difference between the time of arrival of the signals. In fact, if the receiver is moved anywhere along the perpendicular bisector of the line joining the two transmitters, it will indicate a zero time difference between the pairs of recurring signal impulses, since the two transmission paths are equal.

If the receiver is moved to point *A* of Fig. 17·33*b*, a difference in the time of arrival of the two signals is noted. Since the speed of propagation is equal in every direction, the time delay measured by the receiver is directly proportional to the difference in the length of path. At point *B*, the same time difference is obtained as at point *A*. This leads to an ambiguity as to which side of the midpoint line the receiver is located. This ambiguity is resolved by keying one transmitter, designated as the *master,* receiving the signal at the other transmitting location, waiting for a definite time interval, which we might call the *station delay,* and then keying the second, or *slave,* transmitter.

As shown in Fig. 17·33*b*, points *A* and *B*, equidistant from the midpoint, may now be distinguished by the fact that the receiver measures a different time difference at each point. There will be a lesser difference or delay measured between the arrival of the master and slave signals at point *A* than at point *B*. This is because the master signal is received later at *A* than at *B*, and the slave signal is received sooner at *A* than at *B*.

It should be noted that the master signal is always received first, followed a short interval later by the slave signal. At point *B*, the time interval between arrival of the master and slave signals will be greater than the

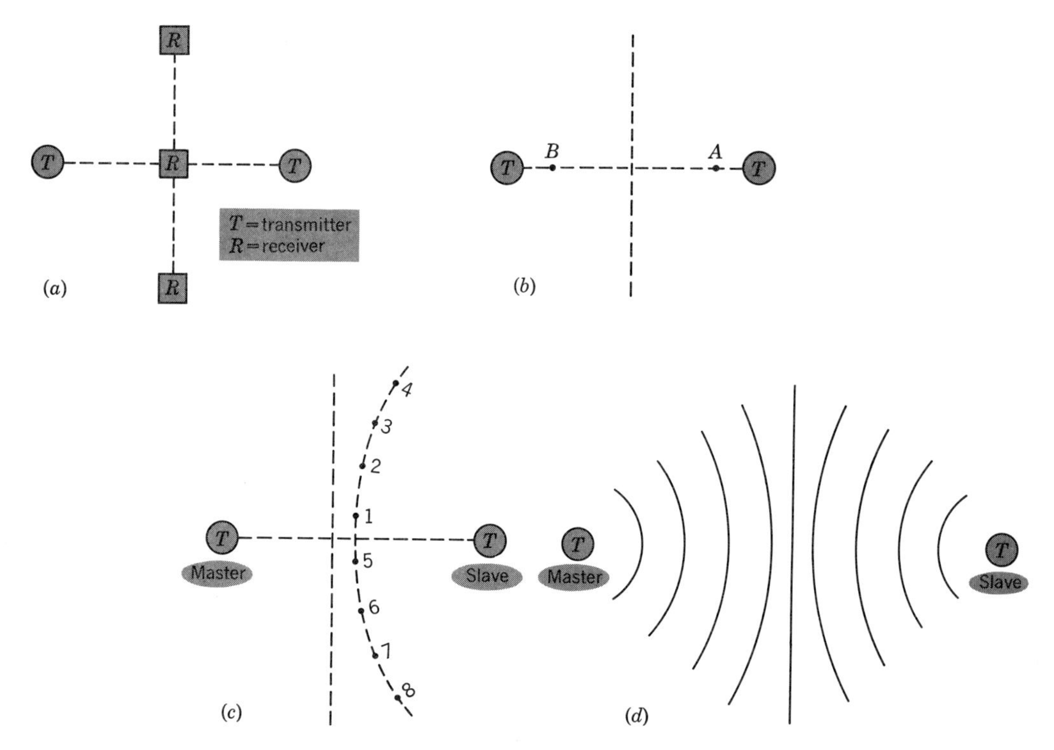


*Fig. 17·33 Basic principles of loran operation. (a) Receivers indicate zero time difference between signals of both transmitters. (b) Time difference is obtained at locations A and B because of pathlengths difference from the transmitter. (c) Points of constant delay. (d) Hyperbolic curves joining points of constant delay.*

interval at point $A$. That is how the two positions are distinguished from each other.

We have seen from Fig. 17·33*a* that a receiver may be moved anywhere along the perpendicular bisector to maintain a constant delay reading. Now, let us take our receiver and move it to a point not on this bisector line. Let us call this point 1 (see Fig. 17·33*c*). At this point, there will be a certain time delay between the arrival of the master and slave pulses. If we move the receiver to point 2, the time delay remains the same. If we keep moving this receiver so that the time delay remains the same, we will find that it describes a hyperbolic[1] arc. This is the dotted line in Fig. 17·33*c*.

In Fig. 17·33*d*, a number of hyperbolic lines are drawn. A receiver, moving along any one line, will record the samc time delay between the arrival of master and slave pulses. For each different line, however, a different time delay is obtained.

Up to this point, we have assumed a flat earth. On the spherical earth,

[1] The hyperbola is a type of geometric curve. Specifically, it includes points whose distance from a fixed point at the focus has a constant ratio to its distance from a fixed line.

these curves of constant time difference are still similar, so that properly modified hyperbola-like lines may be drawn on charts or their values in terms of latitude and longitude tabulated. A pair of stations, as in Fig. 17 · 33*d*, would furnish to any receiver within their range a series of lines of position.

This information alone would prove insufficient for most navigation, since the navigator of a ship or plane would wish to know not only in what lane he was traveling, but also at what point along the lane he happened to be at the moment. Another pair of stations is arranged to give another set of lines of position which intersect the first set. The navigator's receiver then finds the two lines of position along which his ship or plane is traveling at the moment. Their point of intersection tells him where he is at the time of observation. In actual practice, only one transmitter is necessary for sending out both master signals. The navigator then identifies pairs of signals by means other than the difference in their radio frequency. Thus, very little of the radio spectrum is used by the loran system. The transmitters must be keyed or pulsed very rapidly indeed, to furnish a useful continuous service to navigators. Since the velocity of propagation of radio waves is 186,000 statute miles per second, and therefore 0.186 mile per μsec, all our thinking must be done in terms of this small unit of time, if we are to achieve any accuracy at all in the loran system.

In the practical loran system, a pair of shore stations, composed of a master and a slave station, is synchronized by means of ground waves and provides a time-difference pulse signal for the loran receiver. The navigator may select any two pairs to obtain a fix, first reading the time difference of arrival of pulses from one pair, then the other. As many as eight pairs may be operated on one single frequency. The pairs are identified by different recurrence rates at which they operate. Only ground waves are available during the daytime, and these give a coverage of 700 nautical miles over water. At night, owing to reflections from the ionosphere, fixes can be obtained up to 1,500 nautical miles.

*Fig. 17 · 34 Typical airborne loran receiver (left), control box (center), and cathode-ray screen (right).* (Edo Corp.)

The pair of received signals are displayed on a cathode-ray screen, with double traces. Two fast traces are initiated so that one exhibits the signal from the master station and the other exhibits the signal from the slave station. The signals are then superimposed and made equal in amplitude. A line trace, containing markers to give coarse readings in time difference, is switched on in place of the signals. Switching to a double trace gives fine readings of time difference. The procedure is repeated for a second pair of stations. The results are then plotted on a chart to give the navigator his position. Under normal conditions, it usually takes about three minutes to take and plot a fix.

A sky-wave-synchronized loran is a night-time version of the basic loran. The pair of stations are synchronized by reflections from the ionosphere. The navigator follows the same procedure as in standard loran, except that sky waves only are used. Loran operates in the 1,700- to 2,000-kc band and employs high-power pulses of approximately 50 $\mu$sec duration. Figure 17·34 shows a typical loran receiver, control box, and indicator.

## SUMMARY

1. The directivity characteristics of loop and dipole antennas are employed to determine the direction of a distant transmitter in modern radio-direction-finding (RDF) equipment. This is accomplished by detecting the direction of arrival of the signal picked up by the receiving antenna.
2. ADF, or automatic-direction-finding, apparatus provides a continuous bearing indication. It employs a combination of a loop and sensing antenna to provide a unidirectional characteristic.
3. A crossed-loop direction finder (Bellini-Tosi) employs two loop antennas mounted at right angles to each other in conjunction with a goniometer. This is commonly used in ground DF stations.
4. An Adcock antenna is basically a pair of vertical dipoles, crossed over at their centers in order to eliminate the effect of night error.
5. A radio range of beacon makes use of the directional properties of antennas to provide means for marking courses of flight.
6. Radio beams transmitted from range stations produce four courses, including two reciprocal courses. Either loops or vertical dipoles are used for this purpose.
7. The A-N range employs interlocking A and N Morse code signals to provide on- or off-course information. The intensity of the predominating signal is an indication of the off-course distance.
8. The directions of range beams may be shifted by employing antennas with different input phases and different amounts of power. Rectangular, instead of square, antenna arrangements can also be used.
9. Simultaneous range transmission provides weather information as well as course indication.
10. The Alford loop antenna provides a horizontally polarized field, which has very little vertical pickup or radiation. It is useful in VHF range equipment.
11. Omnirange (ODR) is a VHF radio-range system that provides accurate navigational information on any course which the pilot may select. It operates by comparing the phase between two signals.
12. Distance-measuring equipment (DME) permits the pilot to obtain a fix by using only one VHF range station. The time necessary for a wave to travel from the plane to a receiving station and back determines the distance to the range.
13. Tacan (tactical air navigation system) is a navigational aid operating in the 1,000-Mc band, which provides bearing as well as distance indication.

14. Marker beacons replaced the "cone-of-silence" marker with either the Z, or zone, type or the F, or fan, type.
15. ILS (instrument landing system) provides runway direction, distance, and height guidance to permit blind landing of aircraft.
16. Ground-controlled approach (GCA) utilizes a ground-based securing radar for determining the position of plane during the approach procedure. A 10-cm search and a 3-cm sector-scan radar are used in conjunction with a "talk-down" technique.
17. Loran (long-range navigation) operates by measuring the time difference between reception of transmissions from a ground-based master and a synchronized slave station. Accurate fixes are possible for distances of 700 miles during the day and up to 1,500 miles at night.

## SELF-EXAMINATION

Here's a chance to find out how well you have learned the material in this chapter. Work the exercises, then check your answers against the Key at the back of this book. These exercises are for your self-testing only.

1. A common antenna employed in direction-finding apparatus is the (*a*) in-line; (*b*) folded unipole; (*c*) rhombic; (*d*) loop.
2. The directional characteristic of a combined vertical dipole and loop antenna arrangement is a (*a*) circle; (*b*) square; (*c*) cardioid; (*d*) figure eight.
3. ADF systems provide the pilot with (*a*) a continuous bearing indication; (*b*) air traffic-control information; (*c*) blind-landing approach signals; (*d*) hyperbolic navigation lines.
4. The Bellini-Tosi direction finder employs a (*a*) square array of vertical dipoles; (*b*) crossed loops; (*c*) folded horizontal loops; (*d*) parasitic dipoles.
5. The purpose of the Adcock antenna is to (*a*) increase the antenna gain; (*b*) develop a less exact null indication; (*c*) eliminate the effect of night error; (*d*) aid the loop antenna directivity.
6. Aural range equipment makes use of the principle of (*a*) crossed dipole sensitivity; (*b*) loop sensing accuracy; (*c*) interlocking code letters; (*d*) directional switching.
7. Simultaneous radio range provides for (*a*) blind-landing instructions; (*b*) proper approach patterns; (*c*) transmission of weather information; (*d*) accurate distance determination.
8. Omnidirectional range equipment operates on (*a*) very high frequencies; (*b*) two crossed Adcock antennas; (*c*) two radar pulses; (*d*) the principle of "talk-down" technique.
9. Tacan is a navigational aid providing (*a*) bearing and weather information; (*b*) bearing and distance indication; (*c*) instrument-landing glide paths; (*d*) speed and height indication.
10. Loran is a navigation system used primarily for (*a*) blind landing; (*b*) approach control; (*c*) automatic collision warning; (*d*) obtaining fixes over large distances.

## QUESTIONS AND PROBLEMS

1. Explain the action of the loop antenna as employed in direction finders.
2. What is night error? How is it produced?
3. Describe the action and operation of the right-left radio compass.
4. How does the automatic direction finder operate?
5. What is an Adcock antenna, and how is it used in direction finding?
6. Explain the principle of the A-N range.
7. What is the method of operation of DME (distance measuring equipment)?
8. What is the cone of silence, and how is it used as a marker?
9. How does GCA use its two radar units?
10. Describe briefly the taking of a fix in a Loran system.

# Review of chapters 16 and 17

## SUMMARY

1. Radar is designed to measure the distance to a target, which is generally an aircraft or surface ship, by bouncing a high-frequency radio signal off the object and measuring the echo time electronically.
2. A radar system consists of a pulser or transmitter modulator, a transmitter, an antenna, a receiver, a display unit, and a synchronizer.
3. Sonar applies the principle of radar to the detection of underwater objects. Instead of radio frequencies, sonar uses sound waves.
4. Sonar may be active or passive. In the active system, the equipment generates a sound wave that is bounced off another ship, reflected back to a receiver, and then analyzed for information. In passive sonar, the system simply listens to underwater sounds.
5. Another electronic search device is the mine detector. This consists of a search head, a sensitive receiving amplifier, and an audio generator. Detection of a hidden metallic object (such as a mine) occurs when the search head comes close enough to the object to have its characteristics altered.
6. A fire-control system accurately aims and fires guns by electronic controls. Information concerning the target is obtained by radar and fed to a computer which predicts where the target and shell should meet. The gun is then positioned accordingly and fired at the proper moment.
7. Infrared systems locate objects by the heat which these objects radiate. Both passive and active infrared systems are in use, although passive systems are more widely employed.
8. A missile guidance system is able to control flight by computing target and flight information and feeding it to the missile controlling members to steer it toward the target. Servos, radar, infrared, and telemetering are some of the more important sections of a guided missile.
9. The directional characteristics of loops and dipole antennas are used to determine the direction of a distant transmitter in radio-direction-finding equipment.
10. Automatic direction finder (ADF) provides a continuous bearing indication. A loop and a sensing antenna combine in this equipment to provide a unidirectional pattern.
11. Aircraft can be guided through the sky by means of radio ranges or beacons. These radio beams establish fixed courses which can be readily followed with the proper receiving equipment. Simultaneous range transmission provides weather information as well as course indication.
12. Omnirange (ODR) is a VHF radio-range system that provides accurate navigational information on any course which the pilot may select. It operates by comparing the phase between two signals.
13. Distance-measuring equipment (DME) enables a pilot to obtain a "fix" by using only one VHF station. The time necessary for a wave to travel from the plane to a receiving station and back determines the distance to the range.
14. Tactical Air Navigation System (tacan) operates around 1,000 Mc and provides a bearing indication in degrees and a distance indication in miles.
15. Instrument landing systems (ILS) provide runway direction, distance, and height guidance to enable aircraft to land in bad weather for a blind landing.
16. Ground-controlled approach (GCA) utilizes a ground radar system to determine the position of a plane during its approach. A 10-cm search and a 3-cm sector-scan radar are used in conjunction with a "talk-down" technique.

17. Long-range navigation (loran) operates by measuring the time difference between reception of transmissions from a ground-based master and a synchronized slave station. With this system, accurate fixes are possible for distances of 700 miles during the day and up to 1,500 miles at night.

## REFERENCES (*Additional references at back of book.*)

*Books*

Barton, D. K., *Radar Systems Analysis,* Prentice-Hall, Inc.
Lang, D. G., Marine Radar, Pitman Publishing Corporation

*Government Publications*

The following books are published by the U.S. Government Printing Office, Washington 25, D.C., and may be purchased from them:
*Radarman 3 & 2*
*Aviation Electronics Technician 3 and 2*
*Loran*
*Fire Control Technician 1 and C,* Vol. 3
*Fire Control Technician 2*
*Fire Control Technician 3*
*Radar Circuit Analysis*
*Radar Electronic Fundamentals*
*Radar Fundamentals and Surveillance, Precision, and Route Radar*
*Radar System Fundamentals*

## REVIEW SELF-EXAMINATION

Here's another chance to check your progress. Work the exercises just as you did those at the end of each chapter and check your answers.

1. One disadvantage of frequency-shift radar is that it *cannot* (*a*) detect an object moving toward the radar antenna; (*b*) detect an object moving away from the radar antenna; (*c*) detect a stationary object; (*d*) operate at high frequencies.
2. In pulse radar, the transmitted pulses are obtained from a (*a*) synchronizer; (*b*) modulator; (*c*) lighthouse tube; (*d*) magnetron.
3. IFF stands for (*a*) International Frequency Forum; (*b*) Identification, Friend or Foe; (*c*) Intermediate Frequency Follower stage; (*d*) Independent Forecast Forum.
4. The sonar transducer (*a*) serves as the transmitting and receiving antenna of these systems; (*b*) develops the signal to be transmitted; (*c*) receives only incoming signals; (*d*) is equivalent to the cathode-ray tube in a radar system.
5. Infrared detection depends on the fact that all objects (*a*) emit light; (*b*) radiate electromagnetic waves; (*c*) possesses some sort of odor; (*d*) contain some radioactivity.
6. Range beacon systems use the following two letters for guidance: (*a*) C-D; (*b*) A-N; (*c*) G-H; (*d*) X-Z.
7. Three basic instruments are used by the pilot when flying with a VHF ODR. One of these is a (*a*) cross-pointer; (*b*) cathode-ray tube; (*c*) right-left radio compass; (*d*) goniometer.
8. The R-$\theta$ computer is designed to (*a*) act as an automatic pilot; (*b*) take the place of an antenna; (*c*) present a view of the terrain being flown over; (*d*) automatically calculate the flight path.
9. A Z marker is employed in (*a*) tacan; (*b*) radio range systems; (*c*) loran; (*d*) instrument landing systems.
10. GCA operates with (*a*) distance-measuring equipment; (*b*) master and slave stations; (*c*) ground-based radar; (*d*) local radio stations.

Answer true or false

1. Sonar uses high-frequency radio waves.
2. Mine detectors use high-frequency sound waves.

# *Appendix B* *FCC frequency allocations*

*FCC frequency allocations from 30 kc to 300,000 Mc*

| *Band* | *Allocation* | *Remarks* |
|---|---|---|
| 30–535 kc | Includes maritime communications and navigation, international fixed public band, aeronautical radio navigation | Very low, low, and medium radio frequencies |
| 535–1,605 kc | Standard radio broadcast band | AM broadcasting |
| 1,605 kc–30 Mc | Includes amateur radio, loran, government radio, international short-wave broadcast, fixed and mobile communications, radio navigation, industrial, scientific, and medical equipment | Amateur bands 3.5–4.0 Mc and 28–29.7 Mc; industrial, scientific, and medical band 26.95–27.54 Mc; Citizen's band class D for voice is 26.965–27.225 Mc and 27.255 Mc. |
| 30–50 Mc | Government and nongovernment, fixed and mobile | Includes police, fire, forestry, highway, and railroad services; VHF band starts at 30 Mc |
| 50–54 Mc | Amateur | 6-meter band |
| 54–72 Mc | Television broadcast channels 2–4 | Also fixed and mobile services |
| 72–76 | Government and nongovernment services | Aeronautical marker beacon on 75 Mc |
| 76–88 | Television broadcast channels 5 and 6 | Also fixed and mobile services |
| 88–108 Mc | FM broadcast | Also available for facsimile broadcast; 88–92 Mc educational FM broadcast |
| 108–122 Mc | Aeronautical navigation | Localizers, radio range, and airport control |
| 122–174 Mc | Government and nongovernment, fixed and mobile, amateur broadcast | 144–148-Mc amateur band |

*FCC frequency allocations from 30 kc to 300,000 Mc (continued)*

| *Band* | *Allocation* | *Remarks* |
|---|---|---|
| 174–216 Mc | Television broadcast channels 7–13 | Also fixed and mobile services |
| 216–470 Mc | Amateur broadcast, government and nongovernment, fixed and mobile, aeronautical navigation, citizens' radio | Radio altimeter, glide path and meteorological equipment; citizens' radio band 462.55–465 Mc; civil aviation 225–400 Mc; UHF band starts at 300 Mc |
| 470–890 Mc | Television broadcasting | UHF television broadcast channels 14–83 |
| 890–3,000 Mc | Aeronautical radionavigation, amateur broadcast, studio-transmitter relay, government and nongovernment, fixed and mobile | Radar bands 1,300–1,600 Mc |
| 3,000–30,000 Mc | Government and nongovernment, fixed and mobile, amateur broadcast, radio navigation | Super high frequencies (SHF) 8,400 to 8,500 Mc for satellite communications |
| 30,000–300,000 Mc | Experimental, government, amateur | Extremely-high frequencies (EHF) |

# *Appendix C Logarithms*

A logarithm is an exponent. When we express a number in powers of 10, its exponent is the logarithm of the number to base 10. Some examples follow:

$$
\begin{array}{rll}
10 = 10^1 & \text{or} & 1 \text{ is the log of } 10 \\
100 = 10^2 & \text{or} & 2 \text{ is the log of } 100 \\
1{,}000 = 10^3 & \text{or} & 3 \text{ is the log of } 1{,}000 \\
10{,}000 = 10^4 & \text{or} & 4 \text{ is the log of } 10{,}000 \\
100{,}000 = 10^5 & \text{or} & 5 \text{ is the log of } 100{,}000 \\
1{,}000{,}000 = 10^6 & \text{or} & 6 \text{ is the log of } 1{,}000{,}000
\end{array}
$$

With 10 as the base, its exponents are called common logarithms, also known as Briggs logarithms.

There are two reasons for the common use of logarithms. Since they are exponents, logarithms reduce the work of multiplying and dividing numbers to adding or subtracting their logarithms. An interesting example is the slide rule, which has scales with logarithmic spacing. As another advantage, logarithms can be useful in compressing a range of numbers. For example, the wide range of 10,000 to 10 in numbers is a range of only 4 to 1 in their logarithms. This feature is the reason why the decibel is useful as a logarithmic unit. Furthermore, a logarithmic scale is generally used for compressing the range of frequencies in a response curve (see Fig. C·1). This graph is semi-log because one axis is linear. The frequency axis has logarithmic spacing to show the lowest and highest extremes, without extending the graph too far. The three cycles or repeats of log-spacing here are used for tens, hundreds, and thousands.

*Characteristic and mantissa*

Numbers like 10 or 100 have the logarithm of 1 or 2, without any fractions, because these numbers are exact multiples of the base 10. However, any number has a logarithm. The number 32, for instance, is between 10 and 100. Therefore its logarithm is between 1 as the log of 10 and 2 as the log of 100. In fact, the log of 32 happens to be 1.5, approximately. With such logarithms that are not round numbers, the whole-number part is called the characteristic, and the fractional part is the mantissa. In this example of 1.5 the logarithm consists of 1 for the characteristic, plus 0.5 for the mantissa. Determining a logarithm consists of two parts, therefore: (1) decide on the characteristic, (2) find the mantissa in a log table.

*Determining the characteristic*

How the number fits in the scale of 10, 100, 1,000, and so forth, determines the characteristic. Or, the characteristic in the log indicates where the decimal point is in the number. For

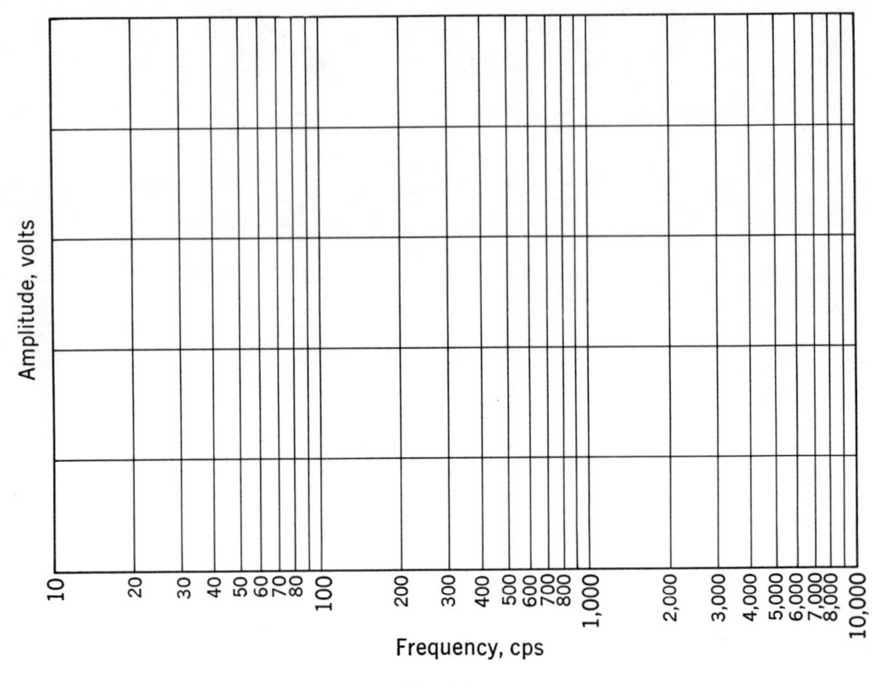


*Fig. C·1*

instance 320 has the same digits as 32, but 320 has a characteristic of 2 because the number is in the hundreds scale between 100 and 1,000. The rule is: For positive numbers more than 1 the characteristic equals one less than the number of places to the left of the decimal point. As examples:

| *Number* | *Characteristic* |
|---|---|
| 1 | 0 |
| 3.2 | 0 |
| 32 | 1 |
| 320 | 2 |
| 3,290 | 3 |

The decimal point is usually omitted for whole numbers, but it is after the last place. Note that the log of 1 is zero.

Because fractions are less than 1, they have a negative characteristic. The rule for decimal fractions is: For positive numbers less than 1, the negative characteristic equals one more than the zeros to the right of the decimal point. Some examples follow:

| *Number* | *Characteristic* |
|---|---|
| 0.1 | −1 |
| 0.32 | −1 |
| 0.032 | −2 |
| 0.0032 | −3 |

*Table of mantissas*

The mantissa, which is the fractional part of the logarithm, is found from the log tables given here. Although they are called log tables, only mantissas are listed. To find the mantissa look for the first two digits of your number in the N column and read this row across, looking for the mantissa in a column under the third digit of the number. Here are some examples of mantissas from the log table:

| *Number* | *Mantissa* |
|---|---|
| 2 | 3010 |
| 3.2 | 5051 |
| 20 | 3010 |
| 32 | 5051 |
| 200 | 3010 |
| 320 | 5051 |
| 329 | 5172 |

When you read the digits from left to right in the number, forget the decimal point. It has no effect on the mantissa. The numbers 3.2, 32, and 320 all have the same mantissa because the digits are the same. Similarly, the mantissa is the same for 2, 20, and 200. These numbers are different, though, in that they have different characteristics.

*Finding the logarithm*

In summary, to find the logarithm of a number, determine its characteristic by inspection of the decimal places; then find the mantissa in the log table, using the digits in the number from left to right regardless of the decimal point. Some examples follow:

| *Number* | *Logarithm* |
|---|---|
| 1 | 0.0000 |
| 2 | 0.3010 |
| 3.2 | 0.5051 |
| 10 | 1.0000 |
| 32 | 1.5051 |
| 54 | 1.7324 |
| 100 | 2.0000 |
| 200 | 2.3010 |
| 320 | 2.5051 |
| 329 | 2.5172 |
| 1,000 | 3.0000 |
| 1,230 | 3.0899 |
| 2,000 | 3.3010 |
| 10,000 | 4.0000 |

For numbers less than 1, the characteristic is negative. However, the mantissa is always positive. Therefore we generally write a negative characteristic, such as $-1$, in the converted form of $9 - 10$, to make the entire logarithm positive. As examples: log 0.1 equals $-1$, or $9.0000 - 10$; log 0.32 equals $9.5051 - 10$; log 0.032 equals $8.5051 - 10$.

*Antilogarithms*

Sometimes it may be necessary to find the number that belongs to a logarithm. The number is the antilog. In the preceding table numbers in the left column are antilogs of their corresponding logs in the right column. As examples, 100 is the antilog of 2; 200 is the antilog of 2.3010. When you want to find the antilog, reverse the process of finding the log: (1) use just the mantissa of the log, look for it in the log table, and find the corresponding numbers; (2) use the characteristic of the log to determine the decimal place in the number.

Suppose you want to find the antilog of 2.5172. The mantissa 5172 in the log table corresponds to the digits 329 for the number. Since the characteristic 2 is one less than the decimal places, the number must have three places. Therefore the number is 329. If the log were 1.5172, its antilog would be 32.9, with two decimal places for the characteristic of 1.

*Table of logarithms (four-place mantissas)*

| No. | 0 | 1 | 2 | 3 | 4 | 5 | 6 | 7 | 8 | 9 |
|---|---|---|---|---|---|---|---|---|---|---|
| **10** | 0000 | 0043 | 0086 | 0128 | 0170 | 0212 | 0253 | 0294 | 0334 | 0374 |
| 11 | 0414 | 0453 | 0492 | 0531 | 0569 | 0607 | 0645 | 0682 | 0719 | 0755 |
| 12 | 0792 | 0828 | 0864 | 0899 | 0934 | 0969 | 1004 | 1038 | 1072 | 1106 |
| 13 | 1139 | 1173 | 1206 | 1239 | 1271 | 1303 | 1335 | 1367 | 1399 | 1430 |
| 14 | 1461 | 1492 | 1523 | 1553 | 1584 | 1614 | 1644 | 1673 | 1703 | 1732 |
| **15** | 1761 | 1790 | 1818 | 1847 | 1875 | 1903 | 1931 | 1959 | 1987 | 2014 |
| 16 | 2041 | 2068 | 2095 | 2122 | 2148 | 2175 | 2201 | 2227 | 2253 | 2279 |
| 17 | 2304 | 2330 | 2355 | 2380 | 2405 | 2430 | 2455 | 2480 | 2504 | 2529 |
| 18 | 2553 | 2577 | 2601 | 2625 | 2648 | 2672 | 2695 | 2718 | 2742 | 2765 |
| 19 | 2788 | 2810 | 2833 | 2856 | 2878 | 2900 | 2923 | 2945 | 2967 | 2989 |
| **20** | 3010 | 3032 | 3054 | 3075 | 3096 | 3118 | 3139 | 3160 | 3181 | 3201 |
| 21 | 3222 | 3243 | 3263 | 3284 | 3304 | 3324 | 3345 | 3365 | 3385 | 3404 |
| 22 | 3424 | 3444 | 3464 | 3483 | 3502 | 3522 | 3541 | 3560 | 3579 | 3598 |
| 23 | 3617 | 3636 | 3655 | 3674 | 3692 | 3711 | 3729 | 3747 | 3766 | 3784 |
| 24 | 3802 | 3820 | 3838 | 3856 | 3874 | 3892 | 3909 | 3927 | 3945 | 3962 |
| **25** | 3979 | 3997 | 4014 | 4031 | 4048 | 4065 | 4082 | 4099 | 4116 | 4133 |
| 26 | 4150 | 4166 | 4183 | 4200 | 4216 | 4232 | 4249 | 4265 | 4281 | 4298 |
| 27 | 4314 | 4330 | 4346 | 4362 | 4378 | 4393 | 4409 | 4425 | 4440 | 4456 |
| 28 | 4472 | 4487 | 4502 | 4518 | 4533 | 4548 | 4564 | 4579 | 4594 | 4609 |
| 29 | 4624 | 4639 | 4654 | 4669 | 4683 | 4698 | 4713 | 4728 | 4742 | 4757 |
| **30** | 4771 | 4786 | 4800 | 4814 | 4829 | 4843 | 4857 | 4871 | 4886 | 4900 |
| 31 | 4914 | 4928 | 4942 | 4955 | 4969 | 4983 | 4997 | 5011 | 5024 | 5038 |
| 32 | 5051 | 5065 | 5079 | 5092 | 5105 | 5119 | 5132 | 5145 | 5159 | 5172 |
| 33 | 5185 | 5198 | 5211 | 5224 | 5237 | 5250 | 5263 | 5276 | 5289 | 5302 |
| 34 | 5315 | 5328 | 5340 | 5353 | 5366 | 5378 | 5391 | 5403 | 5416 | 5428 |
| **35** | 5441 | 5453 | 5465 | 5478 | 5490 | 5502 | 5514 | 5527 | 5539 | 5551 |
| 36 | 5563 | 5575 | 5587 | 5599 | 5611 | 5623 | 5635 | 5647 | 5658 | 5670 |
| 37 | 5682 | 5694 | 5705 | 5717 | 5729 | 5740 | 5752 | 5763 | 5775 | 5786 |
| 38 | 5798 | 5809 | 5821 | 5832 | 5843 | 5855 | 5866 | 5877 | 5888 | 5899 |
| 39 | 5911 | 5922 | 5933 | 5944 | 5955 | 5966 | 5977 | 5988 | 5999 | 6010 |
| **40** | 6021 | 6031 | 6042 | 6053 | 6064 | 6075 | 6085 | 6096 | 6107 | 6117 |
| 41 | 6128 | 6138 | 6149 | 6160 | 6170 | 6180 | 6191 | 6201 | 6212 | 6222 |
| 42 | 6232 | 6243 | 6253 | 6263 | 6274 | 6284 | 6294 | 6304 | 6314 | 6325 |
| 43 | 6335 | 6345 | 6355 | 6365 | 6375 | 6385 | 6395 | 6405 | 6415 | 6425 |
| 44 | 6435 | 6444 | 6454 | 6464 | 6474 | 6484 | 6493 | 6503 | 6513 | 6522 |
| **45** | 6532 | 6542 | 6551 | 6561 | 6571 | 6580 | 6590 | 6599 | 6609 | 6618 |
| 46 | 6628 | 6637 | 6646 | 6656 | 6665 | 6675 | 6684 | 6693 | 6702 | 6712 |
| 47 | 6721 | 6730 | 6739 | 6749 | 6758 | 6767 | 6776 | 6785 | 6794 | 6803 |
| 48 | 6812 | 6821 | 6830 | 6839 | 6848 | 6857 | 6866 | 6875 | 6884 | 6893 |
| 49 | 6902 | 6911 | 6920 | 6928 | 6937 | 6946 | 6955 | 6964 | 6972 | 6981 |
| **50** | 6990 | 6998 | 7007 | 7016 | 7024 | 7033 | 7042 | 7050 | 7059 | 7067 |
| 51 | 7076 | 7084 | 7093 | 7101 | 7110 | 7118 | 7126 | 7135 | 7143 | 7152 |
| 52 | 7160 | 7168 | 7177 | 7185 | 7193 | 7202 | 7210 | 7218 | 7226 | 7235 |
| 53 | 7243 | 7251 | 7259 | 7267 | 7275 | 7284 | 7292 | 7300 | 7308 | 7316 |
| 54 | 7324 | 7332 | 7340 | 7348 | 7356 | 7364 | 7372 | 7380 | 7388 | 7396 |
| *No.* | **0** | **1** | **2** | **3** | **4** | **5** | **6** | **7** | **8** | **9** |

*Table of logarithms (continued)*

| No. | 0 | 1 | 2 | 3 | 4 | 5 | 6 | 7 | 8 | 9 |
|---|---|---|---|---|---|---|---|---|---|---|
| **55** | 7404 | 7412 | 7419 | 7427 | 7435 | 7443 | 7451 | 7459 | 7466 | 7474 |
| 56 | 7482 | 7490 | 7497 | 7505 | 7513 | 7520 | 7528 | 7536 | 7543 | 7551 |
| 57 | 7559 | 7566 | 7574 | 7582 | 7589 | 7597 | 7604 | 7612 | 7619 | 7627 |
| 58 | 7634 | 7642 | 7649 | 7657 | 7664 | 7672 | 7679 | 7686 | 7694 | 7701 |
| 59 | 7709 | 7716 | 7723 | 7731 | 7738 | 7745 | 7752 | 7760 | 7767 | 7774 |
| **60** | 7782 | 7789 | 7796 | 7803 | 7810 | 7818 | 7825 | 7832 | 7839 | 7846 |
| 61 | 7853 | 7860 | 7868 | 7875 | 7882 | 7889 | 7896 | 7903 | 7910 | 7917 |
| 62 | 7924 | 7931 | 7938 | 7945 | 7952 | 7959 | 7966 | 7973 | 7980 | 7987 |
| 63 | 7993 | 8000 | 8007 | 8014 | 8021 | 8028 | 8035 | 8041 | 8048 | 8055 |
| 64 | 8062 | 8069 | 8075 | 8082 | 8089 | 8096 | 8102 | 8109 | 8116 | 8122 |
| **65** | 8129 | 8136 | 8142 | 8149 | 8156 | 8162 | 8169 | 8176 | 8182 | 8189 |
| 66 | 8195 | 8202 | 8209 | 8215 | 8222 | 8228 | 8235 | 8241 | 8248 | 8254 |
| 67 | 8261 | 8267 | 8274 | 8280 | 8287 | 8293 | 8299 | 8306 | 8312 | 8319 |
| 68 | 8325 | 8331 | 8338 | 8344 | 8351 | 8357 | 8363 | 8370 | 8376 | 8382 |
| 69 | 8388 | 8395 | 8401 | 8407 | 8414 | 8420 | 8426 | 8432 | 8439 | 8445 |
| **70** | 8451 | 8457 | 8463 | 8470 | 8476 | 8482 | 8488 | 8494 | 8500 | 8506 |
| 71 | 8513 | 8519 | 8525 | 8531 | 8537 | 8543 | 8549 | 8555 | 8561 | 8567 |
| 72 | 8573 | 8579 | 8585 | 8591 | 8597 | 8603 | 8609 | 8615 | 8621 | 8627 |
| 73 | 8633 | 8639 | 8645 | 8651 | 8657 | 8663 | 8669 | 8675 | 8681 | 8686 |
| 74 | 8692 | 8698 | 8704 | 8710 | 8716 | 8722 | 8727 | 8733 | 8739 | 8745 |
| **75** | 8751 | 8756 | 8762 | 8768 | 8774 | 8779 | 8785 | 8791 | 8797 | 8802 |
| 76 | 8808 | 8814 | 8820 | 8825 | 8831 | 8837 | 8842 | 8848 | 8854 | 8859 |
| 77 | 8865 | 8871 | 8876 | 8882 | 8887 | 8893 | 8899 | 8904 | 8910 | 8915 |
| 78 | 8921 | 8927 | 8932 | 8938 | 8943 | 8949 | 8954 | 8960 | 8965 | 8971 |
| 79 | 8976 | 8982 | 8987 | 8993 | 8998 | 9004 | 9009 | 9015 | 9020 | 9025 |
| **80** | 9031 | 9036 | 9042 | 9047 | 9053 | 9058 | 9063 | 9069 | 9074 | 9079 |
| 81 | 9085 | 9090 | 9096 | 9101 | 9106 | 9112 | 9117 | 9122 | 9128 | 9133 |
| 82 | 9138 | 9143 | 9149 | 9154 | 9159 | 9165 | 9170 | 9175 | 9180 | 9186 |
| 83 | 9191 | 9196 | 9201 | 9206 | 9212 | 9217 | 9222 | 9227 | 9232 | 9238 |
| 84 | 9243 | 9248 | 9253 | 9258 | 9263 | 9269 | 9274 | 9279 | 9284 | 9289 |
| **85** | 9294 | 9299 | 9304 | 9309 | 9315 | 9320 | 9325 | 9330 | 9335 | 9340 |
| 86 | 9345 | 9350 | 9355 | 9360 | 9365 | 9370 | 9375 | 9380 | 9385 | 9390 |
| 87 | 9395 | 9400 | 9405 | 9410 | 9415 | 9420 | 9425 | 9430 | 9435 | 9440 |
| 88 | 9445 | 9450 | 9455 | 9460 | 9465 | 9469 | 9474 | 9479 | 9484 | 9489 |
| 89 | 9494 | 9499 | 9504 | 9509 | 9513 | 9518 | 9523 | 9528 | 9533 | 9538 |
| **90** | 9542 | 9547 | 9552 | 9557 | 9562 | 9566 | 9571 | 9576 | 9581 | 9586 |
| 91 | 9590 | 9595 | 9600 | 9605 | 9609 | 9614 | 9619 | 9624 | 9628 | 9633 |
| 92 | 9638 | 9643 | 9647 | 9652 | 9657 | 9661 | 9666 | 9671 | 9675 | 9680 |
| 93 | 9685 | 9689 | 9694 | 9699 | 9703 | 9708 | 9713 | 9717 | 9722 | 9727 |
| 94 | 9731 | 9736 | 9741 | 9745 | 9750 | 9754 | 9759 | 9763 | 9768 | 9773 |
| **95** | 9777 | 9782 | 9786 | 9791 | 9795 | 9800 | 9805 | 9809 | 9814 | 9818 |
| 96 | 9823 | 9827 | 9832 | 9836 | 9841 | 9845 | 9850 | 9854 | 9859 | 9863 |
| 97 | 9868 | 9872 | 9877 | 9881 | 9886 | 9890 | 9894 | 9899 | 9903 | 9908 |
| 98 | 9912 | 9917 | 9921 | 9926 | 9930 | 9934 | 9939 | 9943 | 9948 | 9952 |
| 99 | 9956 | 9961 | 9965 | 9969 | 9974 | 9978 | 9983 | 9987 | 9991 | 9996 |
| **No.** | **0** | **1** | **2** | **3** | **4** | **5** | **6** | **7** | **8** | **9** |

# *Appendix* **D** *Decibel table*

| *Db* | *Current and voltage ratio* | | *Power ratio* | | *Db* | *Current and voltage ratio* | | *Power ratio* | |
|---|---|---|---|---|---|---|---|---|---|
| | *Gain* | *Loss* | *Gain* | *Loss* | | *Gain* | *Loss* | *Gain* | *Loss* |
| 0.1 | 1.01 | 0.989 | 1.02 | 0.977 | 8.0 | 2.51 | 0.398 | 6.31 | 0.158 |
| 0.2 | 1.02 | 0.977 | 1.05 | 0.955 | 8.5 | 2.66 | 0.376 | 7.08 | 0.141 |
| 0.3 | 1.03 | 0.966 | 1.07 | 0.933 | 9.0 | 2.82 | 0.355 | 7.94 | 0.126 |
| 0.4 | 1.05 | 0.955 | 1.10 | 0.912 | 9.5 | 2.98 | 0.335 | 8.91 | 0.112 |
| 0.5 | 1.06 | 0.944 | 1.12 | 0.891 | 10.0 | 3.16 | 0.316 | 10.00 | 0.100 |
| 0.6 | 1.07 | 0.933 | 1.15 | 0.871 | 11.0 | 3.55 | 0.282 | 12.6 | 0.079 |
| 0.7 | 1.08 | 0.923 | 1.17 | 0.851 | 12.0 | 3.98 | 0.251 | 15.8 | 0.063 |
| 0.8 | 1.10 | 0.912 | 1.20 | 0.832 | 13.0 | 4.47 | 0.224 | 19.9 | 0.050 |
| 0.9 | 1.11 | 0.902 | 1.23 | 0.813 | 14.0 | 5.01 | 0.199 | 25.1 | 0.040 |
| 1.0 | 1.12 | 0.891 | 1.26 | 0.794 | 15.0 | 5.62 | 0.178 | 31.6 | 0.032 |
| 1.1 | 1.13 | 0.881 | 1.29 | 0.776 | 16.0 | 6.31 | 0.158 | 39.8 | 0.025 |
| 1.2 | 1.15 | 0.871 | 1.32 | 0.759 | 17.0 | 7.08 | 0.141 | 50.1 | 0.020 |
| 1.3 | 1.16 | 0.861 | 1.35 | 0.741 | 18.0 | 7.94 | 0.126 | 63.1 | 0.016 |
| 1.4 | 1.17 | 0.851 | 1.38 | 0.724 | 19.0 | 8.91 | 0.112 | 79.4 | 0.013 |
| 1.5 | 1.19 | 0.841 | 1.41 | 0.708 | 20.0 | 10.00 | 0.100 | 100.0 | 0.010 |
| 1.6 | 1.20 | 0.832 | 1.44 | 0.692 | 25.0 | 17.8 | 0.056 | $3.16 \times 10^{2}$ | $3.16 \times 10^{-3}$ |
| 1.7 | 1.22 | 0.822 | 1.48 | 0.676 | 30.0 | 31.6 | 0.032 | $10^{3}$ | $10^{-3}$ |
| 1.8 | 1.23 | 0.813 | 1.51 | 0.661 | 35.0 | 56.2 | 0.018 | $3.16 \times 10^{3}$ | $3.16 \times 10^{-4}$ |
| 1.9 | 1.24 | 0.803 | 1.55 | 0.646 | 40.0 | 100.0 | 0.010 | $10^{4}$ | $10^{-4}$ |
| 2.0 | 1.26 | 0.794 | 1.58 | 0.631 | 45.0 | 177.8 | 0.006 | $3.16 \times 10^{4}$ | $3.16 \times 10^{-5}$ |
| 2.2 | 1.29 | 0.776 | 1.66 | 0.603 | 50.0 | 316 | 0.003 | $10^{5}$ | $10^{-5}$ |
| 2.4 | 1.32 | 0.759 | 1.74 | 0.575 | 55.0 | 562 | 0.002 | $3.16 \times 10^{5}$ | $3.16 \times 10^{-6}$ |
| 2.6 | 1.35 | 0.741 | 1.82 | 0.550 | 60.0 | 1,000 | 0.001 | $10^{6}$ | $10^{-6}$ |
| 2.8 | 1.38 | 0.724 | 1.90 | 0.525 | 65.0 | 1,770 | 0.0006 | $3.16 \times 10^{6}$ | $3.16 \times 10^{-7}$ |
| 3.0 | 1.41 | 0.708 | 1.99 | 0.501 | 70.0 | 3,160 | 0.0003 | $10^{7}$ | $10^{-7}$ |
| 3.2 | 1.44 | 0.692 | 2.09 | 0.479 | 75.0 | 5,620 | 0.0002 | $3.16 \times 10^{7}$ | $3.16 \times 10^{-8}$ |
| 3.4 | 1.48 | 0.676 | 2.19 | 0.457 | 80.0 | 10,000 | 0.0001 | $10^{8}$ | $10^{-8}$ |
| 3.6 | 1.51 | 0.661 | 2.29 | 0.436 | 85.0 | 17,800 | 0.00006 | $3.16 \times 10^{8}$ | $3.16 \times 10^{-9}$ |
| 3.8 | 1.55 | 0.646 | 2.40 | 0.417 | 90.0 | 31,600 | 0.00003 | $10^{9}$ | $10^{-9}$ |
| 4.0 | 1.58 | 0.631 | 2.51 | 0.398 | 95.0 | 56,200 | 0.00002 | $3.16 \times 10^{9}$ | $3.16 \times 10^{-10}$ |
| 4.2 | 1.62 | 0.617 | 2.63 | 0.380 | 100.0 | 100,000 | 0.00001 | $10^{10}$ | $10^{-10}$ |
| 4.4 | 1.66 | 0.603 | 2.75 | 0.363 | 105.0 | 178,000 | 0.000006 | $3.16 \times 10^{10}$ | $3.16 \times 10^{-11}$ |
| 4.6 | 1.70 | 0.589 | 2.88 | 0.347 | 110.0 | 316,000 | 0.000003 | $10^{11}$ | $10^{-11}$ |
| 4.8 | 1.74 | 0.575 | 3.02 | 0.331 | 115.0 | 562,000 | 0.000002 | $3.16 \times 10^{11}$ | $3.16 \times 10^{-12}$ |
| 5.0 | 1.78 | 0.562 | 3.16 | 0.316 | 120.0 | 1,000,000 | 0.000001 | $10^{12}$ | $10^{-12}$ |
| 5.5 | 1.88 | 0.531 | 3.55 | 0.282 | 130.0 | $3.16 \times 10^{6}$ | $3.16 \times 10^{-7}$ | $10^{13}$ | $10^{-13}$ |
| 6.0 | 1.99 | 0.501 | 3.98 | 0.251 | 140.0 | $10^{7}$ | $10^{-7}$ | $10^{14}$ | $10^{-14}$ |
| 6.5 | 2.11 | 0.473 | 4.47 | 0.224 | 150.0 | $3.16 \times 10^{7}$ | $3.16 \times 10^{-8}$ | $10^{15}$ | $10^{-15}$ |
| 7.0 | 2.24 | 0.447 | 5.01 | 0.199 | 160.0 | $10^{8}$ | $10^{-8}$ | $10^{16}$ | $10^{-16}$ |
| 7.5 | 2.37 | 0.422 | 5.62 | 0.178 | 170.0 | $3.16 \times 10^{8}$ | $3.16 \times 10^{-9}$ | $10^{17}$ | $10^{-17}$ |

# *Appendix* **E** *Universal time-constant graph for RC or RL circuits*

## UNIVERSAL TIME-CONSTANT GRAPH FOR RC OR RL CIRCUITS

With the curves in Fig. E · 1 you can determine voltage and current values for any amount of time. The rising curve *a* shows how the voltage builds up across *C* as it charges in an *RC* circuit; the same curve applies to the current increasing in the inductance for an *RL* circuit. The decreasing curve *b* shows how the capacitor voltage declines as *C* discharges in an *RC* circuit or the decay of current in an inductance.

Note that the horizontal axis is in units of time constants, rather than absolute time. The time constant is the *RC* product for capacitive circuits but equals *L*/*R* for inductive circuits. For example, suppose that the time constant of a capacitive circuit is 5 μsec. Therefore, one *RC* unit corresponds to 5 μsec, two *RC* units equals 10 μsec, three *RC* units equals 15 μsec, etc. To find how much voltage is across the capacitor after 10 μsec of charging, take the value

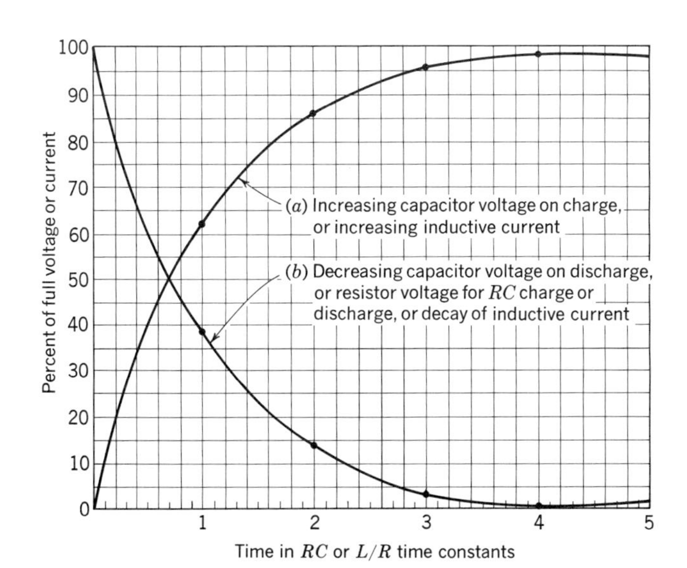


*Fig. E · 1*

*Table E · 1*

| FACTOR | CHANGE, PER CENT |
|---|---|
| 0.5 time constant | 40 |
| 0.7 time constant | 50 |
| 1 time constant | 63 |
| 2 time constants | 86 |
| 3 time constants | 96 |
| 4 time constants | 98 |
| 5 time constants | 99 |

on graph *a* corresponding to two time constants, or approximately 86 per cent of the applied charging voltage. The point where curves *a* and *b* intersect shows that a 50 per cent change is accomplished in 0.7 time constant. The curves can be considered linear within the first 40 per cent of change, or within 0.5 time constant.

In Fig. E · 1 the entire *RC* charge curve actually adds 63 per cent of the net charging voltage for each increment of 1 time constant, although it may not appear so. In the second interval of *RC* time, for instance, $e_c$ adds 63 per cent of the net charging voltage, which is $0.37E$. Then $0.63 \times 0.37$ equals 0.23, which is added to 0.63 to give 0.86 or 86 per cent. This value of 86 per cent is the change from the start, at zero time, as are all the values listed in Table E · 1.

# *Appendix F Color codes*

Included here are color codes for chassis wiring, carbon resistors, and small fixed capacitors with mica or ceramic dielectric. Most of these codes are standardized by the Electronic Industries Association (EIA). Members are not required to follow the codes but it is industry practice to do so where practical.

*Chassis wiring*

Colors for the wires in electronic circuits generally follow the system in Table F·1. By noting the wiring color code, you can often save time in tracing the connections.

## *Table F·1 Chassis wiring color code*

| COLOR | CONNECTED TO |
|---|---|
| Red | B+ voltage supply |
| Blue | Plate of amplifier tube or collector of transistor |
| Green | Control grid of amplifier tube or base of transistor (also for input to diode detector) |
| Yellow | Cathode of amplifier tube or emitter of transistor |
| Orange | Screen grid |
| Brown | Heaters or filaments |
| Black | Chassis ground return |
| White | Return for control grid (AVC bias) |

*Leads or terminals on i-f transformers*

Blue—plate
Red—$B^+$
Green—control grid or diode detector
White—control grid or diode return
Violet—second diode lead for duodiode detector

*A-f transformers*

Blue—plate lead (end of primary winding)
Red—B+ (center-tap on push-pull transformer)
Brown—plate lead (start of primary winding on push-pull transformer)
Green—finish lead of secondary winding
Black—ground return of secondary winding
Yellow—start lead on center-tapped secondary

*Power transformers (Fig. F・1)*

Primary without tap—black
Tapped primary:
    Common—black
    Tap—black and yellow stripes
    Finish—black and red stripes
High-voltage secondary for plates of rectifier—red
    Center tap—red and yellow stripes
Low-voltage secondary for rectifier filament—yellow
Low-voltage secondary for amplifier heaters—green, brown, or slate
    Center tap—same color with yellow stripe

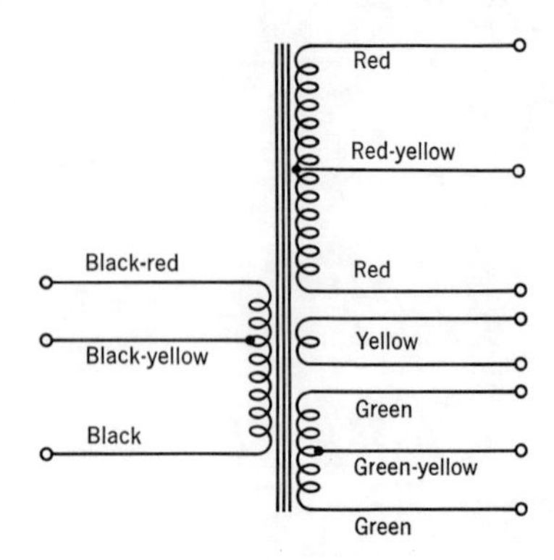


*Fig. F・1 Power transformer color code.*

*Carbon resistors*

For ratings of 2 watts or less, carbon resistors are color-coded with either bands or the body-end-dot system as summarized in Table F・2.

The color values summarized in Table F・3 apply to both resistors and capacitors. However, the colors for voltage ratings apply only to capacitors. Also, only gold or silver is used for carbon-composition resistor tolerance, but all the colors apply to tolerances for capacitors or film resistors.

Similarly, the preferred values in Table F・4 are for resistance values in ohms or capacitor values in picofarads. Only the basic value is listed, from which multiple values are derived. As an example, a 1,500-ohm $R$ or 1,500-pf $C$ is a standard component value, with a tolerance of either 20, 10, or 5 per cent. However, if you want a value of 2,000, this is available only in 5 per cent tolerance.

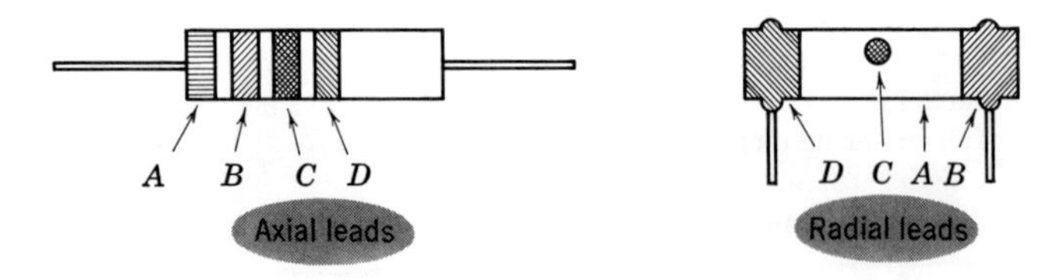


*Table F・2 Color codes for carbon resistors*

| *Axial leads* | *Color* | *Radial leads* |
|---|---|---|
| *B* and *A* | First significant figure | Body *A* |
| *B* and *B* | Second significant figure | End *B* |
| *B* and *C* | Decimal multiplier | Dot *C* |
| *B* and *D* | Tolerance | End *D* |

*Notes:* Band *A* is double width for wirewound resistors with axial leads. Body-end-dot system is a discontinued standard but may still be found on some old resistors. For resistors with color stripes and axial leads, body color is not used for color-coded value. Film resistors have five stripes; fourth stripe is multiplier and fifth is tolerance.

Mica capacitors

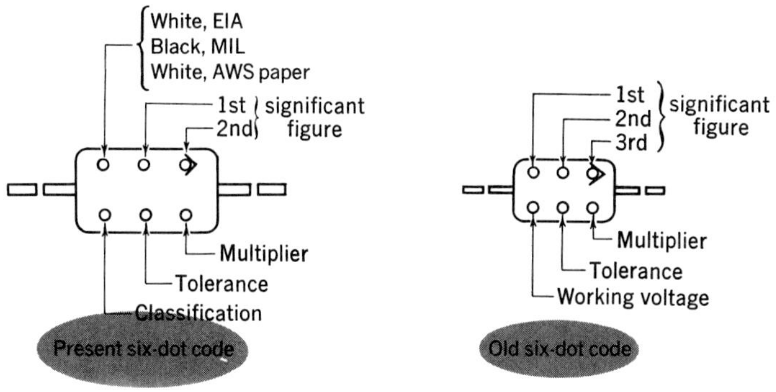

White, EIA
Black, MIL
White, AWS paper
1st } significant
2nd } figure
Multiplier
Tolerance
Classification
Present six-dot code

1st
2nd
3rd } significant figure
Multiplier
Tolerance
Working voltage
Old six-dot code

*Table F · 3 Color values for resistor and capacitor codes*

| *Color* | *Significant figure* | *Decimal multiplier* | *Tolerance,* %* | *Voltage rating** |
|---|---|---|---|---|
| Black | 0 | 1 | 20 | |
| Brown | 1 | 10 | 1 | 100 |
| Red | 2 | $10^2$ | 2 | 200 |
| Orange | 3 | $10^3$ | 3 | 300 |
| Yellow | 4 | $10^4$ | 4 | 400 |
| Green | 5 | $10^5$ | 5 | 500 |
| Blue | 6 | $10^6$ | 6 | 600 |
| Violet | 7 | $10^7$ | 7 | 700 |
| Gray | 8 | $10^8$ | 8 | 800 |
| White | 9 | $10^9$ | 9 | 900 |
| Gold | | 0.1 | 5 | 1,000 |
| Silver | | 0.01 | 10 | 2,000 |
| No color | | | 20 | 500 |

* Tolerance colors other than gold and silver and voltage-rating colors for capacitors only.

*Mica capacitors*

These may be coded with old RMA or new EIA methods, military (MIL), Joint Army Navy (JAN) specifications, or American War Standards (AWS). Actually, for both mica and ceramic capacitors the trend has been for manufacturers to print the values on the capacitor, to avoid confusion with different color codes.

The new EIA six-dot code starts with a white dot. If this dot is black it indicates the MIL six-dot code. Or, if this dot is silver, it indicates a paper capacitor in the AWS code. In all three cases, though, the capacitance in pf units is read from the next three color dots. However, if the first dot has a color, this indicates the old EIA six-dot code, where the first four dots are used for the capacitor value.

The characteristics indicated by the last dot in the new EIA six-dot code specify five classes from $A$ to $E$, according to leakage resistance, temperature coefficient, and other factors.

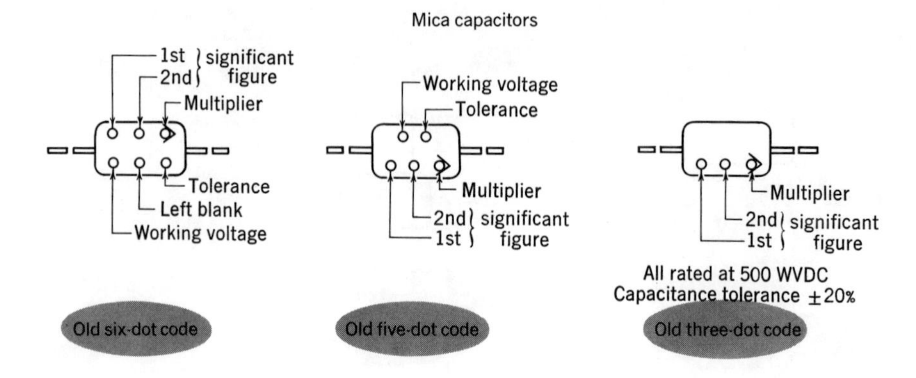


*Table F·4 Preferred values for resistors and capacitors*

| *20% tolerance* | *10% tolerance* | *5% tolerance* |
|---|---|---|
| 10* | 10 | 10 |
| | | 11 |
| | 12 | 12 |
| | | 13 |
| 15 | 15 | 15 |
| | | 16 |
| | 18 | 18 |
| | | 20 |
| 22 | 22 | 22 |
| | | 24 |
| | 27 | 27 |
| | | 30 |
| 33 | 33 | 33 |
| | | 36 |
| | 39 | 39 |
| | | 43 |
| 47 | 47 | 47 |
| | | 51 |
| | 56 | 56 |
| | | 62 |
| 68 | 68 | 68 |
| | | 75 |
| | 82 | 82 |
| | | 91 |
| 100 | 100 | 100 |

* Numbers and decimal multiples for ohms or pf.

*Ceramic capacitors*

These have stripes or dots with three colors or five colors. With five stripes or dots, the first and last colors indicate temperature coefficient and tolerance, as listed in Table F·5. The middle three colors give the capacitance in picofarads, with the same color values as for resistors and mica capacitors. The fourth color is for tolerance, in per cent for sizes larger than 10 pf but in pf units for smaller capacitors.

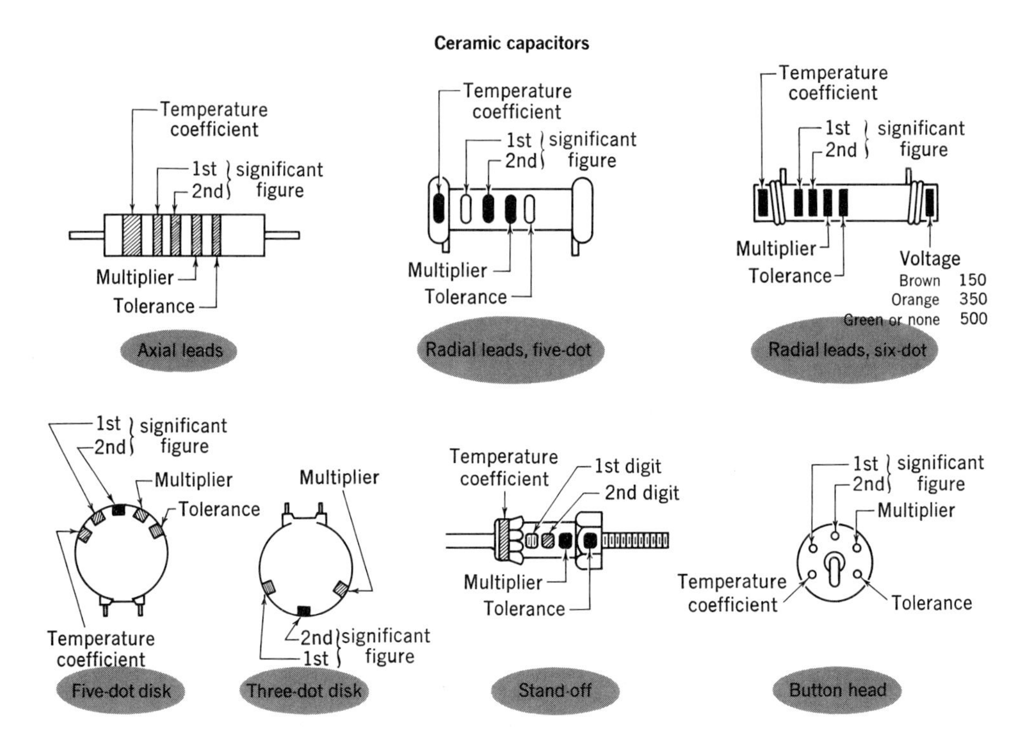


*Table F·5 Color code for ceramic capacitors*

| *Color* | *Decimal multiplier* | *Tolerance* | | *Temp. coeff. ppm per °C* |
|---|---|---|---|---|
| | | *Above 10 pf, %* | *Below 10 pf, in pf* | |
| Black | 1 | 20 | 2.0 | 0 |
| Brown | 10 | 1 | | −30 |
| Red | 100 | 2 | | −80 |
| Orange | 1,000 | | | −150 |
| Yellow | | | | −220 |
| Green | | 5 | 0.5 | −330 |
| Blue | | | | −470 |
| Violet | | | | −750 |
| Gray | 0.01 | | 0.25 | 30 |
| White | 0.1 | 10 | 1.0 | 500 |

# *Appendix G Abbreviations and schematic symbols*

*Table G·1 Greek letters*

| *Letter* | *Meaning* | *Letter* | *Meaning* |
|---|---|---|---|
| $\alpha$ (alpha) | Ratio of collector current to emitter current in transistors | $\Omega$ (omega) | Ohm unit of resistance |
| $\beta$ (beta) | Ratio of collector current to base current in transistors | $\phi$ (phi) | Magnetic flux |
| $\Delta$ (delta) or $d$ | A small change in value | $\pi$ (pi) | Constant of 3.14, generally used as $2\pi$ or 6.28 |
| $\lambda$ (lambda) | Wavelength | $\rho$ (rho) | Resistivity |
| $\mu$ (mu) | Amplification factor of tube, or permeability of magnetic material, or prefix meaning one-millionth | $\theta$ (theta) | Angle, often the phase angle in a-c circuits |

*Table G·2 Electrical units*

| *Quantity* | *Symbol* | *Basic unit* |
|---|---|---|
| Current | $I$ or $i$* | ampere |
| Charge | $Q$ or $q$ | coulomb |
| Voltage | $V$ or $v$, $E$ or $e$ | volt |
| (electromotive force, | emf | |
| potential difference) | PD | |
| Power | $P$ | watt |
| Resistance | $R$† | ohm |
| Reactance | $X$ | ohm |
| Impedance | $Z$ | ohm |
| Conductance | $G$ | mho |
| Admittance | $Y$ | mho |
| Susceptance | $B$ | mho |
| Capacitance | $C$ | farad |
| Inductance | $L$ | henry |
| Frequency | $F$ or $f$ | cps |
| Period | $T$ | second |

* Capital letter for $I$, $Q$, $V$, and $E$ generally used for peak, RMS, or d-c value; small letter for instantaneous values.

† Small $r$ and $g$ usually for internal values, as $r_p$ and $g_m$ of a tube.

*Table G · 3 Multiples and submultiples of units*

| *Value* | *Prefix* | *Symbol* | *Example* |
|---|---|---|---|
| $1\,000\,000\,000\,000 = 10^{12}$ | tera | $T$ | Tc = $10^{12}$ cps |
| $1\,000\,000\,000 = 10^{9}$ | giga | $G$ | Gc = $10^{9}$ cps |
| $1\,000\,000 = 10^{6}$ | mega | $M$ | Mc = $10^{6}$ cps |
| $1\,000 = 10^{3}$ | kilo | $K$ or $k$ | kv = $10^{3}$ volts |
| $100 = 10^{2}$ | hecto | $h$ | hm = $10^{2}$ meters |
| $10 = 10$ | deka | $dk$ | dkm = 10 meters |
| $0.1 = 10^{-1}$ | deci | $d$ | dm = $10^{-1}$ meter |
| $0.01 = 10^{-2}$ | centi | $c$ | cm = $10^{-2}$ meter |
| $0.001 = 10^{-3}$ | milli | $m$ | ma = $10^{-3}$ amp |
| $0.000\,001 = 10^{-6}$ | micro | $\mu$ | $\mu$v = $10^{-6}$ volt |
| $0.000\,000\,001 = 10^{-9}$ | nano | $n$ | nsec = $10^{-9}$ second |
| $0.000\,000\,000\,001 = 10^{-12}$ | pico | $p$ | pf = $10^{-12}$ farad |

SCHEMATIC SYMBOLS

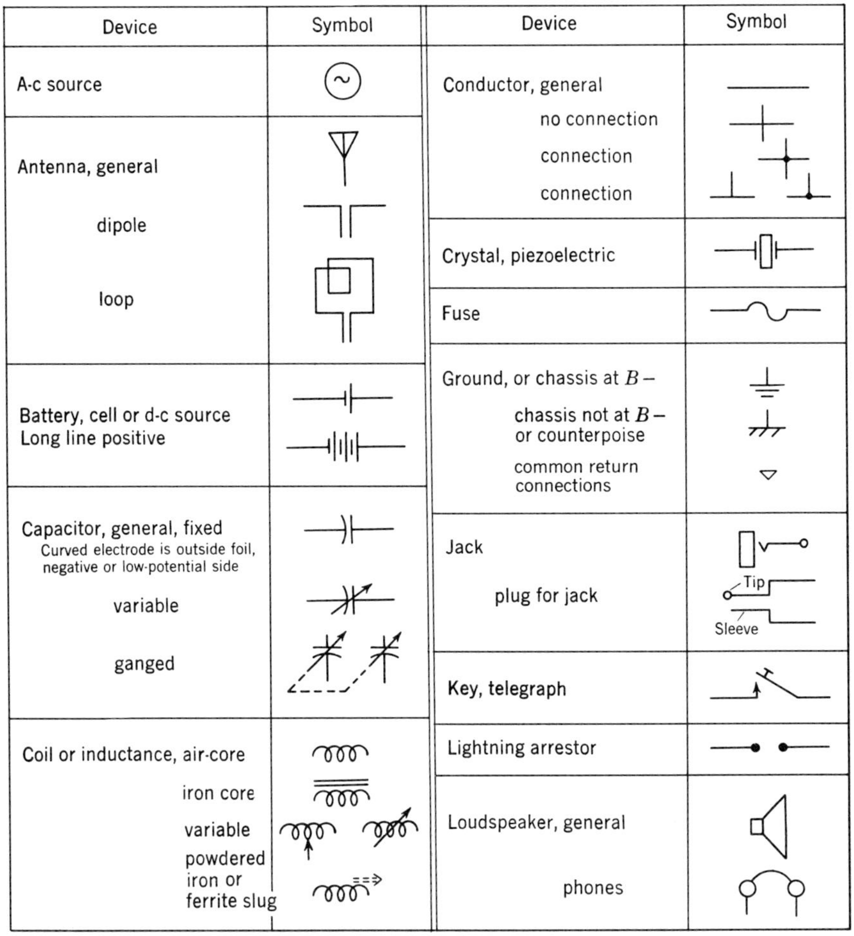

| Device | Symbol | Device | Symbol |
|---|---|---|---|
| A-c source | | Conductor, general | |
| | | no connection | |
| | | connection | |
| | | connection | |
| Antenna, general | | Crystal, piezoelectric | |
| dipole | | Fuse | |
| loop | | Ground, or chassis at $B-$ | |
| Battery, cell or d-c source | | chassis not at $B-$ or counterpoise | |
| Long line positive | | common return connections | |
| Capacitor, general, fixed<br>Curved electrode is outside foil, negative or low-potential side | | Jack | |
| variable | | plug for jack | Tip<br>Sleeve |
| ganged | | Key, telegraph | |
| Coil or inductance, air-core | | Lightning arrestor | |
| iron core | | Loudspeaker, general | |
| variable | | phones | |
| powdered iron or ferrite slug | | | |

SCHEMATIC SYMBOLS

| Device | Symbol | Device | Symbol |
|---|---|---|---|
| Magnet, permanent | 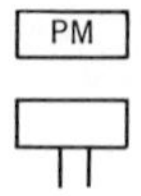 PM | Relay, coil | |
| electromagnet | | contacts | |
| Meters, letter or symbol to indicate range or function | A MA V | Transistors and semiconductors | |
| Resistor, fixed | | *PNP* triode transistor | e c b |
| tapped | | *NPN* triode transistor | e c b |
| variable | | Rectifier (arrow shows hole current) | |
| Switch, general | | Gate-controlled rectifier | |
| toggle | | Zener, avalanche, breakdown or voltage reference diode | |
| 2-pole, double throw | | Tunnel diode or Esaki diode | |
| 3-pole, 3-circuit wafer | | Varactor, varicap, reactance diode or parametric diode | |
| Shielding | | Photodiode or solar cell | |
| shielded conductor | | Tubes, envelope | |
| Thermistor, general | T | gas-filled | |
| Thermocouple | | directly heated cathode | |
| Transformer, air core | | indirectly heated cathode | |
| iron-core | | grid | |
| autotransformer | | | |
| link coupling | | | |

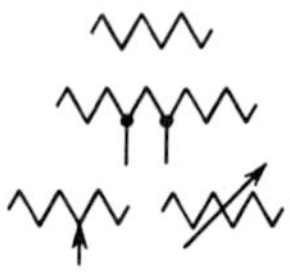

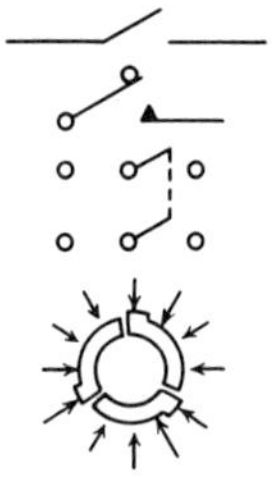

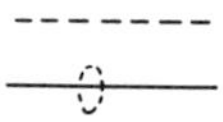

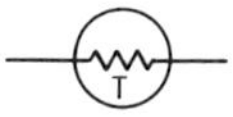

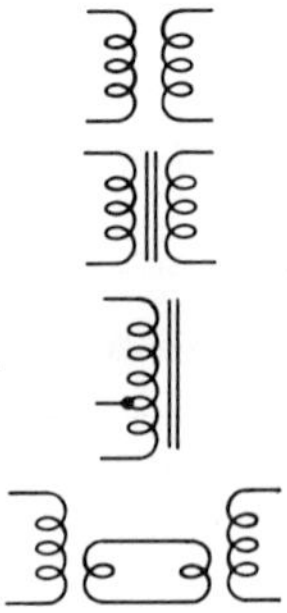

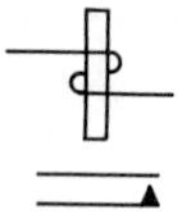

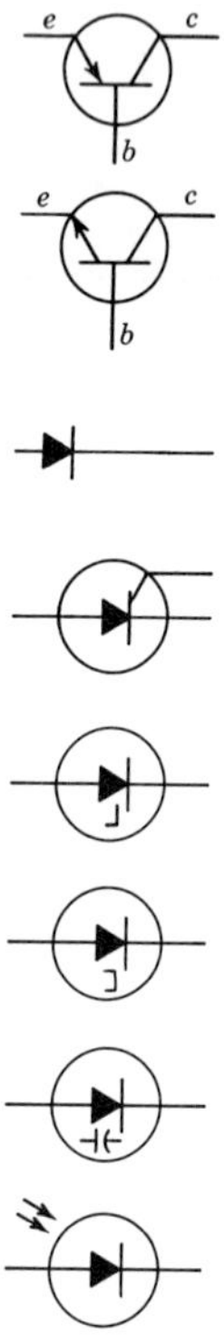

SCHEMATIC SYMBOLS

| Device | Symbol | Device | Symbol |
|---|---|---|---|
| Tubes, contd. | | Tubes, contd. | |
| diode |  | cold-cathode gas diode |  |
| triode | 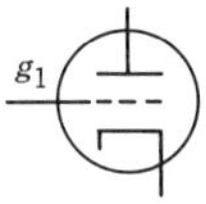 $g_1$ | photototube | |
| tetrode | 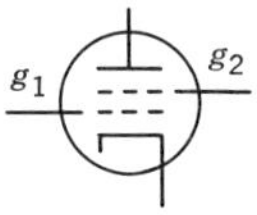 $g_1$ $g_2$ | |  |
| pentode | 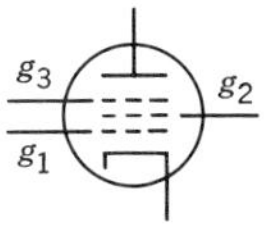 $g_3$ $g_2$ $g_1$ | cathode-ray tube with electrostatic deflection plates | 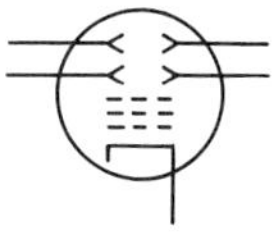 |

# Appendix H International, or Continental, Morse code

*The International, or Continental, Morse Code*

| | | | | | |
|---|---|---|---|---|---|
| A | ·— | N | —· | 1 | ·———— |
| B | —··· | O | ——— | 2 | ··——— |
| C | —·—· | P | ·——· | 3 | ···—— |
| D | —·· | Q | ——·— | 4 | ····— |
| E | · | R | ·—· | 5 | ····· |
| F | ··—· | S | ··· | 6 | —···· |
| G | ——· | T | — | 7 | ——··· |
| H | ···· | U | ··— | 8 | ———·· |
| I | ·· | V | ···— | 9 | ————· |
| J | ·——— | W | ·—— | 0 | ————— |
| K | —·— | X | —··— | | |
| L | ·—·· | Y | —·—— | | |
| M | —— | Z | ——·· | | |

| | | |
|---|---|---|
| . | (period) | ·—·—·— |
| , | (comma) | ——··—— |
| ? | (question mark) ($\overline{\text{IMI}}$) | ··——·· |
| / | (fraction bar) | —··—· |
| : | (colon) | ———··· |
| ; | (semicolon) | —·—·—· |
| ( or ) | (parentheses) | —·——·— |
| ’ | (apostrophe) | ·————· |
| - | (hyphen or dash) | —····— |
| $ | (dollar sign)* | ···—··— |
| “ | (quotation marks) | ·—··— |
| | or, | ·—··—·* |

| | |
|---|---|
| Error sign (8 dots) | ········ |
| Separation indicator also known as $\overline{\text{BT}}$ | —···— |
| End of transmission of a message ($\overline{\text{AR}}$) | ·—·—· |
| Invitation to transmit | —·— |
| Wait ($\overline{\text{AS}}$) | ·—··· |
| End of work ($\overline{\text{SK}}$ or $\overline{\text{VA}}$) | ···—·— |
| Starting signal | —·—·— |

* These characters are not listed internationally but may be heard in domestic communications.

# *Bibliography*

*Books*

American Radio Relay League, *Radio-Amateurs Handbook,* Newington, Connecticut
Annett, F. A., Practical Industrial Electronics, McGraw-Hill Book Company
Barton, D. K., *Radar Systems Analysis,* Prentice-Hall, Inc.
Blitzer, R., *Basic Pulse Circuits,* McGraw-Hill Book Company
Chute, George M., *Electronics in Industry,* 3d ed., McGraw-Hill Book Company
Crowhurst, N. H., *Basic Audio Course,* John F. Rider, Publisher, Inc.
Everitt, W. L., *Fundamentals of Radio and Electronics,* Prentice-Hall, Inc.
Griffith, B. W., Jr., *Radio-Electronic Transmission Fundamentals,* McGraw-Hill Book Company
Grob, B., *Basic Television,* 3d ed., McGraw-Hill Book Company
Hellman, C. I., *Elements of Radio,* 3d ed., D. Van Nostrand Company, Inc.
Henney, Keith, and James D. Fahnestock, *Electron Tubes in Industry,* 3d ed., McGraw-Hill Book Company
Henney, K., and G. A. Richardson, *Principles of Radio,* 6th ed., John Wiley & Sons, Inc.
Hickey, H. V., and W. M. Villines, *Elements of Electronics,* 2d ed., McGraw-Hill Book Company
Johnson, J. Richard, *How to Use Signal and Sweep Generators,* John F. Rider Publishers, Inc.
Kaufman, M., *Radio Operator's Q and A Manual,* John F. Rider Publisher, Inc.
Kiver, M. S., *TV Simplified,* D. Van Nostrand Company, Inc.
Kiver, Milton S., *Transistors,* 3d ed., McGraw-Hill Book Company
Kiver, M. S., *Introduction to UHF Circuits and Components,* D. Van Nostrand Company, Inc.
Lang, D. G., Marine Radar, Pitman Publishing Corporation
Levy, A., and Frankel, M., *Television Servicing,* McGraw-Hill Book Company
Marcus, A., and W. Marcus, *Elements of Radio,* 3d ed., Prentice-Hall, Inc.
Platt, Sidney, *Industrial Control Circuits,* John F. Rider Publisher, Inc.
Philco Technological Center, *Servomechanism Fundamentals and Experiments; Electronic Precision Measurement Techniques and Experiments,* Prentice-Hall, Inc.
Prensky, Sol. D., *Electronic Instrumentation,* Prentice-Hall, Inc.
Ruiter, Jacob H., Jr., *Modern Oscilloscopes and Their Uses,* Rinehart & Company, Inc.
Sheingold, A., *Fundamentals of Radio Communication,* 3d ed., D. Van Nostrand Company, Inc.
Shrader, R. L., *Electronic Communication,* McGraw-Hill Book Company
Slurzberg, M., and W. Osterheld, *Essentials of Radio,* 2d ed., McGraw-Hill Book Company
Slurzberg, M., and W. Osterheld, *Essentials of Radio-Electronics,* 2d ed., McGraw-Hill Book Company

Swallow, K. P., and W. T. Price, *Elements of Computer Programming,* Holt, Rinehart and Winston, Inc.
Turner, Rufus P., *Basic Electronic Test Instruments,* Rinehart & Company, Inc.
Weinstein, S. M., and Keim, A., *Fundamentals of Digital Computers,* Holt, Rinehart and Winston, Inc.
Wheeler, F. J., *Introduction to Microwaves,* Prentice-Hall, Inc.
Zbar, P. B., *Electronic Instruments and Measurements,* McGraw-Hill Book Company
Zbar, P. B., and P. W. Orne, *Advanced Servicing Techniques,* John F. Rider Publishers, Inc.

*Government Publications*

The following books are published by the U.S. Government Printing Office, Washington, D.C., and may be purchased from them.

*Radarman, 3 & 2*
*Aviation Electronics Technicians 3 and 2*
*Loran*
*Fire Control Technician 1 and C,* Vol. 3
*Fire Control Technician 2*
*Fire Control Technician 3*
*Radar Circuit Analysis*
*Radar Electronic Fundamentals*
*Radar Fundamentals and Surveillance, Precision and Route Radar*
*Radar System Fundamentals*

*Magazines*

*Electronics,* McGraw-Hill Publications, New York.
*Electronic Servicing,* Cowan Publishing Corp., New York.
*Electronic Technician,* Ojibway Press Inc., Duluth, Minn.
*PF Reporter,* Howard W. Sams & Co., Inc., Indianapolis, Ind.
*QST,* American Radio Relay League, Newington, Conn.
*Radio-Electronics,* Gernsback Publications, Inc., New York.
*Radio & TV News,* Ziff-Davis Publishing Co., New York.
*Service,* Bryan Davis Publishing Co., Inc., New York.

*Diagrams*

There are also available schematic diagrams of AM, FM, and television receivers, as well as communications receivers and high fidelity equipment from the following sources. These companies publish their material from time to time, as new models become available:

Howard W. Sams & Company, Inc.
John F. Rider, Publisher, Inc.

# *Answers to self-examinations*

### *Chapter 1*

| | | | | |
|---|---|---|---|---|
| 1. (*b*) | 2. (*c*) | 3. (*c*) | 4. (*a*) | 5. (*b*) |
| 6. (*d*) | 7. (*d*) | 8. (*d*) | 9. (*c*) | 10. (*d*) |

### *Chapter 2*

| | | | | |
|---|---|---|---|---|
| 1. (*c*) | 2. (*b*) | 3. (*c*) | 4. (*d*) | 5. (*a*) |
| 6. (*d*) | 7. (*d*) | 8. (*b*) | 9. (*b*) | 10. (*c*) |

### *Chapter 3*

| | | | | |
|---|---|---|---|---|
| 1. (*b*) | 2. (*c*) | 3. (*b*) | 4. (*a*) | 5. (*d*) |
| 6. (*b*) | 7. (*c*) | 8. (*a*) | 9. (*c*) | 10. (*d*) |

### *Chapters 1 to 3 review*

| | | | | |
|---|---|---|---|---|
| 1. (*c*) | 2. (*l*) | 3. (*k*) | 4. (*j*) | 5. (*m*) |
| 6. (*o*) | 7. (*n*) | 8. (*a*) | 9. (*f*) | 10. (*b*) |
| 11. (*d*) | 12. (*g*) | 13. (*h*) | 14. (*e*) | 15. (*p*) |
| 16. (*i*) | 17. (*q*) | 18. (*z*) | 19. (*y*) | 20. (*x*) |
| 21. (*r*) | 22. (*s*) | 23. (*t*) | 24. (*u*) | 25. (*v*) |
| 26. (*w*) | | | | |

### *Chapter 4*

| | | | | |
|---|---|---|---|---|
| 1. (*b*) | 2. (*b*) | 3. (*d*) | 4. (*c*) | 5. (*b*) |
| 6. (*b*) | 7. (*a*) | 8. (*d*) | 9. (*c*) | 10. (*c*) |

### *Chapter 5*

| | | | | |
|---|---|---|---|---|
| 1. (*d*) | 2. (*b*) | 3. (*d*) | 4. (*c*) | 5. (*a*) |
| 6. (*b*) | 7. (*b*) | 8. (*d*) | 9. (*b*) | 10. (*b*) |

### *Chapter 6*

| | | | | |
|---|---|---|---|---|
| 1. (*c*) | 2. (*d*) | 3. (*d*) | 4. (*c*) | 5. (*d*) |
| 6. (*d*) | 7. (*b*) | 8. (*d*) | 9. (*d*) | 10. (*b*) |

### *Chapters 4 to 6 review*

| | | | | |
|---|---|---|---|---|
| 1. T | 2. F | 3. T | 4. T | 5. T |
| 6. T | 7. T | 8. F | 9. T | 10. F |
| 11. T | 12. T | 13. T | 14. T | 15. F |
| 16. F | 17. T | 18. T | 19. T | 20. F |
| 21. T | 22. T | 23. T | 24. T | 25. T |
| 26. T | 27. T | 28. T | 29. T | 30. F |

### *Chapter 7*

| | | | | |
|---|---|---|---|---|
| 1. (*d*) | 2. (*b*) | 3. (*b*) | 4. (*c*) | 5. (*c*) |
| 6. (*b*) | 7. (*d*) | 8. (*c*) | 9. (*c*) | 10. (*a*) |

### *Chapter 8*

| | | | | |
|---|---|---|---|---|
| 1. (*a*) | 2. (*d*) | 3. (*a*) | 4. (*b*) | 5. (*c*) |
| 6. (*c*) | 7. (*c*) | 8. (*c*) | 9. (*a*) | 10. (*d*) |

### *Chapter 9*

| | | | | |
|---|---|---|---|---|
| 1. (*a*) | 2. (*c*) | 3. (*d*) | 4. (*a*) | 5. (*b*) |
| 6. (*c*) | 7. (*a*) | 8. (*b*) | 9. (*d*) | 10. (*b*) |

### *Chapters 7 to 9 review*

| | | | | |
|---|---|---|---|---|
| 1. T | 2. T | 3. T | 4. T | 5. T |
| 6. F | 7. T | 8. F | 9. T | 10. T |
| 11. F | 12. F | 13. T | 14. T | 15. T |
| 16. T | 17. T | 18. F | 19. F | 20. T |
| 21. F | 22. T | 23. T | 24. T | 25. F |
| 26. T | 27. T | 28. F | 29. T | 30. T |

### *Chapter 10*

| | | | | |
|---|---|---|---|---|
| 1. (*c*) | 2. (*a*) | 3. (*a*) | 4. (*a*) | 5. (*b*) |
| 6. (*c*) | 7. (*b*) | 8. (*b*) | 9. (*a*) | 10. (*d*) |

## *Chapter 11*

1. (*b*) 2. (*c*) 3. (*d*) 4. (*c*) 5. (*a*)
6. (*d*) 7. (*b*) 8. (*c*) 9. (*b*) 10. (*c*)

## *Chapter 12*

1. (*c*) 2. (*b*) 3. (*a*) 4. (*b*) 5. (*d*)
6. (*c*) 7. (*b*) 8. (*c*) 9. (*b*) 10. (*d*)

## *Chapters 10 to 12 review*

1. T 2. F 3. T 4. T 5. T
6. T 7. T 8. T 9. F 10. T
11. F 12. T 13. T 14. F 15. T
16. T 17. T 18. T 19. T 20. T
21. T 22. T 23. F 24. F 25. F
26. F 27. T 28. T 29. F 30. T
31. T 32. T 33. T 34. F 35. T
36. T 37. T 38. T 39. T 40. T

## *Chapter 13*

1. (*c*) 2. (*d*) 3. (*a*) 4. (*b*) 5. (*c*)
6. (*a*) 7. (*b*) 8. (*a*) 9. (*c*) 10. (*b*)

## *Chapter 14*

1. (*c*) 2. (*b*) 3. (*c*) 4. (*c*) 5. (*c*)
6. (*a*) 7. (*d*) 8. (*c*) 9. (*d*) 10. (*c*)

## *Chapter 15*

1. (*c*) 2. (*b*) 3. (*c*) 4. (*b*) 5. (*c*)
6. (*d*) 7. (*d*) 8. (*b*) 9. (*b*) 10. (*d*)

## *Chapters 13 to 15 review*

1. (*d*) 2. (*b*) 3. (*d*) 4. (*c*) 5. (*a*)
6. (*c*) 7. (*c*) 8. (*b*) 9. (*c*) 10. (*b*)
11. (*b*) 12. (*c*) 13. (*d*) 14. (*a*) 15. (*b*)
16. (*b*) 17. (*c*) 18. (*c*) 19. (*c*) 20. (*d*)

## *Chapter 16*

1. (*a*) 2. (*c*) 3. (*b*) 4. (*d*) 5. (*b*)
6. (*c*) 7. (*b*) 8. (*b*) 9. (*c*) 10. (*c*)

## *Chapter 17*

1. (*d*) 2. (*c*) 3. (*a*) 4. (*b*) 5. (*c*)
6. (*c*) 7. (*c*) 8. (*a*) 9. (*b*) 10. (*d*)

## *Chapters 16 and 17 review*

1. (*c*) 2. (*d*) 3. (*b*) 4. (*a*) 5. (*b*)
6. (*b*) 7. (*a*) 8. (*d*) 9. (*b*) 10. (*c*)

1. F 2. F 3. T 4. T 5. T
6. F 7. T 8. F 9. T 10. T

# Index